STAR FORMATION
NEAR AND FAR

STAR FORMATION NEAR AND FAR

Seventh Astrophysics Conference

College Park, MD October 1996

EDITORS
Stephen S. Holt
NASA/Goddard Space Flight Center
Greenbelt, Maryland

Lee G. Mundy
University of Maryland
College Park, Maryland

AIP CONFERENCE
PROCEEDINGS 393

American Institute of Physics   **Woodbury, New York**

Authorization to photocopy items for internal or personal use, beyond the free copying permitted under the 1978 U.S. Copyright Law (see statement below), is granted by the American Institute of Physics for users registered with the Copyright Clearance Center (CCC) Transactional Reporting Service, provided that the base fee of $10.00 per copy is paid directly to CCC, 222 Rosewood Drive, Danvers, MA 01923. For those organizations that have been granted a photocopy license by CCC, a separate system of payment has been arranged. The fee code for users of the Transactional Reporting Service is: 1-56396-678-6/ 97 /$10.00.

© 1997 American Institute of Physics

Individual readers of this volume and nonprofit libraries, acting for them, are permitted to make fair use of the material in it, such as copying an article for use in teaching or research. Permission is granted to quote from this volume in scientific work with the customary acknowledgment of the source. To reprint a figure, table, or other excerpt requires the consent of one of the original authors and notification to AIP. Republication or systematic or multiple reproduction of any material in this volume is permitted only under license from AIP. Address inquiries to Office of Rights and Permissions, 500 Sunnyside Boulevard, Woodbury, NY 11797-2999; phone: 516-576-2268; fax: 516-576-2499; e-mail: rights@aip.org.

L.C. Catalog Card No. 97-71978
ISBN 1-56396-678-6
ISSN 0094-243X
DOE CONF- 96-10274

Printed in the United States of America

TABLE OF CONTENTS

PREFACE xiii
    Holt, S. S.

INTRODUCTION TO STAR FORMATION

Current Issues in Star Formation 3
    Silk, J.

The Fourth Day of Creation: The Search for a History of Star Formation 15
    Trimble, V.

CLOSE-UP VIEWS OF STAR FORMATION

Early Star Formation 41
    Myers, P. C.

Turbulence and Magnetic Fields in Star Formation 51
    Ostriker, E. C.

Observations of Circumstellar Disks and Infall 63
    Mundy, L. G.

YSOs in Transition 72
    Magnier, E. A.

Self-similar Evolution of Supercritical Cores 75
    Basu, S.

Virial Balance in Turbulent MHD Two Dimensional
Numerical Simulations of the ISM 81
    Ballesteros-Paredes, J., and Vázquez-Semadeni, E.

Highly Compressible MHD Turbulence and Gravitational Collapse 85
    Vázquez-Semadeni, E., Passot, T., and Pouquet, A.

Numerical Simulations of MHD Turbulence and Gravitational Collapse 89
    Balsara, D. S., Crutcher, R. M., and Pouquet A.

Dynamically Infalling Envelopes Around Low-mass Protostar Candidates 93
    Ohashi, N.

Structure of Dark Clouds from Stellar Extinction 97
    Padoan, P., Jones, B. J. T., and Nordlund, Å.

A Physical Model for the Stellar IMF 101
    Padoan, P., Nordlund, Å. and Jones, B. J. T.

Velocity Coherence in Dense Cores   105
    Goodman, A. A., Barranco, J. A., Wilner, D. J., and Heyer, M. H.

Interferometric Imaging of Dense Gas Tracers
in the Protostellar Collapse Candidate L1527   109
    Wilner, D. J., Mardones, D., and Myers, P.C.

A Statistical Study of Infall Motions in Nearby Young Stellar Objects   113
    Mardones, D., Myers, P. C., Tafalla, M., Wilner, D. J., Bachiller, R., and Garay, G.

Accretion Disks Around Class O Protostars: The Case of VLA 1623   117
    Wilson, C. D., Pudritz, R. E., Carlstrom, J. E., Lay, O. P., Hills, R. E., and Ward-Thompson, D.

A Deflected Jet in the Bipolar Molecular Outflow NGC 2264G   121
    Fich, M., and Lada, C.J.

IRAS 04302+2247: Butterfly Star in Taurus!   125
    Lucas, P., and Roche, P. F.

High Resolution 2.7 mm Observations of L1551 IRS5: A Protobinary System?   129
    Looney, L. W., Mundy, L. G., and Welch, W. J.

Fragmentation of Molecular Clouds with GRAPESPH   133
    Klessen, R. S.

Rotation of Starless Bok Globules   137
    Kane. B. D., and Clemens, D. P.

CIRCUMSTELLAR DISKS

Hubble Space Telescope Observations of the Environments of Young Stars   143
    Hester, J. J.

Observations of the Inner Accretion Disk Around Young Stars   153
    Hartigan, P.

The Theory of Circumstellar Accretion Disks   160
    Stone, J. M.

Axisymmetric MHD Simulations of Stellar
Magnetosphere/Accretion Disk Interaction   171
    Miller, K., and Stone, J. M.

Hard X-ray Emissions from Star Forming Regions   175
    Tsuboi, Y., Koyama, K., and Ueno, S.

X-Ray Emission from Protostars   179
    Feigelson, E. D., Koyama, K., and Montmerle, T.

Dispersed T-Tauri Stars and Galactic Star Formation ... 184
    Feigelson, E. D.

The Influence of Photoevaporation on Star Formation in NGC 6611 ... 189
    de Winter, D., van den Ancker, M. E., and Pérez, M. R.

Extra-Solar Comets Near Young β Pic Analogs ... 193
    Grady, C. A., Pérez, M. R., Bjorkman, K. S., Sitko, M. L., and de Winter, D.

Time Variations of Water Vapor Masers in Star-Forming Regions ... 197
    Mendoza-Torres, J. E., and Lekht, E. E.

Star Formation in M16 with Adonis and Hubble ... 201
    Currie, D., Bonaccini, D., Kissell, K., Shaya, E., Avizonis, P., and Dowling, D.

On the Origin of Narrow, Very Long, Straight Jets from Some Newly Forming Stars ... 205
    Greyber, H. D.

Gas to Dust Ratios in Vega-excess Stars ... 209
    Coulson, I. M., Walther, D. M., and Dent, W. R. F.

THE VERY LOW END OF THE IMF

Formation of Low Mass Stars and Brown Dwarfs ... 217
    Lin, D. N. C.

Early Hints on the Substellar Mass Function ... 228
    Basri, G., and Marcy, G. W.

Hydrogen Flash Divides Accreting Objects into T-Tauri Stars and Embedded Protostars ... 241
    Murai, T.

DYNAMICAL PROCESSES

Dynamical Influences on Star Formation in Sprial Galaxies ... 247
    Kenney, J., and Jogee, S.

Mergers, Interactions, and the Fueling of Starbursts ... 259
    Hibbard, J. E.

Starbursts and Cosmogony ... 271
    Heckman, T. M.

Young Stellar Aggregates Embedded in Expanding Supershells ... 279
    Surdin, V. G., and Moskal', E. V.

Star Formation in Leading/Trailing Single Arm Galaxies: NGC 4378   283
    Byrd, G. G., Purcell, G. B., Buta, R. J., McCormick, D., and Freeman, T.

Orchestration of Starbirth Activity in Disk Galaxies:
New Perspectives from Ultraviolet Imaging   287
    Waller, W. H., Stecher, T. P., and the UIT Science Team

Preliminary Results of the ASU/UGA O-star Project   291
    Scowen, P. A., Hauschildt, P. H., Aufdenberg, J. P., and Sankrit, R.

Early Results from an HST Imaging Survey of the Ultraluminous IR Galaxies   295
    Borne, K. D., Bushouse, H., Colina, L., and Lucas, R. A.

HST Imaging of Sub-Kiloparsec Scale Structure in Markarian Galaxies   299
    Nelson, C. H., MacKenty, J. W., and Simkin, S. M.

Infrared Imaging of the Starburst Galaxy NGC 7469   303
    Jayawardhana, R., Fazio, G., Eikenberry, S., Hughes, D., Hora, J.,
    Hoffmann, W., Dayal, A., and Deutsch, L.

Another Twin Peaks Galaxy - The Barred Spiral NGC 5383   307
    Sheth, K., Regan, M. W., and Vogel, S. N.

Dynamical Stability and Galaxy Evolution in LSB Disk Galaxies   311
    Mihos, C., McGaugh, S., and de Blok, E.

The Age of LSB Discs   315
    Padoan, P., Jimenez, R., and Antonuccio-Delogu, V.

Pre-Starbursts in Luminous IR Galaxies?   319
    Gao, Y., Gruendl, R., Lo, K. Y., Hwang, C. Y., and Veilleux, S.

Starbursts in Our Home Galaxy   323
    Ozernoy, L. M.

Primordial Star Forming Regions in a CDM Universe   329
    Zhang, Y., Norman, M. L., Anninos, P., and Abel, T.

MULTIPLE STAR SYSTEMS

Accretion in Pre-Main-Sequence Binaries   337
    Mathieu, R. D.

Star Formation in Clusters   347
    Clarke, C. J.

The Initial Mass Function: Now and Then   357
    Richer, H. B., and Fahlman, G. G.

Binary Multiplication: The Formation of Close Binaries from Wide Ones   367
    Watkins, S., Bhattal, A., Francis, N., and Whitworth, A.

Disc Formation in Protobinary Systems 371
    Bate, M. R.

A Near-Infrared/Millimeter Study of
Six Southern Hemisphere Star Forming Regions 375
    Megeath, S. T., and Sollins, P. K.

Molecular & Photodissociated Gas in the
Massive Star Formation Region NGC 6334 379
    Kraemer, K. E., and Jackson, J. M.

A Bright, Young Molecular Outflow near Sharpless 302 383
    Shure, M.

Evidence for Core Collapse Toward 3 Young Stellar Clusters 387
    Williams, J. P., and Myers, P.C.

VLA Ammonia (3,3) Observations of Heated and High Velocity Gas in Orion-KL 391
    Wiseman, J. J., Putman, M. E., and Ho, P. T. P.

High-Spatial Resolution Imaging of the NGC 2024 Molecular Ridge 395
    Pound, M. W., Gruendl, R., Lada, E. A., and Mundy, L.

Organization of the Magnetic Field in W3(OH) on Fine Spatial Scales 399
    Bloemhof, E. E.

Observational Evidence for the Present-Day Formation of Globular Clusters 403
    Ho, L. C., and Filippenko, A. V.

GALACTIC STAR FORMATION

Some High Class (or High Mass) Neighborhoods -- the Sites
of the Most Massive Stars in the Milky Way and our Neighbors 409
    Humphreys, R. M.

Observations of the Extragalactic Initial Mass Function
and Modes of Star Formation 414
    Heap, S. R.

Controlling Factors for Global Star Formation 423
    Skillman, E. D.

Star Formation Rates, Efficiencies and Initial Mass Functions in Sprial Galaxies 433
    de Pablos, F., and Cepa, J.

The History of the Star Formation Rate in the Local Disk 437
    Rocha-Pinto, H. J., Maciel, W. J.

Percolating Star Formation in Barred Spirals 441
    Mott, A. S., Alexander, P., and Sleath, J. P.

Turbulent Fragmentation of Interstellar Gas and the Stellar Mass Spectrum   445
    Khersonsky, V.

MHD Turbulence and Scaling Laws in Molecular Clouds   449
    Xie, T.

Star Formation at the Intermediate Distances:
Gravitational Collapse in Massive Cores   453
    Zhang, Q., and Ho, P. T. P.

Origin of the Mass in Massive Star Outflows   457
    Churchwell, E.

Star Formation in Translucent Clouds   461
    Hearty, T., Magnani, L., Caillault, J.-P., Neuhäuser, R.,
    Schmitt, J. H. M. M., and Stauffer, J.

Toward a Better Understanding of the IR
Spectral Energy Distributions of H II Regions   465
    Leisawitz, D., Digel, S. W., and Hanson, M. M.

The Spectacular Ionized Interstellar Medium of NGC 55   469
    Ferguson, A. M. N., Wyse, R. F. G., and Gallagher, J. S.

Far-Ultraviolet Emission from NGC 4038/4039, "The Antennae":
Massive Star Formation in Compact Clusters   473
    Neff, S. G., Hollis, J. E., Hill, J. K., Fanelli, M. N., Smith, D. A.,
    Smith, A. M., Stecher, T. P., Bohlin, R. C., O'Connell, R. W.,
    and Roberts, M. S.

M31 vs. M33: Different Modes of Star Formation   475
    Hodge, P., Wyder, T., and Olsen, K.

STAR FORMATION HISTORY IN SPIRALS

Cosmic Star Formation History   481
    Madau, P.

The Star Formation History of Low Redshift Spiral Galaxies   491
    O'Connell, R. W.

Near-IR Discoveries of Groups of Star-Forming Galaxies at $z > 2$ 502
    Teplitz, H., Malkan, M. A., and McLean, I. S.

The ISO-IRAS Faint Galaxy Survey: Early Results   506
    Hurt, R. L., Lonsdale, C. J., Levine, D. A., Smith, H. E., Helou, G.,
    Van Buren, D., Beichman, C. A., Lord, S. D., Neugebauer, G.,
    Moshir, M., Soifer, B. T., Wehrle, A., Cesarsky, C., Elbaz, D., Klaas, U.,
    Laureijs, R., Lemke, D., McMahon, R. G., and Wolstencroft, R. D.

Gas Content and Star Formation Thresholds in the Evolution of Spiral Galaxies 510
    McGaugh, S., and de Blok, E.

The Evidence for Massive Star Formation in Early-Type Spiral Galaxies 514
    Hameed, S., and Devereux, N.

Hα, Far Infrared and Thermal Radio Continuum Emission
within the Late-Type Spiral Galaxy M33 518
    Devereux, N., Duric, N., and Scowen, P.

STAR FORMATION HISTORY IN ELLIPTICALS

Star Formation History of Elliptical Galaxies from Low-Redshift Evidence 525
    Worthey, G.

HST Observations of Distant Clusters:
Evidence for Old Ellipticals and Younger S0s 535
    Dressler, A.

The K-band Luminosity Function of Galaxies 543
    Gardner, J. P., Sharples, R. M., Frenk, C. S., and Carrasco, B. E.

High Redshift Reflection Nebulae: Implications for Galaxy Formation 547
    Chambers, K. C.

Ultraviolet Imaging Observations of Abell 1795:
Further Evidence for Massive Star Formation 551
    *Smith, E. P., Bohlin, R. C., Bothun, G. D., O'Connell, R. W.,
    Roberts, M. S., Neff, S. G., Smith, A. M., and Stecher, T. P.*

Star Formation Associated with the NE Radio Lobe of NGC 5128 (Cen A) 555
    Graham, J. A.

STAR FORMATION HISTORY IN IRREGULARS AND DWARF SPHEROIDALS

Star Formation History in Irregular Galaxies 561
    Hatzidimitriou, D.

The Star Formation Histories of Dwarf Spheroidal Galaxies 571
    Smecker-Hane, T. A.

The HI Supergiant Shells in the Large Magellanic Cloud 582
    Kim, S., and Staveley-Smith, L.

Induced Star Formation and Chemical Enrichment in NGC 5253 586
    *Kobulnicky, C., Skillman, E., Roy, J.-R., Walsh, J. R.,
    and Rosa, M. R.*

UIT Observations of the SMC 590
    Cornett, R. H., Stecher, T. P., and the UIT Science Team

UIT Astro-2 Observations of NGC 4449 594
    Hill, R. S., Fanelli, M. N., Smith, D. A., Bohlin, R. C., Neff, S. G., O'Connell, R. W., Roberts, M. S., Smith, A. M., and Stecher, T. P.

Luminosities and Star Formation Rates of Galaxies Observed with the Ultraviolet Imaging Telescope: A Comparison of Far-UV, Hα, and Far-IR Diagnostics 598
    Fanelli, M. N., Stecher, T. P., and the UIT Science Team

The Impact of Star Formation on the Interstellar Medium 602
    Martin, C. L.

Dust, Gravitational Lensing and Star Formation 606
    Blain, A. W.

Structural Parameters of Hubble Deep Field Galaxies 610
    Takamiya, M.

SUMMARY

Summary: Star Formation, Near and Far 617
    Larson, R. B.

Appendix A - Conference Programme   627

Appendix B - List of Attendees   631

Appendix C - Physical Constants   639

Author Index   643

Subject Index   649

Preface

This is the seventh in a series of annual October Astrophysics Conferences in Maryland. These conferences are organized by astrophysicists at the Goddard Space Flight Center and the University of Maryland. The topic for each conference is selected by a permanent committee of senior scientific staff with the help of an International Advisory Committee, the current membership of which is:

Marek Abramowicz, Göteborg	*Sir Martin Rees*, Cambridge
Roger Blandford, Pasadena	*Vera Rubin*, Washington
Claude Canizares, MIT	*Joseph Silk*, Berkeley
Arnon Dar, Haifa	*David Spergel*, Princeton
Alan Dressler, Pasadena	*Rashid Sunyaev*, Moscow
Guenther Hasinger, Potsdam	*Alex Szalay*, Budapest
Steve Holt, Greenbelt	*Yasuo Tanaka*, Tokyo
Dick McCray, Boulder	*Scott Tremaine*, Toronto
Jim Peebles, Princeton	*Simon White,* München

The subject chosen for this conference was "Star Formation, Near and Far," with its program developed by the Scientific Organizing Committee:

Eli Dwek	*Nino Panagia*
Sally Heap	*Joe Silk*
Steve Holt	*David Spergel*
Steve Maran	*Virginia Trimble*
Lee Mundy	*Stuart Vogel*
Susan Neff	

Star Formation is perhaps the quintessential subject of modern astronomy. The conference attendance reflected this, in the sense that the fraction of the attendees at the meeting whose early training was in physics rather than astronomy (and who might therefore choose to designate themselves "astrophysicists" rather than "astronomers") was noticeably smaller than has been the case at previous meetings in this series. This is not to say that star formation is not of interest to physicists; on the contrary, this year's subject has simply increased the participation of traditional astronomers in this series.

We began the conference with invited reviews by *Virginia Trimble* on the history of Star Formation research, and by *Joe Silk* on current issues associated with the subject. The conference then proceeded through the next two days with a series of non-paralleled sessions, each devoted to a specific topic and led by a distinguished session chair. Each of these sessions featured two or three invited talks and an extensive discussion period that may have also included one or two short contributions "promoted" from the poster papers by the session chair.

All the chairs should be commended for keeping the activities in their sessions lively, and their speakers to the allotted times:

Bob Brown   *John Graham*   *Susan Neff*
Ed Churchwell   *Mike Hauser*   *Anneila Sargent*
Eli Dwek   *Steve Maran*   *David Spergel*

We are especially indebted to *Gibor Basri, Jeff Kenney, Piero Madau , Bob O'Connell* and *Evan Skillman* for agreeing to substitute for speakers who had to withdraw after having accepted invitations; to *Richard Larson* for a thoughtful and definitive rapporteur paper; to *Kirk Borne* for invaluable help in editing postscript files; and to *John Trasco* and *Susan Lehr* for their usual professionalism in handling all the details necessary to make the conference a success.

As in previous meetings in this series, the banquet at the conclusion of the second day featured a distinguished speaker. This year *Riccardo Giacconi* provided us with an interesting perspective on the future of astronomical observatories. In particular, he delighted us with a whimsical comparison between the next European Southern Observatory facility and that of Tycho Brahe. I'm sure that all of us who were present will remember that one of the essential facilities in Tycho's observatory was a prison.

After a one-year experiment with a different publisher, we are pleased to be returning to the American Institute of Physics Press, which also published the first five Proceedings in this series.

Steve Holt
January 1997

Introduction to Star Formation

Current Issues in Star Formation

Joseph Silk

Departments of Astronomy and Physics, and Center for Particle Astrophysics, University of California, Berkeley, CA 94720

INTRODUCTION

Star formation is the key to understanding many aspects of astrophysics, that span areas as diverse as the solar system, our galaxy, and the distant universe. Yet despite decades of intensive study of regions of nearby star formation, theoretical understanding of star formation has remained elusive.

The parameters that define the essential characteristics of star formation are the efficiency and rate of star formation, and the initial stellar mass function. Detailed observations of these are mostly local. A complete understanding, or even a phenomenological model, for the parameters of star formation is mostly lacking. We need to develop an empirical, or even better, a fundamental, theory before we can succeed in extrapolating from regions such as the Orion Molecular Cloud to the more extreme environments of starbursts and galaxy formation.

The problem of star formation involves complex issues. One has to incorporate interstellar physics and chemistry, as well as galactic dynamics. In order to evaluate how many stars a molecular cloud forms, one presumably has to specify the spectral and temporal dependences of such quantities as the densities of various molecular species as well as that of hydrogen, the ionization, dust extinction, magnetic field strength, the dust and gas temperatures, the turbulent velocity field and the angular momentum distribution. All of this has to be implemented for a three-dimensional gas distribution that is subject to external influences from contiguous regions of star formation as well as, on the more global scale, galactic differential rotation, spiral density waves, tidal torques induced by a central bar or satellites, and satellite accretion. Needless to say, progress has been slow.

There is a diverse range of scales over which one needs to understand star formation, extending (see Table 1) from nearby molecular cores of dark clouds to forming galaxies at high redshift. Star formation dominates the luminosities of objects that are among the most luminous and the most distant in the universe. Presumably there is an underlying theory that we can hope to apply over this range of scales and environments.

I will discuss our current understanding of star formation by proceeding from near to far. I begin with the initial mass function (IMF) and the star formation efficiency (SFE) in nearby regions, and the star formation rate (SFR) in disk galaxies, before turning to starbursts and galaxy evolution.

THE INITIAL MASS FUNCTION

Observations

The initial mass function is defined to be the mass distribution of all stars ever formed in a specified region. One measures the current epoch luminosity function of stars. Since massive stars are short-lived, one has to correct this for the past birthrate of stars that have died in order to derive the IMF. Hence it is customary to represent the uncertainty in the IMF for stars above 1 $M_\odot$ by the ratio b of present to past birth rates. This factor is typically in the range $0.3 \leq b \leq 3$ when averaged over disk galaxies, but must be of order $b \lesssim 0.01$ for ellipticals.

The IMF is relevant for many reasons. One needs to ascertain the IMF in order to derive star formation rates and mass-to-light ratios, interpret galaxy spectra, and model galaxy evolution, gas recycling, chemical evolution and supernova rates. Knowledge of the origin of the IMF and of its possible variations is important for understanding starburst physics, galaxy formation, and the nature of dark matter.

The IMF as presented by Miller and Scalo in 1979 has stood the test of time remarkably well. It is close to a log-normal distribution, rolling over below $\sim 0.5 M_\odot$ and declining below $\sim 0.3 M_\odot$ and above $\sim 1 M_\odot$. Particularly impressive are recent observations of one of the youngest nearby star clusters, the Orion Nebula cluster, centered on the Trapezium, and of several globular clusters. In the case of Orion, a Miller-Scalo IMF fits data over 5.0 to 0.1 $M_\odot$. For the globular clusters, all dynamically unevolved systems have a similar IMF which flattens below about 0.3 $M_\odot$. There is no evidence that the IMF differs between Population I and Population II.

The only slight shadow on the horizon comes from the MACHO gravitational microlensing experiment. The data analyzed thus far for events towards the Large Magellanic Cloud provides evidence that a substantial fraction of the dark halo of our galaxy is in the form of objects of mass $\sim 0.5 M_\odot$ that cannot be hydrogen-burning stars, otherwise one would have seen them in deep star counts at high galactic latitude. If the results survive (at present there are only 6 undisputed events) as new data accumulates, one would be compelled to adopt a highly top-heavy IMF for the halo of the early galaxy in order to form a large enough number of white dwarfs (or possibly of neutron stars).

Theory

Understanding the origins of the IMF is centered on the following questions: what determines the mass of a star? An early idea, formulated originally by Hoyle, is that a collapsing cloud undergoes fragmentation on ever-increasing mass scales. If the cloud is isothermal, as the density increases, the fragmentation scale is given by the Jeans mass M_J. From linear theory, $M_J \propto T^{3/2}\rho^{-1/2}$ and increases until other physical effects intervene. Opacity will eventually result in adiabatic contraction and establish a minimum mass of the Jeans mass. This is found to be too small ($\sim 10^{-3}\,M_\odot$) in spherical collapse to be relevant to star formation.

However the fragmemtation physics is far too simplistic in such analytical calculations. A more realistic treatment would allow for fragment interactions, mergers, and accretions, all of which tend to raise the final mass scale before the onset of formation of Kelvin-Helmholtz contracting protostar cores. Angular momentum will result in disk formation, from which gas can accrete onto the central core. Magnetic fields stabilize sufficiently massive clouds against collapse, but neutral molecular gas undergoes ambipolar diffusion over a time-scale $t_{ad} \sim 10^{14}\,x$ yr, where x is the fractional hydrogen ionization.

Clearly, incorporation of all of the relevant physics presents a challenging task. One could in principle study fragmentation numerically, but the resolution needed is beyond the range of the largest available computers. Hence it is useful to consider the relevant time scales. The gas accretion time for a core of M is $t_{acc} \propto GM\,\Delta V^{-3}$, and the core will grow over this time-scale. What limits the mass of a star?

Nuclear energy release can limit infall, but only once a star of $\sim 10 M_\odot$ has formed. The only possibility for feedback is if the protostellar energy source of gravitational energy can be tapped. The Kelvin-Helmholtz contraction time-scale is $t_K \propto T_c M^{-2}$, where T_c is the central temperature, and is primarily sensitive to the protostellar mass. As the protostar accretes, t_K becomes less than t_{acc}. There is abundant observational evidence that protostellar energy drives outflows, in the form of jets and bipolar molecular flows. Whether or not these outflows deposit sufficient momentum into the ambient molecular cloud gas to inhibit infall is still undemonstrated.

Circumstantial evidence does however suggest an initimate coupling between protostar luminosities (and therefore masses) and cloud linewidths. Moreover the linewidths, which almost invariably are found to be supersonic, are correlated with distance to the nearest young star clusters. It seems plausible that the observed cloud turbulence is driven by protostellar outflows. The turbulence is likely to regulate protostellar masses, as suggested by the mass accretion rate in spherical collapse onto an isothermal core, $\sim \Delta V^3/G$, where ΔV is the linewidth. As protostellar outflow-driven shells interact with neighboring outflows, a network of turbulent remnants develops. The distribution of turbulent velocities in a molecular cloud should be dominated by low ΔV clumps, since it is the many old flows that are volume-filling and are responsible for most of the momentum dissipation.

The oldest, and slowest, flows are supersonic but subAlfvénic. Dissipation of energy via shocks is especially important for the high velocity, younger outflows. If ΔV indeed controls the protostellar mass, then the IMF can be understood as an approximately log-normal distribution. The predominance of low mass stars is due to the predominance of low ΔV or old outflow remnants.

The outflows are responsible for inhibiting core collapse and for explaining the observed supersonic line widths. Important questions remain unanswered in this type of scheme. How are starless molecular clouds explained? How do the first stars in a cloud form? What is clear is that the Jeans mass is irrelevant for star formation if feedback and self-regulation are the dominant processes.

The Jeans scale may however play an important role in accounting for core masses. The generalized Jeans mass is

$$M_{Jeans} \approx (\Delta V)^4 G^{-3/2} p^{-1/2}; \Delta V = \left(V_s^2 + V_A^2 + \Delta V^2\right)^{\frac{1}{2}},$$

where V_s is the sound velocity, V_A is the Alfvén velocity and p is cloud pressure, all appropriately averaged. Fragmentation is likely to occur on scales down to the Jeans mass. Cold clouds are observed over a mass range spanning thousands of solar masses, down to stellar mass scales. This suggests that on small scales the star formation efficiency must be high. In fact, the distribution of core masses is flatter than the IMF by approximately one power in mass at scales above a few solar masses. This is consistent with the observation that stars form in clusters rather than in isolation.

Indications that fragmentation occurs on scales down to $\sim 0.1\,M_\odot$ come from a study of the surface density of stars in the Taurus-Auriga cold clouds. The correlation function of stellar companions changes slope at a length scale of 0.04 pc, corresponding to a mass scale of $\sim 0.1 M_\odot$. The observed correlations imply that only on larger scales are there many companions per star. One could perhaps identify this scale with the transition from fragmentation to a predominance of binary formation.

Star formation is an inefficient process, in that molecular clouds survive for ~ 30 collapse times. Star formation inefficiency is usually attributed to the magnetic Jeans barrier. The magnetic Jeans mass,

$$M_J \approx 10^4 \left(B/10^{-5}\,\text{g}\right) \left(10^3\,\text{cm}^{-3}/n\right)^2\,M_\odot,$$

greatly exceeds a typical stellar mass. The time-scale for flux loss by ambipolar diffusion is of order 10 free-fall times for a magnetically supported cloud in quasistatic equilibrium.. Collapse along field lines is likely to be inefficient, given the turbulent input to molecular clouds. Cloud aggregation is driven by supernova remnants on small scales, and density waves and gravitational instability on large ($\gtrsim$ 100pc) scales. Either flux loss or cloud aggregation above the Jeans mass is necessary to form stars.

One might speculate that the star formation efficiency is inversely proportional to M_J, and therefore varies as $p^{1/2}T^{-2}$, where T is the molecular gas temperature. Since cloud temperatures are unlikely to vary much because of the efficiency of molecular cooling, one could speculate that enhanced star formation efficiencies would occur in high pressure regions, notably in starbursts driven by mergers. Since ΔV is likely to be enhanced in such regions, it is tempting to also speculate that the IMF may be top-heavy, systematically dominated by more massive stars. Conversely, unusually quiescent star forming regions, perhaps typical of star formation that may have occurred in the outermost halo of the galaxy, could be expected to have an unusually large proportion of low mass stars. However there is no definitive evidence for any IMF variations, in starbursts or elsewhere.

STAR FORMATION IN DISK GALAXIES

Spiral galaxies form stars steadily and inefficiently. The star formation rate is relatively constant over time. The star formation efficiency is a few percent, and the gas reservoir, including possible halo infall, lasts over a Hubble time. There appears to be a threshold in gas surface density above which most star formation occurs, as monitored by Hα mapping. The star formation rate is proportional to approximately the 3/2 power of total ($HI + H_2$) gas surface density, and declines abruptly below a critical gas surface density that is interpreted theoretically as $\Sigma_{cr} \approx \kappa \sigma_g / \pi G$, where κ is the epicyclic frequency and σ_g is the gas velocity dispersion. In the solar neighborhood, for example, $\Sigma_{cr} \approx 7 M_\odot$ pc^{-2}, whereas the observed gas surface density is about 15 $M_\odot$ pc^{-2}. Hence, locally the Galaxy is mildly unstable to star formation. In fact a similar value for the ratio of the gas surface density Σ_g to Σ_{cr} applies throughout the star-forming regions of disks.

The physical motivation for Σ_{cr} is based on disk instabilities: a cold self-gravitating disk is gravitationally unstable to local (and also to non-axisymmetric) instabilities provided that $Q \equiv \Sigma_{cr}/\Sigma \lesssim 1$. The Toomre parameter Q demarcates the competition between the stabilizing effects of shear (via κ) and velocity dispersion (via σ_g) versus the destabilizing tendencies of disk self-gravity (via Σ). An unstable disk is subject to both local gravitational instabilities that drive cloud collisions and coalescence, and to bar formation that drives radial gas flows that also lead to cloud build-up in the innermost disk. Star formation is a natural consequence.

A semi-phenomenological theory of disk star formation involves an expression of the form

$$\text{SFR} = \epsilon \Omega \Sigma_{gas} T$$

where ϵ ($\approx 1 - 3$ % over 10^8 yr) is the efficiency of star formation, $\Omega(r)$

is the differential foration rate, Σ_{gas} is the gas surface density and $T(r)$ is an appropriate threshold criterion. The expression for star formation rate incorporates gas cloud aggregation at the instability growth rate $\sim \Omega^{-1}$ of linear perturbation theory. Alternative expressions proposed for the threshold factor include $T = |1 - Q|$ and $T = \Omega(\sigma) - \Omega(p)$, where Ω_p is the (fixed) angular velocity of the spiral density waves. The former expression is based on linear perturbation theory. The latter depiction of T is motivated by the theory of cloud coagulation driven by spiral density waves as a precursor to star formation.

Not only does the expression for the star formation rate acount for the observed radial distributions of star formation and gas, but one can integrate backwards over time to derive the star-forming history of the disk. This integration provides the metallicity distribution of stars in the solar neighborhood, and the runs of stellar metallicity with age and galactic center radius, as well as the metallicity gradient of the gas. All of these can be compared with data, and give acceptable fits.

In fact, the universality of a star formation threshold merits closer examination. Low level star formation occurs well beyond the radius corresponding to the azimuthally-averaged critical surface density. Presumably one can attribute this to local density waves, where the surface density may exceed the critical threshold value.

Two implications of the disk star formation model are worthy of note. One observes $Q \sim 1$ throughout the disk star forming regions. This suggests that disks self-regulate to maintain $\epsilon \ll 1$. One could imagine that disk instabilities drive massive star formation, which heats the gas and thereby provides a negative feedback by helping to quench the instabilities. The mechanism by which feedback occurs could involve driving supernova remnants into the gas, leading to large-scale disturbances such as interstellar bubbles and chimneys.

Disks are inferred to form inside-out since $\Omega \propto r^{-1}$ (for a flat rotation curve) and the star formation rate is therefore proportional to Σ_g/r. There are interesting consequences for disk galaxies at high redshift: such objects should be spheroid-dominated and appear unresolved even at HST resolution. Chemical evolution should also occur inside-out, perhaps helping to explain the surprisingly low metallicities observed for damped lyman alpha clouds seen in absorption towards high redshift quasars. These systems are putative protodisks which are preferentially probed at large impact parameter, or equivalently, radius.

STARBURST GALAXIES

Starbursts are defined to occur when the current star formation rate exceeds the time-averaged star formation rate by an order of magnitude or more. This is accomplished when the initially circular orbits of clouds in the disk develop appreciable non-circular components, due to formation of a central massive bar, a close tidal interaction, or a satellite merger. The clouds react inelastically and undergo greatly enhanced coalescence. Angular momentum transfer occurs via tidal torquing and gas is driven inwards. The presence of a preexisting bar, or formation of a new bar via disk global instability, can enhance the star formation rate by a factor of ~ 10, while a massive merger enhances the star formation rate by ~ 100 or more.

Numerical simulations have provided considerable insights into how the gas becomes concentrated into the central kpc. The starburst timescale is inferred to be the dynamical time scale for gas inflow aggregation, typically $\sim 10^8$ yr. However, detailed knowledge of how gas concentrations lead to star formation is lacking. Precisely what happens within the central kiloparsec of a starburst in terms of triggering star formation is still somewhat of a mystery

Observations do provide some clues. The aftermath of the starburst is readily apparent. The impact on the interstellar medium can be seen with large holes and chimneys in the interstellar gas, and the prevalence of galactic winds within the starburst region, usually concentrated to the inner few hundred parsecs. Several observations are indicative of a gas pressure enhancement by up to ~ 1000 relative to the local interstellar medium. Far infrared lines such as [OI] and [SII] emitted in photodissociation regions surrounding molecular clouds provide the most direct measures of gas pressure. The supernova-injected momentum is inferred from the number of OB stars (and hence supernovae) required to account for the ionizing radiation flux. The inferred cosmic ray pressure, obtained from equipartition arguments applied to the observed diffuse synchrotron radio emission within the starburst region, confirms the enhanced starburst pressure.

Starbursts have a high efficiency of star formation. Perhaps the enhanced pressure helps explain why. Suppose the star formation efficiency is proportional to the number of Jeans mass clumps on the grounds that individual clumps self-destruct once an OV star forms. Ordinarily, the Jeans mass is large, $\sim 10^4 p_{ism}^{-1/2} T^2$ M$_\odot$ for local interstellar medium values. However, in a starburst, the Jeans mass may be reduced by an order of magnitude, suggesting that the star formation efficiency, proportional to the number of massive star-forming clumps, could be similarly enhanced. High pressure also limits the porosity of the interstellar gas to supernova-remnant driven bubbles that would otherwise be disruptive and drive a galactic wind.

There are several uncertain observational aspects to starbursts. Many stars form in star clusters, that appear to be the analogs of young globular star clusters. It is not known whether star formation is instantaneous within a starburst region, or proceeds from the outer to the inner parts as appears to be the case for NGC 5253. There should be many post-starburst galaxies, expected to outnumber starbursts by 10 to 1, that have not been identified. The relation between active galactic nuclei and starbursts remains a mystery:

does one trigger the other?

STARBURSTS AND ELLIPTICAL GALAXY FORMATION

There are intriguing parallels between starbursts and the predicted properties of forming spheroidal galaxies. An ultraluminous starburst has a far infrared luminosity in excess of $10^{12} L_\odot$. If due to massive stars, this corresponds to a star formation rate of several hundred solar masses per year. Over 10^8 yr, star formation sustained at this rate suffices to account for the stellar content of a typical spheroidal component. The star formation history inferred from population synthesis of ellipticals requires a similar star formation rate.

Ultraluminous starbursts invariably are major mergers. Mergers today are rare but increase dramatically at $z \approx 1$, as viewed in deep HST images. Theory predicts a rapid rise in merger rate with redshift. Galaxies form hierarchically via a merging tree that culminates in a major merger to form an elliptical, or in minor mergers to form disk galaxies. Major mergers allow efficient star formation since supernova-driven disruption of the massive merging gas clouds is unlikely, and effective angular momentum transfer, since the dense stellar systems formed in the merger can sink by dynamical friction into the center of the galaxy. The central concentration of gas in a starburst, of order $10^4 M_\odot$ pc^{-2} in H_2 within the central kiloparsec, is sufficient to produce the central surface brightness of an elliptical galaxy. Density profiles of the old star components in mergers often display de Vaucouleurs profiles. This is what is anticipated if mergers form ellipticals and bulges.

If spheroids formed as starbursts, one might expect to find relic signatures of starbursts in the stellar populations. One possible signature may be the fundamental plane for E's and $S0$'s. The fundamental plane is equivalent to a systematic increase in mass-to-light ratio with luminosity, $M/L \propto L^{1/6}$, for luminous galaxies. This may be due to a systematic increase in compact stellar remnants, which could be the product of a progressively changing IMF, either by an increase in the peak mass at the low mass end or a decrease in slope at the high mass end. Another stellar correlation is suggestive of the latter systematic change towards a more top-heavy IMF, namely the Mg/Fe ratio enhancement with decreasing distance from the elliptical galaxy nucleus.

An increase in remnant fraction also implies a higher supernova rate and thereby enhances the prospect of an early galactic wind. The correlation of Mg abundance with local escape velocity in ellipticals suggests that a wind may indeed have systematically depleted enriched gas from lower luminosity spheroids. Indirect evidence in support of this hypothesis comes from the heavy element abundances in the intracluster medium of rich galaxy clusters. The intracluster medium is a dumping ground for wind ejecta. The iron content of the intracluster gas exceeds that in the galaxies and is proportional to the stellar mass in E's and S0's. Moreover there is evidence for enhanced Mg and Si relative to Fe, indicative of a Type II supernova yield. If confirmed, this would provide unambiguous evidence for an early IMF biased towards massive stars.

THE HIGH REDSHIFT UNIVERSE

Distant galaxies provide a glimpse of star formation at early stages of galactic evolution. Spectroscopy of high redshift galaxies has made a major advance via detection of the redshifted Lyman break in deep CFH telescope and HST images to identify candidates at $z = 3 - 4$. These have been confirmed to $I = 24$ with Keck 10m spectroscopy. At present, more than 60 galaxies have confirmed redshifts above 3. Many of these have star formation rates that are similar to those of the current epoch counterparts. The spectra of star-forming galaxies at redshift 3 are indistinguishable from those of nearby late-type spirals. HST imaging reveals a mix of morphologies, ranging from ongoing mergers and irregulars to normal Hubble types.

One can systematically study the universe to greater depths via number counts and morphological studies in deep HST imaging. There seems little doubt that the galaxy counts at the faintest magnitudes (to $I = 27$) are dominated by irregulars. These must be dwarfs rather than giants, according to measured redshift distributions at slightly brighter magnitudes.

Several hypotheses exist as to the origin and nature of these dwarfs. Most conservatively, the distant dwarfs could be a low luminosity upturn in the local luminosity function, where the data on nearby galaxy samples is incomplete. The dwarfs could be a population that formed at high redshift and subsequently disrupted after undergoing starbursts. These hypotheses fall under the rubric of luminosity evolution.

An alternative is number evolution: the distant dwarfs could be a population that subsequently merges to form luminous galaxies. A more extreme version is the possibility that the deep counts are dominated not by whole galaxies, but by giant HII regions within large galaxies. As viewed through the Hubble telescope, many of the compact, barely resolved objects may be parts of larger, lower surface brightness galaxies that are not visible even in deep HST images.

It is difficult to definitively dismiss any of these hypotheses. If in addition we introduce the IMF parametrisation as adjustable with redshift, model discriminants are ever more elusive. Perhaps disks were smaller in the past, as suggested by chemical evolution and star formation models of our galaxy. This could provide an explanation of why the Hubble deep field is dominated by small objects: there seem to be a paucity of large disks in the distant universe. However dynamical modes of secular instability of disks that lead to bar and bulge formation do suggest that disks should be present at the earliest stages of galaxy evolution. The presence of dust might help resolve this apparent discrepancy by helping hide the most distant disks.

Luminosity evolution is measured for disks, or at least for the star-forming disk candidates in the deep redshift surveys. One finds that disks are about 1 magnitude more luminous at redshift unity relative to the present epoch. Ellipticals, or more precisely, their red counterparts, show little evidence of luminosity evolution to $z = 1$. The observed net star formation rate per unit volume appears to rise by about a factor of 10 back to $z = 1$ and is then approximately constant to $z \approx 3$. This is suggestive of the star formation history of disks rather than of ellipticals. Disk formation appears to be occurring, in the sense that the inferred cumulative star formation as well as the measured rate of star formation is consistent with disk models.

However elliptical formation remains a mystery. One does not see the optically luminous counterparts predicted by theory at high redshift. Are the ellipticals forming in highly obscured starbursts at $z = 2 - 3$? Weak evidence in support of the dust-shrouded starburst interpretation of elliptical formation comes from the recent measurement of excess COBE FIRAS residuals that, when combined with a zodiacal light model, yields a tentative detection of a diffuse far infrared background amounting to about $\nu i_\nu \approx 3\,\mathrm{nwm}^{-2}\,\mathrm{sr}^{-1}$ at 400μ. The predicted diffuse background light from forming galaxies can be normalized to metallicity production and amounts to

$$\nu i_\nu = \epsilon_{nuc} Z \rho_0 \frac{c}{4\pi} (1 + z_f)^{-1},$$

where ϵ_{nuc} is the nuclear energy release per gram, Z is the metallicity, ρ_0 is the density in stars, and z_f is the formation epoch. For galaxy disks, one finds $\nu i_\nu \approx 1\,\mathrm{nw\,m}^{-2}\,\mathrm{sr}^{-1}$ at optical wavelengths. For galaxy spheroids, with metallicity normalized to the yields required for the intracluster gas, one finds $\nu i_\nu = 3 - 10\,\mathrm{nw\,m}^{-2}\,\mathrm{sr}-1$. If the associated starbursts are dust-shrouded, most of this emission is expected to peak in the far infrared, at $50 - 100\mu$. Perhaps one is viewing in the diffuse far infrared background the integrated contribution from the elusive forming ellipticals.

UNRESOLVED ISSUES IN STAR FORMATION

I will conclude with a series of questions that I expect will be answered by the next Maryland meeting on star formation. Of course, the answers will in turn generate far more questions, but these may safely be left to the next generation of astronomers to resolve.
- What processes determine the IMF?
- How is the core mass function related to IMF?
- Why is star formation inefficient on GMC scales?
- What was the primordial IMF?
- What is the IMF in extreme starbursts?

- What is the relation of individual SF regions to global processes determining global SFR, SFE?
- What triggers starbursts? What is the role of environment?
- Are starbursts nearby analogs of the galaxy formation process?
- How do starbursts evolve? What is the nature of the host galaxy?
- Do disks grow? Are bulges old?
- Do mergers dominate at early epochs?
- What is the role of AGN and quasars in early galaxy formation?

ACKNOWLEDGEMENTS

I would like to thank Daniela Calzetti, Annette Ferguson, Harry Ferguson, Lynne Hillenbrand, James Lowenthal, Crystal Martin and Rosemary Wyse for allowing me to mention some of their recent results. My research is supported in part by NASA and NSF.

TABLE 1. RANGE OF STAR FORMATION SCALES

	M_{gas} $M_\odot$	R (k)pc	n cm^{-3}	SFE	SFR $M_\odot$/y	number (density)
molecular cores	1 – 10	0.1 pc	10^4	50%/10^7yr	10–6	105
dark clouds	10^3	10	10^3	10%/10^7yr	10^{-5}	10^4
giant molecular clouds	$10^{5.5}$	$10^{1.5}$	$10^{0.7}$	1%/10^7yr	3×10^{-4}	$10^{3.5}$
Milky Way	4×10^9	10 kpc	1	1%/10^8yr	10	10^{-2}Mpc^{-3}
LSB galaxy (giant)	10^{11}	$10^{1.5}$	0.1	0.1%/10^8yr	1	10^{-5}
starburst	10^{10}	$10^{0.7}$	10	30%/10^8yr	10^2	10^{-4}
ultraluminous starburst	10^{11}	1	10^3	50%/10^8yr	10^3	10^{-6}
disk galaxy formation	10^{11}	10	1	1%/10^8yr	10	10^{-2} (comoving) at z ~ 1
elliptical galaxy formation	10^{11}	10	10^{-2}	30%/10^8yr	$10^{2.5}$	10^{-3} at z ~ 3

Fourth Day Of Creation: The Search For A History Of Star Formation

Virginia Trimble

Astronomy Department, University of Maryland, College Park, MD 20742 and Physics Department, University of California, Irvine, CA 92697-4575

Abstract. The history of how our present understanding of star formation was achieved appears not to have been previously investigated at any length. We begin here with the ancients; note the difficulty with which the community accepted the existence of widespread interstellar material; digress briefly on the origins of the solar system; explore the dark ages in which star formation was not regarded as a part of serious astronomy, followed by gradual redawning light; and end with a few notes on the first appearances of some ideas now regarded as important.

INTRODUCTION: THE ANCIENTS FROM CHEOPS AND HERSCHEL TO RITTER AND RUSSELL

Vaya'ash Elohim et shney hamorot hag'dolim.... And God made two great lights, the greater light to rule the day and the lesser light to rule the night. V'et hakochabim. And also the stars... Vayechi erev, vayechi boker, yom r'vi-i. And there was evening, and there was morning. A fourth day. There are three things to notice in this scenario for star formation. First, all the stars will be the same age. Second, star formation is not really a separate process, but is more or less incidental to the creation of the universe and the formation of the solar system. We will encounter both of these again. Third, stars did not appear until the fourth day, and since this was a three day conference, we cannot expect to have solved all the problems. When I started to think seriously about this presentation, in July or thereabouts, I began by asking for advice from colleagues involved more directly with history of science. The most succinct and eventually helpful answer came from Owen Gingerich. He said, "There is no history of star formation." I feared briefly that he meant the subject had no intellectual underpinnings. But what he actually meant was that nobody had written about it at length. This is, therefore, a pioneering effort. Thus I am not sure I have hold of all the right ideas (and additional information and insights would be very welcome). But I am sure I have hold of too many ideas. Thus the oral presentation simply stopped. Quite suddenly. And this one may too. In contrast to the Hebraic view, the Egyptians and Greeks seem to have thought of star formation as an ongoing process, where minor gods and important people (including Pharoah) might take their places among the stars from time to time. Numerous ancient and medieval philosophers speculated on the nature of the stars and possible analogies with the sun. Giordano Bruno, shortly before 1600, was apparently the last person for whom the statement "stars are suns" was a dangerous one to make. Applications of early telescopes by Galileo and others led to the resolution of many fuzzy patches on the sky (nebulae) into stars and to the widespread opinion that all nebulae could be so resolved with sufficient telescopic power. Later, Newton put his considerable scientific weight behind the view that attenuation of starlight on its way to us was always unimportant. William Herschel, in his last three decades, 1784-1814, concluded that the sorts of nebulae that he had been studying would gradually transform themselves into stars. He suggested a sequence from planetary nebulae, to bright nebulae (like Orion), to nebulous stars and clusters (like the Pleiades), with drawings of real objects to illustrate each stage (Hoskin 1963, Berry 1898, Herschel 1784). Herschel's sequence implied a picture of star formation and evolution in which: "Young stars were the hottest and therefore the bluest [the ones seen around the nebulae]. As they cooled and contracted, they became middle age yellow stars like the sun. Still cooler in old age, they reddened. Gradually their light became redder and feebler until they ceased to shine." (from Newcomb and Baker, 1932, describing the situation "until recent time.") Curiously, this picture of stellar evolution lingered in most minds and in the literature long after it had ceased to make sense. Thus we find Astronomer Royal David Gill pondering in 1884: "The nebulae - what are they? Are they, too, condensing into clusters or stars, or will their ghost-like forms remain for ever unchanged among the stars? or do they play some part in the scheme of nature of which we have as yet no conception." And, in 1885, Newcomb and Holden: "We are thus led to the general conclusions that, so far as our knowledge extends, nearly all the bodies of the universe are hot, and are cooling off by radiating their heat into space." They go on to speculate about, and reject, regeneration processes, such as might be caused by accretion (for instance of the planets into the sun), and in this were wiser than a later generation. Simon Newcomb is generally said to have been Walt Whitman's "learned astronomer". That most of us find the stars more beautiful, not less, for knowing how they shine is why we are astronomers and not poets!

The first half of the nineteenth century saw conservation of energy established as a firm principle, requiring an energy source for the stars. Helmholtz (1853) and Kelvin (1862) are universally credited with the idea of contraction under self-gravity as the main energy source for the sun and stars (though agreement on the proper reference is less universal!). The corresponding picture of star formation then begins with "meteoric material" that condenses, heats, vaporizes, and, when it can contract no more, radiates away stored heat, reddening and fading as it goes. S.C. Vogel's 1874 classification of stellar spectra implies such a scenario, and it is more explicitly presented by Ritter (1882), and, especially, Norman Lockyer (1887, 1888, perhaps best known as the discoverer of helium in the sun and the founder and first editor of Nature). Lockyer's paper includes a drawing that looks like water spraying up and out of a hose from right to left, but is meant to be viewed left to right, "On one arm of this we have those stages in the various heavenly bodies in which in each case the temperature is increasing, while on the other arm we have that other condition in which we get first vaporous combination, and then ultimately the formation of a crust due to the gradual cooling of the mass, in dark bodies like, say, the companion of Sirius."

The year 1913 appears to have been something of a watershed, and in the same volume of Observatory, we find Eddington (1913) describing B stars as young, M stars as old, and evolution as a cooling process, while Russell (1913) is enunciating what became known as his "giant and dwarf" theory, essentially the equivalent of Lockyer's picture, in which cool nebular material contracts and heats across the top of the HR diagram (giants), and then stars gradually cool off, moving diagonally downward on the main sequence (dwarfs). The phrases "early" and "late" type stars for hot and cool are left from these early pictures of stellar evolution. By about 1923, belief in the giant and dwarf theory was nearly universal. Eddington (1923) noted correctly that the largest mass stars would get hottest, and Shapley (1923) believed (incorrectly) that the largest masses evolved most slowly and that open clusters were globular clusters that had been disrupted by passage through the galactic plane. It is worth emphasizing that, throughout this period, star formation was regarded as an on-going process, not very well understood (though perhaps connected with spiral nebulae among others), but worthy of the attention of a serious astronomer. This is clear, for instance, in the Newcomb and Holden (1885) volume, in Young's (1904) astronomy text, in the first (1926) edition of Russell, Dugan, and Stewart, in Ball's (1902) Royal Institute lecture, and so forth.

THE GRUDGING ACCEPTANCE OF AN INTERSTELLAR MEDIUM

The view espoused by Galileo, Newton, and others that all nebulae could be resolved into stars held sway for a very long time. For instance, Lord Rosse's original discussion of the Crab Nebula regarded it as a fuzzy patch of imperfectly separated star light, and Agnes M. Clerke, the summarizer of the 19th century, opposed any form of diffuse material until she realized that accepting it would spare her having to consider the ridiculous idea of external galaxies. But in August 1864, William Huggins (1865) put the slit of his spectroscope across "a small but comparatively bright nebula 37 H. iv," and was most surprised to find three "bright, bluish-green lines, separated by dark intervals", indicating the presence of "a luminous gas." He attributed the lines to nitrogen, hydrogen, and an unknown source. (I have not checked whether he was right about N and H, but the unknown line was surely [OII] at 5007 Å, later attributed to nebulium). He was also very firm in saying that this luminous gas (which he later also found in Orion, the dumb-bell nebula, the "annular" nebula in Lyra, and a number of others catalogued by Herschel and Struve) was not Herschel's "nebulous fluid" from which stars could condense. A firm disciple of Kirchhof and Bunsen, Huggins believed that every line of every element seen in absorption in stellar spectra should appear as an emission line in gas able to make stars. That the emitting gas is mostly hydrogen does not seem to have been fully realized until the work of Struve and Elvey (1938) on diffuse H-alpha emission. They derived a ratio by number of H/Ca=100, which is still much too small by modern standards, but did not seem so far out of line at a time when stars were, perhaps, 10-40% hydrogen (Eddington 1923a, Stromgren 1937). In retrospect, we can say that diffuse absorbing gas was discovered in 1904 when Hartmann (1904) described spectra of δ Ori, saying: "The calcium line at λ 3934 does not share the periodic displacement of the lines caused by the orbital motion of the star." These stationary lines had also been reported in the hydrogen spectrum of Nova Persei 1901. Hartmann suggested that the gas responsible might have "some relation to nebular masses seen by Barnard," but also admitted the possibility that the gas might be closely associated with the stars in whose spectra the lines appeared. This remained the majority opinion for a couple of decades, partly because stationary lines were, until quite late, seen only in the light of stars earlier than B3 (this is a matter of crowding and the extent technology). An important argument against "associated lines" (as the quasar people say) was that the line velocities, corrected for solar motion, were very close to the local standard of rest in Hartmann's two examples and, soon after, other stars (Plaskett 1924 and references therein). Slipher (1909) opined that stationary lines were probably interstellar, and by 1926, nearly everybody was sure of this (Eddington 1926). An important confirmation was that the strengths of the lines tended to be proportional to our distances from the stars concerned (i.e. inversely proportional to proper motions, Gerasimovich and Struve 1929). The Hartmann (1904) paper is a translation from a German original prepared by the editors of Astrophysical Journal (a service no longer available, as a quick scan of any recent issue will persuade you). To modern eyes, one of the most interesting aspects of the issue of ApJ in which the "stationary lines" paper appears is the advertizements. Modern scientific journals normally ban ads, or allow only books, telescopes, statistical software, and the like. But ApJ in 1904 featured (a) Pears' Soap ("beau-begging beauty bubbles from..." indicating some uncertainty about the gender of their readership), (b) Vose pianos and Weber pianos, (c) Baker's Breakfast Cocoa (less than one cent a cup), (d) Sapolio hand Kosher soap, recommended to "Gentiles as well as Jews", and (e) most spectacular of all, Buffalo Lithia water, endorsed by a couple of dozen doctors and professors as of use in Bright's disease, gout, renal calculi, rheumatism, and so forth. We wonder whether readers then were more or less likely to believe everything they read in ApJ than readers now. Interstellar matter that obscures our view of the stars behind was also seen early and recognized late. Herschel's first statement (Hoskin 1963) was, "Mein Gott, da ist ein Loch in Himmel" (a hole in the heavens), and the opinion that the coalsack and other dark regions had few or no stars lasted a long time. Real obscuration was suggested repeatedly, by Angelo Secchi in 1853, by A.C. Ranford (1894, not an astronomer), by Max Wolf (1908), and H.D. Curtis (1918) among others. Curtis said that one should study spectra of the stars embedded in the dark nebulosity to learn about it (a very Curtis-like remark). E.E. Barnard was, in some sense, the last to favor true obscuration (because thereafter everybody believed in discrete opaque clouds). He began photographing the Milky Way in 1904 and, 15 years later, said: "I did not at first believe in these dark obscuring masses. The proof was not conclusive. The increase of evidence, however, from my own photographs convinced me later, especially after investigating some of them visually, that many of these markings were not simply due to an actual want of stars, but were really obscuring bodies nearer to us than the distant stars. In this way it has "fallen to my lot to prove this fact." (Barnard 1919, writing in a style no longer fashionable in the main archival journals). He expressed uncertainty about whether the obscurers were gaseous or something else. Looking ahead for a moment, we find Bart Bok and Edith Faith Reilly pointing to some of the more compact of Barnard's dark features as protostars (Bok and Reilly 1947). We now call them Bok globules, and many are indeed known to have infrared-emitting young stellar objects at their cores. As late as Lynds' (1967) article in the Kuiper compendium, it was not certain what the masses of the globules were by more than a factor of 10, because there was no way to be sure whether gas was as over-dense in them as dust. In the same time frame, Minkowski (1949) finally enunciated the idea that absorption and emission nebulae must often be part of a united complex of gas and dust. Why Bok globulaes and not Bok-Reilly globules? Perhaps because, apart from some observations of meteors, comets, and variable stars, this was Reilly's only astronomical paper, and she did not proceed to a PhD, in turn perhaps partly because a physical handicap that interfered both with the clarity of her speech and with walking and manual dexterity (Hoffleit 1996). Meanwhile, Hubble (1922) defined a category of reflection nebulae whose spectra were essentially those of stars in or near them, and Russell (1922) pointed out that these must be made of finely divided dust to produce the observed effect (though he expressed very limited enthusiasm for such quantities of "meteoric material"). The existence of widespread, diffuse interstellar absorption was explicitly denied by pundits from Newton (Hoskin 1982) to Shapley (1918) (though advocated by King 1914 and a few others), with unfortunate consequences for the establishment of distance scales within and outside the Milky Way (see Fernie 1969 and Trimble 1996 for various aspects of the story). The official culture hero here is, of course, Robert Trumpler (1930), whose study of the apparent magnitudes vs. angular diameters of open clusters persuaded him and, soon, everybody else, that the clusters were being systematically dimmed by about 1^m/kpc in the galactic plane as well as by their distances. Molecular interstellar gas comes next, beginning with optical observations of unidentified lines in the period 1937-41 and their gradual recognition as belonging to CH and CN (Dunham 1937, Swings and Rosenfeld 1937, McKellar 1940, 1941, Adams 1941, Douglas and Herzberg 1941). Because the CN lines came from two different lower levels, an excitation temperature could be measured. It was just a little less than 3K, a fact which has (so says Gerhard Herzberg 1951 near the end of his classic treatise on the spectra of diatomic molecules) "only a very restricted meaning."

THE ISM THICKENS

Notice that, through the 1920's, 30's, and 40's, though the existence of several components of interstellar material was generally recognized, it was all pretty tenuous, spotty stuff. There were Huggins's bright and Barnard's dark nebulae in a few isolated regions. The stationary lines, both of atomic Ca, Na, etc. and of molecular CH and CN came from small, isolated clouds with small velocity dispersions, and a typical line of sight would penetrate only a handful per kiloparsec. Trumpler's dust might be quite widespread, but it was clearly very tenuous indeed, with a few dozen grains per cubic kiloparsec. There was, in other words, very little evidence of massive, gravitationally bound clouds of gas and dust, waiting eagerly to collapse into stars. The discovery of 21 cm emission by neutral hydrogen changed the situation substantially. The expected wavelength and line intensity had been predicted by van de Hulst (1945) and Shklovski (1949), and the line was seen in quick succession by observers in the US (Ewen and Purcell 1951), The Netherlands (Muller and Oort 1951), and Australia (Pawsey 1951). These three papers appear in that order in a single issue of Nature. In contrast to the spottiness of stationary lines, bright, and dark nebulae, 21 cm emission was, literally, all over the sky. And it quickly became obvious that neutral gas outweighed ionized gas by a factor of 50-100 and comprised at least 5% of the mass of the Milky Way (Kerr and Westerhout 1965). Here at last was a worthy candidate for the raw material of ongoing star formation. The only trouble was that it didn't seem to be concentrated toward those promising Bok globules (remarked upon by G. de Vaucouleurs in 1957 and undoubtedly many others). The final, missing, phase was, of course molecular hydrogen. Van de Hulst (1948) had said it could form on grains, but, like all the other phases, it seems to have had to endure a period of, at best, marginal respectability. Though Spitzer (1949) thought it might be quite abundant and came back to this view, at least where Bok globules were concerned in his Compendium review (Spitzer 1968), two of the more serious discussions came from Gold (1961) and Zwicky (1959). They have in mind molecular hydrogen as a significant dark matter contributor in the Milky Way ("Oort limit"), clusters of galaxies, or both. Zwicky indicated that he was tooling up to look for the 85μ ortho/para transition, apparently not realizing the extent to which the earth's atmosphere does not want us to do 85μ spectroscopy from the ground! In the same time frame, Bok (1959), also a supporter of H_2 as dark matter, suggested that not being able to detect it was the most significant barrier to understanding the interstellar medium. Like any molecule made of two identical atoms, H_2 has no strong, low-lying lines that are likely to be conspicuous in emission or absorption if the gas is cool (as it clearly had to be, since the HI was checking in at 100-125K). Thus arose the need for tracer molecules, whose lines would be strong enough to see and whose chemistry might be simple enough to reason back from their abundances to that of molecular hydrogen. First Shklovski (1953) and then Townes (1955) recommended the 18 cm, lambda doubling lines of OH as the best bet. Townes' accurate frequency predictions led directly to the detection of interstellar OH, first in absorption against Cas A (Weinreb et al. 1963) and then in emission (Weaver et al. 1965). For better or for worse, OH emission, from the very beginning, displayed the peculiar polarization properties and anomalous ratios of line intensities that are the signatures of its being mased. The numbers of photons you see in its lines are, therefore, a very poor indicator of the amount of gas responsible. And so astronomy waited another five years, for the discovery of widespread CO emission (Wilson et al. 1970), before we had the right tracer. The correct conversion factor from CO line strength to total mass of H_2 and how it varies with conditions in the gas remain under dispute to this day (Trimble and Leonard 1996), but, at long last, astronomers could be sure that there were massive, fairly dense, bound clouds of gas and dust out there ready and willing to form stars. Other components of the interstellar medium that presumably have some bearing on star formation include cosmic rays and the magnetic field. Just when cosmic rays joined the inventory depends on whether you want one person to understand them, most of the community, or all knowledgeable scientists. By this last criterion, the answer is 1994, when Hannes Alfven died (but, of course, he was also among the first to put them in the galactic inventory!). Alfven (1937) was also an early exponent of a galactic magnetic field, whose function was to confine these same cosmic rays. Spitzer (1946) first predicted a field of 10^{-6-5}G, but, within the same 600 word letter, backed off to 10^{-12}G. Fermi (e.g. 1949) argued that one needs the larger value to accelerate cosmic rays as well as confine them. Davis and Greenstein (1951) proposed the same number for the purpose of aligning interstellar grains to polarize starlight (Hall 1949, Hiltner 1949). The field briefly climbed to a few $\times 10^{-5}$G, sufficient to dominate dynamics of the interstellar medium (Woltjer 1965) before direct measurements (Zeeman broadening of radio lines and rotation and dispersion measures of pulsars) pulled it back down to something $\times 10^{-6}$G. Virtually all modern theorists of star formation agree that the magnetic field is important (good vs. bad is another matter). But the cosmic rays seem to have been put there just to make everything more complicated.

A BRIEF DIGRESSION ON FORMATION OF THE SOLAR SYSTEM

Theories of formation of the solar system have a much longer and more distinguished history than those of star formation and have been written about at much greater length (e.g. Brush 1978, 1990, 1996, and many other people and places), frequently under the subject heading "cosmogony". Very crudely, one can distinguished two streams of ideas. In one, often called the nebular hypothesis, the sun and planets (etc.) form together, more or less at the same time, out of a single rotating cloud of stuff (mostly gas we would now say, but this was not obvious when the sun was believed to be made of the same things the earth is). The official propounders were Swedenborg, Kant, and, especially, Laplace (who met Herschel in Paris in 1802, Lubbock 1933). The modern picture of star and planet formation (e.g. Shu et al. 1987) is a remote descendent of the nebular scenario, though squabbling over details has obscured this from time to time. The primary alternative view of solar system formation was tidal expulsion, first put forward by "the celebrated Buffon" (1707-1788, and somewhere in the writings of G.B. Shaw is a story about confusion between the two of them, the latter as "the celebrated Buffoon"). Buffon's version was a comet hitting the sun and kicking out the planets. The early 20th century version of tidal expulsion is closely associated with the names of Thomas Crowder Chamberlain (1863-1928) and Forest Ray Moulton (1872-1952), who envisaged a close encounter between two stars. A late variant, due to Raymond Arthur Lyttleton in about 1938, invoked a single star plus a binary system in an effort to overcome various kinematic and thermal difficulties in the basic Chamberlain-Moulton hypothesis. It is hard from our vantage point to realize just how thoroughly dominant tidal expulsion was at one time. Here, for instance, is Struve (1932) reviewing Moulton's monograph, Astronomy: "The more advanced reader will derive particular satisfaction from chapter xiv, "The Evolution of the Solar System," which in large measure rests on Dr. Moulton's own work. It will be recalled in this connection that in 1899 he proved the Laplacian hypothesis to be untenable–a fact which is now universally recognized." And some discussion among characters in Dorothy Sayers' novels from the 1930's illustrates that tidal expulsion was the accepted explanation of the solar system among educated laymen as well as among professional astronomers. In a general sort of way, one expects star formation to be an ongoing process in the nebular case, while tidal expulsion fits more comfortable in a picture where all stars formed on Jeans' "long" time scale (of which more shortly), and a few experienced encounters when they were already mature. Both camps made use of spiral nebulae as part of their scenarios at one time or another. Ball (1902) and Russell (1913, 1925) are among those who mention spirals as pre-solar-systems in a nebular context. In contrast, Lowell (as in Lowell Observatory), the polymathic Svante Arrhenius, and the creative spirit T.J.J. See saw spirals as the transition stage between tidal expulsion and condensation of planets. It was in this spirit that Lowell set V.M. Slipher to taking spectra of spirals, and only gradually did Slipher come to believe that he was in the process of demonstrating the existence of external galaxies rather than studying the formation of solar systems. Jeans (1928) at one point described spiral nebulae as new material pouring into the observable universe from elsewhere, an idea which has later echoes in the work of Ambartsumian and others on active galactic nuclei.

DARK AGES AND CONFUSION

The period from about 1935 to 1945 is a very important one in this subject because of a sudden epidemic of disbelief in star formation as a continuing process. It is also, unfortunately, a difficult period to investigate systematically, because the advent of WWII meant that many books and journals did not reach the German astronomers who had been compiling annual, very complete abstract volumes since 1899 (found on your shelves with the word Jahresbericht conspicuous), and one volume, covering 1944-45, seems never to have been distributed. This section is correspondingly incomplete and in need of amplification. At the beginning of the period, the decision had not yet completely been made between the "long" and "short" time scales for cosmic evolution. Sir James Jeans (from about 1904 until very shortly before his death; see any of many editions of his Astronomy and Cosmogony) had long advocated a universe that began in some singular event more than 10^{12} years in the past. He had concluded that this much time was needed for an assortment of dynamical processes, including the relaxation of elliptical galaxies, clusters of galaxies, and star clusters, and the modification of binary star orbits by encounters, to bring systems into the conditions we see. (His binaries all began in circular orbits, and the wide ones were gradually eccentricised by passing near other stars, a slow and painful process indeed.) He accepted as a necessary corollary that stars must live on annihilation (presumably of electrons and protons) as their primary energy source, gradually losing mass and moving down the main sequence (a la Russell's giant and dwarf theory) over 10^{12-13} years. The competing, "short" time scale came from measurements of the ages of earth rocks (from decay of uranium and thorium to lead) and carried as a corollary the fueling of stars by some sort of nuclear transformation, which would release only about 1% as much energy as complete annihilation. Eddington's 1926 book, Internal Constitution of the Stars is a standard landmark in establishing the latter view. The balance began to shift decisively in favor of the "short" time scale with Hubble's announcement of a velocity-distance relation for external galaxies (Hubble 1929) that implied a time scale of about 2 Gyr for the age of the universe as a whole. Soon after, new considerations of various dynamical processes suggested that the short time scale was also the right one for spiral galaxies (McVittie 1932), the rotation of the Milky Way (Eddington 1931), and pairs and clusters of stars (Bok 1936, Mineur 1939, Chandrasekhar 1942ff). A 1935 conference, whose proceedings appear in that year's volume of PASP (Epstein 1935 and surrounding papers) showed a clear majority of participants firmly on the side of the short time scale. Unfortunately, it was also a short step from saying that most things seemed to be about 2 billion years old to saying that everything was about two billion years old and that it had all formed from dense, chaotic conditions that no longer exist and hardly deserve study. The quotations that follow are intended to give some feel for the intellectual climate that resulted. The collection is very incomplete and strongly biased toward the British and American literature. I am aware of, but have not been able to get hold of, relevant papers by Bertil Lindblad and G.C. Armellini (who wrote primarily in Italian and who was the only astronomer at the 1957 Vatican conference on stellar populations whose name is not instantly recognizable by most astronomers today) and of papers by Severny and Pariisky that I could not read if I had them! And there are surely others of which I am not even aware and about which information would be very much appreciated. "The trend of thought in recent years has been favorable to the "short" time scale suggested by the Expansion; but practically everyone is uncomfortable about its brevity. It does not seem sufficiently dignified that the uncompromisingly majestic universe measure its duration as scarcely greater than the age of the oldest rocks on this small planet's surface or the age of life in the crannies of the rocks." (Shapley 1943). "The combined evidence rather suggests that the stars, which constitute the observed clusters did not exist as stars until the clusters came into being. The theory of the Expanding Universe indicates that a "catastrophe" took place 3×10^9 years ago, and it is tempting to place the origin of stars and stellar systems tentatively at the epoch of this catastrophe." (Bok 1936) "At a certain time, not many billions of years ago, conditions in the universe were so crowded and presumably so different from those which now prevail, that our backward reckoning cannot safely be extended further. It is tempting to look upon this as a time of turbulence, in which various things happened which are very difficult to account for otherwise–such as the origin of the solar system and of double stars" (Russell et al. 1938). "Bright stars rely on a much more generous source [of energy] than synthesis....It is possible to get around it by supposing that heavy stars (and presumably also light ones) are constantly being made anew from some diffuse material that consists of nearly pure hydrogen; but the idea is not very attractive." (Atkinson 1936, followed by a detailed explanation of what is unattractive). "Might it not be possible that other Hyades or Pleiades might be formed?... it seems very unlikely...It is rather satisfactory to blame it all on conditions as they were three billion years ago, at a time when, according to the hypothesis of the expanding universe, all matter in the physical universe was packed much closer together than it is at present....But, you may well ask, why should we not admit the possibility that stars are still being formed?....There are, however, several reasons why most astronomers are reluctant to take this view." (Bok 1945) "But great difficulties remain for the red giants and the highly luminous blue stars. We do not know what keeps these celestial power houses running. We have yet to learn whether they are recent creations or whether they have existed in some prestellar form since the beginning of time, or operate on unknown principles." (Goldberg and Aller 1943)

CURIOUS CONNECTIONS

Three sorts of ideas arose during the same period, each of which now sounds distinctly odd (as well as wrong). The vociferousness of the contemporaneous objections suggest that they were at the time (especially the idea of significant accretion) somehow perceived as threatening. First was an assortment of unified hypotheses for the combined formation of stars and galaxies, some of which survived well past the time when "everything the same age" was still a possible view. Lindblad (1934) put forward such a scenario, as did Stromberg (1934,1935) and von Weizsacker (1937 to 1951). The last exponent of such unified pictures of star formation appears to have been Layzer (1964 and references therein). The second idea is a sort of inverted picture of stellar evolution in which superdense, prestellar stuff comes first and develops into stars which then expel various sorts of nebulae. Part of the idea goes back at least to 1737, when a chap named Jean-Jacques Dortous de Mairan took the solar corona (extensions of which he thought hit the earth's upper atmosphere, causing aurorae – true, or near enough) as a general model and attributed all nebulous material to streams emanating from stars. A.A. Belopolsky (who attended at least the 11th meeting of the AAS since he appears in the conference photo) seems to have envisaged a sort of balance between attractive forces of gravitation and repulsive forces (like radiation pressure) in the universe. The late Viktor Ambartsumian was apparently much influenced by him (Ozernoi 1996) and this is perhaps part of the origin of Ambartsumian's (1938, 1953, 1960 and beyond) conviction that star clusters, and, later, spiral arms, clusters of galaxies, and active galaxies form from dense, precursor material and are in a state of expansion. There seems to have been a sort of polarization of opinion in the USSR between viewing all nebulae as the precursors of stars and all nebulae as the products of stellar expulsion, and one is surprised to find Shklovsky (in his popular treatment, Stars: Their Birth, Life, and Death) expending a good many words to convince the reader that planetary nebulae mark death not birth. Others mixed up two or more of the ideas just mentioned. Waterfield (1938), for instance, sketches out a picture in which star formation is part and parcel of galaxy formation, but all nebulae are, in turn, expelled from stars, and the hottest stars therefore necessarily the most evolved. The idea of dense, prestellar matter had one very attractive feature. It permitted the synthesis of the heaviest elements. Gamow and his co-workers, of course, located the necessary hot, dense stuff in the early universe, but Chandrasekhar and Henrichs (1942) appear to have had prestellar, but post-big-bang material in mind. Hoyle, even before the advent of steady state, was the earliest exponent of late, rather than early, phases of stellar evolution as a solution to the problem. Third, and at least as surprising as the other two, was the proposal that all stars had been born a few billion years ago, with masses small enough that "nuclear transformations" would fuel them down to the present time, but then a few had wandered into nebulae, where accretion of ambient material turned them into short-lived OB stars and supergiants. There is a hint of this in Eddington (1926), who remarks that accretion is most significant in nebulae and at small relative velocity, and you find B stars with these characteristics because that is where they can "grow". The main advocates of accretion as an alternative to recent star formation were, however, Hoyle and Lyttleton (1939,1940,1941, and beyond). They considered the effects of massive accretion on the orbits of binary stars as well as on the lives of singletons, and supposed that we might be seeing very tenuous accretion on to the sun in the form of its corona. W.H. McCrea considered both star formation and accretion in the same time frame and concluded that "accretion has at least the merit of being known to occur." You may or may not be surprised to hear that some of the Hoyle and Lyttleton papers are still occasionally cited today. This is not because anybody (or at least anybody I know) still either believes that accretion on main sequence stars is important or feels the need to refute it, but because in the process they derived the formula for accretion by a moving point mass that scales as v^{-3}, a formula we all use in a variety of quite different contexts. A last, still unclear, word comes from Greenstein (1951), who considers first accretion and then condensation of new stars and concludes "Therefore only in regions of unusual quiescence can a star condense. Favorable conditions may occur, but, like the accretion hypothesis, the condensation hypothesis now seems suitable only for creation of a few exceptional objects rather than all the stars." But, rather than telling us next where most of the stars actually come from, he moves on immediately to interstellar polarization.

THE REVIVAL OF STAR FORMATION

Even as the lights of peace were going out all over Europe, astronomical light began to dawn again. Two of the swords that had cut down Jeans' "long" time scale still had some bite to their edges. The first was dynamical time scales for star clusters. Most indeed have held together for the few billion years needed by the "all together boys" hypothesis (Spitzer 1940,e.g.),but some OB associations (starting with the Trapezium in Orion) would surely fall apart in only a few million years (Ambartsumian 1938). Similarly, as work led up to Bethe's (1939) synthesis of our understanding of hydrogen burning in stars, it became clear that a few billion years was a perfectly reasonable age for stars of, say, 0.5 - 2.0 $M_\odot$, but that very massive ones could live on hydrogen fusion or similar reaction for only millions of years (Gamow 1938). Bethe (1939) said firmly that the B stars must have been born comparatively recently, "by what process we cannot say." Russell (1939) was not prepared to go quite that far, and said of stars like Gamma Cygni that they "have begun to shine rather late in the history of the galaxy" or that "they have still greater stores of energy to draw on." He made life more difficult than necessary for himself by continuing to consider a sun (and so presumably other stars) that was only 1/3 hydrogen by weight. German astronomers were rapidly reaching similar conclusions, with, perhaps, a bit more confidence. Unsold (1944) wrote "dass solche Sterne [OBs] heute noch fortlaufend neu gebildet werden," though the choice of "gebildet" still sounds like a slight weasel. Von Weiszacker (1947) laid down a set of conditions for recognizing new stars that included rapid rotation ("rotierende"), location in spiral arms ("Spiralarme"), location near the galactic plane ("galaktische Ebene"), and a ratio of luminosity to mass exceeding 100 erg/sec/gram ("Hauptreihe"). Continuing reservations persist, however, in the direction of believing that only stars that had to have formed recently did so. For instance: "the origin of a certain class of stars which may have been created recently" in contrast to "the formation of stars in general is still a closed book, since the explosion of the universe a few billion years ago has so far defied any attempts at detailed analysis "(Spitzer 1948). And "We are thus forced to conclude that some stars may be of recent formation and even that the process of stellar birth is still going on." (Menzel 1949). The proposed site was once again the nebulae that Herschel had in mind, and Bok (1948) described the Harvard Centennial Symposium on interstellar matter as having "a central theme that in some cosmic clouds we are now witnessing the operation of the process of star formation." Struve (1949) in a review of the proceedings of the symposium is apparently still not fully convinced and writes of Joy's groups of T Tauri stars, "But perhaps they are not really 'associations,' in the sense of very wide clusters, but represent accidental groupings of stars which have drifted into the dust cloud and show similar spectroscopic features only because they happen to be subjected to the same physical influence – bombardment by interstellar grains." (We would, of course, say some similar things, but blame gas more than dust and put the source in a residual disk of material left over from the recent condensation of the stars). And (quoting Struve again), "It is, of course, tempting to search for a connection between the T Tauri stars and Bok's 'globules', but we must admit that at present there is no evidence of any objects that could be considered intermediate between the two groups." (A point that remains nearly true in so far as one has in mind protostars whose energy output is dominated by accretion.) And then in 1951, Schwarzschild, Spitzer, and Wildt (1951) step bravely forward with "The suggestion that all type I stars have been formed from the interstellar clouds may, perhaps, be taken as a working hypothesis." But "Type II stars are all old, having been formed by some different process 3 billion years ago." And the subject got what might be claimed as its first review article (Spitzer 1951). The year 1952 could reasonably be said to mark the admission of current star formation to the inventory of respectable problems in astrophysics. Blaauw (1952) studied the expanding star cluster around Zeta Persei and then the Lacertae group (Blaauw and Morgan 1953), finding them to be only 1.3 and 4.2 million years old (much as Ambartsumian had said before, but now not burdened with the weight of dense "prestellar matter"). Herbig (1952) explained in some detail why Joy's (1945,1946) variable, F5-G5, emission line stars near nebulosity (T Tauris we call them now) had to be young. An important point was their position in the HR diagram, firmly above the zero age main sequence. He returned repeatedly to the topic (e.g. Herbig 1960), gradually widening the pre-main-sequence net to take in what we now call Herbig Ae/Be stars as massive analogues of T Tauri variables and M dwarfs with emission lines as the lowest mass examples. Baade (1952), speaking at the Rome General Assembly of the International Astronomical Union, devoted a major review talk to asking the question, "Are we today in a position to point out two groups of stars, one of which in terms of cosmical time scale is young, the other which is old?" And he answered his own question:"The two groups of stars which I want to discuss are the O- and B-type stars of high luminosity as an example of young stars and the stars in globular clusters as an example of old stars." In this context, "young" meant a few million years (as per Blaauw) and "old" meant $3.4 X 10^9$ years, as per calculations of stellar evolution by Sandage and Schwarzschild (1952), who assumed CN cycle hydrogen burning, convective cores, and no other source of mixing over the full range of 1-6 $M_\odot$. And, at that same Rome IAU meeting, Hoyle (1952) was still defending accretion as an important source of massive stars. Incidentally, other astronomers referring back to this period sometimes credit Baade with the old/ young dichotomy, but attribute it to his 1944 stellar populations paper and indicate that Joy (1945) described his T Tauri stars as young. They must have read different versions of those papers from the ones that exist in our library journal collection! Finally, we close the year 1952 by looking at a non-technical book by Cecilia Payne Gaposchkin (1952, Stars in the Making), one prescient passage of which says: "This, I imagine, is why the oldest members of the stellar system (as I suppose the globular clusters to be) are so large and populous, for the available material was richer then. As the layer of dust and gas sank toward the galactic plane, stars continued to form. Dust and gas still lie dense in this layer, and stars are still being formed there." This strikes me as remarkably similar to the ideas we now associate with Eggen, Lynden-Bell, and Sandage (1962).

SOME IMPORTANT (AND UNIMPORTANT) PROCESSES IN STAR FORMATION

We are rapidly approaching the modern era, but it remains to note the first (and sometimes the last) appearance of an assortment of physical processes that have appeared in scenarios of star formation. Important ones include effects that might assemble dense clouds and various kinds of triggers to turn those clouds into stars. There seems to be no rational way to put these into purely chronological order, and I have not tried. It is, however, worth remembering that many of the earlier ones predate the discovery of neutral and molecular hydrogen, so that minds were necessarily being bent to the very difficult problem of trying to turn quite hot, quite tenuous stuff into stars (Spitzer 1941, 1942).

Radiation Pressure. Eddington (1926) was firmly of the opinion that radiation pressure directly regulated the masses of stars. What he meant was that real stars have gas and radiational pressure roughly equal at their centers (which would be true if they were made primarily of elements other than hydrogen). Radiation pressure came to look important again when Whipple (1946), Spitzer (1948a, 1948), Spitzer and Schwarzschild (1951), and Schwarzschild et al.(1951) began to ask how interstellar matter could be concentrated into clouds dense enough to form stars. The idea was that dust grains would shield each other from the general interstellar radiation field and so be pushed together. Later, when a dense enough cloud of grains had assembled, gravitation would kick in and add gas to the mix, resulting in the compositions of stars as they are actually seen. Spitzer and Schwarzschild (1951) and Schwarzschild et al. (1951) introduced the additional refinement that this was relevant only to population I stars, which therefore are metal rich because grain rich. Population II stars had been formed part and parcel with galaxies in the early universe, and so have only their fair share of heavy elements (which, remember,in those days were generally thought to be left over from cosmological or prestellar processes). Only after heavy elements had had time to condense into grains would the current mechanism of star formation kick in. Much more recently, such shielding from radiation pressure (of the cosmic background radiation) has been invoked, somewhat playfully, as a contributor to galaxy formation.

The Parker Instability. A very different way of assembling dense interstellar clouds, this is apparently truly the idea of E.N. Parker (1966) and does not favor dust over gas. Instead, part of the support of the gas layer of the galactic disk is attributed to magnetic fields (and perhaps cosmic rays). These are less dense than the gas and will, from time to time, pop loose, leaving clumps of gas behind (perhaps to be further organized and compressed by spiral density waves).

Contraction. This was, of course, the main energy source of 19th century stars, as per Helmholtz and Kelvin. Later on, the phase of contraction of a gas cloud to a star (which might then live on some other energy source) came to be seen as interesting in its own right. There are very qualitative treatments in Russell (1913, 1925) and Eddington (1926). More quantitative ones (considering also the process of fragmentation) were given by Ebert (1955), Bonner (1956), and McCrea (1957), partly in the context of condensing material in an expanding universe. That normal clouds would have too much angular moment to contract easily had been one of the reasons for doubting the reality of current star formation, and the problem of contraction of clouds with both angular momentum and magnetic fields (and whether the combination was better or worse than one by itself) was discussed by Spitzer (1948a), Mestel and Spitzer (1956), Burbidge and Burbidge (1958), and especially by Mestel (1965) and many others since. The first "modern" treatment of the contraction of an isolated cloud is generally credited to Larson (1972). A very important piece of information was missing from all early calculations of contraction and of what a pre-main-sequence star ought to look like. This is that cold, molecular gas becomes completely convective, leading to a unique equation state for the gas and unique temporal development of luminosity and effective temperature of a gas cloud of a given mass. The resulting Hayashi (1966) tracks continue to provide a reasonable fit to the locations of known pre-main-sequence stars in the HR diagram today.

The Jeans' Mass and Fragmentation. The Jeans' mass (or length) is the smallest one that will contract under its own self gravitation in a homogeneous fluid of given density and temperature. (Expansion of the substrate does not change the critical mass, but greatly slows the contraction.) Credit to Jeans, of course, customarily the 1929 edition of Astronomy and Cosmogony (Jeans 1929). A much cited paper down to the present is Hoyle (1953) on "hierarchical fragmentation," that is, the idea that as a gas mass contracts and radiates, it gets cooler and denser, and the Jeans' mass therefore gets smaller, and what was once the smallest self-gravitating body can then break into additional pieces, the process to continue until the fragments become optically thick or bored. The work of Lifshitz (1946), Ebert, Bonner, and McCrea is also relevant. Chandrasekhar and Fermi (1953) explored the effects of a uniform, oriented magnetic field, and discovered that the Jeans' instability was suppressed for a cylinder of gas representing a spiral arm if B were greater than 7 μG. A early modern treatment of fragmentation is that of Rees (1976), who concluded that stars ought to contain a number of particles equal to the 3/2 power of the Dirac large number (the ratio, e.g. of electromagnetic to gravitational force).

Coagulation. Layzer (1963) believed that hierarchically fragmenting bits would get in each others' way so much as to invalidate the entire process. That the fragments are likely at some point to start sticking back together seems plausible (Arny and Weissman 1973) and is part of many very recent calculations of how dense cloud cores become stars.

Triggered star formation. This is one of the ideas that has been in and out of fashion a number of times in the past 50 years (especially in connection with some event that might have initiated the formation of the solar system and simultaneously endowed it with short-lived radioactive nuclides). The traditional triggers are (a)supernovae (Opik 1953, who was concerned with the then-very-real problem of assembling dense gas clouds, and imagined expanding supernova remnants acting like giant brooms), revived in the radionuclide context by Cameron (1962, Cameron and Truran 1977), (b) HII regions expanding around OB associations (Oort 1954), which, like the SN case, is likely to give you a circle or shell of young, secondary stars, (c)cloud-cloud collisions (Kahn, 1953) (d)shocks in general (Dibai 1958), and (e) especially shocks associated with spiral arms in the density wave picture of spiral disks (Fujimoto, 1966, a paper apparently never published in either English or Japanese; Roberts 1969). The idea of stochastic, self-propagating star formation (Seiden and Gerola 1979, 1982) merges several of the possible triggers, so that star formation, once initiated, can go sailing across a good part of a galactic disk, getting twirled into an apparent spiral by differential rotation as it goes.

Bimodal star formation. This phrase has been used to describe a considerable range of ideas in which the stars that form under different conditions or at different times have different mass distributions or other variations. The particular form in which an early generation of stars left many massive (now very faint) white dwarfs behind and not much else apparently belongs to Schwarzschild (1954). He was thinking of dark matter for elliptical galaxies. Very recently, something similar has been suggested to account for MACHO micro-lensing events in the halo of the Milky Way.

L'ENVOI

The Eocene ("dawn of the recent") period in star formation is clearly marked by Spitzer's (1967) article in the Kuiper compendium (which was in press since 1962 or thereabouts). I began this discussion with the words of one anonymous poet and would like to end with the words of another. That he is unknown is not for lack of trying. J.D.G.M. (1943) signed only his initials. I was able to establish that he was not a contributor either to Observatory or to MNRAS at that time nor an IAU member. The current editors of Observatory and the secretaries of the Royal Astronomical Society and the British Association of Amateur Astronomers have culled their minds and records without providing a firm identification. David Evans who was one of the four 1943 editors has some memory that the poem reached them through his fellow editor, Alan Hunter, and had been sent in by a Chancellor of Melbourne University (where the search is now being continued). The other two editors at the time were George C. McVittie and H.F. Finch (the only one of the four I never met). I quote the poem here for three reasons. It reflects a much higher level of doggerel than we usually see today; it demonstrates a clear belief in star and planet formation as ongoing processes, despite dating from the dark ages when such belief was rare, or at least rarely expressed; and I was born in 1943. It carried the title "III.", being preceded by two shorter meditations on solar and Jovian conditions of life.

> Some time ago my late Papa
> Acquired a spiral nebula.
> He bought it with a guarantee
> Of content and stability.
> What was his undisguised chagrin
> To find his purchase on the spin,
> Receding from his call or beck
> At several million miles per sec,
> And not, according to his friends,
> A likely source of dividends.
> Justly incensed at such a tort
> He hauled the vendor into court,
> Taking his stand on Section 3
> Of Bailey "Sale of Nebulae."
> Contra was cited Volume 4
> Of Eggleston's "Galactic Law"
> That most instructive little tome
> That lies uncut in every home.
> "Cease" said the sage "your quarrel base:
> Lift up your eyes to Outer Space.
> See where the nebulae like buns,
> Encurranted with infant suns,
> Shimmer in incandescent spray
> Millions of miles and years away.
> Think that, provided you will wait,
> Your nebula is Real Estate,
> Sure to provide you wealth and bliss
> Beyond the dreams of avarice.
> Watch as the rolling aeons pass
> New worlds emerging from the gas:

Watch as the brightness slowly clots
To eligible building lots.
What matters a depleted purse
To owners of a Universe?"
My father lost the case and died:
I watch my nebula with pride
But yearly with decreasing hope
I buy a larger telescope.

ACKNOWLEDGEMENTS

I am most grateful to the following for comments, advice, and other assistance: David DeVorkin, Steven Dick, David S. Evans, Owen Gingerich, Richard Larson, Donald Osterbrock, Leonid Ozernoy, Anneila Sargent, and David Stickland. I hope to turn this contribution into a semi-serious discussion of the history of star formation, and input of all kinds would be very much appreciated.

REFERENCES

1. Adams, W.S. 1941, PASP, 53, 209.
2. Adams, W.S. 1942, ApJ, 97, 103.
3. Alfven, H. 1937, Zs. f. Ph., 107, 579.
4. Ambartsumian, V.A. 1938, Ann. Leningrad State Univ. #22 (Astron. Sec. 4), p. 19.
5. Ambartsumian, V.A. 1947, A Zh, 26, 3 and Stellar Evolution and Astrophysics (Acad. Sci. Armenian SSR, Erevan).
6. Ambartsumian, V.A. 1953, Liege Mem. Soc. Roy. Sci. 4th Ser., v. 14, p. 297.
7. Ambartsumian, V.A. 1960, QJRAS, 1, 152.
8. Arny, T., and Weissman, A. 1973, AJ, 78, 309.
9. Atkinson, R. d'E. 1936, ApJ, 84, 84.
10. Baade, W. 1944, ApJ, 100, 131.
11. Baade, W. 1952, Trans. IAU 8, ed. P. Th. Oosterhoff, Cambridge Univ. Press, p. 682.
12. Ball, R. 1902, in B. Lovell ed. Royal Institute Friday Evening Discourses, Astronomy, Vol. 2, Elsevier, 1970, p. 31.
13. Barnard, E.E. 1919, ApJ, 49, 12.
14. Bethe, H.A. 1939, PR, 55, 434.
15. Blaauw, A. 1952, BAN, 11, 459.
16. Blaauw, A., and Morgan, W.W. 1953, ApJ, 117, 256.
17. Bok, B.J. 1934, Harv. Circ., 384.
18. Bok, B.J. 1948, Harvard Observatory Monograph #7, p. 53.
19. Bok, B.J. 1959, IAU Symp., 9, 430.

20. Bok, B.J., and Bok, P.F. 1945, The Milky Way (Blackiston), p. 187 ff.
21. Bok, B.J., and Reilly, E.F. 1947, ApJ, 105, 255.
22. Bonner, W. 1956, MNRAS, 116, 351.
23. Brush, S.G. 1978, J. Hist. Astron., 9, 1 and 77; 1990, RMP, 62, 43.
24. Brush, S.G. 1996, A History of Modern Planetary Physics (Cambridge Univ. Press), 3 volumes.
25. Burbidge, G.R., and Burbidge, E.M. 1958, Handb. der Physik, 51, 134.
26. Cameron, A.G.W. 1962, Icarus, 1, 13.
27. Cameron, A.G.W., and Truran, J.W., Icarus, 30, 447.
28. Chandrasekhar, S. 1942, Principles of Stellar Dynamics (U. Chicago Press), p. 205 ff.
29. Chandrasekhar, S. 1943, AJ, 51, 86 (Yerkes Annual Report).
30. Chandrasekhar, S. 1944, ApJ, 99, 54.
31. Chandrasekhar, S., and Fermi, E. 1953, ApJ, 118, 116.
32. Chandrasekhar, S., and Henrichs, L.R. 1942, ApJ, 95, 288.
33. Curtis, H.D. 1918, PASP, 30, 65.
34. Davis, L.J., and Greenstein, J.L. 1951, ApJ, 114, 206.
35. Dibai, C. 1958, A Zh, 35, 469 (v. 2, 429 in translation).
36. Douglas, A.E., and Herzberg, G. 1941, ApJ, 94, 381.
37. Dunham, T. 1937, PASP, 49, 26.
38. Ebert, R. 1955, Zs. f. ApJ, 37, 217.
39. Eddington, A.S. 1913, Observatory, 36, 467.
40. Eddington, A.S. 1923, in B. Lovell, ed. Royal Institute Friday Evening Discourses, Astronomy, Vol. 2 (Elsevier 1971), P. 256.
41. Eddington, A.S. 1926, Proc. Roy. Soc. A111, 452.
42. Eggen, O.J., Lynden-Bell, D., and Sandage, A. 1962, ApJ, 136, 748.
43. Ewen, H.I., and Purcell, E.M. 1951, Nat., 168, 356.
44. Fermi, E. 1949, PR, 75, 1169.
45. Fernie, J.D. 1969, PASP, 81, 707.
46. Fujimoto, M. 1966, IAU Symp. 29, 453 (in Russian).
47. Gamow, G. 1938, PR, 54, 480.
48. Gerasimovich, B., and Struve, O. 1929, ApJ, 69, 7.
49. Gill, D. 1884, in B. Lovell, ed. Royal Institute Friday Evening Discourses, Astronomy, Vol. 1 (Elsevier, 1971), p. 214.
50. Gold, T. 1961, Mem. Soc. Roy. Sci. Liege. Ser. 15, v. 4, p. 476.
51. Goldberg, L., and Aller, L.H. 1943, Atoms, Stars, and Nebula (Harvard: Blackison), p. 277.
52. Greenstein, J.L. 1951, in J.A. Hynek, ed., Astrophysics McGraw Hill, P. 593.
53. Hall, J.S. 1949, Sci., 109, 166.
54. Hartmann, J. 1904, ApJ, 19, 268.
55. Hayashi, C. 1961, PASJ, 13, 450.
56. Hayashi, C. 1966, ARA&A, 4, 171.
57. Helmholtz, H. von. 1853, Lecture discussed by Kelvin (1863).
58. Herbig, G.H. 1952, JRASC, 46, 233.
59. Herbig, G.H. 1962, Adv. A&A, 1, 63.

60. Herschel, Wm. 1784-1814, Phil. Trans. Roy. Soc., A74, 437 (1784) and see, Hoskin (1963) and A. Berry, 1898, A Short History of Astronomy (Dover 1961), p. 340.
61. Herzberg, G. 1951, Spectra of Diatomic Molecules (Van Nostrand), p. 496.
62. Hiltner, W.A. 1949, Sci, 109, 165.
63. Hoffleit, D. 1996, private communication.
64. Hoskin, M. 1963, William Herschel and the Construction of the Heavens.
65. Hoskin, M. 1982, Stellar Astronomy (Science History Publications), p. 90.
66. Hoyle, F. 1952, IAU Trans. 8, Cambridge University Press.
67. Hoyle, F. 1953, ApJ, 118, 513.
68. Hoyle, F., and Lyttleton, R.A. 1939, Proc. Cam. Phil. Soc., 35, 405, 595, and 608.
69. Hoyle F., and Lyttleton, R.A. 1940, Proc. Cam. Phil. Soc., 36, 424.
70. Hoyle, F., and Lyttleton, R.A. 1941, MNRAS, 101, 227.
71. Hubble, E.P. 1922, ApJ, 56, 162 & 400.
72. Hubble, E.P. 1929, Proc. NAS, 15, 168.
73. Jeans, J.H. 1928, Astronomy and Cosmogony, Cambridge Univ. Press.
74. Jeans, J.H. 1929, Astronomy and Cosmogony, 2nd ed., CUP, p. 345-7.
75. Joy, A. 1945, ApJ, 112, 168.
76. Joy, A. 1946, PASP, 55, 244.
77. J.D.G.M. 1943, Observatory, 65, 88 (reprinted with permission of the current editors).
78. Kahn, F.D. 1953, BAN, 12, 187 and IAU Symp. 2, Gas Dynamics of Cosmic Clouds (North Holland), p. 115.
79. Kelvin, Lord William Thompson 1863, Brit. Assoc. Rep. Part II, p. 27, Les Mondes 3, 472.
80. Kerr, F.J., and Westerhout, G. 1965, in A. Blaauw & M. Schmidt, eds. Galactic Structure (U. Chicago Press), p. 167.
81. King, E.S. 1914, Harvard Obs. Annals, 76, 1.
82. Larson, R.B. 1972, MNRAS, 156, 437 & 157, 121.
83. Layzer, D. 1963, ApJ, 137, 351.
84. Layzer, D. 1964, ARA&A, 2, 341.
85. Lifshitz, E.M. 1946, JETP, 16, 587 (in Russian).
86. Lilley, A.E. 1958, ApJ, 121, 559.
87. Lindblad, B. 1936, MNRAS, 94, 23.
88. Lockyer, N. 1887, Proc. Roy. Soc., 43, 44.
89. Lockyer, N. 1888, Proc. Roy. Soc., 44.
90. Lubbock, C.A. 1937, The Herschel Chronicles (Cambridge Univ. Press).
91. Luyten, W., and Hill, 1937, ApJ, 86, 470.
92. Lynds, B. 1967, in B.M. Middlehurst & L.H. Aller, eds., Nebulae and Interstellar Matter (U. Chicago Press), p. 119.
93. McCrea, W.H. 1957, MNRAS, 117, 562.
94. McKellar, A. 1940, PASP, 52, 187.
95. McKellar, A. 1941, Publ. Dom. Ap. Obs., 7, 251.
96. McVittie, G.C. 1932, MNRAS, 92, 500.

97. Menzel, D.H. 1949, Our Sun (Harvard UP), p. 234
98. Mestel, L. 1965, QJRAS, 6, 161 & 265.
99. Mestel, L., and Spitzer, L. 1956, MNRAS, 116, 503.
100. Minkowski, R. 1949, PASP, 61, 151.
101. Mineur, H. 1939, Ann. d'Ap., 2, 1.
102. Muller, C.A., and Oort, J.H. 1951, Nat., 168, 357.
103. Newcomb, S., and Baker, R.H. 1932, Astronomy for Everybody (The New Home Library), p. 323.
104. Newton, I. see Hoskin (1982), p. 90.
105. Oort, J.H. 1954, BAN, 12, 177.
106. Oort, J.H. 1955, IAU Symp. 2, Gas Dynamics of Cosmic Clouds, p. 157.
107. Opik, E.J. 1953, Irish AJ, 2, 219.
108. Ozernoi, L. 1996, private communication.
109. Parenago, P.P. 1952, A Zh, 30, 249.
110. Parker, E.N. 1965, ApJ, 145, 811.
111. Pawsey, J.L. 1951, Nat., 168, 358.
112. Payne Gaposchkin, C.H. 1952, Stars in the Making (Harvard U.P.), p. 146.
113. Plaskett, J.S. 1924, MNRAS, 84, 80 and Publ. Dom. Obs. Victoria, 1, 163, & 287.
114. Ranford, A.C. 1894, Knowledge, 17, 253.
115. Rees, M.J. 1976, MNRAS, 176, 483.
116. Ritter, H. 1882, Ann. Phys. 34rd Ser., 20, 137.
117. Roberts, W.W. 1969, ApJ, 158, 123.
118. Russell, H.N. 1913, Observatory, 36, 290 & 324.
119. Russell, H.N. 1922, Proc. NAS, 8, 115.
120. Russell, H.N. 1925, Nature, 116, 209.
121. Russell, H.N. 1939, Proc. Am. Phil. Soc., 8, 295.
122. Russell, H.N., Dugan, & Stewart 1938, Astronomy (2nd ed.) (Boston: Ginn and Co.), Vol. 2, p. 924.
123. Sandage, A., and Schwarzschild, M. 1952, ApJ, 116, 467.
124. Schwarzschild, M. 1954, AJ, 59, 273.
125. Schwarzschild, M., Spitzer, L., and Wildt, R. 1951, ApJ, 114, 406.
126. Seiden, P.E. and Gerola, H. 1979, ApJ, 233, 56.
127. Seiden, P.E. and Gerola, H. 1982, Fund. Cosmic Phys., 7, 241.
128. Shapley, H. 1918, ApJ, 48, 89.
129. Shapley, H. 1923, Pop. Astron., 31, 316.
130. Shapley, H. 1943, Galaxies (Harvard), p. 213.
131. Shklovsky, I.S. 1949, A Zh, 26, 10 (in Russian).
132. Shklovsky, I.S. 1953, Dokl. Acad. Nauk USSR, 92, 25 (in Russia).
133. Shu, F., Adams, F.C., & Lizano, S. 1987, ARA&A, 25, 23.
134. Slipher, V.M. 1909, Lowell Obs. Bull., 2, 1.
135. Spitzer, L. 1940, MNRAS, 100, 396.
136. Spitzer, L. 1941, ApJ, 93, 369 & 94, 232.
137. Spitzer, L. 1942, ApJ, 95, 329.
138. Spitzer, L. 1946, PR, 70, 777.

139. Spitzer, L. 1948, Phys. Today, 1, No. 5, 7.
140. Spitzer, L. 1948a, Harvard Observatory Monographs, No. 7.
141. Spitzer, L. 1949, ApJ, 109, 337.
142. Spitzer, L. 1951, J. Wash. Acad. Sci., 41, 313.
143. Spitzer, L. 1967, in B.M. Middlehurst and L.H. Aller, eds. Nebulae and Interstellar Matter (U. Chicago Press).
144. Spitzer, L. 1968, Diffuse Matter in Space.
145. Spitzer, L., and Schwarzschild, M. 1951, ApJ, 114, 383.
146. Stromberg, B. 1937, quoted by Luyten and Hill 1937.
147. Struve, O. 1937, ApJ, 75, 66.
148. Struve, O. 1937, ApJ, 109, 180.
149. Townes, C.H. 1955, IAU Symp., 4, 92 (published 1957).
150. Trimble, V. 1996, PASP, 1007, 1073.
151. Trimble, V., and Leonard, P.J.T. 1996, PASP, 107, 8.
152. Trumpler, R.J. 1930, Lowell Obs. Bull, 14, 154 (No. 420).
153. Unsold, A. 1944, Zs. f. Ap., 24, 278 (published 1947).
154. van de Hulst, H. 1945, Ned. T. Naturk., 11, 201.
155. van de Hulst, H. 1948, Harvard Observatory Monograph No. 7, p. 73.
156. von Weizsacker, C.F. 1943, Zs. f. Ap., 22, 319.
157. von Weizsacker, C.F. 1947, Zs. f. Zp., 24, 181.
158. von Weizsacker, C.F. 1951, ApJ, 114, 165.
159. Waterfield, R.L. 1938, A Hundred Years of Astronomy.
160. Weaver, H.F., et al. 1965, Nature, 208, 440.
161. Weinreb, S., et al. 1963, Nature, 200, 829.
162. Whipple, F. 1946, ApJ, 104, 1.
163. Wilson, R.W., et al. 1970, ApJ, 161, L43.
164. Wolf, M. 1908, Die Milchstrasse und die Kosmische Nebel.
165. Woltjer, L. 1965, in A. Blaauw & M. Schmidt, eds., Galactic Structure (Univ. of Chicago Press), p. 531.
166. Young, C.A. 1904, General Astronomy (Boston: Ginn and Co.), p. 576.
167. Zwicky, F. 1959, PASP, 71, 468.

Close-Up Views of Star Formation

Early Star Formation

P. C. Myers

Harvard-Smithsonian Center for Astrophysics
Cambridge MA 02138

Abstract. We describe recent progress in two areas of the early stages of star formation–initial conditions for making stellar groups, and kinematic evidence for star-forming infall.

INTRODUCTION

The 1980s saw the rise of a remarkably successful picture of low-mass star formation [1]. In this "standard model" a slowly rotating core condenses quasistatically via ambipolar diffusion, on a time scale of a few Myr, until it becomes dynamically unstable and collapses into a star-disk system, on a free-fall timescale of a few $\times$ 0.1 Myr. This theoretical picture incorporates and accounts for numerous observed properties, including the typical size, density, temperature and velocity gradient of dense cores, the spectral energy distributions of highly embedded candidate protostars, and in a few cases the spectral line profiles corresponding to "inside-out" collapse of a dense core. This standard model can be considered a description of the simplest system which displays the basic physics of the problem. In analogy with atomic physics, this system is like the "hydrogen atom" case.

But in the last several years, it has become clear that such simple systems can not be considered "typical" of most stars and the circumstances of their formation. Infrared observations of young clusters [2], young stellar groups [3,4], and pre-main sequence binaries [5] have shown that star formation is a process whose typical outcome is more than one star. Most pre-main sequence stars in Taurus and Ophiuchus are in binary systems, as is also seen in the field star population [6,7]. The number of young stars in groups is at least comparable to the number of stars in isolation, and may be significantly greater, according to a near-infrared imaging study of 164 young stellar objects with molecular outflows [4]. The largest nearby young clusters, such as in Orion and M17, have relatively few massive stars, accompanied by $\sim$ 1000 low-mass stars. The distribution of stellar fluxes at 2 μm in such young clusters is generally consistent with that expected from the initial mass function (IMF) for field stars [8]. This consistency suggests that such clusters can provide most of the stars which are counted to estimate the IMF. Thus the typical star may be a member of a binary system which formed in a group or cluster.

As a result, it is also clear that the typical star cannot be considered to arise solely from the initial conditions derived from quiescent, thermally-dominated, low-mass cores. This is so even though such cores are evidently the birth sites of many young low-mass stars [9,10]. The physical properties of low-mass cores in Taurus, Lupus, and Chamaeleon differ from those of dense cores in cluster-forming regions. "Massive cores" are larger, denser, and more turbulent than low-mass cores, even when they do not contain any embedded young stars, and their molecular spectral line widths are dominated by nonthermal motions [11–13]. Thus we may safely neglect turbulence in modelling the formation of *some* low-mass stars, but we probably cannot neglect turbulence in modelling the formation of the *typical* low-mass star.

The problem of forming a typical star is thus more complex than the problem of forming an isolated low-mass star from a thermal dense core. This issue has been discussed by many other authors, *e.g.* [14,15]. To address this new problem we need to better understand two important features: the physical conditions that lead to stellar groups, and the motions associated with star-forming infall, both in isolation and in groups. This paper presents some recent progress in each of these areas.

INITIAL CONDITIONS FOR STELLAR GROUPS

Dense cores in observed in molecular lines tracing density $> 10^4$ cm^{-3}, such as the $(J, K) = (1, 1)$ line of NH$_3$, have significantly greater size, temperature, and turbulent line width in the Orion A and B complexes, which tend to form stars in groups, than they do in Taurus and other regions of more isolated star formation. For example the core associated with IRAS 05338+0624 in L1641 in Orion has FWHM size 0.2 pc, temperature 17 K, FWHM line width 1.0 km s^{-1}, and some 30 young stellar objects within its NH$_3$ line map contours [11,3]. In contrast the L1527 core associated with IRAS 04368+2557 in Taurus has FWHM size 0.08 pc, temperature 11 K, FWHM line width 0.28 km s^{-1}, and what appears to be a wide protostellar binary [9,16]. How are these representative gas properties of "massive" and "low-mass" dense cores related to the number and mass of the stars which such cores tend to produce?

This general problem of core-star genetics has been discussed by many authors, *e.g.* [17–22] . In this paper we use simple models of the density and ionization structure of massive and low-mass cores. These show that cores tend to have low-ionization zones of weak field-gas coupling. Further, the mass of this zone tends to be comparable to the mass of the young stars observed in the zone, suggesting that a stellar group may arise through a process of "turbulent decoupling" and core fragmentation.

Density Model

We specify the characteristic profile of density with radius in massive and low-mass cores, following the approach of [12]. The model assumes that the nonthermal velocity dispersion observed in the widths of lines of NH$_3$, C^{18}O, and ^{13}CO combines with the thermal velocity dispersion derived from NH$_3$ line observations, assumed constant, and with the static magnetic field to provide supporting pressure against gravity. The physical basis of the nonthermal line widths is presumed to be MHD turbulence and waves. The mass density ρ varies with radius r as

$$\rho = \frac{\sigma_T{}^2 + 2\sigma_{NT}{}^2}{2\pi G r^2} \qquad (1)$$

where σ_T and σ_{NT} are respectively the thermal and nonthermal velocity dispersions for the molecule of mean mass, given by [12], and G is the gravitational constant.

In this model, massive cores are typically about 5 times denser than low-mass cores at the same radius, for radii 0.03 to 1 pc. Further, massive cores are dominated by nonthermal motions over this entire range of radii, while low-mass cores are dominated by thermal motions for radii less than about 0.1 pc. For simplicity we treat these cases as distinct, while in reality they are two samples of a continuous range of properties.

These density profiles are used to predict the typical ionization structure in massive and low-mass cores, using the model of photoionization and cosmic ray ionization of McKee [23,24]. The resulting ionization structure consists of an outer high-ionization zone, where photoionization dominates, and an inner low-ionization zone, where cosmic ray ionization dominates. The transition between these two zones occurs at about $r = 0.2$ pc for massive cores, and at about $r = 0.05$ pc for low-mass cores, corresponding in each case to an external visual extinction of a few magnitudes.

Coupling Parameters

The foregoing model is used to estimate three closely related measures of the coupling between the magnetic field and the neutral gas. For each core type and radius r, we compute the magnetic Reynolds number Re_M, which indicates the ability of the field-fluid system to develop turbulent motions; r/λ_c, the range of wavelengths of MHD waves which can propagate above

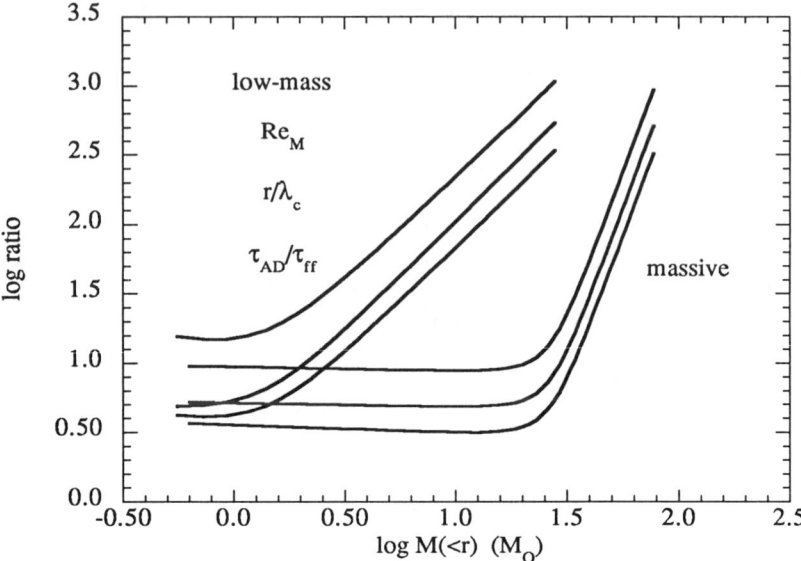

FIGURE 1. Variation of field-gas coupling parameters at radius r in massive and low-mass dense cores, vs. the mass within r. Each type of core has a zone of low ionization, whose mass is similar to that of the stars which form in the core.

cutoff; and τ_{AD}/τ_{ff}, the relative time that the core gas takes to condense via ambipolar diffusion, compared to the time it would take to collapse via free-fall, following [24]. Each of these quantities increases with field-fluid coupling, and decreases toward unity as coupling weakens.

Figure 1 is a log-log plot of each measure of coupling at r, plotted against the core mass within r. For each type of core the field-gas coupling parameters decrease from values of several hundred where photoionization dominates to values 3-10 where cosmic-ray ionization dominates–a region enclosing about 1 $M_\odot$ for low-mass cores, and about 20 $M_\odot$ for massive cores.

The quantities τ_{AD}/τ_{ff} and r/λ_c and have been discussed and interpreted for low-mass cores ionized solely by cosmic rays, by many authors, e.g. [25,1]. These discussions emphasize the quasistatic nature of ambipolar diffusion during core formation, and the negligible role of MHD waves in low-mass systems. In the present case we emphasize the contrast between the massive and low-mass cases, and the dramatic change in field-gas coupling between the zone of high ionization due to UV photons, and the zone of low ionization due to cosmic rays. The numerical values of the coupling parameters are probably uncertain by a factor of order 3, because of uncertainties in the cosmic ray ionization. However the dramatic decrease in the coupling parameters with decreasing core radius, and the value of that radius, appear better determined, and these are the basis of our interpretation.

"Turbulent Decoupling" and Stellar Groups

For both core types, the zone of weak coupling has comparable mass in gas and stars. For low-mass cores, their tendency to produce stars of mass <1 $M_\odot$ is already well-known (*e.g.* [10]). The more massive cores represented in Figure 1 are associated with small groups having 5-40 members, including HH 26IR, L1641-N, L1641-C, and L1641-S3 [3,4]. Nearly all of the sources in these groups have near-infrared brightness typical of red low-mass stars, so a typical mass is probably $\sim$0.5 $M_\odot$. If so, some of these more massive cores have already produced up to $\sim$20 $M_\odot$, and most of them are still in the process of forming new stars, as indicated by their strong IRAS emission.

This similarity in mass between the low-ionization zone and its stellar products may be coincidental, and further studies are needed to establish how robust this connection is. We offer the following physical interpretation. In the outer zone where the field-gas coupling is strong, the MHD turbulent motions are well-developed, with a large range of propagating wavelengths. These motions provide a sufficiently steady and isotropic pressure to support the gas. But in the inner zone where the coupling of the field and gas is weak, the turbulent pressure support is probably more intermittent and anisotropic, since fewer MHD wavelengths can propagate above cutoff. We consider this "turbulent decoupling" as a transition between steady isotropic turbulent support and steady isotropic thermal support. A similar conclusion is based on the increase in the dispersion of the line width in low-mass core maps, from lines tracing higher to lower density [26,27].

The low-ionization zone in a massive core may be unstable against fragmentation and collapse. As its turbulence becomes more intermittent and anisotropic, its effective pressure will decrease, tending toward the thermal pressure along the field lines, and toward the static magnetic pressure across the field lines. If the sound speed is much less than the Alfven speed, as in most massive cores, this decrease can be dynamically significant. When a core with such fragile support is compressed, it should be more vulnerable to collapse and fragmentation than either the well-coupled gas around it, which has steady isotropic turbulent support, or the thermally dominated cores, which have steady isotropic thermal support.

These speculations can be tested by studies of how stellar multiplicity is related to core properties, particularly core ionization. It is also desirable to study the isotropy and steadiness of the pressure exerted by gas as its value of Re_M is reduced toward unity, as in Figure 1. Some simulations at higher Re_M have been carried out, suggesting that appropriate simulations at low Re_M may also be feasible [28,29].

KINEMATIC EVIDENCE FOR INFALL

In the past several years, the prospects for studying star-forming infall with observations have improved, because millimeter-wavelength telescopes have better sensitivity and finer angular and spectral resolution; and because we have better knowledge of infall candidates, through far-infrared and submillimeter continuum observations of highly obscured sources.

Several low-mass systems have recently been studied in detail, including B335, I16293, L1527, and NGC1333-2 [30–33]. In such sources, optically thick and optically thin lines with critical density $>10^4$ cm^{-3} show the "infall asymmetry" signature, where the thick line profile is skewed to the blue of the thin line profile [34], in a significant number of map positions. Radiative transfer and excitation models reproduce these profiles when the density and velocity fields are specified by the well-known model of "inside out" collapse of a singular isothermal sphere [35].

These spectroscopic indications of infall are encouraging, but not yet definitive. We do not understand the basis of reverse infall asymmetry seen in some positions in most "infall" maps. It is not always clear whether the observed infall asymmetry is concentrated on a single star, the members of a binary, or on the multiple components of a fragmenting system. It is not known how to distinguish "young" from "old" infall, or how infall asymmetry should vary between systems which are primarily thermal, and primarily turbulent.

Despite these uncertainties, the observational study of infall is in an exciting period. In this section we present some of the most recent results in this field.

Infall Surveys

Surveys in optically thick and thin lines tracing high density offer a way to define groups of young stellar objects which do and do not show signs of infall. They also offer an advantage over a study of a single object, which is subject to the possibility that its infall asymmetry arises mainly from outflow motions or rotation. If the symmetry axes of rotation and/or outflow are distributed randomly with respect to the line of sight, these motions should contribute equally toward infall asymmetry and toward its reverse asymmetry. In contrast, infall onto a point mass should always yield infall asymmetry, provided the line excitation temperature is centrally elevated, as expected for a centrally condensed core.

A pioneering survey of 23 "Class 0" [36] sources was made in the $J = 4$-3 and 3-2 lines of HCO$^+$, with 19 of these sources also observed in the optically thin isotopic line, the $J = 3$-2 line of H^{13}CO$^+$ [37]. Of the 19 sources observed in both thick and thin lines, 9 have infall asymmetry, 3 have the reverse asymmetry, and 7 have relatively symmetric lines. Application of the requirement that an infall candidate have the same asymmetry in each thick line observed, and that its map does not track the outflow map, yields 3 already known kinematic infall candidates–B335, L1527, and IRAS 16293–and 3 new ones–HH25MMS, Ser SMM4, and IRAS 20050.

A survey of Class 0 sources having bolometric temperature [38] T_{bol} <70 K and also of "Class I" [39] sources having 70 K< T_{bol} <200 K was made in three lines: the optically thick 2_{12}-1_{11} line of H_2CO (52 sources), the optically thick 2-1 line of CS (38 sources), and the optically thin 1-0 line of N_2H^+ (52 sources) [40]. This survey also finds an excess of sources with infall asymmetry. For the CS sample there are 21 Class 0 sources, of which 12 have a velocity shift between the thick CS and thin N_2H^+ peaks of at least one-quarter of the thin FWHM line width. Remarkably, all 12 of these shifts are to the blue. For the 17 Class I sources, the 10 sources with significant shifts are more evenly divided, with 6 to the blue and 4 to the red. These results make a clear kinematic distinction: Class 0 sources have a highly significant excess of sources with infall asymmetry, while Class I sources do not.

Infall on Scales < 1000 AU

The infall motions implied by the single-dish observations in lines of HCO^+, H_2CO, and CS discussed above probably have spatial extent $\sim$0.05 pc, or about 10^4 AU, according to the available maps of their infall asymmetry. To probe smaller scales, approaching the separation of members of a binary system, or the diameter of a circumstellar accretion disk, it is necessary to use interferometers. As millimeter-wavelength interferometers have become more sensitive, it is now possible to make images with resolution of a few hundred AU in lines having excitation energy E/k >20 K and critical density $> 10^6$ cm^{-3}.

A recent example is the observation of the candidate protostar in the dense core L1527, in the optically thick 3_{12}-2_{11} line of H_2CO with the Caltech Owens Valley array, and in the optically thin $J = 11$-10 line of H_3CN with the IRAM Plateau de Bure array [41]. The line emission extends about 600 × 2000 AU, elongated nearly perpendicular to the molecular outflow. These small-scale observations are remarkably consistent with the infall picture. They show strong infall asymmetry over the entire map, in some 15 independent positions. They have no evidence of reversals of the blue-red asymmetry–unlike most larger-scale single-dish maps. They show no alignment of the intensity map with the molecular outflow lobes, unlike some interferometer maps in lines tracing lower density. Further, their spectra show no evidence of outflow wings, unlike single-dish spectra of the same source.

Turbulent Infall

The best-studied kinematic infall candidates, B335 and L1527, have thermally-dominated line widths: the nonthermal component of the observed line width is generally less than the thermal width of the molecule of mean mass, 0.45 km s^{-1} for gas at 10 K. These two sources are also associated with either single or binary stars, as opposed to a stellar group [16]. But most known kinematic infall candidates have significantly broader lines, with FWHM 1-2 km s^{-1}. If they are indeed infall sources, their infall involves primarily turbulent gas, as opposed to primarily thermal gas. Furthermore, many of these candidates are associated with groups or are located in cluster-forming regions. Such "turbulent infall" candidates associated with groups or clusters include IRAS 16293, Serpens SMM4, NGC1333-IRAS4, IRAS 20050, and L1251B. Detailed maps have been made of three of these regions, defining the projected radius of their zones of infall asymmetry to be 0.05 - 0.07 pc [42].

These sources, and others like them, may offer insight into the physics of infall and of turbulence, and into the problem of forming stellar groups, discussed earlier.

Prestellar Infall

The earliest stages of infall should precede the formation of a point source of infrared luminosity. Searching for "prestellar infall" in dense cores without stars appears more difficult than searching for infall toward infrared point sources, because the cores are extended over many beamwidths of the typical millimeter-wavelength telescope. However those regions, once found, should have no confusion from outflows.

A well-studied example of a starless core with extended infall asymmetry is L1544 in Taurus [43,44]. This core has no IRAS point source, no embedded near-infrared source, and no T Tauri star within the contours of its maps in the dense gas tracer lines of NH_3, N_2H^+, and $C^{34}S$. It has no CO wings suggestive of outflow motions. Yet it shows infall asymmetry in lines of CS, H_2CO, and HCO^+, extended over about 0.05 pc. Figure 2 shows this infall asymmetry in optically thick and thin CS and $C^{34}S$ lines. If the infall asymmetry corresponds to infall, the infall seems inconsistent with "inside-out" collapse [35], because the time for an expansion wave to reach the boundary of the observed zone of infall asymmetry is $\sim$0.3 Myr. At this time an infrared point source of >1 $L_\odot$ should have formed according to the standard model, yet none is seen. Further studies of this intriguing source, and others like it, are underway.

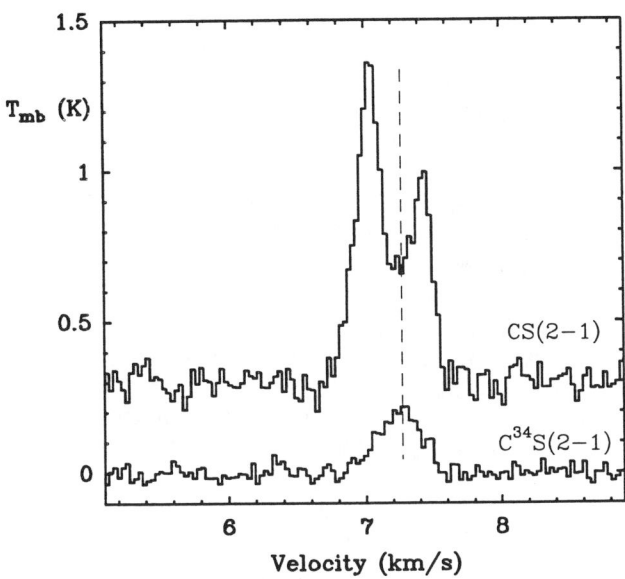

FIGURE 2. CS spectra showing infall asymmetry in the starless dense core L1544.

REFERENCES

1. Shu, F. H., Adams, F. C., & Lizano, S. 1987, ARAA, 25, 23
2. Lada, E. A., & Lada, C. J. 1991, in The Formation and Evolution of Star Clusters, ed. K. Janes (San Francisco: ASP Conf. Ser., 13), 3
3. Chen, H., & Tokunaga, A. T. 1994, ApJS, 90, 149
4. Hodapp, K. 1994, ApJS, 94, 615
5. Mathieu, R. D. 1994, ARAA, 32, 465
6. Ghez, A. M., Neugebauer, G., & Matthews, K. 1993, AJ, 106, 2005
7. Leinert, Ch., Zinnecker, H., Weitzel, N., Christou, J., Ridgway, S. T., Jameson, R., Haas, M., Lenzen, R. 1993, A&A, 278, 129
8. Lada, E. A., & Lada, C. J. 1995, AJ, 109, 16
9. Benson, P. J., & Myers, P. C. 1989, ApJS, 71, 89
10. Beichman, C. A., Myers, P. C., Emerson, J. P., Harris, S., Mathieu, R. D., Benson, P. J., & Jennings, R. E. 1986, ApJ, 307, 337
11. Harju, J., Walmsley, C. M., & Wouterloot, J. G. A. 1993, A&AS, 98, 51
12. Caselli, P., & Myers, P. C. 1995, ApJ, 446, 665
13. Walmsley, C. M. 1995, in Circumstellar Disks, Outflows and Star Formation, eds. S. Lizano & J. Torrelles, (Mexico City: RMA&A Conf. Ser., 1), 137
14. Pringle, J. E. 1989, MNRAS, 239, 361
15. Whitworth, A. P., Bhattal, A. S., Francis, N., & Watkins, S. J. 1996, MNRAS,

in press
16. Fuller, G. A., Ladd, E. F., & Hodapp, K. 1996, ApJ, 463, L97
17. Larson, R. B. 1992, MNRAS, 256, 641
18. Myers, P. C., & Fuller, G. A. 1992, ApJ, 396, 631
19. Zinnecker, H., McCaughrean, M. J., & Wilking, B. A. 1993, in Protostars and Planets III, eds. E. Levy & J. Lunine (Tucson: Univ. Arizona Press), 429
20. Silk, J. 1995, ApJ, 438, L41
21. Adams, F. C., & Fatuzzo, M. 1996, ApJ, 464, 256
22. Elmegreen, B. G., & Falgarone, E. 1996, ApJ, 471, L816
23. McKee, C. F. 1989, ApJ, 345, 782
24. Myers, P. C., & Khersonsky, V. K. 1995, ApJ, 442, 177
25. Mouschovias, T. Ch. 1987, in Physical Processes in Interstellar Clouds, eds. G. Morfill & M. Scholer (Dordrecht: Reidel), 453
26. Barranco, J. A., & Goodman, A. A. 1996, ApJ, submitted
27. Goodman, A. A., Barranco, J. A., Wilner, D. A., & Heyer, M. H. 1996, ApJ, submitted
28. Passot, T., Vazquez-Semadeni, E., & Pouquet, A. 1995, ApJ, 455, 536
29. Gammie, C., & Ostriker, E. 1996, ApJ, 466, 814
30. Zhou, S., Evans, N. J., Kompe, C., & Walmsley, C. M. 1993, ApJ, 404, 232
31. Zhou, S., Evans, N. J., Wang, Y., Peng, R., & Lo, K. Y. 1994, ApJ, 433, 131
32. Myers, P. C., Bachiller, R., Caselli, P., Fuller, G. A., Mardones, D., Tafalla, M., & Wilner, D. J. 1995, ApJ, 449, L65
33. Ward-Thompson, D., Buckley, H. D., Greaves, J. S., Holland, W. S., & Andre, P. 1996, MNRAS, 281, 53
34. Leung, C. M., & Brown, R. B. 1977, ApJ, 214, L73
35. Shu, F. H. 1977, ApJ, 214, 488
36. Andre, P., Ward-Thompson, D., & Barsony, M. 1993, ApJ, 406, 122
37. Gregersen, E. M., Evans, N. J., Zhou, S., & Choi, M. 1997, ApJ, in press
38. Myers, P. C., & Ladd, E. F. 1993, ApJ, 413, L47
39. Lada, C. J., in Star Forming Regions (IAU Symposium 115), eds. M. Peimbert & J. Jugaku (Dordrecht: Reidel), 1
40. Mardones, D. M., Myers, P. C., Tafalla, M., Wilner, D. J., Bachiller, R., & Garay, G. G. 1997, in Star Formation, Near and Far, eds. S. Holt & L. Mundy (New York: AIP Press) [this volume]
41. Wilner, D. J., Mardones, D., & Myers, P. C. 1997, in Star Formation, Near and Far, eds. S. Holt & L. Mundy (New York: AIP Press) [this volume]
42. Williams, J. P., & Myers, P. C. 1997, in Star Formation, Near and Far, eds. S. Holt & L. Mundy (New York: AIP Press) [this volume]
43. Myers, P. C., Mardones, D., Tafalla, M., Williams, J. P., & Wilner, D. J. 1996, ApJ, 465, L133
44. Tafalla, M., Mardones, D., Myers, P. C., Bachiller, R., Benson, P. J., & Caselli, P. 1996, FCRAO Newsletter, August, 6

Turbulence and Magnetic Fields in Star Formation

Eve C. Ostriker

Department of Astronomy
The University of Maryland
College Park, MD 20742-2421

Abstract. Magnetized turbulence is believed to play an important role in the evolution of star-forming clouds in our Galaxy. I summarize the properties of these clouds, and outline the distinctions between various regimes of turbulence. I then present arguments and supporting calculations that in such clouds, MHD turbulence with mean magnetic field strengths of a few $\times 10 \mu G$ is able to (1) preserve clouds against self-gravitating collapse for > 10 Myr, (2) yield velocity power spectra $\sim v(k) \propto k^{-1/2}$ comparable to observations, and (3) generate very inhomogeneous density structure, in which lumps often coincide with discontinuities in velocities and magnetic field directions.

INTRODUCTION

Magnetic fields and turbulence are thought to be important in many stages of star formation as observed "close up" in our Galaxy [1]. In the late stages of star formation, MHD turbulence in the form of a saturated Balbus-Hawley instability may be one of the primary mechanisms driving accretion in the ionized portions of circumstellar disks (see chapter by Stone). Large-scale magnetic fields are also considered to be necessary in accelerating and collimating the jets and outflows that are a natural byproduct of the accretion process [2]. On yet larger scales than individual star/disk/outflow systems, magnetic fields are believed to play an important role in mediating turbulent motions, supporting against gravitational collapse, and encouraging the growth of condensations within the molecular clouds where stars form [3–5]. This paper focuses on the interplay between magnetic fields, turbulence, and gravity on this largest scale, in particular summarizing ideas about what parameters control the physics, and describing some results of numerical studies relevant to this problem.

Cloud properties

Well-observed star-forming regions in our galaxy range from the dark clouds which form low-mass stars in a distributed fashion (typical cloud mass $M \sim 10^4\ M_\odot$, typical cloud size $L \sim 10$ pc), to the giant molecular cloud complexes (GMCs) which form both high- and low- mass stars, primarily in clusters (cloud $M \sim 10^5\ M_\odot$, $L \sim 40$ pc) [6–8]. Strong molecular cooling leads to temperatures in the range $T = 10 - 50$ K (corresponding sound speeds are $c_\text{s} = 0.2 - 0.4$ km s^{-1}). These clouds are quite inhomogeneous, with mean number densities in the range $n = 25 - 100$ cm^{-3}, but volume filling factors of only ~ 0.05 [9,10]; structure is observed down to the smallest resolved scales [11]. Star-forming clouds show a hierarchy of density structure: complex > cloud > clump ($n \sim 10^3$) > core ($n \sim 10^4$) > star. At typical temperatures and mean cloud densities, the thermal Jeans length $L_\text{J} = c_\text{s}\sqrt{\pi/G\rho}$ is in the range $2 - 5$ pc, which is much smaller than size of the whole cloud.

For structures of all scales within molecular clouds, the turbulent component of the linewidth is at least as large as the thermal component, with the former strongly dominating in all but centers of low-mass dense cores [6,12–14]. The larger clumps and clouds (and larger dense cores) have turbulence consistent with virial velocities $v \sim (GM/L)^{1/2}$ (a few × km s^{-1}) implying they are self-gravitating, and also show a linewidth increasing with size as approximately $\sigma_v \propto L^{0.5}$ [15–18]. The lower-mass clumps are not self-gravitating, and are thought to be pressure-confined [10,19,20].

Unfortunately, the magnetic field strengths and directions within molecular clouds are less well-known than other cloud properties, because of the difficulty in making, and sometimes interpreting, the relevant observations [21,22]. Observational studies of magnetic fields in star-forming regions have been thoroughly reviewed by [23]. Field strength observations for star-forming clouds are mostly based on Zeeman splitting in HI or OH, with the result that measured line-of-sight field strengths fall in the range $B = 10 - 30\mu$G [21,24,25]. In several objects, detailed comparisons of mean Alfvén speed $v_\text{A} = B/\sqrt{4\pi\rho}$, turbulent velocity dispersion, and virial speeds suggest approximate equipartition between magnetic, kinetic, and gravitational energies [26]. Optical polarization of background stars traces the averaged plane-of-sky field direction, assuming absorption by spinning dust grains aligned perpendicular to the field. Maps of optical polarization suggest smoothly-varying field directions in the lower-density gas [27,28]. More recently, polarization studies have been extended to the far-infrared [29,30], where polarized emission is believed to come from warm dust grains; these maps also imply large-scale correlations in the mean field orientation (perpendicular in the plane of the sky to the direction of polarized emission).

Regimes of turbulence

Taken together, the observations of star-forming regions (and molecular clouds more generally) suggest that their underlying dynamics lies in a novel regime of magnetohydrodynamic (MHD) turbulence. The basic state in this regime is characterized by (1) highly supersonic motions up to $\sigma_v/c_s \sim 10$, (2) a strongly magnetized, compressible medium with $v_A \sim \sigma_v$, and (3) self-gravity stronger than thermal pressure forces over a wide range of scales within the cloud ($L \gg L_J$). This basic state is far from the regimes of turbulence that have heretofore received most attention; thus, there are few predictions about how spectra of velocity, magnetic field, and density perturbations should develop, how the turbulence should decay in time, etc.

The most familiar regime of turbulence assumes an incompressible, unmagnetized fluid; in the classical Kolmogorov picture, the dynamics consists of a conservative energy cascade from large to small scales such that an eddy's velocity v scales with its size L as $v \propto L^{1/3}$, corresponding in wavenumber space to $v \propto k^{-1/3}$ hence an energy spectrum $dE/dk \propto k^{-5/3}$ [31]. Also well-studied by fluid dynamicists [32,33], but perhaps more familiar to cosmologists (e.g. [34]) than to the interstellar medium community, is Burgers turbulence. Burgers turbulence consists of the development of discontinuities in the velocity and density fields in pressure-free, unmagnetized, flows; the spectral description of the shocks that form is $dE/dk \propto k^{-2}$ (i.e. $v(k) \propto k^{-1/2}$). The Burgers model approximately describes the early supersonic phase seen in numerical simulations of compressible (unmagnetized) turbulence [35,36]. Most of the work on MHD turbulence has concentrated on the case of an incompressible medium, with different energy spectra predicted depending on on the system's dimensionality and the level of nonlinearity in the perturbations [37–39].

Sustained turbulence of any sort requires a "driver" – i.e. a source of energy to replace what is turned into heat (through viscous dissipation at the "bottom" of a cascade, or locally in shocks). In massive molecular clouds, a considerable stock of available energy exists in the gravitational potential. In clouds where stars are already forming, the outflows from young stars are very effective at returning this energy to the cloud, and may act as a feedback element in self-regulated star-formation [40,41]; it is also possible that local instabilities can tap into this gravitational store during the early contraction of a cloud, and that turbulence may be excited by energy (in shear and compressive motions) carried into the cloud from the exterior and partially trapped, cf. [42].

Because of the basic distinction between a conservative, incoherent energy cascade, and a collection of dissipative, coherent, local interactions (in shocks), the dynamics of star-forming regions might end up bearing more resemblance to compressible unmagnetized Burgers turbulence than to incompressible turbulence (magnetized or unmagnetized) [43]. Adding magnetic fields may, however, be expected to alter the dynamics of compressible, supersonic, self-gravitating fluid turbulence significantly, and indeed early MHD simulations showed many effects absent in unmagnetized simulations [44–46] (e.g. reduced shock strengths, strong transient matter compression by magnetic tension and pressure forces, reduction in contraction driven by self-gravity). Further discussion of these and more recent simulations appears below.

THE INTERNAL DYNAMICS OF SELF-GRAVITATING, MAGNETIZED CLOUDS

If we consider an idealized piece of a gas cloud cloud of uniform density n, temperature T, magnetic field B, and linear size L, we can define characteristic timescales which describe the relative importance of gravity, thermal pressure, and magnetic pressure/tension. Respectively, we define

$$t_G \equiv \frac{L_J}{c_s} = \left(\frac{\pi}{G\rho}\right)^{1/2} = 10^7 \text{ yr} \left(\frac{n}{10^2 \text{ cm}^{-3}}\right)^{-1/2}, \quad (1)$$

$$t_T \equiv \frac{L}{c_s} = 5 \times 10^7 \text{ yr} \left(\frac{L}{10 \text{ pc}}\right) \left(\frac{T}{10 \text{ K}}\right)^{-1/2}, \quad (2)$$

and

$$t_M \equiv \frac{L}{v_A} = \frac{L}{B/\sqrt{4\pi\rho}}$$
$$= 8 \times 10^6 \text{ yr} \left(\frac{L}{10 \text{ pc}}\right) \left(\frac{n}{10^2 \text{ cm}^{-3}}\right)^{1/2} \left(\frac{B}{10\mu G}\right)^{-1}. \quad (3)$$

Quiescent cloud stability

A uniform cloud with just thermal pressure is susceptible to gravitational instabilities on scales where $t_T > t_G$. Star-forming molecular clouds have $t_T \gg t_G$, implying they would be unstable by Jeans's original criterion if they were unmagnetized and quiescent. Since the Galactic star formation rate would be much too high if all the cold molecular clouds fragmented and collapsed to make stars in a time $\sim t_G$, the simplest Jeans analysis is clearly insufficient as a prescription for when gas is converted to stars (at least in present-day spiral galaxies like our own).

If we now allow for nonzero magnetic field strengths, we have the result that gravitational instabilities perpendicular to the mean magnetic field direction are suppressed by magnetic pressure on scales where $t_M < t_G$ (in the limit $t_M \ll t_T$) [47]. This criterion is equivalent to having the column density satisfy $N_H < 4 \times 10^{21}$ cm$^{-2}(B/10\mu\text{G})$; thus $B \sim 25\mu\text{G}$ would make observed clouds stable to Jeans modes with $\hat{k} \perp \hat{B}$. Such an initially quiescent cloud would tend to "pancake" along the direction of the uniform magnetic field, however, because there are no magnetic forces parallel to field lines to oppose gravity. We would require a ratio of B/N_H a factor of π larger to prevent fragmentation of a cold "pancake" threaded by uniform fields [48,49]; the critical cases of magnetized cloud equilibria have the same central flux-to-mass ratio [50,51].

Turbulence in magnetized clouds – quasilinear theory

In a time-dependent magnetic field, matter experiences restoring forces due to magnetic pressure and tension; this allows supersonic oscillatory motions without "collisions," hence reducing shock losses and permitting longer-lived turbulence. Arons & Max [52] early argued that the supersonic motions observed in molecular clouds may owe their existence to Alfvén-type (transverse) MHD waves which have $\delta B_\perp/B_0 = \delta v_\perp/v_A$ perpendicular to the mean magnetic field $\mathbf{B_0}$.

Because Alfvén waves have nonzero perturbations perpendicular to $\mathbf{B_0}$ (the time-averaged field), they offer a mechanism to inhibit gravitational collapse along the mean field direction [1,53]. ¿From quasilinear theory, time-averaged Alfvén waves yield a pressure force

$$\mathbf{F}_{wave} = -\frac{\nabla|\delta B_\perp|^2}{8\pi\rho}; \tag{4}$$

the corresponding polytropic wave pressure obeys $P_{wave} \propto \rho^{1/2}$ [54]. A pseudo-Jeans analysis with this quasilinear Alfvén wave pressure would then predict no mean-field collapse in a cloud provided

$$n_J \equiv \frac{L}{L_J} \lesssim \frac{\delta B_\perp}{4c_s\sqrt{\pi\rho}}, \tag{5}$$

i.e.

$$\delta B_{rms} \gtrsim 18\mu\text{G} \left(\frac{L}{20 \text{ pc}}\right) \left(\frac{n}{100 \text{ cm}^{-3}}\right) \tag{6}$$

and

$$\sigma_{v,\ wave} \gtrsim 2.3 \text{ km s}^{-1} \left(\frac{L}{20 \text{ pc}}\right) \left(\frac{n}{100 \text{ cm}^{-3}}\right)^{1/2}. \tag{7}$$

Because the gas in clouds is only partially ionized, ambipolar diffusion acts to damp such waves, especially at short wavelengths [3,55,56]. Unless the background magnetic field were extremely strong (milligauss-level), perturbations of the amplitude in equation (6) would be quite nonlinear, and significant energy losses would be incurred due to the generation of compressive motions (regardless of the initial polarization of the Alfvén wave), which steepen to shocks [57,58]. In summary, analytic arguments (described in this section) suggest that MHD waves should be important to the dynamics of star-forming regions; the actual nonlinear amplitudes of such waves, however, necessitates direct numerical simulations (described in the next section) to investigate the nature of the resulting MHD turbulence, and to quantitatively discriminate among model parameters.

SIMULATIONS OF MAGNETIZED, TURBULENT CLOUDS

In order to understand the physics of star-forming gas clouds, and eventually to quantify how parameters such as the mean magnetic field strength may influence the star formation rate or IMF in a cloud or galaxy, detailed studies are required to characterize the nature of turbulence under a realistic range of conditions. In particular, methodical campaigns of numerical MHD simulations should help address many of the longstanding problems in cloud dynamics for which magnetic fields have been proposed as a solution. Eventually, such MHD simulations will be able to address many aspects of cloud dynamics in a fairly realistic fashion – incorporating, for example, the effects of cosmic-ray heating, radiative cooling, ion-neutral diffusion, and the interface with the surrounding medium. Even very simple cloud models (e.g. isothermal or adiabatic equation of state with periodic boundary conditions and negligible ion-neutral diffusion), however, can be used in the initial attack on such important questions as:

- Can nonlinear Alfvén-like waves prevent or delay cloud collapse? How much wave energy is needed?

- What sorts of velocity and density spectra/structures develop in magnetized, compressible media?

- How do dissipation rates of MHD turbulence depend on σ_v, v_A, and c_s? How much energy input would be required to prevent collapse of a self-graviating cloud?

As mentioned above, questions like these have been addressed in several plane-parallel and two-dimensional MHD simulations in the early 1990's [44–46]. More recently, Gammie & Ostriker [59] performed a survey of plane-parallel simulations over a wide range of parameters describing the relative levels of kinetic, magnetic, thermal, and gravitational energies in a model cloud; these computations are presently being extended to three dimensions

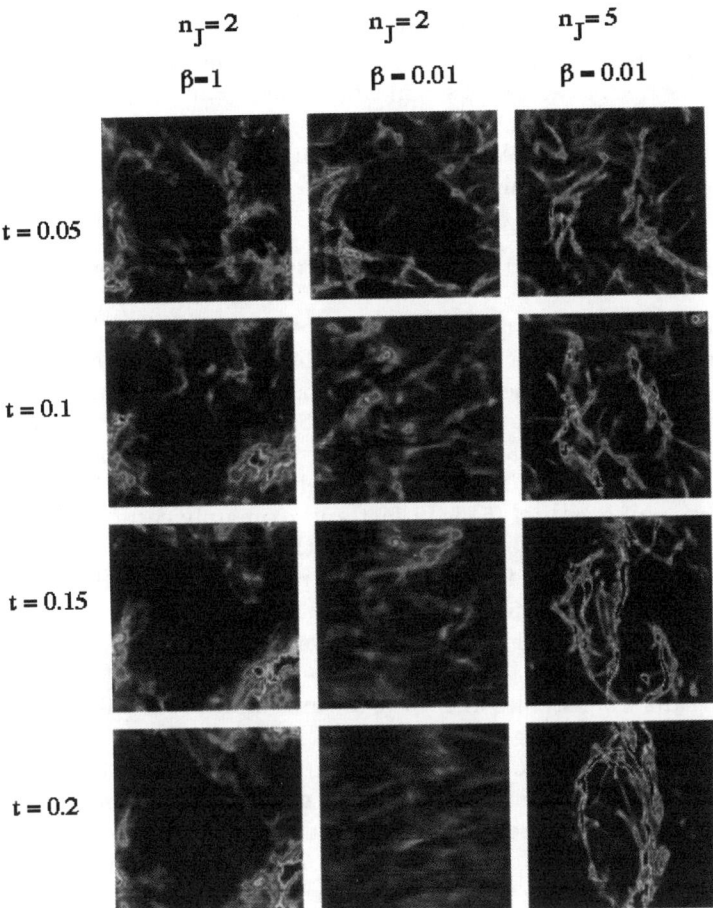

FIGURE 1. Evolution of the integrated line-of-sight density in three simulations with varying values of the mean magnetic field strength (described by $\beta = c_s^2/v_A^2 \equiv 4\pi\rho_0 c_s^2/B_0^2$) and size (described by the number of Jeans lengths $n_J \equiv L/L_J$). Time increases downward in each column from 0.05 to 0.2 sound crossing times, and the mean field runs right to left. All runs have uniform initial density, and initial wave energy $E_W = 100\bar{\rho}c_s^2 L^3$ in random velocity perturbations. The first column has $\beta = 1$ and $n_J = 2$; the second column has $\beta = 0.01$ and $n_J = 2$; the third column has $\beta = 0.01$ and $n_J = 5$. Higher column density regions are brighter (except for the central peaks, which appear dark in this monochromatic representation of a color figure).

[60]. In order to study cloud formation on large scales in a galactic disk, [61] have included parameterized shear flow, diffuse and localized heating, and cooling in a series of two-dimensional simulations. For additional related work on this topic, see the contributions to this Proceedings by Ballestros-Paredes, Balsara, Padoan, and Vazquez-Semadeni. Below I highlight some of the results that have been obtained so far.

Simulation results on cloud support by MHD waves

As described above, Alfvén-type MHD waves are expected to be able to support clouds against collapse along the mean field direction, provided that the energy in the time-dependent, perturbed magnetic field is large enough compared to the gravitational energy of the cloud. Early two-dimensional simulations indeed showed weaker gravitational contraction in models with stronger magnetic fields [46]. The plane-parallel simulations of [59] verified that the prediction of equation (5) is in fact approximately correct quantitatively – marking the boundary between clouds that collapse rapidly (within $\sim t_G$) and those that do not, for the case of initially turbulent clouds. For model clouds in which turbulence is continuously forced at scales smaller than the box, equation (5) marks the boundary between clouds that collapse and those that survive indefinitely [59].

The initial results from studies underway in three dimensions [60] concur in showing that sufficiently strong MHD turbulence can support a cloud against gravitational collapse. Figure 1 shows the time evolution of three different model clouds with different levels of mean magnetic field (measured by $\beta \equiv c_s^2/v_A^2$) and mass (measured by the number of Jeans lengths n_J in each boundary of the simulated region), but all the same initial turbulent energy in random velocities. Comparing the two models with the same mass ($n_J = 2$), but differing magnetization (weak field $\beta = 1$ and strong field $\beta = 0.01$), the figure shows that a strongly self-gravitating condensation forms in only the weakly-magnetized case. Comparing two models with the same magnetization ($\beta = 0.01$) but different gravitational potental energies (weak gravity $n_J = 2$, strong gravity $n_J = 5$), we see that only the $n_J = 5$ cloud that has gravity strong compared to the level of magnetic perturbations ends up collapsing.

Simulation results on density structure

All the simulations cited have discovered highly inhomogeneous/filamentary structure growing in the model clouds as they evolve. For example, in the plane-parallel simulations of [59], typically 70-80% of the volume has density less than the mean, while the preponderance of the mass (75-80%) is in condensations with density above the mean value. The self-gravitating simulations show weak positive correlation between magnetic pressure and gas density, with the correlation increasing over increasing spatial scale [61,59]. On the other hand, non-self-gravitating, isothermal simulations show anticorrelation between magnetic and gas pressure because it is the magnetic field that "confines" the gas clumps [44,59]. In three-dimensional runs [60], strongly magnetized clouds are found to have density structures aligned preferentially perpendicular to the mean field. Models with low magnetization show rounder clumps than those with high magnetization (see Fig. 1). In general, high density regions are correlated with discontinuities of **v** and **B**; low density regions have relatively straight field lines.

Simulation results on velocity/magnetic field perturbation spectra

Independent of the nature of the driving source of the turbulence (i.e. decay of a spectrum of perturbations imposed in the initial conditions or continuous energy injection; fluctuations imposed in just the velocity field, just the magnetic field, or both), the model cold, magnetized clouds evolve toward equipartition of energy in kinetic and fluctuating magnetic fields [61,59,60]. Because compressive motions damp more easily than transverse ones, the latter dominate the velocity field. Also independent of the spatial scale at which perturbations are imposed, the power spectra evolve toward distributions close to $v(k)^2 \propto B(k)^2/(4\pi\bar{\rho}) \propto k^{-2}$ [61,59,60]. The corresponding "linewidth-size" relation $\sigma_v \propto L^{0.5-0.7}$ has been directly verified for the plane-parallel case [59]. Both the plane-parallel and two-dimensional simulations have identified an "inverse cascade" of energy in growing power of δB and δv at large scale (provided that a mean magnetic field threads the box) [61,59].

Simulation results on dissipation rates

Dissipation rates are difficult to measure accurately in high dimensional computations because of the high resolution required. For the plane-parallel simulations [59], dissipation rates have been measured by equating the dissipation rate to the power injected in simulations which have reached a quasi-steady forced equilibrium. Given an supply of energy $\dot{E}_{\text{wave}}$ injected on a scale λ into a box of size L, the dissipation time is computed to be

$$t_{\text{diss}} \equiv \frac{E_{\text{wave}}}{\dot{E}_{\text{wave}}} = 0.5 \left(\frac{\lambda}{L}\right)^{1/2} \left(\frac{v_A}{c_s}\right)^{1/2} \frac{L}{\sigma_v}, \tag{8}$$

larger than the "eddy turnover time" by a factor $\sim (v_A/c_s)^{1/2}$. For typical GMCs, this result would imply that turbulence and support against collapse could be sustained indefinitely given a mechanical luminosity input of tens of

$L_\odot$. How this result will change in three-dimensional simulations with more realistic physics (especially ambipolar diffusion) remains to be seen.

ACKNOWLEGEMENTS

I am grateful to my collaborators Charles Gammie and Jim Stone and for their comments on this manuscript, and for permission to include Figure 1 in this review.

REFERENCES

1. Shu, F.H., Adams, F.C. & Lizano, S. 1987, ARAA, 25, 23
2. Königl, A., & Ruden, S. P., in Protostars & Planets III, ed. E. Levy & J. Lunine (Tucson: University of Arizona Press), p. 641
3. McKee, C. F., Zweibel, E. G., Goodman, A. A., & Heiles, C. 1993, in Protostars and Planets III, ed. E. Levy & J. Lunine (Tucson: University of Arizona Press), p. 327
4. Elmegreen, B. G. 1991, in The Physics of Star Formation and Early Stellar Evolution, Eds. C. J. Lada & N. D. Kylafis (Dordrecht: Kluwer), p. 35
5. Mouschovias, T., in The Physics of Star Formation and Early Stellar Evolution, Eds. C. J. Lada & N. D. Kylafis (Dordrecht: Kluwer), p. 61
6. Blitz, L. 1993, in Protostars and Planets III, ed. E. Levy & J. Lunine (Tucson: University of Arizona Press), p. 125
7. Cernicharo, J. 1991, in The Physics of Star Formation and Early Stellar Evolution, Eds. C. J. Lada & N. D. Kylafis (Dordrecht: Kluwer), p. 287
8. Lada, E. A., Strom, K. M., & Myers, P. C. 1993, in Protostars and Planets III, ed. E. Levy & J. Lunine (Tucson: University of Arizona Press),
9. Pérault, M., Falgarone, E., & Puget J. L. 1985, A&A, 152, 371
10. Williams, J. P., Blitz, L. & Stark, A. A. 1995, ApJ, 451, 252
11. Falgarone, E., Puget, J. L., & Pérault, M. 1992, A&A, 257, 715
12. Benson, P. J., & Myers, P. C. 1989, ApJS, 71, 89
13. Caselli, P., & Myers, P. C., ApJ, 446, 665
14. Goodman, A. A., Barranco, J. A., Wilner, D. J., & Heyer, M., ApJ, submitted
15. Larson, R. B. 1981, MNRAS, 194, 809
16. Dame, T. M., Elmegreen, B. G., Cohen, R. S., & Thaddeus, P. 1986, ApJ, 305, 892
17. Solomon, P. M., Rivolo, A. R., Barrett, J., & Yahil, A. 1987, ApJ, 319, 730
18. Myers, P. C., & Goodman, A. A. 1988b, ApJ, 329, 392
19. Loren, R. B. 1989, ApJ, 338, 925
20. Bertoldi, F. & McKee, C. F. 1992, ApJ, 395, 140
21. Goodman, A. A., & Heiles, C. 1994, ApJ, 424, 208
22. Goodman, A. A., Jones, T. J., Lada, E. A., & Myers, P. C. 1995, ApJ, 448, 748

23. Heiles, C., Goodman, A. A., McKee, C. F.,& Zweibel, E. G. 1993, in Protostars and Planets III, ed. E. Levy & J. Lunine (Tucson: University of Arizona Press), p. 279
24. Crutcher, R. M., Troland, T. H., Goodman, A. A., Heiles, C., Kazes, I., & Myers, P. C., ApJ, 407, 175
25. Troland, T. H., Crutcher, R. M., Goodman, A. A., Heiles, C., Kazes, I., & Myers, P. C. 1996, ApJ, 471, 302
26. Myers, P. C., & Goodman, A. A. 1988a, ApJ, 326, L27
27. Moneti, A., Pipher, J. L., Helfer, H. L., McMillan, R. S., & Perry, M. L., ApJ, 282, 508
28. Goodman, A. A., Bastien, P., Menard, F., & Myers, P. C. 1990, ApJ, 359, 363
29. Davidson, J. A., Schleuning, D., Dotson, J. L., Dowell, C. D., & Hildebrand, R. H. 1995, in Astronomical Society of the Pacific, Airborne Astronomy Symposium on the Galactic Ecosystem: From Gas to Stars to Dust, Volume 73, p 225
30. Hildebrand, R. H., Dotson, J. L., Dowell, C. D., Platt, S. R., Schleuning, D., Davidson, J. A., & Novak, G. 1995, in Astronomical Society of the Pacific, Airborne Astronomy Symposium on the Galactic Ecosystem: From Gas to Stars to Dust, Volume 73, p 97
31. Landau, L. D., & Lifschitz, E. M. 1987, Fluid Mechanics (New York:Pergamon)
32. Burgers, J. M. 1974, The Nonlinear Diffusion Equation (Reidel: Dordrecht)
33. Whitham, G. B. 1974, Linear and Nonlinear Waves (Wiley-Interscience: New York)
34. Gurbatov, S. N., Saichev, A. I., Shandarin, S. F. 1989, MNRAS, 236, 385
35. Porter, D. H., Pouquet, A., & Woodward, P. R. 1992, Phys. Rev. Lett. 68, 21
36. Porter, D. H., Pouquet, A., & Woodward, P. R. 1994, Phys. Fluids 6, 2133
37. Biskamp, D. 1993, Nonlinear Magnetohydrodynamics (Cambridge: Cambridge Univ. Press)
38. Sridhar, S., & Goldreich, P. 1994, ApJ, 432, 612
39. Goldreich, P., & Sridhar, S. 1995, ApJ, 438, 763
40. Norman, C., & Silk, J. 1980, ApJ, 238, 158
41. McKee, C. F. 1989, ApJ, 345, 782
42. Falgarone, E. & Puget, J. L. 1986, A&A, 162, 235
43. Passot, T., Pouquet, A., & Woodward, P. 1988, A&A 197, 228
44. Elmegreen, B. G. 1990, ApJ, 361, L77
45. Pouquet, A., Passot, T., & Leorat, J. 1991, in Fragmentation of Molecular Clouds and Star Formation, eds. E. Falgarone, F. Boulanger, & G. Duvert (Dordrecht: Kluwer), p. 101
46. Pouquet, A., Leorat, J., & Passot, T. 1991, in Advances in Turbulence 3, eds. A. V. Johansson & P. H. Alfredsson (Springer-Verlag: Berlin), p. 343
47. Chandrasekhar, S. & Fermi, E. 1953, ApJ, 118, 116
48. Nakano, T., & Nakamura, T. 1978, PASJ, 30, 671
49. Tomisaka, K., & Ikeuichi, S. 1983, PASJ, 35, 187
50. Mouschovias, T., & Spitzer, L. 1976, ApJ, 210, 326
51. Tomisaka, K., Ikeuchi, S., & Nakamura, T. 1988, ApJ, 335, 239

52. Arons, J. & Max, C. E. 1975, ApJ, 196, L77
53. Fatuzzo, M., & Adams, F. C. 1993, ApJ, 412, 146
54. McKee, C. F. & Zweibel, E. G. 1995, ApJ, 440, 686
55. Kulsrud, R. M., & Pearce, W. P. 1969, ApJ, 156, 445
56. Myers, P. C. & Khersonsky, V. K. 1995, ApJ, 442, 186
57. Zweibel, E. G., & Josafatsson, K. 1983, ApJ, 270, 511
58. Ghosh, S., & Goldstein, M.L. 1994, JGR, 99, 13351
59. Gammie, C. F., & Ostriker, E. C. 1996, ApJ, 466, 814
60. Gammie, C., Stone, J., & Ostriker, E. 1997, in preparation
61. Passot, T., Vázquez-Semadeni, E., & Pouquet, A. 1995, ApJ, 455, 536

Observations of Circumstellar Disks and Infall

Lee G. Mundy*

Astronomy Dept, Univ. of Maryland, College Park, MD 20742

Abstract. This paper presents an overview of the current status of millimeter and submillimeter wavelength observations of circumstellar disk and infall associated with young stars. Particular attention is paid to the progress and promise of high resolution interferometer observations.

INTRODUCTION

Considerable progress has been made in our observational understanding of the formation of stars over the past ten years. This progress has been driven by a wide range of technological developments including major improvements in infrared detectors, the coming of age of millimeter arrays, and the launching of the Hubble Space Telescope. Each of these major wavelength regimes – infrared, millimeter, and optical – has contributed vital information to our current understanding of circumstellar disks and the infall dynamics associated with star formation. In this paper, I will concentrate on recent results at millimeter and submillimeter wavelengths. In particular: how far have we progressed in our observational understanding of disks and infall? What are the important questions for the near future?

Millimeter and submillimeter wavelengths have proven particularly useful in studying the earliest stages of star formation for a number of reasons. First, these wavelengths provide access to a wide variety of molecular transitions which provide information about the physical conditions in the gas and about the chemical state of the gas. Second, the molecular line and dust continuum emission at these wavelengths are good tracers of the cool gas (10 to 200 K) characteristic of molecular clouds and circumstellar environments beyond a few tens of stellar radii. Third, the dust opacity at these wavelengths is small enough to permit direct observation of embedded systems and measurements of the bulk material distribution. And fourth, the resolution necessary to probe size scales from 10's to 1000's of AU is now possible. Thus, these wavelengths provide the opportunity to study the distribution and physical properties of the dominant material in circumstellar environments on linear scales comparable to our solar system.

The first section of this article discusses results which highlight the dynamics of the gas in young stellar systems. The second section looks at the ubiquity of circumstellar disks and new high resolution observations which are beginning to reveal the structure of disks. Known binary systems and the need to incorporate multiple star systems into our basic viewpoint are addressed in the third section. The issue of multiple stars is then extended to the cluster environment in the fourth section. The concluding section sets forth a short presentation of possible directions for future work.

INFALL, OUTFLOW, AND ROTATION

Star formation occurs when a molecular cloud core undergoes gravitational collapse. Such a collapse, if initially spherically symmetric and unimpeded by magnetic fields, rotation or anisotropic turbulence, results in a systematic, spherically symmetric velocity field with speeds characteristic of the gravitational potential. The specific structure of the velocity field depends on the nature of the physical processes that previously supported the cloud and the mechanism by which that support is rendered insufficient to prevent collapse. If the support is removed simultaneously, collapse motions occur throughout the core [16,33]. If the core undergoes an "inside-out" collapse [17,38] the collapse begins at the center and expands outward through the core at the speed of sound (c.f. [39]). Magnetic fields, if present at dynamically significant field strengths, can alter the velocity field by slowing the collapse or by breaking the spherical symmetry [6]. Similarly, a non-spherical initial configuration, rotation, or dynamical interaction between the stellar outflow and surrounding core can, and usually does, break the spherical symmetry of the velocity field for real sources.

The observational signature of an infalling envelope is clear once the envelope is internally heated by the forming star. Discussion and calculation of the red-blue asymmetry expected for optically thick lines can be found in a number of papers (c.f. [50,28]). For optically thick lines, the portion of the line blueward of ambient cloud velocity appears brighter than the redward portion because the gas contributing to blue-shifted velocities is closer to the center of the core, and hence hotter than, the gas at the corresponding red-shifted velocities. For optically thin lines, the red-blue asymmetry does not occur because all of the collapsing gas contributes to the line profile.

There are now a number of detailed observational examples of candidate infall envelopes in the literature. One of the earliest proposed infall systems was IRAS 16293-2422 [44] which showed a clear red-blue line asymmetry in the CS J=2-1 line. Other candidate infall systems with detailed studies include B335 [48], L1527 [51,27], and NGC 1333 IRAS 2 [46]. These systems are considered "candidate" infall systems because it is difficult to prove that their line profiles are not affected by rotational motions or by material interacting with the outflow. The literature on IRAS 16293-2422 provides a good illustration of this problem [24,50].

One approach to resolving this ambiguity between infall and other motions is to conduct surveys of young embedded sources [45,49,22]. If line asymmetries are caused by rotation or outflow interaction rather than infall, then line profiles of embedded YSO's should show no statistical preference for red or blue-shifted emission peaks. If the systems are dominated by infall motions, line profiles with blue-shifted peaks should predominate. The findings to date suggest that blue-shifted peaks are much more common than red-shifted ones, suggesting that infall is present in at least some fraction of the systems.

A second approach to resolving velocity ambiguities is to map the velocity structure of systems at high resolution and look for systematic behavior. For example, if the velocity field of a disk-like distribution is spatially resolved, the line of sight velocity along the minor axis of a rotation system is zero for material in circular orbit; hence, a velocity-position cut along the minor axis can highlight radial infall in a disk [9]. As shown by Ohashi et al. [32], rotation and infall also display distinct signatures in a velocity-position cut along the major axis of a flattened envelope. Unfortunately, kinematics of low velocity material entrained in outflows is not well understood and not likely to be so well behaved. As all well-studied infall "candidates" to date also have outflows, this uncertainty, along with nature's abhorrence of perfect spherical symmetry, leaves room for interpretation of the observations.

The HL Tau system is a good example of a system that has been extensively studied at high resolution and given various interpretations. First studied at high resolution by Sargent and Beckwith [37,36], the HL Tau system shows an elongated gas structure roughly 3000 AU in size, which was interpreted as a rotating disk. Hayashi et al. [9] present more detailed maps of the HL Tau system which they argue support the presence of both infall and rotation, with infall dominant. More recently, Cabrit et al. [2] made additional observations which they interpret as indicating the presence of infall and outflow motions, but not rotation. Overall, the observations provide good evidence for infall motions in the HL Tau system but the connection between some of the small-scale structure and features in the larger scale outflow, as elucidated by Cabrit et al., make detailed measurements of the infall and possible rotation problematic. A number of other T Tauri systems studied at high resolution show rotation, infall, or outflow behavior in various mixtures [31,30,13,35,8].

THE STRUCTURE OF CIRCUMSTELLAR DISKS

Observations accumulated over the past fifteen years allow a number of strong statements to be made about the nature of circumstellar disks. Cir-

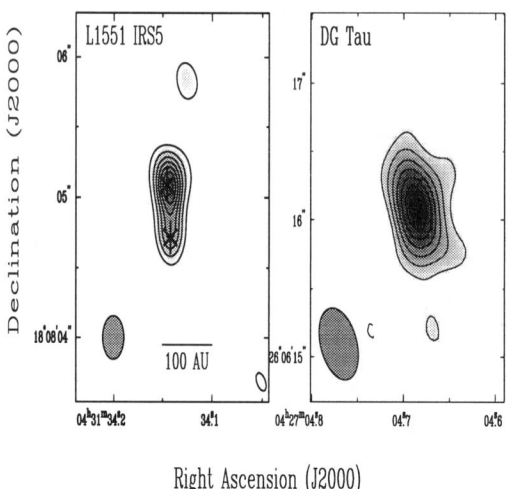

FIGURE 1. $\lambda=2.7$ mm continuum map of the disk around HL Tau overlayed on an HST image of the reflection nebulosity. The $\lambda=2.7$ mm emission (contours) traces the disk. The image was reconstructed with the 0."5 circular beam shown in the lower left corner.

cumstellar disks are common around young stars: most, perhaps all, stars in the T Tauri phase or younger have some form of circumstellar disk. While disk masses around solar type stars range up to ~ 0.1 $M_\odot$, the median mass is closer to ~ 0.02 $M_\odot$, or perhaps somewhat lower. The high column density portions of these disks are generally < 200 AU in radius. Extended thick disks or toroidal envelopes are often associated with embedded young stars. Circumstellar disks appear very early in these systems and often remain until well after the star enters the T Tauri phase.

Recently, progress has been made in resolving circumstellar disks around some nearby systems. For example, Lay et al. [18] used the single baseline CSO-JCMT interferometer to resolve the dust emission from the HL Tau and L1551 IRS5 systems. Assuming that the dust distribution could be characterized by an elliptical Gaussian, Lay et al. found deconvolved major axes of 160 AU (FWHM) for HL Tau and 120 AU for L1551 IRS5; in both cases the minor axis was ≤ 52 AU. These major axes values represent estimates of the circumstellar disk sizes in the systems, although the true disks are not expected to be Gaussians.

A full image of the HL Tau disk, obtained at $\lambda=2.7$mm with the BIMA array by Mundy et al. [26], is displayed in Figure 1. The 0."3 resolution image, which appears as contours overlaid on an HST image of the reflection nebulae [43], clearly shows emission extending over ~ 1 arcsecond along the edge of the reflection and centered on the centimeter radio position for the star (the asterisk). Fitting a standard power-law disk model to the $\lambda=2.7$mm

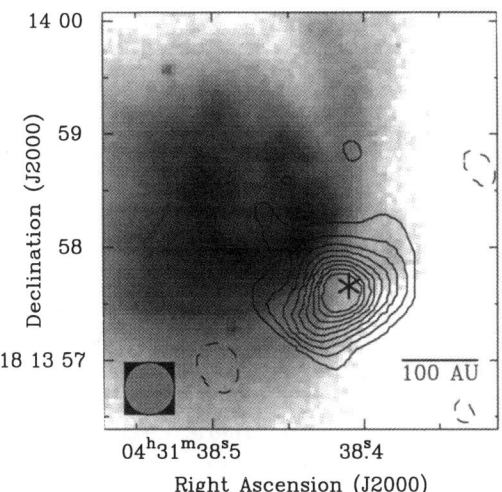

FIGURE 2. $\lambda=2.7$ mm continuum map of the circumstellar material in the L1551 IRS5 (left) and DG Tau (right) systems. The two asterisks in the left panel mark the positions of the $\lambda=2$ cm sources, which are likely the stellar positions in a binary system. In both L1551 IRS5 and DG Tau, the millimeter emission is tracing circumstellar disk material. The L1551 IRS5 image has a 0."3 circular beam; the DG Tau has a 0."64 x 0."45 beam.

maps, as well as the JCMT-CSO interferometer data of Lay et al. and flux measurements at millimeter through mid-infrared wavelengths, yields a disk outer radius of 90-180 AU. Figure 2 shows images of two other systems, L1551 IRS5 and DG Tau, with roughly 0."5 resolution; the dust emission from the L1551 IRS5 system appears to arise from two sources which are coincident with centimeter wavelengths sources, probably indicating that it is a binary system with two small (<30 AU) disks [21]. DG Tau has a disk comparable in size to that of HL Tau. In a broader survey with lower resolution (3" resolution), Dutrey et al. [4] find outer radii of typically ≤ 150 AU for 12 T Tauri type systems.

In addition to estimates of disk sizes, the current resolutions at millimeter and submillimeter wavelengths is becoming sufficient to permit estimates of the radial surface density distribution within disks. In the first such estimates, Mundy et al. [26] find power-law indices between 0 and 0.5 best fit the data for the HL Tau system. Dutrey et al. [4] estimate that their 12 systems are best characterized by surface density distributions flatter than $r^{-1.5}$. And Lay et al. [20] use new CSO-JCMT interferometer data at 220, 345, and 460 GHz to estimate that the most probable power-law index for the HL Tau disk is between 1.0 and 1.25. Clearly, the uncertainty in these estimates is significant, but the ability to measure the surface density distribution in circumstellar disks is coming within reach.

BINARY AND MULTIPLE STAR SYSTEMS

For the past fifteen years, models for star formation have centered around the ideal formation of a single star. Observations, wherever ignorance and lack of spatial resolution permitted, have been interpreted in terms of a single star system. It is now clear, however, that single star systems are not the norm. Surveys of pre-main sequence stars find that binaries are at least as common among these stars as among main sequence stars, which means that roughly two thirds of young stars are in binary systems [40,7]. Thus, most stars must form in multiple star systems. So, where are these systems among the embedded population? What are the properties of circumstellar disks in multiple star systems?

There have been several recent surveys comparing dust emission properties of binary and single star systems where the multiplicity of the system is determined by optical or near infrared observations [40,41,12]. The overall conclusions of these works is that multiple systems generally contain less gas and dust than single star systems. This is particularly true for binary separations of $\leq$100 AU and becomes marginal for wide binaries. The gas and dust in binary systems is generally associated with the individual sources, although circumbinary material is seen.

In addition to broad, low resolution surveys, a growing number of binary systems (or serendipitous binaries) have been imaged with millimeter interferometers. These systems show a variety of behaviors. The close (40 AU separation) binary GG Tau shows strong dust emission which arises from circumbinary material extending out to a radius of roughly 400 AU [5]. UZ Tau shows dust emission associated with both the E and W components (separated by 530 AU) [11]. As shown in Figure 2, the well known L1551 IRS5 system is likely to be a close binary (50 AU separation) with individual circumstellar disks [21]. Imaging more such systems is needed to draw any broad conclusions.

For deeply embedded sources where optical and near infrared techniques can not penetrate, the binarity of even well-studied systems is unknown. The deeply embedded systems NGC 1333 IRAS4 [19] and IRAS 16293-2422 [47] were both found to be binary systems with separations of roughly 400 and 750 AU, respectively. These systems are likely to be the tip of the iceberg since few observations to date have been made with sufficient resolution to detect binaries with separations "typical" of optical stars ($\sim$40 AU [3].

DISKS IN CLUSTER ENVIRONMENTS

On a larger scale than multiple star systems, it is well established that the majority of stars form in dense clusters [14]. How does the cluster environment affect star formation? Does it change the nature of the infall? Do circumstellar disks still play an important role in the star formation process? None of these questions can really be answered yet, but studies at millimeter and submillimeter wavelengths will help find the answers. With the recent improvements in the resolution and dynamic range of millimeter interferometers, the study of star formation within clusters is now becoming possible at millimeter and submillimeter wavelengths. Recent works on the Trapezium cluster illustrate the results that such studies can yield.

Optical and near infrared studies of the Trapezium cluster have shown that $\geq 50\%$ of the stars in the cluster have circumstellar material (c.f. [42]). HST imaging of a number of systems within the central cluster have revealed bright compact nebulae which have been interpreted as protoplanetary disks [29] and dark shadow disks suspected to be circumstellar disks seen in projection against the H II region [23]. One of the key questions about these systems is how much mass is present in the circumstellar material? While optical observations can only provide a lower limit on the mass, high resolution millimeter and sub-millimeter interferometry can image the bulk material distributions and derive disk masses for these cluster stars.

A first attempt at measuring disk masses in the inner portion of the Trapezium was conducted by Mundy *et al* [25] at $\lambda=2.7$mm. They were able to set an upper limit of 0.1 $M_\odot$ on the mass of individual disks and an upper limit of 0.03 $M_\odot$ on the mean disk mass for all previously proposed disk systems in their survey region. These limits eliminate the possibility that cluster stars have massive disks but allow the possibility that cluster stars have disks with the same mass distribution as disks around stars in loose associations or in the field.

The mass limits set in such studies can be improved by a factor of eight or more by observing at $\lambda=1.3$ mm wavelength. Preliminary results from a $\lambda=1.3$mm search of two fields in the Trapezium cluster with the IRAM interferometer [15] suggest that the Trapezium disk systems are significantly less massive than the above limits. Of the fifteen systems in the surveyed region, three were detected and may have disk masses as large as 0.014 $M_\odot$; the twelve undetected systems have masses ≤ 0.005 $M_\odot$. These limits are pushing below the typical masses for disks around T Tauri stars and are low enough to raise questions about the possibility of planet formation in these systems.

ISSUES FOR THE NEAR FUTURE

The future for studies of star formation at millimeter and submillimeter wavelengths is indeed bright as the full capabilities of the existing instrumentation are only beginning to be utilized. Further detailed observations of envelope kinematics for isolated forming stars and for stars forming in small clusters are needed to test current star formation theories. Observations and modeling of infall, outflow, and rotation on the circumstellar scale can be done at factors of 2-4 better resolution than currently published works. Crude measurements of radial surface density and temperature structures of disks in the Taurus and Ophiuchi region should be possible at $\lambda=1.3$ mm with the current millimeter arrays, and even more detailed work will become possible when the SAO Submillimeter Array comes online. Measurements of masses of disks in nearby clusters, and surveys of the cluster age dependence and stellar mass dependence of circumstellar disks, can be done on the existing arrays with sufficient array time. Finally, while modest progress has been made to date on the measurement of magnetic fields in the circumstellar environment [1,10], the instrumentation is in place to push forward.

REFERENCES

1. Akeson, R.L., Carlstrom, J.E., Phillips, J.A., and Woody, D.P., *ApJ* **456**, L45 (1996).
2. Cabrit, S., Guilloteau, S., André, P., Bertout, C., Montmerle, T., and Schuster, K., *A&A* **305**, 527 (1996).
3. Duquennoy, A., and Mayor, M., *A&A* **248** 485 (1991).
4. Dutrey, A., Guilloteau, S., Duvert, G., Prato, L., Simon, M., Schuster, K., and Meńard, F., *A&A* **309**, 493 (1996).
5. Dutrey, A., Guilloteau, S., and Simon, M., *A&A* **286**, 149 (1994).
6. Galli, D., and Shu, F.H., *ApJ* **417**, 243 (1993).
7. Ghez, A.M., Neugebauer, G., Matthews, K., *AJ* **106**, 2005 (1993).
8. Guilloteau, S., and Dutrey, A., *A&A* **291**, L23 (1994).
9. Hayashi, M., Ohashi, N., and Miyama, S.M., *ApJL* **418**, L71 (1993).
10. Holland, W.S., Greaves, J.S., Ward-Thompson, D., and Ph. André, *A&A* **309**, 267 (1996).
11. Jensen E.L.N., Koerner, D.W., and Mathieu, R.D., *AJ* **111**, 2431 (1996).
12. Jensen, E.L.N., Mathieu R.D., and Fuller, G.A., *ApJ* **458**, 312 (1996).
13. Koerner, D.W., and Sargent, A.I., *AJ* **109**, 2138 (1995).
14. Lada, E.A., Strom, K.M. and Myers, P.C., and Lizano, S., *Protostars and Planets III*, Tucson: Arizona Press, 1993, pp. 245-277.
15. Lada, E.A., Mundy, L.G., Guilloteau, S., and Dutrey, A., *ApJ in preparation* (1997).
16. Larson, R.B., *MNRAS* **145**, 271 (1969).
17. Larson, R.B., *MNRAS* **157**, 121 (1972).
18. Lay, O.P., Carlstrom, J.E., Hills, R.E., and Phillips, T.G., *ApJL* **434**, L75 (1994).
19. Lay, O.P., Carlstrom, J.E., and Hills, R.E., *ApJL* **452**, L73 (1995).
20. Lay, O.P., Carlstrom, J.E., and Hills, R.E., *ApJ in preparation* (1997).
21. Looney, L.W., Mundy, L.G., and Welch W.J. *ApJL submitted* (1997).
22. Mardones, D., Bachiller, R., Myers, P.C., Tafalla, M., *ApJ submitted* (1997).
23. McCaughrean, M.J., and O'Dell, C.R., *A&A* **111**, 1977 (1996).

24. Menten, K.M., Serabyn, E., Gusten, R., and Wilson, T.L., *A&A* **177**, L57 (1987).
25. Mundy, L.G., Looney, L.W., and Lada, E.A., *ApJL* **452**, L137 (1995).
26. Mundy, L.G., *et al.*, *ApJL* **464**, L169 (1996).
27. Myers, P.C., Bachiller, R., Caselli, P., Fuller, G.A., Mardones, D., Tafalla, M., and Wilner, D.J., *ApJL* **449**, L65 (1995).
28. Myers, P.C., Mardones, D., Tafalla, M., Williams, J.P., and Wilner, D.J., *ApJL* **465**, L133 (1996).
29. O'Dell, C.R., and Wen, Z., *ApJ* **436**, 194 (1994).
30. Ohashi, N., Hayashi, M., Ho, P.T.P, Momose, M., and Hirano, N., *ApJ* **466**, 957 (1996).
31. Ohashi, N., Hayashi, M., Kawabe, R., and Ishiguro, M., *ApJ* **466**, 317 (1996).
32. Ohashi, N., Hayashi, M., Ho, P.T.P., and Momose, M, *ApJ* **475**, 211 (1997).
33. Penston, M.V., *MNRAS* **144**, 425 (1969).
34. Prosser, C.F., Stauffer, J.R., Hartmann, L., Soderblom, D.R., Jones, B.F., Werner, M.W., and McCaughrean, M.J., *ApJ* **421**, 517 (1994).
35. Saito, M., *et al.*, *ApJ* **453**, 384 (1995).
36. Sargent, A.I., and Beckwith, S.V.W., *ApJL* **382**, L31 (1991).
37. Sargent, A.I., and Beckwith, S., *ApJ* **323**, 294 (1987).
38. Shu, F.H., *ApJ* **214**, 488 (1977).
39. Shu, F.H., Najita, J., Galli, D., Ostriker, E., and Lizano, S., *Protostars and Planets III*, Tucson: Arizona Press, 1993, pp. 3-45.
40. Simon, M., Chen, W.P., Howell, R.R., Benson, J.A., and Slowik, D., *ApJ* **384**, 212 (1992).
41. Simon, M., *et al.*, *ApJ* **443**, 625 (1995).
42. Stauffer, J.R., Prosser, C.F., Hartmann, L., and McCaughrean, M.J., *AJ* **108**, 1375 (1994).
43. Stapelfeldt, K.R., *et al.*, *ApJ* **449**, 888 (1995).
44. Walker, C.K., Lada, C.J., Young, E.T., Maloney, P.R., and Wilking, B.A., *ApJL* **309**, L47 (1986).
45. Wang, Y., Evans, N.J.II, Zhou, S., and Clemens, D.P., *ApJ* **454**, 217 (1995).
46. Ward-Thompson, D., Buckley, H.D., Greaves, J.S., Holland, W.S., and André, P., *MNRAS* **281**, L53 (1996).
47. Wootten, A., *ApJ* **337**, 858 (1989).
48. Zhou, S., Evans, N.J.II, Kompe, C., and Walmsley, C.M., *ApJ* **404**, 232 (1993).
49. Zhou, S., Evans, N.J.II, Wang, Y., Peng, R., and Lo, K.Y., *ApJ* **433**, 131 (1994).
50. Zhou, S., *ApJ* **442**, 685 (1995).
51. Zhou, S., Evans, N.J.II, and Wang, Y., *ApJ* **466**, 96 (1996).

YSOs in Transition

Eugene A. Magnier

Astronomy Dept. 351580, University of Washington, Seattle, WA 98195, USA

Abstract. We are studying YSOs at the transition between the embedded and exposed evolutionary stages. In this paper, we discuss the use of the reflection nebulae to study the inner regions of these systems, where direct imaging of the accretion disk is frequently unfeasible, even with HST.

We are studying a group of young stellar objects (YSOs) which appear to be close to the transition between the embedded and exposed phases of YSO evolution. These sources have flat IRAS spectra somewhat intermediate between the Lada & Wilking [1] Class I (embedded) and Class II (exposed) IRAS sources. These systems have other properties, some of which are typical of Class I and others of Class II objects. For example, many have strong CO emission or outflows, typical of Class I objects, but they also have visible central stars, typical of Class II objects. In these systems, the presence of a reflection nebula and a visible central star allows us to draw some interesting conclusions on the structure of the inner regions of the system, without directly imaging that region. Of these systems, we have concentrated on the source nicknamed "Holoea" (IRAS 05327+3404), and will use that source as an example of our work on the other sources.

Figure 1 shows an I band and a K band image of Holoea. The nebulosity seen in this image is a reflection nebula. There is a clear morphological difference between these two images, and in fact there is a smooth change of the morphology from V to K. We can use the reflection nebula to probe the structure of the accretion disk near the central star.

First, we make the assumption that the bulk of the light from the central source is coming to us directly from the central star, with some amount of extinction by intervening dust, but without scattering. Thus, the observed flux from the central star is:

$$f_\star(\lambda) = f_{\star,0}(\lambda) e^{-0.921 A_{V,1} \frac{A_\lambda}{A_V}} \quad (1)$$

where $f_\star$ is the observed flux of the central star, $f_{\star,0}$ is the unextinguished flux of the central star, $A_{V,1}$ is the extinction to the central star along the

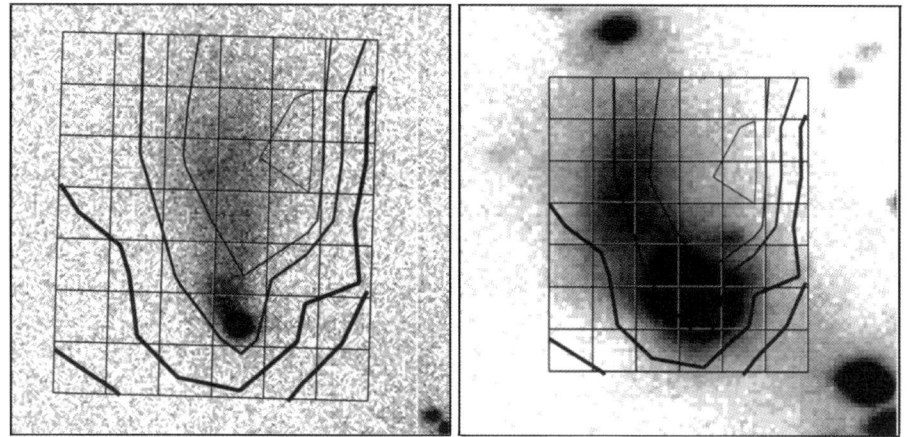

FIGURE 1. I and K images of Holoea (IRAS 05327+3404) with extinction contours overlayed (see text).

direct line of sight, and $\frac{A_\lambda}{A_V}$ is the extinction law. We assume the cannonical extinction law, with $E(B-V) = 3.1 A_V$. Next, we assume that the light we observe from the reflection nebula is the result of scattered light from the central star. The flux we observe from the reflection nebula therefore has both the effects of extinction as well as scattering:

$$f_{neb}(\lambda) = f_{\star,0}(\lambda) e^{-0.921 A_{V,2}(\frac{A_\lambda}{A_V})} \beta (\frac{\lambda}{\lambda_o})^{-\alpha} \qquad (2)$$

Where $f_{neb}(\lambda)$ is the observed flux from the reflection nebula, $A_{V,2}$ is the total extinction between us and the star along the path to the reflection nebula, β is a normalization factor to determine how much light is scattered by the dust particles in the nebula, and α is the slope of the scattering function. For Raleigh scattering, $\alpha = 4$, and for most other types of scattering, α is still quite close to 4 [3].

Dividing these terms, we can remove the uncertainty in the intrinsic spectrum of the central star:

$$\frac{f_n}{f_\star} = e^{-0.921(A_{V,2}-A_{V,1})\frac{A_\lambda}{A_V}} \beta (\frac{\lambda}{\lambda_o})^{-\alpha}. \qquad (3)$$

Thus, by fitting to the flux ratios, we can determine the increased amount of extinction experienced by the light as it travels to different locations in the reflection nebula from the central star, $\Delta A_V = A_{V,2} - A_{V,1}$.

To perform this measurement, we have divided the images into a 6×7 grid of boxes (see Figure 1). For every box, we fit the predicted flux ratio above, with free parameters of ΔA_V and β. In Figure 1 we show contour plots of the measured values of ΔA_V across the image. The trend in this figure is clear: the regions along the optical tail (and slightly to the west) have the lowest total extinction, while the extiction increases gradually as the line of sight from the star to the nebula tilts towards the east or the west. This is what would be expected if the regions in the east and west, with high extinction, were the equatorial regions of the system, and light from the central star must pass through significant amounts of material (i.e., an accretion disk) to reach the reflection nebula. The region towards the north, along the optical tail, is roughly the direction towards which an ionized jet is observed in spectra [2]. It is expected that the ionized jet is travelling perpendicular to the accretion disk, which is consistent with what we observe.

This analysis shows that the asymmetry observed in the reflection nebula at longer wavelengths can be understood as a variation in the amount of dust in these areas available for the scattering process. The extinction contours represent the amount of disk material, which is probably quite close to the central star, perhaps within $0.1''$, corresponding to ~ 100 AU, the typical size of YSO accretion disks. The dust causing the scattering is on a much larger scale and may be left over from the original cloud. The fact that the extinction contours are more or less symmetrical about the axis of lowest extinction (perpendicular to the assumed disk), even though the emission is predominantly on the east, lends credence to this suggestion.

Above, we found that the light which is known to be scattered, i.e., that which comes from the reflection nebula, is well fit by a spectrum incorporating the flux of the central source plus scattering. This reinforces the assumption that the central source is a direct (unscattered) view of the central star. If the central source resulted from scattered light, as in L 1551 IRS 5 then the reflection nebula would have to be the result of *two* scatterings to be consistent with the spectral fits. This seems to be quite implausible. It is very difficult to arrange a model in which the light we observe from the central source arrives after one scatter, and the reflection nebula receives scattered light as well.

We have identified roughly 15 sources which appear to be at a similar evolutionary state as Holoea and which have both a reflection nebula and central source visible. We are currently obtaining optical and near-IR images of these sources to perform similar studies of the accretion disks in a variety of systems.

REFERENCES

1. Lada C.J., Wilking B.A., 1984, ApJ 287, 610.
2. Magnier E.A., Waters L.B.F.M., Kuan Y.-J., et al., 1996, A&A 305, 936
3. Whittet D.C.B., 1992, Dust in the Galactic Environment. IOP Publishing Ltd., London, p. 61

Self-Similar Evolution of Supercritical Cores

Shantanu Basu

Canadian Institute for Theoretical Astrophysics
University of Toronto
60 St. George Street
Toronto, Ontario M5S 3H8, Canada

Abstract. We use a semi-analytic model to examine the collapse of supercritical cores (i.e., cores with a mass-to-flux ratio exceeding a critical value). Recent numerical simulations of the formation and contraction of supercritical cores show that the inner solution tends toward self-similar evolution. We use this feature to develop analytic expressions for quantities such as the density, angular velocity, and magnetic field. All forces involved in the problem (e.g., gravitational, magnetic, thermal, and centrifugal) can be calculated analytically in the thin-disk geometry of the problem. The role of each force during the contraction is analyzed, and we identify the key role of ambipolar diffusion in accelerating the collapse. We find that the collapse is dynamic and supersonic velocities are achieved in the innermost region of the core by the time of protostar formation. The mass accretion rate is significantly greater than the canonical C^3/G at the moment of protostar formation, although we argue that it is time-dependent and will eventually decrease. Comparisons are made with the predictions of existing spherical similarity solutions.

INTRODUCTION

The collapse of gravitationally unstable objects is a central problem in the theory of star formation. Similarity solutions play an important role in understanding gravitational collapse since they provide a simple analytical description of the evolution. Larson (1969) showed that the approach to protostar formation in his numerical simulations could be described by an isothermal similarity solution. This similarity solution, also developed by Penston (1969), is known as the Larson-Penston (LP) solution. A different similarity solution has been found by Shu (1977), which begins at the moment of protostar formation (identified with the presence of a central density singularity and labeled as $t = 0$ in both solutions) and describes the subsequent accretion onto the central point mass. The two similarity solutions differ greatly in their description of the protostellar environment at $t = 0$ (see § 4). However, both solutions assume spherical symmetry and do not include the effects of magnetic fields or rotation. In this paper, we estimate the conditions at $t = 0$ when these sources of support are taken into account. We use the results of detailed magnetohydrodynamic (MHD) numerical simulations (Fiedler & Mouschovias 1993; Ciolek & Mouschovias 1994, hereafter CM94; Basu & Mouschovias 1994, hereafter BM94) to develop analytic expressions for important physical quantities. We find that magnetic fields *cannot* enforce a near-quasistatic approach to protostar formation, due to the effect of ambipolar diffusion, and that the maximum velocity in the innermost region is supersonic. This means that the inner solution is qualitatively more similar to the LP solution for spherical collapse. We find that the mass accretion rate onto the protostar is much higher than in the Shu solution, although we expect that it is time-dependent and will eventually decrease.

SELF-SIMILAR PROFILES

Molecular clouds with initially subcritical mass-to-flux ratios (i.e., magnetically dominated) tend to flatten along the (vertical, or z) direction of the ambient magnetic field, and evolve due to ambipolar diffusion, the drift of neutral particles relative to charged species. When a supercritical region (core) is formed, gravitational contraction proceeds more rapidly, and the inner core

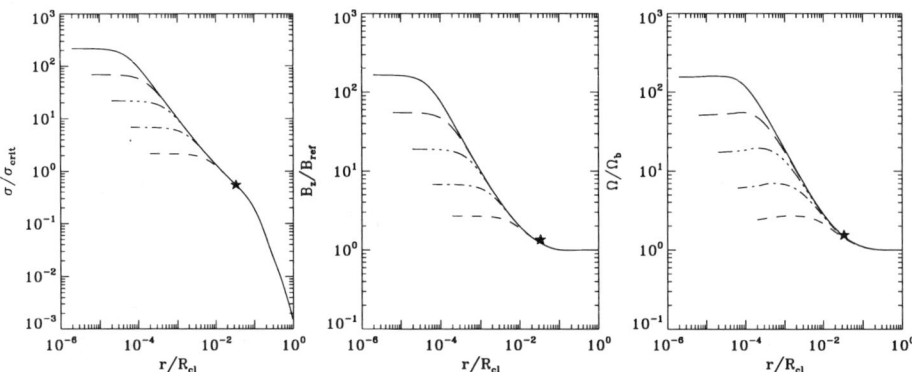

FIGURE 1. Spatial profiles of physical quantities at five different times following the formation of a supercritical core, in the standard model of BM94. The star marks the boundary of the supercritical core. The column density σ is normalized to the value $\sigma_{\rm crit}$ at which a critical central mass-to-flux ratio is achieved, the vertical magnetic field B_z is normalized to the ambient field $B_{\rm ref}$, and the angular velocity Ω is normalized to the background value Ω_b. All are plotted versus the radius r normalized to the initial cloud radius $R_{\rm cl} = 5.76$ pc. The inner profiles tend to evolve in a self-similar manner.

ultimately evolves through a series of self-similar profiles. Figure 1 shows the evolution of the supercritical core in the standard model of BM94.

The self-similarity of the inner core takes the following form. The (vertically integrated) column density σ, vertical component of the magnetic field B_z, and angular velocity Ω are described by

$$\sigma(r,t) = \sigma_c(t) \Big/ \sqrt{1+(r/R)^2}, \tag{1}$$

$$B_z(r,t) = 2\pi\sqrt{G}\,\mu^{-1}\,\sigma_c(t) \Big/ \sqrt{1+(r/R)^2}, \tag{2}$$

$$\Omega(r,t) = 2\Omega_c(t)\left(\frac{R}{r}\right)^2 \left[\sqrt{1+(r/R)^2} - 1\right], \tag{3}$$

where $R = R(t)$ is the scale factor. Equation (2) is obtained from equation (1) since μ, the mass-to-flux ratio in units of the critical value $(2\pi\sqrt{G})^{-1}$ for a uniform disk (Nakano & Nakamura 1978), is spatially uniform in the inner region (CM94; BM94). Equation (3) follows from equation (1) since the specific angular momentum is a linear function of the enclosed mass (BM94). Angular momentum conservation implies $\Omega_c(t) = \Omega_b\,\sigma_c(t)/\sigma_{\rm crit}$, where Ω_b is the rotation rate of the background medium and $\sigma_{\rm crit}$ is the column density at which the central flux tube achieves a critical mass-to-flux ratio.

The scale factor $R(t)$ measures the size of a central near-uniform column density region. Physically, it is the region in which thermal-pressure plays a significant role and helps to smooth out any density inhomogeneities. Therefore, we can relate it to the critical thermal length scale $\lambda_{\rm T,cr}$ for a thin disk (see discussion in CM94 and BM94), i.e.,

$$R(t) = \ell\lambda_{\rm T,cr} \equiv \ell\,C^2/2G\sigma_c(t), \tag{4}$$

where C is the isothermal sound speed. The numerical simulations of BM94 show that $\ell \simeq 2$ in the late stages of collapse.

Equations (1)-(3) give excellent fits to the inner profiles of the supercritical cores calculated by BM94, and allow an analytic study of the central evolution.

APPROACH TO PROTOSTAR FORMATION

The analytic profiles given above allow a calculation of all forces in the inner region of the thin-disk cloud, using the formulas given by CM93 and BM94. We find that the ratio of thermal-pressure to gravitational acceleration $a_{\rm T}/|g_r|$ stays constant during the collapse, and is equal to $1/\pi = 0.32$ in the central region. The central centrifugal support $a_c/|g_r|$ also stays constant to the extent that angular momentum is conserved during the supercritical phase. However, it equals $\sim 10^{-5}$, so that centrifugal support is negligible during this stage. The central magnetic support is $a_{\rm M}/|g_r| = (1+2/\pi)/\mu^2$ and is also constant during collapse if the central mass-to-flux ratio μ is fixed (flux-freezing). However, ambipolar diffusion continues to play a slow but significant role in the supercritical phase, causing an increase in μ of the form

$$\mu \propto \sigma_c^\epsilon, \tag{5}$$

where $\epsilon \simeq 0.05$ in the standard model of BM94. This is significant enough to eventually reduce the once dominant magnetic force to a secondary source of support to thermal-pressure in the inner core (see quantitative discussion in Basu 1997). However, the magnetic forces continue to be the dominant source of support in the outer core and certainly in the subcritical envelope.

We can also estimate the infall velocity v_r (consistent with self-similarity) in the manner originally used by Narita, Hayashi, & Miyama (1984) to analyze self-similar evolution in rotating, nonmagnetic clouds: 1) we use the mass continuity equation with the column density profile of equation (1) to find

$$v_r = \frac{R}{r}\left(\sqrt{1+r^2/R^2}-1\right)\dot{R}, \tag{6}$$

and 2) we differentiate the above expression and equate (to first order in r/R) to the analytically calculated accelerations obtained from equations (1)-(3). Taking the limit of protostar formation ($\sigma_c \to \infty$ or $R \to 0$), when thermal-pressure forces provide the dominant central support, we find

$$\dot{R} = -\sqrt{2(\pi-1)}\, C = -2.07\, C. \tag{7}$$

Equation (6) shows that v_r tends to a constant value $v_r = \dot{R}$ for $r \gg R$. In the limit that $R \to 0$, v_r has this value at all radii. Thus, the mass accretion rate at protostar formation is

$$\dot{M} = -2\pi\sigma r v_r = 2\pi\frac{\sigma_c R}{r} r\, 2.07\, C = 13\,\frac{C^3}{G}, \tag{8}$$

where we have used equation (4) to replace the constant value $\sigma_c R$.

CONCLUSIONS

The study of similarity solutions for spherical, isothermal collapse have found two very different scenarios. The solution of Shu (1977) envisages a slow, quasistatic approach to protostar formation ($t = 0$), followed by a dynamic inside-out collapse of the cloud onto the protostar. There is negligible acceleration before the stellar core forms and the subsequent mass accretion rate is $\dot{M} = 0.975\, C^3/G$. In the LP solution, the contraction is dynamic as $t = 0$ is approached (net central acceleration $a_{\rm tot} = -0.4|g_r|$) and the infall velocity is $v_r = -3.3\, C$ at all radii at $t = 0$. The mass accretion rate is

$\dot{M} = 29\,C^3/G$ at $t = 0$ and rises to $\dot{M} = 47\,C^3/G$ for $t > 0$ (Hunter 1977), when a free-fall flow is established in the vicinity of the protostar.

Our semi-analytic model, which is based on numerical simulations which realistically model the effects of magnetic fields and rotation, allows us to estimate core properties during the approach to protostar formation. Our conclusions are: 1) The inner region achieves considerably dynamic contraction *before* a protostar is formed. The central acceleration tends to the limiting value $a_{\text{tot}} = -0.68|g_r|$ as magnetic support is lost due to the relatively slow but effective magnetic flux diffusion in the inner region. 2) The infall velocity is supersonic, and tends to the limiting value $v_r = -2.07\,C$. 3) The mass accretion rate at $t = 0$ approaches

$$\dot{M} = 13\frac{C^3}{G} = 13(1.6 \times 10^{-6})\left[\frac{C}{0.19\,\text{km s}^{-1}}\right]^3 \frac{M_\odot}{\text{yr}}. \tag{9}$$

Therefore, the collapse in the inner region of our model is qualitatively more similar to the LP solution, though there are quantitative differences.

We believe that our model agrees with observations which suggest that the mass accretion rate around a newly formed protostar (a Class 0 object) is considerably higher than C^3/G (Ward-Thompson 1996; Bontemps et al. 1996). Ward-Thompson (1996) estimates a mass accretion rate of $\sim 10^{-5} M_\odot$ yr^{-1} in the Class 0 stage, in agreement with our estimate in equation (9). The estimated accretion rate at the later Class I phase is $\sim 10^{-6} M_\odot$ yr^{-1} and implies that $\dot{M}$ decreases with time following protostar formation. This is inconsistent with all of the similarity solutions, which require that the infall velocity is the same at all radii at $t = 0$ (e.g., see eq. [6]) and predict constant mass accretion rates for $t > 0$. However, a similarity solution is only valid in a relatively small inner region (where boundary effects are not important), and numerical simulations show that the infall velocity is *not* spatially constant as protostar formation is reached. The MHD simulations of core collapse provide a natural reason; the inner solution has to match onto an outer solution of slow (ambipolar-diffusion regulated) infall through a subcritical envelope. Hence, the magnitude of the infall velocity peaks in the inner core and decreases outward toward the core boundary. The hydrodynamic simulations of Hunter (1977) and Foster & Chevalier (1993), which also had outwardly decreasing speeds due to the presence of a numerical boundary and followed the evolution past $t = 0$, found that $\dot{M}$ achieved a peak value just after $t = 0$ and subsequently decreased as mass shells with successively smaller infall speeds at $t = 0$ reached the protostar. We expect the same pattern in the realistic case with magnetic fields, although quantitative estimates of $\dot{M}$ for $t > 0$ in a supercritical core must await detailed calculations.

REFERENCES

Basu, S. 1997, ApJ, submitted
Basu, S., & Mouschovias, T. Ch. 1994, ApJ, 432, 720 (BM94)
Bontemps, S., André, P., Terebey, S., & Cabrit, S. 1996, A&A, 311, 858
Ciolek, G. E., & Mouschovias, T. Ch. 1993, ApJ, 418, 774 (CM93)
_____. 1994, ApJ, 425, 142 (CM94)
Fiedler, R. A., & Mouschovias, T. Ch. 1993, ApJ, 415, 680
Foster, P. N., & Chevalier, R. A. 1993, 416, 303
Hunter, C. 1977, ApJ, 218, 834
Larson, R. 1969, MNRAS, 145, 271
Nakano, T., & Nakamura, T. 1978, PASJ, 30, 671
Narita, S., Hayashi, C., & Miyama, S. M. 1984, Prog. Theor. Phys., 72, 1118
Penston, M. V. 1969, MNRAS, 144, 425
Shu, F. H. 1977, ApJ, 214, 488
Ward-Thompson, D. 1996, Ap&SS, 239, 151

Virial Balance in Turbulent MHD Two Dimensional Numerical Simulations of the ISM [1]

Javier Ballesteros-Paredes and Enrique Vázquez-Semadeni.

Instituto de Astronomía, UNAM. Apdo. Postal 70-264, México, D.F. 04510.
`javier@astroscu.unam.mx, enro@astroscu.unam.mx`

Abstract. We present results from a virial analysis of fully nonlinear two-dimensional (2D) simulations of the ISM. We discuss the Eulerian Virial Theorem in 2D, and describe preliminary results on the virial budget of clouds in the simulations. The clouds are far from a static equilibrium, and the Virial Theorem is dominated by the time-derivative terms, indicating the importance of flux through the cloud boundaries and mass redistributions. A trend towards greater importance of the gravitational term at larger scales is observed, although a few small clouds are strongly self-gravitating. The magnetic and kinetic terms scale linearly with each other.

INTRODUCTION

Vázquez-Semadeni *et al.* (1995, hereafter Paper I) and Passot *et al.* (1995, hereafter Paper II) have produced a numerical model of the interstellar medium (ISM) including enough physical agents as to render it feasible to perform statistical studies of the clouds formed in the simulations. The simulations include self-gravity, magnetic fields, parameterized cooling and diffuse heating, the Coriolis force, large-scale shear, and localized stellar energy input. In the present work, we discuss the Virial Theorem (VT) as it applies to the simulations, and present preliminary statistical results from a two-dimensional (2D) simulation with a resolution of 800 × 800 grid points, performed specifically for this analysis. In § II we discuss the VT, applying the formalism developed by McKee & Zweibel (1992) to the 2D case. In § III we describe the cloud-identifying algorithm and show preliminary statistical results, and in § IV we present some remarks and future work.

---

[1] This work has received partial finantial support from grants UNAM/CRAY grant SC-002395, UNAM-DGAPA IN105295, UNAM-PADEP grant 003319 and scholarship from UNAM-DGAPA.

VIRIAL THEOREM IN 2D

The VT is obtained by dotting the momentum equation (eq. (1b) in Paper I) with the position vector **x** and integrating over volume. McKee & Zweibel (1992) have discussed an Eulerian form of the VT, which is most appropriate for our simulations, since they are performed with an Eulerian code. Because the simulations are 2D (in order to reach a sufficiently large resolution), we must consider the VT in 2D as well. It reads:

$$\frac{1}{2}\frac{d^2 I}{dt^2} = 2\left(\tau_{\rm kin} + \tau_{\rm int}\right) + M - W - E_{\rm cor} - \frac{1}{2}\frac{d\Phi}{dt} \qquad (1)$$

where $\tau_{\rm kin} = 1/2(\int \rho u^2 dV - \oint_X x_i \rho u_i u_j \hat{n}_j dS)$ is the kinetic term, $\tau_{\rm int} = \int P dV - 1/2 \oint_S P x_i \hat{n}_i dS$ is the thermal term, $M = 1/8\pi \oint_S x_i T_{ij} \hat{n}_j dS$ is the magnetic term, $W = \int x_i \rho\, g_i dV$ is the gravitational term, $E_{\rm cor} = 2 \int x_i (\mathbf{\Omega} \times \mathbf{u})_i dV$ is the Coriolis term, $\Phi = \oint_S \rho u_i r^2 \hat{n}_i dS$ is the flux of moment of inertia through the surface S, and ρ, $\mathbf{u}$, P and g_i are the density, velocity, thermal pressure and self-gravitational acceleration, respectively. Because of two-dimensionality, we must replace volumes by areas and surfaces by contours in (1). However, we retain the above notation for generality. Since in 2D $\nabla \cdot \mathbf{x} = 2$, in equation (1) we note the three following interesting points: *a)* Although magnetic fields are present in the surface term $M = \int_S x_i\, T_{ij}\, \hat{n}_j\, dS$, (where the Maxwell stress tensor is defined as $T_{ij} \equiv 1/4\pi [B_i B_j - 1/2 B^2 \delta_{i,j}]$), the "classical" magnetic energy term $E_{\rm mag} = 1/8\pi \int B^2 dV$ does not enter the virial equation, so it does not provide support against gravity. *b)* The internal energy $E_{\rm int} \equiv \int P dV$ does not contain the 3/2 factor as in 3D. Nevertheless, in 2D this term still coincides with the total internal energy, because there are only two translational degrees of freedom. *c)* Additionally, it can be shown that the gravitational term $\int x_i\, \rho\, g_i\, dV$ does not coincide with the gravitational energy $E_{\rm grav} = 1/2 \int \rho\, \phi\, dV$ as it does in 3D for isolated clouds. Essentially, this is due to the slower distance dependence of the gravitational potential in 2D.

PRELIMINARY STATISTICS

In order to calculate the terms in equation (1), we have performed a 2D simulation similar to the one called "Run 28" in Paper II, but with a resolution of 800×800 grid points. In this run we analyze the data shortly after turning off star formation, in order to allow for the largest possible density gradients (see Vázquez-Semadeni, Ballesteros-Paredes & Rodríguez 1997, hereafter Paper III) while still retaining the structure induced by the stellar energy injection. We have developed a numerical algorithm to identify clouds and evaluate within them the various terms entering the VT, as well as their velocity dispersion and mean density. We define a cloud as a connected set of pixels whose densities are larger than an arbitrary threshold ρ_t. Previous calculations (Paper III) have shown that the simulations exhibit similar scaling properties as those observed in real interstellar clouds (Larson 1981), except for the density-size scaling relation, supporting the possibility that it may be the result of an observational effect (see also Larson 1981, Kegel 1989, Scalo 1990). With this motivation, we have now performed evaluations of the various terms in the VT. We have the following preliminary results: 1.- Both the second derivative of the moment of inertia and the last term in the equation (1) are dominant in the overall virial balance (fig. 1a). 2.-Comparing the remaining terms, the turbulent terms are seen to dominate (fig. 1b). 3.-The surface terms (which are often neglected under the assumption of vanishing fields outside the clouds) are in general of magnitude comparable to that of the volumetric ones (figs. 1c and d). 4.-The gravitational term is most important at large scales (fig. 1e). However, there are a few small (low energy content) clouds which have large values of the gravitational term. These may be the best candidates for collapse and star formation. Their scarcity appears consistent with the low efficiency of star formation. 5.-The magnetic term and the sum of the kinetic terms are proportional to each other (fig. 1f). This suggests there is equipartition between kinetic and magnetic modes, except for a constant factor, which may be due to the fact that clouds have bulk velocities with respect to the integration volume.

FINAL REMARKS

The dominance of the time-derivative and kinetic terms indicates the importance of flow through the volume boundaries, contrary to the cases considered by McKee & Zweibel (1992). In order to minimize this effect, it appears necessary to consider Eulerian volumes instantaneously at rest with respect to the center of mass of the clouds. However, preliminary attempts suggest that the flow through the boundaries cannot be eliminated completely, since the clouds are extremely amorphous and change shape rapidly. This work will be reported in a future paper (Ballesteros-Paredes & Vázquez-Semadeni 1997, in preparation).

REFERENCES

1. Kegel, W. H. 1989. Astron. & Astrophys, **225**, 517.
2. Larson, R. B. 1981. MNRAS **194**, 809.
3. McKee C. F. & Zweibel, E. G. 1992. ApJ **399**, 551.
4. Passot, T., Vázquez-Semadeni E. & Pouquet, A. 1995. ApJ **455**, 536 (**II**).
5. Scalo, J. M. 1990, in Physical Processes in Fragmentation and Star Formation, ed. R. Capuzzo-Dolcetta, C. Chiosi, & A. di Fazio (Dordrecht:Kluwer), 151.
6. Vázquez-Semadeni E., Passot, T. & Pouquet, A. 1995. ApJ, **441**, 702 (**I**).
7. Vázquez-Semadeni E., Ballesteros-Paredes J. & Rodríguez L. F. 1997. ApJ *in press*. January 1. 1997. (**III**)

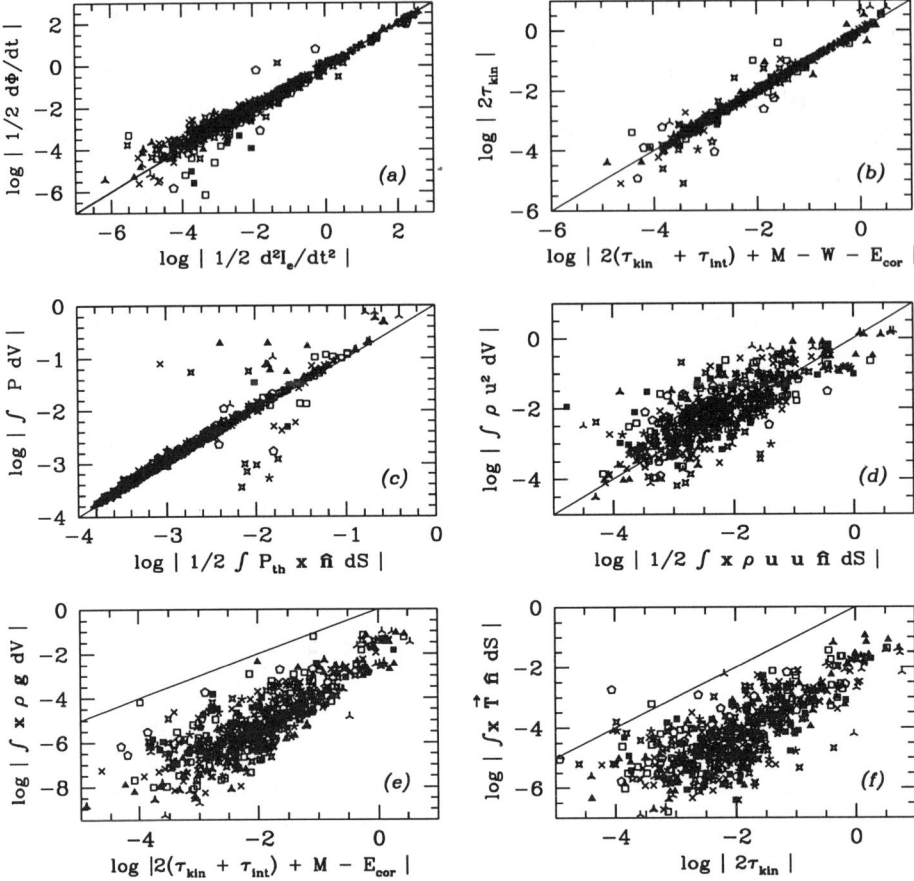

FIGURE 1. In all panels, the solid line is the identity. *(a)* $1/2||d\Phi/dt||$ vs. $1/2||d^2I/dt^2||$. Their near equality shows that the term $1/2d\Phi/dt$ dominates the virial sum, indicating the importance of the variability of the mass flux through the clouds' borders for the total virial balance. *(b)* Volume-plus-surface kinetic terms vs. the virial sum neglecting the $1/2d\Phi/dt$ term. The near equality of both terms indicates the dominance of the kinetic terms over the remaining ones. This effect may be due to cloud bulk motion and should be eliminated by using an instantaneously-at-rest frame of reference for each cloud. *(c)* Volume vs. surface terms for internal energy (pressure) and *(d)* kinetic energy. The surface terms are seen to be comparable to the volume terms in general. The few points with large scatter in *(c)* are likely to correspond to regions of anomalous pressures due to recent star formation. *(e)* The gravitational term W vs. the sum of the remaining virial terms. A trend towards greater importance at larger scales is seen. However, a few points at near balance with gravity are seen at all scales. *(f)* Magnetic term M vs. the sum of the kinetic terms. An almost linear relation is observed. This is consistent with equipartition between kinetic and magnetic modes, if an offset is present, again due to the fact that clouds may have bulk velocities with respect to the integration volume.

Highly Compressible MHD Turbulence and Gravitational Collapse[1]

E. Vázquez-Semadeni*, T. Passot† and A. Pouquet †

*Instituto de Astronomía, UNAM Apdo. Postal 70-264, México, D.F. 04510 †Observatoire de la Côte d'Azur, B.P. 4229, 06304, Nice Cedex 4, France

Abstract. We investigate the properties of highly compressible turbulence and its ability to produce self-gravitating structures. The compressibility is parameterized by an effective polytropic exponent γ_e. In the limit of small γ_e, the density jump at shocks is shown to be of the order of e^{M^2}. In the presence of self-gravity, we suggest that turbulence can produce bound structures for $\gamma_e < 2(1 - 1/n)$, where n is the typical dimensionality of the turbulent compressions. We show, by means of numerical simulations, that, for sufficiently small γ_e, small-scale turbulent density fluctuations eventually collapse even though the medium is globally stable. This result is preserved in the presence of a magnetic field for supercritical mass-to-flux ratios.

INTRODUCTION

In this paper we present a brief discussion on the dynamical properties of highly compressible turbulence, with and without self-gravity. A full-length discussion can be found in Vázquez-Semadeni et al. (1996, hereafter Paper I). The compressibility of the medium is parameterized by an effective polytropic exponent γ_e arising from balance between heating and cooling processes (Elmegreen 1991; Vázquez-Semadeni et al. 1995; Passot et al. 1995), such that the pressure P and the density ρ are related by $P \propto \rho^{\gamma_e}$. We consider the cases where γ_e is either constant, or has a piecewise density dependence (a "piecewise polytropic model" or ppm). We also consider fully thermodynamic cases. Most of the results are based on numerical calculations which solve the full MHD equations in two or three dimensions, at resolutions of 128^2 or 64^3, respectively (Paper I).

---

[1] This work has received partial finantial support from grants UNAM/CRAY grant SC-002395, UNAM-DGAPA IN105295, and a joint CONACYT-CNRS grant.

NO SELF-GRAVITY

Density Structures

It can be easily shown that the density jump $X \equiv \rho_2/\rho_1$ across a shock in a barotropic gas of index γ_e satisfies:

$$X^{1+\gamma_e} - (1 + \gamma_e M^2)X + \gamma_e M^2 = 0, \qquad (1)$$

where M is the upstream Mach number.

In the isothermal case where $\gamma_e = 1$, $X = M^2$, while for $0 < \gamma_e \ll 1$, $X \sim e^{M^2}$. This implies that for small γ_e :
i) The density jump is much larger than in the isothermal case.
ii) Less than supersonic motions (with respect to the isothermal sound speed) are sufficient for producing large density fluctuations.

Vorticity evolution

From the momentum conservation equation (Paper I), we can derive the evolution equation for the *potential vorticity* $\omega_p \equiv \omega/\rho \equiv \nabla \times \mathbf{u}/\rho$:

$$\frac{\partial \omega_p}{\partial t} + \mathbf{u} \cdot \nabla \omega_p = \omega_p \cdot \nabla \mathbf{u} + \nabla \times \mathbf{F}_s + \nabla P \times \nabla \rho / \rho^3 \qquad (2)$$

where $\mathbf{F}_s$ is the solenoidal (or rotational) part of the turbulent energy sources.

While compressive motions can easily be generated from vortical motions, only the "vortex stretching" term $\omega_p \cdot \nabla \mathbf{u}$ and the "baroclinic" term (nonzero only in the fully thermodynamic case) are available as nonlinear sources of ω_p. How effective are they?

Fig. 1 (*left*) shows the evolution of the ratio e_s/e_k of solenoidal to total kinetic energy per unit mass for four 3D runs. Runs 60 and 67 are *ppm* runs with $\gamma_{e\min} = 0.25$ and $\gamma_{e\min} = 0.75$, both with fully solenoidal initial velocity modes but fully compressible forcing ($\mathbf{F}_s \equiv \mathbf{0}$). Run 68 is similar to run 60, but fully thermodynamic. Run 69 is similar to run 60, but with a Coriolis term added. In all cases except run 69, e_s/e_k is seen to decay at roughly the same rate in time. Thus :
i) We have found negligible nonlinear transfer from compressible to rotational kinetic energies. Similar results in the weakly compressible case have been found by Kida & Orszag (1991a,b).
ii) Additional sources of vorticity, such as the Coriolis force (acting on large scales), are necessary for the maintenance of significant amounts of vorticity.

Additionally, the presence of an initially uniform magnetic field also allows for the production of rotational energy from the compressive motions, with an efficiency proportional to the magnetic field strength (fig. 1 (*right*) ; in runs 86, 84, 87, 88, the non-dimensional field intensity is respectively 0.05, 0.3, 1 and 3). This process is more important at small scales.

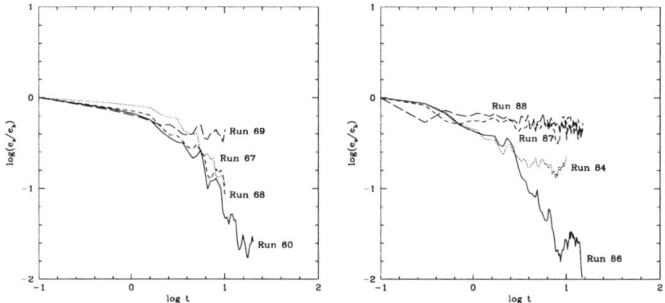

FIGURE 1. Evolution of the ratio of rotational to total specific kinetic energy for various runs. *Left*: non-magnetic runs. *Right*: magnetic runs.

FLOWS WITH SELF-GRAVITY

Non-magnetic case

It can readily be shown that for a barotropic medium with polytropic exponent γ_e, the effective Jeans length is :

$$L_{\text{eff}} = \Big[\frac{\gamma_e \pi c_i^2}{G\rho_0^{2-\gamma_e}}\Big]^{1/2} = \sqrt{\frac{\gamma_e}{\gamma}} \rho^{\frac{\gamma_e-1}{2}} L_J, \tag{3}$$

where c_i is the isothermal sound speed, such that $P = c_i^2 \rho$, and L_J is the Jeans length based on c_i.

The critical density required for destabilizing a length scale L is thus $\rho_J \propto L^{2/(\gamma_e-2)}$. In order to account for turbulent compressions acting on n directions, we consider a volume $V = L^n L_0^{3-n}$, where L is the side of the volume which varies upon compression, and L_0 is the side that remains unaltered. The critical mass to destabilize this volume is thus given by $M_J \propto L^{\frac{n+2}{\gamma_e-2}} L_0^{3-n}$. If M_J is a decreasing function of L, turbulent compression can produce gravitationally unstable structures, i.e, if (see also McKee et al. 1993)

$$\gamma_e < \gamma_{\text{crit}} \equiv 2(1 - 1/n). \tag{4}$$

This result is illustrated in fig. 2 (*left*), which shows the evolution of the global density maximum for three single-γ_e simulations, all with an isothermal Jeans length $L_J = 1.1$ times the box size (i.e., gravitationally stable) and purely compressible random forcing, but with $\gamma_e = 0.9$, 0.3 and 0.1 respectively. While the simulation with $\gamma_e = 0.9$ never develops a gravitational collapse, the other two do, earlier in the case with $\gamma_e = 0.1$.

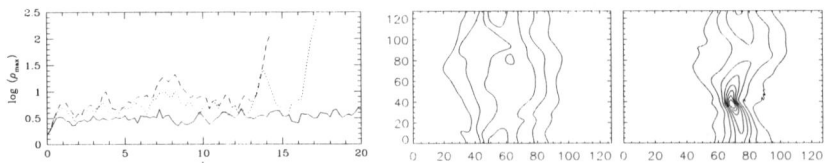

FIGURE 2. *Left*: Evolution of the global density maximum for three runs with $\gamma_e = 0.9$ (*solid line*), $\gamma_e = 0.3$ (*dotted line*) and $\gamma_e = 0.1$ (*dashed line*). *Right*: Final density fields of two magnetic simulations, one with $\gamma_e = 0.9$ (*center*) and one with $\gamma_e = 0.3$ (*right*).

Magnetic case

The above result is preserved in the presence of a magnetic field. Fig. 2 also shows contour plots of the final density field in simulations with an initially uniform magnetic field along the x-direction, a Jeans length $L_J = 0.9$ (i.e., gravitationally unstable with respect to a uniform density configuration), and purely compressible random forcing. One simulation has $\gamma_e = 0.9$ (*center*) and the other has $\gamma_e = 0.3$ (*right*). The latter is seen to have undergone collapse, while the former only contracts to a pancake-type structure supported by thermal pressure against final collapse. However, for large enough magnetic strengths, we have found that turbulence is again not capable of inducing gravitational collapse, suggesting that a subcritical magnetic regime (Mouschovias & Spitzer 1976) cannot be forced to collapse by large-scale turbulence (although the presence of small-scale turbulent modes may affect this result).

REFERENCES

1. Elmegreen, B. G., 1991, ApJ, 378, 139
2. Kida S., Orszag S.A., 1990a, J. of Sci. Comp., 5, 1
3. Kida S., Orszag S.A., 1990b, J. of Sci. Comp., 5, 85
4. McKee, C. F., Zweibel, E. G., Goodman, A. A., & Heiles, C. 1993, in Protostars and Planets III, ed. E. H. Levy & J. I. Lunine (Tucson: Univ. of Arizona Press), 327
5. Mouschovias, T. C., Spitzer, L. Jr. 1976, ApJ, 210, 326
6. Passot T., Vázquez-Semadeni E., Pouquet A., 1995, ApJ, 455, 702
7. Vázquez-Semadeni E., Passot T., Pouquet A., 1995, ApJ, 441, 536
8. Vázquez-Semadeni E., Passot T., Pouquet A., 1996, ApJ, in press (Dec. 20) (Paper I)

Numerical Simulations of MHD Turbulence and Gravitational Collapse

D.S. Balsara[*], R.M. Crutcher[*] and A. Pouquet[†]

[*]*National Center for Supercomputing Applications*
[†]*Observatoire de la Cote d'Azur, CNRS*

Abstract. In this paper we report on our effort to study star formation in turbulent magnetized media. The significance of this study derives from the fact that ram pressure of the turbulent motions in molecular clouds is comparable to the magnetic pressure. We have carried out several one dimensional and three dimensional simulations. The intent was to discover the dominant physical effects regulating star formation in turbulent environments. We find that the turbulent Mach number and the presence of an ordered magnetic field are perhaps the most dominant physical effects. Several other correlations and effects are also mentioned.

INTRODUCTION

Several detailed observational studies of molecular clouds and cores within them have been carried out in recent years, see [5], [11], [8], [6], [7] and references therein. The observed line widths suggest high levels of random internal motion which has been conjectured to be turbulence or waves. From the point of view of numerical simulations this distinction is largely irrelevant. The reason is that the character of these motions, whether turbulent motions or waves, will be borne out by the simulations themselves. It is important, however, to realize that the line widths indicate that the motions are strongly supersonic though perhaps sub-Alfvenic. Thus the turbulent ram pressure is significantly larger than the gas pressure and comparable to the magnetic field pressure, see [1]. An interesting counterpoint is presented in [9]

We present results from a large number of very high resolution one dimensional simulations and a smaller number of medium resolution three dimensional simulations. Because of the ease with which the one dimensional simulations can be carried out we have scoped out several physical effects that might regulate star formation. This is done with a view to picking out the most dominant ones. The three dimensional simulations are done with a view to verify the major conclusions from the one dimensional study and also to help us to understand the interplay of multidimensionality, turbulent cascade in three dimensions, magnetic field structure and collapsed cores. The simulations have been carried out using the RIEMANN code for numerical MHD. It is based on the work described in [13], [2] and [3].

ONE DIMENSIONAL COLLAPSE CALCULATIONS

For our typical cloud we take a region 2 pc across with mean density of 10^{-20} g cm^{-3}. 2048 zones were used. An initial linear density fluctuation was put in to catalyze collapse. Fluctuations were initialized in the velocity and transverse magnetic fields with a spectrum given by $E^V(k) \sim k^2 \exp\left(-2(\frac{k}{k_0})^2\right)$
. The various physical effects included in the calculations are as follows: 1) Varying turbulent Mach numbers. 2) Ambipolar drift. In keeping with the work of [12] we allow for situations where cosmic rays (CR) dominate the ionization in the cloud and situations where far ultra-violet (FUV) radiation dominates the ionization. 3) Varying fractions of compressive motions in the initially impressed turbulence. This mimics different initial input mechanisms for the turbulence (i.e. shocks or large-scale shear). 4) Varying initial length scales on which the turbulence is initially impressed. 5) Varying amounts of magnetic fluctuations.

In view of space constraints we summarize significant results here: 1) The strength of the initial turbulence (as measured by the turbulent Mach number) was the strongest discriminant of the final collapsed object. Thus lower Mach numbers produced cores with strongly peaked density profiles while higher Mach numbers produced broader cores. 2) Shocks do form in the flow and the turbulence rapidly acquires a k^{-2} power spectrum. It does so within one non-linear turnover time scale. 3) The structure of the initially input turbulence is mostly irrelevant. The reason is that mode mixing couples compressive and solenoidal modes very effectively. 4) The length scale on which the turbulence is initially put in is also largely irrelevant. The reason is that the nonlinear turnover time is much smaller than the free fall time for supersonic conditions in molecular clouds. Thus the turbulent cascade puts energy on all length scales with the appropriate power spectrum faster than the collapse time of the collapsing object. 5) The collapsed core aquires a larger scale envelope around it. The turbulence does not penetrate right into the core. The reason for that is that the core has too high a density compared to the ambient. This causes an impedance mismatch for waves in the ambient trying to enter the core. This is consistent with the observations of [8]. 6) Ambipolar drift plays a role at subsonic Mach numbers. But at supersonic Mach numbers its role as a purely dissipative influence is diminished. The reason derives from the realization in [4] that the non-linear terms in the ambipolar diffusion equation can also produce narrow profiles.

THREE DIMENSIONAL COLLAPSE CALCULATIONS

We have carried out medium resolution simulations in three dimensions using initial conditions reasonably similar to the ones described in the previous section. Fig1a shows the logarithm of the density in a plane with the largest density concentration for a simulation with a Mach 2 turbulence. Fig 1b shows the magnetic pressure in the same plane for the same simulation. The initial magnetic pressure fluctuations were random in all three directions and the magnetic pressure was set to be roughly equal to the ram pressure in the velocity field. Figs 2a and 2b show the logarithm of the density and magnetic pressure for a different simulation. Here too the plane in which the core forms is displayed. The difference between Fig 1 and Fig 2 is simply that the x-component of the magnetic field was set to be uniform in the simulation corresponding to Fig 2. Thus Fig 2 corresponds to the situation where the magnetic field has an ordered component on the larger scales. Both figures correspond to about one free-fall time.

The three dimensional simulations lead us to the following results: 1) As before, the turbulent Mach number is one of the strong discriminants. 2) The density peak in Fig 2 corresponds to a density that is more than a factor of ten lower than the density peak in Fig 1. Thus the result of gravitational collapse in a turbulent magnetized medium depends strongly on whether or not that patch of molecular cloud has a large scale magnetic field component. The presence of an ordered large scale magnetic field is, therefore, another key discriminant governing gravitational collapse. 3) Focussing on the lower density parts of the figures shows clearly that magnetosonic shocks do form in the flow. As before, the turbulence rapidly populates all length scales in the problem. 4) A discernable core does form. Filamentary structures form around the core, consistent with observations of the Orion molecular cloud. Thus the notion of a simple envelope around the core might be a simplified (angle averaged) view of the scenario. The actual envelopes around molecular cloud cores may have plentiful substructure. 5) We still lack resolution to map out the variation of the simulated line widths as a function of distance from the core's center but we anticipate that in the next round of simulations we will begin to resolve that question. 6) Consistent with observations (and contrary to the conjecture of [10]) density condensations correlate positively with regions of high magnetic pressure in both the displayed figures. The reason is that the magnetic field is dredged in by the accreting matter.

REFERENCES

1. Balsara, D.S., 1996, ApJ, **465**, 775

2. Balsara, D.S., 1996, ApJ, "Linearized Riemann Solver for Adiabatic and Isothermal MHD", to appear
3. Balsara, D.S., 1996, ApJ, "TVD Scheme for Adiabatic and Isothermal MHD", to appear
4. Brandenburg, A. and Zweibel, E., 1994, ApJ Lett, **427**, L91
5. Crutcher, R.M. *et al*, 1993, ApJ, **407**, 175
6. Blitz, L., 1993, in Protostars and Planets III, eds E.H. Levy and J.I. Lunine, (Univ. of Ariz. Press) 125
7. Falgarone, E. and Puget, J.L., 1986, Astr. Ap. **162**, 235
8. Goodman A.A., Barranco, J.A., Wilner, D.J. and Heyer, M.H., this proceedings
9. Leorat, J., Passot, T. and Pouquet, A., 1990, MNRAS, **243**, 293
10. Lubow, S.A. and Pringle, J., 1996, MNRAS, to appear
11. Myers, P.C. and Goodman, A.A., 1988, ApJ Lett, **326**, L27
12. McKee, C.F., 1989, ApJ, **345**, 782
13. Roe, P. and Balsara, D.S., 1996, SIAM J. Appl. Math., **56**, 57

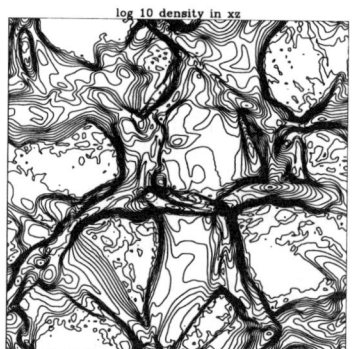 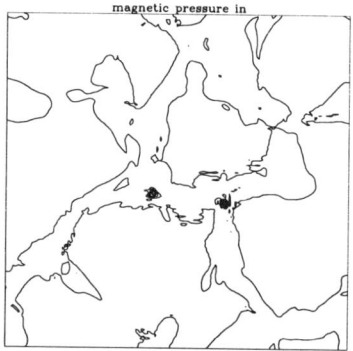

FIGURE 1.

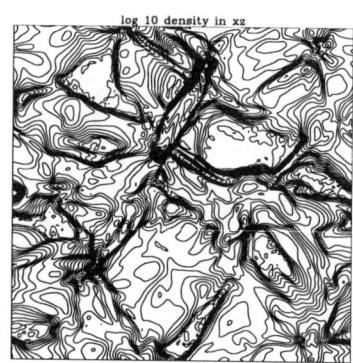 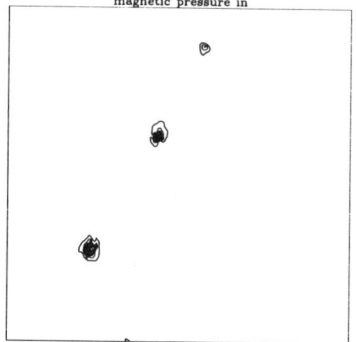

FIGURE 2.

Dynamically Infalling Envelopes around Low-mass Protostar Candidates [1]

Nagayoshi Ohashi

Harvard-Smithsonian Center for Astrophysics, Cambridge, MA 02138

Abstract. Interferometric observations of two protostar candidates, L1551 IRS5 and IRAS 04368+2557(L1527), in Taurus have directly identified dynamically infalling motions in their circumstellar envelopes. The observed infalling envelopes have disklike structures, which are perpendicular to the associated outflows, with a thousand AU scale. In addition to infall, rotation was detected in the envelope around IRAS 04368+2557. This rotation is 5 times slower than the infall at 2000 AU in radius, suggesting that the envelope around IRAS 04368+2557 is not rotationally supported. Mass infall rates were estimated to be 10^{-5} $M_\odot$ yr^{-1} for L1551 IRS5, and 10^{-6} $M_\odot$ yr^{-1} for IRAS 04368+2557.

INTRODUCTION

Dynamical infall is an essential phenomenon in star formation processes. Young stellar objects that are still accreting a large amount of matter are called "protostar". After discovery of low-mass protostar candidates in nearby star forming regions by *IRAS* [1], many efforts have been made to obtain evidence for infall around the candidates [2,3]. Nevertheless, there was no *direct* evidence for infalling motions around them.

In order to identify infalling motions around solar-type protostars directly, both high angular and velocity resolutions are required because low-mass protostars can yield infall motions with detectable velocities only in the vicinity of them. Millimeter wavelength interferometers are unique instruments that provide both high angular and velocity resolutions. Since the successful detection of infall around HL Tau with the Nobeyama Millimeter Array(NMA) [4], we have observed protostar candidates in Taurus with the NMA. In this paper, we will present results of L1551 IRS5 [5] and IRAS 04368+2557 [6].

---

[1] This work has been done under collaboration with M. Hayashi, P. T. P Ho, M. Momose, N. Hirano, & M. Tamura

L1551 IRS5

L1551 IRS5 is a typical protostar candidate with both a strong molecular outflow [7] and a large molecular disk [8]. $C^{18}O$ $(1-0)$ observations with the OVRO interferometer suggested a disklike gaseous condensation around L1551 IRS5 [9]. We have observed ^{13}CO $(1-0)$ with the NMA.

Fig.1a shows the total intensity map of ^{13}CO at $5''\!.1 \times 3''\!.9$ resolution, superimposed on an image of the 2.2 μm reflection nebula [10]. A compact condensation is detected centered on the central star, together with a north-south elongated structure. The compact condensation, which is slightly resolved at the present angular resolution, shows an elongated structure in the northwest-southeast direction with a deconvolved size 1200 AU×670 AU. This elongated structure is almost perpendicular to the outflow axis, suggesting that the compact condensation is a disk around the central star. The detected disklike structure is consistent with the previous $C^{18}O$ results obtained at OVRO [9]. We should note that the infrared reflection nebula is extended to the southwest of the disk, which means that the disk inclines with its northeast part located on the near side and its southwest part on the far side.

Fig.1b, showing the position-velocity (PV) diagram cutting along the minor axis of the disk, demonstrates remarkable velocity structures of the disk along the minor axis: while blueshifted emission is extended to the southwest part

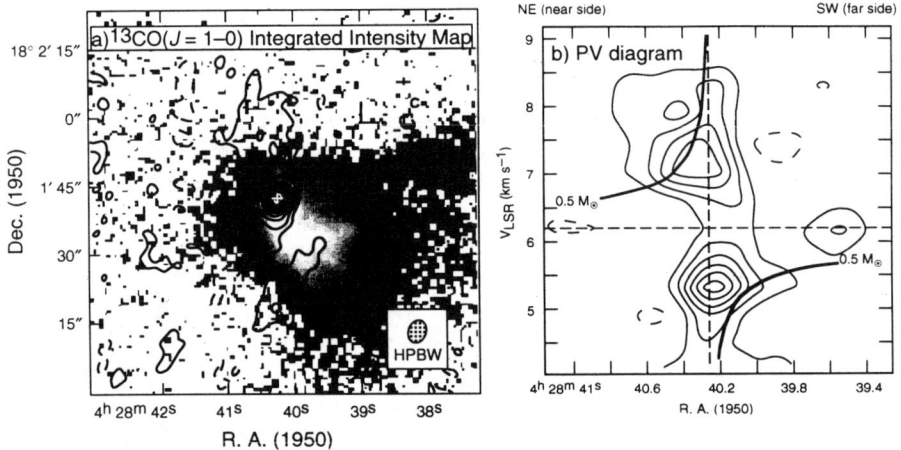

FIGURE 1. (a) ^{13}CO total intensity map of L1551 IRS5 (*contour*) is superimposed on an image of the 2.2 μm infrared reflection nebula (*gray scale*). (b) PV diagram of ^{13}CO cutting along the minor axis of the disk structure through the central star position. The systemic velocity and the stellar position are indicated by a horizontal and a vertical dashed line, respectively. The thick curve shows a radial velocity distribution of a dynamical infall disk around a 0.5 $M_\odot$ star.

(i.e., far side) of the disk, redshifted one is extended to the northeast part (i.e., near side). If these traces radial motions in the disk plane, then this velocity gradient suggests that the disk is contracting toward the central star. In addition to this velocity gradient, it must be noted that emissions with higher blueshifted ($\lesssim 5$ km s^{-1}) and redshifted ($\gtrsim 8$ km s^{-1}) velocities are confined to the vicinity of the central star. This characteristic is well explained in terms of infall because infalling matter is accelerated as it approaches the central star. The entire velocity structures in Fig.1b are consistent with dynamically infalling motions in a disk around a 0.5 $M_\odot$ star.

IRAS 04368+2557(L1527)

IRAS 04368+2557 (hereafter I04368) is an IRAS source, which is very deeply embedded in the L1527 dense core. A molecular outflow with a nearly edge-on configuration is associated with the central star [11], suggesting that any disklike structure around the central star should be edge-on. Asymmetric line profiles were detected from this source [3], suggestive of presence of infalling matter. We have observed $C^{18}O$ (1 − 0) from I04368.

Results are summarized in Fig.2. The $C^{18}O$ total intensity map at $6''\!.0 \times 4''\!.9$ resolution presented in Fig.2a clearly shows a flattened structure in the north-south direction, which is almost perpendicular to the outflow axis, centered on the central star. The east and west sides of the flattened structure are concave. This flattened structure perpendicular to the outflow is naturally considered to be a disk with an edge-on configuration around the central star. We must note, however, that the disk is not spatially thin because the edge-on flattened structure is resolved even along the minor axis. It is also remarkable that the concave boundaries of the disk are well traced by ^{13}CO outflowing shells [6], also observed with the NMA, as shown in Fig.2b.

Fig.2c and 2d suggest a velocity gradient of the disk along the major axis, which can be explained in terms of rotation of the disk. At the same time, however, north-south elongated structures centered on the central star appear at both the blueshifted and redshifted velocities. To explain these velocity features, radial motions along the disk plane, such as infall, as well as rotation are necessary. In fact, the PV diagram of $C^{18}O$ cutting along the major axis of the disk (Fig.2e) can be well reproduced by a model, in which a disk has both rotation and infall(Fig.2f). In contrast, the observed PV diagram cannot be explained by only rotation or only infall(Fig.2g). The infall and rotation velocities at 2000 AU in radius are estimated to be 0.3 km s^{-1} and 0.05 km s^{-1}, respectively: the infall velocity is 5 times larger than the rotation velocity, suggesting that the disk is not centrifugally supported but dynamically contracting. In such a dynamical infalling envelope with rotation, a centrifugally supported (i.e., Keplerian) disk must be formed as a result of angular momentum conservation: the radius of the expected Kepler disk around I04368

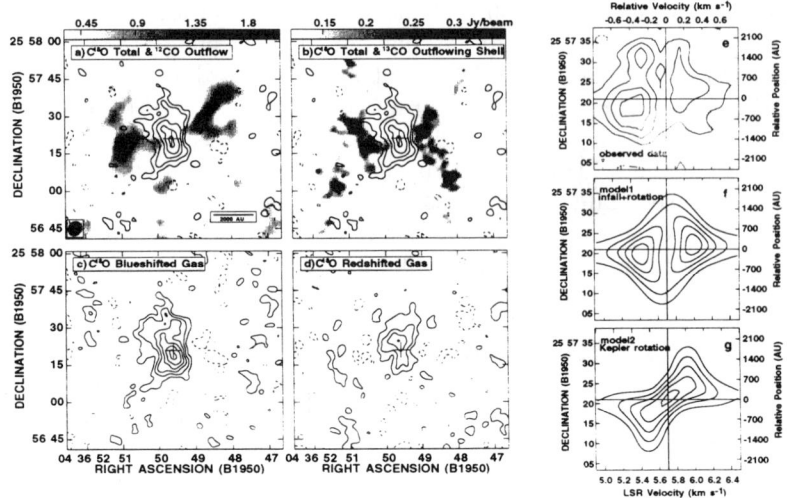

FIGURE 2. (a) $C^{18}O$ total intensity map of I04368 (*contour*) is superimposed on the associated ^{12}CO outflow (*gray scale*) mapped with the NMA. (b) The ^{13}CO outflowing shells (*gray scale*) observed with the NMA is compared with the $C^{18}O$ disk presented in (a). (c) Total intensity map of $C^{18}O$ with redshifted velocities. (d) the same as (c) but the blueshifted component. (e) PV diagram of $C^{18}O$ cutting along the disk major axis, obtained from the observations. (f) the same as (e) but obtained from a model, in which a disk has both infalling and rotation motions with a 0.1 $M_\odot$ central star. (g) the same as (f) but a model with only rotation.

is estimated to be ~100 AU using the infall and rotation velocities.

REFERENCES

1. Beichman et al. 1986, ApJ, 307, 337
2. Adams, F. C., Lada, C. J., & Shu, F. H. 1987, ApJ, 312, 788
3. Myers, P. C. et al. 1995, ApJ, 449, L65
4. Hayashi, M., Ohashi, N., and Miyama, S. M. 1993, ApJ, 418, L71 (1993).
5. Ohashi, N., Hayashi, M., Ho, P. T. P., Momose, M., & Hirano, N. 1996, ApJ, 466, 957
6. Ohashi, N., Hayashi, M., Ho, P. T. P., Momose, M. 1997, ApJ, in press
7. Snell, R. L., Loren, R. B., & Plambeck, R. L. 1980, ApJ, 239, L17
8. Kaifu, N. et al. 1984, A&A, 134, 7
9. Sargent, A. I., Beckwith, S., Keene, J., & Masson, C. R. 1988, ApJ, 333, 936
10. Hodapp, K.-W. 1994, APJS, 94, 615
11. Tamura, M. et al. 1996, AJ, 112, 2076

Structure of Dark Clouds from Stellar Extinction

Paolo Padoan [1], Bernard J. T. Jones [2], & Åke Nordlund [1,3]

[1] *Theoretical Astrophysics Center, Juliane Maries Vej 30, DK-2100 Copenhagen, Denmark*
[2] *Imperial College of Science Technology and Medicine, Blackett Laboratory, Prince Consort Road, London SW7 2BZ, UK*
[3] *Astronomical Observatory and Theoretical Astrophysics Center, Juliane Maries Vej 30, DK-2100 Copenhagen, Denmark*

Abstract. We show that the 3D density field of the cloud IC5146 is well described by a Log-Normal distribution down to very small scales; the power spectrum and the standard deviation of the 3D density field can be constrained; the cloud structure is likely to be determined by the random supersonic motions present in the gas.

INTRODUCTION

Lada et al. (1994) have illustrated the method of mapping the distribution of dust in dark clouds, by using stellar extinction measurements in the near-infrared. Lada et al. plotted the mean extinction, A_V versus its dispersion, σ, measured at any position of a regular grid. They found that the dispersion grows with the average extinction, and realized that this behavior contains information about the structure of the extinction (therefore of the gas mass distribution) in the cloud, on scales smaller than the resolution of the extinction map.

Fig. 1, which is the equivalent of Fig. 7 in Lada et al. (1994), shows the $\sigma - A_V$ plot obtained from the original data, kindly provided to us by the authors. The measurements are taken through the dark cloud IC5146 in Cygnus.

RANDOMLY FORCED SUPERSONIC FLOWS

Nordlund and Padoan (1997) and Padoan, Nordlund and Jones (1997) have recently discussed the importance of supersonic flows in shaping the density distribution in the cold interstellar medium (ISM).

They have run numerical simulations of isothermal flows randomly forced to high Mach numbers. Their experiments are meant to represent a fraction of a giant molecular cloud: $\approx 10pc$ in size and $10^3 - 10^4 M_\odot$ in mass. The simulated random supersonic motions are observed in molecular clouds.

It is found that most of the mass concentrates in a small fraction of the total volume of the simulation, with a very intermittent distribution. The statistic of the density field is well approximated by a Log-Normal distribution, with standard deviation that depends linearly on the rms Mach number of the flow, $\mathcal{M}$:

$$\sigma_x = \beta \mathcal{M} \tag{1}$$

where $x = n/<n>$, and $\beta \approx 0.5$.

It is also found that the power spectrum, $P(k)$, of the density distribution is consistent with a power law:

$$P(k) \sim k^{-2.6} \tag{2}$$

where k is the wavenumber.

STATISTICS AND POWER SPECTRUM OF THE ISM DENSITY FIELD

In order to interpret the extinction data we have generated random 3-D density distributions with given statistics and power spectra, projected them in 2-D, and sampled them randomly as it happen when stars are found through the cloud. We find that the observational plot is well reproduced with a lognormal 3-D density distribution.

Fig.1 shows the observational $\sigma - A_V$ plot. A linear regression analysis gives:

$$\sigma = const + (0.36 \pm 0.02) A_V \tag{3}$$

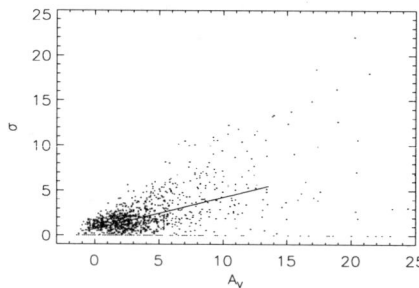

FIGURE 1. The dispersion versus the mean extinction for every bin in the regular grid superposed to the observed region.

Statistics

The slope of the numerical plot could depend on both the power spectrum index, α, and the standard deviation of the 3-D density distribution, $\sigma_{x,3D}$. In Fig. 2 we draw lines of constant $\sigma - A_V$ linear regression coefficient, C_r, on the plane $\alpha - \sigma_{x,3D}$.

Since the value of C_r is known observationally with very small uncertainty (see (3)), its contours on the numerical plane $\alpha - \sigma_{x,3D}$ may in principle be used to constrain the power index of the 3-D density field when its standard deviation is known, or vice-versa.

Given the observational value of C_r and the value of α determined below, one gets:

$$\sigma_{x,3D} = 5.0 \pm 0.5 \tag{4}$$

Power spectrum

The distribution obtained from the 2-D projection of the 3-D distribution, and the 3-D density distribution itself are related in a way that depends on the value of the spectral index. This can be shown using the numerically generated random distributions. In Fig. 3, lines of constant spectral index are plotted in the plane $\sigma_{x,2D} - \sigma_{x,3D}$. For a fixed 3-D standard deviation, the value of the projected 2-D standard deviation decreases towards steeper spectra.

The 2-D standard deviation, $\sigma_{x,2D}$, is measured in the extinction map:

$$\sigma_{x,2D} = 0.7 \pm 0.1 \tag{5}$$

Entering the plane $\sigma_{x,2D} - \sigma_{x,3D}$ with this value and with the previously determined value of $\sigma_{x,3D}$, one gets:

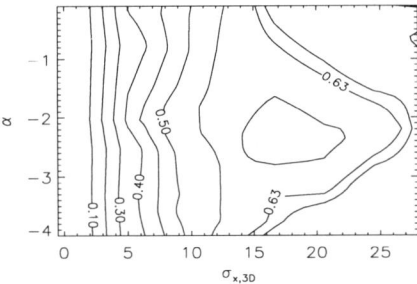

FIGURE 2. Contours of constant value of the slope of the $\sigma - A_V$ plot, C_r. α is the spectral index of the 3-D density distribution, and $\sigma_{x,3D}$ is the standard deviation of the same distribution.

$$\alpha = -2.6 \pm 0.5 \qquad (6)$$

CONCLUSIONS

The connection between the observations and numerical supersonic turbulence is qualitative (Log-Normal shape of the 3-D density distribution), but also quantitative.

In fact, the relation between the standard deviation of the density field and the observed rms Mach number of the flow is consistent with the one predicted in our numerical experiments of supersonic turbulence, and the spectral index estimated from our simulations, ~ -2.6, is consistent with that obtained from the observational data.

We therefore conclude that the dust extinction measurements in the cloud IC5146 are consistent with a scenario, where the cloud structure is shaped primarily by random supersonic motions, as proposed by Padoan (1995), and Padoan, Nordlund, & Jones (1997; also Padoan et al., this volume).

ACKNOWLEDGEMENTS

This work has been supported by the Danish National Research Foundation through its establishment of the Theoretical Astrophysics Center.

REFERENCES

Lada, C. J., Lada, E. A., Clemens, D. P., & Bally, J. 1994, ApJ, 429, 694
Padoan, P., Nordlund, Å. P., & Jones, B. J. T. 1997, MNRAS (in press)
Padoan, P. 1995, MNRAS, 277, 377
Nordlund, Å. P., & Padoan, P. 1997, in preparation

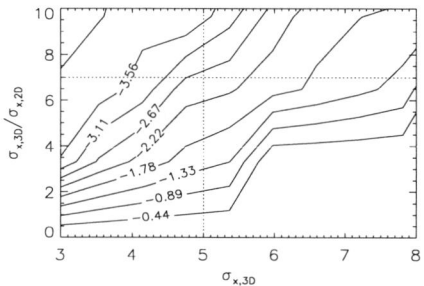

FIGURE 3. Contours of constant spectral index. $\sigma_{x,2D}$ and $\sigma_{x,3D}$ are respectively the 2-D and 3-D standard deviations of the density field.

A physical model for the stellar IMF

Paolo Padoan [1], Åke Nordlund [1,2], & Bernard J. T. Jones [3]

[1] *Theoretical Astrophysics Center, Juliane Maries Vej 30, DK-2100 Copenhagen, Denmark*
[2] *Astronomical Observatory and Theoretical Astrophysics Center, Juliane Maries Vej 30, DK-2100 Copenhagen, Denmark*
[3] *Imperial College of Science Technology and Medicine, Blackett Laboratory, Prince Consort Road, London SW7 2BZ, UK*

Abstract. We propose that the stellar initial mass function (IMF) arises as a consequence of the existence of random supersonic flows in molecular clouds. A Miller-Scalo like IMF is predicted for the typical physical conditions in molecular clouds, and a more "massive" one in star bursts.

INTRODUCTION

We use the idea of gravitational instability, but instead of using the values of temperature and density averaged on a large scale, we define the critical mass using the local value of the density. We assume that the density field is shaped by the random supersonic motions present in the star forming gas, and we predict the density distribution by simulating numerically such random flows. The stellar mass distribution is therefore a consequence of the density distribution.

THE DENSITY FIELD IN RANDOM SUPERSONIC FLOWS

Nordlund and Padoan (1997) have recently discussed the importance of supersonic flows in shaping the density distribution in the cold interstellar medium (ISM).

They have run numerical simulations of isothermal flows randomly forced to high Mach numbers. Their experiments are meant to represent a fraction of a giant molecular cloud, where in fact such random supersonic motions are observed.

It is found that the flow develops a complex system of interacting shocks, and these are able to generate very large density contrasts, up to 5 orders of magnitude, $\rho_{max}/\rho_{min} \approx 10^5$. In fact, most of the mass concentrates in a small fraction of the total volume of the simulation, with a very intermittent distribution. The probability density function of the density field is well approximated by a Log-Normal distribution:

$$p(lnx)dlnx = \frac{1}{(2\pi\sigma^2)^{1/2}} exp\left[-\frac{1}{2}\left(\frac{lnx - \overline{lnx}}{\sigma}\right)^2\right] dlnx \quad (1)$$

where x is the relative number density:

$$x = n/\overline{n} \quad (2)$$

and the standard deviation σ and the mean $\overline{lnx}$ are functions of the rms Mach number of the flow, $\mathcal{M}$:

$$\overline{lnx} = -\frac{\sigma^2}{2} \quad (3)$$

and

$$\sigma^2 = ln(1 + \mathcal{M}^2\beta^2) \quad (4)$$

THE DERIVATION OF THE STELLAR IMF

A simple way to define a mass distribution of protostars is that of identifying each protostar with one local Jeans' mass. In this way the protostar MF is simply a Jeans' mass distribution. Since the gas is cooling rapidly, the temperature is uniform, and the Jeans' mass distribution is just determined by the density distribution.

From the density distribution, we get the protostar MF:

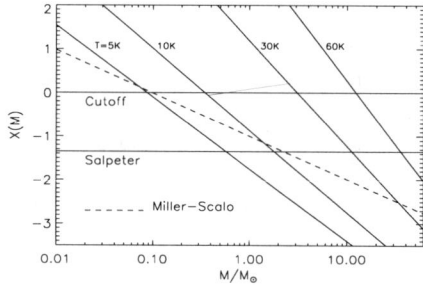

FIGURE 1. The power law exponent of the theoretical MF versus the mass, for different temperatures. The mean density and velocity dispersion have been taken to be $n = 1000 cm^{-3}$, and $\sigma_v = 2.5 km/s$.

$$F(M)dM = \frac{2B^2}{(2\pi\sigma^2)^{0.5}} M^{-3} exp\left[-\frac{1}{2}\left(\frac{2lnM - A}{\sigma}\right)^2\right] dM \qquad (5)$$

where M is in solar masses, and:

$$A = 2lnB - \overline{lnx} \qquad (6)$$

$$B = 1.2 \left(\frac{T}{10K}\right)^{3/2} \left(\frac{\overline{n}}{1000 cm^{-3}}\right)^{-1/2} \qquad (7)$$

This MF has got a long tail at large masses and an exponential cutoff at the smallest masses. The cutoff is an important result, because most other models for the origin of the stellar IMF are not able to explain it.

In Fig. 1 we have plotted the exponent of the power law approximation of the MF for different temperatures. The exponent is defined as:

$$X = \frac{\partial ln(F(lnM))}{\partial lnM} = \left(\frac{2A}{\sigma^2} - 3\right) - \frac{4}{\sigma^2} lnM \qquad (8)$$

The Salpeter MF has $X = -1.35$, and the Miller-Scalo MF (Miller & Scalo 1979) has $X = -1.0 - 0.43 lnM$, where M is in solar masses.

The most probable stellar mass per logarithmic mass interval, is defined by $X(M_{max}) \equiv 0$. One finds:

$$M_{max} = 0.2 M_\odot \left(\frac{n}{1000 cm^{-3}}\right)^{-1/2} \left(\frac{T}{10K}\right)^2 \left(\frac{\sigma_v}{2.5 km/s}\right)^{-1} \qquad (9)$$

This is equivalent to the Jeans' mass at constant external pressure, if turbulent ram pressure is considered. It is an important result because it has been obtained from a realistic statistical description of random supersonic flows, that allows the prediction of the whole shape of the MF.

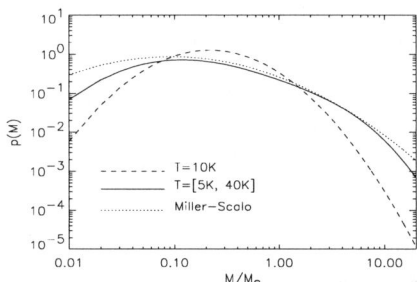

FIGURE 2. Log-log plot of the theoretical MF (dashed line), for a temperature $T = 10K$, and for a temperature distribution between 5 and 40 K (continuous line), compared with the Miller-Scalo.

THE MILLER-SCALO IMF

In Fig. 2 the theoretical MF for $T = 10K$ (dashed line) is compared with the Miller-Scalo MF (MSMF) (dotted line). We also plot the theoretical MF in a case of temperature distribution, between 5 and 40 K. The temperature integration improves the shape of the single temperature MF, making the theoretical MF practically coincident with the MSMF.

The model is therefore consistent with the MSMF, as long as most of the solar neighborhood stars are formed in molecular clouds similar to the ones that are the sites of present day star formation, with temperatures between $5K$ and $40K$.

It has been claimed by many authors, on both theoretical and observational grounds, that the IMF in star-burst regions is more "massive" than in the solar neighborhood (e.g. Doane & Mthews 1993; Riecke et al. 1993; Doyon, Joseph, & Wright 1994).

These "massive" MFs are in agreement with our theoretical prediction. In fact, a value of $4M_\odot$ is predicted for the cutoff in the MF, for $T \approx 60K$, which is reasonable in environments with strong UV and X-ray radiation fields, and with enhanced (even by a factor 100) cosmic ray flux. We also predict a slope of the MF considerably smaller than in the MSMF. For example, $X = -0.9$ (the value found by Malumuth and Heap (1994) in the core of 30 Doradus, R136a, which is a local example of a star-burst event) is predicted by the model for $T = 60K$, or slightly warmer (fig.2).

ACKNOWLEDGEMENTS

This work has been supported by the Danish National Research Foundation through its establishment of the Theoretical Astrophysics Center.

REFERENCES

Doane, J. S., & Mathews, W. G. 1993, ApJ, 419, 573
Doyon, R., Joseph, R. D., & Wright, G. S. 1994, ApJ, 421, 101
Lada, C. J., Lada, E. A., Clemens, D. P., & Bally, J. 1994, ApJ, 429, 694
Malumuth, E. M., & Heap, S. R. 1994, AJ, 107, 1054
Miller, G. E, & Scalo, J. M. 1979, ApJS, 41, 413
Nordlund, Å., & Padoan, P. 1997, in preparation
Padoan, P. 1995, MNRAS, 277, 377
Padoan, P. & Nordlund, Å. 1997, MNRAS (in press)
Rieke, G. H., Loken, K., Rieke, M. J., & Tamblyn, P. 1993, ApJ, 412, 99

Velocity Coherence in Dense Cores

Alyssa A. Goodman[i], Joseph A. Barranco[ii],
David J. Wilner[iii], and Mark H. Heyer[iv]

[i] *Harvard University Department of Astronomy, 60 Garden Street, Cambridge, MA 02138*
[ii] *Astronomy Department, University of California, Berkeley, CA 94720*
[iii] *Harvard-Smithsonian Center for Astrophysics, 60 Garden Street, Cambridge, MA 02138*
[iv] *Five College Radio Astronomy Observatory, University of Massachusetts, Amherst, MA 01003*

Abstract. At the meeting, we presented a summary of two papers which support the hypothesis that the molecular clouds which contain star-forming low-mass dense cores are self-similar in nature on size scales larger than an inner scale, R_{coh}, and that within R_{coh}, the cores are "coherent," in that their filling factor is large and they are characterized by a very small, roughly constant, mildly supersonic velocity dispersion. We expect these two papers, by Barranco & Goodman [1] and Goodman, Barranco, Wilner, & Heyer, to appear in the Astrophysical Journal within the coming year. Here, we present a short summary of our results. The interested reader is urged to consult the on-line version of this work at cfa-www.harvard.edu/~agoodman/vel_coh.html [2].

SUMMARY

After studying how line width depends on spatial scale in low-mass-star-forming regions, we propose that the so-called "dense cores" [3,4] in these regions represent the "inner scale" of a self-similar process which characterizes the larger-scale molecular clouds in which the cores are found.

Before drawing this conclusion, we define four distinct Types of line width-size relation ($\Delta v \propto R^{a_i}$), which have power-law slopes a_1, a_2, a_3, and a_4, as follows: Type 1– multi-tracer, multi-cloud intercomparison; Type 2– single-tracer, multi-cloud intercomparison; Type 3– multi-tracer study of a single cloud; and Type 4– single-tracer study of a single cloud. We discuss the varying meaning and uses of these Types, and we conclude that Type 3 studies are best for measuring the density profile in an individual region, and Type 4 studies are best for studying the dynamics of gas within narrow density ranges. Type 1 studies (of which Larson 1981 [5] is the seminal example) are compendia of Type 3 studies which illustrate the range of variation in the line width-size relation from one region to another.

Using new measurements of the OH and $C^{18}O$ emission emanating from the environs of several of the dense cores studied in NH_3, we show that line width increases with size outside the cores with $a_4 \sim 0.2$. On scales larger than those traced by $C^{18}O$ or OH, ^{12}CO and ^{13}CO observations indicate that a_4 *increases* to ~ 0.5 [6]. By contrast, within the half-power contour of the NH_3 emission from the cores, line width is virtually constant, with $a_4 \sim 0.08$. We interpret the correlation between increasing density and decreasing Type 4 power law slope as a "transition to velocity coherence." Our data indicate that the radius, R_{coh}, at which the gas becomes coherent (i.e. $a_4 \to 0$) is of order 0.1 pc in regions forming primarily low-mass stars. The value of the *non-thermal* line width at which "coherence" is established is typically less than but still of order of the thermal line width of H_2. Thus velocity coherent cores are similar to, but not exactly the same as, isothermal balls of gas.

Two other results bolster our proposal that a "transition to coherence" takes place at ~ 0.1 pc. First, the OH, $C^{18}O$, and NH_3 maps show that the dependence of column density on size is much steeper ($N \propto R^{-0.9}$) inside R_{coh} than outside of it ($N \propto R^{-0.2}$), implying that the volume filling factor of coherent cores is much larger than in their surroundings. Second, Larson [7] has recently found a break in the power-law characterizing the clustering of stars in Taurus at 0.04 pc, just inside of R_{coh}. Larson and we interpret this break in slope as the point where stellar clustering properties change from being determined by the (fractal) gas distribution (on scales > 0.04 pc) to being determined by fragmentation processes within coherent cores (on scales < 0.04 pc).

We expect that the transition to coherence takes place when a "dissipation threshold" for the MHD turbulence which characterizes the larger-scale medium is crossed at a critical inner scale, R_{coh}. We argue that the most likely explanation for this threshold is the marked decline in the coupling of the magnetic field to gas motions due to a decreased ion/neutral ratio in dense, high filling-factor, gas.

FINDINGS

As explained in the previous section, we have undertaken a new investigation into the nature of the line width-size relations often called "Larson's Laws." We find that the power-law slope of single-tracer line width-size relations gets progressively shallower toward higher densities, and we interpret this result as a "transition to velocity coherence." Our specific findings are as follows:

1. Maps of the OH and $C^{18}O$ emission from the $n \sim 10^3$ cm^{-3} gas around dense cores show line width decreasing toward antenna temperature peaks. By fitting antenna temperature as a function of size

and then line width as a function of antenna temperature, we deduce line width-size relations for several maps. The power-law slopes in these single-tracer single-cloud line width size relations are typically ~ 0.2.

2. NH_3 maps of dense cores within the regions mapped in OH and $C^{18}O$ show an even shallower dependence of line width on size, with power-law slope ~ 0.08 (i.e. line width is almost constant inside the cores; see Paper I.) We interpret the decrease in slope between the lower density regions (OH and $C^{18}O$ maps) and the higher density regions (NH_3 maps) as a "transition to velocity coherence."

3. The value of the *non-thermal* line width at which "coherence" is established is typically less than but still of order of the thermal line width of H_2. Thus velocity coherent cores are similar to, but not exactly the same as, isothermal balls of gas.

4. The NH_3 data indicate a much steeper dependence of column density on size than do the OH or the $C^{18}O$ data, which implies that the filling factor within the NH_3 cores is much larger than in their surroundings.

5. Given 2 and 4, we put forth the hypothesis that the molecular clouds which contain low-mass dense cores are self-similar in nature on size scales larger than R_{coh}, and that *within R_{coh}, cores are "coherent," in that their filling factor is large and they are characterized by a very small, roughly constant, velocity dispersion.*

6. We place our and previous studies of line width-size relations into context by defining four Types of line width-size relations, and discussing the operational and physical differences among them. The Types and their most important features are as follows:

    – *Type 1: Multi-tracer, multi-cloud intercomparison.* Shows the (cosmic) scatter about an overall line width-size relation.
    – *Type 2: Single-tracer, multi-cloud intercomparison.* Heavily influenced by the column density-tracing properties of probe used.
    – *Type 3: Multi-tracer study of a single cloud.* Good for studying density profile of individual cores and their environments.
    – *Type 4: Single-tracer study of a single cloud.* Gives best idea of dynamics within a particular density regime.

    Through simulations, we find that a realistic mixture of Type 3 relations can reproduce the original Type 1 relation uncovered by Larson [5]. We show how the data presented in this paper can be used to create each of the four distinct "Types" of line width-size relation. We further illustrate that the slope derived for each Type need not be equal.

7. We point out that the transition to coherence takes place at $R_{coh} \sim$ 0.1 pc, which is just a bit larger than the size scale (0.04 pc) where Larson [7] has found a transition in the power-law slope of the stellar clustering function in Taurus. This result suggests a picture where the distribution of gas and stars on scales $\gg R_{coh}$ is determined by a self-similar process (such as MHD turbulence), and fragmentation dominates on smaller scales.

8. The transition to coherence is likely produced when the conditions in the gas cross a threshold where the forces responsible for maintaining the turbulence and/or waves on larger scales can no longer operate. The forces responsible are likely to rely on the coupling of magnetic fields to the neutral gas, and the critical conditions producing the transition consist of a critical combination of low temperature, high density, and low ionization fraction.

REFERENCES

1. Barranco, J.A. & Goodman, A.A. 1996, ApJ, submitted [Paper I]
2. Goodman, A.A., Barranco, J.A., Wilner, D.A., Heyer, M.H., ApJ, submitted [Paper II]
3. Myers, P.C. & Benson, P.J. 1983, ApJ, 266, 309
4. Benson, P.J. & Myers, P.C. 1989, ApJS, 71, 89
5. Larson, R.B. 1981, MNRAS, 194, 809
6. Heyer, M.H. & Schloerb, F.P. 1996, ApJ, in press
7. Larson, R.B. 1995, MNRAS, 272, 213

Interferometric Imaging of Dense Gas Tracers in the Protostellar Collapse Candidate L1527

D.J. Wilner[1], D. Mardones and P.C. Myers

Harvard-Smithsonian Center for Astrophysics
Cambridge, MA 02138

Abstract. We present $3''$ (~ 400 AU) OVRO and IRAM interferometer observations at $\lambda = 3.0$ mm and 1.3 mm of L1527, one of the closest protostellar collapse candidates, We imaged high excitation lines of H_2CO and HC_3N in an effort to probe the highest densities, temperatures and infall velocities close to the L1527 protostar. The integrated intensity maps show an elongated structure of major axis size $20''$ oriented perpendicular to the bipolar outflow axis. Spectral signatures associated with infall persist to the smallest scales observed.

INTRODUCTION

Despite an abundance of evidence that links star formation to dense cores in dark clouds, convincing evidence for gravitational collapse motions has remained hard to find. The main difficulty is that infall velocities expected at size scales accessible to single dish telescopes are small, only ~ 0.1–0.3 km s^{-1}, comparable to intrinsic line widths and much less than bipolar flow velocities that develop soon after the onset of collapse. Consequently, infall motions are generally inferred from subtle radiative transfer signatures. In an influential study, Zhou et al. (1993) observed a variety of trace species toward the globule B335 and showed that detailed radiative transfer models based on the velocity and density fields of "inside-out" collapse theory reproduced the observed spectra extremely well [1]. In particular, lines of moderate optical depth showed red-shifted self-absorption localized to the protostar. Extensive surveys of deeply embedded young stellar objects are now identifying a growing sample of collapse candidates using this criterion [2,3].

The L1527 region was found to exhibit the spectral signatures characteristic of infall in early high resolution surveys of Taurus cores [4,5]. Recently, several

---

[1] Hubble Fellow

detailed molecular line studies have provided additional support for inside-out collapse in L1527 [6–8,2]. However, neither the single dish observations of dense gas tracers nor the interferometer images of CO isotopes obtained so far have both the sensitivity to high densities *and* the high angular resolution needed to resolve the morphology and kinematics of the suspected infall zone.

OBSERVATIONS AND RESULTS

We imaged the 225.7 GHz H_2CO $3_{12} - 2_{11}$ line and the 100.1 GHz HC_3N $11 - 10$ line toward L1527 at 3″–4″ resolution using the OVRO and IRAM arrays. These high excitation lines were chosen for their (1) range of optical depth (thick and thin) (2) high critical densities ($\sim 10^6$ cm^{-3}), (3) high excitation ($E_l/k > 20$ K), and (4) minimal outflow contamination (line wings weak or absent), with the goal of maximizing sensitivity to the high densities, temperatures and infall velocities close to the protostar.

Figure 1 *(top)* shows the basic geometry of the L1527 core. The solid contours show the integrated intensity of the H_2CO line, revealing dense gas surrounding the embedded source. The 3″ beam (420 AU) resolves the circumstellar environment into an elongated structure oriented perpendicular to a bipolar outflow traced by the near-infrared reflection nebula shown in greyscale. Channel maps provide no evidence for rotational support of the H_2CO structure. The HC_3N results are similar. One possibility is that these lines trace the dense "pseudo-disk" thought to develop in the collapse of an isothermal sphere threaded by a magnetic field [9]. For a core with a 30 μG field and effective sound speed 0.25 km s^{-1} that started collapsing 10^5 years ago, the characteristic radius of the "pseudo-disk" is 500 AU, comparable to the H_2CO and HC_3N emission regions.

Figure 1 *(bottom)* shows the line profiles of the H_2CO and HC_3N lines at the 1.3 mm continuum peak which we associate with a dusty accretion disk surrounding the protostar. The H_2CO line shows pronounced redshifted self-absorption, causing the well-known "infall asymmetry" expected for lines of moderate optical depth arising from a centrally condensed core with radial motions inward. The HC_3N line shows no self-absorption, consistent with low optical depth, thereby identifying the systemic velocity.

Myers et al. (1995) presented a radiative transfer model using the spherical inside-out collapse prescription of Shu (1977) to describe the spectra of two H_2CO lines observed toward L1527 with the IRAM 30 meter telescope [6]. The model assumes a sound speed 0.25 km s^{-1}, a central mass 0.2 $M_\odot$, kinetic temperature profile (30 K)(r/100 AU)$^{-0.3}$ (with a lower bound of 12 K), local microturbulent broadening fwhm 0.15 km s^{-1}, and H_2CO abundance 2.0×10^{-9}. The left panel of Figure 2 reproduces the spectrum of the H_2CO 3_{12}-2_{11} line as observed and modeled. The right panel of Figure 2 shows the OVRO 3″ resolution spectrum and the prediction of the model for this resolution.

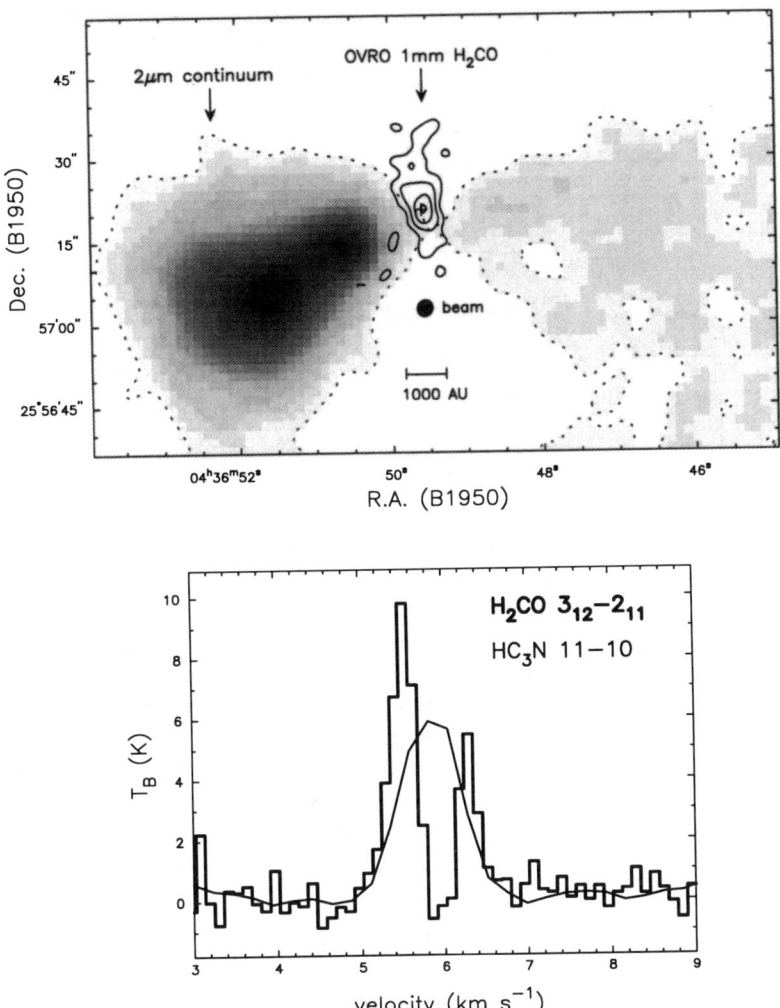

FIGURE 1. *(top)* Images of L1527 in the H_2CO $3_{12} - 2_{11}$ line show a dense core elongated perpendicular to the bipolar outflow direction. *(bottom)* The spectral line profiles of optically thick (H_2CO) and thin (HC_3N) tracers at 400 AU resolution show the characteristic signature of infall with no evidence for high velocity "outflow" wings. Together, these properties suggest a high resolution view of the infall process with surprisingly little contamination from the bipolar flow.

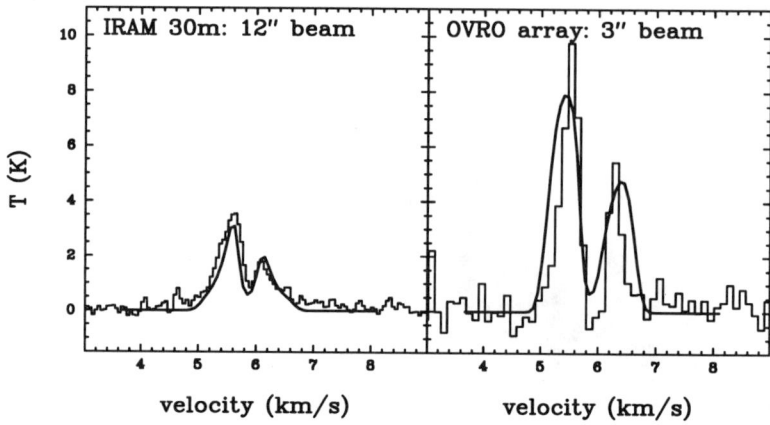

FIGURE 2. H_2CO $3_{12}-2_{11}$ spectra as observed (histogram) and modeled (line).

At the higher resolution, the two peaks of the infall model profile become brighter (the emission is centrally concentrated), broader (less beam dilution for material with high infall velocity) and farther apart (red peak shifts farther red through self-absorption). The OVRO data are generally consistent with these predictions, though the overall broadening is less than predicted. Better agreement may be achieved with a reduction in H_2CO abundance at small radii. Such a reduction might be explained by depletion or destruction of H_2CO molecules at the high densities at close proximity to the protostar. Note also that the weak high velocity "outflow" wings in the IRAM spectrum are virtually absent in the OVRO spectrum, which suggests little high density gas with outflow velocities within $1''.5$ (200 AU) of the protostar.

REFERENCES

1. Zhou, S., Evans, N.J. II, Kompe, C. & Walmsley, C.M. 1993, ApJ, 404, 232
2. Gregerson, E.M., Evans, N.J. II, Zhou, S. & Choi, M. 1996, ApJ, submitted
3. Mardones, D. et al. 1996, this volume (ApJ in preparation)
4. Zhou, S., Wang, Y., Evans, N.J. II, Peng, R. & Lo, K.Y. 1994, ASP v. 65, p. 183
5. Mardones, D., Myers, P., Caselli, P. & Fuller, G. 1994, ASP v. 65, p. 192
6. Myers, P.C., Bachiller, R., Caselli, P., Fuller, G.A., Mardones, D., Tafalla, M. & Wilner, D.J. 1995, ApJ, 449, L65
7. Zhou, S., Evans, N.J. II & Wang, Y. 1996, ApJ, 466, 296
8. Ohashi, N., Hayashi, M., Ho, P.T.P. & Momose, M., 1996, ApJ, in press
9. Galli, D. & Shu, F.H. 1993, ApJ, 417, 220

A Statistical Study of Infall Motions in Nearby Young Stellar Objects

D. Mardones*, P.C. Myers*, M. Tafalla*, D.J. Wilner*,
R. Bachiller† and G. Garay‡

*Harvard-Smithsonian Center for Astrophysics, Cambridge, MA 02138
†OAN, Apartado 143, E-28800, Alcala de Henares, Spain
‡University of Chile, Casilla 36-D, Santiago, Chile

Abstract. We observed 49 nearby low mass pre-main-sequence stars in the $H_2CO\ 2_{12} - 1_{11}$, CS 2–1 (both optically thick) and N_2H^+ 1–0 (optically thin) mm-wavelength lines using the IRAM 30-m, SEST 15-m, and Haystack 37-m radio telescopes. The sources were selected from their spectral energy distributions ($T_{bol} < 200$ K) and distance ($d < 400$ pc). We find an overabundance of sources whose optically thick lines have blue-shifted asymmetry. We quantify the observed line asymmetries by measuring the difference in the peak velocities of the optically thick and optically thin lines. The observed distribution of velocity differences is skewed toward negative values of $V_{thick} - V_{thin}$ velocities. This excess is statistically more significant toward the subsample of the class 0 sources ($T_{bol} < 70$ K). This can be most naturally explained if low T_{bol} sources tend to have inward motions. Kinematics of bipolar outflows or rotation can't reproduce the statistics if the source sample has randomly oriented symmetry axes. The statistical results are insensitive to variations in pointing, and are indistinguishable for the CS and H_2CO line profiles.

INTRODUCTION

Millimeter-wavelength observations have increased the evidence for infall motions onto nearby YSOs in recent years. The main procedure is to observe spectral lines that trace high densities ($n > 10^4$ cm^{-3}) and have red-shifted self-absorption spatially concentrated around an embedded YSO. The spatial concentration and the use of dense gas tracers are important to ensure that the observed kinematic signature is associated with the YSO. Several good infall candidates have been reported in the literature (eg B335 [1], L1527 [2]). However, peculiarities of source structure and kinematics (eg, outflows or rotation) could possibly account for the apparent signatures of infall in such a small sample. These well studied cases are not sufficient by themselves to support a claim for observed infall motions onto a wide class of sources.

Another approach is to observe a large sample of YSOs in optically thick and optically thin lines tracing high density. In such a sample, rotation and bipolar outflow motions having symmetry axes in random orientations would tend to produce an equal number of optically thick line profiles with red- and blue-shifted self-absorption. On the other hand, infall motions in centrally concentrated regions would only produce red-shifted self-absorption in the line profiles. A large spectral line survey is able to reveal whether there is a statistically significant excess of sources with red- or blue-shifted self-absorption. This approach does not depend much on the details of the models, but mainly on whether there is a prevalence of inward or outward motions. At the same time such a study can yield a good list of collapse candidates for further observations.

We present the results of a molecular line survey toward 49 nearby (closer than 400 pc) low mass PMS stars. We observed YSOs with "embedded" colors, ie. spectral class 0 or I [3,4] or $T_{bol} < 200$ K [5] in the H_2CO $2_{12} - 1_{11}$ and CS 2–1 optically thick lines and the N_2H^+ 101–012 optically line at the IRAM 30-m, SEST 15-m and Haystack 37-m radio telescopes.

RESULTS

We measured the velocity of the brightest peak of the H_2CO and CS line profiles and compared it to the velocity fit to the N_2H^+ line. Sources with broad lines tend to have large velocity differences (VD) and fall at the edges of the distribution while sources with asymmetric but narrow line profiles like B335 fall close to the origin. To compensate for this bias, we normalized the VDs dividing by the width of the N_2H^+ line. The observed N_2H^+ velocity widths vary from 0.3 km s^{-1} in narrow sources to $\sim$ 1.3 km s^{-1} in the broad sources. Figure 1 displays histograms of the H_2CO and CS normalized VD distributions.

To characterize the sources responsible for the VD distribution asymmetries, we plot $(V_{CS} - V_{N_2H^+})/\Delta V_{N_2H^+}$ as a function of T_{bol} in figure 2. Sources located in the star forming clusters NGC1333, Ophiuchus or Serpens are indicated by circles, and the other sources by triangles. Sources with asymmetric CS lines are indicated by filled symbols. Note that the observed VD are highly skewed towards negative values for sources of low T_{bol}. To find out whether these results are statistically significant, we evaluate the student's t-test with a null hypothesis that the observed VD distribution is drawn from a normal parent distribution with zero mean. Table 1 lists the number of sources (N), median, standard deviation (σ), 95% confidence limits on the true mean of the parent distribution (μ_{min}, μ_{max}), and the probability of the null hypothesis (p). We can reject the null hypothesis with a 99% or greater confidence level

TABLE 1. Statistics of the $(V_{CS}-V_{N_2H+})/\Delta V_{N_2H+}$ distribution

subsample	N	median	σ	μ_{min}	μ_{max}	p
all	38	-0.14	0.46	-0.32	-0.02	0.03
$T_{bol} \leq 70$ K	21	-0.37	0.43	-0.51	-0.11	0.004
$70 < T_{bol} < 200$ K	17	0.01	0.45	-0.22	0.24	0.95
Asymmetric	22	-0.46	0.44	-0.57	-0.18	0.001
Symmetric	16	0.10	0.34	-0.07	0.29	0.20
isolated	25	0.06	0.43	-0.20	0.16	0.80
clusters[a]	13	-0.42	0.40	-0.69	-0.21	0.002

[a] Sources in NGC1333, Ophiuchus and Serpens

for subsamples of sources: a) with $T_{bol} \leq 70$ K (class 0), b) with asymmetric CS 2–1 line profiles (filled symbols), or c) located in clusters (circles).

DISCUSSION

We can rule out the presence of multiple velocity components to explain the observed VD distribution asymmetries since the observed N_2H^+ line profiles are single in all sources. If the kinematics around a YSO are dominated by infall, then we expect the foreground gas to be preferentially at red-shifted velocities with respect to the source, causing asymmetric line profiles [6] and VD distributions skewed toward negative values. Expansion causes the opposite effect. Spherically symmetric infall alone can, therefore, explain the observed statistics. However, we know that rotation is kinematically important very

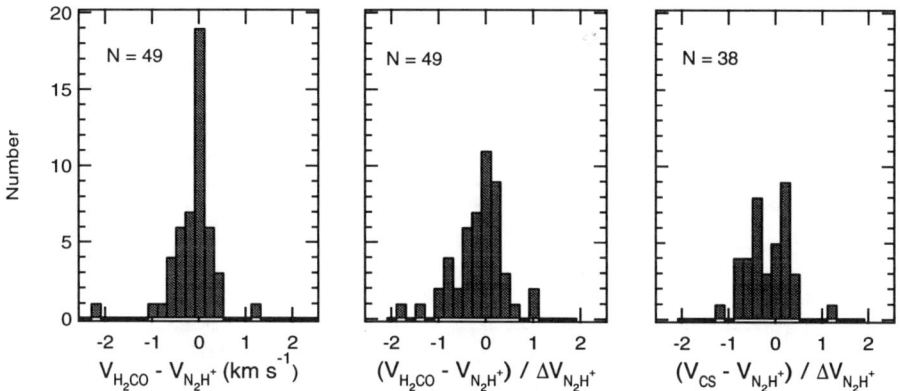

FIGURE 1. a) Distribution of the line peak velocity differences (VD) of $V_{H_2CO} - V_{N_2H+}$ for the full survey. b) Histogram of the VD normalized by the width of the N_2H^+ line $(V_{H_2CO} - V_{N_2H+})/\Delta V_{N_2H+}$. The normalized VD avoids selecting against sources with narrow lines in the histograms. c) same as in b but for the CS and N_2H^+ lines.

close to the YSOs and that bipolar outflow motions are common. Outflow or rotation kinematics would produce infall or expansion signatures on opposite sides of the source [7]. An ensemble of bipolar outflows with random orientation would thus produce a VD distribution symmetric about zero, unlike our results. Our statistical results imply then, the presence of infall motions toward most nearby low mass YSOs. Rotation and outflows are most likely also present in our sample sources, but they can't reproduce the observed line asymmetries without the additional presence of infall.

In summary, the excess of sources with optically thick line profiles having blue-shifted peaks can be naturally explained if most sources in our sample have infalling envelopes. This tendency is more pronounced in subsamples having low T_{bol} (class 0), asymmetric profiles, or sources in clusters.

REFERENCES

1. Zhou, S., Evans, N.J., Kompe, C. and Walmsley, C.M. 1993, ApJ, 404, 232
2. Myers, P.C., Bachiller, R., Caselli, P., Fuller, G.A., Mardones, D., Tafalla, M. and Wilner, D.J. 1995, ApJ, 449, L65.
3. Lada, C.J. and Wilking, B.A. 1984, ApJ, 287, 610.
4. Andre, P., Ward-Thompson, D. and Barsony, M. 1993, ApJ, 406, 122.
5. Myers, P.C. and Ladd, E.D. 1993, ApJ, 413, L47.
6. Leung, C.M. and Brown, R.L. 1977, ApJ, 214, L73.
7. Adelson, L.M. and Leung, C.M. 1988, MNRAS, 235, 349.

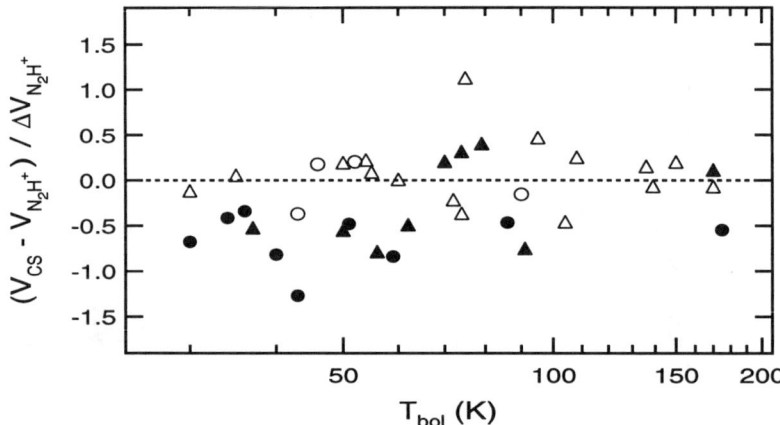

FIGURE 2. Correlation between the normalized velocity differences and T_{bol}. Filled symbols indicate sources with asymmetric CS lines and open symbols indicate sources with symmetric lines. Sources located in the NGC1333, Ophiuchus or Serpens complexes are indicated by circles, all others by triangles.

Accretion Disks Around Class 0 Protostars: The Case of VLA 1623

C. D. Wilson[1], R. E. Pudritz[1], J. E. Carlstrom[2], O. P. Lay[3], R. E. Hills[4], and D. Ward-Thompson[5]

[1] *McMaster University*
[2] *University of Chicago*
[3] *California Institute of Technology*
[4] *Cambridge University*
[5] *Royal Observatory, Edinburgh*

Abstract. We have detected continuum emission at 220 and 355 GHz from the prototype Class 0 source VLA 1623 using the JCMT-CSO interferometer. Gaussian fits to the data place an upper limit of 70 AU on the half-width half-maximum radius of the emission, which implies an upper limit of ~ 175 AU for the cutoff radius of the circumstellar disk in the system. In the context of existing collapse models, this disk could be magnetically supported on the largest scales and have an age of $\sim 6 \times 10^4$ yr, consistent with previous suggestions that Class 0 sources are quite young. Alternatively, if 175 AU corresponds to the centrifugal radius of the disk, the age of the system is $\sim 2 \times 10^5$ yr, closer to age estimates for Class I sources.

INTRODUCTION

Submillimeter observations of deeply embedded star forming regions have identified a small number of strong sources that have virtually no emission at wavelengths below 10 μm and whose spectral energy distributions are characterized by single blackbodies at $T \simeq 15 - 30$ K. These cold "Class 0" sources are good candidates for protostars whose circumstellar envelope masses exceed that of their central protostars [1] in which case their ages may be as low as 2×10^4 yrs [2]. Sources in this class are also defined by the presence of energetic bipolar outflows. Most current models of outflow posit the existence of disks as an essential element of the engine [3].

There is now good evidence that a large fraction of young stellar objects have dense, dusty accretion disks [4]. This is in accord with standard collapse models of star formation [5], which predict that systems older than 10^5 years have centrifugally supported disks extending to 100 AU. Disks around the sources L1551 IRS5 and HL Tau have been resolved with semi-major axes of 80 and 60 AU, respectively [6], [7]. Even larger magnetically supported "pseudo" disks extending out to 1000 AU are predicted for some magnetohydrodynamic collapse models [8].

If the Class 0 sources are young, then the models of Terebey et al. [5] suggest that a disk with radius of 100 AU at 2×10^5 yrs has a centrifugal radius of only $r_c \propto t^3 \simeq 0.8$ AU at 2×10^4 yrs. Similarly a larger magnetically supported outer region of $r_B \propto t^{7/3} \simeq 20$ AU [8] should be present at these early times if 1000 AU structures exist at 2×10^5 yrs.

OBSERVATIONS

VLA 1623 was observed with the JCMT-CSO interferometer on 7 April 1995; we detected both 355 and 220 GHz continuum emission. The source appears to be unresolved and the flux is too strong to be attributed to the inner core of the observed circumstellar envelope. Thus we attribute the observed flux to the presence of an unresolved disk in VLA 1623. We have fit the data with a two-dimensional Gaussian with the position angle fixed at $35°$ to orient the disk perpendicular to the large scale outflow [9]. These fits yield a flux of $0.5^{+0.4}_{-0.1}$ Jy and place an upper limit of 70 AU on the half-width half-maximum radius of the emission, which implies an upper limit of ~ 175 AU for the cutoff radius of the circumstellar disk in the system.

FITTING ACCRETION DISK MODELS

We adopt the approach of Beckwith et al. [4] and assume a geometrically thin disk with power law scalings with radius of the disk temperature and surface density (for details see [10]). For the limiting cases of completely optically thick or thin truncated disks in the Rayleigh-Jeans limit, the cutoff radius R_D can be related to the observed flux F_ν by

$$R_D/r_o = [(\gamma \bar{L}_\nu / x \cos\theta) + 1]^{1/\gamma} \qquad (1)$$

where $\bar{L}_\nu = D^2 F_\nu / 2\pi (2k\nu^2/c^2)$ can be computed from the observations, D is the distance to the source, θ is the inclination of the disk to the line of sight, and r_o is the inner radius of the accretion disk. For the optically thick case, the parameters are $\gamma = 2 - q$ and $x = T_o r_o^2$, while for the optically thin case, $\gamma = 2 - p - q \neq 0$ and $x = \tau_o T_o r_o^2$, where $\tau_o \equiv \kappa_\nu \Sigma_o / \cos\theta$. In these expressions T_o and Σ_o are the temperature and surface density of the disk at radius r_o, p and q are the power laws of the surface density and temperature, and κ_ν is the specific dust opacity. Thus we can simplify our analysis of the model by reducing six free parameters to essentially the two parameters γ and x (with R_D measured in units of r_o). Then a disk model characterized by a given

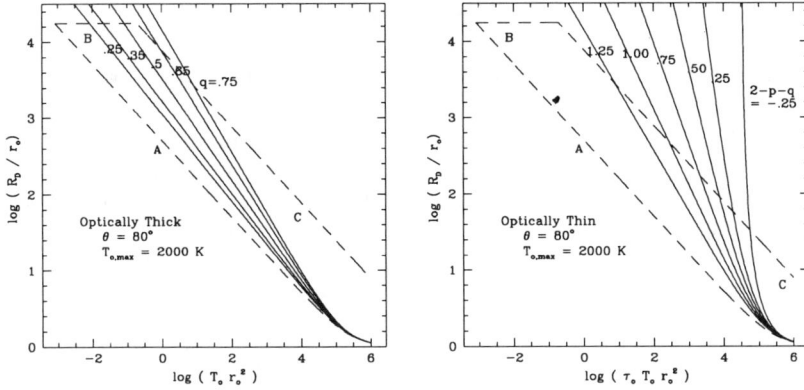

FIGURE 1. (a) Cutoff radius R_D of an optically thick disk in units of the inner disk radius r_o versus the product $T_o r_o^2$, where T_o is the temperature of the disk at radius r_o. Solid lines are model fits for different values of q, the exponent in the temperature gradient $(T(r) = T_o(r/r_o)^{-q})$. Viable accretion disk models lie within the region demarcated by the dashed line segments, A,B,C. The angle between the disk axis and the line of sight is $\theta = 80°$. (b) As in (a), but for optically thin emission. In this case R_D depends on the product $\tau_o T_o r_o^2$.

index γ produces a unique curve in the $R_D/r_o - x$ plane. These curves are plotted in Figure 1 for VLA 1623.

Our observed flux and upper limit to the cutoff radius restrict possible disk models to lie within the region demarcated by the three dashed lines in Figure 1 [10]. Figure 1 shows that while optically thick models with values of $q \simeq 0.25 - 0.5$ fill the largest portion of the permitted parameter space, models with q up to 0.75 are also acceptable. Optically thin models fill a much smaller portion of parameter space, with low values of $2 - p - q$ being less likely. In general, the optically thin solutions have relatively large values for the inner radius, which suggests that they may also have a small optically thick region that contributes relatively little to the total flux of the source. Thus both the spectral index and the models summarized in Figure 1 are consistent with either optically thin or optically thick disks with a cutoff radius not exceeding 175 AU.

DISCUSSION

The constraints on the size of the central source have implications for theoretical models of the evolution of sources such as VLA 1623. If we identify the upper limit to the cutoff radius of 175 AU with the magnetic support radius of the pseudo-disk model of Galli & Shu [8], then this model predicts an age for the system of $t < 5.9 \times 10^4 (B/30\,\mu\mathrm{G})^{-4/7} (a/0.27\,\mathrm{km\,s^{-1}})^{1/7}$ yr, where B is the magnetic field strength and a is the effective sound speed [10]. This age agrees with general estimates for Class 0 ages [2], but is somewhat longer than initial estimates for the age of VLA 1623 [1]. In this picture, the centrifugal balance radius in the pseudo-disk then occurs at $r_c = 0.058 a \Omega_o^2 t^3 < 6.2 (\Omega_o/10^{-13} s^{-1})^2$ AU. This region is large enough either to drive off a disk wind or to interact with a possible stellar magnetosphere, and thus theoretical models developed to explain outflows in Class I sources are likely to be equally valid for Class 0 sources.

Alternatively, if we identify 175 AU with the centrifugal balance radius, then the age of the system is $< 1.8 \times 10^5$ yr, similar to age estimates for Class I sources [11]. In this picture, the magnetic pseudo-disk, if present, would extend to 2300 AU, and must have a very low column density in order to be undetectable with lower resolution measurements [1]. In this case, we may be detecting a Class I source viewed nearly edge-on.

REFERENCES

1. André, P., Ward-Thompson, D., & Barsony, M., 1993, ApJ, 406, 122
2. Barsony, M., 1994, in Clouds, Cores, and Low Mass Stars, eds. D. Clemens & Barvainis, (San Francisco; Astronomical Society of the Pacific), 197
3. Pudritz, R. E., Pelletier, G., & Gomez de Castro, A. I., 1991, in The Physics of Star Formation and Early Stellar Evolution, C.J. Lada & N.D. Kylafis, eds. (Dordrecht; Kluwer), 539
4. Beckwith, S. V. W., Sargent, A. I., Chini, R. S., & Güsten, R., 1990, AJ, 99, 924
5. Terebey, S., Shu, F. H., & Cassen, P., 1984, ApJ, 286, 529
6. Lay, O. P., Carlstrom, J. E., Hills, R. E., & Phillips, T. G., 1994, ApJ, 434, L75
7. Mundy, L. G., et al., 1996, ApJ, 464, L169
8. Galli, D., & Shu, F.H., 1993, ApJ, 417, 243
9. André, P., Martin-Pintado, J., Despois, D., & Montmerle, T., 1990, A&A, 236, 180
10. Pudritz, R. E., Wilson, C. D., Carlstrom, J. E., Lay, O. P., Hills, R. E., & Ward-Thompson, D., 1996, ApJ, 470, L123
11. Wilking, B. A., Lada, C. J., & Young, E. T., 1989, ApJ, 340, 823

A Deflected Jet in the Bipolar Molecular Outflow NGC 2264G

Michel Fich* and Charles J. Lada[†]

*Department of Physics, University of Waterloo, Waterloo, Ont. Canada N2L 3G1
[†]Smithsonian Astrophysical Observatory, 60 Garden Street, Cambridge, MA 02138

Abstract. Recently published H_2 emission data for the bipolar molecular outflow NGC2264G are compared to new CO $(2 \to 1)$ images. The highest velocity CO emission seems to be arise in a jet that is probably the driving mechanism of the outflow. Virtually all of the H_2 emission is found either behind clumps of swept-up ambient material (i.e. closer to the central source of the outflow) or along the axis of the highest velocity CO. The highest velocity CO features are inclined with respect to the general outflow direction, and appear to emerge from H_2 emission knots and generally follow a pattern consistent with a deflection of the jet at these points. Radio continuum observations of the central source are also consistent with this model. Deflections of the underlying jet will inevitably produce an outflow significantly wider than the underlying jet itself. We propose that deflections such as the one observed here play a role in producing the wide outflow lobes seen in many, especially older, outflows.

INTRODUCTION

Material flowing out of star forming regions is observed in a variety of structures: most commonly as bipolar molecular outflows, often as Herbig-Haro objects and associated optical jets and occasionally as small, faint ionized or neutral jets. These structures are probably all manifestations of the same underlying phenomenon. It has been suggested that this phenomenon is some kind of primary wind originating either at the star or at the disk formed by the accreting material. An alternate suggestion that has recently been investigated by several authors (e.g. [1–3]) is that the primary wind is a jet. In this paper we directly compare the ionized jets, H_2 emission, and the structures visible in CO in the outflow NGC 2264G.

NGC2264G is a well collimated, high velocity, but uncomplicated bipolar molecular outflow. In a new study of this outflow [4,5] the entire area has been mapped in CO(2 → 1) with high resolution and sensitivity. The central source of the outflow was recently discovered from VLA NH_3 and radio continuum observations [6] and has been observed in the submm continuum and identified as a Class 0 source [7]. The highest velocity component has been identified in each lobe with the underlying jet, as opposed to the more extended bipolar lobes that primarily consist of swept up ambient material ([5]). This identification of a molecular jet component was based on a break in the integrated line profiles of each wing. This break occurs in both wings at the same velocity where the CO spatial images show a long unresolved feature extending beyond the low and intermediate velocity CO structures.

Recently H_2 in emission has been imaged in NGC 2264G [8]. Using our new, higher resolution and higher signal-to-noise data we show in this paper that the H_2 observations provide a vital clue that leads to the idea that the underlying jets are deflected by the swept up material.

DATA

In Figures 1 and 2 we show contour plots of the high and low velocity CO components of each lobe. The (0,0) position is at the peak of the blue lobe

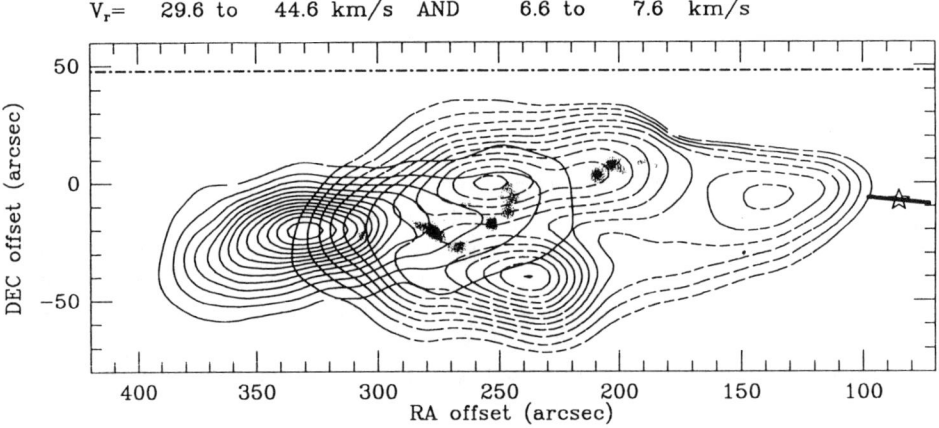

FIGURE 1. Red lobe of NGC 2264G showing CO and H_2.

seen in the earlier CO(1 → 0) maps [9]. The central velocity of the outflow, and the velocity of the central source, is 4.6 km/s. The solid contours show the highest velocity CO emission while the dashed contours show a lower velocity range. The solid contour interval is 0.05 K while the much stronger low velocity emission is shown with contours interval of 0.25 K and the lowest contour is at 2.5 K. The central source is shown with a star and the position angle of the cm wave radio continuum emission is shown (much exaggerated in length) with a heavy solid line. The molecular hydrogen emission is shown in a grey scale.

The red lobe high velocity gas is elongated in the East-West direction with a slight tilt to the South, extends somewhat further from the central source than the low velocity gas but does not point to the central source nor does it extend all of the way inwards to the central source. The blue lobe high velocity gas is also elongated in the East-West direction but with a slight tilt to the North nearly parallel to the red lobe high velocity component. The blue lobe high velocity component extends much further away from the central source than the low velocity gas but, like the high velocity red lobe emission, it does not point to or extend in to the central source.

The highest velocity gas in each lobe, identified with the jet driv-

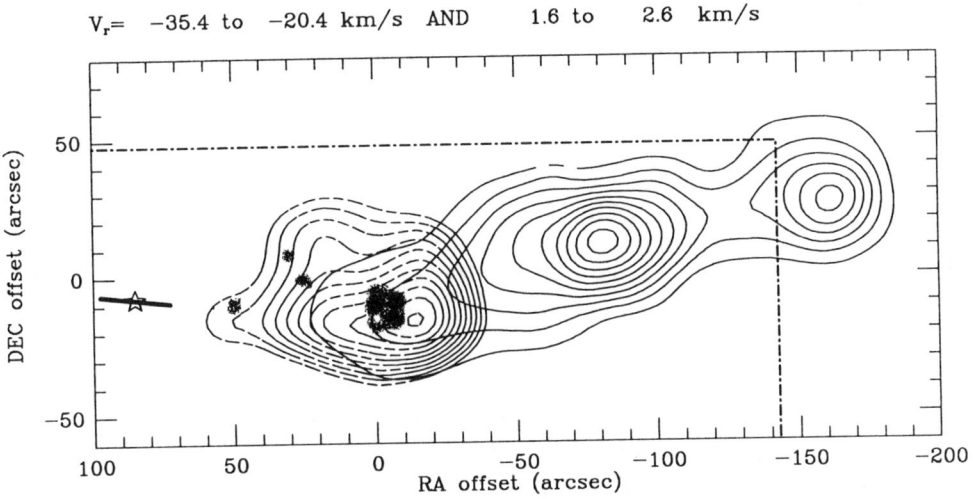

FIGURE 2. Blue lobe of NGC 2264G showing CO and H_2.

ing the outflow, follow axes that are not parallel to the axis defined by the radio continuum at the central source. The brightest H_2 spots occur at the intersection point of the axis defined by the highest velocity CO and the axes of the radio continuum emission.

DISCUSSION

The geometrical relationship between the various components of this object suggests the following interpretation: that the central source emits a jet of material (as seen in the VLA radio continuum data), that deflects from swept up and outflowing material ejected earlier (the low velocity CO components), producing H_2 emission at the deflection points, after which the jet becomes denser, probably slowing and perhaps forming molecules (the highest velocity CO components).

One effect of a deflected jet model is that this deflection of the jet must act to widen the outflow. There has been considerable debate as to the nature of the underlying mechanism producing bipolar molecular outflows. One difficulty of jet models has been there inability to produce wide outflow lobes. However deflections of the underlying jet will inevitably produce an outflow that is significantly wider than the underlying jet itself. If such deflections are a common feature of jets ejected from outflow sources then this process could play an important role in producing the wider outflow lobes seen in many outflows.

REFERENCES

1. Masson, C. R. and Chernin, L. 1993, ApJ, 414, 230
2. Raga, A. and Cabrit, S. 1993, A&A, 278, 267
3. Stahler, S. 1994, ApJ, 422, 616
4. Lada, C. J. and Fich, M. 1995, in Circumstellar Disks, Outflows, and Star Formation, eds. S. Lizano and J.M. Torrelles (Mexico: Rev. Mex. A. A., Ser. de Conf.), 1, 93
5. Lada, C. J. and Fich, M. 1996, ApJ, 459, 638
6. Gómez, J. F., Curiel, S., Torrelles, J. M., Rodríguez, L. F., Anglada, G., Girart, J. M. 1994, ApJ, 436, 749
7. Ward-Thompson, D., Eiroa, C., Casali, M. M. 1995, MNRAS, 273, L25
8. Davis, C. J. and Eislöffel, J. 1995, A&A, 300, 851
9. Margulis, M., Lada, C. J., Hasegawa, T., Hayashi, S. S., Hayashi, M., Kaifu, N., Gatley, I., Greene, T. P., and Young, E. T., 1990, ApJ, 352, 615

IRAS 04302+2247:
Butterfly Star in Taurus !

Philip Lucas* and Patrick F. Roche*

*Astrophysics, Nuclear Physics Building, 1 Keble Road, Oxford OX1 3RH, UK.

Abstract. We present near infrared observations of IRAS 04302+2247, a Class I protostar in the Taurus-Auriga molecular cloud complex whose equatorial plane is inclined precisely edge-on to the line of sight ($i = 90° \pm 4°$). This system displays a unique quadrupolar morphology, which had not been previously predicted in any simulations of a single protostar. We use the Monte Carlo method to tailor a model to our imaging and polarimetric data, and we explain the quadrupolar structure in terms of a dusty jet or outflow which lies perpendicular to the equatorial plane. We constrain the circumstellar structure to the form of an equatorially condensed envelope with a concave bipolar cavity. The circumstellar disk is not seen, which requires that it is physically thin. With its convenient orientation and proximity (d = 140pc) this system should yield many insights into the Class I phase of evolution.

INTRODUCTION

Class I protostars are embedded sources which emit most of their energy at far infrared wavelengths. In the near infrared they are seen primarily in scattered light, the protostar itself being obscured from view. It has been assumed that their appearance is dominated by the structure of the circumstellar envelope and the orientation and shape of the bipolar cavity, which allows scattered light to reach the observer along paths of low optical depth. However, our high resolution images of a small sample of sources in Taurus show that dust in the bipolar outflow can have a prominent effect on the near infrared flux distribution.

High resolution imaging and polarimetry coupled with Monte Carlo modelling allows us to investigate the envelope and cavity structures of nearby Class I sources. The models also constrain the scattering properties of the dust grains. We have been using the Shift and Add technique at the United Kingdom Infrared Telescope (UKIRT) to obtain resolve structures on the scale of the solar system (~ 50 au) and have combined these data with deep imaging polarimetry. Near infrared structures can admit to more than one interpretation, and we have generally found it very useful to incorporate existing millimeter and radio observations into our models of each source.

IRAS 04302+2247

An excellent example of this class of object is IRAS 04302+2247, see fig.1, a protostar whose equatorial plane is oriented precisely edge-on to the line of sight, producing the vertical dark lane. This source has a unique quadrupolar morphology which arises from extinction by dust in the outflow, which is perpendicular to the disk. The 4 flux peaks also exhibit variability on a timescale of months, which is probably due to changing star-envelope extinction caused by irregularities in the outflow. Alternatively, it might be due to the orbital motion of matter within 1 au of the protostar.

Regrettably, the multipolar and diffuse nature of the source prevented the use of Shift and Add, since there is no single bright peak to use as a reference for a wavefront sensor.

Details of the Scattering Model

(1) The density profile of the envelope was modelled using the Ulrich profile [1], with an exponential cut-off in the outer regions. This did not precisely match the wavelength dependence of the east-west separation of the flux peaks, and a more equatorially concentrated profile is indicated. The models shown here employ Rayleigh scattering.

(2) The bipolar cavity has a base radius of 25 au, an opening angle of 150° at base, curving upward in a "concave" manner. The cavity contains an outflow, which is responsible for the dip in intensity along the horizontal axis. The outflow is modelled as being optically thick and having an opening angle of 45°, except within 25 au of the core, where the collimation is weaker and a 90° opening angle is used. The flow adjoins smoothly to the polar regions of the protostar.

(3) There is a thin, flat circumstellar disk of radius 100 au around the star. A physically thick inner disk or toroid can be ruled out since it would be strongly visible in the flux distribution.

(4) The mass of the system is $8.0 \times 10^{-3} M_\odot$, which agrees well with the estimate of $6.6 \times 10^{-3} M_\odot$ made by Moriarty-Schieven et al. [2] from 800 μm photometry of the thermal dust continuum.

(5) The equatorial plane is observed almost edge-on: i=90° ± 4°. At inclinations i< 83° the obscuring effect of the approaching outflow is less apparent and the morphology becomes tripolar at 2.2μm; an increasing contribution from unscattered stellar flux produces a central peak at about i< 79° and at lesser inclinations the morphology becomes bipolar and then monopolar.

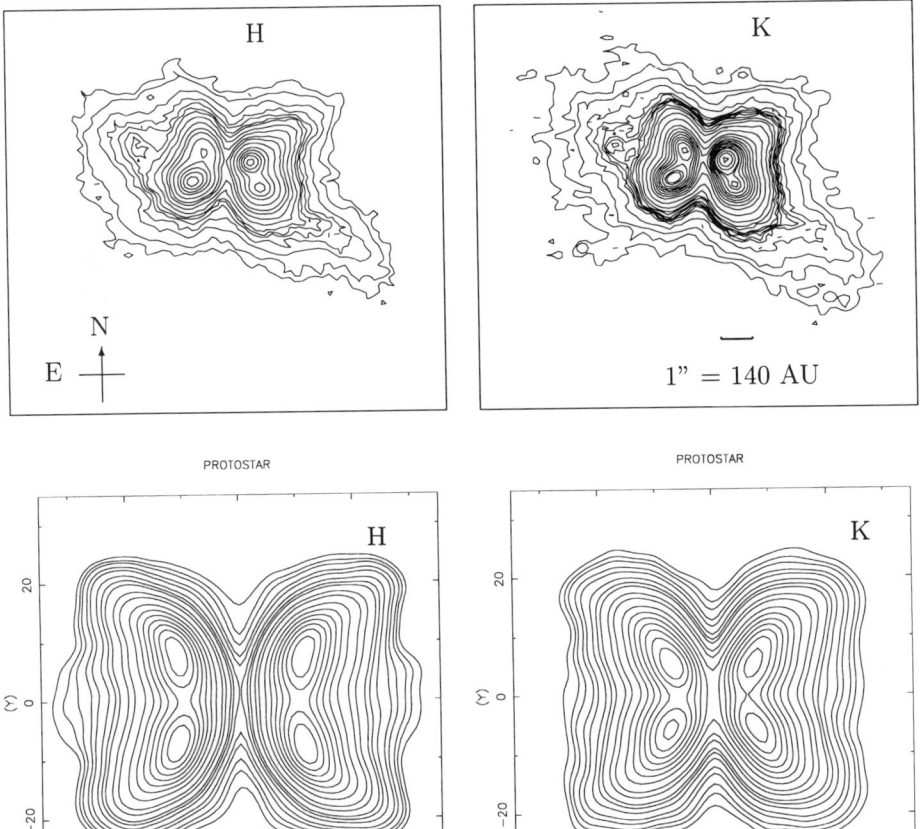

FIGURE 1. IRAS 04302+2247 at H and K. *(above)* Data. *(below)* Models. A knot in the outflow is visible to the east of the core.

The outflow is visible in scattered light in the H band model, but not at K where the scattering efficiency is less. The contribution of line emission has yet to be determined. The polarization data in fig.2 show that the nebulosity is illuminated by a central source. These data were taken in poorer seeing conditions, but the quadrupolar structure is resolved in polarized intensity. Models of the polarization reproduce the "polarization disk" and nulls in the raw data. However, these faint details are drowned out after convolution with the PSF of UKIRT, contrary to observation. Hence we do not rule out a contribution to the polarization from dichroic extinction by aligned grains.

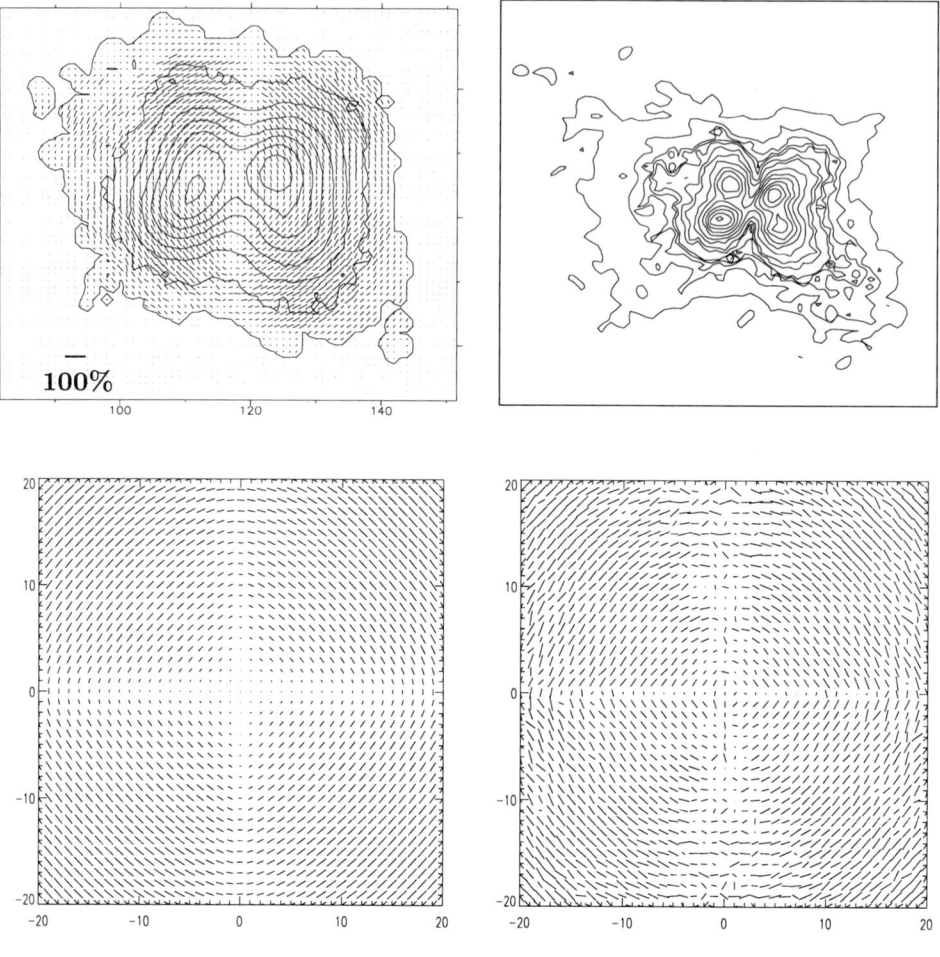

FIGURE 2. IRAS 04302+2247. Polarization data at K band. *(upper left)* Polarization. *(upper right)* Polarized Intensity. *(lower left)* Model of core polarization. *(lower right)* Model, not convolved with the PSF of UKIRT, showing the polarization disk.

REFERENCES

1. Ulrich R.K. 1976,ApJ,210,377
2. Moriarty-Schieven G.H., Wannier P.G., Keene J., & Tamura M. 1994, ApJ,436,800

High Resolution 2.7 mm Observations of L1551 IRS5: A Protobinary System?

Leslie W. Looney*, Lee G. Mundy*, & W.J. Welch[†]

*Department of Astronomy, University of Maryland, College Park
[†]Radio Astronomy Laboratory, University of California, Berkeley

Abstract.
We present sub-arcsecond resolution imaging of the $\lambda = 2.7$ mm continuum emission from the young, embedded system L1551 IRS5 using the new, nine element, high resolution configuration of the BIMA array. The observed emission arises from two compact sources separated by $0\rlap{.}''35$ and coincident with the two sources seen at $\lambda = 2$ cm. The data support the view that IRS 5 is a protobinary system with separation of ~ 50 AU and individual circumstellar disks.

INTRODUCTION

We present $0\rlap{.}''73 \times 0\rlap{.}''31$ imaging of the $\lambda = 2.7$ mm continuum emission from the prototypical deeply embedded, young stellar system L1551 IRS5. Observations at this resolution allow the imaging of suspected circumstellar disks and directly probe the physical environments around young stars, while intrinsically filtering-out emission larger than $\sim 3''$; thus we minimize the emission arising from any large-scale circumstellar envelope and directly measure the emission from the core of L1551 IRS5.

L1551 IRS5 with a strong bipolar molecular outflow, an optical jet, outward moving HH objects, and an envelope-disk structure in the surrounding material, has been used as a standard in the current paradigm for single-star formation. But is it really a single-star system?

In high resolution $\lambda = 2$ cm continuum observations, L1551 IRS5 is resolved into two compact sources with a separation of $\sim 0\rlap{.}''28$ [1,2], which have been interpreted as a protobinary system [1], or as the inner ionized edges of a toroid surrounding a single star [2].

Under the assumption that L1551 IRS5 is a single star system, Keene and Masson (1990) modeled $\lambda = 2.7$ mm interferometric observations to deduce the presence of a 45 AU radius circumstellar disk within an envelope. High resolution JCMT-CSO interferometric observations at $\lambda = 870$ μm resolved the compact central source [3]. The emission was modeled as arising from an 80 AU radius Gaussian source, which was inferred to be the accretion disk around the young star.

BIMA INTERFEROMETRIC OBSERVATIONS

The observation was made in March 1996 using this season's extended baseline array with nine antennas. This configuration had 3 antennas spread over the main array, and 6 antennas located on outrigger stations, connected with fiber optics to the main control room. Baselines ranged from a minimum of 60kλ to a maximum of 480 kλ (1.3 km!). A quick calibrator-source switching technique was utilized to track the atmosphere on 2 minute time-scales. A second calibrator was used to correct for baseline errors on 90 minute timescales.

DISCUSSION

- We detect an elongated structure that is slightly extended in the North-South direction. (Fig. 1a)

- The peak flux is 45 mJy beam^{-1} and the total integrated flux is 75 mJy.

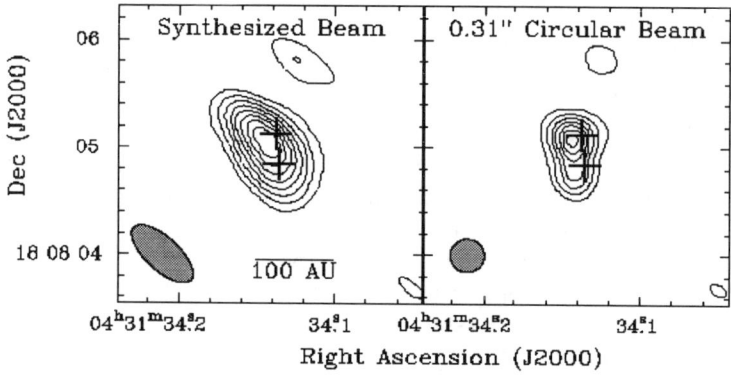

FIGURE 1. L1551 IRS5 continuum emission. The left panel shows the synthesized beam restored map. The right panel shows the map emphasizing the longer baseline information present. Contours are at intervals of 2,3,4,5,6,7,8,9, and 10 times the rms noise, $\sigma = 4.8$ mJy/beam. The inset shows the half-power contour of the clean beam. The two crosses mark the positions of the two point sources seen at $\lambda = 2$ cm by Rodríguez et al. 1986.

- With the interferometer we are sensitive to size structures smaller than the size of the synthesized beam. To emphasize the longer baseline information present in our dataset we have reconstructed our map with a $0\farcs31$ circular beam. (Fig. 1b)
- By fitting two point sources to Fig. 1b., the integrated flux of the northern component is 45 mJy and the integrated flux of the southern source is 23 mJy— a total flux density of 68 mJy.
- By fitting two Gaussians to Fig. 1b., the positions of the two sources agree to within $0\farcs05$ with the $\lambda = 1.3$ cm sources of Koerner and Sargent (1996).
- Radiative transfer modeling of the emission assuming HL Tau typical parameters for density and temperature, gives mass estimates of 0.0024 $M_\odot$ for the northern source and 0.009 $M_\odot$ for the southern source.

VLA COMPARISON

A recent high resolution VLA observation at $\lambda = 1.3$ cm by Koerner and Sargent (1996) also resolves the two components: 2.0 mJy for the North source and 1.5 mJy for the South, yielding a spectral index of $\alpha \sim 1.1$. Extrapolating to our frequency, the expected flux from free-free emission would be 11.5 mJy for the North and 8.6 mJy for the South. The emission at $\lambda = 2.7$ mm is therefore dominated by dust.

COMPARISON TO KEENE AND MASSON OBSERVATIONS

The flux measured in our maps (75 mJy) and the expected flux from the Keene and Masson model (150 mJy 128 AU diameter core) suggests that there is more to understand in this system. By modeling the envelope using submillimeter and far-infrared flux measurements with the core constrained to our observations, we calculated the expected flux values at lower resolution. (Fig. 2.) This type of model does not fit the Keene and Masson data. The downturn seen in the Keene and Masson data can not be explained by a envelope and two disks model, so the excess flux is suggestive of an additional structure, perhaps a circumbinary disk. This could also be the resolved structure in the Lay et al. data.

SUMMARY

- $\lambda=2.7$ mm continuum observations of L1551 IRS5 have resolved a compact central structure.

- The $\lambda=2.7$ mm emission closely resembles the two peaks of ionized emission observed at centimeter wavelengths, both in absolute position and separation of the peaks.

- Our interpretation is that we are observing thermal dust emission from small disks around the individual stars, and the centimeter emission arises from the stellar winds.

- Modeling of the emission from the dust envelope suggests a circumbinary disk that may be the object seen with the JCMT-CSO and deduced by Keene and Masson.

REFERENCES

1. Bieging, J.H., & Cohen, M., *ApJ*, **289**, L5 (1985).
2. Rodríguez, L.F., Cantóo, J, Torrelles, J.M., & Ho, P.T.P.,*ApJ*, **301**, L25 (1986).
3. Lay, O.P., Carlstrom, J.E., Hills, R.E., & Philips, T.G., *ApJ*, **434**, L75 (1994).
4. Koerner, D., & Sargent, A., Private Comm. (1996).
5. Keene, J., & Masson, C.R., *ApJ*, **355**, 635 (1990).
6. Ladd, E.F., Fuller, G.A., Padman, R., Myers, P.C., & Adams, F.C., *ApJ*, **439**, 771 (1995).

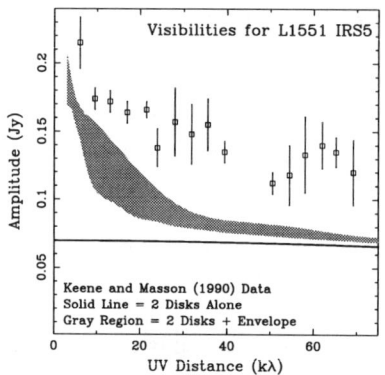

FIGURE 2. Expected flux from a standard envelope and two disks model compared to data from Keene & Masson (1990). Our data was used to constrain the compact source structure and the submillimeter data of Ladd *et al.* (1995) and IRAS $\lambda = 60$ μm and $\lambda = 100$ μm were used to constrain the envelope structure. The solid line is the emission from the two disks alone. The gray area represents the region of permitted amplitudes for the envelope plus disk models.

Fragmentation of Molecular Clouds with GRAPESPH

Ralf S. Klessen[1]

Max-Planck-Institut für Astronomie, Königstuhl 17, 69117 Heidelberg, Germany

Abstract. The dynamical evolution of molecular clouds is very complex. Besides thermal pressure, molecular clouds seem to be supported to a large fraction by 'turbulent motion'. Observations furthermore reveal a hierarchy of clumps and sub-clumps, indicating complicated structure on all scales accessible by todays telescopes.
Here, we present first results from simulations of the dynamical evolution and multiple fragmentation of molecular clouds using smoothed particle hydrodynamics implemented with the special hardware device GRAPE.

NUMERICAL DETAILS

The GRAPE Hardware System

GRAPE is a special purpose hardware device, which calculates the forces and the potential in the gravitational N-body problem by direct summation on a specifically designed chip with high efficiency (Sugimoto et al. 1990). The force law is hardwired to be a Plummer law. GRAPE furthermore returns a list of nearest neighbors for each particle. This feature makes it suitable for use in SPH (Bate et al. 1995, and Steinmetz 1996).

Due to its specific design, GRAPE cannot treat periodic particle distributions directly. However, this limitation can be overcome by computing a periodic correction term on the host computer using a PM-like method.

Periodic Force Correction – The Ewald Method

In 1921, P. Ewald suggested a method to compute the forces in an infinite, periodic particle distribution. Solving Poisson's equation, he realized, that convergence can be improved considerably by splitting the Green's function

---

[1] e-mail: klessen@mpia-he.mpg.de

into a short range and a long range part and treating the first in real space and the latter one in Fourier space (for a recent discussion see Hernquist et al. 1991).

We proceed the following way: Once at the beginning of the calculation, we compute a table of pairwise periodic forces on a grid. Since we are interested in a force *correction*, we subtract the contribution of the isolated pair for each entry. Using FFT, we transform this table into Fourier space and obtain so the Green's function for the correction term. At each time-step, additional to the force calculated with GRAPE for the isolated system, we then apply a PM like scheme using this Green's function to add a periodic force correction to each particle (more details will be presented in Klessen 1997).

The stability of this method is demonstrated in Fig. 1. It shows the time evolution of a system of 32^3 particles, which are initially distributed on a regular grid; this distribution is – in principle – force free within periodic boundaries. The sequence are snapshots of this system initially and after 2, 4 and 6 free-fall times[2]. The whole system remains undistorted for almost 6 free-fall times.

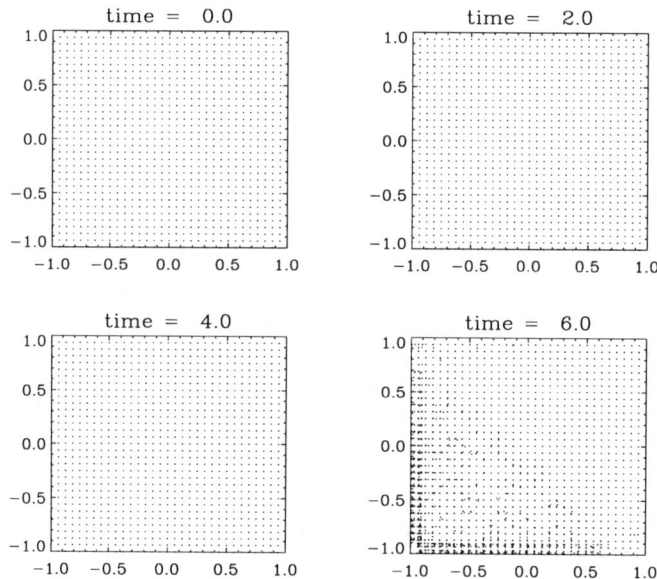

FIGURE 1. 32^3 particles on a regular grid assuming periodic boundaries: Initial condition and at 2, 4 and 6 free-fall times, respectively.

---

[2] An infinite homogeneous medium does not collapse; the free-fall time is per default infinite. Here and in the following we mean the time interval, an *isolated* simulation cube would need to collapse.

FRAGMENTATION OF MOLECULAR CLOUDS

Molecular clouds are highly structured. Observations reveal a hierarchy of filaments and sheets, clumps and sub-clumps, ranging from scales of the size of the whole cloud (see e.g. Bally 1996) down to the smallest objects resolved by todays telescopes (Wisemann & Ho 1996).

Their dynamical evolution is far from being understood in detail. Density-size or linewidth-size relations indicate that typically a cloud is to a large extent supported by 'turbulent motion'; the measured linewidths exceed the thermal line broadening by far. Observations of Zeeman splitting and polarisation indicate the presence of magnetic fields. Furthermore feedback mechanisms from newly formed stars, outflows, stellar winds, ionization fronts and finally supernovae, produce shells and bubbles and deposit huge amounts of energy and momentum into the interstellar medium. The physical processes in molecular clouds are extremely diverse.

However, this is the environment in which stars form. To assess the problem of fragmentation and clump formation that leads to star formation, we start in the most basic way: We follow the dynamical evolution of isothermal gas in a region in the interior of a molecular cloud, starting from an initial density distribution to the formation of self-gravitating clumps. To prevent global collapse, we assume periodic boundary conditions. Ignoring all other effects, we therefore study the interplay between gravity and gas pressure alone. This by itself will produce hierarchical filamentary structure, as was derived by de Vega et al. 1996. Figure 2 illustrates one of our simulations. The left image is the projection of the initial density field. It was assumed to be Gaussian distributed with a $1/k^2$ power spectrum. The right image depicts the same field after one free-fall time. The initial Gaussian fluctuations evolve into a system of filaments and knots, some of which contain collapsing cores.

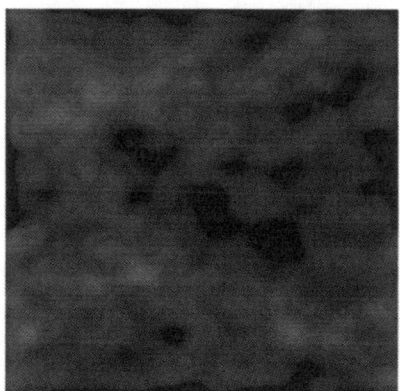 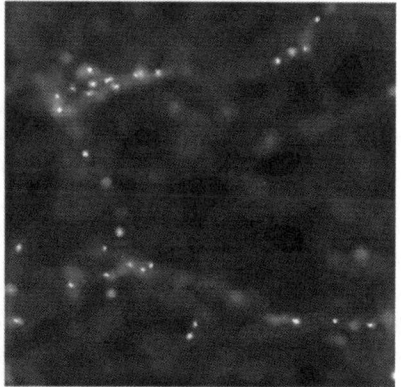

FIGURE 2. Evolution of a Gaussian density field with $P(k) \propto 1/k^2$: Initial condition and after one free-fall time. The density scale is logarithmic and equal in both images.

Figure 3 shows the mass distribution of the identified clumps. With the chosen initial conditions and parameters, this distribution follows a power law, $dN/dm \propto m^n$, with $n \simeq -1.5$, which is indicated by the dashed line. This is in agreement with the observed values (see e.g. Blitz 1993).

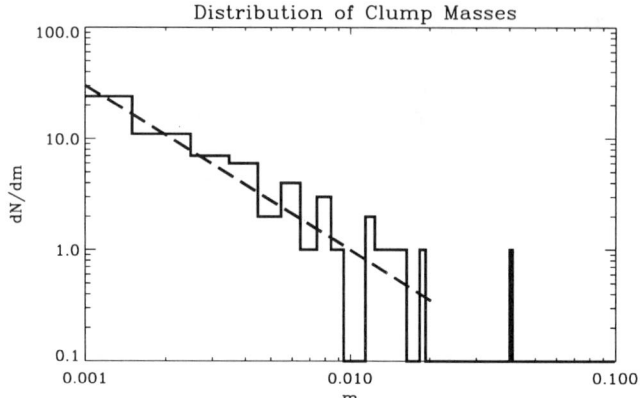

FIGURE 3. Distribution of the masses of the identified clumps, with $m > 0.001$. The dashed line indicates the slope $dN/dm \propto m^{-1.5}$. The total mass in the system is $m = 1$.

Certainly this treatment is very coarse. It neglects the presence of magnetic fields and the input of energy and momentum by young stars. Future numerical studies of the evolution of interstellar medium and the formation of stars have to take these into account. However, the simple isothermal model presented here is able to reproduce some of the observed properties of giant molecular clouds and star forming regions.

REFERENCES

Bally, J.: 1996, *Nature*, **382**, 114
Bate, M.R., Bonnell, I.A., Price, N.M.: 1995, *MNRAS*, **277**, 362
Blitz, L.: 1993, in *Protostars and Planets III*, p. 125, eds. E.H. Levy & J.I. Lunine, University of Arizona Press, Tucson & London
de Vega, H.J., Sanchez, N., Combes, F.: 1996, *Nature*, **383**, 139
Hernquist L., Bouchet, F.R., Suto, Y.: 1991, *ApJSS*, **75**, 231
Klessen, R.: 1997, submitted to *MNRAS*
Steinmetz, M.: 1996, *MNRAS*, **278**, 1005
Sugimoto, D., Chikada, Y., Makino, J., Ito, T., Ebisuzaki, T., Umemura, M.: 1990, *Nature*, **345**, 33
Wiseman, J.J., Ho, P.T.P.: 1996, *Nature*, **382**, 139

Rotation of Starless Bok Globules

Brian D. Kane[*,†] and Dan P. Clemens[†]

Phillips Laboratory/GPOB, Hanscom AFB, MA 01731
†*Boston University Department of Astronomy, Boston, MA 02215*

Abstract.
Fifteen small, apparently starless Bok globules were mapped at high spatial and spectral (0.007 km s^{-1} channel^{-1}) resolution in the ($J = 1 \rightarrow 0$) rotational line of ^{13}CO using the 14 m radio telescope of the Five College Radio Astronomy Observatory, and the fifteen-element 3 mm array receiver QUARRY. From 120 to 360 positions per globule, sampled with half-beam spacing, were observed in the ^{13}CO line.

Gaussian fitting of the emission lines was used to establish mean radial velocities and uncertainties. Each globule radial velocity distribution on the sky was fit to a plane (solid body rotation) to yield mean velocity gradients with position, and rotation axis directions. The globules are rotating at rates about 30 times faster than velocity shifts attributable to local differential Galactic rotation. For globule assumed mean distances of 600 pc, the gradients range in a distinctly *bimodal distribution*, from 0.089 km s^{-1} pc^{-1} ($\omega \sim 3 \times 10^{-15}$ s^{-1}) to 0.950 km s^{-1} pc^{-1} ($\omega \sim 3 \times 10^{-14}$ s^{-1}).

Detailed examination of the globule rotation curves indicated that the kinematics of ten of fifteen globule cores are well-approximated by solid-body rotation. Differential rotation and shearing motions due to external influences (ram pressure stripping and/or bow shocks) are also seen.

INTRODUCTION

We address questions concerning the magnitudes, directions, and natures of the rotational motions of starless Bok globules (SBGs). Past studies have shown SBGs are cold ($T_k \sim 10$ K) and centrally condensed (Arquilla & Goldsmith 1985; Zhou et al. 1990; Turner, Xu, & Rickard 1992; Turner 1994; Lehtinen et al. 1995). However, it has been uncertain whether SBGs possess significant systematic solid-body or differential rotational motion.

Arquilla & Goldsmith (1986) concluded that large Bok globules with strong rotational gradients appear to be uncommon. Other efforts have measured values for dark cloud rotational gradients ranging from 0.3 km s^{-1} pc^{-1} to 2.9 km s^{-1} pc^{-1} (Casali et al. 1987; Lehtinen et al. 1995). All of these studies have lacked sufficient spectral and spatial resolution to characterize the kinematics of globule cores. We summarize a comprehensive study of the rotational motion of fifteen starless small Bok globules, and thereby establish velocity gradients and distinguish between bulk rotational motions and more complex shearing motions.

OBSERVATIONS AND ANALYSIS

Globules were mapped (120 to 360 positions per globule) in the $J = 1 \to 0$ line of ^{13}CO, with the QUARRY fifteen-element 3 mm array receiver and the 14 m FCRAO telescope. Typical system temperatures were ~ 650 K, during integration times of ~ 60 minutes per position. Baselined and folded spectra were fit for single Gaussian profiles, to characterize their velocity centroids. The fit velocity centroid data were employed to find the best solid-body rotation curve, via a matrix inversion method (Heyer 1988; Clemens, Dickman, & Ciardi 1992). After fitting the best solid-body gradient, the velocity residuals γ (normalized by the respective Gaussian fitting uncertainties) were used to characterize four kinematic populations: $\gamma \sim 3$ — clearly solid-body; $\gamma \sim 5$ — mostly solid-body; $\gamma \sim 10$ — kinematically distinct envelopes; and $\gamma > 15$ — differentially rotating.

RESULTS

Assuming a distance of 600 pc (as for CB4; Dickman & Clemens 1983), the velocity gradients range from 0.094 km s^{-1} pc^{-1} ($\omega \sim 3 \times 10^{-15}$ s^{-1}; CB4) to 1.0 km s^{-1} pc^{-1} ($\omega \sim 3 \times 10^{-14}$ s^{-1}; CB161). The mean globule rotation is a factor of 30 larger than the local Galactic value (Clemens 1985). The distribution is bimodal: about half of the SBGs possess mean radial velocity gradients less than 50 m s^{-1} arcmin^{-1}, while the other half of the SBGs have velocity gradients larger than 100 m s^{-1} arcmin^{-1}. A summary of the velocity gradient fits is found in Table 1.

Rotation curves were determined by cataloging observed ^{13}CO mean radial velocities as functions of angular offset perpendicular to the rotation axes through centers of motion for each cloud. By the quality of the solid-body gradient fit to the rotation curve, the dominant type of motion (solid-body rotation, differential rotation, shearing or turbulence) was selected.

Table 2 lists the SBGs and a classification of their shapes, rotation rates, and dominant types of motion. Shearing is noted by the presence or absence of a kinematically distinct envelope. Shearing may also induce turbulence on a scale too small to be directly measured. A statistical measure of the fit (the unnormalized r.m.s. velocity residual [γ']) of the velocity field to the solid-body model may provide an estimate, though.

TABLE 1. SOLID-BODY ROTATION CHARACTERIZATIONS

Cloud ID (1)	VELOCITY GRADIENT			ROTATION DIRECTION	
	∇v [km/s/arcmin] (2)	$\sigma_{\nabla v}$ [km/s/arcmin] (3)	Residuals (γ) [norm. r.m.s.] (4)	Θ_{rot} [°E of N] (5)	$\sigma_{\Theta_{rot}}$ [°] (6)
CB4	0.016	0.002	3.3	44	7
CB16	0.044	0.005	5.2	−5	8
CB17	0.050	0.004	8.1	119	4
CB24	0.103	0.006	5.3	−142	3
CB25	0.030	0.004	4.8	144	6
CB27	0.032	0.001	20.1	−150	2
CB67	0.105	0.002	20.6	112	1
CB148	0.018	0.007	2.1	179	22
CB160	0.113	0.012	6.5	−150	7
CB161	0.171	0.015	4.9	137	5
CB183	0.019	0.005	3.4	65	13
CB195	0.033	0.003	3.4	−108	7
CB202	0.158	0.008	10.1	−86	2
CB228	0.144	0.007	3.2	−109	3
CB246	0.102	0.004	11.1	−165	2
Mean	0.076	0.006	7.6	...	6
Dispersion	0.053	0.004	5.9	...	5

TABLE 2. STARLESS BOK GLOBULE KINEMATIC CLASSIFICATION

Cloud ID (1)	Cloud Shape (2)	Cloud Rotation Rate (3)	Dominant Motion		Kin. Distinct Envelope? (6)	Turbulence(γ') RMS Resid. [km s^{-1}] (7)
			S.B. Rotation? (4)	Diff. Rotation? (5)		
CB4	prolate	Slow	✓	...	...	0.03
CB16	oblate	Slow	✓	...	Red 0.2[a]	0.14
CB17	prolate?	Slow	...	...	Red >0.4	0.13
CB24	prolate	Fast	✓	...	...	0.12
CB25	oblate	Slow	✓	...	Blue 0.1	0.07
CB27	prolate	Slow	...	✓−	Blue 0.05	0.11
CB67	prolate	Fast	...	✓+	...	0.26
CB148	prolate	Slow	✓	...	...	0.05
CB160	oblate	Fast	✓	...	...	0.28
CB161	prolate	Fast	✓	...	...	0.20
CB183	prolate	Slow	✓	...	Blue 0.05	0.05
CB195	oblate	Slow	✓	...	Blue 0.15	0.06
CB202	prolate	Fast	...	...	...	0.21
CB228	prolate	Fast	✓	...	...	0.11
CB246	oblate?	Fast	...	...	...	0.20
Totals	10/15 prol.	...	10/15	2/15	6/15	...

[a] velocity offset in km s^{-1} relative to solid-body gradient fit

GLOBULE ANGULAR MOMENTUM

All SBGs are rotating with velocity gradients too strong to be driven by local differential Galactic rotation. Despite its ubiquity, rotation is not a large reservoir of energy in SBGs. Thermal and shearing motions, and turbulence, are estimated to be much larger. However, the large-scale and generally uniform rotation establishes a minimum angular momentum needed for protostellar disk formation (Stahler et al. 1994). Two-thirds of the SBGs are prolate; the kinematics of the prolate clouds are in general well-fit by solid-body rotation, while the oblate clouds may be tending toward fragmentation or binary formation.

The globules' specific angular momentum J/M, using Kane (1995) J and M estimates ranges from 4×10^{-17} to 2×10^{-15} pc^2 s^{-1}, similar to dense NH$_3$ cores in molecular clouds (Goodman et al. 1993) and to binary protostars embedded in Bok globules (Yun & Clemens 1994a; 1994b). Numerous T Tauri binaries observed towards the Ophiuchus and Taurus star-formation regions (Simon et al. 1995) on average contained about 50 times less J/M than the typical Goodman et al. dense core. J/M for main sequence open cluster stars (Prosser et al. 1995) ranges from 10^{-22} to 10^{-20} pc^2 s^{-1}—100,000 times smaller than the typical SBG and 2,000 times smaller than the typical T Tauri binary system.

This research was funded by NSF Grant AST 92-21194 to D. P. C., as part of Ph. D. dissertation work by B. D. K.

REFERENCES

1. Arquilla, R., & Goldsmith, P. F. 1985, ApJ, 297, 436
2. Arquilla, R., & Goldsmith, P. F. 1986, ApJ, 303, 356
3. Casali, M. M., & Edgar, M. L. 1987, MNRAS, 225, 481
4. Clemens, D. P. 1985, ApJ, 295, 422
5. Clemens, D. P., Dickman, R. L., & Ciardi, D. R. 1992, AJ, 104, 2165
6. Dickman, R. L, & Clemens, D. P. 1983, ApJ, 271, 143
7. Goodman, A. A., et al. 1993, ApJ, 406, 528
8. Heyer, M. H. 1988, ApJ, 359, 363
9. Kane, B. D. 1995, Ph. D. Dissertation, Boston University
10. Lehtinen, K., Matilla, L., Schnur, G. F. O., & Prusti, J. 1995, A&A, 295, 487
11. Prosser, C. F. et al. 1995, PASP, 107, 211
12. Simon, M. et al. 1995, ApJ, 443, 625
13. Stahler, S. W., et al. 1994, ApJ, 431, 341
14. Turner, B. E., Xu, L., & Rickard, L. 1992, ApJ, 391, 158
15. Turner, B. E. 1994 ApJ, 420, 661
16. Yun, J. L., & Clemens, D. P. 1994a, AJ, 108, 612
17. Yun, J. L., & Clemens, D. P. 1994b, ApJS, 92, 145
18. Zhou, S., Evans, N. J., & Butner, H. M. 1990, ApJ, 363, 168

Circumstellar Disks

Hubble Space Telescope Observations of the Environments of Young Stars

J. Jeff Hester

Department of Physics and Astronomy, Arizona State University

Abstract.
Since the installation of the WFPC2 in the *Hubble Space Telescope* in late 1993, the *HST* has proven itself to be a powerful tool for the study of the immediate environments of young stellar objects. In regions of low mass star formation, *HST* images of objects such as HH30 and HL *Tau* have effectively replaced the classical textbook "artist's rendition" of disks, jets, and outflow cavities with images that clearly show the structure of this environment. Together, these images support a view in which the young stars themselves are responsible for limiting infall and clearing away the dense molecular material from which they formed. In short, these stars determine their own masses. Regions in which high mass stars are forming show a very different picture. *HST* images still show disks and dense envelopes surrounding young stars, but in regions such as M16 and M42 these objects are immersed within the evacuated interiors of the H II regions. Intense UV and winds from massive stars have disrupted the environment of star formation, cutting these objects off from the material from which they formed or were forming. The timing of this event has nothing to do with the young stellar object itself, and can happen at any arbitrary stage in the growth of the YSO, effectively "freezing" the mass of the object at that point. The *HST* images provide strong circumstantial evidence that the masses of many low mass stars may be determined, not by the process of star formation itself, but instead by the effects of nearby massive stars.

INTRODUCTION

Ten years ago, had you asked a star formation theorist to describe the immediate environment of a forming star, the picture would have included a number of components. Of prime importance in this list would be a flared disk, serving as a necessary intermediary in the collapse of a dense core of molecular material with far too much angular momentum to collapse directly onto the central star. The star itself would be found embedded at the center of the disk, which in turn provides the material for nascent planetary systems. A final component to the sketch would have been motivated not by theoretical understanding, but instead by observational imperative. This is an outflow of material, possibly highly collimated, moving away along the axis of rotation of the system. A version of this picture, taken from the classic paper by Shu, Adams, & Liszano [1] is still a common feature at meetings such as this one. More artistic representations of the same picture have become a staple of astronomy texts.

The circumstellar disk and accompanying structure is typically a few hundred AU in size, which means that at the distance to the nearest regions of ongoing star formation (140 pc or so), the typical disk should have a size of order a few arcseconds. This is just the size scale where ground based observations could, under exceptional conditions, yield intriguing hints about the structure present. It is also a size scale where a significant improvement in spatial resolution might be expected to reveal a wealth of structure. It was for this reason that the soon-to-fly *Hubble Space Telescope* held great promise and excitement for those working on star formation.

Unfortunately, studies of circumstellar material were also among the programs which suffered most dramatically from the spherical aberration of the *HST* primary mirror. The point spread function of the aberrated *HST* spread approximately 85% of the light from a star into a halo with a radius of about an arcsecond – comparable to the expected size of disk-like structures. As a result, light from a bright central star effectively obliterated the view of the circumstellar environment. While there were a few valiant efforts to conduct this type of work, these efforts met with little success.

WFPC2 OBSERVATIONS OF REGIONS OF LOW MASS STAR FORMATION

With the installation of the second generation Wide Field and Planetary Camera (WFPC2) during the first *HST* Servicing Mission in December 1993, the *HST* became capable of obtaining the sort of images that had originally justified its construction. Observations of regions of star formation took their place among the high priority programs pursued with *HST* during the first year of post-servicing operations. One of the early and most spectacular successes of this program were observations of the source of the HH30 outflow. These observations [2,3] show a shield-shaped, centrally brightened reflection nebulosity approximately $2''\!.5$ long by $0''\!.5$ wide (~ 350 AU $\times$ 70 AU). Offset from this by about $0''\!.5$ is a second reflection nebulosity similar in shape but somewhat fainter than the first. These observations (Figure) are remarkably well fit by calculations of scattering from a 350 AU disk which flares steeply beyond about 50 AU and which is viewed at an inclination of about $7°$ away from edge on. The star itself remains buried behind the disk.

The HH30 observations are also noteworthy for the two highly collimated jets of emission extending out perpendicular to the disk in both directions.

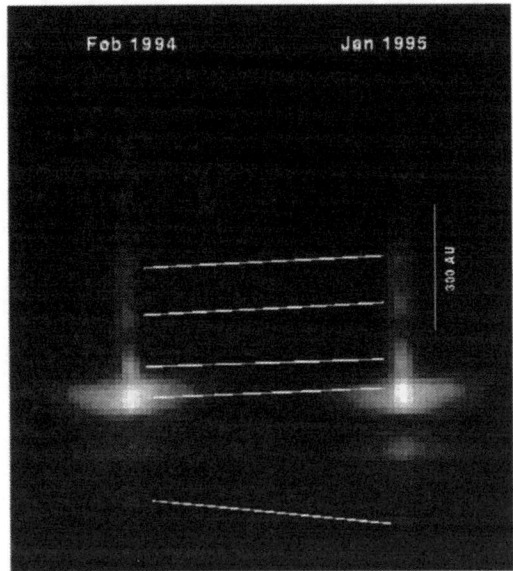

FIGURE 1. WFPC2 images of HH30 taken from Burrows et al. show the disk around a young stellar object viewed at an inclination of about 7° from edge-on. The star itself is hidden behind the dark obscuration of the disk. The two images, taken 11 months apart, show the proper motion of knots in the jet moving out perpendicular to the disk.

The jets are very narrow at their base, and originate from the inner parts of the larger disk. Observations taken over a 1 year period show that knots in the jet originate at the source and move out at velocities of several hundred km/sec. These knots fade, consistent with radiative cooling, but later their relative velocities will cause them to catch up with each other, giving rise to the shocks responsible for bow-shock shaped knots seen in *HST* observations of a number of HH objects jets, including HH 1, HH34 [4], and HH 111 [5].

Subsequent observations show that HH 30 is not the only source for which *HST* observations provide clear evidence of a circumstellar disk. Other sources observed include GM *Aur*, which is surrounded by an inclined flattened disk with an inclination of about 20°. DG *Tau* B shows the reflection nebulosity associated with a thick disk, together with two jets which are inclined by $\sim 25°$ with respect to the plane of the sky (see Krist et al., this volume [13]). In addition to a spectacular jet and bow shock, observations of HH 34 also show a conical reflection nebula with what may be the point source at its apex (see Figure 2). While not as obviously disk-like, this structure is actually in remarkably good agreement with calculations of the appearance of a disk viewed from more nearly face-on. The HH 34 image also provides a somewhat broader perspective on the star forming environment. In the WF chip containing the HH 34 source are four young stellar objects, including both a binary system and what appears to be a disk viewed from more nearly face-on and its associated outflow.

The importance of the HH 30 observations and other similar results can be stated quite simply. The textbook artist's conception of what the disk around a young star *should* look like has been replaced by *HST* images of what the

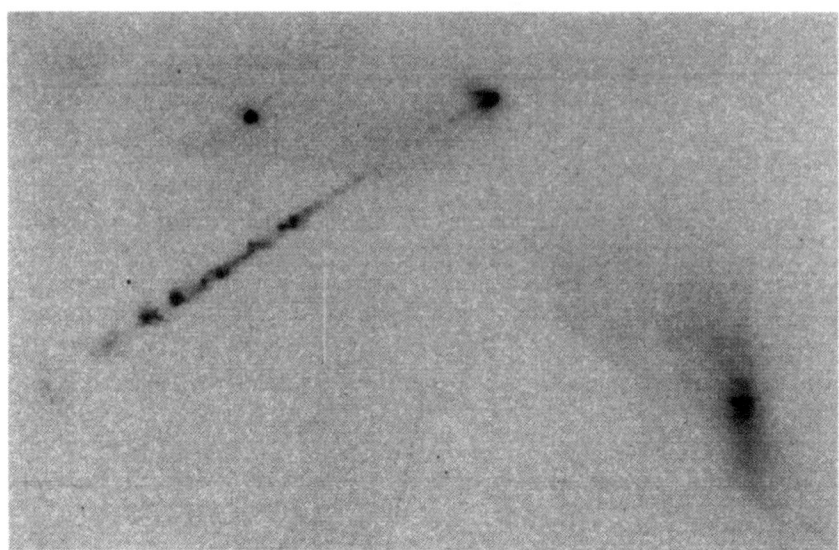

FIGURE 2. A WFPC2 F702W image of the HH34 source and jet. The appearance of the conical reflection nebula at the base of the jet is in good agreement with calculations of reflection from a disk viewed at an angle of approximately 30°. Note the material surrounding the YSO in the lower right part of the frame. A fourth system – a binary – is present off of the upper edge of this frame.

disk around a young star *does* look like.

HST observations of young stellar objects do not always show structure which can readily be interpreted as direct observations of disks. A case in point is the *HST* observation of HL *Tau* [6]. HL *Tau* was long considered to be a prototypical classical T *Tauri* star. Yet when observed with *HST* the object thought to be the star itself was instead resolved into a clumpy C-shaped reflection nebulosity about 1″ across, but which contained *no* visible point source. Careful astrometry indicates that the star itself, seen at radio wavelengths, was located just on the other side of a sharp, straight edge along the side of the reflection nebula. This sharp edge lies along the direction of a elongated disk seen at mm wavelengths, and the dark center of the reflection nebulosity is aligned with an optical jet. Rather than being optically visible, Stapelfeldt et al. concluded that the point source lay obscured behind over 22 magnitudes of visual extinction, more typical of embedded YSO's such as L1551 than of T *Tauri* stars. They also revised the photospheric luminosity of HL *Tau* upward to greater than $3L_\odot$, which in turn indicated a young age for HL *Tau* ($\sim 10^5$ years). In this view the nebulosity was interpreted as reflection from the walls of an outflow cavity, and it was concluded that HL *Tau* is intermediate between a protostar and a classical T *Tauri* star, caught as the outflow from the source is clearing out its circumstellar environment. Other objects, such as DG *Tau* and T *Tau*, also show circumstellar environments that have been shaped by the star formation process, either through outflows or through the dynamics of the accretion process itself.

Overall, *HST* observations of the environments of low mass YSO's can be plausibly interpreted as a sequence in which powerful outflows are responsible for dispersing material in the circumstellar environment and limiting the process of accretion – a picture suggested by Shu, Adams, & Liszano [1] and others. An especially nice example of this is the *HST* observation of HH 47 [7], which shows a spectacular outflow cavity surrounding the source, together with ample evidence of the role of the outflow in disrupting the environment of the star. Regardless of whether this interpretation is correct in detail, it is clear that in regions such as the Taurus-Auriga molecular the process of star formation is able to "run to completion." A star forming in such a region generally "does what it does." While there is disagreement over the details of the process, we can say with some certainty that the final mass of the forming star is set by the characteristics of a star's immediate environment and the characteristics of the accretion process, itself.

HST OBSERVATIONS OF REGIONS OF HIGH MASS STAR FORMATION

In the previous section we argue that *HST* images of regions of low mass star formation show environments that have been shaped by the star formation process itself. These stars are forming more or less in isolation, and are responsible for their own fate. In stark contrast, *HST* observations of star formation in regions containing massive stars show a very different picture. Here the structure of what is seen is dominated instead by the interaction between massive stars and the star forming environment.

WFPC2 Images of M16

WFPC2 images of elephant trunks in M16 [8] offer a look at what happens as an H II region ionization front advances into dense molecular material, characteristic of star-forming environments. These images were taken as part of a program to study the process of photoevaporation. UV radiation incident on the elephant trunks photoevaporates material from the surfaces of these dense structures. The same UV is responsible for photoionizing the resulting flow. The result is a sharply stratified ionization structure in which low ionization lines such as [S II] are concentrated within a few hundred AU of the surface of the dense material, and even high ionization lines such as [O III] are formed predominately within about 10^{17} cm of the interface. Emission from

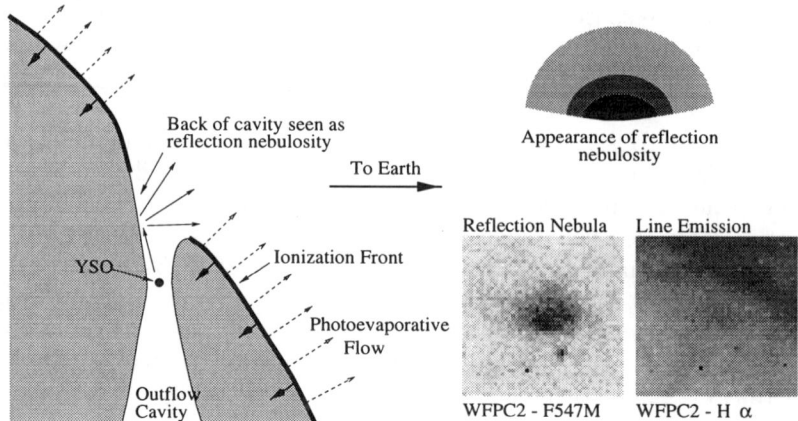

FIGURE 3. An outflow cavity associated with a young stellar object, possibly similar to that seen in HL Tau, is uncovered by an advancing photoevaporation front. At the time that the ionization front cuts across the cavity, the back wall of the cavity will be seen as a reflection nebulosity. One such object is seen in the WFPC2 images of M16.

H II regions may be dominated by such structures, which are well-defined physical systems. As a result, a good understanding of photoionized photoevaporative flows might remove much of the ambiguity that plagues current efforts to model and interpret emission from H II regions.

M16 was chosen for this study because the convex geometry of the elephant trunks in M16 allows the stratified profile of the flow to be clearly seen. Our study of photoionization in M16 was very successful, with calculations of the ionization structure closely matching the observations. At the same time the M16 images offer a clear look at what happens as the photoevaporation front progresses into the molecular cloud, uncovering the structure buried there.

One noteworthy object in M16 is a small reflection nebula seen on the surface of the central elephant trunk. Along one side this reflection nebula has a fairly straight, sharp edge. To the northwest of this edge the reflection nebula is centrally concentrated, but does not contain a strong point source. The nebula fades away with distance from its core. The overall size of the structure is about 1000 AU. The bright, concentrated reflection nebulosity and the lack of any corresponding emission from ionized gas indicates that this object is the result of a local source of nonionizing radiation – a low mass star which is not seen because it is hidden from our line of sight by intervening molecular material. In fact, this refection nebula is reminiscent of HL *Tau* in many respects, and can be easily understood as an outflow cavity that has been "cut open" by the advancing ionization front. Figure 3 shows the proposed geometry of the object. The reflection nebulosity is the illuminated back face of the outflow cavity, which is brighter nearer to the star. The sharp edge at the bottom marks where our view is cut off by the front "lip" of the uncovered cavity, or perhaps by the circumstellar disk itself.

This object is familiar in a sense. It is just what we would expect to see if a photoionization front were to move through a region such as Taurus-Auriga, slicing through and uncovering the structures that are seen there. However it is also the exception rather than the rule in M16. Far more common in the WFPC2 images of M16 are many oval or cometary structures protruding from the surface of the dense clouds (see Figure 4). These globules range in size from a radius of about 150 AU (a limit set by the resolution of the data) to over 1000 AU. The morphology of the interface leaves little doubt as to the basic physical nature of these objects. They are denser than average condensations that formed within the interior of the elephant trunk, then were uncovered as the advancing ionization front moved past, dispersing the lower density material surrounding them. Shadowing of ionizing radiation and the flow of material past the globules are responsible for the formation of "tails" which connect the globules with the main body of the cloud. Hester et al. [8] dubbed these finger-like and tear-drop-shaped objects "evaporating gaseous globules," or "EGGs." On the basis of the proximity of these objects to the surface of the elephant trunk they estimate that most had been uncovered by photoevaporation within the last 30,000 years or so.

The original *HST* images of M16 showed that a few EGGs contain visible stars, establishing a plausible link between the EGGs and star formation. While it is unclear exactly what fraction of EGGs contain young stellar objects, preliminary analysis of ground based near infrared images of these objects by McCaughrean [9] suggest that about 10% of the EGGs contain stellar objects that are visible at 2μm and are more massive than about 0.5 $M_\odot$.

Comparison with M42

WFPC2 images of M42 [14,10,11] show that perhaps 25-50% of YSO's in M42 are surrounded by some type of circumstellar matter. These structures include dark oval-shaped structures, neutral globules with ionized surfaces, or combinations of the two. There is disagreement over some aspects of the interpretation of these objects. O'Dell et al. have proposed a picture in which the neutral globules consist of material that has been photoevaporated from the surfaces of disks, while Hester et al. [8] favor a picture in which the neutral globules are the remains of dense material surrounding the forming stars, more like the EGGs in M16. The objection to the evaporating disk picture is motivated in part by a comparison between the objects seen in M42 and the objects seen in M16. The presence of disks notwithstanding, could not the globules seen in M42 simply be the same sort of recently uncovered envelopes

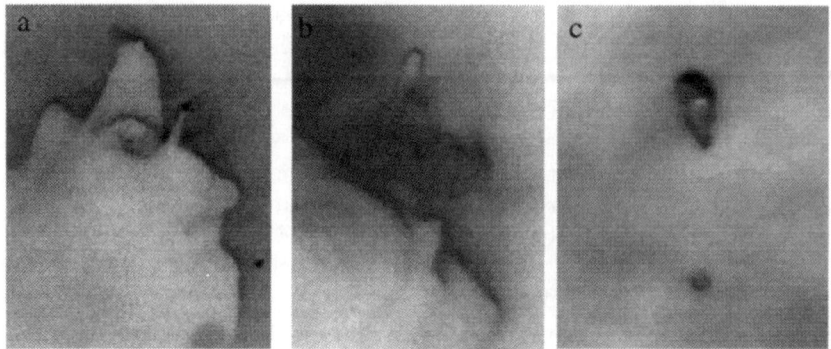

FIGURE 4. (a) Evaporating gaseous globules, or EGGs in M16. Note in particular the finger-shaped EGG with the visible star at its tip. (b) M16 EGGs showing details of the interaction of the photoionization front with small scale structure in the molecular cloud. (c) YSOs embedded within globules in M42. While the M42 objects are significantly closer than the M16 EGGs, they are similar in structure.

around young stars seen as EGGs in M16? There may also be a theoretical difficulty with the evaporating disk picture of the objects in M42. This picture requires that enough nonionizing UV penetrate through an extended neutral globule to evaporate material from the embedded disk at a rate that matches the photoevaporation rate of the globule itself. The globule is large and feels the full brunt of the radiation from the O star. The embedded disk is smaller and feels only the highly attenuated nonionizing radiation. It may be difficult to balance the two resulting mass loss rates.

While this disagreement over the detailed interpretation of the origin of the envelopes seen in M42 is interesting, it has probably drawn attention away from a more obvious and significant respect in which images of M16 and M42 are clearly telling the *same* basic story. In both objects we see young stellar objects, still surrounded by the immediate trappings of star formation and accretion, but adrift within the evacuated interiors of the H II regions. We stress that these objects were separated from the dense environment of star formation *not* as a result of their own evolution, but because of the UV radiation and winds from the massive stars which are responsible for shaping the H II region. This is fundamentally different from regions such as Taurus-Auriga, where YSO remain surrounded by dense molecular material until they "dig themselves out."

Limiting the Masses of Stars

In the case of regions of low mass star formation it is unclear what cuts off accretion, thereby setting the mass of the resulting star. While the author favors a view in which outflows disrupt the star forming environment, most workers in the field would probably agree that regardless of the details, the process is limited by something having to do with the star itself. In the case of regions of high mass star formation the *HST* images tell a different story. The M16 EGGs and the PIGs (partially ionized globules) in M42 were dense condensations buried within dense molecular gas up until a few times 10^4 or 10^5 years ago. At that time they were in contact with the reservoir of material from which they formed and from which they may still have been accreting. However when these objects were uncovered by the advancing ionization front they were abruptly cut off from any source of additional material.

The fact that some of the EGGs contain stars that are seen at 2 μm while others do not makes them especially interesting objects, because it provides strong circumstantial evidence that *ongoing* star formation in this region has been *interrupted* by photoevaporation. At the time they were uncovered some EGGs contained objects that may have already accreted the majority of the material that they would have if left within the molecular cloud. Yet it seems likely that others had not. We can only speculate as to the number of EGGs that contain Class 0 objects, but a few should.

We have argued that we are seeing a process in M16 in which disruption of the star forming environment by massive stars can determine the final masses of other stars. The next obvious question is to ask how important a process is this overall. What fraction of the stars in our Galaxy had their growth truncated in this way, and what implications does that have for questions such as our understanding of the IMF? We do not yet have the data needed to give firm answers to this question, but consideration of the numbers for M16 and M42 suggest that this process might be very significant. Each of these H II regions contains hundreds of young stellar objects located inside H II region volumes of several tens of cubic parsecs. Within a few $\times$ 10^5 or 10^6 years ago all of these YSOs – or at least the cores from which they formed – were buried within the molecular cloud which was to become the H II region. Today they are instead surrounded by tenuous 10^4 K gas, and have accreted all of the material that they are going to.

If the time scale for evolution of the H II region were either much longer or much shorter than the time scale associated with accretion then it would be easy to understand the relationship between the two. In one case we might imagine that if a star begins to form, it will probably be able to complete the formation process before being uncovered. In the other case we might imagine that when massive stars turn on they "instantaneously" destroy their environment, bringing star formation to a halt and leaving behind the population of stars that existed prior to the ignition of the massive stars. However, neither of these is the case. Instead, the time scale over which H II regions evolve is comparable to the time scale over which low mass star formation takes place, which maximizes the diverse range of possibilities for interaction between the two processes.

The possibility that the effects of massive stars may be directly responsible for determining the masses of other stars forming nearby might lead to some interesting theoretical approaches to understanding the IMF, as well as to some interesting predictions. For example, if the events which shut off infall onto a star forming in isolation are coupled to the onset of early stages of nuclear burning, as some have supposed, then it might be very difficult to form brown dwarfs in such environments. In contrast, the truncation of infall by the action of massive stars can occur at any stage in the growth of a star. One might expect that if both of these ideas are correct, then isolated brown dwarfs (i.e., brown dwarfs outside of binary systems) should be found preferentially in regions which contain (or at one time contained) massive stars. This is a prediction for which answers might come in the next few years.

In closing, we note that there has been a good deal of discussion over the years of the possibility that the energy released by massive stars might be important in triggering the formation of new stars (e.g., [12]). *What the HST images of M16 and M42 demand is that we recognize that the events which initiate star formation (whether "triggered" or not) may be closely followed by events which separate forming stars from their reservoir of mass.* Such a process may be responsible for determining the final masses of a significant fraction of stars.

REFERENCES

1. Shu, F. H., Adams, F. C. & Liszano, S. 1987, ARA& A, 25, 23
2. Stapelfeldt, K. R., et al. 1994, BAAS, 185, 4802
3. Burrows, C. J., et al. 1996, Ap.J., 473, 437
4. Hester, J. J. et al. 1994, BAAS, 26, 1386
5. Hartigan, P., Reipurth, B., Morse, J., Bally, J., Heathcote, S., Schwartz, R. 1996, BAAS, 188, 4103
6. Stapelfeldt, K. R., et al. 1995, Ap.J., 449, 888
7. Heathcote, S. et al. 1996, AJ, 112, 1141
8. Hester, J. J., et al. 1996, AJ, 111, 2349
9. McCaughrean, M. 1997, in preparation
10. McCaughrean, M. J., & O'Dell, C. R. 1996, AJ, 111, 1977
11. Bally, J. et al. 1997, in preparation
12. Elmegreen, B. G. 1992 in Proceeding III Canary Island Winter School of Astrophysics, Star Formation in Stellar Systems, edited by G. Tenorio-Tagle, M. Prieto, and F. Sanchez (Cambridge University Press, Cambridge), p. 381
13. Krist, J. et al. 1997, this volume
14. O'Dell, C. R., & Wong, K. 1996, AJ, 111, 8460

Observations of the Inner Accretion Disk Around Young Stars

Patrick Hartigan

Space Physics and Astronomy Dept., Rice University
6100 S. Main, Houston TX 77005-1892

Abstract. Many of the spectroscopic signatures characteristic of low mass young stars, including near-infrared, visible, and ultraviolet excess emission, inverse P-Cygni absorption, and a variety of emission lines, originate from the inner regions of accretion disks. The area where the disk interacts with the stellar photosphere is particularly important to understand because it controls how angular momentum transfers between the disk and star, and because stellar jets originate there. This review summarizes some of the recent observational and theoretical work of the inner regions of T Tauri accretion disks.

INTRODUCTION

During the past few decades it has become clear that accretion disks surround a large fraction of the youngest low mass stars in our galaxy. Unfortunately, the closest regions of star formation are too distant to allow us to spatially resolve the region of greatest interest, where material from the disk falls onto the stellar photosphere. Radii of T Tauri stars are $\sim 3 R_\odot$, or $\sim 10^{-4}$ arcseconds for the closest star formation regions. This angular size is some three orders of magnitude smaller than can currently be resolved with the best telescopes. To date, no eclipsing binaries have been discovered among the sample of young stars with accretion disks. Hence, astronomers have not yet been able to exploit eclipse mapping, which has been so successful in clarifying the structure of CV disks.

In light of the above constraints, we must employ some sort of remote sensing to understand more about T Tauri accretion disks. Studies of T Tauri disks typically focus on either separating the spectral energy distribution (SED) of the disk from that of the photosphere, or observing the kinematics of emission lines produced as material falls onto the star and is ejected into a jet. These approaches are complimentary, as the spectral energy studies give estimates of mass accretion rates, while the emission line work provides some insight into the geometry and dynamics of the accretion process near the star.

In this article I will briefly summarize some of the techniques that have been used by various researchers in the last decade or so to study inner ($\lesssim$ 0.1 AU) accretion disks around young stars, and I will also highlight some recent theoretical work that has been put forth to explain these observations.

SPECTRAL ENERGY DISTRIBUTIONS (SEDS)

It has been known for decades [1,2] that the youngest stars often exhibit excess emission above photospheric levels at both near-infrared, optical, and ultraviolet wavelengths. Though there is some disagreement about the statistics, it seems that at least half of the youngest stars are surrounded by accretion disks (classical T Tauri stars; cTTs) [3]. The fraction of young stars which lack opaque disks (weak-lined T Tauri stars; wTTs) increases substantially at ages $\gtrsim 3 \times 10^6$ yr [4]. The cTTs continuum excesses at near-infrared wavelengths might plausibly arise from either a passive disk, which simply absorbs radiation from the star and re-emits at longer wavelengths characteristic of the disk's cooler temperature [5], or from an accretion disk, where the loss of gravitational potential energy during accretion heats the disk, which then radiates this energy into space [6]. The infrared luminosity of a passive disk is limited to a fraction of the stellar luminosity, while that of an accretion disk is proportional to the mass accretion rate. Both accretion disks and passive disks give rise to similar SEDs in the IR, so it is difficult to distinguish between them unless the accretion rate is large enough to be inconsistent with a reprocessing origin.

A steady accretion disk deposits as much energy in the 'boundary layer' where material falls onto the star as it does throughout the entire disk. Because the boundary layer has a much smaller surface area than does the disk, its temperature is correspondingly higher, and the radiation emerges in the optical and ultraviolet rather than in the IR and sub-mm [7]. The UV/optical excesses add extra continua to the photospheric spectra and thereby reduce the observed equivalent widths of the photospheric absorption lines, a phenomenon called 'veiling' [8,9]. Absorption lines in the blue part of the spectrum are more heavily veiled than those in the red because the excess continuum rises slowly toward the blue, while the photospheric fluxes of K and M stars (the spectral type of low mass $\lesssim 1 M_\odot$ stars on the Hayashi track) drop sharply [10].

One can measure the amount of veiling at each wavelength by comparing the depths of the photospheric lines of the object with those of a template star (typically a weak-lined T Tauri star that lacks veiling). For example, if an absorption line in a cTTs is only half as deep as that of a wTTs with the same spectral type, we know that the ratio r_λ of the excess continuum to the photospheric continuum must be ~ 1 at that wavelength. By measuring r_λ at different wavelengths and knowing the true SED of the photosphere (*e.g.*, a K star), we can construct an SED for the excess emission. Because this excess SED is constructed from the relative depth of an absorption line with respect to the adjacent continuum, and any intervening reddening reduces the flux in both the line and the continuum equally, the SED of the veiling measured in this manner is independent of the amount of reddening and of the reddening law. This independence is particularly useful because the reddening toward young stars is difficult to measure accurately.

The total luminosity of the UV/optical excess is proportional to the rate that mass accretes onto the star. Typical accretion rates for classical T Tauri stars range between $\sim 10^{-6} M_\odot \mathrm{yr}^{-1}$, and $10^{-8} M_\odot \mathrm{yr}^{-1}$ [11]. The T Tauri phase lasts $\sim 10^6$ yr, so these stars accrete a significant fraction of their total mass through the disk [10,9]. Some disk accretion rates may be high enough to affect how cTTs evolve in the HR diagram [12].

Temperatures of the UV/optical excesses are high ($\sim 10^4$ K) and the filling factors low ($\sim$ a few percent of the stellar surface area), consistent with the accretion disk scenario [13,14,10,15]. Objects that have veiling also have significant near-IR excesses, as expected. Remarkably, the converse is also true; objects with near-IR excesses all show veiling or inverse P-Cygni profiles indicative of accretion [11]. Hence, any passive disks around young stars must rapidly become optically thin, perhaps as a result of dust coagulation into larger rocky bodies.

The effective temperature of the disk at each radius determines the shape of the SED in the IR and sub-mm [16]. Similarly, the behavior of the near-IR colors is governed by the inner accretion disk. Studies of the near-IR SEDs of intermediate mass (1 – 3 $M_\odot$) young stars have shown that the disk becomes transparent within a few stellar radii [17]. It is more difficult to observe this phenomenon among lower mass stars because the photospheric colors are close to those of the disk for late-type stars, but spectroscopic observations (see below) suggest that gaps also occur close to the star within disks that surround low mass young stars.

EMISSION LINE DIAGNOSTICS

Balmer Lines

Classical T Tauri stars are often discovered because of their strong Balmer emission lines, which typically exhibit linewidths of several hundred km s^{-1} [18,19]. Metallic lines such as Fe, Na, and Ca are also prominent in the spectra of cTTs as are forbidden lines like [O I]λ6300. In contrast, wTTs show only weak, narrow emission lines that can be explained as arising in an active chromosphere, and these stars always lack forbidden lines (*e.g.* [20]). Hence, it is natural to associate the emission lines of cTTs with the accretion process.

When the emission lines of cTTs are studied at high ($\lesssim 20$ km s^{-1}) resolution, it is possible to measure the veiling accurately and subtract photospheric absorption lines from the observed spectrum, leaving only residual emission or absorption features. This method has proved to be an extremely powerful way to measure weak emission lines and also weak absorption within emission line profiles [11]. It has been known for some time that a small fraction of cTTs, known as YY Ori stars, exhibit redshifted absorption in their Balmer lines. However, new studies of the line profiles that correct for photospheric absorption have shown that the vast majority of cTTs are, in fact, YY Ori stars [19]. The absorption features are typically redshifted by 100 km s^{-1} – 250 km s^{-1}; the orbital velocity at the surface of a typical T Tauri star is $\sim$ 250 km s^{-1}.

To produce an inverse P-Cygni profile, material which flows from the disk to the star must be redshifted by several hundred km s^{-1} along the line of sight, and must lie between the observer and a bright source, either the stellar photosphere or the hot spot where material from the disk impacts the star. These observations rule out a model where a planar accretion disk simply intersects a spherical star, as such geometry would never produce inverse P-Cygni profiles. One way to explain these observations is if a stellar magnetic field funnels material from the disk onto the star [21]. In such a model, emission lines and inverse P-Cygni profiles arise from the accretion columns, and the veiling comes from the hot spots at the base of the column [22]. Recent theoretical models of the accretion columns now include detailed cooling, and predict line profiles like those observed [23]. There is also some evidence for hot spots from analysis of periodic light curves of cTTs [24,25].

Rotational periods have been measured from the light curves of many cTTs and wTTs [26]. One might expect that the accretion of angular momentum from the disk would cause cTTs to rotate more rapidly than their wTTs counterparts; in fact, the *opposite* occurs: rotation periods of cTTs are always $\gtrsim 10$ days, while those for wTTs range from 1.5 – 10 days [27,26]. Equatorial rotational velocities for both cTTs and wTTs are substantially lower than breakup velocity, which occurs for rotational periods $\lesssim 0.7$ days. Hence, it appears that somehow accretion disks regulate the angular momentum of young stars, perhaps by ejecting high angular momentum material in a wind. The means by which accretion disks accelerate and collimate jets is a topic of much current theoretical research [28,29].

Forbidden Lines

There is a one-to-one correspondence between the presence of forbidden line emission and infrared excess among cTTs and wTTs [11]. This remarkable fact ties the presence of an optically thick disk to the heating of cTTs forbidden lines. Forbidden lines in cTTs arise from two very distinct regions. There is a high-velocity component that is typically blueshifted by several hundred km s^{-1} and resembles a stellar jet which is unresolved spatially (the disk probably blocks the receding portion of the jet) [30]. This component is present most often in systems that accrete rapidly, but not all cTTs with high accretion rates have bright high-velocity forbidden line emission. A second low-velocity component appears to be much denser than the high-velocity component, and has a small blueshift of a few km s^{-1}. This material may arise in a disk wind, though the method of heating remains unknown. A low-velocity component forbidden line always seems to be present in all cTTs [11].

Studies of the high-velocity component of the [O I] and [S II] lines in cTTs have focussed on estimating mass loss rates from the line luminosities [11,31]. There is a strong correlation between mass loss rates found in this manner and mass accretion rates measured from veiling. Hence, accretion appears to drive outflows from young stars. The spectacular stellar jets that have been discovered within the last two decades are the best examples of such outflows (*e.g.* [32]); jets emanate from very young systems that have high accretion luminosities. In all cases the accretion luminosity exceeds the mechanical luminosity in the jet, though in some objects both these exceed the photospheric luminosity [33].

DISK PHOTOSPHERES

Roughly a dozen objects, known as FU Ori stars, have been discovered in dark clouds that have similar spectral line shapes and SEDs. Some FU Oris have been observed to be eruptive variables, suddenly increasing in brightness by ~ 5 magnitudes in less than a year, and then fading slowly on a timescale of ~ 100 years. These objects are thought to represent accretion disks that undergo a sudden rise in the mass accretion rate, perhaps brought on by an instability in the disk [34].

When the mass accretion rate increases high enough, heating at the disk midplane causes the disk to resemble a photosphere, with a hot layer obscured by a cooler layer. The result of a vertical temperature gradient in an optically thick material is to form absorption lines. These absorption lines should partake of the rotation of the disk, and become broadened or even double-peaked. Also, because the outer parts of the disk are cooler, these regions should dominate the observed spectrum at longer wavelengths and cause the rotational broadening to decrease as one moves from the optical to the infrared. Both the dependence of line broadening with wavelength and double-peaked absorption line profiles have been observed in several FU Ori objects [35].

Theoretical models of FU Ori accretion disks have met with considerable success in reproducing the observed line shapes [36]. Mass accretion rates in FU Oris are as high as $10^{-4} M_\odot \mathrm{yr}^{-1}$, which means that $\gtrsim 0.01$ $M_\odot$ may accrete in a single event [37]. A typical T Tauri star probably experiences at least one, and perhaps several FU Ori events before the disk dissipates. Hence, a significant fraction of the final stellar mass derives from steady and episodic disk accretion. The close connection between accretion and outflow suggests that massive FU Ori accretion events are likely to be responsible for the multiple bow shocks observed in stellar jets [34].

High-resolution emission line profiles of FU Ori objects have been analyzed by subtracting disk models from the data and examining the residuals in much the same manner as has been done for cTTs [38]. The resulting line profiles show a massive outflowing wind which models suggest must have a *rotational* component [39]. This indication of rotation in outflows is particularly important in light of the angular momentum issues discussed in the previous section.

Another potentially powerful means to study the inner disks of T Tauri stars comes from observations and models of the CO emission feature at $2.3\mu m$ [40,41]. When convolved with the rotational broadening expected from a disk, the CO emission bandhead matches very well with the observations [41]. These observations indicate the temperature of the molecular gas, and may someday constrain the vertical temperature structure in the disk.

CONCLUDING REMARKS

Though astronomers have made great strides forward in understanding how stars form, the current paradigm will remain fundamentally incomplete until we develop a better idea of how the inner accretion disk operates. Because this region cannot be imaged, models of the interaction region between the disk and the star must connect with observation *via* increasingly sophisticated spectroscopic studies of cTTs. The most obvious extension of the current work is to address accretion disks in binaries. Binaries are common among young stars [42], and studies of these systems could provide some much needed insight into the accretion process. Much could also be learned about the onset of disk instabilities if we could detect and study the early phases of an FU Ori outburst. Any further information about rotation in outflows is clearly of great importance to understanding the angular momentum of accretion disks, though observation of rotation in outflows will likely remain a difficult challenge for the near future.

REFERENCES

1. Rydgren, A.E., and Vrba, F.J. 1983, *A.J.* 88, 1017.
2. Joy, A.H. 1949, *Ap.J.* 110, 424.
3. Kenyon, S.J., & Hartmann, L. 1995, *Ap.J. Supp.* 101, 117.
4. Strom, K.M., Strom, S.E., Edwards, S., Cabrit, S., & Skrutskie, M.F.
5. Adams, F.C., & Shu, F. 1986, *Ap.J.* 308, 836.

6. Lynden-Bell, D., & Pringle, J. 1974, *MNRAS* 168, 603.
7. Bertout, C. 1989, *Ann. Rev. Astr. Ap.* 27, 351.
8. Hartigan, P., Hartmann, L., Kenyon, S.J., Hewett, R.,
9. Basri, G., & Batalha, C. 1990, *Ap.J.* 363, 654.
10. Hartigan, P., Kenyon, S.J., Hartmann, L., Strom, S.E., Edwards, S., Welty, A.D., & Stauffer, J. 1991, *Ap.J.* 382, 617.
11. Hartigan, P., Edwards, S., & Ghandour, L. 1995, *Ap.J.* 452, 736.
12. Hartmann, L., & Kenyon, S.J. 1990, *Ap.J.* 349, 190.
13. Bertout, C., Basri, G., & Bouvier, J. 1988, *Ap.J.* 330, 350.
14. Basri, G., & Bertout, C. 1989, *Ap.J.* 341, 340.
15. Valenti, J., Basri, G., and Johns, C. 1993, *A.J.* 88, 1017.
16. Osterloh, M. & Beckwith, S. 1995, *Ap.J.* 439, 288.
17. Hillenbrand, L.A., Strom, S.E., Vrba, F.J., & Keene, J. 1992, *Ap.J.* 397, 613.
18. Hamann, F. & Persson, S.E. 1992, *Ap.J. Supp.* 82, 247.
19. Edwards, S., Hartigan, P., Ghandour, L. & Andrulis, C. 1994, *A.J.* 108, 1056.
20. Walter, F. 1992, *A.J.* 104, 758.
21. Königl, A. 1991, *Ap.J. Lett.* 370, L39.
22. Calvet, N., & Hartmann, L. 1992, *Ap.J.* 386, 239.
23. Martin, S. 1996, *Ap.J.* 470, 537.
24. Kenyon, S.J., et al. 1994, *A.J.* 107, 2153.
25. Bouvier, J., & Bertout, C. 1989, *Astr. Ap.* 211, 99.
26. Bouvier, J., Cabrit, S., Fernandez, M., Martin, E.L., & Matthews, J.M. 1993, *Astr. Ap. Supp.* 101, 485.
27. Edwards, S. et al. 1993, *A.J.* 106, 372.
28. Shu, F., Najita, J., Ostriker, E., Wilkin, F., Ruden, S., and Lizano, S. 1994, Ap.J. 429, 781. 1989, *A.J.* 97, 1451.
29. Pelletier, G., and Pudritz, R. 1992, *Ap.J.* 394, 117.
30. Hirth, G., Mundt, R., and Solf, J. 1994, *Astr. Ap.* 285, 929.
31. Cabrit, S., Edwards, S., Strom, S.E., & Strom, K.M. 1990, *Ap.J.* 354, 687.
32. Heathcote, S., Morse, J.A., Hartigan, P., Reipurth, B., Schwartz, R.D., Bally, J., & Stone, J.M. 1996, *A.J.* 112, 1141.
33. Hartigan, P., Morse, J.A., & Raymond, J.C. 1994, *Ap.J.* 436, 125.
34. Hartmann, L., Kenyon, S.J. & Hartigan, P. *Protostars and Planets III*, Tucson: University of Arizona Press, 1993, p497.
35. Hartmann, L., & Kenyon, S.J. 1987, *Ap.J.* 312, 243.
36. Popham, R., Kenyon, S., Hartmann, L., & Narayan, R. 1996, *Ap.J.* 473, 422.
37. Hartmann, L., & Kenyon, S.J. 1985, *Ap.J.* 299, 462.
38. Welty, A., Strom, S.E., Edwards, S., Kenyon, S.J., & Hartmann, L.W. 1992, *Ap.J.* 397, 260.
39. Calvet, N., Hartmann, L., & Kenyon, S. 1993, *Ap.J.* 402, 623. & Stauffer, J. 1989, *Ap.J. Supp.* 70, 899.
40. Kenyon, S.J., Hartmann, L., Gomez, M., Carr, J., & Tokunaga, A.T. 1993, *A.J.* 105, 1505.
41. Najita, J., Carr, J. & Tokunaga, A.T. 1996, *Ap.J.* 456, 292.
42. Matheiu, R. 1997, this volume.

The Theory of Circumstellar Accretion Disks

James M. Stone*

Department of Astronomy, University of Maryland, College Park, MD 20742

Abstract. Recent theoretical advances in our understanding of the dynamics of circumstellar accretion disks are reviewed. Of particular importance are developments regarding angular momentum transport processes in disks. It has recently been shown through direct numerical simulation that vertical convection results in inward transport, and thus is unlikely to serve as a source of anomalous viscosity. More importantly, these results have been generalized to demonstrate that any form of hydrodynamic turbulence in which velocity fluctuations do not extract *both* energy and angular momentum from the mean shear flow will be associated with inward angular momentum transport. In fact, the analysis predicts that nonlinear shear instabilities are absent in Keplerian disks, a result which has been confirmed through direct numerical simulation. The important role that MHD turbulence driven by the Balbus-Hawley (BH) instability plays in outward angular momentum transport is emphasized, especially regarding the implications for the production of MHD winds from disks. A detailed understanding of the dynamical interaction between an accretion disk and the magnetosphere of the central object will be crucial to understanding accretion from disks in general.

INTRODUCTION

In the current paradigm, the formation of a single star is a four step process [1,2], consisting of the formation of cloud cores, infall, outflow which terminates accretion, and the T Tauri phase. The first step is controlled by the internal dynamics of molecular clouds, which is thought to be best described by MHD turbulence in a self-gravitating gas (see the review by Ostriker, these proceedings). On the other hand accretion disks are an important ingredient in all of the last three stages, thus the study of all but the earliest stages of star formation necessarily involves the study of accretion disks. Moreover, as reviewed in this conference, there is ample observational evidence of the association of accretion disks with protostars (see Mundy, and Sargent, these proceedings).

The theory of circumstellar accretion disks has been the subject of a number of recent reviews (e.g. see [3–5], and references therein). Rather than providing another general review of the subject, in this talk I will instead focus on recent developments in three specific areas: (1) transport processes in accretion disks, (2) the production of winds and outflows from disks via magnetohydrodynamical (MHD) processes, and (3) the interaction between an accretion disk and the magnetosphere of the central object.

TRANSPORT PROCESSES

Inward accretion of mass in a disk requires outward transport of angular momentum. However, it has long been known that classical microphysical viscosity can only produce an angular momentum transport (and therefore mass accretion) rate which is many orders of magnitude smaller than that inferred from observations of a variety of disk systems. Recognizing the need for an "anomalous viscosity" to explain the observations, Shakura & Sunyaev [6] introduced the ansatz, based on dimensional arguments, that the magnitude of kinematic viscosity could be scaled with the gas pressure according to

$$\nu_T \sim \alpha C_s H \tag{1}$$

where α is a dimensionless constant, C_s the sound speed in the disk, and H the vertical scale height. The subscript reflects the implicit assumption that some sort of subsonic turbulence may be the root cause of the anomalous viscosity. Observations suggest a magnitude for $\alpha \sim 0.01 - 1.0$. The parametrization (1) has served as a useful concept for the study of the global characteristics of accretion disks. Nonetheless, the fundamental question remains: what is the source of this anomalous (turbulent) viscosity? Or equivalently: what is the source of α?

Hydrodynamic Turbulence

It has long been thought that purely hydrodynamic processes may lead to the turbulence that is the source of anomalous viscosity. However, since there are no linear shear instabilities in hydrodynamic disks in Keplerian rotation (although there are other modes of instability, see below), identifying these processes has been a challenge. In the case of protostellar disks, vertical convection driven by steep temperature gradients in the disk has been suggested as one possibility. In addition, since it is known on experimental grounds that Cartesian shear layers are subject to nonlinear instabilities, it has been suggested that, by analogy, such nonlinear shear instabilities may be present in accretion disks (although this has never been shown). Still, even if such processes operate to generate turbulence, the fundamental question of whether, in the nonlinear regime, this turbulence will transport angular momentum *outward* is still open. It is important to emphasize that while turbulence is effective at transporting passive scalars, the angular momentum is a fundamental dynamical variable which plays a crucial role in determining the flow field itself. Simply generating turbulence is not enough. It must be a very specific kind of turbulence, one in which the radial and azimuthal velocity fluctuations are correlated in such a way as to give outward angular momentum transport.

In fact, partially for these reasons, the direction of angular momentum transport produced by convection in disks is somewhat controversial even in the linear regime. For example, [7] conducted a linear analysis of axisymmetric modes in polytropic disks. Although such modes cannot transport angular momentum in and of themselves, these authors estimated that nonaxisymmetric dissipation of the modes will result in $\alpha \sim 10^{-2}$ to 10^{-3} in the most favorable cases. However, a linear analysis of nonaxisymmetric modes (which can result in direct angular momentum transport) by [8] showed that these modes actually transport angular momentum *inward*. [9] argued that there is nothing in principle to preclude outward transport by convection, and found evidence for such in selected modes in three-dimensional *linear* calculations. Two-dimensional axisymmetric simulations described by [10] gave rise to inward transport, but the authors were careful to note that nonaxisymmetric structure might lead to something very different.

Recently, it has become feasible to study the fully three-dimensional and nonlinear nature of convective turbulence in disks via direct numerical simulation. Results have been presented by both Cabot & Pollack [11,12], and Stone & Balbus [13]. Both groups study vertical convection in a local patch of a shearing disk. The simulations in [13] begin with an analytic vertical structure for the disk which is convectively unstable in the core. The growth and saturation of convective modes is followed, and the angular momentum transport rate measured directly from the simulation. It is found that convective motions in the disk are not self-sustaining: without an internal source of heat which drives the convection, the unstable gradients are simply smoothed away and all motions damp after a few overturn times. Figure 3 of [13] shows an image of the structure of convective cells deep in the nonlinear regime of a simulation in which an external source of heat was placed at the disk midplane to drive steady convection. As one might expect, in the $R - Z$ plane, the convective cells look much the same as those in a non-rotating fluid. However, in the $\phi - Z$ plane, the cells are stretched by shear into long sheets. Figure 1 shows the angular momentum transport rate (characterized by the α-parameter) over 40 orbits in disk in which vertical convective motions are sustained. Two results are immediately obvious. The first is that the magnitude of the transport rate is very small, generally $\alpha < 10^{-4}$. The second is that the average value of α is negative, implying inward angular momentum transport. Using a Navier-Stokes solver, [11,12] also find inward transport in

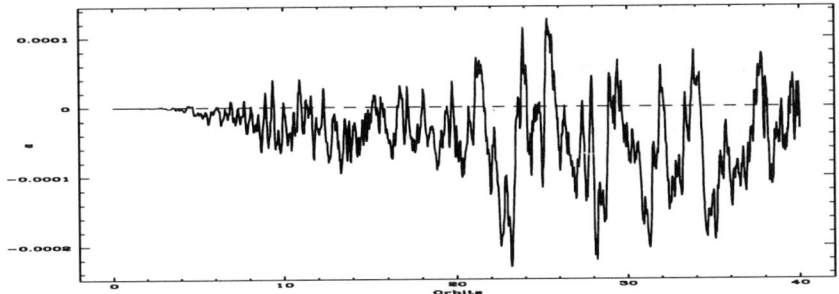

FIGURE 1. Evolution of the Reynolds stress normalized by the pressure at the disk midplane (equivalent to α) over 40 orbits in a disk in which vertical convection is driven. Note α is both very small and, on average, negative.

convectively unstable disks consistent with the above results, although these authors argue higher-resolution simulations are required to settle the issue.

[14] have used simple analytic arguments to explain why convection results in inward transport. In fact, the argument can be extended to predict that *any* source of hydrodynamical turbulence will drive inward transport. Decomposing each dynamical variable into a mean and fluctuating part, [14] demonstrate that in an accretion disk, the time rate of change of kinetic energy associated with fluctuations in the angular velocity can be written as (see also [11])

$$\frac{\partial}{\partial t}\langle\frac{\rho\delta v_\phi^2}{2}\rangle = (q-2)\Omega\langle\rho v_R \delta v_\phi\rangle + PF - VL. \qquad (2)$$

Here, v_R and δv_ϕ are the fluctuating part of the radial and azimuthal components of velocity, $\langle\rho v_R \delta v_\phi\rangle$ is the Reynolds stress, and PF and VL denote terms which represent forcing by coherent azimuthal pressure gradients, and losses due to viscous dissipation, respectively. The parameter q describes the shear profile via $\Omega = \Omega_0 R^{-q}$. Provided the last two terms on the RHS are negligible, equation (2) predicts that for $q \geq 2$, outward transport leads to growth of the kinetic energy, i.e. instability. In fact, it is well known that disks with $q \geq 2$ are subject to the Papaloizou-Pringle instability. In stark contrast, however, if $q < 2$, the only way to produce unstable growth is through inward transport. Outward transport in a Keplerian disk (with $q = 3/2$) is associated with stability, i.e. damping of nonlinear perturbations. Hydrodynamic simulations of disks with various q values confirm these predictions (e.g. see Figure 1 of [14]). Disks with $q \geq 2$ which are given perturbations with nonlinear amplitudes become unstable, whereas the same perturbations introduced in disks with $q < 2$ simply damp away. In fact, these latter simulations are a direct demonstration that Keplerian accretion disks ($q = 3/2$) do *not* contain any nonlinear shear instabilities. It is important to note that simulations of Cartesian shear layers otherwise identical to those of differentially rotating disks described above capture the nonlinear instabilities known to exist in these systems, and moreover, an analytic analysis of the streamwise kinetic energy similar to equation (2) above also predicts this markedly different behavior.

It should be noted that the failure of hydrodynamic turbulence to provide for outward angular momentum transport does not preclude all hydrodynamic transport mechanisms. Coherent pressure forcing by, e.g., spiral density waves (represented by the PF term in equation 2), or gravitational torques in either a self-gravitating disk or due to companions, remain viable options.

Magnetohydrodynamic Turbulence

Unlike the purely hydrodynamic case, accretion disks which contain a weak magnetic field are subject to a powerful local linear shear instability [15,16], herein referred to as the BH instability. Understanding the properties of the nonlinear, saturated state of the instability, and how it affects the structure and evolution of disks are important problems.

The nonlinear evolution of the instability has been studied via direct MHD simulations in both two- [17,18] and three-dimensions [21,22,19,20]. In three-dimensions, the instability results in MHD turbulence with a power spectrum similar to Kolmogorov. Moreover, vigorous *outward* angular momentum transport is associated with this turbulence. Most simulations to date have used the local approximation, i.e. only a small patch of the disk with a radial extent much smaller than the radial distance from the center is studied. The results presented in [19,20] are quasi-global in nature in that vertical structure is incorporated into the model; the computational domain represents a local patch in the plane of the disk but encompasses several scale heights in the vertical direction. Stratification does not change most of the properties of the instability reported for homogeneous boxes (e.g. MHD turbulence still results). In addition, it is found that buoyancy is not an important saturation mechanism; instead local dissipation dominates. However, buoyancy does affect the resulting vertical structure of a BH unstable disk. For example, figure 2 is a "space-time" plot of various quantities during the evolution of a 3D simulations of a stratified disk with an initially vertical field. The plot is created by averaging each quantity over a horizontal plane, and plotting the result (which is still a function of vertical coordinate Z) versus time. For times $t < 2$ orbits, the linear growth and saturation of the instability as the "channel solution" [18] is evident. The channel solution clearly breaks up at about 3 orbits, and thereafter each variable shows complex structure. However, diagonal streaks which begin at the disk midplane are evident in several variables, especially the magnetic energy. These streaks represent regions of high magnetic field

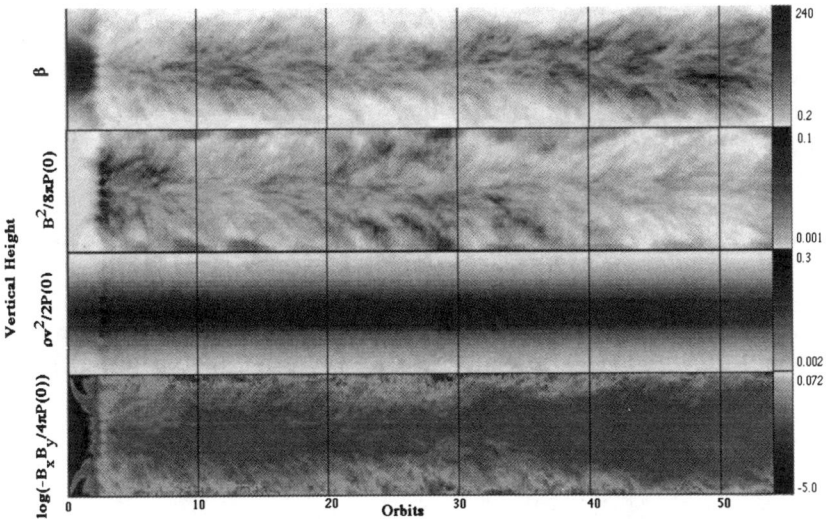

FIGURE 2. Space-time plot of the plasma-β parameter, the magnetic pressure, the kinetic energy, and the logarithm of the Maxwell stress during the evolution of a BH unstable stratified disk. Diagonal stripes, especially evident in the magnetic pressure, represent buoyantly rising flux.

strength which are rising vertically through the disk. Interestingly, while this buoyancy rise of magnetic field has no discernible effect on the saturation amplitude of the instability, it does modify the vertical structure of the disk to produce a strongly magnetized corona which surrounds a weakly magnetized core. Notice that the plasma-β parameter demonstrates this core-corona structure very clearly; near the midplane of the disk $\beta \sim 100$, whereas higher up $\beta < 1$. In [19,20] a variety of simulations are reported with different vertical boundary conditions in which the same structure is evident, thus it is unlikely the effect is produced by the boundary conditions alone.

The results discussed above indicate the BH instability is a promising mechanism for outward angular momentum transport in disks; values of $\alpha \sim 0.01 - 0.1$ are produced in current simulations. However, nearly all of the studies to date use the assumption of ideal MHD, i.e. that the magnetic field is frozen-in to the fluid. Thus, are these results applicable to protostellar disks, which are only weakly ionized? [23] showed that if the ion-neutral collision time is shorter than the orbital period, the BH instability is still present in the disk. Even at ionization fractions as small as 10^{-12}, this condition should be satisfied [24]. However, at the dense cores of protostellar disks, which are shielded from interstellar UV photons and ionizing cosmic rays, the ionization fraction may fall to values as low as 10^{-16}, and at this point the field is no longer coupled to the disk. Gammie [24] has therefore proposed a model for protostellar accretion disks in which beyond $r > 0.2$ AU, only a thin surface layer of the disk is ionized (primarily by cosmic rays) and therefore kept turbulent by the BH instability (within 0.2 AU, the disk is hot enough that collisional ionization of potassium keeps the ionization fraction above 10^{-12}). The central core of the disk is composed of a quiescent "dead zone". Because the surface density of the actively accreting surface layers of the disk are constant with radius, the model cannot be steady (i.e. $\dot{M} = \dot{M}(r)$), leading to the speculation that mass which accumulates in the dead zone is periodically flushed in FU Orionis like events. Moreover, the radial dependence of the mass accretion rate will lead to a temperature structure less steep than the canonical $T \propto R^{-3/4}$ law expected for α-disks, giving rise to a significant infrared excess in the thermal emission from the disk, as observed in many protostellar systems. While there are many aspects of the model which warrant further, more detailed calculations, at the very least the model serves to illustrate that magnetic fields may still be essential to the evolution of even weakly ionized disks.

MHD WINDS FROM DISKS

Magnetic forces have long been suspected as producing outflows from disks either by gradients in toroidal magnetic pressure, or by the "magnetocentrifugal effect" [25]. The former has been studied primarily through time-dependent numerical simulations [26-28]. In the latter case, a number of workers have presented models for the *steady state* structure of the wind accelerated by magnetocentrifugal effects (e.g., see the review in [5]).

Because of the complex evolution of axisymmetric MHD simulations of accretion disks (no simulations yet show relaxation to a steady-state), it is difficult to identify and investigate steady-state wind mechanisms, such as the magnetocentrifugal effect discussed by [25]. Recently, time-dependent numerical studies have begun to focus on the dynamics of the wind region only. In this new approach, the disk is treated as a fixed boundary condition at which the magnetic and velocity components, pressure, and density are specified and held fixed. The advantage to such an approach is that the wind can be evolved to a steady-state, and direct comparison to analytic (self-similar) theory can be made. The disadvantage is, of course, that back reaction of the wind on the disk is not accounted for, and moreover, dynamic processes within the disk itself, which may be important for the production of the wind, cannot be studied.

Results from this approach have now been presented by a variety of authors [29-32]. In every study, the wind is observed to reach a steady state for some region of parameter space. Transitions through each of the three critical points available to an MHD wind are noted in the simulations, and collimation of the flow near or beyond the fast magnetosonic point is observed. Generally such simulations begin with a corona above the disk which is in magnetohydrostatic equilibrium. However, the magnetocentrifugal effect operating along field lines inclined at angles of greater than 30° to the vertical produces strong outflow, and eventually the flow relaxes to a steady state. In some of the models of [30], the wind remains episodic for long dynamical times, indicating that no steady-state is possible for the assumed boundary conditions. These episodic solutions may have direct relevance to time-dependent jets, depending on the extent that the conditions at the base of the wind are determined by internal processes in the disk, as opposed to the wind acceleration mechanism itself. Generally good agreement between the predicted location of the Alfven critical point along magnetic field lines from steady-state theory, and that actually observed in the simulations, is found. Typical terminal speed for the flow are at an Alfvenic Mach number of ~ 5.

It would seem such simulations will be an important tool for studying properties of steady-state winds which are difficult to examine analytically. At the same time, it will be important to merge these simulations with those which treat the internal dynamics of the disk as well. The production of a strongly magnetized corona by dynamo action in the disk implies that the field which drives an outflow may be self-generated. Because current simulations of the BH instability in disks have been confined to local patches of the disk, there is still no evidence that large scale, global fields which are sufficiently strong to produce outflows result from the instability. However, because the power spectrum of the turbulence is a decreasing power law, most of the energy in the field is on the largest scales. Because the core of the disk is turbulent, however, random motions of the footpoints may introduce significant Poynting flux, which might affect the acceleration of the flow near the disk. Examining the large scale structure of the field, and whether it still contains significant fluctuating component at high-Z are both vital questions to be addressed by further studies.

DISK-MAGNETOSPHERE INTERACTION

In many circumstances accretion occurs onto a strongly magnetized central object. The magnetosphere of this object can truncate the inner disk far from the surface, and the resulting dynamical interaction of the disk and magnetosphere can determine not only the resulting accretion flow, but also the rate at which angular momentum is deposited on (or extracted from) the star [33–35]. Moreover, the observation that in many case outflows from accretion disks reach terminal speeds which are comparable to the escape velocity of the central object indicates that at least some fraction of the outflow arises from the disk-magnetosphere interaction region [36]. Thus, studies of the structure and dynamics of accretion disks which focus on the production of disk winds must be merged with studies of the disk-magnetosphere interaction region to build a complete picture of outflows from accreting systems.

One of many critical questions to be addressed by studies of the star-disk interaction region is: to what extent does the magnetospheric field thread the disk? Current models rely on MHD instabilities to mix the disk and magnetospheric plasma. For example, it has been proposed that the mixing rate is controlled by a combination of the Kelvin-Helmholtz instability operating in the shear layer between the magnetospheric and disk plasma [37,38], the Rayleigh-Taylor instability in disk plasma which is supported against gravity by magnetic pressure in the magnetosphere [33], and the BH instability which may keep the disk turbulent [36]. However, there is little agreement as to the efficacy of these processes: in some models the magnetospheric field is confined to a thin boundary layer at the surface of the disk, while in others turbulence in the disk keeps the field well-mixed. Because the nonlinear regime of these instabilities control these processes, detailed answers to these questions will probably require numerical methods.

Fortunately, numerical algorithms and computational resources are sufficient that detailed numerical studies of the star-disk interaction region are now underway. For example, Miller & Stone (these proceedings, and also [39]) have presented the results of a series of axisymmetric simulations which explore a number of assumed initial magnetic field geometries, and include resistivity in the disk to allow investigation of dense protostellar disks. The results indicate that in axisymmetry, polar cap accretion occurs in a very limited regime of field geometries, with the most interesting results occurring when the disk carries magnetic field with opposite sign from the dipole moment of the star. In fact, a number of groups are pursuing similar studies (see the review of [40]). The results of [39] are being extended to three-dimensional studies which focus in each of the processes thought to mix magnetospheric and disk plasma; we can expect more results from all of these groups in the near future.

SUMMARY

There has been much progress in our understanding of transport processes in accretion disks in the last five years. It is now clear that purely hydrodynamical turbulence is not an effective source of angular momentum transport. Thus, processes such as convection, which merely serve as mechanisms to drive turbulence, are probably not important as the source of anomalous viscosity. Furthermore, Keplerian disks do not appear to be subject to the nonlinear shear instabilities known to be present in Cartesian shear layers, a result which has recently been demonstrated via direct numerical simulation. On the other hand, MHD turbulence driven by the BH instability is an effective source of outward angular momentum transport. Models of very weakly ionized protostellar disks indicate they may have a layered structure, with only the surface regions well coupled to the field and therefore turbulent. The nature of this MHD turbulence may have important consequences for the production of MHD winds from disks (for example, through a disk dynamo which generates large scale fields). Moreover, the interaction of a magnetized disk with the magnetosphere of the central object will likely have important consequences for the accretion process as a whole.

REFERENCES

1. Shu, F.H., Adams, F.C. and Lizano, S. 1987, ARAA, 25, 23.
2. Shu. F.H., et al, 1993, in Protostars and Planets III, ed. by E. Levy & J. Lunine (Tucson:Univ. of Arizona Press), p3.
3. Adams, F.C., and Lin, D., 1993, *ibid*, p721.
4. Morfill, G., Spruit, H., & Levy, E.H., 1993, *ibid*, p939.
5. Konigl, A., & Ruden, S., 1993, *ibid*, p641.
6. Shakura, N.J., & Sunyaev, R.A. 1973, A&A, 24, 337.
7. Ruden, S.P., Papaloizou, J.C.B., & Lin, D.N.C. 1988, Ap.J., 329, 739.
8. Ryu, D., & Goodman, J. 1992, Ap. J., 388, 438.
9. Lin, D.N.C., Papaloizou, J.C.B., & Kley, W., 1993, ApJ, 416, 689.
10. Kley, W., Papaloizou, J.C.B., & Lin, D.N.C. 1993, ApJ, 416, 679.
11. Cabot, W., & Pollack, J., 1992, Geophys. Astrophys. Fluid Dyn., 64, 97.
12. Cabot, W., 1996, Ap.J., 465, 874.
13. Stone, J.M., & Balbus, S.A. 1996, Ap.J., 464, 364.
14. Balbus, S.A., Hawley, J.F., & Stone, J.M., 1996, Ap.J., 467, 76.
15. Balbus, S.A., & Hawley, J.F. 1991, Ap.J., 376, 214.
16. Balbus, S.A., & Hawley, J.F. 1992, Ap.J., 400, 610.
17. Hawley, J.F., & Balbus, S.A. 1991, Ap.J., 376, 223.
18. Hawley, J.F., & Balbus, S.A. 1992, Ap.J., 400, 595.
19. Brandenburg, A., Nordlund, A., Stein, R., & Torkelsson, U. 1995, Ap.J., 446, 741.
20. Stone, J.M., Hawley, J., Gammie, C., & Balbus, S. 1996, Ap.J., 463, 656.
21. Hawley, J.F., Gammie, C.F., & Balbus, S.A., 1995, Ap.J., 440, 742.
22. Hawley, J.F., Gammie, C.F., & Balbus, S.A., 1996, Ap.J., 464, 690.
23. Blaes, O.M., & Balbus, S.A. 1994, Ap.J., 421, 163.
24. Gammie, C.F., 1996, Ap.J., 457, 355.
25. Blandford, R.D., & Payne, D.G. 1982, MNRAS, 199, 883.
26. Uchida, Y., & Shibata, K. 1985, PASJ, 37, 515.
27. Shibata, K., & Uchida, Y. 1986, PASJ, 38, 631.
28. Stone, J.M., & Norman, M.L. 1994, Ap.J., 433, 746.
29. Ustyugova, G.V., et al 1995, Ap.J., 439, L39.
30. Ouyed, R., Pudritz, R.E., & Stone, J.M., 1996, Nature, in press.

31. Romanova, M., 1996, in Accretion Phenomena and Related Outflows, proceedings of IAU Coll. 163, ed. by D. Wickramasinghe, L. Ferrario, & G. Bicknell, in press.
32. Meier, D., 1996, *ibid.*
33. Ghosh, P. & Lamb, F.K., 1979, Ap.J. 232, 259.
34. Lovelace, R.V.E., Romanova, M.M., & Bisnovatyi-Kogan, G.S., 1995, MNRAS, in press.
35. Konigl, A. 1991, Ap.J., 370, L39.
36. Shu, F., Najita, J., Ruden, S., & Lizano, S., 1994, Ap.J. 429, 797.
37. Scharlemann, E.T., 1978, Ap.J., 219, 617.
38. Aly, J.J., 1980, A&A, 86, 192.
39. Miller, K., & Stone, J.M., 1997, in preparation.
40. King, A., in Accretion Phenomena and Related Outflows, proceedings of IAU Coll. 163, ed. by D. Wickramasinghe, L. Ferrario, & G. Bicknell, in press.

Axisymmetric MHD Simulations of Stellar Magnetosphere/Accretion Disk Interaction

Kristen Miller* and James Stone*

*University of Maryland, Department of Astronomy
College Park, Maryland 20742

Abstract. The evolution of the inner disk region of magnetized accreting systems is studied by means of numerical simulations. Two different initial magnetic field topologies are studied: a pure dipole field which threads the disk continuously and a dipole field excluded from the disk by surface currents. When present, the Balbus-Hawley (BH) instability provides efficient angular momentum transport, and accretion is equatorial. Otherwise, magnetic coupling between the disk and star surfaces as the dominant mechanism for transport and nonequatorial accretion results. In all of the simulations, low density, collimated disk-driven winds lead to mass loss.

INTRODUCTION

Central to an understanding of the accretion process is a quantitative analysis of the disk-star interaction and, in particular, the role magnetic fields play in this interaction. Such analyses require time-dependent, nonlinear calculations to properly determine the dominant mechanisms of angular momentum transport and the resulting accretion flow pattern. The current work consists of time-dependent, nonlinear, axisymmetric MHD simulations of the interaction between a magnetized star and a Keplerian disk in the inner disk region. Specifically, it summarizes the results for two initial configurations: Model I depicts a dipolar stellar field which threads the disk continuously, and Model II examines the case where the dipolar stellar field is completely excluded from the disk by surface screening currents. In this model, the stellar field topology bends inwards (toward the star) at the equator (the component of the magnetic field normal to the disk being completely cancelled by the field of the disk surface currents, leaving only a component parallel to the disk surface). (See Ghosh & Lamb [1] (1979), Aly [2] (1980), and Shu et. al. [3] (1994) for analytical descriptions of similar topologies.) Using these models, we examine the roles of magnetic braking, the BH instability, and resistivity in driving angular momentum transport and accretion.

RESULTS

A number of exploratory simulations are performed by varying the field strength and the disk density (Miller & Stone, [4] 1996). Resistivity is set to zero in Model I since the field already threads the disk everywhere; thus, adding resistivity in this model would only damp the magnetic instabilities which drive evolution. It is varied in Model II to allow the field to penetrate the disk and to control the size of the interaction region. The evolution is computed by solution of the equations of ideal magnetohydrodynamics with resistivity, assuming equatorial and axisymmetry. We use free flow boundary conditions at the outer boundary of the disk and magnetosphere. A hard stellar surface is assumed at the inner radial boundary. The midplane disk density is uniform in r and Gaussian in $\hat{z}$. The magnetosphere and stellar magnetic field are in solid body rotation at the stellar velocity. The density of the magnetosphere is $\approx 10^{-2} \rho_{disk}$; it is in pressure balance with the disk.

Model I. In Model I, for all values of R_c (corotation radius), density, and magnetic field strength used, angular momentum transfer is so efficient that the ram pressure of the accreting material is able to easily overcome the equatorial magnetic pressure and thus accrete onto the equator of the star. This transport is provided either by the BH instability (when present) or by magnetic coupling between the disk and star (in simulations with a higher initial Alfvén speed). We find no combination of the input parameters such that polar accretion results. In all cases, reconnection events develop at large radii in the magnetospheric region, suggesting the formation of a wind. Figure 1 shows the end state of a representative simulation in which magnetic coupling dominates the inner disk region. Note how the disk plasma has overpowered the magnetic tension and pulled the field lines in as it moved toward the stellar equator. Note also the multiple reconnections which have occured in the magnetosphere.

Model II. Resistivity controls the evolution of Model II. Because the initial state is not in radial force balance, a large resistivity is required to allow the field to diffuse into the disk, alleviating the radial magnetic tension force at the inner disk edge. However, a small resistivity is needed to ensure the field remains sufficiently coupled to the disk material to provide angular momentum transport once it has entered the disk. The optimum choice is to have the magnetic Reynolds number, $R_m, \approx 1$.

Nonequatorial accretion ($\theta \approx 10° - 20°$ above the equator) is achieved with this value of the resistivity for relatively weak fields and (likely) strong fields as well. In these simulations, the accreting plasma moves with subKeplerian speeds and does not follow the field lines, but, instead, areas of high magnetic field strengths. This suggests the accreting disk material is at least partially supported by magnetic pressure. Accretion takes place at a single hot spot on the stellar surface and appears to be steady. If the footpoints of the field are held fixed on the stellar surface, episodic outflows originate at the hot spot and move radially out from the star. As in Model I, winds are also suggested during the evolution; the winds in this model appear to be strongly collimated in the polar direction. Figure 2 shows the evolution of a simulation with $R_m \approx 1$. Note that the accreting plasma creates a hot spot on the stellar surface, just above the equator, and then spreads out along the stellar surface toward the equator and the pole as accretion continues.

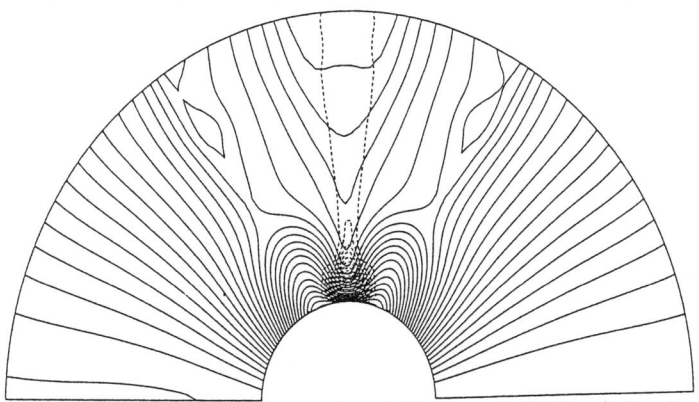

FIGURE 1. Density contours (dotted lines) and field lines (solid lines) showing equatorial accretion after 7 orbits.

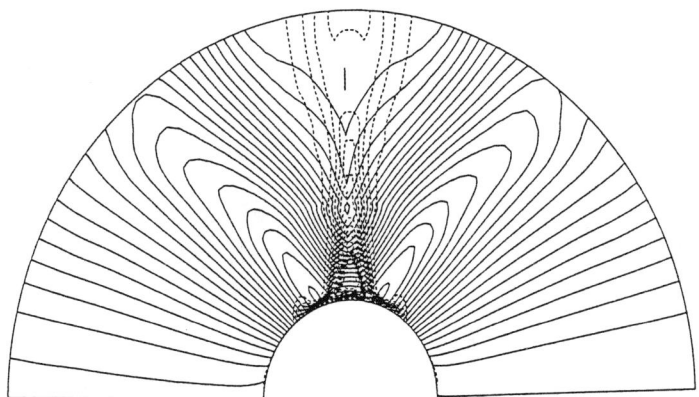

FIGURE 2. Density contours (dotted lines) and field lines (solid lines) showing nonequatorial accretion after 3 orbits. Note that the accreting material does not follow the field lines.

DISCUSSION

In both models, small values of the resistivity are needed to allow sufficient coupling between the field and the disk for angular momentum transport to occur via magnetic instabilities. When such coupling exists, this transport, driven by either the BH instability or magnetic braking, dominates the evolution. The occurrence of nonequatorial accretion is highly dependent on both the initial magnetic field configuration and the diffusion rate of the field into the disk (i.e., the value of the resistivity). For the purely dipolar field with negligible resistivity, we find no combination of the input parameters which will result in polar accretion. In this model, the pressure of the accreting disk plasma inevitably overwhelms the magnetic pressure and equatorial accretion results. For the dipole field screened from the disk, a value of $R_m \approx 1$ results in nonequatorial accretion which does not follow the field lines but instead areas of high magnetic pressure. Here, accretion takes place at a single hot spot on the stellar surface and appears to be steady.

In general, none of the simulations of either model result in dynamical flow (i.e., $v \approx v_{kep}$) of material along the field lines from the midplane of the disk to the stellar poles. In both models, reconnection events are present in the field lines at large radii, suggestive of the formation of a disk driven wind. In Model II, the winds appear to be collimated in the polar direction, and radially directed outflows occur.

REFERENCES

1. Ghosh, P., and Lamb, F.K. 1979, Ap.J., 232, 259.
2. Aly, J.J. 1980, A & A., 86, 192.
3. Shu, F.H., Najita, J., Ostriker, E., Wilkin, F., Ruden, S., and Lizano, S. 1994, Ap. J., 429, 781.
4. Miller, K., and Stone, J. 1996, In Preparation.

Hard X-ray emissions from Star Forming Regions

Tsuboi, Y., Koyama, K., Ueno, S.

Cosmic Ray Group, Dept. of Physics, Kyoto Univ. Kitashirakawa-Oiwake-Cho, Sakyo, Kyoto 606-01, Japan

Abstract. ASCA is the first satellite which has an imaging capability in high energies up to 10 keV, hence can peer deeply into the molecular cloud where protostar candidates may reside. We have carried out a systematic survey of hard X-rays from the core of molecular clouds, and discovered hard X-rays from protostar candidates (Class I objects) in the core of Rho Oph dark cloud. Two of them showed clear time variability; one (EL29) is flare-like and the other (WL6) is sinusoidal flux variation.
On the other hand, in Taurus dark cloud, we have detected a giant flare of a weak-lined TTS, V773 Tau in hard X-ray band. The plasma temperature was extremely high, about 8 keV, at the maximum flux, then it decreased as the flux decreased. The decay time and estimated electron density were found to be comparable to those of the sun, but the geometrical flare size was equal or even larger than the stellar size. The absorption column during the flare was large, about 3.6×10^{22} H cm^{-3}, and it makes impossible to detect the flare in soft X-ray band. Interestingly, in the decay phase, we found the apparent change of the metal abundance.

X-RAYS FROM CLASS I CANDIDATES IN ρ OPH DARK CLOUD

The ρ Oph dark cloud is one of the well known nearest star forming region (d$\sim$ 165 pc). The previous observations in the soft X-ray band (0.3-4 keV) using imaging telescope in this region were made by the Einstein satellite [1]. These observations established that low mass pre-main sequence stars (T Tauri stars) emit strong soft X-rays. The spectra can be fitted by temperatures of about 1 keV suffering absorption column densities ranging 10^{20-21} cm^{-2} from source to source. Some of X-ray sources show strong flare activities in the Einstein IPC band pass [1]. No X-ray was found from any Class I sources. Casanova et al. [2], with the recent deep ROSAT observation, reported 55 reliable detection of X-ray sources including several Class I sources in the error region of X-ray peaks in the 1.0-2.4 keV energy band. X-ray from one of the Class I sources was confirmed with the HRI observation (Casanova et al. private comm.).

On the other hand, the wide energy band imaging spectroscopy, including hard band(> 4 keV), was also made by ASCA for the first time. In the early report [3], 11 hard X-ray sources were detected in the core region including new X-ray sources. We have done detailed data analysis of the ASCA observation, then report newly identified X-ray sources, especially focusing Class I objects.

To demonstrate the higher capability of the present source finding procedure than previous one [3], we show two-energy band SIS contour map (combined image SIS 0+1), together with a molecular cloud contour map [4] and cataloged infrared sources [5] in Figure 1. Most of the point sources near the core are visible only in the hard X-ray band. A remarkable fact is that we found X-rays from 3 Class I and candidates (EL29 and possibly from WLY16A and WL6) in the ASCA field of view. The two of them (EL29 and WL6) showed clear flux variability.

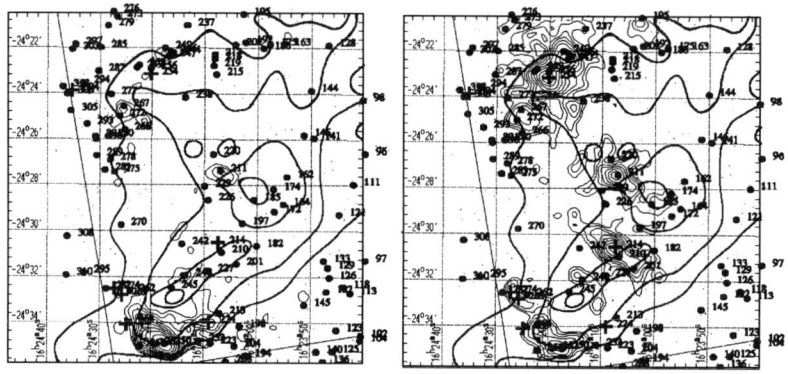

FIGURE 1. ASCA SIS 0+1 contour map near the core of L1681 cloud in the 0.5-2 keV and 2-10 keV energy bands. $C^{18}O$ contour and IR sources are also indicated. Class I is indicated by "plus". Refer the number of the IR sources to [5].

EL29(=YLW7) is an infrared (K band) variable, Class I protostar deeply embedded in the Lynds 1681 core. This source shows hard X-ray flare visible only in the 2-10 keV energy band.

WL6 is a Class I object located at the north position from the $C^{18}O$ peak of Lynds1681. This source shows sinusoidal temporal variability by a factor of 2. The hardness ratio shows no significant variability, which implies that no significant difference is seen between the spectra of high and low state.

X-ray emission mechanisms of Class I objects would be due to a solar-type flare activity, because these often showed an X-ray flare similar to the solar flare with the fast rise and slow decay, as seen in EL29.

The time variability seen in WL6 is very suggestive, because this shows no flare-like characteristics rather shows sinusoidal flux change with no change of the hardness ratio. Thus a native idea is that the X-ray variability is due to a simple geometrical effect, possibly the spinning of the protostar. If this idea is true, we can roughly estimate the spin period to be about 1 day, which make this source to be one of the fastest spinning protostar.

GIANT FLARE FROM WTTS, V773 TAU

As mentioned above, Einstein and ROSAT revealed the T Tauri stars (TTS) are strong X-ray emitters, with prominent X-ray flares. However, the detailed flare study of TTSs have been lacking, due to the limited energy band and resolution. Recently, we detected giant flare from WTTS, V773 Tau with ASCA. Here, we present for the first time, high quality X-ray spectra during the decay phase of an X-ray flare, and discuss the emission mechanism based on the precise physical parameters of the TTS flare.

The light curve of the V773 Tau is shown in Figure 2 (left). Due to earth occultation, however, the flare on-set was not detected. We note that the flare is found only in the hard X-ray band. This indicates that the flare emission is largely absorbed. The flare decayed with an e-folding time of about 7.5 ksec. We then made four X-ray spectra using the time-sliced data set 1-4 as is given in the light curve (Figure 2 left). We also made two X-ray spectra; that from the pre-flare state and the other taken from the post-flare state. The best-fit parameters are given in Table 2, while the best-fit flare spectra are shown in Figure 2 (right). With the spectral analysis, we obtained three major results.

The first important result is that the temperature at the data set 1 is extremely high, about 8 keV and it decreased as the flux decreased (Table 2). Assuming that the temperature and luminosity decrease are due to radiative cooling, we can estimated the electron density to be 2.1×10^{11} cm^{-3}, where the cooling time τ is expressed by $\tau = 3n_e kT/n_e^2 \Lambda(T)$. The loop size is calculated to be 4.6×10^{11} cm with the assumptions that the shape of the loop is similar to that of the solar and $\Gamma = 1$ [6]. This flare loop size is comparable to or even larger than the V773 Tau radius of 2.9×10^{11} cm [7].

The second important finding is that the apparent abundance (iron K-line at 6.7 keV) (Table 2) changed during the flare. The abundance at the data set 1 showed about 0.6 solar, as the flux decreased, it decreased to less than 0.1 solar. Since the plasma density is 2.1×10^{11} cm^{-3}, the flare could reach the ionization equilibrium within 100 sec; negligibly shorter than the flare time scale of $\sim$ 7.5 ksec. Thus ionization non-equilibrium cannot solve this puzzle. The abundance of star forming regions are reported to be about 0.3 solar with the ASCA observations (ex. Kamata private comm.). Therefore the apparent abundance change may be a key to solve the "low abundance problem".

The last notable finding is that the flare X-rays exhibited large absorption of 3.6×10^{22} H cm^{-3}, which was constant during the flare. In the quiescent states, on the other hand, we found no large absorption (Table 2). V773 Tau has a disk with mass about 0.01 $M_\odot$ estimated from millimeter measurements [8]. The flare probably occurred behind this disk.

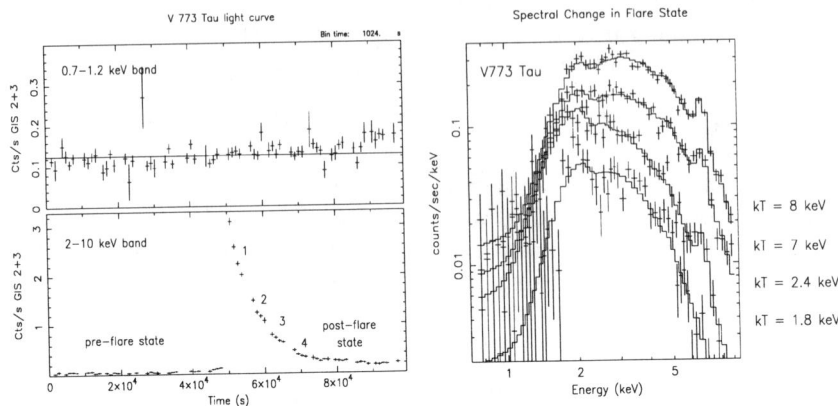

FIGURE 2. The light curve of the V773 Tau (left) and Spectral change of the flare (right).

TABLE 1. The best fit parameters of V773 Tau (GIS2+3)

Parameter	pre-flare	post-flare	1	2	3	4
kT (keV)	2.1±0.2	3.0±0.2	8±1	7^{+1}_{-2}	$2.4^{+0.3}_{-0.2}$	$1.8^{+0.2}_{-0.3}$
Abundance	< 0.2	$0.1^{+0.2}_{-0.1}$	0.6±0.1	0.4±0.1	0.3±0.2	<0.1
N_H (10^{22} H cm^{-2})	<5.5×10^{-2}	0.4±0.1	$3.6^{+0.2}_{-0.1}$	3.1±0.2	$3.0^{+0.4}_{-0.3}$	$4.3^{+0.8}_{-0.6}$
Lx (ergs s^{-1})^a	7.4×10^{30}	3.1×10^{31}	6.0×10^{32}	2.9×10^{32}	1.6×10^{32}	1.1×10^{32}
E.M.(10^{55}cm^{-3})^b	0.09±0.01	0.29±0.02	3.2±0.2	1.7±0.2	$1.6^{+0.3}_{-0.2}$	$1.5^{+0.8}_{-0.3}$

Note.— The errors and upper limits show one-parameter 90% confidence level. The background for the flare data (1~4) is the pre-flare data. The background for the pre- and post-flare data is taken form the annular region surrounding the source.

^{a} in 0.5~10 keV band (corrected for absorption), for d = 140 kpc.
^{b} in 0.5~10 keV band.

REFERENCES

1. Montmerle, T. et al. 1983, *ApJ* **269**, 182.
2. Casanova, S., Montmerle, T., Feigelson, E. D., Andre, P. 1995, *ApJ* **439**, 752.
3. Koyama, K. et al. 1994, *PASJ* **46**, L125.
4. Wilking, B. A., Lada, C. J. 1983, *ApJ* **274**, 698.
5. Greene, T. P., Young, E. T. 1992, *ApJ* **198**, 95.
6. van den Oord, G. H. J., Mewe, R., Brinkman, A. C. 1988, *A&A* **205**, 181.
7. Bouvier, J. et al. 1995, *A&A* **299**, 89.
8. Beckwith, S. V. et al. 1990, *AJ* **99**, 924.

X-ray Emission from Protostars

Eric D. Feigelson[1], Katsuji Koyama[2], and Thierry Montmerle[3]

[1] Department of Astronomy and Astrophysics, Pennsylvania State University, University Park PA 16802
[2] Department of Physics, Faculty of Science, Kyoto University, Sakoy-ku, Kyoto 606-01 Japan
[3] Service d'Astrophysique, CES/DAPNIA/SAp Centre d'études de Saclay 91191, Gif-sur-Yvette Cedex France

Abstract. Strong and flaring X-ray emission, 2 − 5 orders of magnitude above main sequence levels, is well-established for low-mass T Tauri stars (ages $10^6 - 10^7$ yr). This has been attributed to enhanced solar-type magnetic activity. We summarize here recent reports for X-ray emission from earlier phases of star formation, corresponding to Class I/II (ages $\sim 10^5$ yr) and Class I (ages $\sim 10^4 - 10^5$ yr) infrared sources. These include XZ Tau in the Taurus L1551 cloud, the Coronet Cluster of infrared sources in the R Corona Australis cloud, and IRS 43 in the Ophiuchi cloud core. Data are obtained from the PSPC and HRI detectors on *ROSAT*, and the SIS and GIS detectors on *ASCA*. Two of the sources, R1 in CrA and IRS 43 in Oph, exhibited powerful flares during the observations. The presence of X-rays in protostars may involve non-solar-type magnetic flaring events involving the circumstellar disk. Protostellar X-rays should have a significant effect on circumstellar material, such as producing ionization for magnetic coupling of disks and outflows, and promoting disk accretion through photoionization.

INTRODUCTION

Extensive observations of nearby star forming regions with the *Einstein* and *ROSAT* imaging X-ray observatories showed that low-mass ($\simeq 0.3 - 2$ $M_\odot$) pre-main sequence stars along the Hayashi tracks are copious emitters of soft X-rays. The majority of classical T Tauri stars, and a large population of previously unrecognized weak-lined T Tauri stars, were detected with X-ray luminosities ranging from the sensitivity limit around $10^{28.5}$ erg/s to $10^{31} - 10^{32}$ erg/s during the most powerful flares. (The contemporary Sun, for comparison, typically ranges between 10^{26} and 10^{27} erg/s in this band.)

The temporal behavior − moderate variability of timescales of days or longer with occasional rapid flares on timescales of $10^2 - 10^4$ seconds − is similar to the Sun and similar magnetically active late-type stars with multipolar fields undergoing sudden reconnection events. X-ray spectra correspond to plasmas with temperatures around 10^7 K. X-ray properties of classical T Tauri stars (CTTs, with strong infrared excesses and hot winds or accretion attributed to star-disk interactions) and weak-lined T Tauri stars (WTTs, with little or no circumstellar disk and associated activity) show similar X-ray properties. Optical photometric and spectroscopic study confirmed that WTTs have strong chromospheric emission and large cool starspots. WTT properties are very similar to those of magnetically active RS CVn binary systems.

All evidence thus points to enhanced solar-type magnetic activity in stars as they descend the Hayashi tracks. This phenomenon can be attributed to an enhanced magnetic dynamo, since T Tauri stars simultaneously have deeper convection zones and more rapid rotation than main sequence stars. The evidence and arguments for magnetic activity in pre-main sequence stars are reviewed in Feigelson et al. (1991) and Montmerle et al. (1993).

However, until 1996 there was no clear evidence for X-ray emission from the earlier, protostellar phases of low-mass stellar evolution. Earlier studies (Casanova et al. 1995; Kamata et al. 1996) found X-ray sources spatially coincident with Class I infrared sources, but the identifications were uncertain due to the crowded field. We summarize here three recent reports of highly variable X-ray emission from younger pre-main sequence stars, and speculate briefly regarding its possible implications. A closely related study is reported by Tsuboi et al. in this volume.

XZ TAU

XZ Tau is an unusually unobscured very young star with a strong emission line optical spectrum, a flat-spectrum infrared excess, and powers a molecular outflow and possible Herbig-Haro objects. Its IR classification is Class I/II, and its age is estimated to be $\simeq 5 \times 10^4$ yr. It was a relatively weak source in a 1993 *ROSAT* exposure ($L_x \simeq 8 \times 10^{29}$ erg/s) but was the strongest source in a 1994 *ASCA* exposure ($L_x \simeq 6 \times 10^{30}$ erg/s) (Carkner et al. 1996). In addition to being the most variable star in the L1551 cloud, it also showed the hardest *ASCA* spectrum with $kT \simeq 2.3$ keV. The high state was not due to a rapid flare, as it was constant during the 26-hour exposure. A similar Class I/II source in the Taurus L1495W cloud, IRAS 04113+2758, has also been detected (Carkner et al., in preparation). However, neither the *ROSAT* nor *ASCA* images detected any emission from three Class 0/I sources in the L1551 cloud, including the very luminous sources L1551-IRS5 and L1551-NE.

CORONET CLUSTER

The core of the nearby R Corona Australis (RCr A) molecular cloud contains a dense grouping of very young stars called the Coronet Cluster, consisting of Herbig Ae/Be stars, embedded Class I, and a few Class II and III sources. Images from a 40 ks exposure with the *ASCA* satellite are very dramatic. The low energy (0.5 − 2 keV) band shows the expected CTTs and WTTs sources. At higher energies (4 − 10 keV), these sources have disappeared and a bright complex structure about 4′ in extent appears near the densest molecular gas. It consists of 5 local peaks which coincide with the positions of Class I sources within an error of 20″. Spectral analysis indicates that the Class I sources have an unusually high temperature (kT∼7 keV) and absorption ($N_H \sim 4 \times 10^{22}$ cm^{-2}). Each source emits at levels exceeding ∼10^{30} erg/s, greater than ∼90% of CTTs and WTTs.

The brightest source within the cluster, tentatively associated with the protostellar infrared source R1, exhibited a flare during the observation with a fast rise and slower decay on a timescale around 5 hours. The flare spectrum shows a surprising feature: a strong and broad emission line complex at 6-7 keV, attributable to unresolved iron-like lines at 6.2 and 6.8 keV. A full description of this study is given by Koyama et al. (1996).

IRS 43

A 40 ks exposure of the Core F in the ρ Ophiuchi cloud was obtained with the *ROSAT* High Resolution Imager. This instrument is sensitive to the 0.1 − 2.4 keV energy range aith 5″ FWHM spatial resolution, but has no spectral resolution. After alignment with stellar sources in the field, astrometric precision of ±2″ is achieved. The image clearly shows a distinct source associated with IRS43 (= YLW15), a bright and well-studied infrared source in the cloud. It has properties characteristic of a Class I protostar: an infared spectrum rising in the near- infrared, peaking around 80μm, falling but detectable in the millimeter band ($L_{bol} \simeq 10$ L$_\odot$); a bipolar carbon monoxide outflow; and radio continuum emission. Models indicate a dusty envelope $\simeq$ 2000 A.U. in extent with mass ≥ 0.05 M$_\odot$.

IRS 43 displayed a strong flare during the X-ray exposure; indeed, the source is undetected except during the event. The peak HRI count rate jumped to a very high level 16.7 ± 5 cts/ksec, at least a factor 20 above quiescent levels, and then decreased with an e-folding time around 5 hours.

The uncertain line-of-sight column density to IRS 43 and the lack of spectral capability in the HRI precludes a definitive measurement of flare properties. Assuming a reasonable range of plasma temperatures (kT $\simeq 1 - 10$ keV) and absorptions ($A_V \simeq 10 - 40$), the peak flare luminosity is estimated to be $10^{32} - 10^{34}$ erg/s in the 0.1 − 2.4 keV HRI band, and $10^{33} - 10^{36}$ erg/s in the full X-ray band. These are extraordinarily high values, matching the most powerful flares ever detected in a T Tauri (or in fact any late-type) star. For a simple spherical or loop geometry, the X-ray emitting region is likely to be of order 10^{-1} A.U., far larger than the stellar radius. This study is described by Grosso et al. (1996).

DISCUSSION

These results clearly show that X-ray emission, common among $10^6 - 10^7$ yr old low-mass stars, can be present in $10^4 - 10^5$ yr old stars as well. However, such emission is not ubiquitous, and the patterns of occurrence are not understood. The three Class 0/I sources in L1551 are not detected at levels seen in all five Class I sources in the Coronet Cluster, assuming their $A_V \leq 50$. A survey of archival *ROSAT* fields with very young stellar objects reveals that X-ray emission with $L_x \geq 10^{30}$ erg/s is relatively rare (Carkner et al., in preparation). We can not distinguish between the possibilities that: all protostars have powerful X-ray flares with small duty cycles; all protostars have powerful X-ray emission but some have $A_V \gg 50$ along the line-of-sight; or that only a small fraction of protostars are X-ray emitters.

Although T Tauri flares are traditionally attributed to solar-type magnetic reconnection events on the surface of the star, there are reasons to consider more exotic mechanisms for these events in protostars (and possibly CTTs). These include magnetic field reconnection: within the infalling envelope; at the Alfven surface between a stellar magnetosphere and an accretion disk; around the corotation radius in the disk where shearing will occur; at the X-point above the corotation radius where magnetic geometries change from closed to open; at the interface between a stellar wind and the disk (see references in Grosso et al. 1996). The data outlined above give tantalizing suggestions that the X-ray emitting region may be associated with the circumstellar disk. The broad emission line complex in the Coronet source might be iron K-shell emission from a hot plasma loop (6.8 keV) and fluorescence from a cold disk (6.4 keV). The large loop size inferred for the IRS 43 flare may represent magnetic fields between the star and disk corotation region.

Whether the emission processes in IRS43 are solar-like or involve disk interactions, our result shows that large numbers of X-ray photons can be present in protostellar environments. They will efficiently irradiate the circumstellar envelopes and, if source geometries are favorable, the accretion disk of protostars. New physical effects will emerge, such as photoionization-induced accretion (Glassgold & Najita 1997) or flare melting of chondrules (Cameron 1994) which may significantly further our understanding of the earliest stages of low-mass star formation and their proto-planetary disks.

Acknowledgements: EDF was partially supported by grants NAGW-2120 and NAS 8-38252.

REFERENCES

1. Cameron, A.G.W. 1994, Meteoritics, 29, 454
2. Carkner, L., Feigelson, E. D., Koyama, K., Montmerle, T. and Reid, I. N., 1996, ApJ 464, 286
3. Casanova S. , Montmerle T., Feigelson E. D., Andre Ph. 1995, ApJ 439, 752
4. Feigelson, E. D., Giampapa, M. S. & Vrba, F. J. 1991, in *The Sun in Time*, eds. C. Sonett & M. S. Giampapa, (Tucson:Univ. Az. Press), p. 658
5. Glassgold, A.E., & Najita, J. 1996, ApJ, in press
6. Grosso, N., Montmerle, T., Feigelson, E. D., Casanova, S., Gregorio- Hetem, J. & André, P. 1996, submitted
7. Kamata Y., Koyama K., Maeda Y., Ozaki, M., Ueno S., Tawara Y., Yamauchi S. 1996, MNRAS, submitted
8. Koyama, K., Hamaguchi, K., Ueno, S., Kobayashi, N. & Feigelson, E. D. 1996, PASJ 48, L87
9. Montmerle, T., Feigelson, E.D., Bouvier, J., & André, P. 1993, in *Protostars and Planets III*, eds. E.H. Levy & J.I. Lunine, (Tucson : University of Arizona Press), p. 689

Dispersed T Tauri Stars and Galactic Star Formation

Eric D. Feigelson

Department of Astronomy & Astrophysics, Pennsylvania State University, University Park PA 16802

Abstract. The spatial distribution, age distribution and kinematics of T Tauri stars, both close to and widely distributed around active clouds, are considered using simple models of T Tauri dispersal. Models are compared to observations in and around the nearby cloud complexes, in particular the recent discovery of widely scattered young stars from the *ROSAT* All-Sky Survey. We suggest the dispersal of T Tauri stars has two major causes: slow isotropic drifting of stars away from long-lived star forming clouds, and star formation in short-lived rapidly moving cloudlets associated with large-scale turbulent motions of molecular cloud complexes. A third mechanism for dispersal, dynamical ejection of high velocity T Tauri stars, appears to be less important. Other implications include: star formation in at least one cloud (Chamaeleon I) has been continuous for $\simeq 20$ Myr; star formation efficiencies of clouds may often be 20% or higher; a large fraction of low-mass stars may form in small shoft-lived cloudlets each producing no more than a few stars; and T Tauri kinematics support molecular evidence for large-scale turbulence in molecular clouds. A full presentation of this study appears in Feigelson (1996).

INTRODUCTION

Unbiased and complete samples of T Tauri stars are essential for addressing a number of important issues concerning star formation in the Galaxy and early phases of low mass stellar evolution. Does a molecular cloud produce star continuously or episodically? What is a cloud's star formation efficiency and stellar initial mass function (IMF)? What fraction of stars are formed in small cloudlets *vs.* giant molecular clouds? How do cloud dynamics affect stellar dynamics and bound cluster formation? What is the evolution of circumstellar disks from which planetary systems may condense? The samples of T Tauri stars from which these questions have been addressed have grown considerably due to Hα emission, infrared array, radio continuum and particularly X-ray surveys. But despite these advances, T Tauri census is far from approaching completeness: there is a steep decline in the numbers of stars known with ages older than ~2 Myr in all nearby clouds.

The mystery of the missing older T Tauri stars is more dramatic today than when Herbig (1978) first raised the issue. Several possible explanations, upon detailed consideration, are not viable. Short-lived star formation is contradicted by the presence of some older stars near the clouds, and the absence of many inactive molecular clouds in the solar vicinity. Episodic star formation should have produced multiple bumps in the age distributions, which are not seen. Although theoretical isochrones are sensitive to the details of stellar interior models, this uncertainty is not sufficiently large to solve the essential age distribution problem. Accretion luminosity is also unlikely to solve the problem, as it does not affect weak-lined T Tauri (WTT) stars. Flux limits of current T Tauri surveys may be responsible for poor sampling of stars older than ~20 Myr, but does not appear to account for the sharp decline in population in the 2 − 20 Myr age range.

The most likely explanation is that older T Tauri stars do exist in large numbers, but lie outside of well-surveyed regions. T Tauri stars should inherit velocity dispersions of 1 km s^{-1} or greater, so that older T Tauri stars are expected to travel tens of parsecs (i.e. many degrees for nearby regions) from their parent cloud, where careful surveys have not been conducted.

DISPERSED OF T TAURI STARS

This idea receives considerable support from the large numbers of dispersed lithium-rich magnetically active late-type stars recently reported, most prominently in the *ROSAT* All-Sky Survey studies of the environs of the Taurus-Auriga and Chamaeleon clouds (e.g. Alcalá et al. 1995; Neuhäuser et al. 1995; Wichmann et al. 1996). It has been argued that these stars are mainly $10^7 - 10^8$ yr ZAMS stars, rather than $10^6 - 10^7$ pre-main sequence stars (Micela et al. 1993; Briceño et al. 1997). This is possible, though a recent radio continuum survey indicates strong magnetic activity comparable to WTT stars (Carkner et al. 1997). In either case, the *RASS* samples represent only the brighter end of a large underlying population of dispersed young stars, which may solve the puzzle of the missing older T Tauri stars. Other cases of T Tauri stars far from molecular clouds, such as the TW Hya group, are also known. We therefore investigate the properties expected of T Tauri populations that are dispersed from molecular clouds by different mechanisms.

Dispersal with slow velocity dispersion

We first consider the simple model wherein stars form at a single location from gas with an isotropic Gaussian velocity dispersion of $\Delta v = 1$ km/s due to thermal motions in the parent cloud core. The result is radial dispersion of the population, with younger T Tauri stars concentrated in the inner region and older stars drifting up to several degrees away (assuming $d \simeq 150$ pc). Studies of ages and kinematics restricted to the inner region will exhibit significant biases: undersampling of older stars, underestimation of the proper motion and radial velocity dispersions. These effects may explain discrepancies found between radial velocity and proper motion distributions in Taurus-Auriga T Tauri stars. Counting the number of older stars ejected along the line-of-sight can measure the longevity of star formation in a cloud; in this way, the Chamaeleon I cloud is estimated to have been producing stars for ~ 20 Myr.

Dispersal by dynamical ejection

It is possible that some stars acquire significantly higher velocities, either from the complex dynamics of core collapse, or gravitational interactions with other stars after formation (Sterzik & Durisen 1995; Gorti & Bhatt 1996). Ejection velocities around $5-10$ km/s from scattering by hard binaries, for example, is astrophysically quite reasonable. However, such a mechanism is unlikely to explain the dispersed young stars found by ROSAT because of the large number of bright very young (<2 Myr if they lie at the distance of the cloud) stars very far from the cloud. This mechanism could produce a small number of high velocity stars as far as ≥ 100 pc ($\geq 50°$) from nearby clouds, and may explain isolated T Tauri stars like FK Ser.

Cloudlets in turbulent molecular clouds

Molecular gas in low-mass star forming molecular regions have line structures which can be divided into thermal motion with $\Delta v_t \simeq 0.5$ km/s and a larger nonthermal motion Δv_{nt} within or between individual clouds. The amplitude of the latter component increases with the size of the region R under consideration as $\Delta v \propto R^p$, where p ranges from 0.2 to 0.7, and reaches $10-20$ km/s (e.g. Larson 1981, Myers & Goodman 1988, Falgarone 1991). These findings, combined with the fractal geometry and self-similar density structure of molecular clouds, are widely interpreted as manifestations of Kolmogorov-type turbulence. Star formation within a hierarchical cascade of a turbulent cloud is likely to be unsynchronized in space and time (Henriksen 1991).

A broad agreement between the integrated line widths of cloud complexes and T Tauri kinematics was first noted by Larson (1981). However, since most of the gas velocity dispersion is due to intercloud rather than intracloud motions, clouds should gradually separate from each other as each produces a comoving cluster of T Tauri stars. To explain the poor correspondance between dispersed *RASS* stars and existing molecular clouds, the cloudlets either rapidly dissipate after active star formation, or are decelerated by interaction with the intercloud medium. In either case, the cloudlets leave behind isolated T Tauri stars or comoving groups of T Tauri stars such as the TW Hya group. A simulation of such a star formation model for a cloud complex forming stars for ∼20 Myr predicts groups of stars distributed over tens of parsecs with large radial velocity vectors and randomly oriented proper motion vectors. Reasonable cloud turbulence parameters reproduce the distribution of dispersed young stars seen by ROSAT.

CONCLUSIONS

Herbig's (1978) question 'Where are the post-T Tauri stars?' (or, in modern terminology, 'Where are the post-classical T Tauri stars?') can thus be answered: some are WTT stars found close to the active cloud cores, but most are WTT stars have drifted several parsecs, or were born tens of parsecs away, from currently active clouds.

These interpretations must be considered tentative. It is possible, for example, that the high-lithium stars found in the *ROSAT* All-Sky Survey are unrelated to nearby star forming regions, but rather represent the magnetically active portion of a large post-T Tauri or ZAMS population. However, our models make clear testable predictions concerning the spatial, age and velocity distributions of T Tauri stars. For example, the thermal drift model predicts that the known population of 117 stars within 1° of the Chamaeleon I cores should be surrounded by $\simeq$ 200 mostly older T Tauri stars spread over several degrees, with proper motions oriented away from the cloud. This population around long-lived clouds can be located with unbiased Li 6707 Å spectroscopic surveys. Our models further predict that the more widely dispersed T Tauri stars found in the *RASS* were formed *in situ* over a 20 − 50 pc region in small molecular cloudlets that have quickly dissipated, and should be clustered in comoving groups having low internal motions but high relative proper motions between groups. Proper motion surveys of $V \simeq 10 - 16$ stars should be able to investigate this prediction.

ACKNOWLEDGMENTS

This work was supported by NASA grant NAGW-2120.

REFERENCES

1. Alcalá, J. M., Krautter, J., Schmitt, J. H., Covino, E., Wichmann, R. & Mundt, R., 1995, As. Ap. 114, 109
2. Briceño, C., Hartmann, L. W., Stauffer, J. R., Gagné, M., Stern R. A. and Caillault, J.-P., 1997, A. J., in press

3. Carkner, L., Feigelson, E. D., Koyama, K., Montmerle, T., and Reid, I. N., 1996, Ap. J. 464, 286
4. Falgarone, E. & Phillips, T. 1991, in Fragmentation of Molecular Clouds and Star Formation, eds. E. Falgarone et al. (Dordrecht:Kluwer), p. 119.
5. Feigelson, E. D., 1996, Ap. J. 468, 306
6. Gorti, U. & Bhatt, H. C. 1996, MNRAS 278, 611
7. Henriksen, R. N., 1991, Ap. J. 377, 500
8. Herbig, G. H. 1978, in Problems of Physics and Evolution of the Universe, ed. L. V. Mirozan (Yervan:Acad. Sci. Armenian SSR)
9. Larson, R. B. 1981, MNRAS 194, 809
10. Micela, G., Sciortino, S. & Favata, F., 1993, Ap. J. 412, 618
11. Myers, P.C. & Goodman, A. A.1988, Ap. J. 413, 593
12. Neuhäuser R., Sterzik M.F., Schmitt J.H., Wichmann R., Krautter J., 1995, As. Ap. 295, L5
13. Sterzik, M. F. & Durisen, R. H., 1995, As. Ap. 304, L9
14. Wichmann, R., Krautter, J., Schmitt,. J. H. M. M., Neuhäuser, R., Alcalá, J. M., Zinnecker, H., Wagner, R. M., Mundt, R. & Sterzik, M. F., 1996, As. Ap. 312, 439

The influence of photoevaporation on star formation in NGC 6611

D. de Winter[1], M.E. van den Ancker[2], and M.R. Pérez[3]

[1] Dpto. Física Teórica, C–XI, UAM, Cantoblanco, E-28049 Madrid, Spain
[2] Astronomical Institute "Anton Pannekoek", University of Amsterdam, Kruislaan 403, 1098 SJ Amsterdam, The Netherlands
[3] Applied Research Corporation, Suite 1120, 8201 Corp. Dr., Landover, MD 20785, USA

Abstract. The role of photoevaporation on gaseous globules was shortly ago presented as a possible new mode of star formation (SF) in NGC 6611. In this paper we study this process by the distribution of the population of young stars and strong near-IR sources in the field of NGC 6611. We show that the locations of these sources do not coincide with those of the EGGs, but are found in regions shielded from the photoevaporation by the surrounding clouds. We therefore conclude that SF in NGC 6611 takes place at the other side of dark clouds to where the EGGs can be found.

INTRODUCTION

The recently detected evaporating gaseous globules (EGGs) in the very young open cluster NGC 6611 (M16, the Eagle nebula) [2], were proposed to be a possible site of star formation (SF) if the EGG contains a young stellar object (YSO). The influence of the photoevaporation mechanism by hot stars on the circumstellar material of neighbouring extremely young stars has been discussed by us in a previous study [5]. Here we will further investigate the possibility of this model as a possible SF mode and we study the relation of the location of the EGGs in NGC 6611 with the population of embedded near-IR sources [3] and with the optical stellar population of NGC 6611.

THE STELLAR POPULATION OF NGC 6611

It is illustrative to see where the photoevaporation of molecular material in NGC 6611 originates. In [2] it is mentioned that the total ionizing flux that accounts for the detected photoevaporation flow possibly comes from one O3–4 star and three objects with spectral type O5–6, located in the cluster.

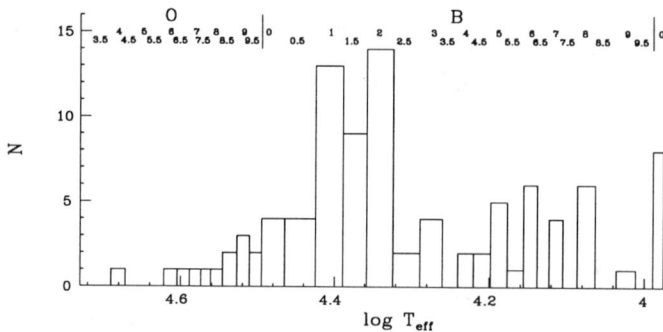

FIGURE 1. Histogram of a sample of the stellar content of NGC 6611 for which reliable spectral types exist.

However, there are many other early type stars in NGC 6611 with sufficient UV flux to play a significant role in this process. It is therefore interesting to investigate the interplay of their location and the evaporating material as well. For these purposes we first sorted a sample of true cluster members of [6,7,3] by spectral type, see Fig. 1, and then determined the location of the O and B sources, Fig. 2.

Fig. 1 shows that most stars are early B-type and that the number of O-type stars is not very large. The earliest type star is one of O4 with an increasing number of stars from O6 up to B2.5. In Fig. 2 all stars with a temperature of 30,000 K or more appear to be located far northwest and west of the central region. Only two O-type objects are located near the dark columns identified [2], being W 367 (O9.5) and W 401 (O8.5). All other sources within the temperature range from 20,000 to 30,000 K appear in or around the central region. Note that most objects cooler than B2.5 are located east, northeast, south and southwest of the central region. Furthermore, Fig. 8 in [7] shows that A_V rapidly increases towards the northwest.

THE YOUNGEST OBJECTS IN NGC 6611

As the different stages of EGG formation are visible and because of the short evolutionary time scale of stars such as in NGC 6611, the location of the "oldest" EGGs and the youngest stars must be correlated.

SF is still in process in NGC 6611, which is evident from the population of 113 embedded, strong JHK, sources [3], from the ages of the youngest stars visible [7] and from the heated giant molecular clouds [1,4]. The correlation between EGGs and the molecular material is not questioned, which seems to be the prime reason to suggest that EGGs can contain YSOs.

The location of the population of embedded and other strong near-IR sources in NGC 6611 is discussed by [3]. Most of these objects are located at the northeast of the cluster center, as the near-IR sources in the south and southwest are supposed to be extremely reddened background objects. In more detail, Fig. 2 shows that most of these candidate YSOs are located at the edges of the bright nebula and only about 1/5th in the more central region of which only 4 are located close to the "elephant trunks". However, none of these sources seem to be near one of Hester's 73 EGGs.

To localize the youngest visible objects is more difficult. Many of the stars in our sample have masses higher than 10 M$_\odot$, and can be pre- or post-ZAMS within a few Myrs. Following the HRD in [7], we assume that the objects younger than one Myr are: their group II objects, plus the most massive stars (spectral type O6 or earlier) plus the ones with masses less than 10 M$_\odot$ that are located right on the MS. This latter group is either extremely young or about 10 Myr, or more for the less massive ones, which is not to be expected

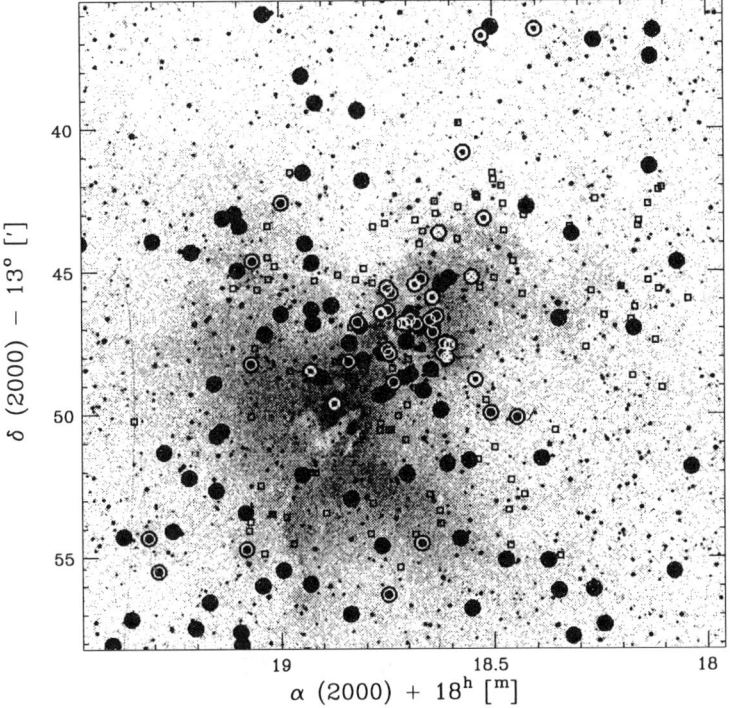

FIGURE 2. The region of NGC 6611, taken from the Digital Sky Survey. The embedded population of [3] are indicated by boxes. The studied stars are indicated by black dots. The different effective temperatures of our sample of objects is also given, see Fig. 1 and text. Stars with 30000 K > T_{eff} > 20000 K are indicated with a white circle, the ones with 40000 K > T_{eff} > 30000 K with a thick circle and the three objects with T_{eff} > 40000 K by filled circles.

[5]. The remaining objects in the sample are a few Myrs old and located mostly in the central region, see Fig. 2. Although the number of objects around the central part in our sample is relatively low, we do detect that this is the location of the youngest visible objects. From these objects the most massive ones are located northwest and the ones with 10 $M_\odot$ or less more to the east and southwest.

CONCLUSIONS

From the stellar population of NGC 6611 we conclude that the highest temperature sources are located in the central region of NGC 6611 with an increasing temperature distribution opposite to the dark pillars. These stars also suffer from the highest intracluster absorption, further weakening their UV-radiation field and are therefore of less importance in ionizing and evaporating the "elephant trunks".

The youngest visible stars have a tendency to be located around the central region, with the most massive, probably youngest ones located at the northwest. The population of embedded sources also shows a tendency to be located further out from the center with most of them placed near the edges of the molecular regions, especially toward the northern cloud. These are evidences that most young and probable protostars do not coincide with the central region of NGC 6611 and with ionizing fronts at the southern molecular cloud. The high number of detected embedded sources in the rather isolated regions, without traces of the presence of YSOs, show that if they are protostars they must be of low mass. Their pre-main sequence timescale exceeds the age of the cluster for masses less than 2 $M_\odot$.

This leads us to believe that the bulk of the stellar population of NGC 6611 has been formed or is in formation in regions shielded from the photoevaporation by the surrounding molecular material, in contradiction with a possible mechanism of SF by EGGs.

Acknowledgements D. de Winter is supported in part by Spanish grant DGI-CYT PB94-0165.

REFERENCES

1. Goudis, C., *Ap&SS* **41**, 105 (1976).
2. Hester, J.J., Scowen, P.A., Sankrit, R., et al., *AJ* **111**, 2349 (1996).
3. Hillenbrand, L.A., Massey, P., Strom, S.E., Merrill, K.M., *AJ* **106**, 1906 (1993).
4. McBreen, B., Fazio, G.G., Jaffe, D.T., *ApJ* **254**, 126 (1982).
5. Pérez, M.R., de Winter, D., van den Ancker, M.E. et al., in preparation (1996).
6. Thé, P.S., de Winter, D., Pérez, M.R., *A&AS* **104**, 315 (1994).
7. de Winter, D., Koulis, C., Thé, P.S., et al., *A&A*, in press (1996).

Extra-Solar Comets Near Young β Pic Analogs

C.A. Grady[1], M.R. Pérez[2], K.S. Bjorkman[3], M.L. Sitko[4] and D. de Winter[5]

[1] *Eureka Scientific, 2452 Delmer St., Suite 100, Oakland CA 94602*
[2] *Applied Research Corporation, Landover MD 20785*
[3] *Ritter Observatory, University of Toledo, Toledo OH 43606-3390*
[4] *Department of Physics, University of Cincinnati, Cincinnati, OH 45221-0011*
[5] *Dpto. Física Teórica, C-XI, Facultad de Ciencias, Universidad Autónoma de Madrid, Cantoblanco, E-28049 Madrid, Spain*

Abstract.
Studies of the evolutionary precursors of β Pic have demonstrated that high velocity, accreting gas features are routinely detectable when the viewing geometry is similar to the β Pic system. We present a first exploration of the composition of the accreting gas toward 4 nearby, UV-luminous Herbig Ae/Be stars viewed through their circumstellar dust disks. We find that in addition to refractory materials, plausibly interpreted as sublimates of silicate grains, the accreting gas features in these stars contain carbon, mildly refractory species such as Zn II, and volatiles. Overall, the composition of the accreting gas is consistent with detection of the gaseous comae of star-grazing planetesimals of either cometary or carbonaceous chondritic composition.

INTRODUCTION

These are exciting times for the study of extra-solar planetary systems and their evolutionary precursors, with reflex motion studies yeilding the first indirect detections of Jupiter-mass planets, and expanding observations of circumstellar disks. Such studies emphasize either the largest bodies in a system or small dust grains. Detection of minor bodies, such as planetesimals, in the disk systems is critical in demonstrating that circumstellar disks routinely evolve into planetary systems [1]. Planetesimals are normally exceedingly difficult to detect. However, if such bodies are on star-grazing orbits, cometary activity, induced either by vaporization of volatile ices, as in the Solar System case, or by tidal and thermal disruption of more refractory-rich bodies [2], may make a gaseous coma surrounding the nucleus detectable. The first indication of the presence of such activity was provided by the nearby β Pictoris system, which routinely exhibits high velocity, accreting circumstellar gas visible in transitions of atomic ions of refractory metals ([3],[4] and references therein). Accreting gas in this system is routinely detected to +200 km s^{-1} with more sporadic detections of material visible to +300 to +400 km s^{-1}, velocities consistent with free-falling material within a few tenths of an AU of the star. The accreting gas profiles have been successfully modelled as absorption from the gaseous comae of swarms of star-grazing comets [5].

Systems like β Pic do not just happen. The evolutionary precursors of stars like β Pic are intermediate-mass pre-Main Sequence stars such as the Herbig Ae/Be (HAeBe) stars which include both objects in high density star-forming regions and isolated objects [8]. Like β Pic, these stars have IR excesses indicative of the presence of circumstellar gas and dust [6] which is apparently distributed in disks [7][8][9] extending for several hundred AU [10][11]. UV and optical studies of the nearest and minimally embedded HAeBe stars demonstrate that accreting, circumstellar gas is routinely detected in the line of sight to those stars [12] for which polarimetric, spectroscopic, and photometric studies indicate that the disk is viewed close to equator-on. The available IUE and optical data indicate that the accretion is clumpy. When high accretion state spectra are normalized by either low accretion state spectra or main sequence comparison star spectra, high velocity accreting gas profiles similar to those seen toward β Pic are produced (figure 1).

THE ACCRETING, CIRCUMSTELLAR GAS

Typically the accreting gas has flat-bottomed absorption profiles consistent with the presence of optically thick absorption which does not fully obscure the star[13]. The % coverage, or covering factor of such material, gives us a measure of the spatial distribution of the various gas-phase species. The relative prominence of species with comparable f-value transitions and similar ionization potentials enables us to place some coarse limits on the composition of the accreting gas. In the case of HD 100546 (B9 Ve, L=90 L$_\odot$) the Fe II features are anomalously weak compared to both the Si II and Mg II profiles, which have covering factors of 50-60%, suggesting that iron is underabundant in the accreting gas. Similarly, prominent transitions of siderophiles such as Mn II and Cr II are not detected toward this star. ISO SWS spectra of HD 100546 confirm that the silicates in this isolated Herbig Be star are predominantly Mg-rich, crystalline olivene, e.g. forsterite [14][15]. The accreting gas seen toward HD 95881, in contrast, appears to be underabundant in Mg II relative to both Si II and Fe II transitions. The other nearby, suitably oriented Herbig Ae stars, HD 163296 and HR 5999, have comparable covering factors for Fe II, Si II, and Mg II suggesting compositions similar to β Pic for the silicates.

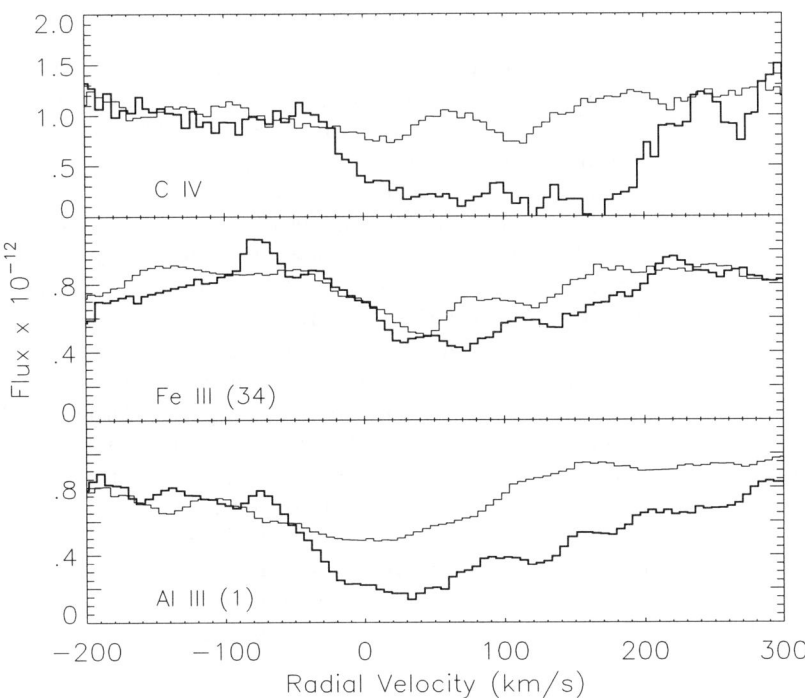

FIGURE 1. Accreting, circumstellar gas viewed in transitions of C IV, Fe III (34), and Al III toward HD 95881 and the comparison field A-shell star HD 15253 (shallower absorption features.

The UV data provide some important compositional clues for species which do not have transitions in either the mid-UV or optical. Accreting carbon ions are routinely detected in the line of sight toward these stars, including species up to C IV. Neutral atomic carbon is firmly detected in the accreting gas, and *only in the accreting gas*, for HD 100546 and HD 95881 with covering factors comparable to that of Si II. C I is clearly present in spectra of both HR 5999 and HD 163296, but saturation of C I (2) in the limited IUE data precludes more detailed analysis. Mildly refractory species such as Zn II and for the hottest stars, S II, are detected in the accreting gas toward the HAeBe stars, unlike β Pic. Typical covering factors for these species are smaller than for the silicate constitutents. We also detect N I absorption, typically with covering factors of 25% or smaller, and in the optical He I with covering factors near 10%.

The available data for these HAeBe stars suggest that accreting gas detections are not primarily dictated by nominal "cosmic" abundances, since we detect, in HD 100546, trace elements such as zinc and not similarly abundant, but more refractory elements such as manganese. Overall, the more refractory elements have larger covering factors than species such as Zn II or S II which are considered to be mildly refractory, and species such as He I and N I. Our data, therefore, indicate that the accreting gas is most likely to be the vaporized remnants of solid bodies, rather than accretion streams of material which have remained in the gas phase since the formation of the circumstellar disks. The IUE data also indicate that the carbon, while nominally considered a volatile, has a spatial distribution similar to the most refractory material sampled, and thus may originate in refractory grains rather than volatile ices. Coupled with the velocities achieved by the accreting gas (0 to +400 km s^{-1} with neutral species seen to +200 km s^{-1}) which are inconsistent with production of the gas from grains gradually spiralling in under Poynting-Robertson drag [9], and with small grain blow-out radii of several microns [16], the covering factor data indicate that we are likely to be viewing, as in the case of β Pic, the gaseous comae of larger, unseen star-grazing bodies. The spectral data, coupled with the emerging ISO data suggest that these bodies have compositions similar to either carbon-rich meteorites or to comets, in agreement with theoretical expectations for planetesimals.

REFERENCES

1. Beckwith, S. V., and Sargent, A. I., *Nature* 383, 139 (1996).
2. Grinin, V.P., Natta, A., and Tambovtseva, L., *A&A*(in press), (1996b).
3. Kondo, Y. and Bruhweiler, F.C. *Ap.J.*291, L1 (1985).
4. Lagrange, A.M., et al. *A&A*296 499 (1995).
5. Beust, H., et al., *A&A*. 310, 181 (1996).
6. Thé, P.S., et al. *A&AS* 104, 315 (1994).
7. Hillenbrand, L.A., et al., *Ap.J.*397, 613 (1992).
8. Grinin, V.P. et al., *APSS* 186, 283 (1991).
9. Grinin, V.P., et al., *A&A*292, 165 (1994).
10. Grinin, V.P., and Rostopchina, A. *Astr. Rep.*40, 1 (1996a).
11. Mannings, V.G., and Sargent, A.I. *From Stardust to Planetesimals*, ed. Y. Pendleton, NASA CP (in press) (1996).
12. Henning, Th., et al. *A&A*291, 546 (1994).
13. Grady, C.A., et al., *A&AS* 120, in press (1996).
14. Vidal-Madjar, A. et al. in *Circumstellar Dust and Planet Formation*, eds. A. Vidal-Madjar and R. Ferlet, Paris: Editions Frontiéres, p. 7 (1995).
15. Waelkens, C. et al., *A&A*, in press (1996).
16. Sitko, M.L. et al., *From Stardust to Planetesimals*, ed. Y. Pendleton NASA CP, in press (1996).
17. Sitko, M.L. et al. 1994, *Ap.J.*432, 753 (1994).

Time Variations of Water Vapor Masers in Star-Forming Regions

J.E. Mendoza-Torres and E.E. Lekht

Instituto Nacional de Astrofisica Optica y Electronica
Apdo. Postal 51 y 216, Puebla, Pue. Z.P. 72000, Mexico.

Abstract. The results of observations of water vapor maser at 22.2 GHz made during of about 15 years are presented. The results show that anticorrelation between the fluxes of different spectral ranges as well as of different spectral features is a common phenomenon in Star-Forming Regions (SFR) masers. We interpret these results as indicating that partial saturation commonly occurs in SFR water vapor masers.

INTRODUCTION

Water vapor maser sources are frequently found in Star-Forming Regions (SFR). In this environment the maser condensations may undergo the action of dynamic phenomena like stellar wind and bipolar outflow (1, 2). Thus, the understanding of the time variations of maser emission represent a potential tool to have a better insight into the SFR dynamics.

The time variations of the observed maser radiation may be due basically to three different causes: a) Time variations of pumping b) Variations of the seed radiation c) Beaming.

a) The input energy for pumping may present time fluctuations. For example, it has been analyzed the possibility of modulation of the water vapor maser emission through variations of the inversion level by stellar wind (3) or shock waves (4). Another source of energy for pumping (and consequently of the maser emission variations) may be a background optical source (3). The possibility of maser emission fluctuations due to inhomogeneities in the inversion level, inside the vapor water condensation itself, also has been studied (5).

b) The maser emission may be just the amplification of a background source (6). Hence, the maser emission may undergo variations due to fluctuations of the intensity of the incident radiation.

c) The maser beam and the observed radiation intensity are close related (7). So, a relative change in the maser beam and the line of sight of the observer may lead to variations of the recorded flux. Moreover, spontaneous emission take place inside the condensation itself as well as in other condensations. The competition for pumping of this radiation with the radiation going out in the main beam may not be significant for high inversion levels but become important when there are low pumping rates.

The evolution of the observed flux shows two types of variations (8): short time enhancements (outbursts) and long-term fluctuations. The short time enhancements are usually denominated outburst since the observed flux rises suddenly (9). However, is not yet clear at which mere the observed flux variations are fluctuations of the output energy and at which mere - are the result of the beaming.

Another parameter to evaluate the maser evolution is the centroid velocity. The centroid velocity of the entire spectrum shows, as a rule, regular velocity drifts, which may be explained in some cases as a result of condensations in an expanding shell (10).

OBSERVATIONS

The observations were carried out with the 22-m Radiotelescope of the Radio Astronomical Station of the Lebedev Physical Institute in Pushchino, Russia at 22.2 MHz. The noise temperature was about 200 K and the beam width of 2.6 arc minutes. A 96 channel filter type spectrograph, which has a spectral resolution of 7.5 KHz (0.1 km/sec), was used. The time integration was about 15 minutes and an antenna temperature, for a point source, of 1K corresponded to a flux density of 25 Jy. The observations were made from the beginning of the eighties with intervals of about 1-2 months (observing all the sources each time).

DATA REDUCTION

The integrated flux was computed for the entire spectrum of each source and it was also computed for different spectral ranges depending on the variations observed in the spectra at different times (11). The centroid velocity was also computed for the entire spectrum and for different spectral intervals. To analyze separate features each spectrum was considered as the superposition of gaussian curves. For each curve we determine its velocity and flux.

RESULTS

Below we describe the main results of the observed water maser fluctuations of the sources: W31A (RA 18 07 30.3, Dec -19 56 38), W75S (RA 20 37 14, Dec 42 12 11), S128 (RA 21 30 37, Dec 55 40 36), ON1 (RA 20 08 10, Dec 31 22 39), and S252A (RA 6 05 36.5, Dec 20 39 34).

Most of the observed outbursts lasted from some months to about one year. In some cases two outbursts separated by about one year were observed. Also were observed a series of consecutively occurring outbursts (Fig.1).

As a rule, an outburst take place preferentially in a given spectral range of the spectrum. For example, during 1984-1987 were observed water maser outbursts in S128 in the velocity range $V < -75$ km/sec, while at velocities $-75 < V < -73$ km/sec the obsereved flux was very low.

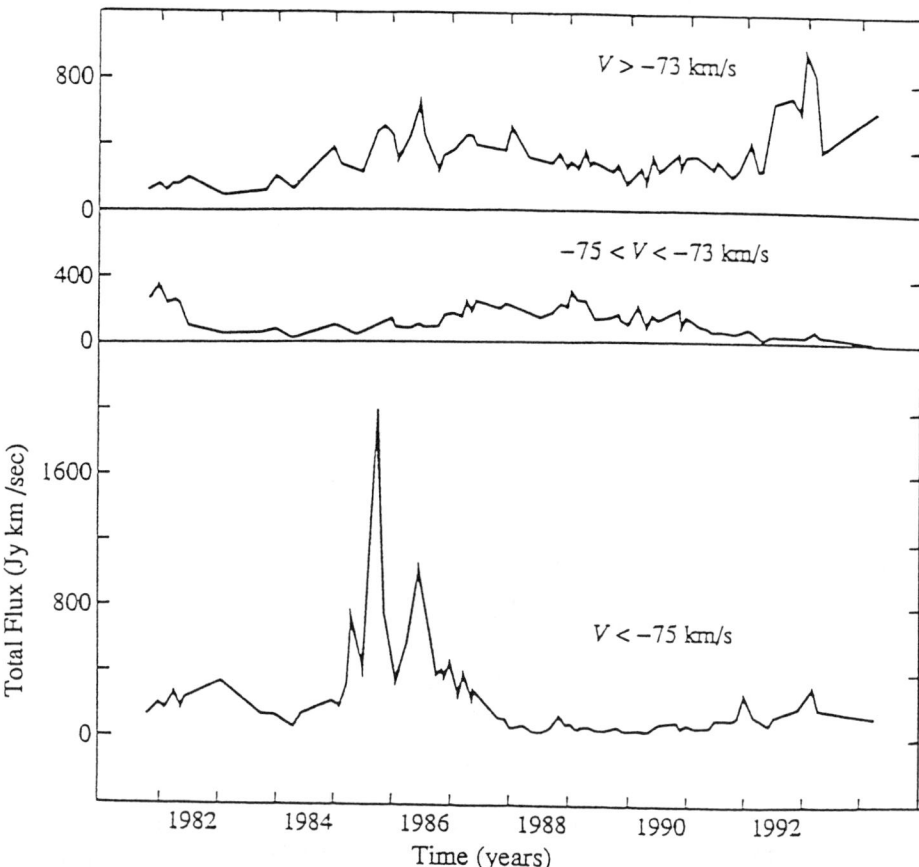

FIGURE 1. The total flux and the flux integrated over three different spectral ranges for the maser source S128.

In the ON2 water vapor maser the spectral features appeared in two spectral ranges. The integrated flux of these groups of features changed mostly in counterphase, including the period of outbursts. The flux of some ON2 spectral features also showed time variations in counterphase (Fig.2). Similar situations were observed in other sources.

The amplitude of the centroid velocity fluctuations for the entire spectrum, for most of the sources, are about 5-10 times the amplitude of the variations of the velocity for separate spectral features. This fact is possible due to the transmission of energy to the maser condensations in time dependence releases.

In the S252A spectrum two features approximated during an outburst. The initial difference in radial velocity was about 0.6 km/sec, and during the flux maximum the difference decreased to a value of 0.3 km/sec. The approach most likely was originated by partial saturation.

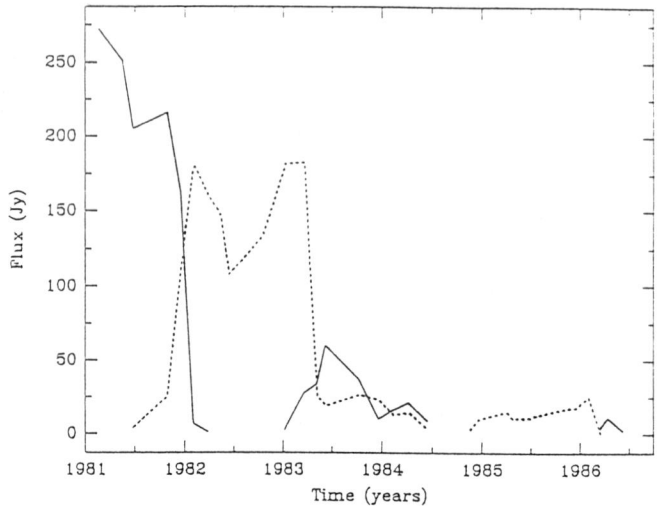

FIGURE 2. Curves for the flux of two neighboring features in the spectrum of ON2. Dashed line represents the curve for the spectral feature of velocity 4.1 km/sec while continued line the flux of the feature of 3.5 km/sec.

CONCLUSIONS

The variations appeared in counterphase when low fluxes were observed. At these times the pumping could not be enough to maintain high inversion levels. However, during the strongest outbursts the observed flux increased for wide spectral ranges. So, it seems that the input energy was large enough to maintain high inversion levels during such outbursts. We believe that the observed variations are real variations of the output energy from the maser condensations and since their large amplitude, most probably, they are due to variations of the input energy for pumping.

REFERENCES

1. Chernin, M.L. 1995, Ap.J. 444, L97
2. Felli, M., Palagi, F. and Tofani G. 1992, Astron. and Astrophys. 255, 293
3. Strelnitskii, V.S. 1974, Sov. Astron. 17, 717
4. Gomez-Balboa, A.M. and Lepine, J.R.D. 1986, Astron. and Astrophys. 159, 166
5. Elitzur, M. 1991, Ap.J. 370, L45
6. Kylafis, N.D. and Pavlakis, K.D. 1992, Ap.J. 400, 344
7. Elitzur, M. 1992, Ann. Rev. Astr. Astroph. 30, 75
8. Lekht, E.E., Mendoza-Torres, J.E. and Sorochenko, R.L. 1995, Ap.J. 443, 222
9. Matveenko, L.I. 1986, Sov. Astron. 63, 996
10. Lekht, E.E., Mendoza-Torres, J.E. and Sorochenko, R.L. 1995, Astron. Rep. 72, 34
11. Berulis, I.I., Lekht, E.E., and Mendoza-Torres, J.E. 1995, Astron. Rep. 39, 411

STAR FORMATION IN M16 WITH ADONIS AND HUBBLE

D. Currie*, D. Bonaccini†, K. Kissell*, E. Shaya*, P. Avizonis*, D. Dowling*

*Department of Physics, University of Maryland, College Park, MD 20742
†European Southern Observatory, D-85748 Garching bei Muenchen, Germany

Abstract. New details of the star formation process have been revealed by a coordinated use of the ADONIS system on the 3.6 meter telescope at La Silla with data from the WFPC2 Camera of the Hubble Space Telescope. In this very preliminary report, we illustrate some of the unique capabilities of the ADONIS system for high-resolution observations of the stellar formation processes in the near infrared region.

INTRODUCTION

These observations address M16 (NGC 6611, and the Eagle Nebula). This is a molecular cloud in which a small cluster of high mass stars has recently formed. The very high ultraviolet flux emitted by these early-type stars has dispersed or "photo-eroded" surrounding regions of the molecular cloud. Variations in the density of the gas and dust of the cloud have resulted in an uneven irregular erosion, forming the well known "elephant trunks" seen in the ground-based images. Investigation of the details of the resulting structure gives a measure of the resistance of the cloud to the photo-erosion process, which, in turn is a measure of the density of the various regions of the cloud. While other properties, such as the magnetic field and local temperature may also affect the rate of erosion, this may be one of the most direct methods of measuring the density profile about pre-protostellar objects.

This data was taken under ESO Proposal 57.C-0796 by D. Currie with K. Kissell of the University of Maryland and D. Bonaccini of ESO. Preliminary image processing has been conducted by E. Shaya, P. Avizonis, and D. Dowling of the University of Maryland. The analysis of this data will be conducted by this group, in collaboration with other members on the WPFC2 IDT team (i.e., J. Hester, P. Scowen and others) and other individuals at the University of Maryland. The observations described here were conducted in May, 1996 on

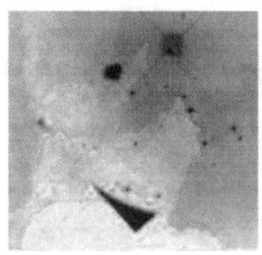

FIGURE 1. Observations of the region near Walker Star #367. This is a greyscale of the K band image, with a background image (a "ghost") from the HST Hα data superimposed for reference. A color image is available at *http://www.physics.umd.edu/rgroups/astro-metro/eagle.html*. This region has a large number of bumps where the photo-erosion process is slowed by positive density fluctuations of different sizes. The ADONIS image is 37" x 37" which represents four pointings of the 25.6" x 25.6" FoV of the SHARP II Camera. The resolution (before deconvolution) is about 280 mas FWHM for the K-band. This is within a factor of two of that which can be achieved in the visible on the Hubble Space Telescope. The region at the edge of the image is the annular mount which holds the beam-splitter, illustrating the low (300 K) temperature emission of this room-temperature object. The high-resolution information obtained by the dual use of Hubble and ADONIS allows a unique probe into the star forming region. Multiple systems are now resolved.

the 3.6 meter telescope at La Silla using the ADONIS adaptive optics system and the SHARP II NICMOS Camera.

OBJECTIVES

The scientific objectives of these observations is to provide observational data for the understanding of star formation processes. The following discusions describe areas of interest to our group at the University of Maryland and define the objectives of these observations, as well as the target of the succeeding analysis.

Density Profiles as a Function of Stellar Formation Stage. The dimensions of the features revealed in the Hubble data yield information related to the density profile of the in-falling dust and gas. For example, we can obtain a "characteristic size" related to the current stage of the formation process for a given object. The ESO data allows one to place the object within the normal classification schemes for star formation, i.e., from the classes described by Hillenbrand to the very early class described by Andre (1995).

Evidence of Pre-Main Sequence Objects in M16. M16 has long been in interesting region for the search for "pre-main sequence" objects. This work has been developed by Walker, et al. (1988), by Chini, et al. (1990), and by Hillenbrand, et al. (1993). In each case, the primary limitations have been resolution, (with overlap of objects due to the extreme crowding), and limiting magnitude in various bands. The combination of the increased resolution and the deeper exposures of the ADONIS data and the capabilities of Hubble would appear to permit a significant extension of the analysis of Hillenbrand. Figures 1 and 2 show far more objects than are visible in the HST data due to the increased infrared sensitivity of the ADONIS system.

Investigation of the Very Red source 367 B. About 10" away from Walker Star 367, there is a very strong infrared source which does not appear in our V filter data. Because far infrared data is needed in order to distinguish the early stage of the individual star formation regions, the program at the University of Maryland is currently using data which we have obtained using Hubble WFPC2 data, 350μm data from Caltech Submillimeter Observatory using the GSFC bolometer array, and the millimeter interferometric observations obtained with the Berkeley-Illinois-Maryland (BIMA) array in addition to the data from the observations at ESO.

It is critical to obtain the best resolution in order to understand the extended nature of the objects. To this end, we are using the Lucy algorithm for the deconvolution. This can remove the extended, one arc second skirt of the Adaptive Optics PSF due to the correction residuals, as illustrated in Figure 3. The presence in adaptive optic corrected images of an extended halo around the central high resolution object is typical. This has been observed and its cause well understood. Astronomers who are looking for faint structures underneath this skirt will have to familiarize themselves with deconvolution,

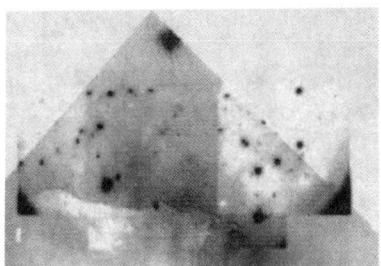

FIGURE 2. ADONIS Observations illustrate the region around the "TIP" of Column III. As in Figure 1, this figure is composed of both WFPC2 and ADONIS observations. The region contains a number of protrusions of different sizes and very high activity due to the proximity to the blue stars. Many of the objects detected in the ADONIS data are above and outside of the molecular cloud of the tip, indicating that they are not directly associated with the column. The outline of this image is 60" x 32" which consists of the data from six individual pointings in K band.

as well as take care that enough SNR is present in the observed object and structure.

REFERENCES

1. Andre, P.,*Astrophysics and Space Science* **24**, Pp. 29 (1995).
2. Chini, R., et al, *Astronomy and Astrophysics* **227**, Pp. 213 (1990).
3. Hillenbrand, L., et al, *The Astronomical Society of the Pacific* **100**, Pp. 1509-1521 (1988).
4. Hester, J., et al, *The Astronomical Journal* **111**, 6(1996).
5. Currie, D., et al, *In Proceedings, Topical Meeting on Adaptive Optics*, Garching bei Munchen, Germany, by the European Southern Observatory, October 2-6, 1995.

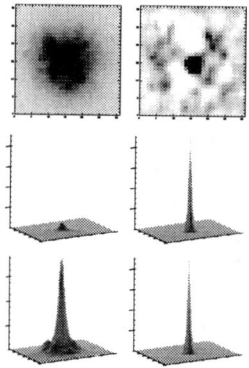

FIGURE 3. In order to utilize fully the higher frequency components available in the ADONIS data, we need to perform an image deconvolution and/or image reconstruction. For the present data, we use our Lucy-Richardson procedures developed for HST, modified for adaptive optics images (Currie, et al.1995). This allows an improved understanding and identification of the extended and double sources. The upper left figure is the image of a (nominally) point source in our K-band data set. The small patch is 3" square and the FWHM of the image from ADONIS is 280 mas. This same image is shown in a surface representation in the two lower left images. The deconvolved images are shown along the right side. In this case, the FWHM of the deconvolved image is 130 mas. The center image pair is normalized to illustrate the improvement obtained in the Strehl ratio (about a factor of 12, one third of the gain coming from the reduction of the width of the core and the remainder coming from the redistribution of the energy in the extended base of the image) The bottom image pair is normalized to illustrate the narrowing of the peak and the extended structure at the base of the ADONIS image. The wings out to about 1" are normal for an adaptive optics system. The sub-peaks are due to aliasing caused by the regular pattern of actuators in the deformable mirror. This same feature can be seen in the StarFire Optical Range data described in an earlier publication (Currie, et al., 1995).

On the Origin of Narrow, Very Long, Straight Jets from Some Newly Forming Stars

Howard D. Greyber
10123 Falls Road, Potomac, MD 20854, U.S.A.

Abstract: Observations have shown the existence of narrow, very long, straight jets emitted by some newly forming stars (1). It is highly likely that stars forming in the plane of a spiral galaxy do so in the presence of an almost uniform magnetic field. In the Strong Magnetic Field model (SMF), gravitational collapse of a highly conducting plasma in the presence of such a field will result in the formation of a stable, highly relativistic current loop (storage ring) around the central object. The concept was first described by Greyber (2-14). In the figures in Mestel & Strittmatter (15), one can see such a storage ring beginning to form. Such an increasing dipole magnetic field (formed temporarily for 10^4 to 10^6 years) will produce, accelerate and confine a narrow, very long, straight jet. When the density becomes too high, either the loop is destroyed, or the current-carrying plasma ring is buried inside the newly forming star and is the source of primordial stellar magnetism.

INTRODUCTION

The famed physicist Enrico Fermi introduced equipartition into astrophysics in the Forties, and did it right. He discussed shock waves, and obviously turbulence close to the shock made equipartition a reasonable assumption there. However astronomers soon forgot the caveat, i.e. that there was no reason to assume plasma turbulence producing equipartition must exist everywhere in astrophysics. Thus, the very small magnetic fields, often deduced and published, from assuming equipartition, are irrelevant in many situations.
  Actually equipartition applies only when the physics demands it does!
  Thus, for almost four decades, there has been a widespread misconception that "equipartition" between the particle energy and the magnetic field energy was absolutely necessary in most astrophysical situations. Very strong cosmic magnetic fields *are accepted* as real in white dwarfs and neutron stars. However until recently it has been alleged that strong magnetic fields could not exist in the cores of quasars and active galactic nuclei (AGN), nor in newly forming stars.
  For 35 years, the Strong Magnetic Field model, Greyber (2-14), has argued forcefully for what Chi and Wolfendale (16) wrote recently. "there is, however, no compelling justification for this assumption of equipartition". Originally created to explain spiral arms and answer Oort's famous questions, SMF has since been applied to the physical model of the central engine of AGNs, jet formation, galactic energetics and morphology, gamma ray bursts, etc.

The Strong Magnetic Field Model (SMF)

Greyber (17) has proposed an original model, within the Big Bang hypothesis, involving cosmical magnetism as well as gravitation, that explains the origin of large-scale primordial magnetic fields, as well as the origin of the observed highly structured nature (thin sheets of galaxies and voids) of matter in the Universe. Note that Pietronero et al (18) conclude, " - galaxy correlations are fractal and not homogeneous up to the limits of the available catalogues".

Kosowsky and Loeb (19) recently emphasized "The origin of the primordial field is still a subject of speculation. In the past, various indirect theoretical arguments were used to favor the dynamo amplification mechanism over the primordial origin alternative. However recent studies argue that a galactic dynamo should saturate due to the rapid growth of a fluctuating small-scale field before it can actually result in a coherent large-scale field of the type observed in galactic disks. The view that the galactic field may, in fact, be primordial gains additional support from observations of damped Lyman α absorption systems in QSO spectra at $Z \sim 2$. - - - The potential existence of a primordial magnetic field is also consistent with observations of clusters of galaxies. Faraday rotation measurements of radio sources inside and behind clusters indicate strong magnetic fields in many of them."

When one considers a galaxy or quasar forming by gravitational collapse of a giant, highly conducting, plasma cloud containing an almost uniform primordial magnetic field, SMF argues that a new physical construct, a storage ring, is created. Since the topology is very similar, the very same process occurs during the formation of a star under gravity. The gravitationally bound current loop, or *storage ring,* is extremely intense and highly relativistic. The bursting force of this very strong unified magnetic field system is in equilibrium, balancing the gravitational force between the slender toroidal plasma (bound to the current loop inside the toroid by the Maxwell "frozen-field" condition), and the central massive object.

The morphology and energetics of objects of galactic dimension are determined in SMF by the ratio of magnetic field energy to rotational energy in that particular object. The ratio is extremely high for quasars and blazars, and decreases steadily for giant elliptical and radio galaxies, Seyferts, Markarians, is low in ordinary spirals and is close to zero for the ordinary elliptical galaxies. However it is important to note that the AGN activity *we now observe* is a function of the accretion of matter into the central engine of the object.

A diagram of the SMF central engine for AGN is in references (10) and (14). It is the same for around a newly forming star, except that the central mass concentration is the protostar. High energy particles in this completely coherent, relativistic current loop store a significant fraction of the huge energy of gravitational collapse. The dipole magnetic field bound to the protostar probably contributes to the loss of angular momentum of the contracting protostar.

One can see such a storage ring forming in the figures in a brilliant, pioneering paper by Mestel and Strittmatter (15). They analyzed the effect of Ohmic diffusion on the magnetic field distribution of a gravitationally bound magnetic gas cloud, illustrating how the magnetic field topology changes as the cloud field detaches from the background field.

A storage ring, once formed, is *uniquely stable*. Due to coherence, one part of the loop does not radiate in the magnetic field of another part. However, if a fluctuation or "bump" occurs somewhere along the loop, the electrons in the "bump" will suddenly radiate furiously in the immensely strong local magnetic field, the energy in the fluctuation will dissipate rapidly, and the storage ring will return quickly to its undisturbed configuration. A relevant point is that the largest external perturbation to a storage ring is limited to solid matter, i.e. objects not much larger than the planet Jupiter, since stars, made of plasma, would break up before penetrating close to the intense magnetic field of the loop.

Klein and Brueckner (20) investigated the motion of a plasma under the action of an increasing magnetic field from a stationary coil. They found that the efficiency of conversion of stored energy into kinetic energy was about 5-10%. A similar or greater efficiency is then expected in SMF for the expulsion of the extremely high conductivity plasma from around a newly forming star, forming a jet. The increasing dipole magnetic field, as the contraction under gravity forming the star continues, accelerates and keeps the plasma jet narrow and confined for extremely long distances from the protostar.

The SMF model predicts that jets are formed by the successive emission of blobs of plasma from the AGN or stellar "central engine". Jets composed of blobs of plasama are just what is observed, both in jets from galaxies and quasars, and also observed as well in some newly forming stars like HH-30, which has been observed recently with HST.

CONCLUSION

The fact that jets are not well defined in many newly forming stars is understandable considering the relatively high density of the plasma in the vicinity of the protostar, and the disruption to the storage ring that may occur from binary or multiple protostars forming in close proximity. When the plasma density becomes too high, either the storage ring is destroyed, or, in some instances, the current-carrying plasma loop is buried inside the newly forming star and is the source of primordial stellar magnetism.

Evidence for the storage ring of current around some newly forming stars, with narrow, straight, long jets, hopefully will be found as observational resolution improves. This would be very important because it would validate the same topology as applied to the physics of the central engine of quasars and galaxies. So far, the SMF model appears to fit the observations.

REFERENCES

1. DeYoung, D., 1991, Science 252, 389
2. Greyber, H. D., Trans. of the I.A.U. XIB, 332; Report of Commission 33
3. Greyber, H. D., 1962, U.S.A.F.O.S.R. Research Report No. 2958, "On the Steady State Dynamics of Spiral Galaxies"
4. Greyber, H. D., 1963, Astron. J. 68, 536
5. Greyber, H. D., 1964, Chapter 31, in "Quasistellar Sources and Gravitational Collapse". ed. Ivor Robinson et al, University of Chicago Press (First Texas Symposium on Relativistic Astrophysics)
6. Greyber, H. D., 1967, in "Instabilitie Gravitationelle et Formation des Etoiles, des Galaxies, de Leurs Structures Caracteristique", Memoirs Royal Society of Sciences of Liege, XV, 189-196
7. Greyber, H. D., 1967, Publications Astron. Soc. of the Pacific, 79, 341
8. Greyber, H. D., 1984, 11th Texas Symposium, Annals of the New York Acad. of Sciences, 422, 353
9. Greyber, H. D., 1988 in "Supermassive Black Holes", ed. M. Kafatos, Kluwer Acad. Press, 360
10. Greyber, H. D., 1989, Comments on Astrophysics, 13, 201
11. Greyber, H. D., 1989, in "The Center of the Galaxy", ed. Mark Morris, Kluwer Acad. Press, 335
12. Greyber, H. D., 1990, 14th Texas Symposium, Annals of the New York Acad. of Sciences, 571, 239
13. Greyber, H. D., 1993, in "Compton Gamma Ray Observatory", A.I.P. Conference Proceedings 280, 569
14. Greyber, H. D., 1994, in "COSMICAL MAGNETISM Contributed Papers", edited by D. Lynden-Bell, Institute of Astronomy, Cambridge, England, 110-118, NATO Advanced Study Institute, in honour of Prof. L. Mestel, FRS
15. Mestel, L. & Strittmatter, P., 1967, M.N.R.A.S., 137, 95
16. Chi, X. & Wolfendale, A. W., 1993, Nature, 362, 610
17. Greyber, H. D., 1996, in "Clusters, Lensing and the Future of the Universe", A.S.P. Conf. Series Vol. 88, 298
18. Pietronero, L. et al, 1997, to be published in "Critical Dialogues in Cosmology", World Scientific, editor N. G. Turok
19. Kosowsky A. & Loeb, A., 1996, Ap.J. 469, no.1, 1
20. Klein, M. M. & Brueckner, K. A., 1960, Journal of Applied Physics, vol. 31, no.8, 1437

Gas to Dust ratios in Vega-excess stars

I.M. Coulson[†], D.M. Walther and W.R.F. Dent

Joint Astronomy Centre, 660 N.A. A'ohoku Place, Hilo HI 96720, U.S.A.
† *e-mail address   imc@jach.hawaii.edu*

Abstract. We report new infrared photometry of a sample of Vega-excess systems, and the results of a search for submillimeter CO emission. The resulting gas to dust ratios are compared with those for the class archetypes, and are found to be similarly reduced from the canonical molecular cloud value.

INTRODUCTION

Vega-excess systems are main-sequence stars with FIR continuum excess [1] arising from circumstellar dust. The archetypes (Vega, Fomalhaut, β Pictoris, ϵ Eridani, β Leonis) show no emission at CO:J=1-0 [12], J=2-1 or J=3-2 [6]. Most other stars in the category seem dustier [10], yet have little or no CO emission [13,5].

Here we report measurements of the dust in the NIR continuum, and the results of a search for submm CO emission. These yield gas to dust ratios that may be compared with those for the archetypes [12,6]. The apparent depletion of CO in these systems may be due to photodissociation [12] or, if there is a proportional depletion of H_2, to a mechanism related to the formation of a disk. The degree of depletion, as measured by the gas to dust ratio, may then provide a timescale for disk evolution in these systems.

OBSERVATIONS

Vega-excess candidates are identified from their (IRAS) far-infrared fluxes [11,10,13], and our sample is shown in Table 1. NIR photometry was obtained with IRCAM3 on UKIRT, and the CO:J=3-2 observations were made with JCMT; both telescopes are on Mauna Kea. Some candidates have already been eliminated because of nearby galaxies or NIR sources, and others show strong and/or extended CO emission that foil analysis. Here we present (V-K) or (V-L) colors from our new photometry, the dust parameters derived from the photometry, the CO fluxes, and, where applicable, the derived gas masses.

ANALYSIS OF THE PHOTOMETRY

We have fit blackbodies to the available optical, IR and submm photometry in a manner described elsewhere [5]. Where submillimeter photometry exists [10] the good fits support the use of unmodified blackbodies to represent the emission from the cold dust, and imply large (>1mm) grains. By contrast, the archetypes show emission deficits at 1mm that imply [4] grain sizes <<1mm. The NIR photometry often suggests a 'warm' ($\sim$1000K) body, perhaps due to very small particles being rapidly heated by the stellar radiation [8], and we have fit this when appropriate.

The fitting algorithm yields the temperature and effective surface area (πd^2_{dust}) of the dust grains. An assumption about the size of the dust particles yields their total mass, and we have adopted 1mm as the grain size.

CO COLUMN DENSITIES AND THE MASS OF GAS

We have detected CO J=3-2 lines from SAO 206462 and SAO 112630 and give upper limits for the remaining stars. The observed integrated line intensities, $\int T_{mb} dv$, are given in Table 1. The CO column densities and the total mass of gas follow assuming that the gas occupies the same space as the cold dust, and that the CO:H_2 ratio is 2×10^{-4}.

DISCUSSION

Gas to dust ratios in these Vega-excess systems are significantly less than 1, confirming the results for the archetypes [12,6]; the canonical molecular cloud value is 100. However, the systems studied here are significantly different from the archetypes :

- they have >10× the amount of cold dust

- their cold dust components comprise large (1mm) grains

- they often show near-IR excess

- some of them show CO emission.

If disks of Vega-excess systems are products of solar nebula evolution they should have ages of 10^7 to 10^8 yr [7], and so these differences might imply that the archetypes are more evolved. However, a dusty, gas rich system may persist if there are enough dust grains to shield the gas from photodissociation by stellar ultraviolet radiation [12].

The evolutionary link between Vega-excess and T Tauri systems is inviting. T Tauris are dustier than the systems examined here ($M_{dust} = 10^{-5}$ to $10^{-2} M_\odot$, [3]) and have gas to dust ratios between 10^{-3} and 10^1. While there is no clear correlation between the gas to dust ratio and age for T Tauri systems, there are also no Vega-excess systems with gas to dust ratios >1, suggesting a possible evolutionary sequence and a general decline of gas content with age.

An independent estimate of age is essential to define any such sequence. Both the position in the H-R diagram and the atmospheric lithium abundance may be unreliable indicators for such young stars [9]. In addition, emission lines may arise either from a disk or from the stellar chromosphere, confusing any interpretation of the observed Brγ emission from SAO 206462 [5] and the Hα emission from SAO 183986 and SAO 112630 [13].

If the disks are *not* an evolutionary feature the dust and lack of CO do *not* constrain their age. The primary age constraint for A-stars remains their main-sequence nature. Photodissociation can preferentially remove the CO [12] and this argument has been extended [6] to explain how CO replenishment could occur by the continuous infall of comets. For the two stars for which we have definitive CO measures, similar calculations yield about 150 comets/yr. A comet supply mechanism was also suggested [2] to explain the continued presence of small dust particles orbiting the archetypes. Large particles (>1mm) may survive throughout either scenario.

CONCLUSIONS

We have presented new NIR photometry and CO J=3-2 results for a sample of Vega-excess stars. Two definitive gas to dust ratios and 10 upper limits confirm the depletion of CO seen in the archetypes. The gas to dust ratios of Vega-excess systems may be correlated with their ages but this needs confirmation using a reliable independent age estimate.

REFERENCES

1. Aumann, H.H., *et al, ApJ* **278**, L23 (1984).

2. Becklin, E.E., Zuckerman, B., *Submillimetre Astronomy* eds Watt & Webster, Kluwer, Dordrecht, p147 (1990).
3. Beckwith, S.V.W., Sargent, A.I., Chini, R.S., Güsten, R., *AJ* **99**, 924, (1990).
4. Chini, R., *et al*, *A&A* **252**, 220, (1991).
5. Coulson, I.M., Walther D.M., *MNRAS*, **274**, 977, (1995).
6. Dent, W.R.F., *et al*, *MNRAS*, **277**, L25, (1995).
7. Lissauer, J.J., *ARA&A* **31**, 129, (1993).
8. Sellgren, K., *ApJ* **277**, 623, (1984).
9. Strom, K.M., Wilkin, F.P., Strom, S.E., Seaman, R.L., *AJ* **98**, 1444, (1989).
10. Sylvester, R.J., Barlow, M.J., Skinner, C.J., *Planetary Systems . . .*, (*Ap&Sp.Sci*, **212**), eds Burke *et al*, Kluwer, p261, (1994).
11. Walker, H., Wolstencroft, R.D., it PASP, **100**, 1509, (1988).
12. Yamashita, T., *et al*, *ApJ*, **402**, L65, (1993).
13. Zuckerman, B, Forveille, T., Kastner, J.H., *Nature* **373**, 494, (1995).

TABLE 1. NIR photometry, Dust measures and CO:J=3-2 data for Vega-excess stars

SAO	Sp	V-K (V-L)	dist pc	T_{dust} K	d_{dust} au	M_{dust} $M_\odot$	$T_{mb}dv$ K.km/s	M_{gas} $M_\odot$	$\dfrac{Gas}{Dust}$
12469	B9V		420	65	210	1.3×10^{-4}	>1.5		
20126	G5		27	60	5	8.7×10^{-8}	<2.0	$<7\times10^{-9}$	<0.08
21171	B7V	0.6	480	60	290	2.5×10^{-4}			
21775	B9V	(0.2)	150	55	52	8.3×10^{-6}	<1.1	$<4\times10^{-6}$	<0.5
21910	B5	1.2	650	60	290	2.6×10^{-4}			
21938	A3V	0.3	95	40	33	3.3×10^{-6}			
34421	B9	0.2	345	40	138	5.7×10^{-5}			
35963	B9	(-0.1)	230	60	70	1.4×10^{-5}	>0.5		
50172	B0V	(1.0)	985	80	395	4.7×10^{-4}	>1.5		
52339	A0	0.3	270	50	160	7.9×10^{-5}			
76159	B8V	-0.1	150	45	82	2.0×10^{-5}	<0.9	$<9\times10^{-8}$	<0.01
76945	A2		180	70	36	3.9×10^{-6}			
77144	A2		185	180	15	6.6×10^{-7}	<0.8	$<1\times10^{-6}$	<0.6
91022	A5V	(0.7)	62	80	3	2.7×10^{-8}	>0.5		
112630	G0	1.0	70	95	15	6.7×10^{-7}	0.8	4×10^{-8}	0.06
131926	A0	1.2	450	75	120	4.3×10^{-5}	<0.2	$<3\times10^{-7}$	<0.01
132384	A0	1.4	385	65	175	9.0×10^{-5}			
132389	A0	-0.3	320	55	225	1.5×10^{-4}			
132393	B9V	2.4	630	170	23	1.5×10^{-6}			
132483	B9Ve	1.9	360	220	25	1.9×10^{-6}	<0.6	$<1\times10^{-6}$	<0.6
134141	A0	1.3	280	80	56	9.4×10^{-6}	<1.1	$<8\times10^{-7}$	<0.09
147886	A1V		80	80	9	2.3×10^{-7}	<1.3	$<6\times10^{-8}$	<0.3
151723	A	-0.5	280	40	85	2.1×10^{-5}			
160017	B9IV	1.0	185	40	120	4.3×10^{-5}			
179815	K5V	3.4	20	160	2	1.2×10^{-8}	<0.6	$<4\times10^{-9}$	<0.3
183956	A7V		193	50	58	1.0×10^{-5}			
183986	G5V	2.9	75	95	12	4.3×10^{-7}	<0.3	$<1\times10^{-8}$	<0.03
206462	F8V	2.7	80	95	23	1.6×10^{-6}	0.9	5×10^{-8}	0.03

The Very Low End
of the IMF

Formation of Low Mass Stars and Brown Dwarfs

Douglas N.C. Lin*

Lick Observatory, University of California, Santa Cruz, CA 95064

Abstract.
Despite the recent identification of a few brown dwarfs and extremely low-mass stars, their abundance does not appear to be sufficiently large for them to be the major contributors of the Galactic halo potential. We examine here the formation of brown dwarf and low mass stars under three different conditions. We show that the first generation metal poor stars, formed from the fragmentation of the protogalactic clouds, are likely to be massive and short lived. The formation efficiency of brown dwarfs is also likely to be low in present-day molecular cloud cores. Finally, we show that the formation of brown dwarfs may require the fragmentation of protostellar disks which differs from the path of protoplanetary formation.

INTRODUCTION

The recent discoveries of planetary companions with masses up to a few times that of Jupiter marks a milestone in the search for extrasolar planetary systems (Mayor & Queloz 1995, Marcy *et al.* 1996). Today, there are over a dozen planets have been detected (Butler *et al.* 1996). In contrast, brown dwarfs and extremely low mass stars are less frequently sighted despite some intense searches (Mayor *et al.* 1992, Marcy & Butler 1994). The discovery of Gl 229 B (Nakajima *et al.* 1996, Kulkarni in this volume) and a few low-mass companions around stars in the recent ELODIE and CORAVEL surveys (Mayor *et al.* 1996) appears to be consistent with a general decline of the mass function in the brown-dwarf range (Basri in this volume). In the MACHO search, an important realization is that brown dwarfs are unlikely to contribute significantly to the potential of the Galactic halo (Alcock *et al.* 1996)

In this contribution, we address some theoretical issues associated with the formation of brown dwarfs and extremely low-mass stars. We discuss the physics of low mass star formation in three contexts. We show that in a metal deficient protogalactic cloud, brown dwarfs are unlikely to form in significant numbers. In the dense cores of molecular clouds, isolated low-mass stars may be formed as a consequence of ambipolar diffusion of the interstellar magnetic field. But the emergence of brown dwarfs requires the fragmentation of collapsing clouds and the avoidance of subsequent capture of other uncollapsed clouds. Finally, we show that the formation of brown dwarf companions around main sequence stars is also difficult to realize although it remains an unsolved issue.

FIRST GENERATION STARS

Our basic conjecture is that the first generation stars are formed within infalling protogalactic clouds (PGC's) (Lin & Murray 1992). During the collapse of a PGC, density inhomogeneities and velocity variations lead to shocks which heat the gas to the virial temperature (T_{vir}) of the galactic halo (Binney 1977; Rees & Ostriker 1977; White & Rees 1978). In order for a PGC to collapse, its cooling time scale, τ_c, must be shorter than the dynamical time scale, τ_d, on which it can contract. For PGC's with masses comparable to the Galaxy, this condition is satisfied when their characteristic length scale $D < 100$ kpc (Blumenthal et al. 1984).

Subsequent fragmentation of the PGC requires the growth of density inhomogeneities on a time scale, $\tau_g < \tau_d$. If the PGC is cold, gravitational instability causes perturbations with initial amplitude δ_0 to become nonlinear when the system collapses by a factor $\sim \delta_0^{2/3}$ (Hunter 1962). Starting with an initial size $D = 100$ kpc, the amplitude of the perturbations must be nearly nonlinear for gravitational instability to trigger fragmentation of the PGC as it contracts to a size of a few kpc.

In the limit that $\tau_c < \tau_d$, thermal instability can lead to the rapid growth of perturbations from infinitesimal δ_0 to nonlinear amplitudes (Field 1965). For T_{vir} in the range of a few 10^6 K, the dominant cooling mechanisms are bremsstrahlung and recombination processes (Dalgarno & McCray 1972) for which τ_c increases with temperature. In this case, any small temperature difference between the cooler perturbed regions and the background is amplified. Across the interface between the two-phase medium, differential cooling leads to a pressure gradient which induces gas flow from the hot background towards the cooler perturbed regions. The density enhancement in the cooler region further reduces τ_c from $\sim \tau_d$ and that of the background.

Fragmentation of a cloud requires instabilities for which the growth time scale, τ_g, increases with wavelength, λ. For thermal instability associated with local cooling, τ_g is independent of λ. During the growth of thermal instability, however, quasi hydrostatic equilibrium is maintained on scales smaller than that (l_s) traveled by sound on the time scale τ_c. Consequently, compact ($< l_s$), cool, dense regions contract as entropy is lost at a much faster rate than the background. Because τ_c increases with temperature, the interface separating the cool regions and the background retreats at an accelerating pace, leading to the growth of Rayleigh-Taylor instability, for which τ_g is an increasing function of the wavelength (Burkert & Lin in preparation). The acceleration of the interface increases as the temperature in the perturbed region decreases. The long-wavelength disturbances become stabilized when their growth time scales become long compared with that over which the acceleration of the interface is modified. Nevertheless, they grow to nonlinearity due to the Richtmyer-Meshkoff instability. As the temperature in the perturbed regions continues to plummet, l_s decreases rapidly to values below the length scale of the perturbed regions, such that most parts of the perturbed region cool without any significant change in density. Pressure balance is, however, enforced and Rayleigh-Taylor instability grows within l_s from the interface.

The cooling efficiency of Bremsstrahlung, recombination, and atomic hydrogen emission decrease rapidly below $\sim 10^4$ K. In metal-free PGC's, however, non-equilibrium recombination leads to the formation of a small amount of H^- ions which combine with neutral H to form H_2. Radiative emission by H_2 reduces the gas temperature (T) to $\sim 10^2$ K (Murray & Lin 1990). If $[Fe/H] > -3$, a lower T (~ 10 K) is attainable due to additional cooling by heavy elements (e.g. CI, CO, and grains). Equilibrium $T \sim 10^4$ K may be maintained in the presence of external heat sources. Once the final T is attained, pressure balance is re-established on all scales and the PGC becomes a two-phase medium with density contrast inversely proportional to the temperature difference. The residual halo gas (RHG) in the background remains at the virial temperature, with a density (n) such that its thermal energy is lost on a time scale $\sim \tau_d$. Energy is loss from the RHG through both radiative cooling and thermal conduction between the RHG and the cold clouds (McKee & Cowie 1977). At ~ 10 kpc, the energy balance implies $nT \sim 10^{3-4}$.

The cool dense clouds are pressure confined by the RHG, and so have similar nT. The RHG also exerts a drag on the motion of the clouds as they are accelerated by the gravity of the Galactic halo. The terminal speed of clouds with size L is $V_t \sim (f_n L/D)^{1/2} V_k$ where f_n is the density ratio of the clouds to the RHG, and V_k and D are the velocity dispersion and size of the halo, respectively. The motion of the clouds through the RHG also leads to mass loss due to the Kelvin-Helmholtz instability (Murray *et al.* 1993), whose growth time scale (τ_{KH}) is a few times L/V_t. Because τ_{KH} increases with λ, the KH instability leads to fragmentation. The break down of the clouds increases their collective area filling factor and collision frequency. A balance between disruption and coagulation establishes an equilibrium size distribution.

A lower limit on their size distribution is set by the clouds' evaporation by the hot RHG. In the high mass limit, the clouds' self-gravity increases the central density and suppresses the Kelvin-Helmholtz instability. But at a critical mass $M_c \sim T^2/(nT)^{1/2} M_\odot$, thermal pressure can no longer support the weight of the envelope (Bonner 1956), and the clouds undergo inside-out collapse (Shu 1977). During the collapse, although the Jean's mass decreases with density, it is larger than the mass contained inside any radius. The collapse is stable and does not lead to fragmentation without any further unstable cooling. Thus, contrary to the opacity-limited fragmentation scenario (Hoyle 1953; Low & Lynden-Bell 1976), M_c represents the minimum mass for isothermal collapsing clouds (Tsai, in preparation).

In a metal-free environment, $T \sim 10^2$ K and $M_c \sim 10 - 10^2 M_\odot$. The resulting massive stars are, however, copious sources of UV radiation. A population of $\sim 10^4$ O5 stars is adequate to photoionize the entire PGC out to 100 kpc. Photoionization raises $T \sim 10^4$ K and $M_c \sim 10^6 M_\odot$. Small (a few $M_\odot$) heated clouds are stable and star formation is quenched. As the massive stars evolve off the main sequence, the UV flux diminishes, cooling again leads to $T \sim 10^2$ K in some sheltered regions, and spontaneous star formation is resumed. This self-regulated star formation scenario has three implications: 1) Stars formed in a metal-poor environment are massive and short-lived, consistent with their rarity today. 2) The elemental abundance distribution are produced by type II supernovae, consistent with that observed among stars with [Fe/H]< -1 (Wheeler et al. 1989). 3) The self-regulated star formation rate naturally contaminate the PGC with [Fe/H] ~ 0.1 on the collapse time scale $\tau_d \sim 1$ Gyr, consistent with the observed metallicity of the halo stars. Only with the present metallicity of the globular clusters, the cold dense clouds can cool to ~ 10 K so that M_c may be reduced to $0.1 M_\odot$. Thus, we do not anticipate the prolific production of brown dwarfs in the outer halo of the Galaxy and we do not expect most of the dark matter out there to be made of substellar objects.

PRESENT-DAY STAR FORMATION

Recent observations find that most stars in the Galaxy today form in clusters in which the time scale for star formation is remarkably short (Lada et al. 1991; Lada 1992). The central density of some young clusters are comparable to that of some globular clusters. In the Trapezium cluster, all the stars appear to have an age $< 10^6$ yr (Prosser et al. 1994) which is comparable to the crossing timescale (τ_{dc}) of the cluster. These time scales are consistent with the model in which star formation proceeds through a sequence of initial gas fragmentation, coagulation, protostellar collapse, and the clearing of residual gas. (There are exceptions such as IC 348, where star formation has persisted for many τ_{dc}, Lada & Lada, 1995).

Molecular clouds are clumpy on all scales (Scalo 1985). Complex cloud substructure may also be inferred from the large dispersion, over a small field, in the observed extinction of the stars in the background cluster IC 5146 (Lada et al. 1994). (The extinction is equivalent to the surface density of the intervening clouds). These observations suggest that fragmentation occurs in the clouds prior to the gravitational collapse of individual protostellar cores as we have postulated above.

Magnetic fields, neglected in our first generation star formation scenario, are observed to be important in regulating the structure of star forming regions today. In these regions, the velocity dispersions, which is correlated with the length scale of the substructures (Larson 1981), are often larger than the sound speed inferred from the transition temperature, but are comparable to the Alfvén speed (Heiles et al. 1993; Caselli & Myers 1995). In some regions, the dispersive motion of the clouds may be regulated by the interstellar magnetic fields. There are also, however, magnetic supercritical regions, where the magnetic field can no longer balance gravity. These are the regions where massive stars and small clusters are formed. The lack of polarization in the densest cores of molecular clouds (Goodman et al. 1995) suggest that magnetic fields may be excluded from these regions. The decoupling of the field from the protostellar clouds is equivalent to the loss of thermal support during a cooling instability, and may also lead to complex substructures (Terquem & Lin in preparation).

The mass function of dark cloudlets in star forming complexes such as Ophiuchus, Taurus, Orion (Scalo 1985), and L1630 (Lada et al. 1991) has a similar power-law distribution which is flatter than that of the stellar IMF. The extrapolated collisional time scale for the small cloudlets is comparable to a few local dynamical time scales, and so their size distribution could arise naturally from a collisional equilibrium. The relatively flat mass spectrum is then consistent with that obtained from numerical simulations of the coagulation processes among protostellar cores (Murray & Lin 1996), and implies that the physical cross section determines the merger rate among the cloudlets.

During their motion through the ambient medium, the cloudlets lose mass through ram pressure stripping and gain mass through coagulation. In a collisional equilibrium, a power-law size distribution is established. At the upper limit of the mass spectrum, the cloudlets' masses exceed M_c and they collapse to form protostellar cores. Since no additional fragmentation is expected, the minimum stellar mass is $\sim M_c$. Based on the typical observed values of the density, temperature, and field strength, we estimate M_c to be a fraction of $M_\odot$ such that very low mass brown dwarfs are unlikely to be formed.

Protostellar cores can continue to acquire additional mass as they merge with residual cloudlets. The observed number of stars in the range m to $m + dm$ is usually approximated as $\frac{dN_*}{dm} \propto m^{-(1+x)}$. Such a power-law mass distribution is expected in the idealized coagulation models (Nakano 1966; Kwan 1979; Silk & Takahashi 1979). Approximate asymptotic solutions of the coagulation equations indicate $x \approx 0$ if the collisional cross-sections are determined by the geometric cross-sections, whereas $x \approx 1$ if encounters are strongly gravitationally focussed. These values of x encompass much of the observed range for both open and globular clusters (Salpeter 1955; Miller & Scalo 1979; Scalo 1986; Francic 1989; Capaccioli, Ortolani, & Piotto 1991).

Numerical simulations of the coagulation of cloudlets in a protocluster cloud

(PCC) (Murray & Lin 1996) shows that the geometric cross section determines the collisional frequency between uncollapsed cloudlets leading to a relatively flat mass distribution. But the protostellar cores have compact sizes and their capture rate of residual cloudlets is strongly affected by gravitational focusing. In this case, the growth time scale is a decreasing function of the mass. A few massive cores rapidly grow prior to any significant mass increase among most other cores, leading to a steep initial mass function (IMF) for the protostellar cores. Collisions between the residual cloudlets and protostellar cores lead to the growth of the cores' mass and the dissipation of their relative motion, while global instabilities increase their velocity dispersion (Aarseth *et al.* 1988). In the final stage of a PCC's collapse, the velocity dispersion increases more rapidly than does the mass of protostellar cores, and the collisional frequency is once again determined by the geometrical cross section of the cloudlets, leading to a flattening of the IMF during the collapse. The final slope of the IMF is determined by whether or not most cloudlets have undergone collisions to form protostellar cores before gravitational focusing ceases to be important.

The growth of the most massive cores is terminated when their UV emission heats and ionizes their nearby cloudlets. The most massive stars require many dissipative mergers, and so are preferentially formed in the cluster center. This expectation is in contrast to the implication of opacity-limited fragmentation scenario, in which the Jean's criterion suggests that the high mass stars are preferentially formed in low density environments. The UV flux from these massive stars is sufficiently intense to photoionize the residual gas throughout a PCC. In the shallow potential of PCC's, the resultant internal heating leads to the formation of an expanding ionization wave which is efficient in clearing out the residual gas within a few 10^6 yr (Tenorio-Tagle *et al* 1986).

Unless more than half of the gas in PCC's is converted into stars, the removal of the gas on a time scale comparable to the dynamical time scale rapidly reduces the depth of the cluster's potential. In most cases, the disposal of residual gas would then lead to the disruption of the cluster (Lada *al.* 1984). If, however, young stars form through the coagulation of low mass cloudlets, energy dissipation resulting from the mergers leads to the resulting star clusters having radii much smaller than the original PCC's. It is therefore more likely that the newly formed clusters will remain gravitationally bound even in the limit of inefficient star formation. Numerical simulations show an order of magnitude reduction in the half mass radius of the star cluster formed through coagulation of a system of collapsing cloudlets (Murray & Lin 1996).

Shortly after the formation of protostellar cores, they are surrounded by the residual gas and appear as embedded sources. IR observations indicate that the embedded sources are strongly clustered (Greene *et. al.* 1994). These clusters are much more centrally condensed than the host cloud complex. Furthermore, the brightest embedded sources are usually found at the center of the clusters (Lada 1992). The luminosity segregation is consistent with the concept that protostellar cores form through dissipative mergers of small cloudlets, such that the most massive stars preferentially form in regions of high density where the collision frequency is greatest.

Finally, in older star forming regions, the young stellar objects emerge as T Tauri stars. In the Orion Nebula, T Tauri stars also appear to be clustered. The higher luminosity T Tauri stars are more centrally condensed than those with low luminosity (Prosser et al. 1994). In these regions, there is not sufficient time for post-formation dynamical evolution toward mass segregation. These observations are consistent with the more massive stars forming in dense, central regions. Since most of their kinetic energy is dissipated during the coagulation, these massive stars remain near PCC's center after the residual gas is cleared.

BROWN DWARFS AND PLANETARY COMPANIONS IN PROTOSTELLAR DISKS

We now turn our discussion to the formation of brown dwarf companions in protostellar disks. Recent observations indicate that typical star-forming dense cores have specific angular momentum $> 10^{21}$ cm^2 s^{-1} (Goodman et al 1993) such that their collapse leads to the formation of rotationally supported disks analogous to the primordial solar nebula (Terebey et al. 1984). During its collapse, gas around one young stellar objects (YSO), HL Tau, is observed (Sargent & Beckwith 1987, Hayashi et al. 1993) to have an adequate amount of angular momentum to form a 50-100 AU disk (Lin et al. 1993). This inner disk has been resolved (Lay et al. 1994, Mundy et al. 1996).

Around the classical T Tauri stars (CTTS), the presence of AU-size disks is inferred from the spectral energy distribution over more than 2 orders of magnitude in the IR wavelength. The observed power index is consistent with that expected from either the viscous dissipation of accretion disks with mass transfer rate $\dot{M} \sim 10^{-7} M_\odot$ yr^{-1}, or the reprocessed stellar radiation emitted by circumstellar geometrically-thin opaque disks (Adams et al. 1987).

The common existence of protostellar disks around YSOs, with properties similar to those of the solar nebula, suggests that the necessary conditions for planetary formation are generally satisfied. If their formation is a robust process, planets would be ubiquitous in nature. In the past year, several short-period low-mass and long-period eccentric planets have been discovered (Butler, Marcy, & Williams 1996).

In the conventional planetary formation models (Pollack et al. 1996), the first stage of protoplanetary formation is the rapid build up of solid cores through the coagulation of planetesimals (Safronov 1969, Wetherill & Stewart, 1989). After acquiring a fraction of an Earth mass ($M_\oplus$), these cores begin to accret gas and to attain a quasi static atmosphere which is heated by the bombardment of solid particles onto the cores. When the cores' mass M_{core} increases above a critical value ($M_{\text{crit}} \sim$ a few $M_\oplus$), the planetary atmosphere becomes unstable, undergoes collapse, and the cores dynamically accret the gas (Mizuno 1980, Bodenheimer & Pollack 1986).

The protoplanet's accretion rate ($\dot{M}$) increase rapidly with M_p. For a planet with $M_p \sim M_J$ (the mass of Jupiter), $\dot{M} \sim 10^{-6} M_\odot$ yr^{-1}. During this phase of rapid growth, the protoplanet also exerts a tidal perturbation on the disk (Papaloizou & Lin 1984). In the limit that the disk has a relatively low viscosity (ν), the protoplanet induces the formation of a gap in the disk near its orbit when its Roche radius ($r_R = (M_p/3M_*)^{1/3} a$ where M_* is the mass of the central star) increases beyond the disk thickness H (Lin & Papaloizou 1993). After the gap formation, M_p no longer increases since the disk gas cannot reach the protoplanet. At $a < 0.1$ AU, the acquisition of $1 M_J$ requires $H \sim 0.1a$ and a midplane temperature (T_c) $> 2 \times 10^3$ K. Since refractory material cannot condense at such high T_c (Palme & Boynton 1993), the short-period planets are unlikely to be formed *in situ*.

Probably all types of planets were formed several AU away from their host stars through the processes outlined in the standard scenario (Lin et al. 1996). After the termination of their growth through gap formation, protoplanets continue to tidally remove/supply angular momentum from/to the disk interior/exterior to the gap (Goldreich & Tremaine 1980, Lin & Papaloizou 1986a). Since gas flow across the gap is quenched, both the surface density (Σ) and the flux of tidal transfer of angular momentum ($\dot{J}_t$) interior to the gap decrease as gas diffuses toward the star. In the outer disk, Σ and $\dot{J}_t$ maintain their value due to the prevention of the inward viscous diffusion by the protoplanet's tidal torque. The imbalance of $\dot{J}_t$ between the inner and outer disk leads to the inward orbital migration of the protoplanet. If $\Sigma >> M_p/a^2$, the orbital migration of the protoplanet would be coupled to the viscous evolution of the disk (Lin & Papaloizou 1986b, Takeuchi et al. 1996).

The truncation condition can be violated if the companion has a large orbital eccentricity (e). Large e may be excited by the tidal interaction between the protoplanetary disk and relatively massive ($> 10^{-2} M_\odot$) companions. In this limit, the gap may be sufficiently wide to exclude the e-damping contribution from corotation resonances and large e may also be excited by the tidal interaction through Lindblad resonances (Goldreich & Tremaine 1979, 1980, Artymowicz 1993, Lin & Papaloizou 1993, Artymowicz & Lubow 1996). The large radial excursion associated with a large e enables the protoplanet to overrun the gap and accret additional amount of mass. Mazeh et al. (1996) suggest that these companions may have $M_p >> M_J$ and they belong to a brown dwarf population. But in the tidal potential of the binary system, mass in the outer disk regions is primarily accreted onto the companion. The rapid acquisition of additional disk mass has a tendency to quickly increase M_p beyond the brown dwarf limit. Thus, the gap overrun process also leads to the reduction of low-mass stars.

Note that in the low-mass limit ($M_p <$ a few M_J), the companion-disk tidal interaction leads to eccentricity damping such that the gap formation criterion is well preserved. Thus, a companion embedded in a protoplanetary disk cannot acquire a mass in excess of a few M_J unless $H \gg 0.1a$. Large H/a requires the midplane temperature to exceed that due to the stellar irradiation. Viscous dissipation can provide the necessary heating source to sustain $H/a > 0.2$, but it requires a relatively large $\dot{M}$ and Σ. In this limit, the disk may become gravitationally unstable. In principle, gravitational instability may lead to the growth of non axisymmetric perturbations and the formation of companions (Adams et al. 1989, Shu et al. 1990). But, numerical simulations indicate that during the growth of non axisymmetric disturbances, a strong tidal torque induces the redistribution of mass and angular momentum and places a limit on their amplitude (Laughlin & Rozyczka 1996). The growth to nonlinearity is particularly problematic in optically thick regions of the disk where the cooling timescale is longer than the dynamical timescale.

Despite these difficulties in the formation of companions through gradual growth or gravitational instabilities in protostellar disks, binary T Tauri stars are commonly found (Ghez et al. 1993). The most likely mechanism for their formation may be rotational fragmentation of a collapsing protostellar cloud when it is first settled into an optically thin disk (Bodenheimer 1995). The most natural outcome of such fragmentation is the formation of binaries with similar masses. The formation of systems with extreme mass ratio requires the preferential collapse around the more massive component which also has a tendency to stablize the growth of the less massive component. Nevertheless, recent numerical simulations (Burkert & Bodenheimer 1996) indicate that multiple systems with extreme mass ratio may be formed but they have a general tendency to merge as a consequence of their interaction with both the infalling envelope and the circumbinary disk. Unless the infall rate decreases and the disk depletes shortly after their formation, the low mass companions may either grow rapidly or merge with the massive components. Thus, it is difficult not only to form but also to retain brown dwarf as companions. These obstacles may account for the relative rarity of brown dwarfs.

We thank Drs. P.H. Bodenheimer, G. Bryden, A. Burkert, G. Marcy, S.D. Murray, & F. Shu for useful conversations. This work is supported by NSF and NASA through grants AST-9315578 and NAGW-4967.

REFERENCES

1. Aarseth, S.J., Lin, D.N.C., and Papaloizou, J.C.B., *Ap.J.* **324**, 288 (1988).
2. Adams, F.C., Lada, C.J., and Shu, F.H., *Ap.J.* **312**, 788 (1987).
3. Alcock, C., Allsman, R.A., Axelrod, T.S., and other *Ap.J.* **461**, 84 (1996).
4. Adams, F.C., Ruden, S.P., and Shu, F.H., *Ap.J.* **347**, 959 (1989).
5. Artymowicz, P., *P.A.S.P.* **105** 1032 (1993).
6. Artymowicz, P., and Lubow, H., *Ap.J.Lett.* **467**, L77 (1996).

7. Binney, J.J., *Ap.J.*, **215**, 483 (1977).
8. Blumenthal, G.R., Faber, S.M., Primack, J.R., and Rees, M.J., *Nature* **311**, 517 (1984).
9. Bodenheimer, P.H. *Ann.Rev.A.A.* **33**, 199 (1995).
10. Bodenheimer, P. and Pollack, J.B., *Icaurs* **67**, 391 (1986).
11. Bonner, W.B., *MNRAS* **116**, 356 (1956).
12. Butler, R.P., Marcy, G.W., and Williams, E., *Ap.J.* in press (1996).
13. Capaccioli, M., Ortolani, S., and Piotto, G., *A.A.* **244**, 298 (1991).
14. Caselli, P. and Myers, P.C., *Ap.J.* **446**, 665 (1995).
15. Dalgarno, A., and McCray, R.A., *ARAA* **10**, 375 (1972).
16. Field, G.B., *Ap.J.* **142**, 531 (1965).
17. Francic, S.P., *A.J.* **98**, 888 (1989)
18. Ghez, A.M., Neugebauer, G., and Matthews, K., *A.J.* **106**. 2005 (1993).
19. Goldreich, P., and Tremaine, S., *Ap.J.* **233**, 857 (1979).
20. Goldreich P., and Tremaine, S. *Ap.J.* **241**, 425 (1980).
21. Goodman, A.A., Benson, P.J., Fuller, G.A., and Myers, P.C., *Ap.J.* **406**, 528 (1993).
22. Goodman, A.A., Jones, T.J., Lada, E.A., and Myers, P.C., *Ap.J.* **448**, 748 (1995).
23. Greene, T.P., Wilking, B.A., André, P., Young, E., and Lada, C.J., *Ap.J.* **434**, 614 (1994).
24. Hayashi, M., Ohashi, N., and Miyama, S.M., *Ap.J.Lett.* **418**, L71 (1993).
25. Heiles, C., Goodman, A.A., and McKee, C.F. in *Protostars and planet III*, eds. E. H. Levy and J. I. Lunine Tucson: Univ. Arizona Press, 279 (1993).
26. Hoyle, F., *Ap.J.* **118**, 513 (1953).
27. Hunter, C., *Ap.J.* **136**, 594 (1962).
28. Kwan, J., *Ap.J.* **229**, 567 (1979).
29. Lada, E.A., *Ap.J.Lett.* **393**, L25 (1992).
30. Lada, E.A., DePoy, D.L., Evans, N.J., and Gatley, I., *Ap.J.* **371**, 171 (1991).
31. Lada, E.A., and Lada, C.J., *A.J.* **109**, 1684 (1995).
32. Lada, C.J., Lada, E.A., Clemens, D.P., and Bally, J. *Ap.J.* **429**, 694 (1994).
33. Lada, C.J., Margulis, M., and Dearborn, D., *Ap.J.* **285**, 141 (1984).
34. Larson, R.B., *MNRAS* **194**, 809 (1981).
35. Laughlin, G., and Rozyczka, M., *Ap.J.* **456** 279 (1996).
36. Lay, O.P., Carlstrom, J., Hills, R.J., and Phillips, T.G., *Ap.J.Lett.* **434**, L75 (1994).
37. Lin, D.N.C., Bodenheimer, P.H., and Richardson, D., *Nature* **380**, 607 (1996).
38. Lin, D.N.C., Hayashi, M., Bell, K.R., and Ohashi, N. *Ap.J.*, **435**, 821 (1994).
39. Lin, D.N.C., and Murray, S.D., *Ap.J.* **394**, 523 (1992).
40. Lin, D.N.C., and Papaloizou, J.C.B., *Ap.J.* **307**, 395 (1986a).
41. Lin, D.N.C., and Papaloizou, J.C.B., *Ap.J.* **309**, 846 (1986b).
42. Lin, D.N.C., and Papaloizou, J., *Protostars and Planets III*, eds. G.H. Levy, J.I. Lunine, Univ. Arizona Press: Tucson, 749 (1993).
43. Low, C., and Lynden-Bell, D. *MNRAS* **176**, 367 (1976).
44. Marcy, G.W., Bulter, R.P., Williams, E., Bildsten, L., and Graham, J., *Ap.J.*

preprint (1996).
45. Marcy, G.W., and Bulter, R.P., *Ap.J.Lett.* **464**, L147 (1996).
46. Mayor, M., Duquennoy, A., Halbwachs, J.L., and Mermilliod, J.C., *Complementary approaches to double and multiple star research*, eds W. Hartkopf & H.A. McAlister, 73 (1992).
47. Mayor, M., and Queloz, D., *Nature* **378**, 355 (1995).
48. Mayor, M., Queloz, D., Udry, S., and Halbwachs, J.L., preprint (1996).
49. Mazeh, T., Latham, D., and Mayor, M., *Ap.J.* in press (1996).
50. McKee, C.F., and Cowie, L.L., *Ap.J.* **215**, 213 (1977).
51. Miller, G. E., and Scalo, J.M., *Ap.J.Suppl.* **41**, 513 (1979).
52. Mizuno, H., *Prog.Theor.Phys.* **64**, 544 (1980).
53. Mundy, L., Looney, L., Erickson, W. *et al.*, *Ap.J.Lett.* **464**, L16 (1996).
54. Murray, S. D., and Lin, D. N. C., *Ap.J.* **467** 728 (1996).
55. Murray, S.D., White, S.D.M., Blondin, J.M., Lin, D.N.C., *Ap.J.* **407**, 588 (1993).
56. Nakajima, T., Oppenheimer, B.R., Kulkarni, S.R., Golimowski, D.A., Mathews, K., Durrance, S.T., *Nature* **378** 463 (1995).
57. Nakano, T., *Prog. Theor. Phys.* **36**, 515 (1966).
58. Palme, H., and Boynton, W.V., *Protostars and Planets III*, eds. G.H. Levy, J.I. Lunine, Univ. Arizona Press: Tucson, 979 (1993).
59. Papaliozou, J. and Lin, D.N.C., *Ap.J.* **285**, 818 (1984).
60. Pollack, J.B., Hubickyj, O., Bodenheimer, P., Lissauer J.J., Podolak, M., and Greenzweig, Y., *Icarus*, in press (1996).
61. Prosser, C.F. *et al.*, *Ap.J.* **421**, 517 (1994).
62. Rees, M.J., and Ostriker, J.P., *MNRAS* **179**, 541 (1977).
63. Safronov, V., *Evolution of the Protoplanetary Cloud and Formation of the Earth and Planets* Nauka: Moscow, (1969).
64. Salpeter, E. E., *Ap.J.* **121**, 161 (1955).
65. Sargent, A.I., and Beckwith, S., *Ap.J.* **323**, 294 (1987).
66. Scalo, J.M., *Protostars and Planets II*, eds. D.C. Black and M. Matthews, Univ. of Arizona Press: Tucson, 201 (1985).
67. Scalo, J.M., *Fundam. Cosmic Phys.* **11**, 1 (1986).
68. Shu, F., *Ap.J.* **214**, 488 (1977).
69. Shu, F.H., Tremaine, S., Adams, F.C., Ruden, S.P., *Ap.J.*, **358**, 495 (1990).
70. Silk, J., and Takahashi, T., *Ap.J.* **229**, 242 (1979).
71. Takeuchi, T., Miyama, S., and Lin, D.N.C., *Ap.J.* 460, 832 (1996).
72. Tereby, S., Shu, F.H., and Cassen, P., *Ap.J.* **286**, 529 (1984).
73. Tenorio-Tagle, G., Bodenheimer, P., Lin, D.N.C., and Noriega-Crespo, A., *MNRAS* **221**, 635 (1986).
74. Wetherill, G.W., and Stewart, G.R., *Icarus* **77**, 330 (1989).
75. Wheeler, J.C., Sneden, C., and Truran, J.W., *ARAA* **27**, 279 (1989).
76. White, S.D.M., and Rees, M.J., *MNRAS* **183**, 341 (1978).

EARLY HINTS ON THE SUBSTELLAR MASS FUNCTION

Gibor Basri and Geoffrey W. Marcy[†]

Astronomy Dept., Univ. of California, Berkeley, California 94720
[†] *also at San Francisco State University*

Abstract. The study of substellar objects, which had no solid empirical basis as little as two years ago, has now achieved remarkable progress. Three indisputable brown dwarfs have been discovered, and six Doppler companions to solar-type stars appear very convincingly to be extrasolar planets. A number of other Doppler companions are almost certainly substellar. We begin by defining the classes of substellar objects on a physical basis. Then we discuss the discoveries of the past two years, and the methods which led to success. We make a preliminary estimation of the substellar mass function, based on the Doppler results and the cluster searches. We use these to predict the success rate of various continuing searches. We must emphasize that these predictions are based on very few objects, and are likely to change in the coming years. Nonetheless, the number of substellar objects is already reasonably constrained, and it is unlikely that they are a major constituent of the baryonic dark matter.

INTRODUCTION

We know that Nature prefers to make low mass stars over high mass stars, though we don't know exactly why. It has been a longstanding mystery how far this preference extends - whether and where the initial mass function (IMF) turns over at very low masses. Does it continue to rise into the substellar domain, first to brown dwarfs and then even to planets? We believe that planet formation occurs in protoplanetary disks, while star formation is a result of molecular cloud fragmentation and collapse. But it turns out that disks are intimately involved in star formation. Is there any real difference between the two modes, and if so, how is it manifested in the IMF? Finally, if there are large enough numbers of substellar objects, they could provide a major constituent of the baryonic dark matter. How many are there? These are some of the reasons for strong interest in the substellar mass function.

Let us first define what we mean by "brown dwarf" (BD) and by "planet". These terms may seem straightforward, but there is some confusion at the boundary between BDs and stars, and much confusion at the boundary between brown dwarfs and planets. The definitions offered here are not accepted by everyone, but we feel they are clear and physically based. By "brown dwarf" we mean an object which in its youth produces a non-negligible luminosity by nuclear fusion, but which never stabilizes its luminosity by hydrogen burning. The reaction engaged in by all BDs is deuterium burning, which occurs early in the evolution of the object. The highest mass BDs also engage in some hydrogen burning for a period of time (in that way they are similar to stars). Fusion may even dominate over gravitational contraction as a source of luminosity early on, but the crucial point is that the luminosity derivative will never be zero or positive after deuterium burning is finished. The object will not reach a stable luminosity or temperature and will continually grow cooler and dimmer. At later stages, gravitational contraction is the only power source.

It should be noted that the lowest mass stars, while they eventually stabilize their luminosity with hydrogen burning, do not do so for a few billion years. Thus they cool very much like the highest mass BDs for a long time [4]. Only now are the Galaxy's oldest very low mass (VLM) stars finally distinguishing themselves from its oldest high mass BDs by reaching the main sequence. If one finds an object which is fainter than the lowest mass star will be on the main sequence, that by itself is proof of the substellar nature of the object. Since the size of objects at and below the substellar boundary is quite similar at all times, this also means that there is a minimum effective temperature below which it too can provide clear proof of substellar status (cf. Kulkarni). While the minimum stellar mass is still subject to some modeling uncertainties, it is somewhere near 75 times the mass of Jupiter (M_J).

For lower mass BDs, the temperature and luminosity are lower at all ages. Below about $60 M_J$ there is never sufficient core temperature to completely burn even the next most fragile element after deuterium, namely lithium. Deuterium itself will experience non-negligible fusion down to about $13 M_J$ [23]. It is here we place the boundary between BDs and planets. Such objects are still supported primarily by free electron degeneracy, while Jupiter itself has a substantial support from electrostatic pressure [22]. The maximum size of a cold object is obtained near $2 M_J$, beyond which the gravitational energy dominates the degeneracy energy. This is the reason one may wish to refer to objects between 2 and 13 M_J as "superplanets", while Jovian-type objects are called "giant" planets. Note that the superplanets are actually smaller (and denser) than the giant planets.

We have not tried to distinguish BDs from planets by their mode of formation, or orbital eccentricity (as a proxy for mode of formation). Too little is known about the formation of low mass BDs, or about the possible masses or final orbital eccentricities of objects formed in a protoplanetary disk, to make this a currently useful means of distinction. Indeed, the evidence is already building that giant planets can be found both near to a star and with high orbital eccentricity [7]. This system (16 Cyg B) emphasizes that its binary nature (or history, since wide binaries can be stripped) can influence the eccentricity of a closer companion.

Unfortunately, it is inherent in the character of substellar objects that they are faint to begin with, and grow fainter with time. Their gravitational effects are also smaller than those of stars. Thus there has been a long and frustrating quest to even demonstrate the empirical existence of these objects. This has only recently (though spectacularly) borne fruit. BDs can be searched for either using their direct luminosity (very low) and color (very red) as diagnostics, or by their indirect gravitational effect on binary companions (using precision radial velocities or astrometry). Imaging searches have proceeded in the field, in clusters, and near faint stars. It is advantageous for these to look for BDs when they are young, because their luminosities are highest then.

DISCOVERY OF SUBSTELLAR OBJECTS

The Lithium Test

The most visible (higher mass) young BDs closely resemble VLM stars in both luminosity and temperature. While in principle one could distinguish them by dynamical mass in a binary system, such mass determinations are not as accurate as one would like. So far this is moot in any case, as we do not know a binary system including a BD where the mass can currently be unambiguously dynamically determined. This has meant that many promising brown dwarf candidates have been found, but certifying them as brown dwarfs has until recently been unsuccessful.

One means of doing so which has finally met with success is the "lithium test" [14]. This relies on the fact that lithium will never be destroyed in BDs with less than about $60M_J$, coupled with the fact that it will always eventually be destroyed in VLM stars. The timescale for this destruction is about 100 Myr in the lowest mass stars. The test can only be applied with knowledge of the object's age, since it is also true that young objects will not have had time to destroy lithium. Thus the best place to use the lithium test is in clusters. It is not entirely useless in the field because an object's luminosity does jointly constrain its mass and age, as does the presence or absence of lithium. An ambitious initial effort to use this test to certify various BD candidates resulted in none of them surviving [17].

The most successful application of the lithium test to date has been for the Pleiades cluster. Although thought to be a little younger (75 Myr) than required for the lowest mass stars to deplete lithium , this cluster is about the right age and relatively close and compact. A number of searches had turned up good candidate objects [24] [25]. Two of the best initial objects, HHJ 10 [17] and HHJ 3 [16] were found to have depleted lithium, however, despite a mass based on its luminosity and presumed age of about $60M_J$ for the latter. This mystery was resolved by us [2], when we found partially depleted lithium in an object just a little fainter: PPL 15. The discovery was announced at the June 1995 AAS meeting, and headed up a rapid series of further announcements of substellar objects which have survived further scrutiny.

We suggested (in [2]) that the reason the previous objects had failed the lithium test is that the age of the cluster had been underestimated. We propose an age of 120 Myr for the Pleiades, which places the mass of PPL 15 right on the substellar boundary. The cluster age had been based on the upper main sequence turnoff. We pointed out (in [2]) that there has actually been a controversy about such ages, and that if there is core convective overshoot in the B stars (as proposed by eg. [19]) then the high and low mass age determinations can be brought into agreement. Indeed, there are already preliminary indications [30] that the same problem is present in the α Per cluster.

By September 1995, the Canarias group [26] announced the discovery of an even fainter Pleiad called Teide 1. If the implications of PPL 15 were correct, this object *had* to be an incontrovertible BD, and show undepleted lithium. We confirm [27] that Teide 1, and a very similar object discovered later (Calar 3) both satisfy the test, and are BDs in the cluster with masses just under $60M_J$. These masses assume the older age for the cluster; if it is younger then their masses are even lower (but then it would be hard to understand the lithium depletion in HHJ 3). The Canarias group has recently discovered even fainter and redder candidates; they are confirmed by our Keck program to be dwarfs and have roughly the radial velocity of the cluster. A similar object has been located by S. Hodgkin and R. Jameson. While none of these are yet confirmed BDs, it is clear that we can discover BDs to quite low mass in the Pleiades (with spectral type M9 or so).

Other Substellar Objects

Only a month after the publication of Teide 1 came the announcement of the discovery of the sort of object everyone was really looking for: a BD old enough that it is well below the minimum temperature and luminosity of a star. Gl 229B has been described by Kulkarni. There can be no question that this is a substellar object. It is probably not too much less massive than Teide 1, perhaps 40 instead of $55-60M_J$. The difference is that it is perhaps 3 Gyr old (instead of 120 Myr for the Pleiades objects). This is the middle range of possible masses which depend on its age, for which we have only indirect arguments with poor precision [1]. For the first time we can study an object truly intermediate in temperature between stars and planets. There has developed, however, an unfortunate tendency in the community to dismiss young brown dwarfs as less convincing, just because they have stellar temperatures. In fact there are no reasons we know of that make the substellar nature of Teide 1 or Calar 3 any less convincing than for Gl 229B.

The excitement over Gl 229B was unfortunately diluted by the announcement at the same meeting of the first extrasolar planet around a solar-type star. Though 51 Peg is quite mundane, its planet has completely surprising properties. It is a giant planet located only a few solar radii above the surface of the solar-type star. 51 Peg turns out to be merely the first example of what are already 4 such planets, announced in a continuing flurry by Marcy and Butler. These low eccentricity giants with very short orbital periods and masses greater than $0.5\text{-}4 M_J$ are also joined by 2 superplanets and 1 giant with high eccentricity, and an intermediate object with low eccentricity, all at longer orbital periods. It is quite likely that there are more objects with longer periods and perhaps lower masses still to be mined out of the high precision surveys.

In fact, several candidate objects in the BD mass range had already been uncovered in medium precision Doppler surveys. What is in retrospect probably the first BD found was discovered by [11] as a Doppler companion to HD 114762. The difficulty with this announcement was that it was the only such detection in a large survey, so the odds that the companion was actually stellar with a very low orbital inclination were uncomfortably high. Only now, when much higher precision Doppler surveys have uncovered very few similar cases, can we say that it is likely to be a substellar object. This same objection held for the low mass objects in [20].

Finally, the best current candidates for free-floating field BDs were announced by the British group. Their survey employs stacking up to 100 photographic plates together to find very faint objects with high proper motion and parallax. These are found as faint red objects with parallaxes which place their luminosities just below the minimum stellar value. The object 269A has already been confirmed to pass the lithium test [28]. Although in this case the ages are not known, the presence of lithium coupled with the luminosity constrains the object to be similar to PPL 15. Thus, in less than one year we went from a situation where there were no confirmed substellar objects of any mass to a suite of objects with masses throughout the substellar regime.

PRELIMINARY ESTIMATION OF THE NUMBER OF SUBSTELLAR OBJECTS

Results from Doppler Searches

One of the best sources of information on the substellar IMF comes from the Doppler searches for low mass companions. These are helpful because they have now actually found objects, because the selection effects are relatively well-defined and understood, and because the situation for low mass stars is also well understood. Thus one can fairly directly compare the detection rates for very low mass stars and brown dwarfs, given that the detection limits have turned up objects well below the lowest brown dwarf. The main searches with large samples have been the moderate precision surveys at CfA [21] and Geneva [18], and the high precision surveys at Lick [5] and Geneva [18].

A given velocity precision implies a lower limit on the $m \sin i$ of the companions which can be detected at a given separation. Closer companions are favored both because they have a larger effect on the primary and because their orbital period is more likely to fit into the survey epochs. In principle the orbital inclination is a random variable. Orbital eccentricity will also bias the sample, because high eccentricity means that the companion spends most of its time at low relative velocities. A careful study of the biases in Doppler samples has been given by [20]. Once these are understood, the next question is to what extent stellar companions will appear as possible substellar objects. This will occur largely because of the correction for orbital inclination, which means that a nearly pole-on system produces a low velocity signal even for a high mass companion. One needs to understand the stellar distribution of companion masses for a given set of primaries, and then calculate how many such stellar companions will show up as possible substellar objects in a survey with given sensitivities.

Such an analysis was given by [15]. They use the observed binary distribution from [8] to specify the shape and normalization of companions. The number of stellar companions of all masses is fixed at 64% of the sample size, but any number of BD companions is allowed (given a specified mass function). This analysis is subject to a number of assumptions (as discussed by [21]) which could be lifted in principle, but which seem adequate for the present. We consider here anew the mass distribution of substellar companions to FGK main sequence stars, with special attention to two questions: 1) Do the lowest values of $m \sin i$ for companions definitively demand the existence of a substellar population or can they be interpreted simply as caused by the tail of the distribution of $\sin i$? 2) What constraints can be placed on the mass distribution of substellar companions from the Doppler searches to date? Probably the most powerful constraint on the nature of the companions having low $m \sin i$ comes from those having $m \sin i < 40 M_J$ (following [18]). These low values can only be dismissed as H-burning companions if the orbital inclinations are extreme, carrying a probability sufficiently low that modeling can probe its likelihood. We will also place a bound of $m \sin i > 5 M_J$ with the justification that yet lower mass objects are more likely to be planets or superplanets. We call objects found in Doppler surveys between the above bounds on $m \sin i$ Doppler brown dwarf candidates (DBDCs).

We set the velocity thresholds in our modeling to correspond to statistically significant detections in each survey. [18] report that their medium-precision Doppler survey of 570 G and K dwarfs reveals 8 DBDCs. We set the threshold

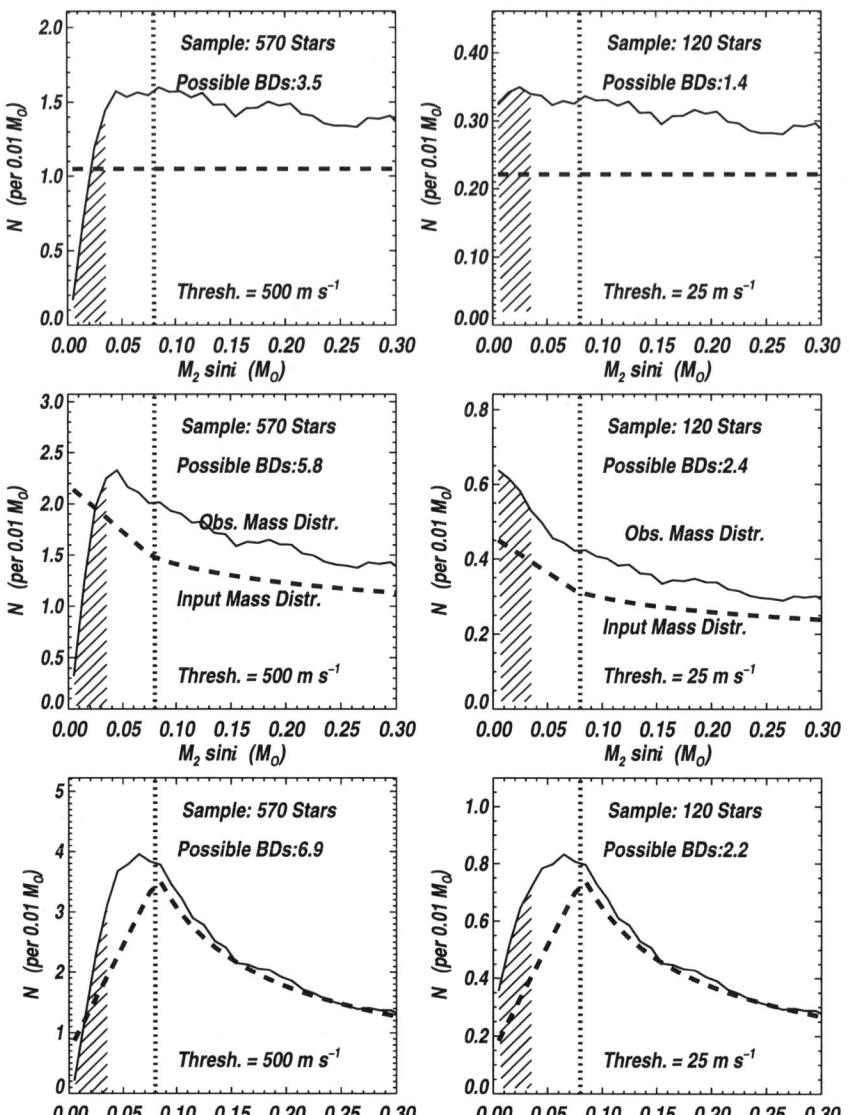

FIGURE 1. A set of predicted companion distributions for Doppler surveys. The vertical dotted line marks the substellar boundary. In each row a distribution of $m \sin i$ has been assumed (dashed line), and an "observed" distribution of companion masses is calculated (solid line). Each column has a different sample size and precision (labeled). The companions with $m \sin i < 40 M_J$ are shown with hatching.

for this work at 500 m^{-1}. They give preliminary results from the initial 2 years of their high precision (threshold 50 m s^{-1}) survey with ELODIE, suggesting they may have found 3 DBDCs. [5] have surveyed 120 stars with a threshold of 25 m s^{-1} for 9 years. They found *no* companions exhibiting $m \sin i > 10 M_J$, but they did find one companion having $m \sin i > 5 M_J$ which is also in an eccentric orbit, namely 70 Vir. Here we will consider this a DBDC (though it could a be superplanet that was perturbed into an eccentric orbit). The other object they found with an eccentric orbit (16 Cyg B) has such a low $m \sin i$ that it is not likely to be a BD.

Can we explain the [18] result of 8 DBDCs by using H–burning companions alone? We begin with a "null" test as in [15] and use an input mass distribution of companions that is constant per unit mass (dN/dM = const), but which is set to zero for all masses less than the H–burning limit (about 80M_J). The predicted number of DBDCs is only 0.6, as opposed to the 8 actually seen. We next consider another "null" possibility that again there are no substellar companions but that the H–burning mass function rises as a power law: dN/dM = $M^{-\alpha}$ with $\alpha = 1$, similar to that suspected for field M dwarfs at the bottom of the main sequence [10]. This model mass function yields only 2.0 predicted DBDCs (all caused by extreme $\sin i$.), still clearly discrepant with the 8 observed. The above two models strongly indicate that some, and probably most, of the companions discovered by [18] which have $m \sin i < 40 M_J$ are indeed substellar.

We now attempt to explain the actual number of detections by including substellar companions in the input mass distribution. We start by assuming a constant mass function for all companions masses, as shown in the top row of the Figure. Such a flat mass distribution would yield 3.5 DBDCs, compared with the observed 8. Of these, 0.8 could be stars masquerading as a BD due to orbital inclination. Given the Poisson statistics of the detected number, it appears that the flat mass distribution is marginally inconsistent with the 8 companions observed by [18]. On the other hand, this model predicts about the right number (1) of DBDCs found by [5]. It also predicts the ELODIE survey should find 2 (true) BDs, consistent with their preliminary result of 3 possible candidates.

We next consider a model which contains a gently rising mass function ($\alpha = 0.2$) within the stellar domain, then rises linearly by a factor of 1.5 from the H–burning limit to zero mass. This model (middle row of the Figure) yields a predicted 5.8 DBDCs, compared with 8 observed for the medium precision survey, constituting reasonable agreement within Poisson statistics. Of these, about 0.9 are really stars. The predicted number of substellar companions could be increased by simply increasing the slope in the substellar domain. But such an increase would violate the constraint coming from the high–precision Doppler work. Here the same model predicts 2.4 (2.75) companions within our mass limit (which are likely to really be substellar), compared with the 1 (3) found by [5] ([18]).

To constrain the slope of the substellar mass function we consider another model. We try to increase the number of companions found at medium precision by increasing α to 0.8 (as suggested by [9]), while suppressing the number of BDs to 0.2 rather than 1.5 times the H–burning limit as before. This model (bottom row of the Figure) predicts 6.9 DBDCs in the medium precision survey (of which 1.5 are really stellar), in good agreement with observations. The high precision surveys also don't change much, with 2.2 (2.7) DBDCs predicted (and about 0.5 of these would be stellar). Apparently either of the last two scenarios are consistent with both the medium and high precision surveys.

We cannot say much about the shape of the substellar mass function beyond that it cannot be empty or rising very fast. Interestingly, the detected companion in the [5] survey has a very low mass, more consistent with the assumption of a rising BD mass function. Note that the high precision survey is *more* sensitive to higher mass companions than the medium precision survey. The reason they were not found by the current high precision surveys is that the survey samples are too small. As more stars are examined, such surveys should detect DBDCs of all masses. The bottom of the stellar mass function (at least for companions within 5 AU) cannot be rising with an α much greater than unity given current observations. Finally, we note that our modeling shows unequivocally that the detections of objects below $m \sin i$ of $5M_J$ cannot be brown dwarfs (as we have defined them). There are low chances for one such fake in the Lick survey; to have 6 is completely out of the question. They must be planets or superplanets.

Results from Cluster Searches

Another good source of information on the substellar mass function at present comes from the lithium studies of the Pleiades cluster. While this work is also in its early phases, there is a clear advantage to studying a cluster for which the stellar mass function is well known. The Canarias survey [31] for brown dwarf candidates is capable of finding brown dwarfs down to $50M_J$ or less based on the luminosity limit at the age of the Pleiades. This is not a proper motion survey and relies on color information to identify candidates. Confirming them as brown dwarfs is difficult and slow (requiring medium resolution spectra of objects that are very faint at the wavelength of interest), since BDs still look like stars at the age of the Pleiades. They have found that there is a significant population of background M dwarfs which can only be ferreted out with the lithium test, so this project will require a few years.

Already, however, one can ask how many brown dwarfs should have been detected in the fraction of the cluster surveyed (about 2%) compared to the number of very low mass stars found by a previous surveys (in particular the work of [25]). They found a little over 50 stars with proper motion membership in the cluster having I magnitudes fainter than 15.5, which are VLM Pleiads. If there are comparable numbers of free-floating brown dwarfs in the cluster, then the Canarias survey should have turned up 1 or 2 so far given the small survey area. That is in fact the case. These surveys suggest there should be of order 100 VLM Pleiads, and as many as 200 BDs. They are compatible with the results so far from the the Doppler searches, with perhaps a gently rising mass function. Note that many of these stars may actually be binary.

The Hyades search of [13] is another current source of information. They used high spatial resolution infrared imaging, and examined the faintest members of the cluster for companions within 20 arcsec. They did not find a single good brown dwarf candidate in a sample of almost 200 primaries. Because of the low detection rate of VLM stars, a reasonable conclusion is that the wide binaries (which they are most sensitive to) have been stripped in the cluster, as is theoretically expected. It is also true that this cluster is old enough to have evaporated many of its lowest mass members (though this is somewhat mitigated for binary systems). Even so, at the inner detectability limit where stripping is less important, their result constrains the low end of the mass function. It is certainly compatible with the proposition that there are not many more BD companions than VLM stellar companions.

Other Searches

It is worth mentioning here that there have been identifications of extremely young brown dwarfs in the ρ Oph star formation region [29]. Such objects can only be claimed as BDs on the basis of their age and luminosity. This is a reasonable way to do it, except that the theory of pre-main sequence tracks is still somewhat unsettled. Unfortunately the lithium test is of no help here, because the true stars have not had time to deplete their lithium either. When the models have been agreed upon, this will be a powerful way to test the production of BDs, because these clusters are not yet evolved away from their IMF (by evaporation or stripping of low mass objects).

The other searches for brown dwarfs are much harder to translate to a mass function so far. Any searches for brown dwarfs which rely on their luminosity for detection but where the age of the objects searched is not strictly controlled suffer from the problem that brown dwarfs fade out as they get older. This limitation will hold for any general search in the field. The survey of Kulkarni et al. made an attempt to study "younger" primaries, but the precision with which this can be done is insufficient to allow strong statements about the mass function. This is illustrated by our ignorance of the mass of Gl 229B. Until we study its (very slow) orbit long enough to get a dynamical mass, its mass will not be known to better than $20M_J$ or so. It is very hard to say how many brown dwarfs have been missed because they are too faint. The separation range probed is also a function of the distance of the target.

Imaging searches for faint companions of selected stars (for example, using white dwarfs as companions) generally suffer the same difficulty, unless they are conducted in a cluster. In particular, an early contender for brown dwarf status, GD 165B [3], has remained uncertified despite being very promising because of the uncertainty in its age and the fact that it resides right at the minimum stellar temperature. An attempt to determine the luminosity function from imaging survey results [12] concludes there is reasonable evidence that the IMF flattens out at around the substellar limit. Obviously the fading of BDs with age does not impair Doppler or astrometric searches, which do not rely on being able to actually see the companion. Astrometric searches are confined to relatively nearby stars. They have not yet sampled enough stars to have found a BD.

THE BOTTOM LINE

Each of the surveys discussed above samples a limited part of the full parameter space in which BDs can be found. The Doppler surveys are only relevant to companions within 5 AU or so of the primary star. They further are relevant at the moment mostly to GK stars; we don't know for example whether the same frequency of companions holds for the more common (in the Galaxy) M star primaries. The period distribution of companions found by [8] resembles a Gaussian in the log of the period (in days), with a peak at 4.8 and width of 2.3. Thus, the current Doppler results imply that less than 2% of GK stars have a BD companion within 5 AU, but the total number of BD companions at *all* separations might be as high as 4-5%. General imaging surveys, which sample the outer half of possible separations for example, should expect a to find at most 2.5% of targets with BD companions, reduced by the fraction of all possible ages to which the survey is sensitive. Thus, such a survey which can see BDs only for the first 2 Gyr should require something like 200 stars in order to find one BD. This is indeed similar to the results which have been achieved so far. If one can choose the primaries so they are all preferentially very young, the success rate should increase accordingly.

The number of free floating BDs should be similar to the number of field M7-10 main sequence stars. Of course, most field BDs should be cooler than M10, and extremely faint and hard to find. Only the most local ones will be found in imaging surveys. The exception is the young ones (younger than 0.5 Gyr or so). These can be found preferentially in young clusters. In addition, perhaps 5% of M7-10 objects identified in the field will prove to be BDs (since very young ones have spectral types in this range). These should be found preferentially in the Galactic plane, and there may also be a surfeit of them in the general direction of Gould's belt, where there is a known excess of young stars. Proof that they are BDs generally requires the lithium test (unless we find a suitable binary system). At low luminosities, the objection that sufficiently young stars will also have retained their lithium is countered by the fact that for an object to be both faint and young, its mass must be very low.

It is really too early to say what the shape of the substellar mass function looks like. What seems safe to say already, though, is that there are nothing like the number of substellar objects required to make them a significant constituent of baryonic dark matter. That would require the mass function to rise like $M^{-\alpha}$ where $\alpha = 3$ through the BD domain. We have argued above that the results from both the Doppler searches and the imaging searches of young clusters are consistent with the number of BDs being comparable to the number of VLM stars. In particular, we find it unlikely that the low end of the stellar IMF has α much over unity. While we know even less about the shape of the mass function for BDs, there is a slight hint that it may continue to rise gently in the BD domain. Coupled with the results from the MACHO project [6], a population of substellar objects which could significantly contribute to the dark matter seems unlikely.

It is clear that there is no reason to speak of a "brown dwarf desert" or "mass gap". These expressions came about because of the lack of BD detections by the high precision Doppler searches. We have shown that is due to insufficient sample size in those surveys rather than a "lack" of BDs. What is found instead is a "planetary jungle" (especially given the wild nature of the objects found so far). The number of substellar objects at very low mass (below BDs) is rising quickly. Even with the limited sensitivity of current surveys it is already much easier to find planetary rather than BD companions to stars. Although these objects cannot contribute significantly to the dark matter, they are obviously of tremendous interest in the ultimate search for other life in the Universe.

REFERENCES

1. Allard, F., Hauschildt, P.H., Baraffe, I. and Chabrier, G. 1996, Ap.J., 465, L123
2. Basri, G., Marcy, G.W., and Graham, J.R. 1996, Ap.J., 458, 600
3. Becklin, E.E. and Zuckerman, B. 1988, Nature, 336, 656
4. Burrows, A.S., Hubbard, W.B., Saumon, D. and Lunine, J.I. 1993, Ap.J., 406, 158
5. Butler, R.P. and Marcy, G.W. 1997, *Astronomical and Biochemical Origins and Search for Life in the Universe*, IAU Coll. 161, in press
6. Chabrier, G., Segretain, L. and Mera, D. 1996, Ap.J., 468, L21
7. Cochran, W.D., Hatzes, A.P., Butler, R.P. and Marcy, G.W. 1997, submitted to Ap.J.
8. Duquennoy, A., and Mayor, M. 1991, A.& A., 248, 485
9. Kirkpatrick, J.D., McGraw, J.T., Hess, T.R, Liebert, J. and McCarthy, D.W. 1994, Ap.J.Supp., 94, 749

10. Kroupa, P., Tout, C.A. and Gilmore, G. 1993, M.N.R.A.S., 262, 545
11. Latham, D.W., Mazeh, T., Stefanik, R.P., Mayor, M. and Burki, G. 1989, Nature, 339, 38
12. Laughlin, G. and Bodenheimer, P. 1993, Ap.J., 403, 303
13. Macintosh, B., Zuckerman, B., Becklin, E., and McLean, I. 1997, preprint
14. Magazzù, A., Martín, E.L., and Rebolo, R. 1993, Ap.J., 404, L17
15. Marcy, G.W. & Butler R.P. 1995, *The Bottom of the Main Sequence - and Beyond*, (Tinney, ed.), ESO Astrophysics Symposia, Springer-Verlag, p. 98
16. Marcy, G.W., Basri, G., and Graham, J.R. 1994, Ap.J., 428, L57
17. Martín, E.L., Rebolo, R., Magazzù, A. 1994, Ap.J., 436, 202
18. Mayor, M., Queloz, D. and Udry, S. 1997, *Astronomical and Biochemical Origins and Search for Life in the Universe*, IAU Coll. 161, in press
19. Meynet, G., Mermilliod, J.-C., and Maeder, A. 1993, A.& A. Supp., 98, 477
20. Mazeh, T., Goldberg, D., Duquennoy, A., and Mayor, M. 1992, Ap. J., 401, 265
21. Mazeh, T., Latham, D.W. and Stefanik, R.P. 1996, Ap.J., 466, 415
22. Nelson, L.A. 1990, *Baryonic Dark Matter*, (Lynden-Bell and Gilmore, eds.), NATO ASI Series, Kluwer Academic, p. 67
23. Saumon, D., Hubbard, W.B., Burrows, A., Guillot, T., Lunine, J.I. and Chabrier, G. 1996, Ap.J. 460, 993
24. Stauffer, J.R., Hamilton, D. and Probst, R. 1994, A.J., 108, 155
25. Steele, I.A., Jameson,R.F. and Hambly, N.C. 1993, M.N.R.A.S., 263, 647
26. Rebolo, R., Zapatero-Osorio, M.R. and Martín, E.L. 1995, Nature, 377, 129
27. Rebolo, R., Martín, E.L., Basri, G., Marcy, G.W., and Zapatero-Osorio, M.R. 1996, Ap. J., 469, L53
28. Thackragh, A., Jones, H. and Hawkins, M. 1996, M.N.R.A.S., in press
29. Williams, D.M., Comeron, F., Rieke, G.H. and Rieke, M.J. 1995, Ap.J., 454, 144
30. Zapatero-Osorio, M.R., Rebolo, R., Martín, E.L. and García Lopez, R.J. 1996, A.& A., 305, 519
31. Zapatero-Osorio, M.R., Martín, E.L. and Rebolo, R. 1997, A.& A., preprint

Hydrogen Flash Divides Accreting Objects into T-Tauri Stars and Embedded Protostars

Tadayuki Murai

Department of Physics, Nagoya University, Nagoya 464-01, Japan

Abstract. The birth of stars at the low mass end of main sequence is investigated. Possible deuterium-burning shells slightly increase the critical mass for H burning which is about 0.08 solar masses for homogeneous stellar models with pop I composition, since the degree of central condensation is higher for inhomogeneous models than for those of homogeneous one. When hydrostatic protostellar core reaches the critical mass, H-flash takes place, provided that the accretion rate is so small that cores can radiate enough entropy for electron gas to become degenerate. H-flash develops vigorous convective motion of magnetized rotating medium, resulting in strong stellar wind phase, and stellar winds break surrounding molecular clouds mainly in the direction of the rotation axis. Stars flare up, settle on the Hayashi track and then descend as T-Tauri stars. Some of bipolar flows in star forming regions may be initiated at the onset of H-flash while progenitors(protostars or accreting brown-dwarf cores) are embedded within dense molecular clouds. Brown dwarfs form rarely and are not the main constituent of galactic dark matter, since protostars continues to receive surrounding matter. Low mass end of initial mass function depends critically on whether H-flash takes place frequently or not.

INTRODUCTION

According to the inside-out collapse scenario of low mass star formation by Shu, Adams and Lizano [1] [2], strong stellar wind due to turbulent convection initiated by the onset of deuterium burning is envisaged to help protostars reveal themselves out of dense molecular clouds. They are able to interpret the first appearance of T-Tauri stars within the dense molecular clouds by simply delineating a locus in the HR diagram for wholy convective stellar models with the central temperatures of 10^6K. The locus is that of a linear series of varying mass for deuterium burning main-sequence. The entropy generation of deuterium burning is the primary mechanism which bring protostars in a wholy convective configuration and in a stage of stellar winds which presumably break accreting circumstellar matter in the direction of rotation axes of the central bodies. Therefore, central bodies could be observed as T-Tauri stars descending along the Hayashi track toward the main sequence if the solar system is located in a cone where circumstellar matter is expelled, or as bipolar-flow objects if the line-of-sight optical depth is still high.

The D-burning sequence in the HR diagram, logically, extends downward to the end point at 0.01 solar masses where central conditions plunge into the electron degenerate regime. It has not been clarified yet whether accreting protostellar cores within cold dense molecular cloud cores evolve to the low mass end of D-burning main sequence or not. Observational data, however, have been accumulated and revealed that there exists a region in the HR diagram right to the Hayashi track [3] corresponding to 0.2 solar masses where T-Tauri stars have scarecely been found, even though theory expects some T-Tauri stars [4](but see also [5]). This is because the low mass end of the zero-age H-burning main sequence, theoretically, extends down to about 0.08 solar masses, the critical mass for H-burning stars with homogeneous chemical composition, which is a natural assumption after the evolution via wholy convective phase(Hayashi phase).

This paper investigates a plausible reason why above mentioned Hayashi limit divides accreting objects in the molecular clouds into two categories; one consists of observed T-Tauri stars or bipolar-flow objects and the other consists of visualy unseen protostars(IRAS point sources) without bipolar flow phenomena.

SUMMARY

Let us consider the role of D-burning shell which presumably located in the surface region where fresh unprocessed materials are accreting onto the central body in quasistatic equilibrium. If the D-burning shell is active, then it plays a similar role of H-burnig shells in stars ascending the red giant branch in the HR diagram of globular clusters just before He-flash, and of He-burning shells in stars in advanced phases approaching toward C-detonation/deflagration supernovae. These nuclear burning shells change stellar structure drastically from a homogeneous configuration to inhomogeneous one and increases the critical mass for nuclear burning(see, e.g. [7]) appreciably, i.e. from 0.3 to 0.5 solar masses in the case of He-flash, and from 1.1 to 1.39 solar masses [8] [9] in the case of C-flash. Therefore, the critical mass for H-burning would increase from 0.08 to a value between 0.1 to 0.2 solar masses, the precise value depending upon accretion rate onto the D-depleted cores dM_{core}/dt and physical conditions of stellar matter in partially degenerate state(screening factor and opacity).

At the onset stage of any nuclear burning phase when gravitational energy generation rate are gradually decreasing due to the increase of nuclear energy release, nuclear energy generation rate overshoot for a brief period of readjustment of the change of stellar structure from a uniform source model to a model of centrally concentrated sources (extreme case is a point source model, see e.g. [3] and [6]]. If the degree of electron degeneracy is strong, this overshoot of nuclear energy release is conspicuous and is called as nuclear flash, therefore, convective motion becomes violent, leading to turbulence. Rotating turbulent cores generates magnetic fields, energy of which might exceed the energy of turbulent motion as anticipated from a recent self-consistent dynamo simulation by Kageyama and Sato [10]. In the simulation, vortex tubes alligned along the rotation axis generates multiple magnetic poles around the polar regions. Resulting stellar wind activity powered by central nuclear energy release would be high enough to expel outer molecular cloud as seen in the phenomena of bipolar flows.

The inside-out scenario does not have a mechanism to halt the accretion process inherently, and moreover, a close inspection into a realistic situation prevailing in star forming regions and also theoretical understanding of supernova collapse phase presents us with a question whether circumstellar accreting matter or stellar envelope gets its outward momentum without a central nuclear energy release, because dynamical time scale of fluid elements in gravitational field is shortest at the center than any other part of accreting matter. Outer part does not respond dynamical developments in the central part quickly, and the accretion onto the central bodies continue until the onset of the central nuclear energy release or some outer mechanisms leading to evaporation of molecular gas as seen in a series of photographs taken by Hubble Space Telescope. Then, brown dwarfs form rarely and do not contribute to galactic dark matter. Low mass end of initial mass function depends critically on whether H-flash takes place frequently or not.

ACKNOWLEDGEMENTS

The author would like to thank Tadashi Nakajima and Mrs. Nakajima for a brief conversation on brown dwarfs in last December at their home in Pasadena, which stimulated his imagination and knowledge acquired in Kyoto in 1965-70 from C. Hayashi, R, Hoshi and D. Sugimoto as a graduate student and result in this work. He is grateful to Charles Barnes, G.J. Wasserburg, and G. Papanastassiou for their warm hospitality during his brief visit at Caltech at the occasion of the Symposium on Nuclear Astrophysics, A Celebration of Willy Fowler. He would like to express his gratitude to Charles Barnes and Juliana Sackmann for discussions and their interests to this work when he visited Caltech after this Conference. The travel expences were due to generosity of Setsu Murai and Yuri Murai who have supported his activities, especially three foreign travels, Varna, Bulgaria, Berlin, Gernany, Cambridge, England in 1990, Pasadena, CA in 1996, and Pisa, Italy, Berlin, Germany,

NY, College Park, Pasadena in 1996.

REFERENCES

1. Shu F.H., Adams F.C., and Lizano S., 1987, *Ann.Rev.A.Ap.*, **25**, 23.
2. Stahler S.W., 1983, *ApJ*, **274**, 822-829.
3. Hayashi C., Hoshi R., and Sugimoto D., 1962, *Progr. Theor. Phys.*, **22**, 1.
4. Cohen M., and Kuhi L.V., 1979, *ApJS*, **41**, 743-843.
5. Kenyon S.J., and Hartmann L., 1995, *ApJ*, **101**, 117-171.
6. Schwarzschild M., 1958, *Structure and Evolution of Star*, Princeton University Press, 1.
7. Takarada K., Sato H., Hayashi C., 1966, *Progr. Theor. Phys.*, **36**, 504-514.
8. Murai T., Sugimoto D, Hoshi R., and Hayashi C., 1968, *Progr. Theor. Phys.*, **43**, 639.
9. Arnett D., 1968, *Ap.Sp.Sci.*, **5**, 180.
10. Kageyama A., and Sato T., 1996, preprint of National Institute For Fusion Science, ISSN 0915-633X, NIFS-458.

Dynamical Processes

Dynamical Influences on Star Formation in Spiral Galaxies

Jeffrey Kenney and Shardha Jogee

Astronomy Dept., Yale University, P.O. Box 208101, New Haven, CT 06520-8101 USA
kenney@astro.yale.edu, jogee@astro.yale.edu

Abstract. The principal dynamical influences on large-scale star formation in relatively undisturbed spiral galaxies are reviewed, using recent observations on flocculent galaxies, spiral arms, bars, resonance rings, and circumnuclear starbursts. Non-axisymmetric features in the gravitational potential like bars and spiral arms impact star formation in at least two ways. They cause radial flows of gas, influencing where gas concentrates and therefore where star formation is likely. They also affect the kinematics and density of gas, and therefore its susceptibility to instabilities. The local gravitational instability theory is relatively successful in predicting whether or not gas in a given location undergoes star formation, on scales ranging from outer disks to starbursting central regions. Much about galaxy evolution depends on the relative values of the radial flow rate and the star formation rate, both of which are strongly influenced by galaxy dynamics.

INTRODUCTION

What aspects of star formation can be understood from the perspective of global galaxy dynamics? A casual glance at the Hα image of a galaxy with a strong bar or spiral arms is enough to demonstrate that certain large-scale, dynamical features orchestrate where stars form. A perhaps less obvious, but equally compelling comparison of theory to observed gas densities and kinematics shows that gravitational instabilities are relevant for understanding where star formation occurs. Less obvious still, yet just as important for galaxy evolution, are observations and modelling suggesting that radial flows of gas driven largely by gravitational torques, concentrate gas in certain regions which become the favored sites of star formation.

The star formation process is initiated when gas on large scales becomes self-gravitating. The local gravitational instability picture [1] [2] [3] [4] for this first stage in star formation is undoubtedly oversimplistic and incomplete. Yet there is mounting evidence that it is relatively successful in predicting, on sufficiently large scales, where gas forms stars. The simplest version of this theory applies only for uniform rings of gas with circular orbits, and therefore is not strictly valid for gas in bars and spiral arms. However, it is in principle straightforward to extend the gravitational instability theory to the non-circular local velocity fields of spiral arms and bars. The effect of shocks acting on the gas at these sites is more difficult to handle with this theory. Shocks may impact star formation in complex ways, dictated by local conditions in the shocked interstellar medium (ISM). Nonetheless, much about the origin, location, and strength of shocks are the direct result of large-scale dynamical processes. In this paper we address these aspects of star formation using recent observational results on flocculent galaxies, spiral arms, bars, resonance rings, and circumnuclear starbursts.

THE SIMPLEST SPIRAL DISKS: FLOCCULENT GALAXIES

Among the simplest galaxies for studying star formation are the flocculent spirals. These are spiral galaxies whose optical disks are dominated by a multitude of short, spiral fragments [5]. Weak spiral features are also seen in flocculent spirals at near-infrared (NIR) wavelengths [6], where light is a better tracer of the stellar mass distribution [7]. There may be weak density waves in flocculent spirals, but the arm-to-interarm contrast in NIR light is only ∼0.1-0.3, compared with at least a factor of ∼2 in strong grand design spirals like M51 [6]. In flocculent spirals, one can study star formation with minimal complications from spiral density waves, bars, and non-circular motions.

NGC 4414, a nearby flocculent spiral with particularly weak NIR arms, has been recently mapped in CO at 4″ (190 pc) resolution [8] [9]. The molecular gas in the disk is clumpy, exhibits no clear symmetries or rings, and has predominantly circular motions. A central hole exists in the distributions of both molecular gas and HII regions (Fig. 1a).

With CO and HI interferometry data, one can measure the gas surface density Σ_{gas}, the critical surface density for the onset of gravitational instabilities $\Sigma_{\text{crit}} = \kappa \sigma / \pi G$, and their ratio $Q = \Sigma_{\text{crit}} / \Sigma_{\text{gas}}$ [1] [3] [10]. The epicyclic frequency κ is closely linked to the rotation curve V(R) through $\kappa = \sqrt{\frac{2V}{R}\left(\frac{V}{R} + \frac{dV}{dR}\right)}$ and is related to the centripetal and coriolis forces which can support gas against gravitational collapse. The gas velocity dispersion σ is related to the pressure. The observed rotation curve and linewidths need to be corrected for beam-smearing in order to determine κ, σ, and hence Q. The H_2 surface density is often inferred from the CO surface brightness, although the applicability of a "standard" CO-H_2 conversion factor remains uncertain.

In the H_2-dominated inner disk of NGC 4414, Q is close to 1 beyond r≃16″ (800 pc) (Fig. 1b), showing that the molecular disk is marginally unstable where stars are forming [9]. Further evidence in NGC 4414 for the instability picture comes from the good agreement between the observed masses of 10^7 $M_\odot$ for the largest molecular gas complexes and the predicted masses of gas lumps whose diameter is the fastest growing mode of a gravitational instability [9]. Since Q$\simeq$1 rather than much less than 1 in the region of star formation, the gas disk is near neutral stability rather than being highly unstable, implying that there is a regulation mechanism to maintain $\Sigma_{gas} \simeq \Sigma_{crit}$. This regulation mechanism is likely to be feedback from star formation itself. Most of the CO complexes in NGC 4414 appear gravitationally bound, but 2 are not, since they have large velocity dispersions for their diameters and masses. These could be complexes in the process of being disrupted by star formation. In NGC 4414 we may be witnessing 3 parts of the star formation cycle: formation of giant molecular associations through gravitational instabilities, star formation in clumps, and clump disruption by star formation [9].

One of the interesting unanswered questions is whether spiral-like chains of molecular complexes in galaxies like NGC 4414 are due to weak density waves [6] or swing amplification [9]. While gravitational instabilities require Q$\leq$1, swing amplification can amplify shearing disturbances even for 1<Q<2 [11] [12]. For Q between 1.5 to 2, amplifications factors ranging from 5 to 30 are typically achieved [13] [14]. Swing amplification can be particularly effective in amplifying the growth of density fluctuations in a clumpy ISM on large (kpc) scales in regions of differential rotation.

Although gravitational instabilities seem to account for the location of star formation (on sufficiently large scales), it cannot predict the star formation rate (SFR) without significant input from local physics. Gas consumption timescales due to star formation are 2-3 orders of magnitude longer than grav-

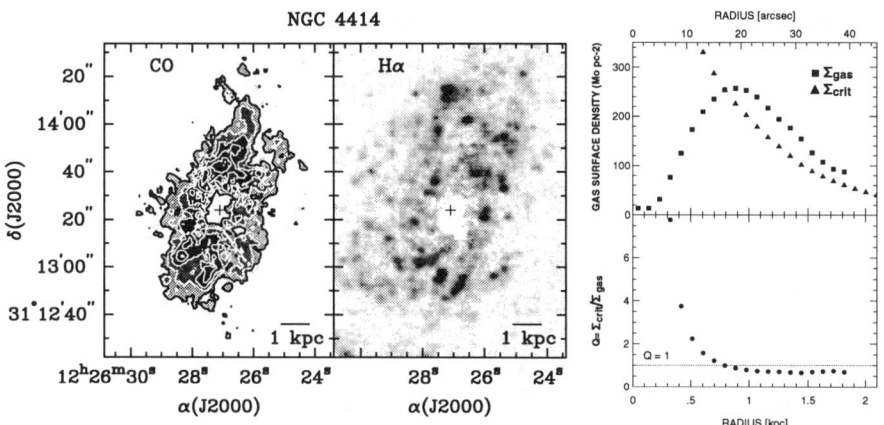

FIGURE 1. a.) CO and Hα images of the flocculent galaxy NGC 4414 [8]. b.) Measured and critical gas surface densities (top) and Q (bottom) versus radius in NGC 4414 [9]. Q$\leq$1 in the regions where gas is forming stars.

itational collapse timescales [15] [16]. This is probably because (1) molecular clouds are supported by magnetic fields and/or turbulence, so their collapse is slower than the free-fall rate, and (2) The energy input from OB stars and supernovae ionizes and disrupts most of the molecular cloud, so that only a small fraction of the cloud mass turns into stars. More extensive studies are needed to determine how the ratio of the cloud formation rate (from the gravitational instability picture) to the SFR, often called the efficiency factor, varies within and among galaxies.

STAR FORMATION IN SPIRAL ARMS

In galaxies with strong spiral arms like M51, virtually all the massive star formation occurs in the arms (Fig. 2a). There are many factors, including shocks, cloud collisions, magnetic fields, and inhomogeneities which may strongly impact the ISM as a strong spiral density wave passes through it. However, even in such galaxies an azimuthally averaged analysis in the framework of the gravitational instability picture can bring surprisingly useful insight into the process of star formation. The classic Kennicutt [3] paper, which provided fairly compelling evidence that this theory is relevant for large scale star formation in galaxies, contained M51 and other galaxies with strong density wave spiral arms. The existence of density waves violates the assumption of azimuthal symmetry. However, if the azimuthally-averaged value of Q is close to 1 in a disk with non-axisymmetric structures, then both stable and unstable regions likely exist in the disk.

While this explanation is overly simplistic, an examination of the conditions within spiral arms shows that arms are indeed favored sites for gravitational instabilities. Even in the absence of dissipative processes like shocks or cloud collisions, the gas density would be increased in the spiral arms as a consequence of orbit crowding [17]. In fact, observations and modelling of the CO intensity distribution and velocity field in the grand design spiral galaxy M51 suggest that orbit crowding is the dominant effect [18], although some dissipation is clearly required [19]. Furthermore, the amount of shear and the value of κ (which are related) decrease strongly in the first half of the spiral arm with respect to both the back half of the arm and the interarm region [20] [21]. The cumulative effect of a high gas surface density and a low value of κ in the first part of the spiral arm makes gas in that region particularly unstable to gravitational collapse. This is consistent with studies of individual giant molecular associations in M51 [21] [22] (Fig. 2a), where gravitationally bound complexes as large as a few 10^7 $M_\odot$ exist in the spiral arms, whereas some of the largest complexes in the interarm regions appear unbound.

The SFR does not vary smoothly along all spiral arms. In M51, both the molecular gas [23] and the SFR [24] are enhanced at symmetric, regularly spaced intervals along the arms, at locations likely related to resonances.

STAR FORMATION ALONG BARS

Recent Hα imaging surveys reveal a wide range of star formation properties along bars [25] [26]. Many barred galaxies have very little star formation along their bars. Observations indicate at least 2 reasons for this. In some galaxies this is due to a relative absence of gas along the bar. Stellar bars drive gas radially inwards, by a combination of dissipation and torques [27] [28], and this inflow will ultimately make the bar region gas-poor. In other cases star formation is weak even though the bar is gas-rich. For instance, the bar of M83 has strong CO emission, suggesting an abundance of molecular gas, yet the ratio of ultraviolet to CO luminosity is several times lower along the bar than in the outer disk or circumnuclear region [29]. Several authors have suggested that the large amount of shear along bars may make star formation inefficient [30] [31] [32] [33]. All of these results need better observational confirmation.

The HII regions which do occur along bars are generally offset in a leading sense from the stellar bars, and are located close to the dust lanes which are associated with shock fronts [25]. A particularly clear illustration of a bar shockfront comes from the beautiful Fabry-Perot Hα observations of NGC 1530 citeRegan 1997 [34] showing the ionized gas velocities throughout the entire region swept by the bar (Fig. 2b). Note that there are HII regions on both

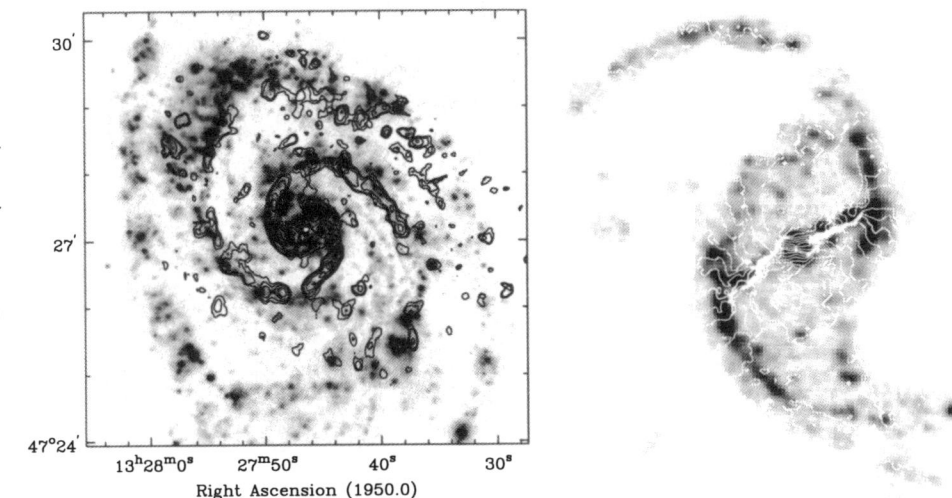

FIGURE 2. (a) CO (contour) on Hα (greyscale) map of the grand design spiral galaxy M51 [22]. (b) Hα intensities (greyscale) and isovelocity contours in the strongly barred galaxy NGC 1530 [28][34]. Note the strong velocity jump at the leading edge of the bar in NGC 1530, and the HII regions on either side of this shockfront, including many on the upstream side In the spiral arms of NGC 1530 and M51, virtually all the HII regions are located downstream from the velocity discontinuity (NGC 1530) and CO peaks (M51).

sides of the shockfront in NGC 1530, including intense star formation on the preshock side. This pattern of star formation has not yet been explained, but probably depends on the gas flow pattern. Regan *et al.* [28] find good agreement between the observed velocity field, including the strong velocity discontinuity at the shockfront, and a model based on ideal gas hydrodynamics. These gas kinematics along the bar indicate a stronger shock and more dissipation than the spiral arms of M51, where the gas velocity field appears intermediate between the ballistic particle case and the hydrodynamic case [18] [21]. The relationship between the pattern of gas flow and the location of star formation along bars deserves further study because the orbits of newly formed stars affect the growth and evolution of bars.

Star formation properties along bars exhibit a trend along the Hubble sequence [25] [26]. Sa's and Sb's generally have little star formation along the bar, but strong star formation in the center. Sc's exhibit a wide variety of star formation morphologies, but many have strong star formation along the bar and weak star formation in the center. Part of these observed trends may be due to the structural differences between flat and exponential stellar bars. Most late type (Sc) barred galaxies have small bars with exponential light distributions along their major axis, whereas most early type (Sa-Sb) barred galaxies have larger bars with flat light distributions [35] [36]. Late type galaxies with exponential bars might have strong star formation along their bars because exponential bars are typically weaker than flat bars, so they produce smaller gas inflow rates, and may also have less shear. Recent simulations [37] suggest that small exponential bars form through internal global instabilities, and large flat bars via tidal interactions or mergers.

The correlation between Hubble type, bar profiles, and star formation properties does not hold for some galaxies. For example, the late type galaxies

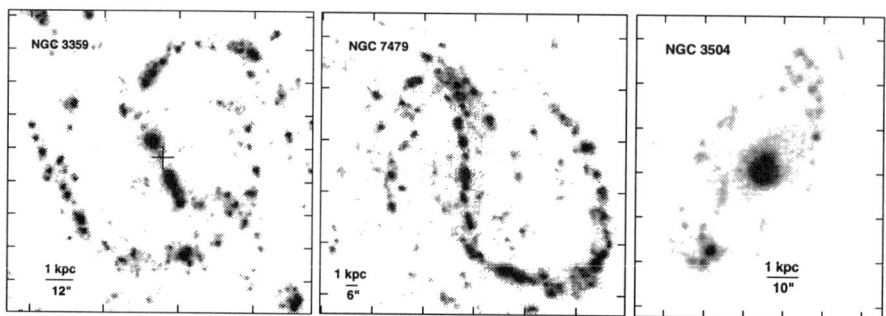

FIGURE 3. Hα images of 3 spirals with flat bar profiles, yet different distributions of star formation along the bar. a.) NGC 3359 (Sc) has strong star formation along its bar but none in the center (marked by a cross). b.) NGC 7479 (Sbc) has strong star formation along the bar and in the center c.) NGC 3504 (Sb) has very weak star formation along the bar but an intense central starburst. Do these galaxies outline an evolutionary sequence?

NGC 3359 (Sc) and NGC 7479 (Sbc) contain large flat bars similar to those in earlier type spirals [35], and have strong star formation along their bars [26] [38], as shown in Fig. 3. These galaxies are in a short-lived evolutionary state, and may be evolving towards earlier Hubble types [39]. By comparing the SFR to the gas inflow rate, one can estimate the growth rate of the central mass concentration. The extinction corrected Hα luminosity in NGC 7479 indicates a SFR along the bar of 0.5 M$_\odot$ yr^{-1} [26], whereas the inflow rate calculated from the gravitational torque is >4 M$_\odot$ yr^{-1} [38]. Thus, most of the gas presently in the bar region should reach the circumnuclear region before it turns into stars. The net inflow of 4×10^9 M$_\odot$ of H$_2$ from the bar will triple the total mass in the central $\sim$200 pc in less than 1 Gyr.

If most of the gas reaching the circumnuclear region turns into stars, then the central stellar concentration would increase significantly. While the stars tend to form in a thin compact circumnuclear disk [40], several mechanisms have been proposed to drive stars to large scale heights to enhance or build a spheroidal bulge [41] [42] [43] [44]. Thus, this galaxy previously classified as Sc or Sbc, would now bear greater similarity to a barred Sb, with a flat bar profile, a large central stellar concentration and little star formation along the bar. This may be one way in which galaxies evolve along the Hubble sequence from late to early type. Figure 3 schematically illustrates this possible evolutionary sequence with the Hα morphologies of 3 spiral galaxies with flat bars: NGC 3359 (Sc), NGC 7479 (Sbc) and NGC 3504 (Sb).

STAR FORMATION IN RESONANCE RINGS

Nowhere are the dynamical effects on large-scale star formation more obvious than in strongly barred galaxies. In these galaxies the existence of strong morphological features like rings makes it is easier to see how large-scale dynamics affect the location of gas concentrations and star formation. The recognition of clear resonance features in strongly barred galaxies is an important step towards recognizing less prominent features related to resonances in galaxies with weaker bars.

Figure 4 shows Hα and off-line continuum images of the barred, ringed galaxy NGC 6782 [45]. Virtually all the star formation in this galaxy is concentrated into 3 ringlike features which are associated with dynamical resonances of the bar. The outer ringlike feature is composed of 2 spiral arms with a small pitch angle, which nearly close on themselves. These 'pseudorings' [46] are located near the outer Lindblad (outer 2:1) resonance (OLR) of the bar. The middle ring (commonly termed 'inner ring') is located just beyond the end of the bar, near the inner 4:1 or ultraharmonic resonance, which itself is $\sim$10-20% inside the corotation (1:1) resonance. The 4-sided shape to the inner ring in NGC 6782 is evidence for the ring being associated with the 4:1 resonance. The smallest ring (commonly termed 'nuclear' or 'circumnuclear' ring) is located deep within the bar, and is associated with the inner Lindblad (inner 2:1) resonances (ILRs) of NGC 6782.

The formation of the OLR outer rings may be understood in terms of gravitational torques exerted by the bar on the spiral arms beyond corotation. These torques push the outer spiral arms outwards towards the OLR, forming a pseudo-ring [47]. Both gas and stars are concentrated here, and star formation may be more concentrated than the gas. In the OLR ring of NGC 1300, the HI surface density contrast is modest between the outer ring and the region just inside, whereas massive star formation occurs only in the ring [48]. This pattern of star formation may be understood by the gravitational instability theory. $Q\simeq1$ in the ring, indicating that the gas there is marginally unstable to gravitational collapse, whereas $Q>1$ interior to the ring, indicating that the gas there is stable.

Between corotation and the ILR, the gravitational torque of the stellar bar drives gas inwards. The torques can approach zero at the inner 4:1 resonance because the orbits there are symmetrically shaped and aligned with the bar. Thus inflow can stall at this resonance, forming an inner ring [49]. The shapes of inner rings vary from pointy ovals (as in NGC 6782) to nearly round, with the distribution of star formation intensity around the rings being correlated with ring shape [45]. Round inner rings have their HII regions distributed nearly uniformly around the ring, while extremely oval inner rings have their HII regions strongly concentrated near the major axis of the ring and bar. Gas trapped in inner rings follows oval streamlines, and moves slower near the ring major axis, leading to enhanced gas densities there. The increase in star formation at that location may be due to higher gas densities, and/or other dynamical effects associated with the bar.

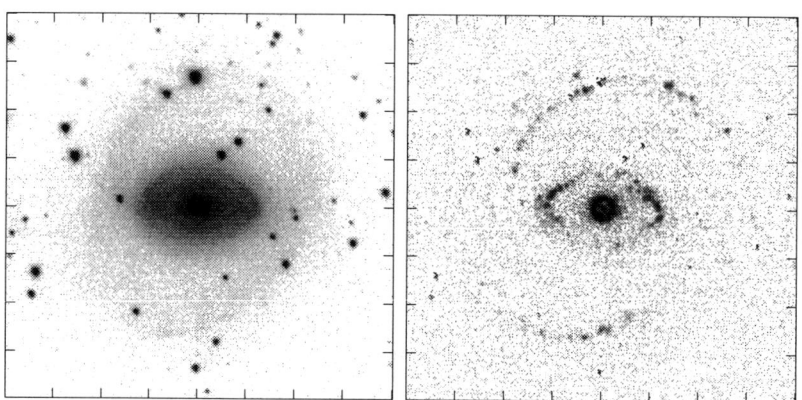

FIGURE 4. Red continuum and Hα images of the barred, ringed galaxy NGC 6782 [45]. Star formation is concentrated in an outer 'pseudoring', an inner oval ring and a circumnuclear ring. All three rings are associated with dynamical resonances of the stellar bar.

CIRCUMNUCLEAR RINGS AND STARBURSTS

Gas is driven inwards in galaxies as a result of dissipation and gravitational torques exerted by bars, spiral arms, or tidal disturbances. Most of the gas driven inwards probably does not reach the nucleus, but is consumed by star formation along the way or blown out in starburst winds. In a galaxy having 2 ILRs, theory shows that oppositely directed gravitational torques can act in the region of the resonances, causing the gas to pile up, at least temporarily, between the ILRs [27] [49].

There is good evidence that the "hot spot" star-forming rings and the maximum molecular gas surface densities in many barred galaxies are located between the 2 ILRs [10] [49] [50] [51]. In many galaxies, including NGC 6951 [50], the concentration of gas and star formation peak near the OILR, while in others they peak well inside it. In the intense starbursts NGC 3504 [16] and NGC 4102 [53] the peak gas density (at $\sim$100 pc resolution) is located well inside the OILR, and perhaps even inside the IILR. The behavior of gas in the vicinity of ILRs is complex and time-dependent. In simulations including star formation [54] [55], gas distributions in the vicinity of a double ILR evolve rapidly, and gas concentrations can form near both the OILR and IILR, and even get inside the IILR. However at present, few simulations can realistically model the effects of star formation, and more observations of the circumnuclear regions are needed to better understand the interplay between gas properties and star formation.

Star formation rates appear to be related to gas morphologies. Galaxies with the most centrally concentrated CO distributions tend to have the most luminous and shortest-lived starbursts [16] [50] [51] [52]. The HII ring ("hot spot") galaxy NGC 4314 [57], provides a striking comparison with the centrally peaked starburst galaxy NGC 3504 [16] (Fig. 5), and illustrates how there is much more contrast in the distributions of star formation than in the molecular gas. In the central 200 pc of these galaxies, the surface brightnesses in CO differ by only a factor of 6, while those in Hα and radio continuum differ by more than a factor of 100 [56]. Inside the star-forming ring of NGC 4314 there is virtually no massive star formation despite inferred molecular gas surface densities of 600 M$_\odot$ pc^{-2} [57]. This likely indicates a high threshold gas surface density required for star formation in the circumnuclear regions of galaxies. These circumnuclear gas surface density thresholds are $\sim$2 orders of magnitude higher than those in the outer disks of spiral galaxies, where the thresholds are typically $\sim$1-10 M$_\odot$ pc^{-2} [3] [58]. One possible reason for this difference is that the epicyclic frequency κ is much higher in galaxy centers than outer disks. A gas element must overcome much higher centrifugal and coriolis forces in order to gravitationally collapse near galaxy centers.

While there is evidence for the existence of a threshold gas surface density for star formation in circumnuclear regions, we still need to establish if it is the same as the critical gas density for cloud formation by gravitational instabilities. The Toomre Q parameter [1] has been measured in the circumnuclear regions of a few galaxies. $Q \simeq 1$ throughout the starburst regions of NGC 3504 [16] and NGC 4102 [53], and in the star-forming ring of NGC 4314. On the other hand, $Q>1$ inside the star-forming ring of NGC 4314 (Fig. 5), which may explain why there is virtually no star formation inside the ring despite significant quantities of molecular gas [56]. However, Q appears to be different from 1 in the star-forming regions of at least 2 other starburst galaxies. CO observations indicate $Q \simeq 0.2$-0.6 in Maffei 2 [59], and $Q \simeq 4$-1 (in adequately resolved regions) in NGC 3628 [60][61]. Given the uncertainties in the CO-H_2 conversion factor and the gas velocity dispersions, it is not clear if the above deviations of Q from 1 are significant. A larger sample of galaxies with Q values measured in a consistent way is needed to address this issue further.

One of the principal questions about starbursts is why they happen preferentially in circumnuclear regions. This may be partially understood in the framework of the gravitational instability picture. The gravitational collapse timescale goes as $\tau \sim \sigma/\Sigma_{gas} \sim Q/\kappa \sim QR$ [15][62]. It is much shorter near galaxy centers than outer disks if Q does not change significantly. Since the critical surface density is so high near galaxy centers, the gas might accumulate to high densities there without undergoing star formation. Once the gas surface density reaches the critical value, it is so high that the associated star formation timescale is very short. Although a convincing correlation between

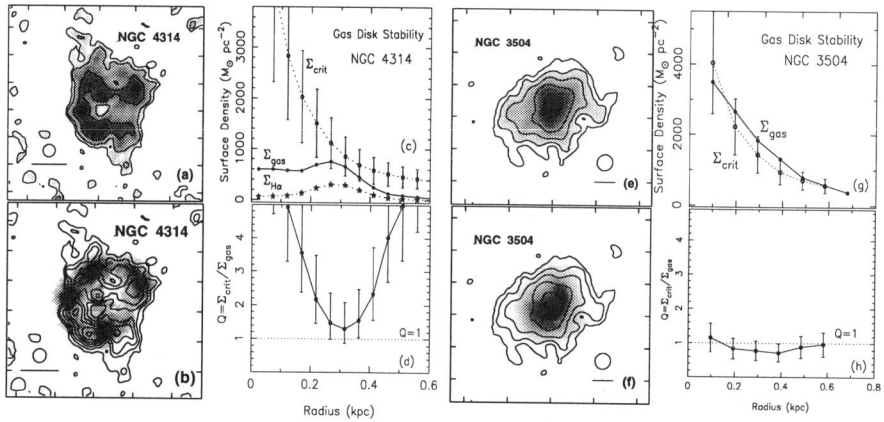

FIGURE 5. A comparison of molecular gas distributions, star formation distributions, and critical surface densities in the HII ring galaxy NGC 4314 and the central starburst galaxy NGC 3504. (a,e) CO maps. (b,f) Hα (greyscale) on CO (contour) maps. The horizontal lines in the maps represent 300 pc, and the circles indicate the beamsizes of $\sim 2''$. Radial profiles of gas and critical surface densities (c,g), and their ratio Q (d,h). $Q \simeq 1$ throughout the starburst regions of NGC 3504 and in the star-forming ring of NGC 4314. On the other hand, $Q>1$ inside the star-forming ring of NGC 4314, where there is hardly any star formation despite 600 $M_\odot$ pc^{-2} of molecular gas [16][56][57].

the local SFR and the value of κ remains lacking, this could well be the reason why starbursts are the common mode of star formation in galaxy centers.

We thank Ron Buta, Deb Crocker, Mike Regan, Kazushi Sakamoto, Michele Thornley, and all-around swell guy Rich Rand, for kindly providing figures.

REFERENCES

1. Toomre, A. 1964, ApJ, 139, 1217
2. Quirk, W. J. 1972, ApJL 176 L9
3. Kennicutt, R. C., Jr. 1989, ApJ, 344, 685
4. Jog, C. J. 1996, MNRAS, 278, 209
5. Elmegreen, D. M., & Elmegreen, B. G. 1984, ApJS, 54, 127
6. Thornley, M. D. 1996, ApJ, 469, L45
7. Rix, H.-W., & Rieke, M. J. 1993, ApJ, 418 123
8. Thornley, M. D. 1997, Ph.D. Dissertation, University of Maryland
9. Sakamoto, K. 1996, ApJ, 471, 173
10. Elmegreen, B. G. 1994, ApJ, 425, L73
11. Toomre, A. 1981, in The Structure and Evolution of Normal Galaxies, eds. S. M. Fall and Lynden-Bell, p. 111. Cambridge University Press, Cambridge.
12. Larson, R. B., 1984, MNRAS, 206, 197
13. Goldreich, P., & Lynden-Bell, D. 1965, MNRAS, 130, 125
14. Julian, W. H., & Toomre, A. 1966, ApJ, 146, 810.
15. Larson, R. B. 1987, in Starbursts and Galaxy Evolution, ed. T. X. Thuan, T. Montmerle & J. Tran Thanh Van (Gif Sur Yvette: Editions Frontieres), p. 467
16. Kenney, J. D. P., Carlstrom, J. C., & Young, J. S. 1993, ApJ, 418, 687
17. Roberts, W. W., Jr., & Stewart, G. R. 1987, ApJ, 314, 10
18. Garcia-Burillo, S., Combes, F., & Gerin, M. 1993, AA, 274, 148
19. Rand, R. J. 1993, ApJ, 410, 68
20. Elmegreen, B. G. 1987 ApJ 312, 626
21. Rand, R. J. 1993, ApJ, 404, 593
22. Rand, R. J., & Kulkarni, S. 1990, ApJ, 349, L43
23. Garcia-Burillo, S., Guelin, M., & Cernicharo, J. 1993, A&A, 274, 123
24. Knapen, J. H., Beckman, J. E., Cepa, J., van der Hulst, T., & Rand, R. 1992, ApJL, 385, L37
25. Phillips, A. C. 1996, in Proceedings of IAU Colloquium 157: Barred Galaxies, eds. R. Buta, D. Crocker, & B. Elmegreen, (ASP Conference Series), p. 44
26. Martin, P. 1996, in Proceedings of IAU Colloquium 157: Barred Galaxies, eds. R. Buta, D. Crocker, & B. Elmegreen, (ASP Conference Series), p. 70
27. Combes, F. 1988, in Galactic and Extragalactic Star Formation, eds. R. E. Pudritz and M. Fich, (Dordrecht: Kluwer), p. 475
28. Regan, M. W., Vogel, S. N., & Teuben, P. J. 1997, ApJ submitted
29. Handa, T., Sofue, Y., and Nakai, N. 1991, in IAU Symposium 146: Dynamics of Galaxies and Their Molecular Cloud Distributions, ed. F. Combes and F. Casoli (Dordrecht: Kluwer), p. 156

30. Elmegreen, B. G. 1979, ApJ, 231, 372
31. Kenney, J. D. P., & Lord, S. D. 1991, ApJ, 381, 118
32. Athanassoula, E. 1992, A&A, 259, 328
33. Das, M. & Jog, C. J. 1995, ApJ, 451, 167
34. Regan, M., Teuben, P., Vogel, S., & van der Hulst, T. 1996, AJ, 112, 2549
35. Elmegreen, B. G. & Elmegreen, D. M. 1985, ApJ, 288, 438
36. Combes, F. & Elmegreen, B. G. 1992, A&A, 267, 17
37. Noguchi, M. 1996, ApJ, 469, 605
38. Quillen, A. C., Frogel, J. A., Kenney, J. D. P., Pogge, R. W., & DePoy, D. L. 1995, ApJ, 441, 549
39. Friedli, D. & Benz, W. 1995, A&A, 301, 649
40. Kormendy, J. 1993, in Galactic Bulges, ed. H. Dejonge & H. J. Habing, (Dordrecht: Kluwer), p. 209
41. Combes, F., Debasch, F., Friedli, D., & Pfenniger, D. 1990, A&A, 233, 82
42. Hasan, H., Pfenniger, D., & Norman, C. 1993, ApJ, 409, 91.
43. Pfenniger, D. 1993 in Galactic Bulges, ed. H Dejonghe & H. J. Habing, p. 387.
44. Norman, C. A., Sellwood, J. A., & Hasan, H. 1996, ApJ, 462, 114.
45. Crocker, D. A., Baugus, P. D., & Buta, R. 1996 ApJS 105 353
46. Buta, R., & Crocker, D. A. 1991, AJ, 102, 1715
47. Schwarz, M. 1981, ApJ 247 77
48. Elmegreen, D. M., Elmegreen, B. G., Chromey, F. R., & Hasselbacher, D. A. 1996, ApJ, 469, 131
49. Combes, F. 1996, in Proceedings of IAU Colloquium 157: Barred Galaxies, eds. R. Buta, D. Crocker, & B. Elmegreen, (ASP Conference Series), p. 286
50. Kenney, J. D. P., Wilson, C. D., Scoville, N. Z., Devereux, N. A., & Young, J. S. 1992, ApJ, 395, L79
51. Kenney, J. D. P. 1997, in The Interstellar Medium in Galaxies, ed. T. van der Hulst, (Dordrecht: Kluwer), in press
52. Kenney, J. D. P. 1997, in Starburst Activity in Galaxies, eds. J. Franco, R. Terlevich & G. Tenorio-Tagle, Rev.Mex.Astron.Ap. (Conf. Ser.), in press
53. Jogee, S., & Kenney, J. D. P. 1996, in Proceedings of IAU Colloquium 157: Barred Galaxies, eds. R. Buta, D. A. Crocker, & B. G. Elmegreen, (ASP Conference Series), p. 230
54. Knapen, J. H., Beckman, J. E., Heller, C. H., Shlosman, I., de Jong, R. S. 1995, ApJ 454, 623
55. Shlosman, I. 1996, in Barred Galaxies and Circumnuclear Activity, eds. A. Sandquist et al., Springer-Verlag
56. Kenney, J. D. P., & Jogee, S. 1997, in prep
57. Benedict, G. F., Smith, B. J., and Kenney, J. D. P., 1996, AJ, 111, 1861
58. Skillman, E. 1997, these proceedings
59. Hurt, R. L., & Turner, J. L. 1991, ApJ, 377, 434.
60. Irwin, J., & Sofue, Y. 1996, ApJ, 464, 738
61. Irwin, J., private communication
62. Silk, J. 1988, in Galactic and Extragalactic Star Formation, eds. R. E. Pudritz and M. Fich, (Dordrecht: Kluwer), p. 503

Mergers, Interactions, and The Fueling of Starbursts

John E. Hibbard

Institute for Astronomy
2680 Woodlawn Drive
Honolulu, Hawai'i 96822

Abstract. The most active starbursts are found in galaxies with the highest IR luminosities, with peak star formation rates and efficiencies that are over an order of magnitude higher than in normal disk systems. These systems are almost exclusively on-going mergers. In this review I explore the conditions needed for interactions to experience such a phase by comparing two systems at similar stages of merging but quite different IR luminosities. These observations show that the most intense starbursts occur at the sites with the highest gas densities, which is a general result for IR luminous mergers. Observations and theory both suggest that the strength of the merger induced starburst depends on the internal structure of the progenitors, the amount and distribution of the gas, and the violence of the interaction. In particular, interactions involving progenitors with dense bulges, gas-rich disks, and/or a retrograde spin are expected to preferentially lead to large amounts of gaseous dissipation, although the interplay between these parameters is unknown. A major outstanding question is how the effects of feedback alter these conclusions.

INTRODUCTION

While galaxies showing both peculiar morphologies and signs of global youth have been known for decades (e.g. [1–3]), it was not until a classic study on the *UBV* colors of peculiar galaxies by Larson & Tinsley in 1978 [4] that the study of starbursts in interacting galaxies began in earnest. In that seminal work the authors showed that the disturbed systems from the Arp *Atlas of Peculiar Galaxies* [5] had a larger spread in colors and significantly bluer colors than a comparison sample of normal Hubble types. The Toomres [6] had recently demonstrated quite convincingly that gravitational interactions provide a natural explanation for many of the types of peculiarities exhibited by the systems in Arps atlas, and Larson & Tinsley posited that such interactions induce small bursts ($\sim$ 1-5% by mass) of star formation within the host(s). They were able to explain the color distribution of the Arp systems

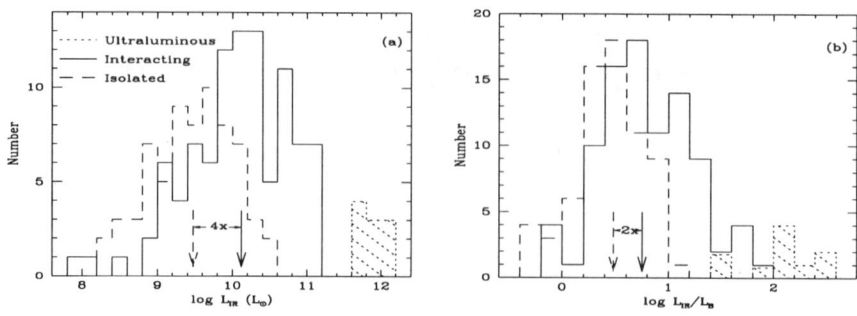

FIGURE 1. Histograms of L_{IR} and L_{IR}/L_B for isolated, interacting, and infrared bright systems. The isolated and interacting samples were chosen optically, independent of their infrared properties. Arrows indicate their means. These plots show that while the brightest IR emitting systems are interacting, many interacting systems are not IR bright. From Bushouse, Lamb & Werner 1988 [8].

with a grid of burst models of varying strength and age superimposed upon an underlying host of normal colors.

Much subsequent work has supported this suggestion [7]. The consensus is that interacting galaxies as a class have mean levels of star formation that are factors of 2–5 higher than normal spirals for optically selected samples, or 2–20 times higher for samples selected on the basis of IR luminosity. This point is illustrated in Figure 1 [8], which compares the IR luminosities of an optically selected isolated and interacting sample as well as a sample of ultraluminous infrared (ULIR; $L_{IR} > 3 \times 10^{11} L_\odot$ in this case, although other definitions are used) galaxies which are known to be on-going mergers [9]. Similar results have been derived using other measures of star formation, finding that interactions serve primarily to concentrate moderately enhanced star formation into the central regions of galaxies rather than to globally raise the star formation rate (e.g. [10–13]).

Figure 1 illustrates several other points. The first is that none of the isolated systems have $L_{IR} > 3 \times 10^{10} L_\odot$ or $L_{IR}/L_B > 15$. At these levels almost all galaxies are interacting or merging [9,14,15]. In fact, IR luminosity appears to be the most efficient way to select interacting systems: while the overall peculiar fraction of optically selected samples is around 9% [16], the fraction of morphologically peculiar galaxies approaches 90% or higher at IR luminosities above $5 \times 10^{11} L_\odot$ (see [17] and references therein). The second is that many systems in the interacting sample do *not* exhibit enhanced levels of star formation. So while the most luminous IR sources are almost invariably mergers, not all mergers are luminous IR sources.

IR luminosity is believed to be directly related to the number of hot stars present, as the dust absorbs the UV photons from these stars and reradiates them in the IR [14,15]. It can therefore be used as a measure of the massive

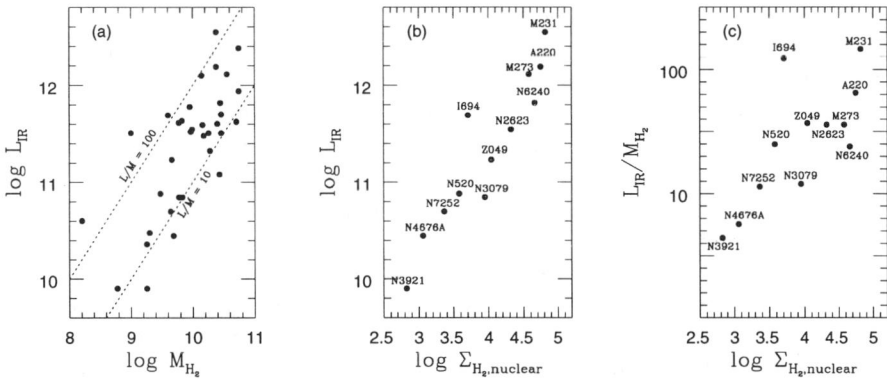

FIGURE 2. (a) total IR luminosity vs. derived H_2 mass for all objects observed at OVRO. Normal spirals fall below the L/M=10 line. (b) L_{IR} vs. the molecular gas column density averaged over a 1 kpc diameter region ($\Sigma_{H2,nuclear}$) for the 12 objects observed with sufficient resolution. (c) SFE vs. $\Sigma_{H2,nuclear}$. From Yun & Hibbard, in preparation.

star formation rate (MSFR; $M > 5M_\odot$) [18]. The IR luminous systems are known to be both gas-rich and dusty [9,19,20], and most of their bolometric luminosity emerges in the far IR [9,17]. While there has been a long-standing debate over whether the luminosity of the most luminous systems is instead powered by an obscured AGN (see e.g. [13,21]), this ambiguity is mostly at the highest IR luminosities, and the majority are believed to be predominantly powered by starbursts (see reviews by [7,22]). The popular picture which has emerged is that two gas rich systems undergo a close interaction, leading to orbital decay and eventual merging. The gas dissipates and moves inward, stimulating a circumnuclear starburst. Dust, which is coupled to the gas, absorbs much of the UV radiation and re-radiates it in the IR. This high level of star formation quickly subsides as the burst consumes the available gas.

But do all mergers go through a strong burst of star formation? If not, what are the deciding factors? These are the issues I will address in this review.

Quantifying the Star Formation Activity

Although star formation activity has historically been defined by some absolute value such as IR luminosity, it would be preferable to use a relative measure which compares the total SFR to the the size of the system. In this way we can address whether star formation proceeds in an inherently different manner at these luminosities, rather than being a scaled up version of processes taking place at lower luminosities. Blue luminosity is frequently used as a substitute for the stellar mass (e.g. Fig. 1b), although this interpretation is wrought with uncertainties due to the presence of young stars and large amounts of dust. L_{IR}/L_K is a more useful measure, although L_K is still affected by the presence of red AGB stars.

A popular normalization introduced by millimeter astronomers is the ratio of the IR luminosity to the molecular gas mass (L_{IR}/M_{H2}), where M_{H2} is estimated from measurements of the CO line. This ratio is termed the "star formation efficiency" (SFE) [20,23,24] as it represents in some global sense the number of massive stars formed per giant molecular cloud. Since $L_{IR} \sim$ MSFR [18], the SFE is inversely related to the gas depletion time, and is thus equivalent to the more classical definition of a starburst ($i.e\ \Delta t_{burst} << H_o^{-1}$).

Many studies have evaluated the SFE in merging galaxies concluding that ULIR mergers are forming stars up to an order of magnitude more efficiently than normal spirals (see Figure 2 and [17]). While there is some question as to whether the Galactic conversion factor between CO and H_2 applies in merging galaxies [19,25,26], the expected variation leads to an overestimate of the molecular gas and would result in even higher intrinsic SFEs. It therefore seems hard to escape the conclusion that massive stars are being formed at a higher rate for a given amount of cold gas than in quiescent systems, although the exact level of the enhancement is uncertain.

The Tails of Two Mergers

To understand how this efficient mode of star formation is triggered, it is instructive to compare two nearby mergers at apparently similar stages of merging but at two quite different levels of star formation activity. The mergers I will examine are "The Antennae" (NGC 4038/9, Arp 244; V_o=1630 km s^{-1}) and Arp 299 (NGC 3690/IC 694; V_o=3080 km s^{-1}). Both the tidal structure and inner disks of these systems are shown in Figure 3 along with distribution of the cold gas components.

The evidence that these systems are at a similar stage of merging is the following: both have a long tidal tails (130 kpc for NGC 4038/9 assuming a distance of 25 Mpc; 180 kpc for Arp 299 assuming a distance of 48 Mpc), suggesting several hundred Myr since first orbital periapse ($\gtrsim$500 Myr for NGC 4038/9; $\gtrsim$700 Myr for Arp 299); the disks of their progenitors are highly distorted and in physical contact, yet still distinct; and their nuclei are still well separated (8 kpc and 4 kpc, respectively). Both systems have similar CO distributions (Fig. 3c-d), with concentrations of cold molecular gas near both nuclei and a significant concentration at the region of disk overlap. Despite these similarities, Arp 299 is almost an order of magnitude more luminous in the infrared than NGC 4038/9, indicating a much higher MSFR (see Table 1).

In Table 1 we use NIR flux measurements [27,28] to divide the IR luminosity among the various components. This comparisons shows that a major difference between the systems is in the amount of nuclear star forming activity: in NGC 4038/9, the overlap region is the most active star forming region, both in terms of the MSFR and the SFE. In Arp 299, on the other hand, the overlap region has similar properties as that in NGC 4038/9, but both nuclei outshine

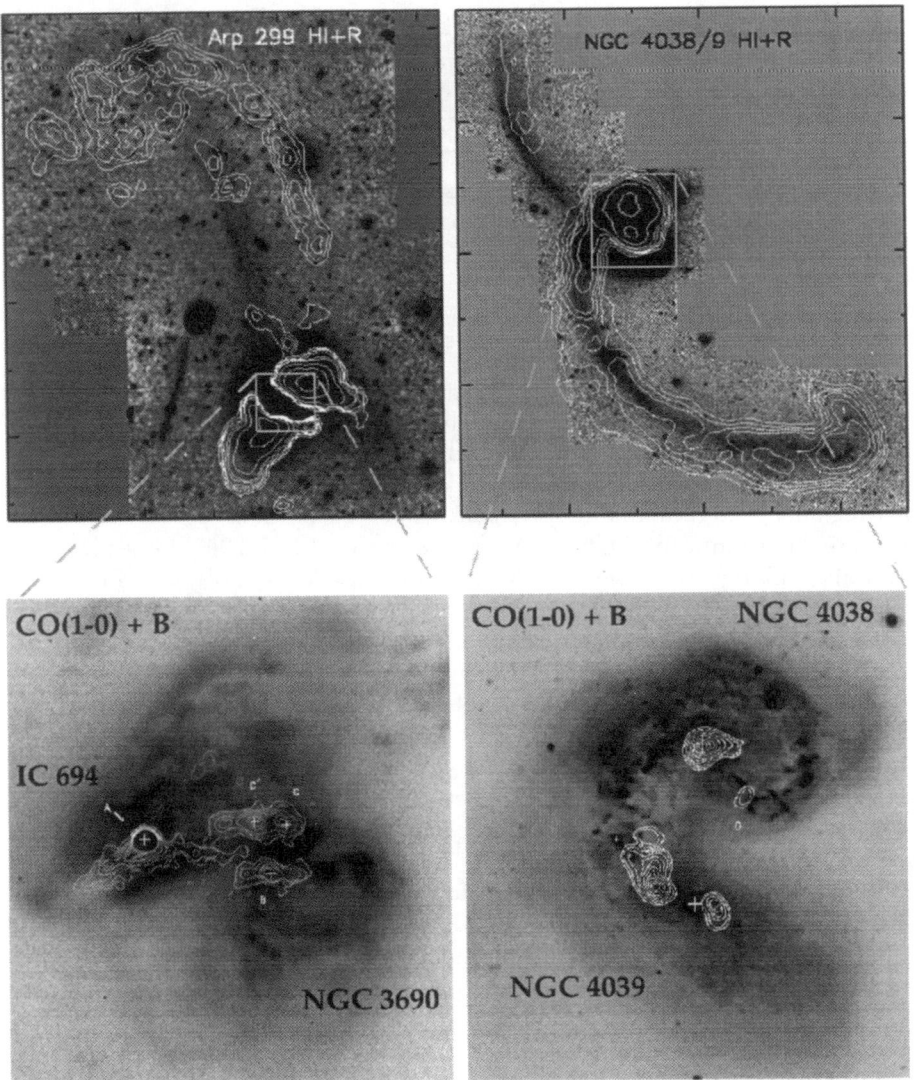

FIGURE 3. Two on-going mergers. (a)&(c) NGC 4038/9 and (b)&(d) Arp 299. In the upper two panels, a deep R-band image of the entire system is shown with VLA H I contours overlaid, while in the lower two panels a B-band image of the inner regions is shown with OVRO ^{12}CO(1-0) contours overlain. The optical data were obtained by the author; the VLA data are from Hibbard & Yun 1996 [31] (Arp 299) and Hibbard & van der Hulst in preparation (NGC 4038/9); the OVRO data are from Aalto et al. 1997 [32] (Arp 299) and Stanford et al. 1990 [33] (NGC 4038/9).

TABLE 1. Properties of Star Forming Regions in Arp 299 & NGC 4038/9.

	L_{IR}[a] ($\times 10^{10} L_\odot$)	M_{H2}[b] ($\times 10^9 M_\odot$)	L_{IR}/L_K	MSFR ($M_\odot \text{yr}^{-1}$)	Σ_{H2} ($M_\odot \text{pc}^{-2}$)	SFE ($L_\odot M_\odot^{-1}$)
IC694	49.8	4.0	—	31	27,000	124
N3690	24.5	1.0	—	15	4,900	245
overlap	4.7	2.0	—	3	3,400	24
Arp 299	79.0	7.0	31	49	27,000	113
N4038	2.0	0.8	—	1	1,200	25
N4039	0.04	0.2	—	0	540	2
overlap	7.8	1.2	—	5	1,350	65
N4038/9	9.8	2.0	13	6	1,350	49

[a] L_{IR} is split between components using 15μm ISOCAM measurements for NGC 4038/9 [27], and 32μm measurement for Arp 299 [28].
[b] From OVRO CO observations of NGC 4038/9 [33] and Arp 299 [34].

this region by large amounts. We note that observations argue against an energetically significant AGN in any of the nuclei [13,27–29] (but see [30]).

Similar differences are seen in the column densities of molecular gas: there is five times as much gas in the central kpc of the nuclei of Arp 299 as in NGC 4038/9. As a result, the peak column densities are over an order of magnitude higher in the nuclei of Arp 299 as in the nuclei of NGC 4038/9 or the overlap regions. The tight correlation between Σ_{H2} and L_{IR} shown in Fig. 2 shows this to be a general result for mergers over a broad range of IR luminosity [20]. While it is still possible that M_{H2} has been overestimated [25,26] a similarly tight correlation exists between L_{IR} and L_{HCN}, as well as between SFE and L_{HCN}/L_{CO} [35], showing that the fraction of gas at very high densities ($n > 10^5 \text{cm}^{-3}$) is greatly increased in ULIR systems. As these fractions are much higher than those found in normal spirals, this indicates that MSFR is intricately linked to the amount of gaseous dissipation [19,26,36].

It therefore seems that in order to understand how such efficient periods of star formation are induced, we need to understand how so much gas attains such high gas densities. Most phenomenological scenarios of star formation predict that the MSFR and the SFE are proportional to the gas density to some power (e.g. [37,38]), although they do not address in detail when and where such densities are attained. For this, we turn now to results from numerical simulations.

Lessons from Simulations

Ever since a dissipational component was first included in numerical simulation of interactions it has been clear that large amounts of gas can be efficiently driven into the central regions [39–41]. In this section I examine individual models for clues as to what parameters play the largest role in driving gas to high densities. This is not intended to be a review of all the simulations (for

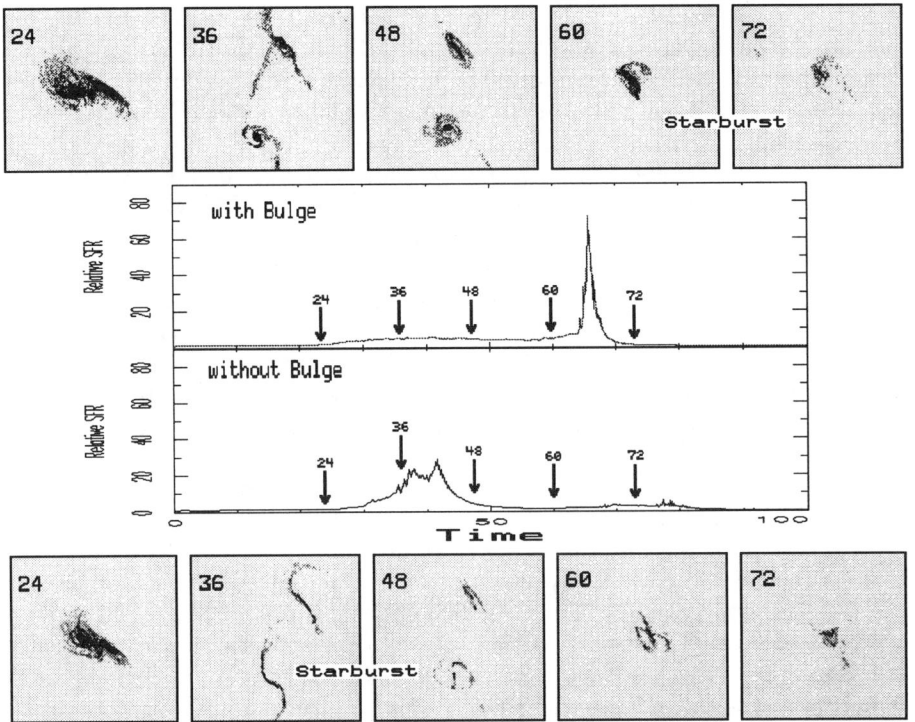

FIGURE 4. Time history of two interaction induced starbursts. In the bottom sequence, the lack of a dense bulge allows strong bar-induced inflows to develop early in the encounter. The relatively mild starburst which ensues consumes much of the available gas supply, leading to a less extreme burst at the final merger. Adapted from Mihos & Hernquist [43].

that see [42]), but rather an assessment of the results which might have the most relevance for understanding the observations discussed above.

Progenitors: A very interesting result to emerge from the numerical studies is that the progenitor structure ought to play a dominant role in the star formation history (SFH) [43,44]. This is illustrated in Figure 4. In the upper sequence the presence of a bulge stabilizes the disks against bar formation, lowering the pre-merger star formation levels, but leaving more gas in the progenitors to fuel a strong starburst when they finally merge. Because of the peaked appearance, it is tempting to associated this SFH with the ULIR systems, but it is important to realize that the vertical axis is a *relative* SFR, and the absolute SFR between the two curves could differ due to other factors (see below). Indeed, the fact that Arp 299 is so luminous while the nuclei are still well separated argues that it follows the lower SFH in Fig. 4. This interpretation is supported by the H I observations, which suggests two late type progenitors [45]. NGC 4038/9 on the other hand appears to have one late type (NGC 4038) and one earlier type (NGC 4039) progenitor [46], and may have a SFH more like the upper curve in Fig. 4, with its most active star forming period still in the future. These considerations suggest that while a dense bulge may push certain systems into the ULIR regime, it need not be a strict requirement. It also points to the need for better constraints on the past and future star forming histories to facilitate comparisons with models.

Gas content: There is over three times as much molecular gas in Arp 299 than NGC 4038/9, although the relative gas contents are similar (M_{H2}/L_K=0.5, 0.4). Could this be responsible for the different levels of activity? Clearly the level of star formation must be closely connected to the amount of fuel available, but plots of SFE or L_{IR} vs. M_{H2} show notoriously large amounts of scatter (e.g. Fig. 2a). So while the gas content may help shift the SFR upward for some systems (see Gao et al. these proceedings), high gas mass alone is not a sufficient condition for fueling an ULIR phase. Olson & Kwan [47] conducted one of the few simulations to explore the effects of different gas masses and distributions. Using a code in which the SFR is proportional to the cloud collision rate, they find that doubling the gas content in one of the disks in a disk-disk merger doubles the SFR, therefore keeping the SFE (=SFR/M_{gas}) constant, while splitting the same amount of gas between two galaxies leads to a 30% higher SFR and SFE than putting it all into one of the system. This suggests that how the gas is distributed and how it collides is as important as how much gas is present. These studies should be continued with a wider range of encounter parameters.

Spin geometries: The two systems under consideration here differ in their spin geometries, with Arp 299 undergoing a prograde-retrograde encounter [45,48] and NGC 4038/9 undergoing a prograde-prograde encounter [46]. Since retrograde encounters fail to raise strong tails [49], there will be more gas left in the inner regions at the late stages of merging where it will be violently perturbed as the encounter progresses. The simulations of Barnes & Hernquist [49] show that the fraction of gas at high densities depends on encounter geometry, with retrograde encounters leading to higher quantities of dense gas than prograde encounters. In Arp 299, the highest gas column densities are found in the nucleus of the disk experiencing the retrograde encounter (IC 694). Mihos & Hernquist [43,44] also evaluated the effects of spin geometry, and while they showed that it had a much less dramatic effect on the SFRs than progenitor structure, geometry still made a difference of about a factor of two in the relative SFRs. If there is either a threshold to the onset of very efficient star formation activity, or if SFE is a function of the fraction of the gas at very high densities, it is possible that spin geometry plays a role beyond that attributed to it in these studies.

The reality is that all of the factors are probably playing a part, although in what combination and order of importance is not clear. Continued parameter studies are needed in concert with detailed observations of individual systems in order to discriminate between the different processes. It is also important

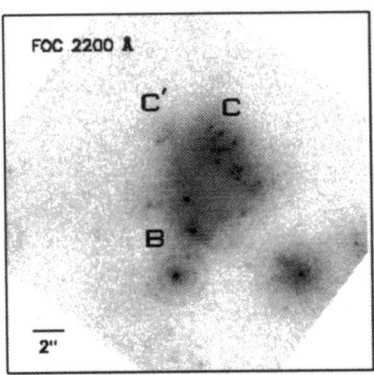

FIGURE 5. *HST FOC* 2200Å observation of the B-C-C' complex in NGC 3690 (Arp 299 West) from Vacca 1995 [51]. These observations show that many of the massive stars in this system are confined to very bright, compact knots.

to compare the actual distributions and dynamics of the star forming regions and dense gas with the predictions of the simulations in order to discriminate between different numerical formalisms for star formation and gas dynamics. Only by doing this will we know how far to trust the models or how to better conduct our observations.

Other Aspects of Interaction Induced Starbursts

There are many other interesting aspects of interaction induced starbursts, and in the remaining space I simply mention a few.

Star Formation Knots: When viewed with sufficient resolution the star-forming regions in Arp 299 and NGC 4038/9 are found to break up into many distinct knots along with a diffuse component. Figure 5 shows an *HST FOC* image of NGC 3690 [50,51]. These knots have typical diameters of less than 10pc and may evolve into globular clusters [50,52]. However these knots are not unique to mergers [50,53–55], and it is not clear if this mode of star formation is increased in such interactions and if it is related to the enhanced SFE. It may just be an important mode of star formation in general. A detailed comparison of the luminosity functions of such knots from different types of environments would help resolve this question.

The Return of Tidal Debris: Due to the strong tides experienced during merging encounters, appreciable amounts of stars and gas are lifted high above the merging systems into tidal features. These features frequently exhibit significant substructure, the largest having observational properties typical of dwarf galaxies [56–59]. Much of this material remains bound to the remnant on long-period orbits, and will take several Gyr to fall-back [60,61]. Tidal clumps that are far out along the tails may be able to avoid tidal stripping when they fall back towards the remnant and should become long-lived dwarf companions [61], while the more tightly bound material will fall back into the remnant. The stellar component of these tails will wrap coherently in the central potential, giving rise to fine structure features [60] while the gas may feed a prolonged period of low-level star formation. The largest clumps have as much as a few $10^8 M_\odot$ of H I [58,59], and their return may give rise to smaller bursts. The overall star formation history in merger events should be similar to that illustrated by Worthy (this volume).

Because of the timescales and amounts of gas involved, we therefore do not expect two merging spirals to turn quickly into an elliptical, but rather for there to be a series of transitions, e.g. to an S0pec, to an S0, to a dust lane elliptical, etc. [59]. By the time the more obvious signs of its merger origin have faded and the remnant has evolved into a *bona fide* elliptical, the stars formed during the merger induced starburst will have aged 2-5 Gyr, leaving very little indication of a merger origin in the remnants broad band colors. This picture is quite similar to the one emerging from studies of cluster populations at high redshift (see Dressler, these proceedings).

Outstanding Questions

The main outstanding question for interaction induced star formation is the same as for any field of star formation: is the IMF universal, and if not how is it different in these violent environments? Are there upper and lower mass limits? Is the IMF top heavy? All of these effects have been claimed [30,62,63], but the observations allow for a wide range of parameters. A major problem with these determinations is that the observered bursts are both temporally and spatially variable, and as a result there will be a mix of burst populations of varying strength and ages spread throughout the merger. An IMF measured globally will reflect the luminosity averaged IMF over the region observed, which will be different in different wavebands. Careful UV and/or NIR spectroscopy of individual knots in conjunction with dynamical modeling should help separate these effects. Such observations should also help constrain the past star formation history in the systems, allowing one to map the age distribution of the star forming episodes.

Another major question is the how does the energy injected back into the surrounding gas via the winds from massive stars and SNe change the gas dynamics and burst properties? This process, referred to as "feedback", is seen in its most extreme form in mergers, as evidenced by the galactic scale superwinds emerging from many ULIR systems (see Heckman, these proceedings). It may simply be an interesting side-effect of the circumnuclear star bursts, or it may dramatically affect the physics of star formation, for example by raising the lower-mass cutoff of the IMF or changing its high-end slope [64], or by regulating the star formation rate at some critical value [65].

There is much hope for further progress to be made in these areas in the future. It is becoming feasible to model a multi-phase ISM, which should help assess the importance of feedback. Further numerical trade studies should help decide how the different parameters interact and predict relationships between gaseous and star forming regions. Careful observations can test these predictions, and in this way we can discriminate between the different numerical formalisms. Not only will this provide a deeper understanding of interaction induced star formation, but ultimately it should be possible to decide which of the two histories depicted in Fig. 4 a merger follows, and which of these lead to an ULIR phase. Since the ULIR systems may be the closest analogs to the star forming galaxies seen at high redshift (see Madau, these proceedings), this understanding will provide valuable insight into the major processes at work during the epoch of galaxy formation.

I wish to thank J. van Gorkom and W. Vacca for comments on this manuscript; to M. Yun, C. Mihos, S. Aalto and W. Vacca for permission to reproduce their figures; and to the organizing comittee for setting up such an interesting meeting. This work is supported by Grant HF–1059.01–94A from STScI, which is operated by AURA, Inc., under NASA contract NAS5–26555.

REFERENCES

1. Zwicky, F. 1950, *Experientia*, 6, 441.
2. Burbidge, E.M., Burbidge, G.R., & Hoyle, F. 1963, ApJ, 138, 873.
3. Searle, L. Sargent, W.L.W., Bagnuolo, W.G. 1973, ApJ, 179, 427.
4. Larson, R.B., Tinsley, B.M. 1978 ApJ, 219, 46.
5. Arp, H.C. 1966, ApJ (Supp.), 14, 1.
6. Toomre, A., Toomre, J. 1972, ApJ, 178, 623.
7. Kennicutt, R.C.Jr. 1990, in "Paired and Interacting Galaxies", IAU Colloq. No. 124, Sulentic, Keel, & Telesco eds. (NASA, Washington), p. 269.
8. Bushouse, H.A., Lamb, S.A., & Werner, M.W. 1988, ApJ, 335, 74.
9. Sanders, D.B., et al. 1988, ApJ, 325, 74.
10. Keel, W.C., et al. 1985, AJ, 90, 708.
11. Bushouse, H.A. 1987, ApJ, 320, 49.
12. Kennicutt, R.C. Jr., et al. 1987, AJ, 93, 1011.
13. Condon, J.J., Huang, Z.-P, Yin, Q.F., Thuan, T.X. 1991, ApJ, 378, 65.
14. Lonsdale, C.J., Persson, S.E., Matthews, K. 1984, ApJ, 287, 95.
15. Joseph, R.D., Wright, G.S. 1985, MNRAS, 214, 87.
16. Arp, H., & Madore, B.F. 1975, Observatory, 95, 212.
17. Sanders, D.B., Mirabel, I.F. 1996, ARAA, 34, 749.
18. Condon, J.J. 1992, ARA&A, 30, 575.
19. Scoville, N.Z., et al. 1989, ApJ (Lett.), 345, L25.
20. Scoville, N.Z., Sargent, A.I., Sanders, D.B., Soifer, B.T. 1991, ApJ, 366, L5.
21. Lonsdale, C.J., Smith, H.E., & Lonsdale, C.J. 1995, ApJ, 438, 632.
22. Heckman, T.M. 1990, in "Paired and Interacting Galaxies", IAU Colloq. No. 124, Sulentic, Keel, & Telesco eds. (NASA, Washington), p. 357.
23. Young, J.S., Schloerb, F.P., Kenney, J.D., Lord, S.D. 1986, ApJ, 304, 443.
24. Solomon, P.M., Sage, L.J. 1988, ApJ, 334, 613.
25. Maloney, P. & Black, J.H. 1988, ApJ, 325, 389.

26. Downes, D., Solomon, P.M. & Radford, S.J.E. 1993, ApJ (Lett.), 414, L13.
27. Vigroux, L. et al. 1996, A&A, 314, L93.
28. Wynn-Williams, C.G., et al. 1991, ApJ, 377, 426.
29. Ridgeway, S.E., Wynn-Williams, C.G. & Becklin, E.E. 1994, ApJ, 428, 609.
30. Shier, L.M., Rieke, M.J. & Rieke, G.H. 1996, ApJ, 470, 222.
31. Hibbard, J.E., & Yun, M.S. 1996, in "Cold Gas at High Redshift", Bremer, Rottgering, van der Werf & Carilli eds. (Kluwer, Dordrecht), p. 47.
32. Aalto, S., Radford, S.J.E., Scoville, N.Z. & Sargent, A.I. 1997, ApJ, in press.
33. Stanford, S.A., Sargent, A.I., Sanders, D.B., Scoville, N.Z. 1990, ApJ, 349, 492.
34. Sargent, A.I., Scoville, N.Z. 1991, ApJ (Lett.), 366, L1.
35. Solomon, P.M., Downes, D., & Radford, S.J.E. 1992, ApJ (Lett.), 387, L55.
36. Kormendy, J. & Sanders, D.B. 1992, ApJ (Lett.), 390, L53.
37. Silk, J. 1996, ApJ, in press (astro-ph/9612117)
38. Elmegreen, B. 1994, in "Violent Star Formation from 30 Doradus to QSOs", ed. G. Tenorio-Tagel (Cambridge University Press, Cambridge).
39. Negroponte, J., White, S.D.M. 1983, MNRAS, 205, 1009.
40. Noguchi, M., Ishibashi, S. 1986, MNRAS, 219, 305.
41. Barnes, J.E., & Hernquist, L. 1991, ApJ (Lett.), 370, L65.
42. Barnes, J.E., & Hernquist, L. 1992, ARA&A, 30, 705.
43. Mihos, J.C., Hernquist, L. 1994, ApJ (Lett.), 431, L9.
44. Mihos, J.C., Hernquist, L. 1996, ApJ, 464, 641.
45. Hibbard, J.E., & Yun, M.S. in preparation.
46. van der Hulst, J.M. 1979, A&A, 155, 151.
47. Olson, K.M., Kwan, J. 1990, ApJ, 361, 426.
48. Augarde, R., and Lequeux, J. 1985, A&A, 147, 273.
49. Barnes, J.E., & Hernquist, L. 1996, ApJ, 471, 115.
50. Meurer et al. 1995, AJ, 110, 2665.
51. Vacca, W.D. 1995, "The Interplay between Massive Star Formation, the ISM, and Galaxy Evolution", Kunth et al. eds. (Ed. Frontieres) p. 321.
52. Whitmore, B.C. & Schweizer, F. 1995, AJ, 109, 960.
53. Conti, P.S. & Vacca, W.D. 1994, ApJ (Lett.), 423, L97.
54. Holtzman etal 1996, AJ, 112, 416.
55. Maoz, D., et al. 1996, AJ, 111, 2248.
56. Schweizer, F. 1978, "The Structure and Properties of Nearby Galaxies", IAU Symp. No. 77, Berkhuijsen and Wielebinski eds. (Reidel, Dordrecht), p. 279.
57. Mirabel, I.F., Duc, P.-A., & Dottori, H. 1994, in "Dwarf Galaxies", Meylan & Prugniel eds. (ESO) p. 371.
58. Hibbard, Guhathakurta, P., van Gorkom & Schweizer, F. 1994, AJ, 107, 67.
59. Hibbard, J.E., & van Gorkom, J.H. 1996, AJ, 111, 655.
60. Hernquist, L. & Spergel, D.N. 1992, ApJ (Lett.), 399, L117.
61. Hibbard, J.E., & Mihos, J.C. 1995, AJ, 110, 140.
62. Doyon, R., Puxley, P.J. & Joseph, R.D. 1992, ApJ, 397, 117.
63. Lançon, A. & Rocca-Volmerange, B. 1996, New Astronomy, 1, 215.
64. Zepf, S. & Silk, J. 1996, ApJ, 446, 114.
65. Lehnert, M.D. & Heckman, T.M. 1996b, ApJ, 472, 546.

Starbursts and Cosmogony

Timothy M. Heckman

Department of Physics & Astronomy, The Johns Hopkins University, Baltimore, MD 21218 USA

Abstract. Starbursts are an energetically significant component of the universe and the sites for at least 25% of the high-mass star-formation in the local universe. They offer unique laboratories for testing our ideas about star-formation, the evolution of massive stars, and the physics of the interstellar medium. They may also serve as local analogs of the processes that were important for the formation and early evolution of galaxies and the heating and chemical enrichment of the inter- galactic medium. In this talk I review starbursts from this cosmogonical perspective, and attempt to relate starbursts to some of the key issues addressed during this conference.

INTRODUCTION

The theme of my talk is to relate starbursts to the 'big picture' issues that are of most relevance to this conference. In other words, while starbursts are fascinating objects in their own right, they are even more important because they also provide 'laboratories' for the study of: 1) the evolution of high-mass stars 2) the physics of the interstellar medium, and especially of star-formation, under extreme conditions 3) processes relating to the formation and early evolution of galaxies 4) the outflows likely responsible for heating and chemically enriching the inter-galactic medium.

Before discussing these specific issues, I would like to begin my talk by trying to place starbursts into context. How energetically significant are starbursts in the present-day universe? What fraction of the (high mass) stars being formed in the local universe are in starbursts? Are starbursts just interesting curiosities or a fundamental astrophysical phenomenon?

I will follow convention and define a starburst to be a galaxy whose luminosity primarily arises in a region in which stars are forming at rate that can be sustained for only a small fraction of a Hubble time (e.g. < Gyr). The most complete and well-characterized sample of starbursts is the far-IR-selected sample drawn from the IRAS survey (e.g. Soifer et al 1989). The rather strong inverse correlation between gas-depletion timescales and far-IR luminosity (Sanders & Mirabel 1996 and references therein) means that we may consider that starbursts dominate the set of galaxies lying above the 'knee' in the far-IR galaxy luminosity function. By this reckoning, starbursts provide about 10% of the bolometric emissivity of the local universe. Thus indeed, starbursts are an energetically-significant phenomenon.

It is also instructive to estimate the rate of high-mass star- formation in starbursts, compared to that in the disks of normal galaxies in the local universe. To do so, I have used the extinction-corrected luminosity of Hydrogen recombination lines as an estimator of the rate of formation of high-mass stars, and applied this technique to the distance-limited Kraan-Korteweg & Tammann (1979) catalog of the nearest galaxies ($D < 10$ h_{50}^{-1} Mpc). I then find that four most luminous circumnuclear starbursts (M 82, NGC 253, NGC 4945, and M 83) together comprise about 25% of the total star- formation rate in this volume and the six most-luminous normal galactic disks contribute a comparable amount. In specific terms, the rate of high-mass star-formation in the few-hundred-pc- scale nucleus of M 82 actually exceeds the rate in the disk of a giant Sc I galaxy like M 101!. Thus, circumnuclear starbursts are comparable to the disks of spiral galaxies in the production of high-mass stars. Gallego et al (1995) have reached a similar conclusion based on their $H\alpha$ objective prism survey.

If the local universe is typical of the universe as-a-whole, it would appear that either: 1) starbursts (unlike galactic disks) make only massive stars and therefore leave little long-term residue, or 2) starbursts are responsible for building the inner disks and possibly the bulges of spiral galaxies and (in extreme cases of merger-driven starbursts) elliptical galaxies. This possibility leads to the idea that secular evolution of disk galaxies may be important even at the current epoch (e.g. Norman, Hasan, & Sellwood 1996), and that what are called 'bulges' may be a diverse class of objects including post-starbursts (M. Stiavelli, private communication).

DISK DYNAMICS, BARS, & THE STARBURST PHENOMENON

Matt Lehnert and I have recently published the results of an optical spectroscopic and imaging survey of the ionized gas in a far-IR flux-limited sample of about 50 starbursting spiral galaxies whose disks were viewed from within 30 degrees of edge-on. In Lehnert & Heckman (1996) we have used the rotation curves to study empirically the relationship between the disk kinematics and the size and strength of the starburst. I would like to highlight some results from this paper that I think are of particular relevance to this meeting:

1. The maximum star-formation rate allowed by causality is one in which all the gas in a starburst is converted into stars in one dynamical time-scale. This

leads to a prediction that the maximum starburst luminosity will scale like the cube of the starburst velocity dispersion (Heckman 1993). Comparing the far-IR luminosities and rotation speeds implies that most of the starbursts in our sample are forming stars at a rate that is typically an order-of-magnitude below this limit (even assuming a normal Initial Mass Function and a normal gas fraction). However, several extreme objects have such high luminosities for their rotation speeds that they violate this limit unless they are forming only massive stars and/or gas dominates the mass of the starburst. Both of these hypotheses have been offered on other grounds for starbursts (cf. Scoville & Soifer 1991; Rieke et al 1993).

2. The rather normal, symmetric rotation curves exhibited by most of our galaxies imply that fueling a circumnuclear starburst does *not* require severe dynamical disruption of the galactic disk, at least for the moderately luminous starbursts in our sample. This is a markedly different situation than for the rare but spectacular 'ultraluminous' IR galaxies in which major mergers are probably occuring (cf. Scoville & Soifer 1991; Mihos & Hernquist 1994a). Thus, the mechanism by which *typical* starbursts are triggered is not yet clear. Perhaps the capture of dwarf galaxies or mild grazing encounters with neighbors are responsible (cf. Mihos & Hernquist 1994b). Secular evolution of the disk with a bar-driven inflow (possibly resulting from a mild tidal interaction) is suggested by the empirical links between bars and circum-nuclear starbursts (e.g. Devereux 1989; Ho, Filippenko, & Sargent 1996).

3. The rotation curves of the starburst host galaxies generically have an inner solid body portion changing at the 'turnover radius' (r_{to}) to an outer flat portion. We find that r_{to} agrees very well with the starburst size as estimated by the Hα half-light radius. Thus, starbursts occur within the inner region of solid body rotation. One possibility is that the competition between tidal shear and gravitational instability favors cloud growth in the region of solid body rotation (Kenney, Carlstrom, & Young 1993). Models of bar-driven inflow (cf. Piner, Stone, & Teuben 1995) also predict that rings of gas will accumulate near r_{to} (between two Inner Lindblad Resonances, if present).

4. Using the rotation speed as a surrogate for galaxy mass, the typical rotation speeds of 100 to 220 km s^{-1} in our sample imply that the starburst host galaxies have typical masses ranging from about 5% to 100% of the mass (M_*) of a fiducial L_* spiral galaxy. We find very little correlation between the host galaxy rotation speed and the IR ($\sim$ bolometric) luminosity of the starburst. Even quite low mass galaxies (10% M_*) can apparently have quite powerful starbursts (L $\sim 10^{11}$ L$_\odot$).

5. Weedman (1983) pointed out that the emission-line widths in starburst nuclei were surprisingly small compared to expectations for gas in virial equilibrium in the bulge of a typical disk galaxy. He proposed that the starburst was therefore not in dynamical equlibrium, and would rapidly evolve into a highly compact configuration (as a precursor to the formation of a Seyfert nucleus). We confirm Weedman's result (by comparing v_{rot} to the width of the

nuclear emission-line profiles). However, the narrowness of the nuclear lines arises in our sample because the gas in the nucleus does not sample the full range of the rotation curve (which is solid body in the central region).

In the long run, these survey results can hopefully be combined with detailed investigations of a small sample of the nearest and brightest starburst galaxies (e.g. Kenney, this conference) to shed some light on the 'care and feeding' of starbursts.

STARBURSTS AND THE IGM

Starbursts can interact with the inter-galactic medium in a variety of ways:

First, some of the ionizing radiation produced by massive stars in the starburst can leak out into the IGM and photoionize it. For star formation with a normal Salpeter IMF from 0.1 to 100 $M_\odot$, every proton that participates in star-formation results in enough Lyman continuum photons to ionize (once!) an average of 3000 H atoms. Thus, in principle even a small amount of early star-formation could have played an important role in photoionizing the very early (pre-QSO) IGM. Even at latter times, the adequacy of QSOs alone in providing the observed metagalactic ionizing background is still a matter of debate (cf. Madau & Shull 1996). However, recent observations below the Lyman edge of a small sample of local starburst galaxies by Leitherer et al (1995) implies that these objects are quite opaque to their own ionizing radiation (less than about 1% of the Lyman continuum escapes).

Secondly, the global outflows of interstellar matter from starburst galaxies driven by the collective effect of the supernovae and stellar wind ("superwinds") are now a well-established phenomenon (cf. Heckman, Lehnert, & Armus 1993). These may play an important role in heating and chemically-enriching the IGM.

It has long been known that the mass of metals contained in the intracluster medium in galaxy clusters is similar to the mass of the metals locked up in the stars inside the cluster galaxies. Theories to explain this have invoked ram pressure stripping, 'late' galactic winds driven by type Ia supernovae and 'early' winds driven by type II supernovae associated with a starburst/formative phase of an early type galaxy. The recent discovery with ASCA (Loewenstein & Mushotzky 1996) that the abundances of α-peak (e.g. O, Si, Ne) elements relative to Fe are several times the solar value implicates type II supernovae and supports the 'early wind' (starburst-driven outflow) scenario.

If galaxies inside clusters can eject as much metals as they retain, why wouldn't this be true globally (e.g. in the field)? While a truly diffuse IGM has not been detected in X-rays, recent spectroscopic investigations of the Lyman-α 'forest' (the cloudy component of the IGM at high redshift) shows that this material is also enriched with metals (cf. Cowie et al 1995). It is tempting to speculate that starburst-driven outflows in the early universe polluted the IGM with metals.

Thirdly, the galaxy interactions that help 'fuel' powerful starbursts, may tidally-liberate significant quantities of interstellar gas. Hibbard (1995 and this conference) has shown some spectacular HI maps that document this process.

Both the outflow driven by the starburst and the tidally-dispersed gas may significantly increase the cross-sectional area of a starburst galaxy, and it is interesting to speculate whether these episodic events may contribute substantially to the population of QSO absorption-line systems. One particularly interesting case-in-point is the strongly interacting starburst system NGC 520. Norman et al (1996) have detected strong and possibly complex UV absorption arising from the envelope of this galaxy along the line-of-sight to a background QSO with an impact parameter of about 50 h_{50}^{-1} kpc. Possible absorption is also seen along a sight-line to a second QSO with an impact parameter twice as large. While tidally-deployed HI can be seen in a deep VLA map of this system (Hibbard 1995), the UV- detected gas is seen well beyond where the HI 21cm line can be detected. Further observations, especially of the highly-ionized gas probed by the CIV and NV ions, will be needed to establish whether the absorbing gas is relatively cool tidal debris or hot (shocked) wind material.

STARBURSTS AS 'YOUNG GALAXIES'

One of the major discoveries in astrophysics during the past several years has been of the existence of a genuine field population of star-forming galaxies at high-redshift (e.g Steidel et al 1996a,b; Lowenthal et al 1997; Madau, this conference). For such galaxies, almost everything we know comes from the rest-frame far-UV spectral region. It is therefore interesting to examine the far-UV properties of *local* starbursts. Do their UV properties exhibit systematic dependences on other important galaxian parameters? If so, what inferences might we make on this basis about the high-z population?

We have recently undertaken a multiwavelength study of the statistical properties of a sample of about 50 local starburst galaxies observed in the UV with the IUE. In agreement with the important pioneering work by Calzetti et al (1994, 1995) and Storchi-Bergmann et al (1994), we find that the UV properties have been strongly modified by the effects of dust. If we parameterize the UV spectral slopes as a power-law $F_\nu \sim \nu^\alpha$, we measure values for α ranging from the value +0.5 predicted for a dust-free starburst population (cf. Leitherer & Heckman 1995) to -2.5. This UV spectral slope correlates strongly with the ratio of far-IR to UV flux (cf. Figure 4 in Heckman et al 1995). The most straightforward interpretation of this is that the dustier the starburst, the greater the reddening suffered by the UV continuum and the correspondingly larger fraction of the UV continuum is absorbed and reradiated in the far-IR. As discussed by Meurer et al (1995) and Calzetti et al (1995) this idea even seems to work quantitatively using an empirical starburst UV-extinction law and a simple foreground screen model as a first approximation. Interestingly, the UV spectral slope and FIR/UV flux ratio both correlate strongly with the metallicity of the starburst: the higher the metallicity, the greater the inferred amount of dust extinction (see also Calzetti et al 1994; Storchi-Bergmann et al 1994). This is not too surprising since the dust-to-gas ratio should roughly track the metallicity.

Taking these results at face value, and naively applying them to the high-redshift samples implies that:

1 The high-redshift samples (which are selected on the basis of UV colors and fluxes) are likely to be biased towards systems with low-metallicity, and may be a quite incomplete census of high-z star-formation.

2) Even among the UV-selected starbursts at high-z, the measured UV luminosity may under-represent the intrinsic (reddening-corrected) value by factors typically ranging from two to ten.

3) The reddest (in the UV) members of the high-redshift samples would have roughly solar metallicity.

We (Meurer et al 1997) have also recently analysed the rest-frame UV photometric properties of the high-redshift star-forming galaxies and compared them to local starbursts. After correcting for the effects of extinction (following the precepts described by Meurer et al 1995), we find that starbursts at all redshifts seem to have the same 'limiting' bolometric surface brightness of a few x 10^{11} $L_\odot$ kpc^{-2}. The physical meaning is unclear (are starbursts self-regulating in some way?), but the similarity of the high- and low-redshift starbursts is striking. The implied surface-mass densities in the starbursts (roughly 10^9 $M_\odot$ kpc^{-2}) are similar to the corresponding values in the inner parts of present-day galaxies.

One of the most striking properties of local starbursts are their retinue of 'super star clusters' (SSC's) - cf. O'Connell et al (1994); Meurer et al (1995). The photometric properties of the SSCs imply that they have sizes and masses similar to present-day globular clusters, and recent dynamical measures of the masses confirm this in several nearby cases (Ho & Filippenko 1996a,b). Whether classical globular clusters were formed under conditions similar to the starburst SSCs is not clear. One challenge is to understand how the power-law luminosity function of the SSCs can be reconciled with the peaked (Gaussian) luminosity function of globular clusters (van den Bergh 1995; Meurer 1995).

Finally, if (as I have argued in the section above) starburst-driven winds have enriched the IGM, the corrollary is that these winds must also have a significant impact on the chemical evolution of galaxies. In the best-studied starbursts studied to date, the temperature of the hot outflowing gas is always in the range of a few million to ten million K (Dahlem, Weaver, & Heckman 1997; Della Ceca et al 1996; 1997). This temperature is presumably set by thermal and gas dynamical processes in the ISM of the starburst, independent of the depth of the gravitational potential provided by the host galaxy. This observed temperature range corresponds roughly to the 'escape temperature' for a L_* galaxy. This then implies that starburst-driven outflows can escape much more readily from low-mass galaxies (and in so-doing, carry away the newly synthesized metals). Such a mechanism provides a physical basis for the mass-metallicity relation among galaxies and the radial metallicity gradients within galaxies (cf. Lynden-Bell 1992; Franx & Illingworth 1990).

ACKNOWLEDGMENTS

I would like to thank the organizers of this conference and acknowledge the support of NASA grant NAGW-3138. I would also like to thank my collaborators (especially M. Dahlem, R. Della Ceca, M. Lehnert, G. Meurer, C. Norman, C. Robert, and K. Weaver, who have played the leading roles in much of the unpublished or just-published work discussed above).

REFERENCES

1. Calzetti, D., Kinney, A., & Storchi-Bergmann, T. 1994, ApJ, 429, 582
2. Calzetti, D., Bohlin. R., Kinney, A., Storchi-Bergmann, T., & Heckman, T., 1995, ApJ, 443, 136
3. Cowie, L., Songaila, A., Kim, T.-S., & Hu, E. 1995, AJ, 109, 1522
4. Dahlem, M., Weaver, K., & Heckman, T. 1997, submitted to ApJS
5. Della Ceca, R., Griffiths, R., Heckman, T., and MacKenty, J. 1996, ApJ, 469, 662
6. Della Ceca, R., Griffiths, R., & Heckman, T. 1997, ApJ, in press
7. Devereux, N. 1989, ApJ, 346, 126
8. Franx, M., & Illingworth, G. 1990, ApJL, 359, L41
9. Gallego, J., Zamorano, J., Aragon-Salamanca, A., & Rego, M. 1995, ApJL, 455, L1
10. Heckman, T. 1993, in *Mass-Transfer-Induced Activity in Galaxies*, ed. I. Shlosman, (Cambridge University Press), 234
11. Heckman, T., Lehnert, M. & Armus, L. 1993, in *The Evolution of Galaxies and their Environments*, ed. S.M. Shull and H. Thronson (Kluwer: Dordrecht), 455
12. Heckman, T., Krolik, J, Meurer, G., Calzetti, D., Kinney, A., Koratkar, A., Leitherer, C., Robert, C., & Wilson, A. 1995, ApJ, 452, 549
13. Hibbard, J. 1995, Ph.D. Dissertation, Columbia University
14. Ho, L., Filippenko, A., & Sargent, W. 1996, in *IAU Colloquim 157: Barred Galaxies*, ed. R. Buta, B. Elmegreen, & D. Crocker (San Francisco: ASP), in press
15. Ho, L.C., & Filippenko, A.V. 1996a, ApJ, 466, L83
16. Ho, L.C., & Filippenko, A.V. 1996b, ApJ, 472, 600
17. Kenney, J., Carlstrom, J., & Young, J. 1993, ApJ, 418, 687

18. Kraan-Korteweg, R., & Tammann, G. 1979, AstronNach, 300, 181
19. Lehnert, M., & Heckman, T. 1996, ApJ, 472, 546
20. Leitherer, C. & Heckman, T. 1995, ApJS, 96, 9
21. Leitherer, C., Ferguson, H., Heckman, T., & Lowenthal, J. 1995, ApJL, 454, L19
22. Loewenstein, M., & Mushotzky, R. 1996, ApJ, 466, 686
23. Lowenthal, J., Koo, D., Guzman, R., Gallego, J., Phillips, A., Faber, S., Vogt, N., Illingworth, G., & Gronwall, C. 1997, ApJ, in press
24. Lynden-Bell, D. 1992, in *Elements and the Cosmos*, M. Edmunds and R. Terlevich (eds), (Cambridge University Press: Cambridge), 270
25. Madau, P., & Shull, J.M. 1996, ApJ, 457, 551
26. Meurer, G. 1995, Nature, 375, 742
27. Meurer, G., Heckman, T., Leitherer, C., Kinney, A., Robert, C., & Garnett, D. 1995, AJ, 110, 2665
28. Meurer, G., Heckman, T., Leitherer, C., Lowenthal, J., & Lehnert, M. 1997, submitted to ApJ
29. Mihos, J.C., & Hernquist, L. 1994a, ApJL, 425, L13
30. Mihos, J.C., & Hernquist, L. 1994b, ApJL, 431, L9
31. Norman, C., Bowen, D., Heckman, T., Blades, C., & Danly, L. 1996, ApJ, 472, 73
32. Norman, C., Hasan, H., & Sellwood, J. 1996, ApJ, 462, 114
33. O'Connell, R., Gallagher, J., & Hunter, D. 1994, ApJ, 433, 65
34. Piner, B., Stone, J, & Teuben, P. 1995, ApJ, 449, 508
35. Rieke, G., Loken, L., Rieke, M., & Tamblyn, P. 1993, ApJ, 412, 99
36. Sanders,D. & Mirabel, I.F. 1996, ARAA, 34, 749
37. Scoville, N., & Soifer, B.T., 1991, in *Massive Stars in Starburst Galaxies*, ed.C. Leitherer, N. Walborn, T. Heckman, & C. Norman (Cambridge: Cambridge University Press), 233
38. Soifer, B.T., Boehmer, L., Neugebauer, G., & Sanders, D. 1989, AJ, 98, 766
39. Steidel,C., Giavalisco, M., Pettini, M., Dickinson, M., & Adelberger,K., 1996a, ApJL, 462, L17
40. Steidel,C., Giavalisco, M., Dickinson, M., & Adelberger,K., 1996b, ApJ, in press
41. Storchi-Bergmann, T., Calzetti, D., & Kinney, A. 1994, ApJ, 429, 572
42. van den Bergh, S. 1995, Nature, 374, 215
43. Weedman, D. 1983, ApJ, 266, 479

Young Stellar Aggregates Embedded in Expanding Supershells

Vladimir G. Surdin and Evgenij V. Moskal'

Sternberg Astronomical Institute, Moscow, 119899, Russia

Abstract. Numerical N-body investigation of dynamical evolution of young stellar aggregate embedded into massive interstellar cloud is presented. We consider gravitational interaction of stars with unprocessed gas during cloud disruption and expanding envelope formation. This interaction leads to dynamical cooling of the stars and its comovement with the envelope. This effect can mimic of stimulated star formation in the supershell. The dynamical cooling also leads to gravitational bound cluster formation in spite of low star formation efficiency.

INTRODUCTION

OB associations are unbound, expanding stellar systems. Most all stars that form in the Galaxy likely form in these associations. But their origin remains a mystery: there is, as yet, no successful theory of the association formation.

Most extremely young stars embedded in GMCs. This stars apparently form in rich temporal star clusters from massive cores of dense molecular gas. Explanation of the transition from this temporal clusters to unbound associations is challenging task for stellar dynamics. Here, we discuss some mechanisms of the stellar association and star cluster formation, and represent some new results obtained with a numerical model. The main goal of the investigation is to clarify how the unprocessed gas influences on the dynamics of stellar aggregate.

HISTORY

Öpik [5] created fruitful idea of star formation in cooling gas around supernova remnants. He suggested that expanding shells could sweep up and compress interstellar material from which new stars are formed. These newly formed stars would retain the outward motion of the supernova shell and form an unbound association which was expanding away from the original site of the supernova explosion.

Oort [4] proposed a similar scenario for association formation based upon the expansion of the H II region into neutral cloud material. According to this scenario, clumpy structure of the cloud leads to the unbound stellar association formation around of progenitor O-type star.

In a few decades, this Öpik–Oort idea of stimulated star formation was applied to H I supershells, which are formed under the collective action of many young massive stars. According to this paradigm, any example of the young stars concentration near the shell was interpreted as evidence of the stimulated star formation.

Another idea for association formation was developed simultaneously with the Öpik–Oort scenario: according to Zwicky [11] and McCrea [3], formation of massive stars in a temporal cluster leads unprocessed gas to escape from this cluster, which increases the cluster energy and forces its expansion. If the energy reaches a positive value, the cluster loses it bound state and forms an association. This idea did not attract much attention and was "rediscovered" a quarter of a century later, just after first observations of dense protostellar groups situated inside molecular cloud nuclei.

The low value of star formation efficiency (SFE) in GMCs is of central importance for understanding the dynamical nature of OB associations. The stars, which were originally bustling in virial equilibrium with the deep potential well of the GMC's core, respond to the rapid removal of the majority of the binding mass by freely expanding into space with their initial virial velocities. From a gas release time which is rapid ($\tau_R \leq 10^6$ yr for GMC's cores), a value of $SFE \geq 50\%$ is necessary to form a bound cluster. For $\tau_R > (4 \div 5) \times 10^6$ yr (corresponding to 4 to 5 crossing times) the numerical models are reduced to analytic models for adiabatic gas release and SFE of only 20 to 30% are required to result in tidally stable bound stellar systems [10].

Theoretical evaluation predicts the large scale $SFE \sim 1\%$ [8,9]. Indeed, observations indicate overall SFE in GMCs is low: from 0.1 to 5% of the total gaseous mass is ever converted to stars [2,1]. Consequently, the inevitable destruction of a GMC by massive stars results in the dispersal of the majority of the initial binding mass of the star-forming complex. Therefore, an unbound system of stars is left behind, expanding into the field with a characteristic of the velocity dispersion in the original cloud core.

Still missing from consideration, however, is the next evolutionary step: that of modelling the gravitational interaction of stars with gas after the temporal cluster disruption. First of all, it is interesting to check numericaly the concentration of stars near a massive gaseous shell, which is analytically predictable. This goal is one of the prime motivations for our investigation.

MODEL

We carried out N-body simulations of the dynamical evolution of a stellar aggregate embedded in a massive spherical cloud with initially isothermal density distribution. The gas was represented in Aarseth's code by an additional term in the gravitational potential function.

Initially, a good relaxed velocity distribution of stars ($N \approx 10^3$ is assumed. We included rotation of stellar component in some models. The initial half-mass radius of stellar component is about 0.5 in arbitrary units. After a few dynamical times, the cloud begins to be disrupted. The gas of the cloud forms the expanding envelope of the radius $R_{sh}(t)$. All the gas inside this radius is sweeping up to the absolutely thin envelope. We used a few analytical approximations for the shell radius: from $R_{sh}(t) \propto t$ to $R_{sh}(t) \propto t^{1/2}$ according to the results of numerical simulations [7].

All parameters of our models are dimensionless. The time intervals are expressed in the units of initial central dynamical time t_{dyn}. The initial cloud radius is $R_g \approx 30$. The model is represented in Figure 1 has very small value of $SFE = 1\%$. It shows the dynamical effect of the massive envelope which leads to the star concentration near the shell. Models with value of $SFE \approx 10\%$ display the cluster formation from stars which dynamically cooled in the massive shell.

We included the stimulated star formation in some models. All the new stars are formed randomly during some period of time, localized in the random positions on the expanding envelope and have space velocities are equal to

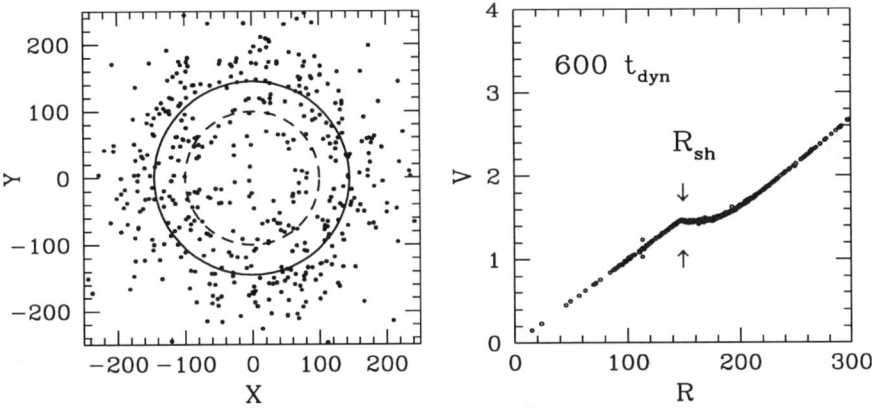

FIGURE 1. Stellar aggregate embedded into massive shell. The model is represented by the sky projection (left) and velocity–radius plot (right). The initial cloud radius is marked by the dash circle; the shell radius is marked by a solid circle. The time is indicated in the right frame.

the current envelope velocity. After termination of the star formation in the shell we calculated the "observable" radial velocities distribution of all stars which are projected on the central region of the shell. These data lets us to distinguish between new and old stars.

RESULTS

Interaction of stars with massive supershell gives a few dynamical effects, one of which is **dynamical cooling** of the association. It leads to comovement newborn stars and envelope, i.e. mimic stimulated star formation in the shell. The structures like this are observed, for example, around NGC 2070 in LMC [6].

Some stars may cross the system a few times, decreasing their energy any time. Finally, from 20% to 50% of stars may form a gravitationally bound cluster under the action of dynamical cooling. In spite of low value of SFE ($\approx 10\%$), this cluster can survive.

For systems with significant rotation and low SFE the dichotomy between the formation of bound cluster and unbound association is very sharp. It is difficult to find the initial conditions that lead to formation of bound cluster together with unbound association of approximately the equal masses.

Taking into account an effect of stimulated formation of the second generation of stars in the expanding envelope we compare the relative spatial and velocity distributions of the both star populations. In a general case, we are able to distinguish between this two populations by the radial velocity distribution only.

REFERENCES

1. Carlberg, R.G., *The Milky Way Galaxy*, van Woerden, H., et. al. (eds.), Dordrecht: Reidel Publ. Comp., 1985, p. 615.
2. Duerr, R., Imhoff, C.L., and Lada, C.J., *Astrophys. J.* **261**, 135 (1982).
3. McCrea, W.H., *Observatory*. **75**, 206 (1955).
4. Oort, J.H., *Bull. Astron. Inst. Netherlands*. **12**, 177 (1954).
5. Öpik, E.J., *Irish Astron. J.* **2**, 219 (1953).
6. Seleznev, A., *Astron. Lett.* **21**, 663 (1995).
7. Silich, S.A., *Astrophys. Space Sci.* **195**, 317 (1992).
8. Surdin, V.G., *Astron. Nachr.* **310**, 381 (1989).
9. Surdin, V.G., *Astron. Lett.* **20**, 318 (1994).
10. Wilking, B.A., and Lada, Ch.J., *Protostars and Planets. II.* Black, D.C., and Matthews, M.S. (eds.), Tucson: Univ. Arizona Press, 1985, p. 297.
11. Zwicky, F., *Publ. Astron. Soc. Pacific.* **65**, 205 (1953).

Star Formation in Leading/Trailing Single Arm Galaxies: NGC 4378

Gene G. Byrd[1], Guy B. Purcell[1], Ronald J. Buta[1], Demetrius McCormick[2] and Tarsh Freeman[1]

[1] Dept. of Physics and Astronomy, Univ. of Alabama, Tuscaloosa, AL 35487-0324
[2] Tuskegee University, Tuskegee Institute, AL 36088

Abstract. de Vaucouleurs (1958) found that all members of a sample of nearly edge-on spirals had trailing arms relative to disk rotation. An exception was provided by the identification of a single arm in the galaxy NGC 4622 by Byrd et al. (1989). Buta, Crocker and Byrd (1992) observationally verified that an inner single arm occurs in the stellar disk of NGC 4622 and that it also shows the color/age position angle sequence of an $m = 1$ leading density wave. Byrd, Freeman and Howard (1993) simulated NGC 4622's pattern with a small mass companion plunging close to the center in a retrograde orbit in the disk plane. A single long-lived leading arm is created in the central regions. NGC 4378 is also a single armed galaxy (Rubin et al. 1978, Sandage 1961). Byrd, Freeman and Howard (1994) simulated the arm pattern of NGC 4378 by a direct planar grazing encounter with a small galaxy. In contrast to NGC 4622, the simulation arm for NGC 4378 trailed. We use new B and I images and models to determine that NGC 4378's arm trails and show its lack of a star formation/age sequence. Thus NGC 4378 has a tidal arm not a one-armed density wave.

A LEADING OR TRAILING SINGLE ARM?

As described above, NGC 4378 contains a single arm. Contrary to the single density wave arm in NGC 4622, the Byrd et al. (1993) simulations predicted that the NGC 4378 single arm is not a density wave but instead is a tidal arm in the side of the disk originally closest to the grazing passage of a perturber. This is supported by new computer simulations of the NGC 4378 rotation curve irregularities with the encounter parameters of Byrd, et al. (1994). These results show a close similarity to Rubin et al's observed rotation curve. We verify this "trailing" prediction observationally. We revive de Vaucouleurs' (1958) method in which color differences are used to determine which edge of the disk is closer using Figures 1 and 2 on the following page.

Preferential scattering of blue light by disk plane dust near the nucleus will

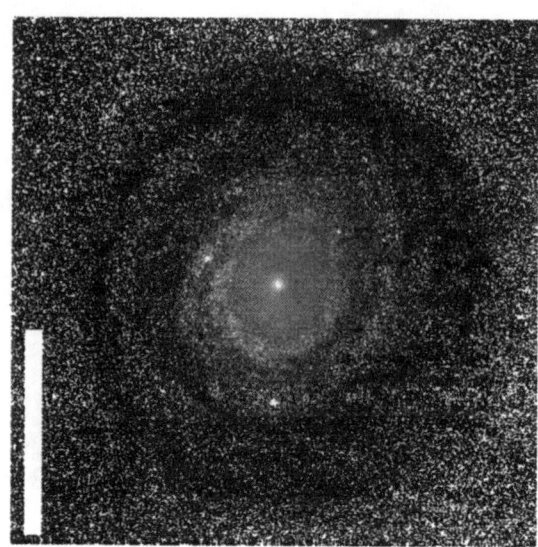

FIGURE 1. Color Index Image of NGC 4378 of deep CCD images obtained with the CTIO 1.5m telescope using a 1024 square CCD and Johnson B and Cousins I filters. Blue ($B - I \sim 1.0$) is dark shading. Red ($B - I \sim 2.5$) is light. Vertical bar = 90 arc sec.

make the side of the disk which is closer to us appear redder. As can be seen in Figures 1 and 2, the east side is redder and thus is the nearer side. Since the arm unwinds counter-clockwise and the northern edge has a red shift relative to the nucleus, the arm trails. Not all single arms lead! The color method is useful in determining disk orientation and arm sense even for the moderate $\sim 35°$ inclination of NGC 4378.

DENSITY WAVE OR TIDAL ARM?

The position of the peak in I band intensity across an arm is the point in the stellar disk where gas clouds are crowded together and may be triggered into star formation. The B peak represents the "lighting up" of the protostars as they reach the main sequence about 10 million years later. For a flat rotation curve like that of NGC 4378, these peaks may or may not be separated in position angle around the nucleus according to the following three possibilities shown on the right hand side of Figure 3 on the next page, top to bottom. The stellar disk stars I line is dotted. The new stars B line is solid. Single leading-arm density wave is the top. The pattern turns opposite the disk orbital motion. The two opposite angular motions result in a large angular separation of the B and I intensity peaks. Trailing arm density wave is the middle. Orbital motion and pattern speed are in same direction. The B and I

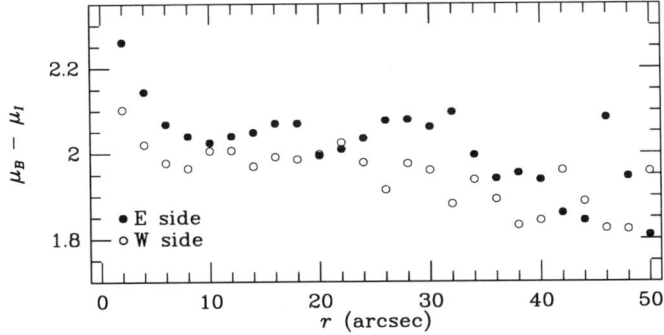

FIGURE 2. Color Index versus radius on east and west sides of disk minor axis.

peaks are less separated than in the leading arm density wave and are switched in

As the only known example of a leading-arm density wave, the position angles of B and I peaks versus radius are shown for NGC 4622 from Buta, Crocker and Byrd (1992) in the left hand side of Figure 3 on the next page. The star-forming single arm is the upward sloping line between 40 and 60 arc sec. The B and I peaks are separated by about $40°$ in position angle! NGC 4622 provides the clearest case we know of a density wave star formation/age sequence.

In contrast, bottom left of Figure 3 on the next page shows the NGC 4378 B and I arm position angle versus radius. The arm extends from 50 to 90 arc sec. Our observations show no systematic azimuthal color separation. This matches the tidal arm computer prediction for NGC 4378 diagramed on the lower right. Our determination that the NGC 4378 single arm trails relative to disk orbital motion and its lack of a star formation color separation are all consistent with its being a single trailing tidal arm. The inner arm of NGC 4622 retains its unique distinction of being the only galaxy determined to have a leading arm density wave.

Acknowledgements. Support was provided by NSF grants REU AST-9424226, EPSCoR RII8996152, and AST 9014137. We thank CTIO for use of their observational facilities.

REFERENCES

1. Buta, R., Crocker, D. and Byrd, G. 1992, AJ 103, 1526.
2. Byrd, G., Freeman, T. and Howard, S. 1994 AJ 108, 2078
3. Byrd, G., Freeman, T. and Howard, S. 1993 AJ 105, 477
4. Byrd, G. G., Thomasson, M., Donner, K. J., Sundelius, B., Huang, T.-Y. and Valtonen, M. J. 1989 Celestial Mechanics 45, 31.

5. de Vaucouleurs, G. 1958 Astrophysical Journal 127, 487.
6. Rubin, V. C., et al. 1978 ApJ 224, 782.
7. Sandage, A. 1961, *Hubble Atlas of Galaxies*, Carnegie Pub. No. 618.

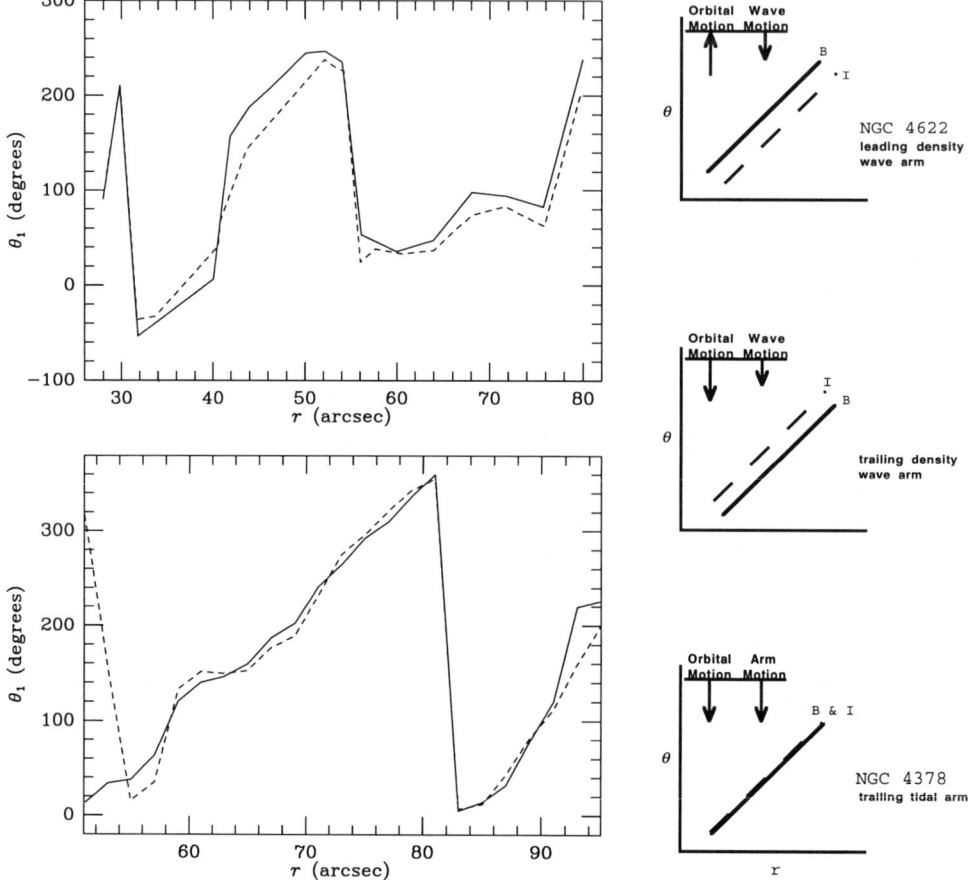

FIGURE 3. B (solid) and I (dashed) arm peaks. Right: in leading density wave (top), trailing density wave (middle), and trailing tidal arm (bottom). Left: Observed peaks for NGC 4622 (top) and NGC 4378 (bottom).

ORCHESTRATION OF STARBIRTH ACTIVITY IN DISK GALAXIES: New Perspectives from Ultraviolet Imaging

William H. Waller[†*], Theodore P. Stecher[*], and the
Ultraviolet Imaging Telescope (UIT) Science Team[*‡]

[†]*Hughes STX Corporation (waller@stars.gsfc.nasa.gov)*

[*]*NASA Goddard Space Flight Center
Laboratory for Astronomy and Solar Physics (LASP)
Code 680, Greenbelt, MD 20771*

[‡] *Website = http://fondue.gsfc.nasa.gov/UIT/UIT_Homepage.html*[1]

Abstract. Ultraviolet imaging of nearby disk galaxies reveals the star-forming activity in these systems with unprecedented clarity. UV images recently obtained with the Shuttle-borne Ultraviolet Imaging Telescope (UIT) reveal a remarkable variety of star-forming morphologies. The respective roles of tides, waves, and resonances in orchestrating the observed patterns of starbirth activity are discussed in terms of the extant UV data.

WHY ULTRAVIOLET IMAGING?

Despite the potential for obscuration by dust, a significant fraction of a disk galaxy's UV emission manages to escape and thus be detected by instruments located beyond the Earth's atmosphere. Even in disk galaxies of high inclination, UV imaging can reveal widespread emission (cf. Fanelli et al., Neff et al., & O'Connell, these Proceedings).

Ultraviolet imaging of nearby disk galaxies reveals the star-forming activity with unprecedented clarity. Unlike imaging at Hα, UV imaging *directly* traces the full range of OBA stellar spectral types, thereby sampling the recent-epoch

---

[1)] UIT research is funded through the Spacelab Office at NASA Headquarters under Project number 440-51.

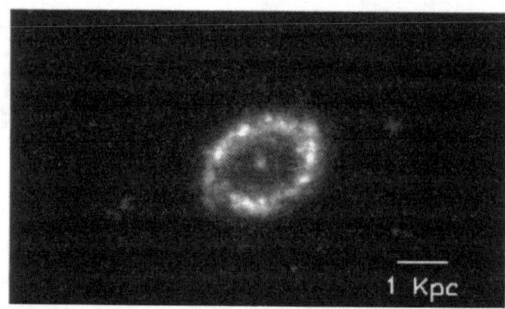

FIGURE 1. FUV ($\lambda 1520$) image of the Sab galaxy M94 (inner disk), showing bisymmetric star-forming knots, a resonant ring of starburst activity and diffuse FUV emission interior to the ring — where Hα is *in absorption* due to the underlying B & A-type stellar populations.

"Population I" component of each galaxy more completely. UV imaging also provides a cleaner separation of the hot star component in regions dominated by cooler stars (e.g. central disks & bulges). Moreover, UV imaging is unaffected by line absorption in the atmospheres of B & A-type stars — unlike imaging at Hα and Hβ — thereby providing a truer representation, where these populations are concentrated (see FIGURE 1). Finally, the UV colors of OB/HII regions can be used to derive extinction-free UV luminosities [1] [2], whereas few constraints exist for deriving extinction-free EUV (e.g. Lyman continuum) luminosities based on H-line, radio-continuum, or other indirect indices.

UV MORPHOLOGIES OF DISK GALAXIES

UV images of nearby disk galaxies obtained with the Shuttle-borne *Ultraviolet Imaging Telescope* (UIT) reveal a remarkable variety of star-forming morphologies. These Pop I patterns yield important insights to the respective roles of *tides, waves, and resonances* in orchestrating starbirth activity in disk galaxies.

A comparison of M33 (Scd), M74 (Sc), and M81 (Sb) at UV and visible wavelengths highlights the Pop I character of the UV imagery [3]. *Flatter radial distributions* are evident in the UV — with exponential scalelengths that are 20–45% larger. The effects of reddening, abundance, and IMF variations do not fully explain the differences. The flatter UV profiles most likely indicate that the median radius of star-forming activity has migrated outward over the past several Gyrs [4].

The UV morphologies also show *narrower arms* — delineated by a combination of direct starlight from OBA associations and indirect (scattered) radiation from dust associated with the massive young stars [1] [5]. The narrower UV features indicate that star formation over the past $\sim$10 Myrs

FIGURE 2. FUV image of the giant ScI galaxy M101, showing multiple "crooked arms" throughout the disk and a faint outer arm terminating in a "curly tail" feature.

occupies a significantly smaller areal domain than the ~1–1000 Myr legacy of star formation that is traced at longer wavelengths.

Tides

In the giant ScI spiral M101, multiple linear arm segments (*"crooked arms"*) can be traced throughout the disk (see FIGURE 2). These features, along with a faint spiral arm and *"curly tail"* feature that links the outermost supergiant HII region with the rest of the galaxy, indicate that *tidal processes of both external and internal origin* are directing the current starbirth activity [5].

Numerical simulations of isolated disk galaxies show that the "crooked arm" behavior can arise through the action of "massive disturbers" orbiting within the disks. The outermost supergiant HII region, NGC 5471 may represent one of these massive disturbers. Larger-scale morphological and kinematic anomalies in M101, including the faint arm and "curly tail" feature, require external interactions with companion galaxies. Such interactions can induce the formation of massive condensations at the ends of the tidal tails, perhaps explaining the origin of NGC 5471 at the terminus of M101's "curly tail" [5]. Similar behavior can be found in other giant Sc galaxies with "companions," including M51, NGC 1232, NGC2805, and NGC 4303.

Waves

In the "grand-design" ScI galaxy M74 (NGC 628), reflection of the UV-emitting disk upon itself shows the spiral structure to be more symmetric than is observed at visible wavelengths — thus arguing for large-scale dynamics (e.g. density waves) governing the current-epoch star formation [6].

Evidence for *spatio-temporal sequences* of molecular-cloud aggregation, massive star formation, cluster evolution, and cloud disruption can be found in M74, M51 (O'Connell, these Proceedings), and the *inner disk* of M101 [5]. In M101, far-UV emission is often found on the outer (downstream) side of the inner-disk CO arms. Modeling the FUV–CO displacements according to density-wave dynamics results in a wave pattern speed and co-rotation radius remarkably similar to those derived from a multi-mode analysis of the optical spiral structure [5]. Similar downstream displacements between the FUV and Hα emission is evident in the SE arm of M74, again indicating density-wave dynamics at work.

Resonances

In the Sab spiral M94 (NGC 4736), UV imaging reveals an inner starbursting ring and bi-symmetric outer knots in high contrast against the underlying visible bulge and disk (see FIGURE 1). Dynamical resonances seem to best explain these transient features. Similar UV rings are evident in the inner disks of NGC 1317 (SBa), NGC 1512 (SBb), NGC 3351 (SBb), and M100 (NGC 4321) (Sc) — most of which are also of "early" morphological type.

Resonances may also explain the dearth of UV emission interior to M81 and M31's ring-like spiral arms [7], whereby star-forming gas migrates outwards from the interior and piles up near the Inner Lindblad Resonances (Kenney, these Proceedings). Because the longer-wavelength emission traces the older stellar populations and is so prominent interior to the ringlike arms, we conclude that the resonant locations and their constructive/inhibiting effects must have evolved with time.

REFERENCES

1. Hill, J. K. et al. 1995, ApJ, 438, 181
2. Hill, J. K. et al. 1996, ApJ, in press
3. Waller, W. H. & Offenberg, J. 1994, Beyond the Blue: Greatest Hits of the Ultraviolet Imaging Telescope, ASP slide set and booklet (San Francisco: ASP)
4. Cornett, R. H. et al. 1994, ApJ, 426, 553
5. Waller, W. H. et al. 1996, ApJ, in press
6. Chen, P. C. et al. 1992, ApJL, 395, L41
7. Hill, J. K. et al. 1992, ApJL, 335, L37

Preliminary Results of the ASU/UGA O-star Project

P.A. Scowen

Dept. of Physics & Astronomy, Arizona State University, Box 871504, Tempe, AZ 85287-1504

P.H. Hauschildt

Dept. of Physics & Astronomy, University of Georgia, Athens, GA 30602-2451

J.P. Aufdenberg and R. Sankrit

Dept. of Physics & Astronomy, Arizona State University, Box 871504, Tempe, AZ 85287-1504

Abstract.
Understanding the physics of the photodissociation regions (PDRs) which occur between the ionized HII regions and the surrounding molecular clouds is critical in assessing the role that primary star formation has on secondary events. It is also pivotal to understanding how HII regions affect their local environment. Modelling of these interfaces, and of the photons that escape to ionize the local ISM, has been hampered by a lack of modern models of the O-star atmospheres. We present preliminary results from a program intended to better model the physics of the extended atmospheres of O stars and thus derive more physically accurate photon fluxes and therefore better estimates of the actual ionization rates. The models include features such as winds, spherical symmetry, accurate and up-to-date opacites, and the effect of metallicity on the extended envelope of material around O stars and the spectra that emerge from them.

INTRODUCTION: MOTIVATION FOR THE PROJECT

The initial motivation for the project was rooted in a desire to better model the ionization structure in HII regions, and hence to obtain a better feel for the nature of the escaping ionizing photons that affect the local ISM around HII regions. It became immediately apparent that current models of O-star atmospheres were physically unrealistic, and that we were in a position to generate models that did a much better job of producing a physically meaningful model of the extended atmospheres of O-stars.

The naive impression of O and B stellar atmospheres is that they must be rather easy to model since the typical spectra we're used to seeing are quite smooth and not etched by large numbers of absorption lines. However once you get into the business of understanding what an O star is and what it is doing, the whole subject becomes a lot more complicated.

TOWARDS BETTER MODELS OF HII REGIONS

We originally wanted to make better estimates of the number of ionizing stars in extragalactic HII regions, in addition to estimating the H ionizing flux from known Galactic HII regions. This well known mapping was described in some detail by Vacca [1] where he attempted to make better estimates of the ionizing fluxes put out by O stars of different spectral types and luminosity classes. Our current work is directed to modelling the ionization front boundaries of HII regions - to understand the process of photoevaporation. It was the combination of wanting to better estimate the distribution of the ionizing photons from OB stars for structure modelling, together with the need for better estimates of the bolometric H ionizing flux for OB stars that the ASU/UGA project was conceived. The most recent efforts [2] to calculate q0 for a series of OB stars used Kurucz atmospheres [3]: but these models suffer severe problems in that a lot of physics is left out of them.

We demonstrate below the effect that q0 and the shape of the incident stellar photon energy distribution have on the ionization stucture at the edge of an HII region. We have made a grid that uses four Kurucz stellar atmosphere models with Teff ranging from 42500 to 50000 and three values of q0 - the central value (0%) is 4.5×10^{49} s^{-1}, with a star-interface distance of 6×10^{18} cm. This is the type of variations in energy distribution and q0 between the existing and our new stellar atmosphere models. The separation between the [S II] and Hα ionization zones for the models is shown in the following table.

TABLE 1. Table showing the change in the separation ($\times 10^{18}$ cm) of the Hα-[SII] peaks when varying q0 and T_{eff}.

Relative Flux	T_{eff}=42,500	T_{eff}=45,000	T_{eff}=47,500	T_{eff}=50,000
+20%	1.9	2.4	3.0	3.8
0%	2.7	3.0	3.5	4.1
-20%	3.6	4.0	5.6	6.2

Apart from the effects on PDRs, changes in the character of the ionizing photons for a given spectral type have far-reaching implications for a variety of problems, including estimates of ionizing populations in extragalactic HII regions, as well as calculations about the expected ionization field in the diffuse interstellar medium (DIM).

PHOENIX: A MORE PHYSICAL MODEL

Hauschildt et al [4–6] have developed over the past 5 years a very sophisticated atmosphere code called PHOENIX. Originally intended for use with nova, SNe and M-dwarf projects we decided to attempt to use the code in the OB star domain with remarkable success. Comparisons with good IUE spectra in the SWP region revealed that we were doing a decent job of replicating important features as well as general energy distribution shape.

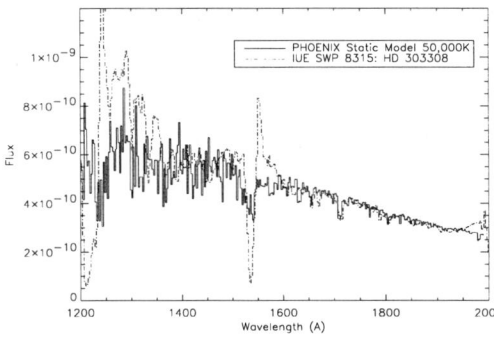

FIGURE 1: *IUE spectrum SWP 8315, HD 303308: Spec. Type O3 V, is compared with a PHOENIX 50,000K LTE static model. The IUE spectrum is unreddened using the Savage and Mathis (1979) extinction curve with an E(B-V) = 0.45. The obvious wind features at CIV and Lyα are not fit by this hydrostatic model. We have not yet attempted to fit this spectrum with our wind models.*

There are numerous differences between ATLAS and PHOENIX. Most importantly, PHOENIX uses spherical geometry while ATLAS uses a plane-parallel static model. In addition, PHOENIX can include NLTE effects on both the structure and the spectrum, while ATLAS is a pure LTE code. We have computed a few pure LTE models with PHOENIX to be able to more directly compare to the Kurucz 1991 model grid. PHOENIX uses the same line list that Kurucz is using but it uses the Verner et al [7] b-f cross sections for a large number of ions to improve the quality of the opacities at higher electron temperatures.

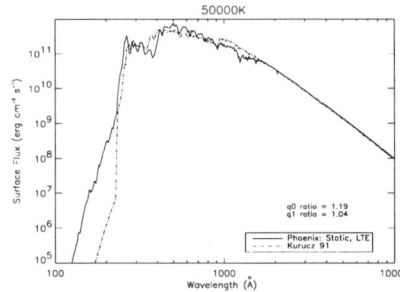

FIGURE 2: *PHOENIX includes by default line scattering in the atmosphere: K91 does not, so we need to turn this off to compare. For the appropriate case of T=50,000 K (ie. an early-type O star, typical of bright HII regions) we see that a static PHOENIX model predicts 20% more H ionizing photons than do K91 models of the same star.*

In addition to hydrostatic atmospheres, PHOENIX can include the effects of the wind velocity on the radiative transfer and the structure of the atmosphere/wind. Current wind models take as input the effective temperature T_{eff}, photospheric radius $R_\star$, gravity $g(R_\star)$ and mass-loss rate $\dot{M}$. Parameters v_∞ and β specify the velocity field above $R_\star$. Below $R_\star$ a hydrostatic structure is calculated. The special relativistic equation of radiative transfer is solved in the co-moving frame (CMF). Temperature corrections are made in accordance with radiative equilibrium. The structure is iterated to convergence.

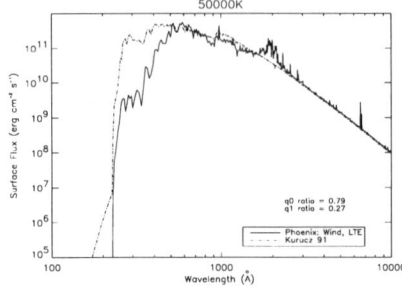

FIGURE 3: Comparison of PHOENIX wind model ($\dot{M} = 10^{-7}$, $v_\infty = 1450$ km/s, $\beta = 0.8$) spectrum to K91 static spectrum for $T_{\text{eff}} = 50{,}000K$. The line blanketing effect of the wind reduces the number of hydrogen and helium ionizing photons compared with the static model.

PRELIMINARY CONCLUSIONS AND SUBSEQUENT DIRECTIONS

With the application of PHOENIX to the regime of OB stellar atmospheres we have made a major step forward in better modelling not only the emission properties of O-stars but also the physics that produce the photon fluxes we observe. Despite excellent work by other researchers in the field it is only now with the advent of advanced numerical methods that we can really start to perform calculations at the level of complexity needed to properly model the ionizing fluxes that etch out the bright nebulae we see as HII regions.

Standard practice has been to use either K91 or K93 atmosphere models for most work that investigates ionizing flux in HII regions and the surrounding ISM. With the new models we are generating this work will be made more accurate. We will also be able to do even better calculations with regards to the structure of ionization fronts in HII regions as was done with the Eagle Nebula. This work should lead to a better understanding of the physics of the boundaries of HII regions and lead to a better insight into how much ionizing radiation can leak out of an HII region.

REFERENCES

1. Vacca, W.D. 1994, ApJ, 421, 140
2. Vacca, W., Garmany, C., and Shull, J.M. 1996, ApJ, 460, 914
3. Kurucz, R.L. 1992, in IAU Symp. 149, "The Stellar Populations of Galaxies", ed. R. Barbuy & A. Renzini (Dordrecht: Kluwer), 225 (K91)
4. Hauschildt, P.H., Baron, E., and Allard, F. 1996, ApJ, *submitted*
5. Hauschildt, P.H., Baron, E., Starrfield, S., and Allard, F. 1996, ApJ, 462, 386
6. Hauschildt, P.H., Starrfield, S., Shore, S. N., Allard, F., and Baron, E. 1995, ApJ, 447, 829
7. Verner, D.A., Ferland, G.J., Korista, K.T. and Yakovlev, D.G. 1996, ApJ, 465, 487

Early Results from an HST Imaging Survey of the Ultraluminous IR Galaxies

Kirk D. Borne*, H.Bushouse[†], L.Colina[†], and R.A.Lucas[†]

Hughes STX, 7701 Greenbelt Road, Greenbelt, MD 20770
[†]*Space Telescope Science Institute, Baltimore, MD 21218*

Abstract. We present results derived from a recent Hubble Space Telescope (HST) imaging survey of the Ultraluminous IR Galaxies. The most significant issue related to this sample of galaxies has been identifying the nature of the primary energy source: starburst or AGN? Through a high-resolution imaging survey of a large sample, some properties of this class of objects are now being better defined, which is helping to clarify the nature of the energy source. Fine structure is seen within the central arcsecond of each galaxy in the sample. In some cases, the structure is smooth and centrally concentrated, suggestive of a bright nuclear energy source (AGN?). In many other cases, the sub-arcsecond morphology is chaotic and extended, suggestive of strong starburst activity. The peculiar, disturbed morphologies that are seen on large (kiloparsec) scales among this sample of galaxies are continued down to the smallest scales in the cores of these strongly starbursting systems. A rich variety of morphological features are noted, many of which are related to the recent intense interaction-induced star formation episode. These starburst-related features (e.g., numerous bright clumps of star formation, shells, and bubbles) are similar to those seen in previous HST imaging observations of other strongly interacting and merging galaxies.

INTRODUCTION

The intense study of interacting galaxies originated in large part with the discovery by IRAS that the most IR–luminous galaxies are nearly all products of collisions and may be the missing link in the chain of evolution from quasars to normal quiescent galaxies [1,2]. These galaxies (with $L_{IR} > 10^{12} L_\odot$) are considered to be the most strongly starbursting of all galaxies in the local universe, have a higher space density than quasars, emit >90% of their power in the IR, are rich in the raw materials of star formation, and to a large extent owe their peculiar morphologies to encounters with other galaxies. The particular importance of IR–luminous galaxies in the grand scheme of cosmology and galaxy evolution has been underscored by the luminosity function studies of Soifer et al. (1986) [3], which indicated that most galaxies have gone through a high–IR luminosity stage. The ultraluminous IR (ULIR) galaxies may also represent an important stage in the formation of elliptical galaxy cores, the formation of globular clusters, and the metal enrichment of the intergalactic medium [4].

We are using the HST to survey the fine–scale features that are associated with the interaction– and activity–related processes that are at work within the ULIR Galaxy Sample. It is widely believed that these galaxies are undergoing star formation at a prodigious rate and are abnormally dust-enshrouded. An alternative to the starburst hypothesis is that these galaxies' IR luminosity is powered by a dust-hidden quasar at its center [1]. *It is important for our understanding of the evolution of galaxies and quasars to determine which of these two hypotheses is valid, or in which objects they are separately valid.*

THE ULTRALUMINOUS IR GALAXY SAMPLE

We have formed our survey sample from a combination of sources. The first "bright" sample of 10 galaxies that satisfied the $L_{IR} > 10^{12} L_\odot$ constraint was compiled by Sanders et al. (1988a) [1]. A second, partially overlapping, "warm" sample of 12 galaxies (9 new ones) was also identified at that time by Sanders et al. (1988b) [2]. An additional, partially overlapping, "bright" sample of 17 galaxies in the south was later compiled by Melnick & Mirabel (1990) [5]. Lawrence et al. (1996) [6] found another 126 low-flux objects in the QDOT all-sky redshift survey of IRAS galaxies, while Kim et al. (1995) [7] and Clements et al. (1996) [8] have added to the numbers at low flux levels. We have selected from these samples most of the objects with redshift $z < 0.20$.

EARLY RESULTS FROM THE HST SURVEY

To date, we have received HST WFPC2 I-band (F814W) images for about 60 galaxies from our total combined sample of 160 ULIR galaxies. With the help of such high-resolution imaging, the properties of this class of objects are now being better defined, including a clarification of the nature of the energy source. *Fine structure is seen within a radius < 2" for each galaxy* (Fig. 1). In ~10% of the galaxies, the structure is smooth and centrally concentrated, suggestive of a bright nuclear energy source (AGN?). *In the other cases, the sub-arcsecond morphology is chaotic and extended, suggestive of strong starburst activity.* Ground-based studies have indicated that ~95% of the ULIR galaxies show morphological evidence for a merger, and ~35% show spectroscopic evidence for a Seyfert nucleus [4]. Clearly the strong buildup of molecular gas in the cores of these systems that results from the merger event is related both to starburst activity and to AGN activity (either through the fueling or the creation of the black hole engine).

The peculiar, disturbed morphologies that are seen on large (kiloparsec) scales among this sample of galaxies are continued down to the smallest (HST resolution) scales in the cores of these strongly starbursting systems. A rich variety of morphological features are seen; these are probably related to the recent interaction-induced starburst episode. These starburst-related features (e.g., numerous bright clumps of star formation, shells, and bubbles) are similar to those seen in previous HST imaging observations of strongly interacting and merging galaxies, particularly the large star-forming regions (super star clusters?) seen in the Cartwheel Ring Galaxy [9,10], the Antennae (NGC 4038/4039) [11], and NGC 7252 [12]. The HST images of the ULIR galaxies are not able to resolve similar super star clusters because the minimum and mean redshifts for this sample are substantially higher than the redshifts of the nearby mergers that have been studied by HST. Nevertheless, the clumpy nature of the luminosity distribution (independent of obvious dust mottling) suggest that super star-forming complexes represent the typical mode for massive star formation among merger-induced starburst systems. It is anticipated that further analysis of our complete HST sample of ULIR galaxy images will substantiate this claim, at least qualitatively, out to redshift $z = 0.20$.

REFERENCES

1. Sanders, D. B., Soifer, B. T., Elias, J. H., Madore, B. F., Matthews, K., Neugebauer, G., & Scoville, N. Z. 1988a, ApJ, 325, 74
2. Sanders, D. B., Soifer, B. T., Elias, J. H., Neugebauer, G., & Matthews, K. 1988b, ApJ, 328, L35
3. Soifer, B. T., Sanders, D. B., Neugebauer, G., Danielson, G. E., Lonsdale, C. J., Madore, B. F., & Persson, S. E. 1986, ApJ, 303, L41
4. Sanders, D. B., & Mirabel, I. F. 1996, ARA&A, 34, 749
5. Melnick, J., & Mirabel, F. 1990, AA, 231, L19
6. Lawrence, A., Rowan-Robinson, M., Saunders, W., Parry, I. R., Xiaoyang, X., Ellis, R. S., Frenk, C. S., Efstathiou, G., Kaiser, N., & Crawford, J. 1996, MNRAS, submitted
7. Kim, D.-C., Sanders, D. B., Veilleux, S., Mazzarella, J. M., & Soifer, B. T. 1995, ApJS, 98, 129
8. Clements, D. L., et al. 1996, MNRAS, 279, 459
9. Borne, K. D., Lucas, R. A., Appleton, P., Struck, C., Schultz, A. B., & Spight, L. 1995, in "Science with the Hubble Space Telescope – II" edited by P. Benvenuti, F. D. Macchetto, & E. J. Schreier, 239
10. Struck, C., Appleton, P. N., Borne, K. D., & Lucas, R. A. 1996, AJ, 112, 1868
11. Whitmore, B. C., & Schweizer, F. 1995, AJ, 109, 960
12. Whitmore, B. C., Schweizer, F., Leitherer, C., Borne, K., & Robert, C. 1993, AJ, 106, 1354

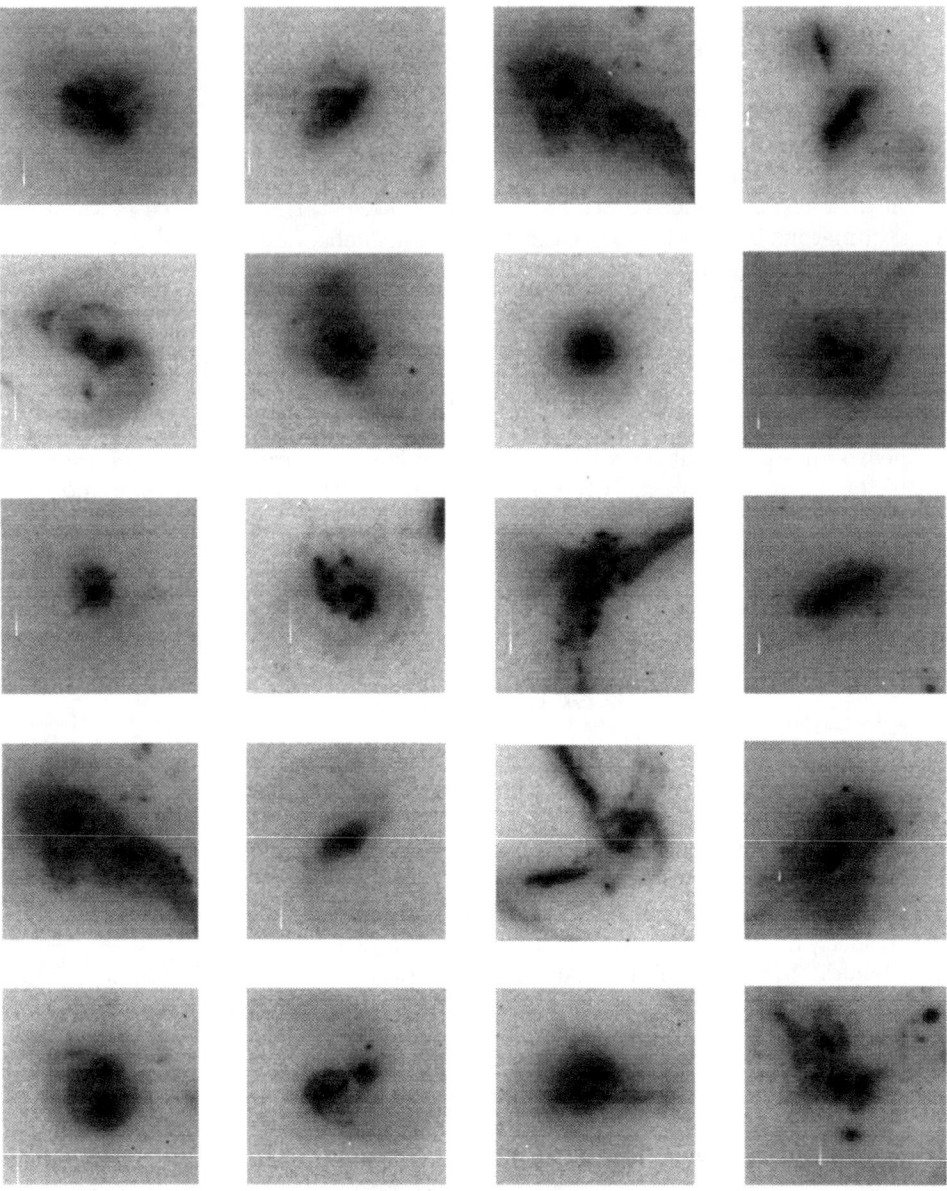

FIGURE 1. Selected HST WFPC2 I-band images (10″ square) for 20 ULIR galaxies. Note the clear interaction/merger morphology for many of the galaxies, but also note the AGN-like appearance of at least 2 of them: #3 in second row, and #1 in third row.

HST Imaging of Sub-Kiloparsec Scale Structure in Markarian Galaxies

Charles H. Nelson*, John W. MacKenty[†], and Susan M. Simkin[#]

*Dept. of Physics, University of Nevada, 4505 Maryland Pkwy., Box 454002, Las Vegas, NV 89154-4002

[†]Space Telescope Science Institute, 3700 San Martin Dr., Baltimore, MD 21218

[#]Michigan State University, Dept. of Physics and Astronomy, East Lansing, MI 48824-1116

Abstract.
As part of our HST WF/PC-1 near-IR survey of Markarian galaxies we have identified several non-Seyferts which display peculiar morphologies on sub-kiloparsec scales. These include double nuclei, spiral structures, and twisting isophotes. One explanation is that we are seeing asymmetric disturbances to the gravitational potential which drives gas inflow providing fuel for the nuclear starburst. Alternatively, these features may be tracing the distribution of massive stars produced by the starburst.

INTRODUCTION

Galaxy interactions and mergers are often cited as important mechanisms for inducing bursts of star formation and activity in galactic nuclei. Theoretical studies and numerical simulations of interacting galaxies have shown that large fractions of the system's interstellar medium can be deposited within the central kiloparsec in a few dynamical times (*e.g.* Barnes & Hernquist 1991). To complete this picture, however, additional gravitational torquing is required for the gas to reach the smallest nuclear scales. This can result from bisymmetric disturbances on scales < 1 kpc such as small bars or double nuclei in a merger of two colliding galaxies. By examining the near-nuclear morphology of Markarian galaxies we can determine if such structures are present and their relationship to the star-formation history.

Table. Small-Scale Structure in Markarian Galaxies

Name	cz (km/s)	M^*_{pg} (mag)	Scale* (kpc/″)	Morphology
Mrk 38	10820	−19.8	0.70	Nuclear arm & oval misaligned with main bar.
Mrk 96	6600	−19.2	0.43	Double nucleus.
Mrk 289	12000	−19.5	0.78	Double nucleus, distorted large scale morphology.
Mrk 420	12860	−20.7	0.83	Inner spiral; twisting isophotes.
Mrk 434	9840	−20.1	0.64	Nuclear arm, twisting isophotes.
Mrk 542	7500	−19.5	0.48	Nuclear arm; extended large scale arm.
Mrk 623	12562	−19.6	0.81	Nuclear arm.

* Assumes $H_o = 75$ km s^{-1}Mpc^{-1}.

OBSERVATIONS AND RESULTS

We have obtained HST WF/PC-1 "snapshots" (0.043 ″/pix) with F785LP filter, ($\sim$ I band) for 52 Seyfert and 50 non-Seyfert Markarian galaxies. The sample is described in MacKenty (1989). Here we concentrate on the non-Seyferts in the sample. The passband contains virtually no emission lines and therefore produces images which trace stellar and nuclear continuum sources. Each image was deconvolved using the accelerated Lucy algorithm (Lucy 1974).

We present images of 7 sample galaxies which display near-nuclear morphological peculiarities. Results are summarized in the Table. These features are found less than 1 kpc from the nucleus in all cases. Recurrent features are nuclear extensions or "arms" (4), double nuclei (2), twisting isophotes (all?). Similar features are seen in our Seyfert sample (MacKenty et al. 1997). We emphasize that these are near-IR continuum images and therefore represent the distribution of starlight and *not* emission line gas.

DISCUSSION

Small-scale, non-axisymmetric structures such as those shown here, can induce the additional gas inflow required to fuel nuclear star formation. (*e.g.* Shlosman, Frank, & Begelman 1989). The two double nuclei objects may be in the late stages of a merger when the most vigorous star formation is likely to occur (Mihos & Hernquist, 1996). The detection of similar structures in the Seyfert galaxies supports the idea that nuclear activity can be fueled by the same mechanism. Alternatively, since numerical simulations find that secondary bars, which provide the dynamical mechanism for inflow, form not

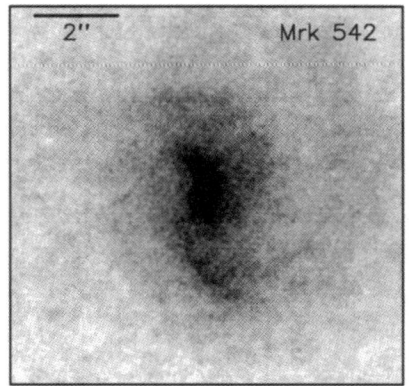 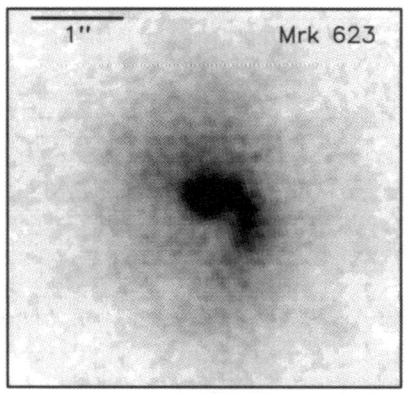

FIGURE 1. Markarian non-Seyfert galaxies with peculiar morphologies on sub-kiloparsec scales.

in the stars but in the gas, the near-IR continuum we observe may come from supergiants produced by star-formation in the gas bar. In this case we are seeing the *effects* rather than the *cause* of the nuclear starburst.

SUMMARY

We present images of Markarian galaxies which display small-scale structure (< 1 kpc) suggesting dynamical mechanisms for producing gas inflow. These features take the form of nuclear arms, double nuclei and isophote twists. It is also possible that these features track massive, evolved stars produced by the starburst rather than the underlying structure of the host galaxy.

This paper is based on observations with the NASA/ESA Hubble Space Telescope obtained at the Space Telescope Science Institute (STScI), which is operated by the Association of Universities for Research in Astronomy, Incorporated, under NASA contract NAS5-26555. Support for this work was provided by NASA through grant numbers GO05-70600 and GO05-97900 from the STSCI.

REFERENCES

1. Barnes, J. E. & Hernquist, L. 1991, ApJ, 370, L65
2. Lucy, L. B. 1974, AJ, 79, 745
3. MacKenty, J. W. 1989, ApJ, 343, 125
4. MacKenty, J. W., Nelson, C. H., Simkin, S. M., Griffiths, R. E., & Jones, G. 1997, submitted to ApJ
5. Mihos, J. C. & Hernquist, L. 1996, ApJ, 464, 641
6. Shlosman, I., Frank, J., Begelman, M. C. 1989, Nature, 338, 45

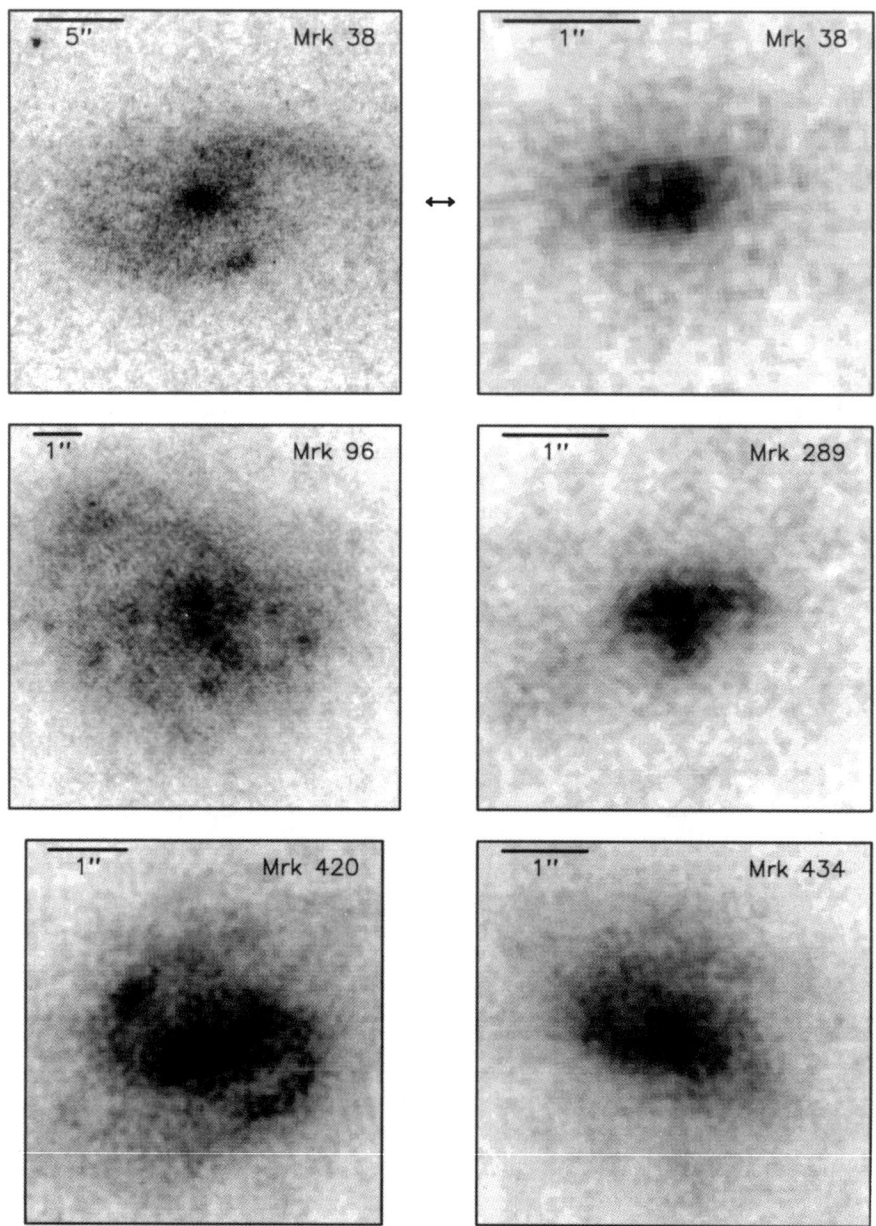

FIGURE 2. Markarian non-Seyfert galaxies with peculiar morphologies on sub-kiloparsec scales.

Infrared Imaging of the Starburst Galaxy NGC 7469

Ray Jayawardhana[1], Giovanni Fazio[1], Stephen Eikenberry[1], David Hughes[2], Joseph Hora[3], William Hoffmann[4], Aditya Dayal[4], Lynne Deutsch[5]

[1] *Harvard-Smithsonian Center for Astrophysics, Cambridge, MA 02138*
[2] *University of Edinburgh, Edinburgh, EH9 3HJ, United Kingdom*
[3] *University of Hawaii, Honolulu, HI 96822*
[4] *University of Arizona, Tucson, AZ 85721*
[5] *University of Massachusetts, Amherst, MA 01003*

Abstract. We have obtained high-resolution mid-infrared and near-infrared images of the Seyfert-starburst hybrid galaxy NGC 7469. Although our direct K-band image is relatively featureless, the residual image created by subtracting a smooth model based on best-fitting ellipses reveals what appear to be inner spiral arm structures. The location of these features is in fair agreement with the inner spiral arms first reported by Mazzarella et al., and the high-surface brightness portions of the spiral features correspond to the previously detected 3" (1 kpc) diameter ring of radio continuum emission. Our deconvolved mid-infrared image also shows lumps of emission corresponding to the reported circumnuclear starburst ring. We compare our images to published infrared and radio observations, and discuss the mid-infrared excess in NGC 7469 and other starburst galaxies.

INTRODUCTION

The *Infrared Astronomy Satellite (IRAS)* identified infrared-luminous galaxies as an interesting class of objects which may be significant in the context of the evolution of starburst galaxies, active galactic nuclei, and quasars (e.g., Soifer et al. 1987). The Seyfert-starburst hybrid galaxy NGC 7469 is one of the most luminous members of this class, with an infrared luminosity $\simeq 3.4 \times 10^{11} L_\odot$. A significant fraction of this luminosity originates from a circumnuclear starburst, seen as a broken ring structure in high-resolution radio (Wilson et al. 1991), optical (Mauder et al. 1994) and infrared (Mazzarella et al. 1994; Miles et al. 1994; Genzel et al. 1995) observations. Infrared features at 3.3μm and 8-13μm, presumably from polycyclic aromatic hydrocarbons (PAHs), have also been seen (Aitken et al. 1981; Cutri et al. 1984).

OBSERVATIONS

We have observed NGC 7469 with sub-arcsecond resolution in the K-band using the COB camera on the 2.1-meter telescope at the Kitt Peak National Observatory, and in the N-band using the Mid-Infrared Array Camera 2 (MIRAC2) on the 3.8-meter United Kingdom Infrared Telescope (UKIRT).

K-band image

Even the direct K-band image shows differences between the position angles and ellipticities of the inner and outer isophotes, as well as slight deviations from ellipticity, suggesting that non-axisymmetric structures may be present. Therefore, we subtracted a smooth model of the galaxy, based on best-fitting ellipses, to obtain the residual image shown in Figure 1. This image shows a small central spiral structure, with a diamter $\sim 3''$, similar to that reported by Mazzarella et al. (1994). However, the presence of an inner bar is not confirmed.

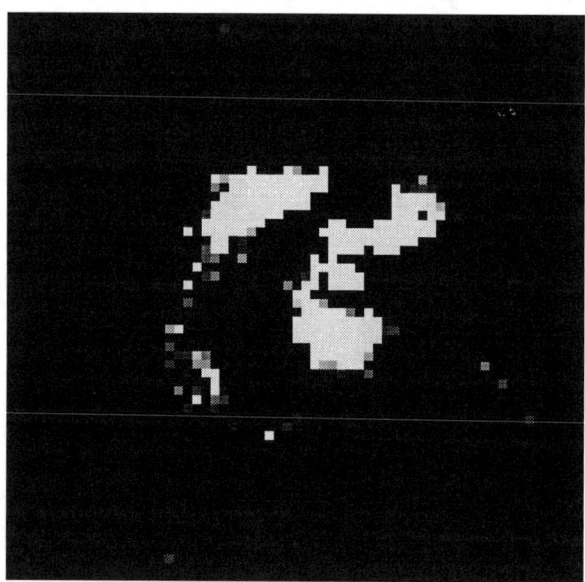

FIGURE 1. Residual K-band image of the central region

N-band image

To look for spatially extended structure, we deconvolved the N-band image using the Wiener deconvolution algorithm provided in the IRAF data analysis package. The resulting contour plot, shown in Figure 2, contains all the major features seen in the deconvolved 11.7μm image of Miles et al. (1994). Thus we are confident that the "knotty" ring structure seen here is real.

DISCUSSION

By comparing our deconvolved N-band image to the 6 cm radio map of Wilson et al. (1991), we confirm that mid-infrared emission in NGC 7469 is spatially correlated with radio continuum emission.

The starburst "knots" in the ring have a typical N-band flux of $\sim$ 60 mJy. Since Wilson et al. measured a spectral index of $\alpha = 0.95 \pm 0.1$ between 2cm and 6cm, the 2cm flux gives an upper limit to the thermal flux density at 6cm. Therefore, the mid-infrared to *thermal* radio flux ratio must be $\geq$ 90. Similarly, we find a ratio $\geq$ 70 for the unresolved nulear source, in fair agreement with Miles et al. (1994). These flux ratios are much greater than those observed in Galactic HII regions (Lebofsky et al. 1978; Thronson et al. 1978).

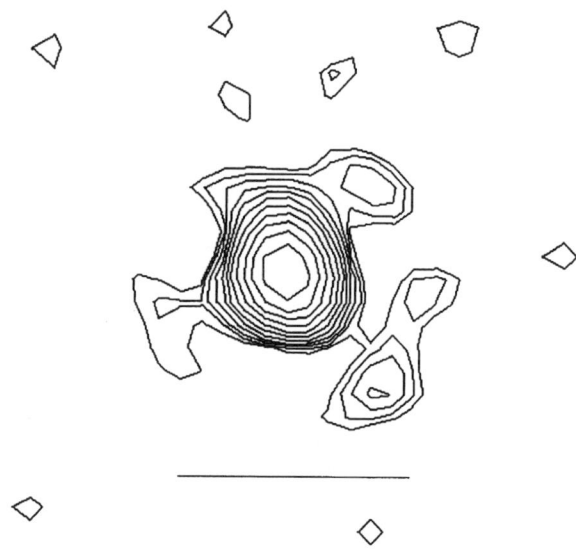

FIGURE 2. Deconvolved N-band image. Horizontal bar is 3" across.

As Ho et al. (1989) have suggested, a population of small grains may account for the observed infrared excess in starburst galaxies. The detection of PAHs in NGC 7469 (Aitken et al. 1981; Cutri et al. 1984) provides strong support for this hypothesis. However, spectral observations by Miles et al. (1994) show that PAHs are not present in the nucleus of NGC 7469, implying that some other explanation is needed for the infrared excess in the nucleus.

REFERENCES

1. Soifer, B.T., Sanders, D.B., Madore, B.I., Neugebauer, G, Lonsdale, C.J., & Rice, W.L. 1987, ApJ, 320, 238
2. Wilson, A.S., Helfer, T.T., Haniff, C.A., & Ward, M.J. 1991, ApJ, 381, 79
3. Mauder, W., Weigelt, G., Appenzeller, I. & Wagner, S.J. 1994, A&A, 285, 44
4. Mazzarella, J.M., Voit, G.M., Soifer, B.T., Matthews, K, Graham, J.R., Armus, L., & Shupe, D. 1994, AJ, 107, 1274
5. Miles, J.W., Houck, J.R., & Hayward, T.L. 1994, ApJ,425, L37
6. Genzel, R., Weitzel, L., Tacconi-Garman, L.E., Blietz, M., Cameron, M., Krabbe, A., Lutz, D. & Sternberg, A. 1995, ApJ, 444, 129
7. Aitken, D.K., Roche, P.F., & Phillips, M.M. 1981, MNRAS, 196, 101
8. Cutri, R.M., Rudy, R.J., Rieke, G.H., Tokunaga, A.T., Wilner, S.P. 1984, ApJ, 280, 521
9. Lebofsky, M.J., Sargent, D.G., Kleinmann, S.B. & Rieke, G.H. 1978, ApJ, 219, 487
10. Thronson, H.A., Campbell, M.F., & Harvey, P.M., 1978, AJ, 83, 1581
11. Ho, P.T.P., Turner, J.L., Fazio, G.G., & Willner, S.P., 1989, ApJ, 344, 135

Another Twin Peaks galaxy - The Barred Spiral NGC 5383

Kartik Sheth, Michael W. Regan, Stuart N. Vogel*

Department of Astronomy, University of Maryland, College Park, MD 20742

Abstract.

We present Berkeley-Illinois-Maryland Array interferometric CO (J = 1-0), KPNO near infrared and H α observations of the barred spiral galaxy NGC 5383. We find that the CO emission is strongly concentrated in the nuclear region with three distinct peaks coinciding with peaks of dust emission observed via our color maps. Two of the peaks clearly correspond to the intersection of the bar dust lanes with the nuclear ring. Significant CO emission extending up to 500 pc along the dust lane is also detected.

The total molecular gas content in the nuclear region provides an important constraint on the star formation rate in the nucleus. We discuss the total gas content as derived from our CO and dust maps. Comparison of the observed H α emission, which is a measure of the star formation rate, to the total gas content then yields a lifetime for the nuclear ring.

INTRODUCTION

Barred spiral galaxies constitute a large fraction of all disc galaxies. The kinematics and morphology of bars provide valuable information about the density distribution in centers of galaxies, the inflow rate of gas to the nucleus and small and large scale spiral structures.
We present BIMA (CO J=1-0), near infrared and optical observations of the prototypical barred spiral NGC 5383.

Conclusions

- CO emission is largely confined to the nucleus in the familiar "twin peaks" pattern.
- Low density gas outside of nucleus and dust lane is suggestive of gas spray from the dust lane-ILR collision point.
- The nucleus and the bar ends are active star forming regions while the bar itself is quiescent.

FIGURE 1. R-band image showing offset dustlanes curving into the nucleus. Bifurcation of the dustlane at various points along the bar is also observed.

TABLE 1. Global Properties of NGC 5383

Property	Value	Reference[a]
R.A. (J2000)	13h 57m 04.61s	(1) K-Band peak
Dec. (J2000)	41d 50d 46.03s	(1)
D25	2.75 arcmin	(2)
Inclination	50 degrees	(2)
Sys. radial velocity	2250 km/s	(3)
Adopted Distance	30 Mpc	..
Bar Length	110 arcsecs	(1)
Bar P.A.	135 degrees	(2)

[a] (1) Our data; (2) Duvall and Athanassoula A.A., 121,279; (3) Becker et. al. ApJ, 450, 559B.

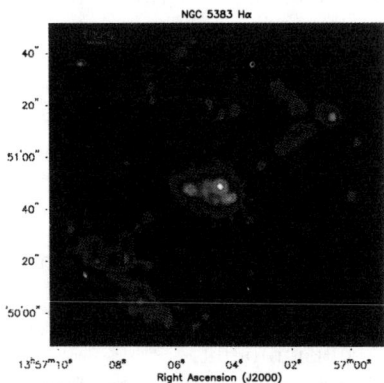

FIGURE 2. Continuum subtracted Hα image showing active star forming regions in the nucleus and bar ends. The total star formation rate (SFR) is 0.12 M☉/yr in the nucleus, 0.21 M☉/yr in the northern bar end and 0.15 M☉/yr in the southern bar end. There is a striking lack of star formation in the bar itself.

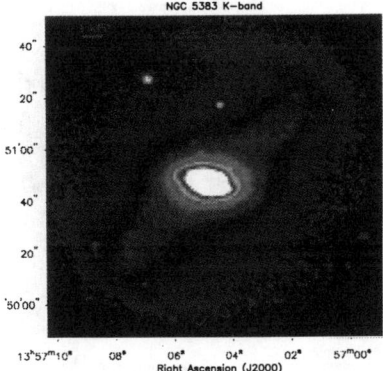

FIGURE 3. K band image depicting the relatively smooth bar potential. The dense gas lies at an obtuse angle from the bar major axis as expected from the interaction of the gas flow in the dust lane and the inner Lindblaad resonance (ILR).

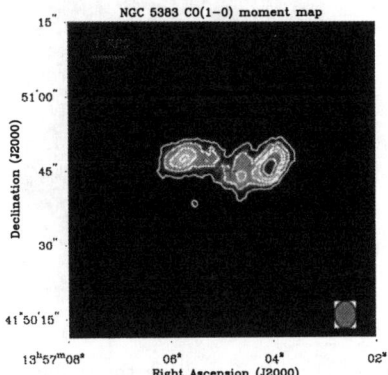

FIGURE 4. Berkeley-Illinois-Maryland Array (BIMA) CO (J=1-0) total intensity map showing the "twin peaks" structure at the intersection of the dust lanes with the ILR.

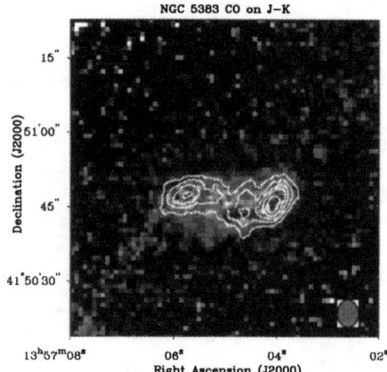

FIGURE 5. CO overlaid on J-K shows that the dense gas is well traced by the dense dust. In addition to the "twin peaks" in CO, we see a third peak in CO which is also coincident with another J-K peak.

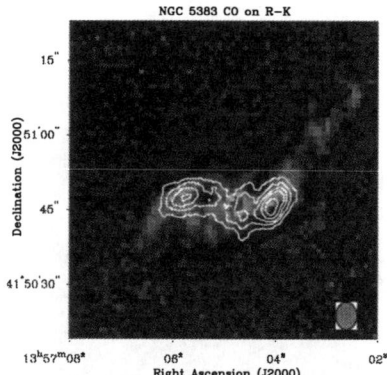

FIGURE 6. Overlay of the CO on R-K showing the striking extension of CO along the low column gas/dust lane traced by the R-K color. Surprising enhancement of low column gas outside the dust lane and the nucleus may be suggestive of gas spray from the collision point of the dust lane and the nuclear ring. Bifurcation of the dust lane seen in the R-band image is also well traced by the R-K color.

Dynamical Stability and Galaxy Evolution in LSB Disk Galaxies

Chris Mihos,* Stacy McGaugh,† and Erwin de Blok‡

Hubble Fellow, Department of Physics and Astronomy, Johns Hopkins University
†*Carnegie Institute of Washington, Department of Terrestrial Magnetism*
‡*Kapteyn Astronomical Institute, University of Groningen*

Abstract. We demonstrate that, due to their low surface mass density and large dark matter content, LSB disks are quite stable against the growth of global bar modes. However, they may be only marginally stable against local disk instabilities. We simulate a collision between an LSB and HSB galaxy and find that, while the HSB galaxy forms a strong bar, the response of the LSB disk is milder, in the form of spiral features and an oval distortion. Unlike its HSB counterpart, the LSB disk does not suffer strong inflow of gas into the central regions. The lack of sufficient disk self-gravity to amplify dynamical instabilities makes it difficult to explain strong interaction-driven starbursts in LSB galaxies without invoking mergers.

The lack of companions around low surface brightness (LSB) disk galaxies [1,2] has led to the suggestion that, without the well-established dynamical trigger provided by interactions, LSB galaxies may simply evolve passively due to their low surface densities [3], and never experience any strong star-forming era in their lifetimes. Indeed, sufficient tidally induced star formation in LSB disks may drive evolution from LSB to high surface brightness (HSB) galaxies. This has been suggested as the cause of the observed isolation of LSB galaxies: interactions in denser environments transform them into HSB or HII galaxies or perhaps even destroy them entirely.

However, the ability for interactions to trigger evolution and starburst activity is linked to instabilities in the stellar disk. As LSB disk galaxies have lower disk mass densities and a greater fraction of dark to visible matter than do HSB galaxies [4], the stability of LSB disks – and their response to tidal interactions – may be quite different than that of "normal" HSB galaxies. In this study, we use analytic stability criteria and numerical simulation to investigate the stability of LSB disks in the context of galaxy interactions.

STABILITY CRITERIA

To study disk stability, we use the structural properties of the LSB disk galaxy UGC 128 and the HSB galaxy NGC 2403, derived by de Blok & McGaugh [4] from HI rotation curve decompositions. UGC 128 has a disk mass density nearly an order of magnitude below that of NGC 2403, and is more dark matter dominated: the mass-to-light ratio within 6 scale lengths is $\Upsilon_B = 30$ for UGC 128 and $\Upsilon_B = 7.4$ for NGC 2403 (see [4] for details). The rotation curves for UGC 128 and NGC 2403 are shown in Figure 1a.

One measure of the susceptibility of galactic disks to global bar instabilities is the X_2 parameter [5]: $X_{m=2} = \frac{\kappa^2 R}{4\pi G \Sigma_d}$, where κ is the epicyclic frequency, R is the radius, and Σ_d is the disk surface density. For flat rotation curves, disks prove stable against growing modes if $X_2 > 3$, while for linearly rising rotation curves $X_2 > 1$ is a sufficient condition for stability. Figure 1b shows X_2 as a function of scale length for our representative galaxies. The HSB galaxy NGC 2403 is only marginally stable over a large range of radius, while the LSB galaxy UGC 128 proves stable throughout the disk, due to its lower mass surface density. We point out that the rotation curve modeling assumed maximum disk models; if LSBs are less than maximal disks, they will be even *more* stable.

If LSB disks are stable against the growth of global instabilities in the disks, are they also stable against *local* instabilities? The growth of local axisymmetric instabilities is measured by the Toomre Q parameter [6]: $Q = \frac{\sigma_r \kappa}{3.36 G \Sigma_d}$, where σ_r is the radial velocity dispersion of the disk stars. Lacking information on σ_r in LSB disks, we use two alternatives: 1) that σ_r is like that in the Milky Way (~ 30 km s^{-1}) or 2) that $\sigma_r^2 \sim \Sigma_d$ (so that $\sigma_r \sim 11$ km s^{-1}). Figure 1c shows Q in each disk; if velocity dispersion drops with surface density as might be expected from energy arguments, LSB and HSB disks may have similar *local* stability properties, such that local instabilities might grow in LSB disks where global modes cannot.

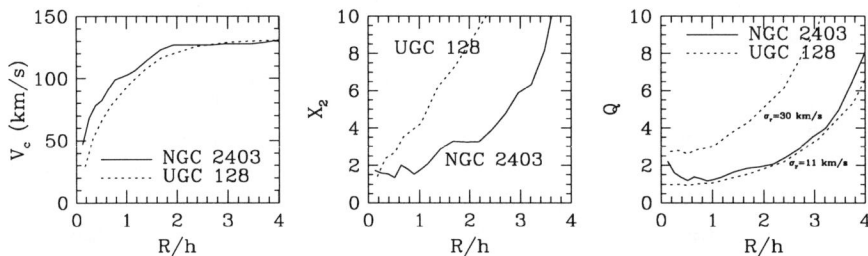

FIGURE 1. Left: Rotation curves of NGC 2403 (HSB) and UGC 128 (LSB), as a function of disk scale length (R/h). Middle: X_2 stability parameter. Right: Toomre Q parameter. The two curves for UGC 128 reflect two choices for σ_r.

NUMERICAL MODELS

To examine how LSB disks will respond to a close interaction, we simulate a grazing encounter between an LSB galaxy and an HSB companion. We choose a prograde, parabolic orbit with a perigalactic separation of $R_p = 10$ disk scale lengths.

Rather than build galaxy models which differ in a number of structural parameters, we focus on variations in disk surface density to define the difference between HSB and LSB disk galaxies. We construct two model galaxies with disk surface densities which differ by a factor of eight, similar to the difference between NGC 2403 and UGC 128. The dark halos have identical mass distributions (as a function of R/h) in both galaxies, resulting in our LSB being very dark matter dominated. We initialize velocities in *both* galaxy disks such that Q=1.5, implying lower velocity dispersion in the LSB disk; the simulation is thus a conservative test of LSB stability. In models which include gas, the gas comprises 10% of the total disk mass in each galaxy.

Figure 2 shows the evolution of the HSB and LSB disks in the stellar dynamical interaction model. Both galaxies respond strongly during the close passage (at T=24). In the HSB disk, the self-gravity of the disk amplifies the perturbation such that by T=44 the galaxy has developed a very strong bar. By contrast, the LSB disk displays a persistent oval distortion and long-lived spiral arms in the disk. Without adequate disk self-gravity no strong bar develops in the LSB disk. Figure 3a shows the strength of the $m = 2$ mode in the inner half mass of each disk. The peak strength is more than twice that of the LSB disk, and declines at late time, probably due to disk heating by the bar. We emphasize that the $m = 2$ mode is not only different in strength between the disks, but also in character: the HSB sports a strong bar, while the LSB displays a milder oval distortion. The bar in the HSB galaxy drives strong inflow (Figure 3b): the gas surface density in the center of the HSB disk has risen significantly by T=36.[1] By contrast, the relatively weak response of the LSB disk results in very little change in the gas mass distribution in the disk, even much later after the encounter at T=72 (Figure 3c).

LSBS AND GALAXY EVOLUTION

Both analytic arguments and numerical simulation indicate that, despite their seemingly fragile nature, LSB disks are quite stable, and resistant against the growth of bars and bar-driven inflows. These results present a problem for the otherwise appealing notion that interacting LSB dwarfs are the progenitors of HII galaxies experiencing central starbursts [2]. Even the relatively

---

[1] At this point, the gas was "switched off" in the HSB to save computational expense; however, inflow was ongoing, and the final gas density at the center of the HSB would be even higher than shown here.

close, strong interaction we have presented will not result in a strong central starburst, nor will it drive strong structural evolution in the galaxy; in order to provoke a violent enough response in the LSB disk, a bona-fide merger may be necessary.

REFERENCES

1. Bothun, G.D., et al. 1993, AJ, 106, 530
2. Taylor, C.L. 1997, ApJ, in press
3. van der Hulst, J.M., et al. 1993, AJ, 106, 548
4. de Blok, W.J.G., & McGaugh, S.S. 1996, ApJ, 469, L89
5. Toomre, A. 1981, in The Structure and Evolution of Normal Galaxies, eds. S.M. Fall & D. Lynden-Bell (London: Cambridge University Press), 111
6. Toomre, A. 1964, ApJ, 139, 1217

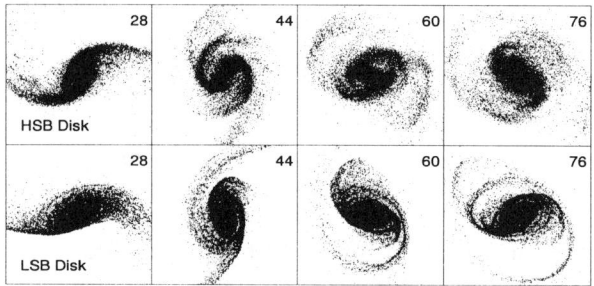

FIGURE 2. Post collision disk evolution. Top: HSB disk. Bottom: LSB disk. Each frame is 10 scale lengths on a side, and time is given in the upper right. One rotation period is approximately 13 time units.

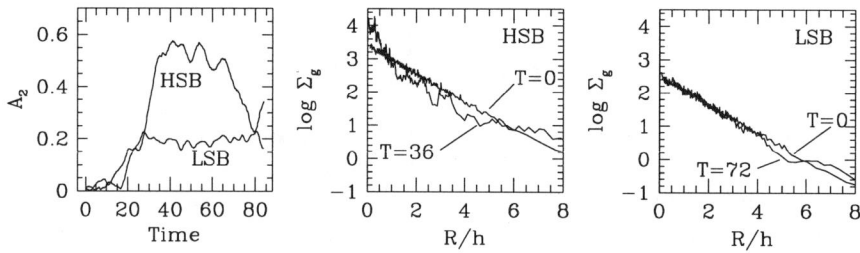

FIGURE 3. Right: Growth of m=2 modes in stellar-dynamical simulation. Middle: Gas mass profile in HSB disk in stellar+hydro simulation. Right: Gas mass profile in LSB disk in stellar+hydro simulation.

The age of LSB discs

Paolo Padoan [1], Raul Jimenez [2], & Vincenzo Antonuccio-Delogu [3]

[1] *Theoretical Astrophysics Center, Juliane Maries Vej 30, DK-2100 Copenhagen, Denmark*
[2] *Royal Observatory, Blackford Hill EH9-3HJ, Edinburgh, UK*
[3] *Osservatorio Astrofisico di Catania, Città Universitaria - Viale A. Doria 6, 95125 Catania, IT*

Abstract.
The UBVRI colors of LSB discs seem to indicate that these galaxies are older than 7 Gyr, and their that mean age is probably about 9 Gyr.

INTRODUCTION

Late-type Low Surface Brightness galaxies (LSBs) are considered to be very young stellar systems, because of their blue colors (de Blok, van der Hulst & Bothun 1995, McGaugh & Bothun 1996). Dalcanton, Spergel, & Summers (1996), and Mo, McGaugh, & Bothun (1994) have recently suggested that LSBs are formed inside dark matter halos that collapsed very recently, at $z \leq 1$, from density fluctuations of small amplitude.

In this work we study the colors of LSBs from the point of view of synthetic stellar populations (SSP), and show that LSBs could not be as young as claimed in the quoted literature.

SINTHETIC STELLAR POPULATIONS

In the following when we will refer to LSBs' we will always mean the sample of late-type disc galaxies observed by de Blok, van der Hulst & Bothun (1995).

De Blok, van der Hulst & Bothun (1995) noted that it is difficult to understand the colors of LSBs, if their stellar population is old or forming at a declining rate. McGaugh and Bothun (1996), from the analysis of their sample, concluded that the stellar populations in LSBs must be very young, because of the very blue colors and of the very low metallicity.

In order to check the validity of these statements, we have produced synthetic stellar population models of LSBs, assuming that their stellar discs have exponential density profiles, with the stellar IMF as predicted by Padoan, Nordlund & Jones (1997; see also Padoan et al., this volume).

We apply the P-IMF to a simple exponential disc model, with height-scale equal to 100 pc, length-scale equal to 5 Kpc, and total mass equal to $M_D = 3 \times 10^9 M_\odot$. The value of the mass is chosen to be representative of the HI content of the galaxies (de Blok, McGaugh, & van der Hulst, 1996).

To compute the synthetic colors we used the latest version of our synthetic stellar population code (Jimenez et al. 1996). The code uses the library of stellar tracks computed with JMSTAR9 and the set of atmospheric models calculated by Kurucz (Kurucz 1992). A careful treatment of all evolutionary stages has been done following the prescriptions in Jimenez et al. (1995), and Jimenez et al. (1996).

We find that the star formation in LSBs can be adequatley described with an initial burst, followed by a quiescent evolution up to the present time. It has been already remarked (van der Hulst et al., 1993) that LSBs' gas surface densities are too low to allow efficient star formation according to Kennicut criterion (Kennicut 1989).

THE AGE OF LSBS

Fig. 1 shows the time evolution of the colors for a very low metallicity ($Z = 0.0002$).

The diamonds are the mean values of the luminosity-weighted colors listed in Table 4 of de Block, van der Hulst, & Bothun (1995). The error bars represent the dispersion around the mean, rather than the photometric uncertainty. For an age younger than 7 Gyr, the theoretical $B - V$ and $B - R$ colors are too blue, compared with the data.

Since the colors $B-V$ and $B-R$ are the best ones to constrain the age of the galaxies, we show in Fig. 2 the color-color diagram (B-V,B-R). The continuous

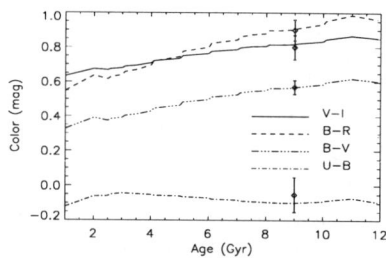

FIGURE 1. The time evolution of the colors in a model with metallicity $Z = 0.0002$ and star formation in a initial burst of 5×10^7 yr. The diamonds are the mean luminosity-weighted values for the sample of LSBs (de Blok et al. 1995).

line is the time trajectory of our model, from about 3 Gyr (left-hand side) to 12 Gyr (right-hand side). This plot shows that the model reproduces well the colors of each galaxies (not only their mean value), and that most galaxies are probably older than 7 Gyr, as it was concluded from the values of the mean colors of the sample (Fig. 2).

TOTAL MAGNITUDE AND OXYGEN ABUNDANCE

Fig. 3 we shows the B-V color of our model versus its absolute B magnitude, at different times. The continuous line is for a star formation efficiency (SFE) of 0.3, the dashed line for a SFE= 0.6 (therefore more luminous). The diamonds are the observed galaxies. One can see that the total magnitude of the model is not inconsistent with the observations, if LSBs' discs have turned into stars from 30% to 60% of their initial mass.

McGaugh (1994) has measured the oxygen abundances in the HII regions of LSB disc galaxies. He has found $log(O/H) < -3.6$, excluding the two galaxies

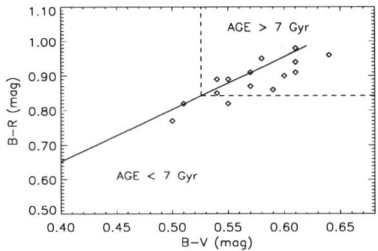

FIGURE 2. The time trajectory of the model in the (B-V,B-R) diagram. The diamonds are the observed luminosity weighted colors of LSB discs, from de Block et al. (1995).

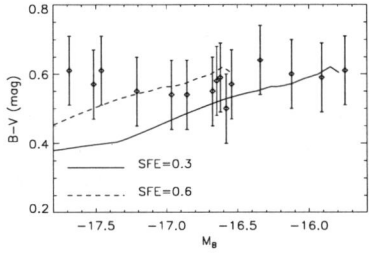

FIGURE 3. The B-V color of the model versus its absolute B magnitude, at different times. The diamonds are the observed galaxies.

with the largest abundance. Since the Padoan et al. IMF, that we use in the model, produces the oxygen abundance $log(O/H) = -3.6$, if about 30% of the initial disc mass is turned into stars, we can say that the SFE, imposed to the model by the observed magnitudes of the galaxies, is not inconsistent with the constraint provided by the oxygen abundance in the gas.

DISCUSSION AND CONCLUSIONS

Our burst model is the most conservative one, as far as the lower limit for the age of the beginning of the star formation process in the discs is considered. Any different star formation history could only make LSBs' discs older than our estimate. Therefore, we can conclude that the colors seem to indicate that LSB discs are older than 7 Gyr, and that *they cannot be formed in primordial density fluctuations of low amplitude, collapsed at $z \leq 1$*.

ACKNOWLEDGEMENTS

This work has been supported by the Danish National Research Foundation through its establishment of the Theoretical Astrophysics Center.

REFERENCES

1. Dalcanton, J. J., Spergel, D. N., Summers, F. J. 1996, preprint
2. de Blok, W. J. G., van der Hulst, J. M., Bothun, G. D. 1995, MNRAS, 274, 235
3. de Blok, W. J. G., McGaugh, S. S., van der Hulst, J. M. 1996, MNRAS, 283, 18
4. Jimenez et. al. 1996, in preparation
5. Jimenez, R. & MacDonald, J., 1997, in preparation
6. Kennicut, R. C. 19080, ApJ, 344, 685
7. Kurucz, R., 1992 ATLAS9 Stellar Atmosphere Programs and 2km/s Grid CDROM Vol. 13
8. McGaugh, S. S. 1994, ApJ, 426, 135
9. McGaugh, S. S., Bothun, G. D. 1996, preprint
10. Miller, G. E, Scalo, J. M. 1979, Ap. J. Suppl., 41, 413
11. Mo, H. J., McGaugh, S. S., Bothun, G. D. 1994, MNRAS, 267, 129
12. Nordlund, & Padoan, *in preparation*, 1997
13. Padoan, P., Nordlund, Å., Jones, B. J. T. 1997, MNRAS (in press)
14. van der Hulst, J.M., Skillman, E.D., Smith, T.R., Bothun, G.D., NcCahugh, S.S. and de Blok, W.J.G 1993, AJ, 106, 548

Pre-Starbursts in Luminous IR Galaxies ?

Y. Gao*, R. Gruendl*, K.Y. Lo*, C.Y. Hwang[+], & S. Veilleux[†]

*Lab. for Astronomical Imaging, Dept. of Astronomy, Univ. of Illinois
[+]IAA, Academia Sinica, Taiwan, ROC   [†]Dept. of Astronomy, Univ. of Maryland

Abstract. We present first results of our on-going BIMA Key Project: imaging the CO(1-0) emission from a sample of 10 LIRGs that are at various merging stages, with special emphasis on systems apparently in the early/intermediate stages of merging. We present here CO images with $\sim 5''$ resolution. An important result is the recognition of a plausible *pre-starburst* phase in some early LIRG mergers (e.g., Arp 302 and NGC 6670). Our initial analysis suggests that a merger-induced starburst phase may not begin before the nuclear separation between the merging galaxies reaches roughly 10 kpc. The surface gas density seems to increase from a few times $10^2 M_\odot pc^{-2}$ to $> 10^3 M_\odot pc^{-2}$ while the prominent CO extent systematically decreases as merging progresses.

INTRODUCTION

Luminous IR galaxies (LIRGs, $L_{IR} \gtrsim 2 \times 10^{11} L_\odot$, $H_0 = 75 \text{km s}^{-1} \text{ Mpc}^{-1}$), emit most of their bolometric luminosity in the far-IR (up to $\gtrsim 90\%$), and are the dominant class of galaxies in the local universe at these high luminosities [10]. Many LIRGs are interacting/merging galaxies [7,3,6] rich in molecular gas [8,2,11]. It is not well understood whether starbursts produce most of the IR luminosity, how the starbursts are initiated and what role galaxy-galaxy interactions might play in triggering these starbursts.

A study of the molecular gas properties at various phases of the merging process in LIRGs would help identify the key physical processes involved. Previous CO imaging studies have concentrated on relatively advanced merger systems [9] in which the interstellar medium (ISM) has already been highly disrupted by the interaction and starbursts. In order to isolate the conditions in the ISM *leading* to starbursts, we have started a program to study a sample of LIRGs chosen to represent different phases of the interacting/merging process, using the newly expanded Berkeley-Illinois-Maryland Association (BIMA) millimeter-wave array [12]. The goal is to sample statistically the

TABLE 1. Luminous Infrared Galaxies in a Merger Sequence.

Source	cz km s^{-1}	R_{Sep}[a] kpc	L_{IR} $10^{11}L_\odot$	M(H$_2$)[b] $10^{10}M_\odot$	Beam "	CO[c]	Σ_{H_2}[d] $M_\odot pc^{-2}$	SFE[e]
ARP302	10166	25.8	4.1	8.0	6.0	u+e	2100	5.0 / 6.0,2.0
N6670	8684	14.6	3.8	5.5	4.9	u+e	1100	6.9 / 8.7,6.9
U2369	9475	13.1	3.9	3.4	6.1	u+e	1800	11.5 / 15.4
ARP55	11773	10.7	4.7	5.8	4.4	u+e	1900	8.1
IZw107	12043	4.8	7.2	3.4	3.9	u+e	1400	21.2 / 30.
N5256	8239	4.7	3.1	2.7	4.6	u+e	1800	11.5
N6090	8830	3.5	3.0	2.4	5.8	u	1300	12.5 / 17.2

[a] Projected separation between the two galaxy nuclei.
[b] M(H$_2$)=4.78 L$_{CO}$, where L$_{CO}$ is the single-dish CO luminosity in Kkm s^{-1} pc^2.
[c] CO morphology, u$\equiv$ unresolved peak; e$\equiv$ extended structures resovled by the beam.
[d] Observed peak surface gas density uncorrected for inclination. Note Arp 302 and NGC 6670 have inclination angles larger than 75°
[e] Given by L$_{IR}$/M(H$_2$)[L$_\odot$/M$_\odot$]. The far-IR luminosities are estimated by scaling the far-IR emission and extent following that of the radio continuum emission. First number is a global value while the second row shows each peak value.

evolution of physical conditions of the molecular material in LIRGs as compared with the properties of the IR emission along the merger sequence.

SAMPLE AND OBSERVATIONS

Our sample emphasizes LIRGs which appear to be in the early and intermediate stages of merging with large nuclear separation (Table 1). These LIRGs are potentially the most molecular gas-rich systems since the CO luminosity is found to increase with increasing separation of the merging nuclei in a sample of $\sim$ 50 LIRG mergers [2].

The BIMA array is ideally suited to study these early/intermediate LIRG mergers given its large primary beam and wide spectral bandwidth. The observations presented here were all made with the 9-element BIMA array in the H/C configurations in 1996.

RESULTS AND DISCUSSIONS

Fig. 1 shows integrated CO intensity (contours) overlayed on broad-band images in 4 LIRGs in an order of decreasing nuclei separation. Although LIRGs in our sample have very small ranges in L$_{IR}$ and L$_{CO}$, the apparent differences in CO morphology and gas properties are clearly seen along the merger sequence:

- The morphology of the molecular gas in LIRGs changes, along the merger sequence, from the weakly disturbed two separated gas disks, e.g., early or pre-mergers like Arp 302 [4] and NGC 6670 → the disturbed or merged-common-envelope gas disks (intermediate mergers like Arp 55 and Mrk 848) → a single common gas disk for the double nuclei of the two galaxies (close to the advanced mergers like NGC 6090).
- The total spatial CO extent drops from ~ 20 kpc for the early mergers to a few kpc for the intermediate and advanced mergers. Very advanced mergers like Arp 220 have typical nuclear CO concentration $\lesssim 1$ kpc [9].
- The corrected face-on central gas surface density (lower-limits due to

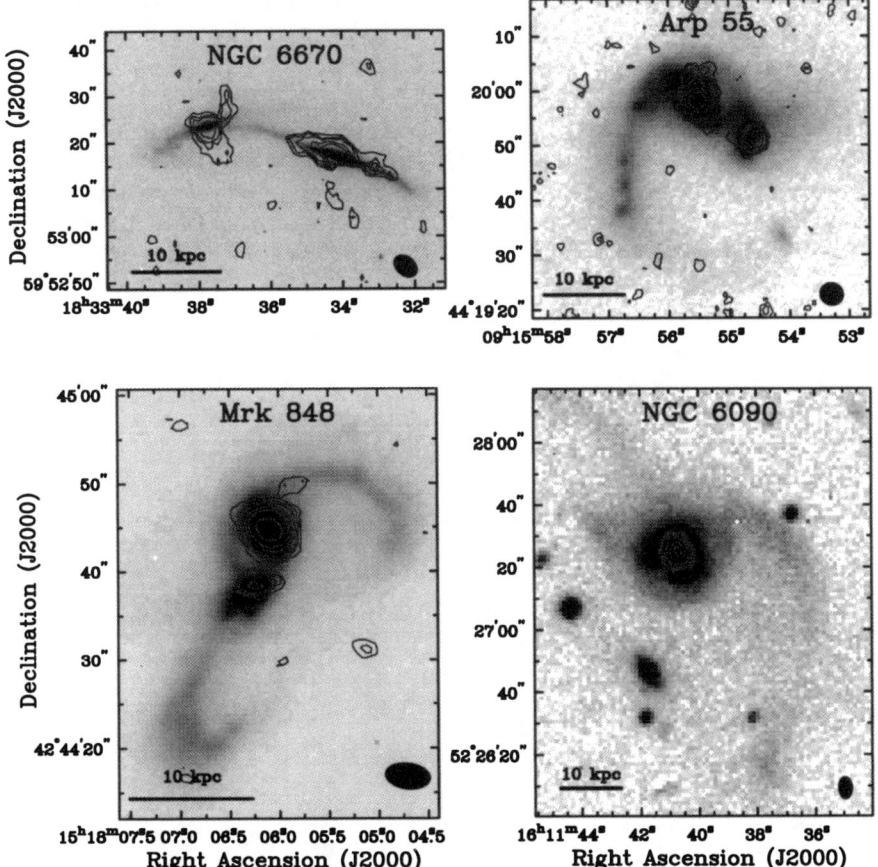

FIGURE 1. CO contours overlayed on CCD images in four typical LIRGs in order of decreasing nuclear separation. The contours plotted are 2,3,4,6,8,12,16,24,32 σ levels.

resolution) increases from a few times 10^2 $M_\odot pc^{-2}$ to $> 10^3$ $M_\odot pc^{-2}$ along the sequence. Whereas advanced mergers such as Arp 220 and Mrk 273 have typical values $> 10^4$ $M_\odot pc^{-2}$.

- The $L_{IR}/M(H_2)$ ratio (star formation efficiency, SFE) increases by roughly a factor of two from the early mergers to the intermediate/advanced mergers. When we scale the IR luminosity and extent with those of the radio continuum emission [1] using the well-known correlation between far-IR and radio continuum flux densities [5], we can estimate the central SFE ratio which tends to increase more drastically than the global SFE along the sequence (see Table 1).
- The SFE ratio usually ranges from 20 to 100 $L_\odot/M_\odot$ in LIRGs. However, we found that early mergers like Arp 302 [4] and NGC 6670 appear to have much smaller SFE throughout the entire interacting/merging disks, comparable to that of GMCs in the Milky Way disk. The level of star formation activity in early mergers is therefore remarkably similar to that of GMCs. This strongly suggests that LIRGs in the early stage of merging are in a pre-starburst phase.
- In intermediate stage LIRG mergers the starbursts appear to have "turned-on", exhibited by higher SFE ratios especially in the nuclei. This may imply that starburst phase does not begin before the separation between the merging galaxies reaches roughly 10 kpc characterizing these intermediate stage LIRG mergers.

REFERENCES

1. Condon, J.J., Helou, G., Sanders, D.B., & Soifer, B.T. 1990, ApJS, 73, 359
2. Gao, Y. 1996, Ph.D. thesis, SUNY at Stony Brook
3. Leech, K.J. et al. 1994, MNRAS, 267, 253
4. Lo, K.Y., Gao, Y., & Gruendl, R.A. ApJL, in press
5. Marsh, K.A., & Helou, G. 1995, ApJ, 445, 599
6. Murphy, T.W.Jr., et al. 1996, AJ, 111, 1025
7. Sanders, D.B. et al. 1988, ApJ, 325, 74
8. Sanders, D.B. et al. 1991, ApJ, 370, 158
9. Scoville, N.Z. et al. 1991, ApJ, 370, 158
10. Soifer, B.T. et al. 1987, ApJ, 320, 238
11. Solomon, P.M. et al. 1997, ApJ, in press
12. Welch, W.J. et al. 1996, PASP, 108, 93

Starbursts in Our Home Galaxy

Leonid M. Ozernoy

Computational Sciences Institute and Department of Physics & Astronomy, George Mason University, Fairfax, VA 22030-4444 [1]
and Laboratory for Astronomy and Solar Physics, Goddard Space Flight Center, Greenbelt, MD 20771[2]

Abstract. I discuss a scenario of a recent (5 to 8 Myr ago) modest starburst in a central region of the Milky Way galaxy. By incorporating a collision of two giant molecular clouds within the central molecular zone ($r \lesssim 0.6$ kpc), this scenario combines formation of massive stars close to the center of the Galaxy with an opportunity to have X-ray massive binary successors at very large distances from the center.

MOTIVATION

Although there is a general agreement that the bulk of the stars in the central stellar bulge of the Milky Way was formed 10^{10} yrs ago, the origin of intermediate-aged and young stellar populations in the bulge is a matter of debates as yet. The current star formation rate in the central molecular zone, about 0.5 $M_\odot$ yr^{-1}, is tens of times lower than that in starburst galaxies, seemingly indicating the Galactic nucleus to be a "slow-burn nucleus" and the star formation process in it to be sustained [1].

While a slow, continuous process of star formation throughout the Galaxy's lifetime might, indeed, be responsible for the augmenting of the bulge, it is at odds with a piece of evidence for a recent (5 to 8 Myr ago) starburst at the Galactic center. The arguments strongly favoring such starburst include: (i) starburst models are able to account for a young stellar population in the central parsec [2–4]; (ii) ultra hot gas in the central 200 pc could be produced by multiple supernova explosions [5]; and (iii) population synthesis of X-ray sources within the central kiloparsec (in particular, the ratio of the number of systems containing a black hole candidate to the number of X-ray transients with a neutron star) supports both the occurence of the starburst and its age [6].

---

[1] E-mail: ozernoy@hubble.gmu.edu
[2] E-mail: ozernoy@stars.gsfc.nasa.gov

In contrast to an enhanced distribution of X-ray sources, which was found extended within the central kiloparsec, the observed distribution of their progenitors, the massive stars, is apparently seen much more concentrated towards the center, although it is not restricted just by the innermost 1 pc site. In addition to this well known central cluster, two more clusters with young massive stars located near the Galactic center Arc have recently been revealed and studied: 'Quintuplet' and 'Object 17' (for detail, see [7]).

In what follows, I discuss a scenario able to explain the origin of the Galactic center starburst, including drastic differences in locations of the bulk of hot, massive stars and the X-ray inheritors.

STARBURST SCENARIO

Although evidence in favor of a recent starburst in the central region of the Galaxy is continued to mount, details of both its origin and initial location are still obscure. Recently I have discussed *collisions between giant molecular clouds* (GMCs) as a feasible mechanism able to trigger (recurrent) starbursts in the central region of the Galaxy [8,9]. Each collision between two GMCs occurs with an average time interval of $\sim$ 200 Myr and has a duration of about 0.8-2 Myr, assuming the radius of each cloud $\sim$20 pc and their relative velocity $\sim$20-50 km/s. The cloud-cloud collision results in the formation of a shocked layer. The gas is compressed there, cools radiatively, and then fragments, making possible for protostars to start condensing out. Due to a high Mach number in the colliding GMCs, a very high number of young stars is expected to be produced in each collision even it is grazing. A 'wave of star formation' could start at comparatively large distances from the Galactic center and gradually propagate towards the center, accompanied by the fall of the leftovers of the clouds onto the center.

In order to quantify this scenario, let us assume that the collision between two GMCs occurs at a large distance (say, 500-750 pc) from the center and produces a shock that initiates an instantaneous starburst. Suppose that the stars formed inherit the initial internal velocity dispersion within the molecular clouds (say, 3 km/s). Since a substantial part of the transverse velocities of the clouds is lost in the collision, the remnant (stars + gas) will be falling towards the Galactic center. Due to the conservation of angular momentum, the velocity dispersion of the stars will be growing as r^{-1} and will reach its maximum when the cloud passes at its minimum approach from the Galactic center. If this distance is $\sim$ 10 pc (i.e. comparable to the initial radii of the clouds), the velocity dispersion of stars reaches $\sim 150 - 225$ km/s, while the systematic velocity of the cloud acquired at the central potential well turns out to be $v_0 \sim 300 - 400$ km/s. A combination of these large systematic and chaotic velocities is expected to be the major factor leading to the scattering of the formed stars in the area of about 750 pc around the center.

For binary stars able to produce X-ray sources, two more factors could contribute to this scattering: (i) ejection of mass during the supernova explosion even if the ejection was spherically-symmetric relative to the exploding star, and (ii) a 'kick' that the binary acquires as a result of an asymmetry of the supernova explosion. As for (i), an estimation for the acquired velocity ranges between 20 and 100 km/s (e.g. [10]). As for (ii), even if we use a rather large estimate of 400 km/s for 'kick' velocity, which is being currently under discussion in literature [12], then a massive binary acquires a recoil velocity by a factor of 10 less unless it is disrupted. Therefore both effects (i) and (ii), while occuring for binaries during their infall onto the Galactic center, could result even in a larger scattering than that for the single stars.

SPATIAL DISTRIBUTION OF YOUNG, MASSIVE STARS AFTER THE STARBURST

As is known, the observed distribution of massive stars in the direction of the Galactic center looks very peculiar: the vast majority of all massive stars are concentrated towards the central 1 pc or so [11]. The observed X-ray sources demonstrate concentration towards the center, although significantly less pronounced: more than half of them are occupying a region of $\approx$ 750 pc $\times$ 750 pc in size.

In ref. [6], we confronted the results of model simulations with the distribution of the *GRANAT* sources and NS+Be systems (the latter are more numerous than BH-sources both in the *GRANAT* observations and in our calculations). Initially all the binaries have small, stochastically oriented peculiar velocities with a Maxwellian distribution and dispersion of 3 km/s. At the moment of the SN explosion they acquire, due to a 'kick' and mass ejection from the system, an additional velocity of about 75 km/s, also stochastically oriented in space. We have explored several variants of the resulting (by T =6-8 Myr) spatial distribution of the binaries formed in an instantaneous starburst for different initial locations of the starburst and different rotational velocities about the Galactic center, v_ϕ, ranging from zero to the circular velocity, v_{circ}.

If the starburst's distance from the center is $r_0 \sim 500$ pc and initial, after cloud–cloud collision, rotational velocity of the cloud about the Galactic center is taken to be $v_{\phi_0} = 50$ km/s, i.e. not too large, the resulting spatial distribution turns out to be quite extended and more or less symmetric, consistent with the observed distribution. Similar computations by Lipunov et al. performed for a range of r_0 and v_{ϕ_0} indicate that, by $T \simeq 7$ Myr, a starburst would produce a quasi-isotropic projected distribution of X-ray sources occupying a large region around the center with the size of several hundred parsecs, consistent with the data, for a rather wide range of initial conditions (initial distance from the Galactic center $r_0 \sim 500$ pc, not too large v_{ϕ_0}).

DISCUSSION AND CONCLUSIONS

Since a lot of young, massive stars are located within the central parsec, one could explore whether the starburst region occured just there, i.e., the progenitors of X-ray binaries were formed in a region of ~ 1 pc in size and then the resulting X-ray systems were ejected into, and scattered within, the central 1 kpc or so due to the 'kick' that accompanied the formation of those systems. However, this alternative scenario turns out to be highly improbable [6]. This is due to the fact that the mass distribution at the Galactic center $M(r) = 3.4 \times 10^6$ $M_\odot$ $(1 + r/1\,\text{pc})$ (ref. [13]) results in such a high escape velocity that the distance reached by the star and its velocity are related by $\ln(r/r_0) = 0.34\,(v_0^2 - v^2)$, where v_0 (in 100 km/s) is the initial ejection velocity at a radius r_0. This implies that, in order to reach $r = 1$ kpc from $r_0 = 1$ pc even with a zero velocity, v_0 needs to be as high as 450 km/s. For $r_0 = 10$ pc, the required $v_0 = 368$ km/s is less but not by a substantial factor. Such high velocities cannot be reached by imposing a 'kick' onto an initial velocity dispersion of the newly formed stars without destroying binaries. This is supported by the absence of fast moving massive X-ray binary systems in our Galaxy (say, with $v > 50$ km/s). Therefore, explaining the observed wide distribution of X-ray binaries at the Galactic center by ejection of them from the central parsec looks unlikely.

Meanwhile in the scenario described above, a GMC collision episode could start at a comparatively large distance from the Galactic center and result in the infall of the products toward the center. The high velocities acquired by the infalling gas in the galactic potential well are inherited by the forming massive stars, which further enables them to be scattered up to 1 kpc or so.

One may argue that the region of 750×750 pc in size around the Galactic center is broad enough so as to be contaminated by X-ray binaries originating in the adjacent regions. However, the fraction of X-ray binaries of such type among the 'field' stars (i.e. not associated with the starburst of interest) is much lower. Yet, one could imagine in the Galactic center another starburst of a similar age but on a much larger scale, compared to what is considered above, which would of course somewhat change our results; however, no evidence for such a burst is known so far. The proposed scenario of the origin of X-ray sources in a comparatively compact starburst has a clear signature: the velocity of an X-ray source is (statistically) expected to be the larger the closer the source is located to the Galactic center.

Unlike to the observed extended distribution of X-ray sources within the inner 1 kpc, the progenitors, massive young stars, are seen much more concentrated towards the Galactic center being assembled in the very center ($r \lesssim 1$ pc) and around, in the three clusters mentioned in the introductionary section. Naturally, the origin of those clusters is associated with the material, which, after the dissipation of an excessive momentum in the cloud collision, could fall into the center and produce massive stars whose velocity dispersion would not exceed 100 km/s. Thus the above scenario combines formation of massive stars close to the center of the Galaxy *and* the presence of massive binary successors, the X-ray systems, at very large distances from the center.

Two more aspects of the starburst at the Galactic center, which concern the central parsec, are worth mentioning:

(i) *Origin of counter-rotation of the ensemble of early-type stars.* The central cluster of ealy-type stars has been found rotating with a best fit rotational velocity of 120 km/s and the sense of rotation opposite to that of galactic rotation [14]. The specific angular momentum of this young stellar population is estimated to be $J \sim 3 \cdot 10^{24}$ cm^{2}s^{-1}. Meanwhile for an ordinary giant molecular cloud $J \lesssim 5 \cdot 10^{24}$ cm^{2}s^{-1} [15]. The fact that those quantities are remarkably close to each other is consistent with our scenario of cloud collision as the the starburst trigger. An alternative conjecture that the early-type stars in the central parsec were born from a counter-rotating cloud or a gas streamer belonging to the Circumnuclear Ring [14] turns out to be in contradiction with its specific angular momentum, which is estimated to be $J \sim (1-3) \cdot 10^{26}$ cm^{2}s^{-1}.

(ii) *Population of compact stellar remnants at the Galactic center.* The proposed starburst mechanism is recurrent, and the time interval between two successive GMC collisions, $T_c \simeq 200$ Myr [9], defines the characteristic cycle of starburst recurrency. Currently, there are several dozens of massive stars in the central parsec [14]. Since up to 10^2 generations of massive stars might have been formed during the lifetime of the Galaxy at its center, they might have produced by now at least 10^4 compact stellar remnants such as neutron stars and black holes and even much more in form of white dwarfs. In fact, the rate of GMC collisions adopted above might be just a lower limit since the galactic bar can substantially increase that rate near the resonances. The latter is supported by the fact that the two most prominent high-mass star forming regions in the Galactic nuclear disk, Sgr B2 and Sgr C, seem to be closely associated with the inner Lindblad resonance [16]. Therefore, is not excluded that the total number of stellar remnants at the Galactic center approaches $10^5 - 10^6$, consistent with the the amount of 'dark matter' in the form of compact stellar remnants advocated in [4].

The results of this paper seem to be also relevant for studying the star formation regions in other galaxies, including starburst galaxies. Even the formation of primeval galaxies has been proposed to involve collisions of protogalactic clouds [17]. In starburst galaxies, short episodes of violent star formation with a time scale of ~10 Myr have been suggested to occur recurrently some billion years. Although the causative agent for the starburst at the Galactic center is a cloud collision driven by internal factors, and not by an encounter with a companion galaxy, starbursts as such have some common features. As follows from our modelling of the population synthesis of X-ray sources, production of a few 10^5 stars in a starburst has to be accompanied by the formation of about 10 hard X-ray sources at the starburst age of 6-8 Myr (and a larger number of X-ray sources at earlier times). It is of interest to compare these figures with the recent data from an X-ray study of starburst galaxies [18].

To conclude, a recent starburst at the Galactic center could be triggered by a collision of giant molecular clouds, which makes possible to combine the formation of massive stars both in and not far from the center of the Galaxy with the presence of massive binary successors, the X-ray systems, at very large distances from the center. Although the Milky Way starburst looks rather modest if compared with the typical starburst galaxy, its consequences observable in more detail may be helpful in our understanding of such fundamental processes in the universe as violent star formation.

Acknowledgement. I am grateful to V. Lipunov and S. Popov for close cooperation in the population synthesis work and related topics.

REFERENCES

1. Serabyn, E. & Morris, M. 1996, Nature 382, 602
2. Tamblyn, P. & Rieke, G.H. 1993, ApJ 414, 573
3. Schaerer, D. 1996, in "Unsolved Problems of the Milky Way". Eds. L. Blitz & P. Teyben. Kluwer Acad. Publ.
4. Tamblyn, P., Rieke, G.H., Hanson, M.M., Close, L.M., McCarthy, D.W., & Rieke, M.J. 1996, ApJ 456, 206
5. Ozernoy, L., Titarchuk, L. & Ramaty, R. 1993, in "Back to the Galaxy", ed. S.S. Holt & F. Verter. AIP Conf. Proc. 278, 73
6. Lipunov, V.M, Ozernoy, L.M., Popov, S.B., Postnov, K.A., & Prokhorov, M.E. 1996, ApJ 466, 234
7. Morris, M. & Serabyn, E. 1996, ARA&A 34, 645
8. Ozernoy, L.M. 1994, BAAS 26, 1420
9. Ozernoy, L.M. 1996, in "Unsolved Problems of the Milky Way". Eds. L. Blitz & P. Teyben. Kluwer Acad. Publ., p. 189
10. Shore, S.N., Livio, M., & van den Heuvel, E.P.J. 1994, "Interacting Binaries". Springer Verlag, Berlin, Ch. 3
11. Genzel, R., Hollenbach, D, & Townes, C.H. 1994, Rep. Prog. Phys. 57, 417
12. Lyne, A.G. & Lorimer, D.R. 1994, Nature 369, 127
13. Lacy, J.H., Achtermann, J.M., & Serabyn, E. 1991, ApJ 380, L71
14. Genzel, R., Thatte, N., Krabbe, A., Kroker, H., & Tacconi-Garman, L. 1996, MPE Preprint 362 = ApJ (submitted)
15. Blitz, L. 1993, in "Protostars and Planets III". Eds. E.H. Levy & J.I. Lunine (Univ. of Arizona Press), p.125
16. Lis, D.C. & Carlstrom, J.E. 1994, ApJ 424, 189
17. Ozernoy, L.M. 1989, Ann. New York Acad. Sci. 571, 219
18. Rephaeli, Y., Gruber, D., & Persic, M. 1995, A&A 300, 91

PRIMORDIAL STAR FORMING REGIONS IN A CDM UNIVERSE

Yu Zhang, Michael L. Norman, Peter Anninos, Tom Abel

Laboratory for Computational Astrophysics
University of Illinois at Urbana–Champaign
405 N. Mathews Ave., Urbana, IL 61801

Abstract.
We developed a three–dimensional 2–level hierarchical cosmological code with a realistic and robust treatment of multi–species non–equilibrium ionization and radiative cooling processes, and use it to investigate primordial star forming regions that originate from high-σ perturbations in a standard CDM dominated cosmology. We find it is possible to produce gravitationally bound and cooled structures at very high redshift ($z \sim 40$) with baryonic masses as small as $\sim 10^3 \, M_\odot$. The molecular hydrogen formation in these small scale structures follows very well the analytical predictions of Abel (1995) and Tegmark et al. (1996). We also discuss the minimum mass that cosmological structures must have in order to be able to cool and collapse.

INTRODUCTION

Models for structure formation are based on the growth of small primordial density fluctuations by gravitational instability on a homogeneously expanding background universe. Depending on the nature of the dark matter and whether the primordial fluctuations were adiabatic or isothermal, the first mass scale to collapse could be as small as one solar mass (very heavy cold dark matter particles) or as high as $\sim 10^{12} M_\odot$ (hot dark matter scenarios). In CDM cosmogonies the fluctuation spectrum at small wavelengths has only a logarithmic dependence for mass scales smaller than $\sim 10^8 M_\odot$, which indicates that the small scale fluctuations in this model collapse nearly simultaneously in time. This leads to very complex dynamics during the formation of these structures, that can be studied accurately only by using realistic numerical computations to model the fluid motion and micro–physical processes as well as the dark matter component.

We have recently been able to develop methods that allow us to study the problem in three dimensions [5]. We describe this code briefly in section but first review the process of H$_2$ formation during small scale structure collapse in the following section. A more extensive discussion of our results is given in a separate publication [3].

MOLECULAR CHEMISTRY AND COOLING

The cooling in small scale fluctuations is dominated by the rotational/vibrational modes of hydrogen molecules. In primordial gas at low temperatures ($\lesssim$ 6000K) molecular hydrogen can not be destroyed efficiently unless there is a radiation flux higher than $\sim 3 \times 10^{-26}$ erg cm^{-2} s^{-1} in the Lyman Werner Bands. Once self-shielding is important, even higher fluxes would be needed. The dominating H$_2$ producing gas phase reaction is the dissociative attachment reaction: H$^-$ + H $\rightarrow$ H$_2$ + e$^-$. In the absence of an external UV background one can integrate the rate equations to find the molecular hydrogen fraction formed during the collapse of primordial gas clouds with neutral hydrogen number density n_H, temperature T, and initial free electron fraction x_0 to be [1,10]:

$$f_{H_2}(t) - f_{H_2}(0) = \frac{k_{PA}}{k_r} \ln(x_0 n_H k_r t + 1) = 10^{-8} T^{1.53} \ln(t/t_r^0 + 1), \quad (1)$$

where k_{PA}, and k_r denote the rates for photo-attachment to H$^-$ and recombination of hydrogen, respectively. The production of H$_2$ only depends logarithmically on time with a typical time scale of one initial recombination time. The temperature dependence is due to the ratio of recombination and H$^-$ formation time scales, which is a measure of the number of electrons available to produce H$^-$. A typical H$_2$ fraction of $\sim 10^{-3}$ is produced during the collapse of structures with virial temperatures greater than 10^3 K. For initial (virial) temperatures higher than 6000 K the charge exchange with protons will efficiently destroy H$_2$, and equation (1) will not be applicable. However, during the collapse of clouds with such high virial temperatures the final H$_2$ fraction will, nevertheless, be $f_{H_2}(T \sim 6000 \text{ K}) \sim few \times 10^{-3}$ [2].

NUMERICAL RESULTS AND DISCUSSION

We achieve high spatial and mass resolution with the two–level hierarchical three–dimensional code (HERCULES) that we have developed for cosmology [4,5]. This code is designed to simulate structure formation in an expanding dark matter dominated universe with Newtonian gravity, multi–fluid hydrodynamics, radiative cooling, non–equilibrium chemistry and external radiation fields. Furthermore, the code independently evolves the following nine species: neutral hydrogen H, ionized hydrogen H^+, negatively–charged hydrogen H^-, hydrogen molecules H_2, ionized hydrogen molecules H_2^+, neutral helium He, singly–ionized helium He^+, doubly–ionized helium He^{++} and free electrons e^-. The 28 most important chemical rate equations (including radiation processes) are solved in non–equilibrium for the abundances of each of the nine species. The reaction rates and an extensive discussion of the chemistry model are provided in [2]. We have also implemented a comprehensive model for the radiative cooling of the gas that includes atomic line excitation, recombination, collisional ionization, free–free transitions, molecular line excitations, and Compton scattering of the cosmic background radiation (CBR) by electrons.

We apply our code to high redshift pre–galactic structure formation and evolution, investigating specifically the collapse of the first high–σ bound objects with total masses in the range $10^5 - 10^9 M_\odot$. Our model background spacetime is a flat ($\Omega_0 = 1$) cold dark matter dominated universe with Hubble constant $H_0 = 50$ km s^{-1} Mpc^{-1}, baryonic fraction $\Omega_B = 0.06$, and a hydrogen mass fraction of 76%. The baryonic matter is composed of hydrogen and helium in cosmic abundance with a hydrogen mass fraction of 76% and ratio of specific heats $\gamma = 5/3$. The initial data for the baryonic and dark matter perturbations is the Harrison–Zel'dovich power spectrum modulated with a CDM transfer function and normalized to the cluster scale $\sigma_{8h^{-1}} = 0.7$. The data is initialized at redshift $z = 100$ using Bertschinger's [6] constrained realization procedure to construct 3 and 4σ fluctuations in cubes of comoving length 1024kpc, 512kpc, and 128kpc, with total masses of 7.5×10^{10} $M_\odot$, 9.3×10^9 $M_\odot$, and 1.5×10^8 $M_\odot$, respectively.

We reproduced the work of Tegmark et al. 1996 with the same cooling function [9] we have used in our cosmological hydrodynamics code so that we can compare their findings directly to our 3D numerical results. (We note that Tegmark et al. used a modified form of the Hollenbach and McKee (1979) H_2 cooling function.) Figure 1a is analogous to Fig. 6 of Tegmark et al. 1996 but includes the mass evolution in our numerical results for the 4σ perturbations. The dotted lines are found by adding up the total mass M_{200} found in cells with dark matter overdensities exceeding 200. It is obvious that the use of a different cooling function has a very strong influence on the predicted mass that can collapse. Although, the quantitative results are very different, the shape for these two different $M_c(z)$ curves is rather similar at redshifts $30 < z < 100$. The slopes are consistent with $M_c \propto (1+z)^{-3/2}$ indicating a constant virial temperature since $T_{vir} \propto M^{2/3}(1+z)$. A constant virial temperature in turn implies a roughly constant final H_2 fraction given by equation (1). For the case in Tegmark et al., the virial temperature in that regime is ~ 1000 K which yields $f_{H_2} \sim 4 \times 10^{-4}$ which they argued to be roughly a constant universal value which, if exceeded, allows the cloud to collapse at its free fall rate. Using the Lepp and Shull H_2 cooling function we find that the virial temperature needed to fulfill the Tegmark et al. requirement for collapse is ≈ 200 K for redshifts > 30. This translates to a molecular fraction of only $\sim 3 \times 10^{-5}$. Our simulations, however, show that, although we used the Lepp and Shull cooling function, the H_2 fraction at the time when the baryons are

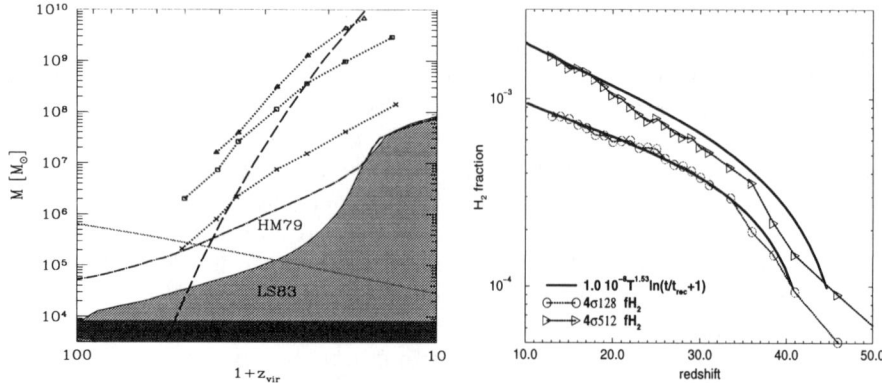

FIGURE 1. (a) Collapsed mass vs. redshift. The dark shaded region depicts the mass scale for which the virial temperature equals the CMB temperature. Only above the light shaded area labeled with LS83 are structures able to collapse. The dotted lines show M_{200} from our 4σ 128, 512, and 1024kpc (crosses, squares, triangles) runs. The dashed upward sloping line is the BBKS CDM spectrum scaled appropriately for 4σ peaks. The gray solid downward sloping line is the Jeans Mass at $18\pi^2$ the background density. The dot–dashed line shows the original delimiting line computed by Tegmark et al. (1996) which was based on a modified form of the Hollenbach and McKee (1979) H_2 cooling function. (b) The molecular hydrogen fraction vs. redshift in the densest zone of the $4\sigma128$ (open circles) and the $4\sigma512$ simulations (triangles). The thick solid lines is the solution (1) initialized with the appropriate free electron densities, initial H_2 fraction, and temperature.

collapsing into the DM potential wells is also about $\sim 5 \times 10^{-4}$. Hence, we find roughly the same critical H_2 fraction as derived by Tegmark et al. (1996) even with a very different cooling function.

In Figure 1b we test equation (1) against results from the $4\sigma128$ kpc (open circles) and the $4\sigma512$ kpc simulations. The fit is astonishingly good, although the initial temperature (or redshift) is somewhat difficult to pick out in this case since the heating due to adiabatic compression is slow and the initial (virial) temperature is not as well defined as in the case of collapse on larger mass scales. Analyzing the time derivative of equation (1) one finds for large times that $f_{H_2} \propto T^{1.53}/t$ which, for the spherical collapse model, translates to $f_{H_2} \propto H_0 M (1 + z_{vir})^{1.53}$ when we compare the slopes for different mass scales at the present time. This explains that the divergence of the two graphs in Figure 1b for low redshifts is due to differences in collapse mass and redshift.

It has been stressed by various authors that early small scale structure might influence the entire pregalactic medium and subsequently play an important role for structure formation on larger mass scales (e.g. [7]). They are in principle capable of ionizing a large fraction of the pregalactic medium as well as to enrich it with metals. We hope to achieve the required dynamical range in future work to estimate the IMF using adaptive mesh refinement techniques and so be able to quantify the feedback of collapsing small scale structure.

We happily acknowledge discussions with M.J. Rees and Max Tegmark. This work is done under the auspices of the Grand Challenge Cosmology Consortium (GC3) and supported in part by NSF grant ASC-9318185. The simulations were performed on the CRAY-C90 at the PSC, and the CONVEX-3880 at the NCSA.

REFERENCES

1. Abel, T., thesis, Univ. Regensburg (1995).
2. Abel, T., Anninos, P., Zhang, Y., and Norman, M.L., submitted to *NewA* (astro-PH 9608040) (1996).
3. Abel, T., Anninos, P., Norman, M.L., and Zhang, Y., in preparation (1996).
4. Anninos, P., Norman, M.L. and Clarke, D.A., *ApJ* **436**, 11 (1994).
5. Anninos, P., Zhang, Y., Abel, T. and Norman, M.L., submitted to *NewA* (astro-PH 9608041) (1996).
6. Bertschinger, E., Private Communication (1994).
7. Couchman, H.M.P. and Rees, M.J., *MNRAS*, **221**, 53 (1986).
8. Hollenbach, D. & McKee, C.F., *ApJS* **41**, 555-592 (1979).
9. Lepp, S., and Shull, J.M., *ApJS* **270**, 578-582 (1983).
10. Tegmark, M., Silk, J., Rees, M.J., Blanchard, A., Abel, T., Palla, F., to appear in *ApJ* **473** (1996).

Multiple Star Systems

Accretion in Pre-Main-Sequence Binaries

Robert D. Mathieu

Department of Astronomy, University of Wisconsin - Madison, Madison, WI 53706

Abstract. Mass accretion at stellar surfaces occurs in pre-main-sequence binaries of all separations. The accretion rates are comparable to those found for single stars. This paper describes several recent observations which shed light on the accretion process in the binary environment. 1) Most classical T Tauri binaries include two classical T Tauri stars, suggesting that the lifetimes of their circumstellar disks are correlated. 2) The classical T Tauri binary DQ Tau shows evidence for enhanced accretion rates at periastron passage, as predicted for accretion streams from circumbinary disks. 3) Two infrared companions show evidence for shocked molecular hydrogen emission, indicative of continued infall onto these stars or associated disks. Timescale arguments indicate that for many binaries existing circumstellar material cannot supply the observed accretion rates for as long as the ages of the binaries. Presuming steady accretion rates, this requires replenishment. Infall from circumbinary envelopes or accretion streams from circumbinary disks have been suggested, although such circumbinary envelopes or disks have not yet been directly detected around most actively accreting binaries.

INTRODUCTION

During the last decade we have gone from knowing that most pre-main-sequence (PMS) stars *must* be binary stars to knowing that most PMS stars *are* binary stars. Extensive surveys for binaries among PMS stars have revealed numerous pairs. Indeed in some star-forming regions such surveys have yielded an excess of riches, with PMS binary frequencies greater than main-sequence binary frequencies (for reviews see [1-3]; also [4]).

At the same time the role of mass accretion through disks has become central to our picture of low-mass single-star formation [5]. In the present paradigm disk accretion is both the means by which most of the stellar mass is accumulated and the origin of many properties of classical T Tauri stars (CTTS), including infrared excesses, ultraviolet excesses and spectral veiling, Balmer and forbidden emission lines, polarization, etc.

The combination of these two lines of study creates an interesting conundrum, which rests on the gravitational interaction of stellar companions with associated disks [6]. Consider a binary located within a coplanar disk. On orbital timescales the binary clears a gap in the disk, creating in the process three distinct disks - two circumstellar and one circumbinary. The circumstellar disks are free to continue accreting onto the stars. However, the expectation has been that the balance of viscous and resonant forces at the inner edge of the circumbinary disk would prevent any flow of circumbinary material across the gap. The consequence of continued accretion at the stellar surfaces would thus ultimately be exhaustion of the circumstellar disks and cessation of the accretion. As such, the formation of a stellar companion within a typical disk radius (e.g., 100 AU) may both preclude long-lived accretion and isolate the stars from the mass reservoir in a circumbinary disk.

In fact observational diagnostics for mass accretion at stellar surfaces are found among PMS binaries of all separations. This paper will review recent observations which reflect on accretion processes in young binaries and present several conjectures as to how circumstellar disks may be replenished. All of these discussions presume that circumstellar and/or circumbinary disks are present in young binary systems, the evidence for which has been reviewed elsewhere [1, 7].

OBSERVATIONS OF DISK ACCRETION

Among PMS stars diagnostics for active accretion include strong Hα emission, inverse-P-Cygni profiles, ultraviolet excesses, and spectral veiling. These phenomena have been found in binaries with separations ranging from many thousand AU to less than 0.1 AU. As a case in point, DF Tau is a canonical example of a PMS star with an actively accreting disk. It has large ultraviolet and infrared excesses, heavy spectral veiling, and an Hα equivalent width of 54 Å. Modeling DF Tau as a single star, Bertout, Basri, and Bouvier derived an accretion rate of 10^{-7} M_O yr^{-1}, typical among T Tauri stars [8]. In fact, DF Tau is a binary system with a projected separation of only 12 AU. Evidently this PMS binary retains at least one circumstellar disk which is actively accreting onto a stellar surface.

Classical T Tauri Stars and Binary Frequencies

Hα emission, the defining property of CTTS, is closely correlated with other diagnostics of accretion at stellar surfaces. Thus by examining the distributions of CTTS among PMS binaries we might hope to find trends that shed light on the nature of disk accretion in young binary stars.

In the scenario described in the Introduction, the circumstellar disk exhaustion timescale will depend, among other things, on the masses of the circumstellar disks. A reasonable presumption is that the reservoir of circumstellar disk material tends to decrease with decreasing separation. Consequently the duration of active accretion at stellar surfaces and the lifetimes of circumstellar disks might tend to decrease with decreasing binary separation. An associated observational prediction is that the frequency of CTTS will decrease among closer binaries.

The recent high-angular-resolution surveys for PMS binaries provide the largest samples for investigating the distribution of CTTS with binary separation. Ghez and collaborators found that within a range of projected separation from 16 AU to 252 AU the frequency of strong Hα emission among closer binaries was significantly less than among wider binaries [9]. However this effect was not found in two other studies of comparable sample size [10, 11]. In addition, among 11 PMS binaries with projected separations of less than 16 AU there are at least 5 with strong Hα emission [1]. Similarly it has been speculated that spectroscopic binaries may be infrequent among CTTS [12], but existing surveys do not support this conclusion [13], and recently two CTTS spectroscopic binaries have been found in the Taurus-Auriga region [14, 15].

Lately attention has shifted to the joint frequency of CTTS among stars *within* binary pairs. Prato and Simon have employed both angularly resolved near-infrared spectroscopy and photometry to identify CTTS among the primaries and secondaries of 12 PMS binaries with separations between 40 AU and 360 AU [16]. In particular they observe Br γ emission as a surrogate for Hα emission in 4 binaries, and consider near-infrared colors as diagnostics for accretion in the remaining 8 systems. They find that in all 12 systems *both* components show, or have shown in the history of their observations, CTTS diagnostics. At the same time they find that commonly one star in each pair exhibits more pronounced CTTS behavior. This star can be either the brighter or fainter in continuum emission.

Prato and Simon argue that this association of CTTS in binaries cannot be the result of chance [16]. Specifically they find that if PMS binaries resulted from a random association of CTTS and weak-lined T Tauri stars (WTTS), then roughly 2/3 of CTTS binaries would in fact be mixed CTTS-WTTS pairs, in marked contrast to their observations. They also argue that the circumstellar disks feeding the CTTS activity must be replenished, since the measured accretion rates would deplete the measured disk masses in less than the stellar ages. Prato and Simon suggest that the disks are being replenished by infall from circumbinary, optically thin, residual dust envelopes. Given such envelopes, the replenishment of the disks around both stars - and consequently their accretion lifetimes - would be correlated, as observed. In support of this conjecture they note that the binaries studied have relatively flat spectral energy distributions, which have been suggested to originate from reprocessing in such envelopes.

For comparison, in an optical study of 13 binaries with similar separations Brandner and Zinnecker find 4 binaries which include one CTTS and one WTTS [17]. They conclude that "chromospheric activity and/or accretion rates of primary and secondary appear not to be related". They too find that occasionally the secondary star (in terms of optical brightness) has the stronger Hα emission.

Time Dependent Accretion: The Case of DQ Tau

Both circumstellar disks and mass accretion at stellar surfaces are found in PMS binaries of all separations. This observational result is particularly notable in the shortest period binaries where periastron separations do not permit the presence of substantial circumstellar disks. Two cases in point are the PMS spectroscopic binaries UZ Tau E and DQ Tau [14, 15]. Both binaries are model CTTS, showing Hα equivalent widths of order 100 Å, the entire Balmer series in emission, spectral veiling indicative of mass accretion rates in excess of 10^{-8} $M_\odot$ yr^{-1}, etc. [18]. Yet with orbital periods of less than 20 days and periastron separations of less than 0.1 AU, these accretion rates cannot be maintained by unreplenished circumstellar disks. Both binaries have potential sources of material in massive circumbinary disks [15, 19]. The issue is whether the circumbinary material can be tapped.

Photometric and spectroscopic monitoring of DQ Tau, in conjunction with recent theoretical work of Artymowicz and Lubow [20], may have provided a clue to the tapping mechanism. As indicated by its name, DQ Tau is photometrically variable. Optical photometric monitoring has shown this variability to be

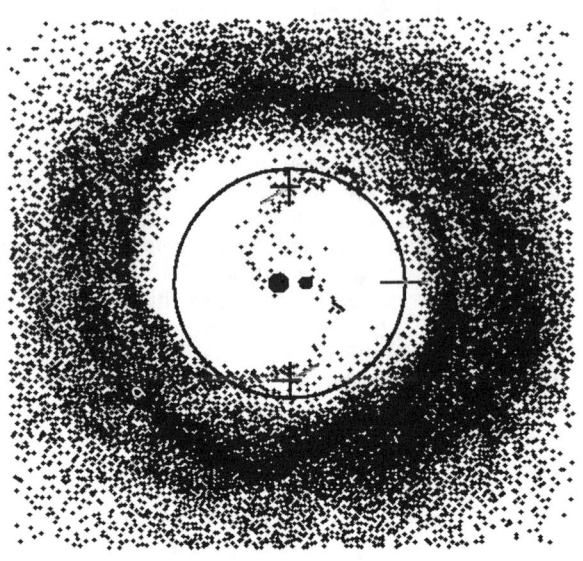

Figure 1. SPH simulation of a binary system (e=0.5, q=0.9) surrounded by a circumbinary disk [20]. Material flows from the circumbinary disk to the binary in two quasi-spiral accretion streams. The stream formation is periodic, with the flow terminating at the stars when they are at periastron (as shown).

characterized by regular brightenings of as much as 0.7 mag at V. Most importantly, these brightenings are periodic with a period of 15.8 days. This is precisely the same as the orbital period, with the brightenings occurring at periastron passage. During these brightenings the system becomes bluer, the veiling increases, and emission line strengths increase [15, 21]. Together, all of these results point toward an increased mass accretion rate at periastron passage.

While DQ Tau was under study, Artymowicz and Lubow [20] suggested that in certain situations accretion streams can flow from a circumbinary disk across a dynamically cleared gap to the stars. In particular they ran simulations of a binary very much like DQ Tau (orbital eccentricity e=0.5 and mass ratio q=0.8) surrounded by a circumbinary disk. In this case two accretion streams develop, spiral-like in morphology, roughly symmetric about the center of mass, and each terminating at a different star (Figure 1). Both streams are narrow, so that the gap region remains almost empty. Nonetheless, the streams carry approximately as much mass flux as would a similar disk extending to the surface of a (single) star.

Most significantly with respect to DQ Tau, Artymowicz and Lubow found the accretion flow to be time variable with a periodicity equal to the binary orbital period. In particular they find the maximum gas accretion rate near the stars to occur just prior to periastron, just as do the brightenings of DQ Tau. In Figure 2 are shown the theoretical gas accretion rate and the photometric observations of DQ Tau, both phased against orbital period. The agreement in phase is striking. The agreement between the width of the accretion pulse and the phase interval of

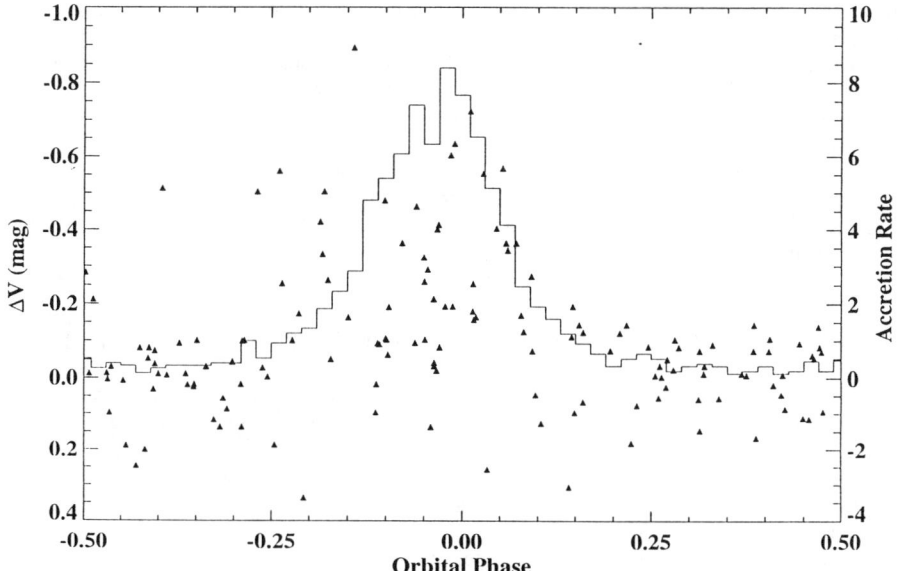

Figure 2. Comparison of observed brightenings of DQ Tau (triangles) with predicted accretion rate (solid curve) due to accretion streams from a circumbinary disk. Both are phased to the binary orbit, with phase 0.0 being at periastron. Note the agreement in phase of the predicted bursts in accretion with the observed brightenings.

observed brightenings is also encouraging, but the reader should note that the durations of individual brightening events vary, as do their amplitudes.

In this scenario, the increase in luminosity is due to the deposition of kinetic energy from the infalling streams and the consequent heating of a region in the system. However, the nature of the interaction of such streams with either circumstellar disks, magnetospheres, or the stellar surfaces is not known; the resolution of the Artymowicz and Lubow simulations is not adequate to study the mass flows near the stellar surfaces. Thus whether the heated regions are on the stellar surfaces or circumstellar disks, whether they are localized in hot spots or more broadly distributed, and whether they are heated directly by the stream or indirectly through provision of circumstellar material which then accretes, is unknown.

In addition, DQ Tau shows spectral veiling at all orbital phases, indicating a continuous low-level flow of material onto the stars [21]. If accretion streams are the origin for this material as well, then they must replenish a reservoir that exists throughout the binary orbit. Such a reservoir might contribute to the observed near-infrared excess. The infrared excess nught also derive in part from the streams themselves, although it remains to be seen whether the streams can subtend an adequate solid angle at the star to power the observed near-infrared emission in themselves.

While the agreement in phasing of the brightenings of DQ Tau with the predicted arrival of accretion streams is promising, there remain additional issues to be considered. First, Artymowicz and Lubow find that accretion streams only develop in warm, moderately viscous circumbinary disks. It is unclear whether the circumbinary disk of DQ Tau satisfies these conditions. Second, the periastron separation of DQ Tau is smaller than the inferred stellar magnetospheric radii of classical T Tauri stars, so that such magnetospheres would interact at each periastron passage. While it is unlikely that the observed brightenings are due to flaring alone given the concurrent increase in veiling emission, it remains possible that the enhanced accretion rates at periastron are the result of reconfiguration of the magnetic fields rather than or in addition to modulation of the accretion flow from the circumbinary disk.

Near-Infrared Spectroscopy of Infrared Companions

Infrared companions have proven to be one of the most challenging puzzles of PMS binary stars (see reviews in [1, 22, 23]). Companions to optically visible PMS stars, infrared companions typically are not visible in the optical and have spectral energy distributions similar to those of embedded sources. Their luminosities range widely; most significantly they can exceed the luminosities of their associated optical "primary" stars.

Suggested explanations for infrared companions vary greatly, reflecting our puzzlement [1, 22, 23]. One class of hypotheses places infrared companions at earlier phases of evolution than their associated optical stars, with their spectral energy distributions being the result of continued infall onto the infrared companion. Other hypotheses resort to geometry to provide large extinctions (for example, lines of sight through an edge-on disk about the infrared companion, through the circumprimary disk, or through a circumbinary envelope centered on the primary).

Recently Herbst, Koresko, and Leinert have used 2-μm spectroscopy to search for shocked molecular hydrogen emission indicative of infalling material [24]. Such emission has previously been detected in the combined light from systems with infrared companions. The essential technical challenge has been to spatially resolve the system so as to properly associate the emission with the optical and/or infrared stars. Herbst et al. successfully resolved two binaries having infrared companions, Haro 6-10 and UY Aur. Their spectra are shown in Figure 3. The spectrum of the optical ("primary") star of Haro 6-10 is typical of a stellar photosphere with spectral type later than K5. The infrared companion, however, shows a very red continuum with no photospheric absorption lines. Also present in the companion spectrum is an emission feature at 2.12 μm associated with the $v = 1 - 0\ S(1)$ transition of molecular hydrogen. The spectrum of the UY Aur infrared companion is notably different, showing a photospheric spectrum appropriate for an early-M star. Again 2.12 μm emission from molecular hydrogen is seen

Herbst et al. argue that the absence of molecular hydrogen emission at 2.248 μm establishes a shock origin for the detected emission, and in particular suggests the presence of material with velocities of order 20 km s^{-1} at which shocked molecular hydrogen emission is efficiently produced. Rejecting a wind origin, they conclude that the emission most likely derives from infall. Using a simple spherical infall model they derive accretion rates of $2 \times 10^{-8}\ M_O\ yr^{-1}$ for Haro 6-10 and $6 \times 10^{-8}\ M_O\ yr^{-1}$ for UY Aur.

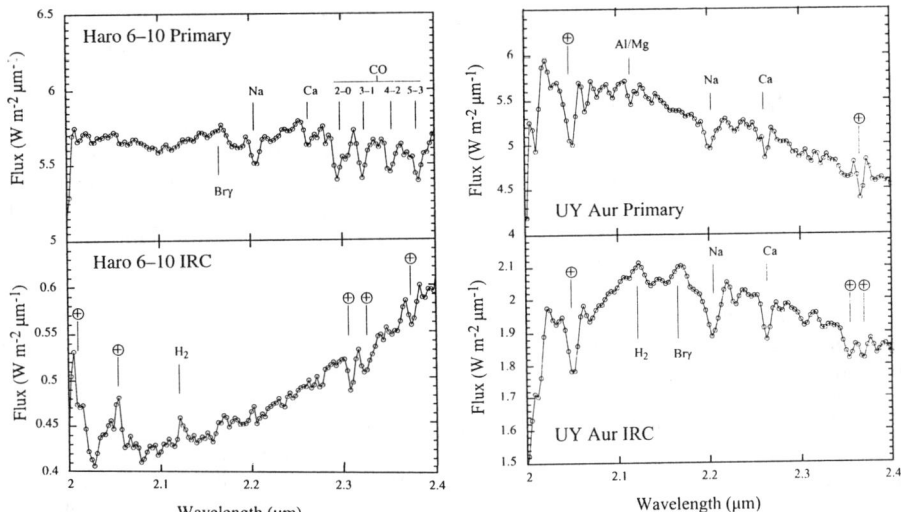

Figure 3. Near-infrared spectra of two binaries with infrared companions [24]. These observations resolved the binaries and provided the spectra of the optical stars and infrared companions separately. Note the presence of molecular hydrogen emission in the spectra of both infrared companions, suggestive of hydrogen shocked in the course of infall.

The extinctions corresponding to these accretion rates are large, indeed implausibly large for UY Aur. Consequently Herbst et al. argue that a flattened distribution is necessary around UY Aur, possibly with infall onto a disk. However in the case of Haro 6-10 even at 2 μm only a dust photosphere is visible and the geometry of the extincting material remains unclear. In any case, the important point is that the evidence for infalling material near both infrared companions strengthens the view that their unusual spectral energy distributions are due to the environments near to the companions.

Again, the essential issue is the origin of the accreting material. For spherical circumstellar envelopes with radii less than the binary separations (of order 100 AU for these two binaries), the infall timescales are of order 1000 yr. If such envelopes are the explanation for infrared companions, then they must be replenished. Arguably replenishment might derive from a circumbinary envelope [25], but such an envelope must not extinct the primary as well. Certain angular momentum distributions in the parent cloud or the clearing of polar cavities by outflows from the primary might establish the required geometry for the envelope [26, 27]. Alternatively, circumstellar envelopes may be replenished from circumbinary disks, perhaps in an episodic mode related to accretion streams [23].

CONCLUSION

Presuming that the observational diagnostics are being correctly interpreted, mass accretion at stellar surfaces is occurring in PMS binaries with separations ranging from tenths of an AU to thousands of AU. A key question is the origin of the circumstellar material that feeds the accretion. Ultimately the origins are likely to be some combination of three possibilities: circumstellar matter remaining from formation, circumbinary envelopes, and circumbinary disks.

For the very widest binaries the accretion may simply be fed from circumstellar disks created during the formation process; the distant companions are unlikely to influence such circumstellar disks once formed. This is consistent with the observation that the widest binaries have circumstellar disk masses similar to the disks of single stars [28, 29]. Even so, such disks may have insufficient mass to support the observed accretion rates for the present ages of some binaries (e.g., [28]), and replenishment from infalling envelopes may still be required.

With decreasing binary separation the companions will truncate the circumstellar disks, which at some point must find themselves with insufficient mass to maintain the observed accretion rates. At this point a replenishment mechanism is demanded. This replenishment may also come from continued infall from circumbinary envelopes, as has been suggested to explain the predominance of CTTS pairs. Alternatively circumbinary disks, if they can generate accretion streams, might also replenish the circumstellar disks of both stars in binaries. For binaries at the smallest separations, replenishment of circumstellar disks by infall becomes unlikely, if only because of the small circumstellar surface area. In at least these cases the replenishment must come from circumbinary disks.

Clearly a direct approach to this issue would be to determine whether circumbinary disks and/or envelopes are present around binaries showing active accretion. In fact, the closest binaries (separations less than $\approx$ 1 AU) with ongoing accretion all have massive circumbinary disks [15, 19, 30, 31]. At intermediate separations ($\approx$10 AU to $\approx$ 1000 AU) a few binaries show evidence for circumbinary disks [1]. Most notable is GG Tau, which not only has a massive circumbinary disk but also shows intriguing suggestions of material in the gap with stream-like structures [32, 33]. However surveys of millimeter emission from PMS binaries suggest that massive circumbinary disks are not common around binaries at intermediate separations [29, 34], many of which nonetheless show evidence for active accretion. The existence of circumbinary envelopes is also not well established for most CTTS binaries, although flat SEDs may be indicative of such envelopes in some cases [16].

In closing, the essential physics underlying the interactions of disks and stellar companions also applies to disks and massive planets. Thus PMS binaries represent laboratories for observational study of processes that may be significant in the formation of planetary systems as well. For example, the possibility of accretion streams onto companions may alter our ideas regarding the processes which set planet masses [20]. Establishing that accretion streams exist may only be possible through observations of young binary systems surrounded by circumbinary disks. Similarly, studies of circumstellar disks and accretion in binary environments will shed light on the survival of circumstellar disks in the presence of a gap-clearing planet and the possibilities for further planet formation. Given that gravitationally bound companions and disks arise in many contexts throughout the universe, the observational study of disk evolution in young binary environments promises to yield insights into binary star formation, planetary formation, and beyond.

ACKNOWLEDGMENTS

I am very grateful to the organizers for their invitation to take part in this excellent meeting. I would like to thank C. Koresko and S. Lubow for assistance in providing figures, and E. Jensen for a critical reading. Funding was provided by the National Science Foundation (AST-941715).

REFERENCES

1. Mathieu, R.D. 1994, ARAA, 32, 465
2. Ghez, A.M. 1996, in Evolutionary Processes in Binary Stars, eds. R.A.M.J. Wijers, M.B. Davies, & C.A. Tout (Dordrecht:Kluwer), 1
3. Mathieu, R.D. 1996, in The Origins, Evolution and Destinies of Binary Stars in Clusters, eds. E.F. Milone & J.-C. Mermilliod (San Francisco:ASP), 231
4. Brandner, W., Alcala, J.M., Kunkel, M., Moneti, A., & Zinnecker, H. 1996, A&A, 307, 121
5. Shu, F.H., Adams, F.C., & Lizano, S. 1987, ARAA, 25, 23
6. Lubow, S.H. & Artymowicz, P. 1996, in Evolutionary Processes in Binary Stars, eds. R.A.M.J. Wijers, M.B. Davies, & C.A. Tout (Dordrecht:Kluwer), 53
7. Mathieu, R.D. 1996, in Evolutionary Processes in Binary Stars, eds. R.A.M.J. Wijers, M.B. Davies, & C.A. Tout (Dordrecht:Kluwer), 11
8. Bertout, C., Basri, G., & Bouvier, J. 1988, ApJ, 330, 350
9. Ghez, A.M., Neugebauer, G., & Matthews, K. 1993, AJ, 106, 2005
10. Leinert, Ch., Zinnecker, H., Weitzel, N., Christou, J., Ridgway, S.T., Jameson, R., Haas, M., & Lenzen, R. 1993, A&A, 278, 129
11. Simon, M., Ghez, A.M., Leinert, Ch., Cassar, L., Chen, W.P., Howell, R.R., Jameson, R.F., Matthews, K., Neugebauer, G., & Richichi, A. 1995, ApJ, 450, 824
12. Mathieu, R.D., Walter, F.M., & Myers, P.C. 1989, AJ, 98, 987
13. Mathieu, R.D. 1992, in Complementary Approaches to Double and Multiple Star Research, eds. H.A. McAlister and W.I. Hartkopf (San Francisco:ASP), 30
14. Mathieu, R.D., Martin, E.L., & Maguzzu, A. 1996, BAAS, 188, #60.05
15. Mathieu, R.D., Stassun, K., Basri, G., Jensen, E.L.N., Johns-Krull, C.M., Valenti, J.A., & Hartmann, L.W. 1997, AJ, in press
16. Prato, L. & Simon, M. 1997, ApJ, in press
17. Brandner, W. & Zinnecker, H. 1997, A&A, in press
18. Valenti, J. A., Basri, G., & Johns, C. M. 1993, AJ, 106, 2024
19. Jensen, E.L.N., Koerner, D.W., & Mathieu, R.D. 1996, AJ, 111, 2431
20. Artymowicz, P. & Lubow, S.H. 1996, ApJ, 467, L77
21. Basri, G., Johns-Krull, C.M., & Mathieu, R.D. 1997, in preparation
22. Zinnecker, H. & Wilking, B.A. 1992, in Binaries as Tracers of Stellar Formation, eds. A. Duquennoy & M. Mayor (Cambridge:Cambridge University Press), 269
23. Koresko, C.D., Herbst, T.M., & Leinert, Ch. 1997, AJ, in press
24. Herbst, T.M., Koresko, C.D., & Leinert, Ch. 1995, ApJ, 444, L93
25. Van Langevelde, H.J., Van Dishoeck, E.F., & Blake, G.A. 1994, ApJ, 425, L45

26. Bate, M.R. & Bonnell, I.A. 1997, in preparation
27. Whitney, B. & Hartmann, L.W. 1993, ApJ, 402, 605
28. Beckwith, S.V.W., Sargent, A.I., Chini, R., & Guesten, R. 1990, AJ, 99, 249
29. Jensen, E.L.N., Mathieu, R.D., & Fuller, G.A. 1996, ApJ, 458, 312
30. Jensen, E.L.N. & Mathieu, R.D. 1997, AJ, in press
31. Mathieu, R.D., Adams, F.C., Fuller, G.A., Jensen, E.L.N., Koerner, D.W., & Sargent, A.I. 1995, AJ, 109, 2655
32. Dutrey, A., Guilloteau, S., & Simon, M. 1994, A&A, 286, 149
33. Roddier, C., Roddier, F., Northcott, M.J., Graves, J.E., & Jim, K. 1996, ApJ, 463, 326
34. Dutrey, A., Guilloteau, S., Duvert, G., Prato, L., Simon, M., Schuster, K., & Menard, F. 1996, A&A, 309, 493

Star Formation in Clusters

C.J. Clarke

Institute of Astronomy, Cambridge, ENGLAND, CB3 0HA.

Abstract. We consider cluster formation in a variety of contexts (i.e. globular clusters, open clusters and small N groups) examining the similarities and differences in the process in different environments and on different scales. We also highlight the most recent theoretical work on the subject, driven by the availability of detailed observations of clustering in nearby star forming regions.

INTRODUCTION

A number of statistics indicate that clustering is an important feature of stellar distributions at birth over a wide range of epochs, scales and star formation conditions. This conclusion is probably best quantified in the case of the solar neighbourhood, where star count analysis indicates that roughly 10% of disc stars form in (open) clusters with lifetimes of the order of 3×10^8 years [1] with most of the remainder existing on timescales of order 10^6 to 10^7 years in unbound associations. At one extreme of the longevity scale, an albeit small fraction ($\sim 0.1\%$) of the stars in the Galactic spheroid formed in (globular) clusters with lifetimes in excess of 10^{10} years, whilst, at the opposite extreme, there is growing circumstantial evidence that a high fraction of stars may originate in mini-clusters for which the lifetime as bound systems is extremely short ($\sim 10^5$ years) This contribution seeks to identify the principal similarities and differences between cluster formation in these diverse environments and to review recent progress on the subject.

Several obvious questions spring to mind when confronted with the problem of cluster formation. Firstly, what are the necessary conditions for cluster formation, and what environmental factors determine whether stars form in clustered or distributed mode? A second question, easier to answer than the first and upon which there is therefore a larger literature, concerns the factors that determine for how long a cluster, once formed, remains bound. A third question, recently prompted by the realization that young clusters can constitute extremely dense stellar systems, is of how the cluster environment affects the properties of the stars it contains.

SOME PRELIMINARIES

We consider the formation of clusters via top-down fragmentation of gravitationally unstable gas. If thermal pressure is the main agent opposing gravitational collapse, the minimum mass scale for collapse is the Jeans limit:

$$M_J = 2 \times 10^6 M_\odot T_4^{3/2} n^{-1/2} \qquad (1)$$

for a medium with temperature $10^4 T_4$K, number density n cm^{-3}. Although this criterion is derived for the case of a uniform, infinite, *static* medium, it has also been applied to dynamically collapsing systems in order to argue that the Jeans mass is progressively reduced during collapse, leading to hierarchical fragmentation down to an 'opacity limited' mass limit where isothermal conditions no longer hold [2].

Numerical simulations [3] however indicate that the ultimate number of fragments in a gravitationally unstable system is chiefly governed by the number of Jeans masses in the system *at the onset of collapse*, this result being explicable in that it is only prior to collapse that the uniform static conditions, invoked in the Jeans analysis, apply. If, by contrast, a Jeans stable sphere is cooled over many dynamical timescales, its quasi-static contraction involves an increasing degree of central concentration, until the Bonnor-Ebert limit [4] is reached, whereupon it collapses dynamically. Since the system is already centrally condensed as it enters this dynamical collapse phase, the collapse is focused on the cloud center, a situation that is not conducive to the fragmentation of the cloud.

Such considerations have led to a widespread belief that the key to cluster formation lies in rapid cooling at the onset of collapse [5], this cooling picking out a new Jeans mass that is small compared with that of the parent cloud. Cooling is required, even though naive application of equation (1) indicates that isothermal compression should also reduce the Jeans mass. However, isothermal compression of a cloud into a layer changes neither the gravitational collapse timescale, nor the sound crossing timescale, in the plane of the slab [6,7] so that the Jeans mass is unchanged. Thus whilst cloud-cloud collisions, and resultant shocks, are often invoked as triggers for cluster formation, their role is to trigger net cooling (i.e. 'better than isothermal' shocks).

THE FORMATION OF GLOBULAR CLUSTERS.

Globular cluster formation is a case where in which collisions between galactic sub-structure are often invoked as a trigger. In galactic halos with characteristic free-fall velocities of the order of hundreds of km/s, collisions between over-dense regions generate strong shocks, provided that the internal temperature of such regions is less than the virial temperature of the galaxy ($\sim 10^6$K). In the absence of external radiation fields, primordial gas subject to strong shock compression cools rapidly to temperatures of order 100K [8,9]. (Note that metal-free gas *in ionization equilibrium* cools very weakly below 10^4K, due to the lack of free electrons to catalyze the formation of the main low temperature coolant, molecular hydrogen; in the shock calculations, however, the shocked gas cools faster than it can recombine, thus providing a supply of free electrons to assist in molecular hydrogen formation). Such calculations therefore indicate that galactic shocks can indeed provide the required rapid cooling, and reduction of Jeans mass, that has been identified above as a pre-requisite for cluster formation. Note that the additional concern of Kang et al [9] (that the cooling should hold up at around 10^4 K for some period) is necessary only if it is desired to identify the masses of globular clusters with the Jeans mass for gas at 10^4K, and that Kang et al achieved this effect by invoking external irradiation to delay molecular hydrogen formation until the shocked material has become self-shielding. It is notable in this regard, that whilst irradiation by hard photons from AGN produces a temporary temperature plateau at 10^4K, followed by rapid cooling on a less than dynamical timescale, the corresponding effect of irradiation by early type stars is quasi-static cooling, which by the arguments presented in Section 2, is problematical for the fragmentation of the cluster into stars. If external irradiation is not invoked, the shocked gas cools quickly through 10^4 K, without imprinting a particular mass scale on the cluster. In this case, the masses of globular clusters would then reflect the scale of inhomogeneity (and the coherence of the velocity field) in the proto-galaxy, modified by mass-dependent effects determining the survival of clusters as bound entities [10]

Once fragmentation of a proto-cluster occurs, the fragmented ensemble collapses, whilst each individual fragment also collapses under its self-gravity. Provided that collapse is isothermal, the latter effect keeps modestly ahead of the over-all collapse but there are potential problems once fragments are held up (at a radius of about an A.U.) by the inefficiency of cooling at high densities [11].The question of whether the fragments then merge thus hinges on what are the processes that cause a reversal in the global collapse. If, for example, the turn-round is caused by two-body scattering of fragment trajectories, then the minimum radius to which the cluster collapses is $N^{-2/3}$ of the initial radius [12] (for N fragments); for large N globular clusters this implied collapse factor of 10^4 would allow fragments to merge before this minimum radius is achieved. More recent analysis [13] however demonstrates that the chief cause of the turn-round of global collapse is the amplification of global collisionless modes, whose seed is provided by Poisson noise in the initial distribution of fragment positions. The resultant collapse factor is then $N^{-1/3}$, which allows the ensemble to collapse, 'bounce' and rebound into a state of approximate virial equilibrium without the constituent fragments merging. Thus fragmentation of a cloud into an initially cold dynamical system remains a viable way of forming globular clusters. (Note that in a variant on this picture [14],fragmentation is not gravitationally driven but derives instead from Rayleigh-Taylor instabilities resulting from shocks driven into the cooled, proto-cluster gas. Such fragments are sub-Jeans mass, so the formation of stars involves the coalescence of fragments during violent relaxation: the larger filling factor for objects that are not collapsing under their own self-gravity means that merging is then significant.)

Another issue is the initial shape of proto-globular clusters. Although the sphericity of old, Galactic globular clusters has encouraged dynamical studies in spherical geometry, there is a well known tendency for young globular clusters (such as those in the LMC) to be pronouncedly aspherical [15,16]. This is a rational expectation if globular cluster formation is triggered by cooling in shocked *layers* and thus suggests that the issue of cluster violent relaxation should be addressed in flattened geometry. Pilot studies (Boily, Clarke and Murray in preparation) show that, due to the preservation of an adiabatic invariant during the radial collapse, such clusters 'remember' their initial shapes during violent relaxation, in contrast to earlier work [17] where this memory was erased by two-body effects during the collapse. This conclusion suggests that the young globular clusters in the LMC (with ages of around 10 dynamical times) should provide good tracers of the initial geometry of the proto-globular cloud. Note that on longer timescales this memory is lost by two-body relaxation, thus explaining the spherical shape of Galactic globular clusters, which are considerably older than their central relaxation timescales.

Finally, it should be noted that this picture (where gas fragments and forms stars on a free-fall timescale) implies that magnetic fields are not dynamically significant in the support of proto-globular clouds. Additional arguments for rapid - i.e. dynamical - formation timescales derive from the observed narrow width of the giant branch in globular clusters [19], which sets limits on self-enrichment, plus the requirement (if the cluster is to remain bound) that most of the gas is incorporated in stars before the first supernova explodes [18].

CLUSTER FORMATION IN NEARBY MOLECULAR CLOUDS.

In contrast to what is inferred in the case of globular cluster formation, star formation at current epochs, in nearby Giant Molecular Clouds (GMCs), is *not* proceeding on the cloud's dynamical timescale, even though GMCs contain a large number of Jeans masses [20]. This apparent contradiction is removed by the postulate that GMCs contain a dynamically significant magnetic field which supports the cloud, both by the cushioning of otherwise dissipative shocks between cloud sub-structure and also by introducing an additional, magnetic, condition for gravitational collapse (see Ostriker, this volume). This latter limit imposes a minimum mass to flux ratio that must be satisfied by a clump in the cloud that can collapse under its own self-gravity. Thus, in this picture, star formation in molecular clouds is controlled not primarily by thermal processes, but by the local decoupling of field and gas.

The prospects for cluster formation thus relate to the question of how, and on what timescale, this local decoupling is achieved. For example, a popular mechanism for decoupling matter and field on the level of individual stars is that of ambipolar diffusion, occurring over a large number of dynamical timescales of the clump involved [21]. Such a quasi-static process however runs into qualitatively similar problems, when it comes to generating clusters, as the scenario of quasi-static cooling outlined above: again, the collapse becomes increasingly centrally condensed, thus focusing the collapse onto the clump center and inhibiting multiple fragmentation. A more promising scenario is one in which regions of the cloud have succeeded in overcoming their magnetic flux problem without becoming centrally condensed in the process (i.e. where self-gravity is not the primary mechanism driving gas-field separation). Such a scenario either requires that pockets of the cloud remain field free or else that some mechanism that is not gravitational in origin removes the field support locally (e.g. X-type reconnection of tangled field lines followed by O-type reconnection of closed flux loops [22]). If parcels of gas can indeed come off field lines without becoming centrally condensed in the process then the possibilities for cluster formation depend on the number of thermal Jeans masses contained in each parcel. If this number is high, then the gas freed from the field is susceptible to prompt initial fragmentation into a number of pieces controlled by the initial Jeans number, analogous to models for globular cluster formation. If, conversely, gas is released in blobs that are highly sub-Jeans in mass (which is a possibility if the field loss mechanism is not gravitationally driven) then the formation of stars requires the coalescence of many such blobs. Thus one might envisage the cloud as a two-phase medium, with the de-magnetized portion trickling through, and interacting with, material still joined to the field lines. In this case, cluster formation could result if a large numbers of Jeans masses of de-magnetized material were able to accumulate in a local minimum of the cloud potential.

Although the above arguments suggest a dominant role for magnetic fields in controlling the rate of star formation in local GMCs, there are observed structures in molecular clouds - the dense low mass cores observed in NH_3 - which are primarily supported against gravity by thermal pressure, and in which additional, non-thermal, contributions to the molecular line widths are relatively small [23,24] (see also Myers, this volume). These cores have thus largely overcome the 'magnetic flux problem' that holds up lower density phases of molecular clouds, so that in their eventual collapse (and possible fragmentation) thermal processes (i.e. further cooling) may play an important role. A possible trigger for core cooling may, for example, be provided by core-core collisions and consequent 'better than isothermal' cooling in the shocked layer thus formed [25]. Alternatively, Clarke and Pringle [26] have pointed out the possible role of 'shear induced cooling' in cooling and destabilizing dense molecular gas: since the thermal equilibrium of such cores involves cooling through optically thick molecular line radiation [27], it follows that any disturbance of the core that increases the velocity width of lines to superthermal values promotes enhanced cooling. Since the thermal and dynamical timescales for the equilibrium core are comparable, order unity perturbations to the cooling rate produce net cooling on a dynamical timescale and therefore set up the conditions required for 'prompt initial fragmentation'. Therefore dynamical disturbances to dense cores (as from interactions with sub-structure in the cloud or from the close passage of a passing star) may induce dense molecular gas to fragment into small clusters of stars.

ISSUES IN EARLY CLUSTER EVOLUTION.

Open Clusters

Survival as bound clusters.

The number of open clusters (age $\sim 10^8$ years) provides a lower limit to the fraction of stars formed in clusters if most clusters dissolve on timescales much shorter than this. One way in which clusters become unbound is through mass loss, simple considerations indicating that a virialized cluster is unbound if subject to the instantaneous loss of 50% or more of its mass. Since the over-all mass fraction of GMCs that is converted into stars over their star forming lifetimes (i.e. the 'star formation efficiency') is only a few per cent, there is a possible contradiction between such inefficient star formation and the formation of bound clusters. The required efficiency for bound cluster formation is lowered if gas is lost over timescales of a few million years [18]. Thus the formation of bound open clusters requires *either* that they form in regions of localized high star formation efficiency *or* else that they are bereft of O stars and therefore retain their gas over the required period. In the former case, the need for high efficiency locally implies considerable coordination for collapse and star formation, with implications for the manner in which magnetic support is lost from regions destined to become open clusters.

Cluster Dynamics and Binarity.

The cluster environment can affect the population of binary stars, both constructively and destructively. Thus soft binaries are prone to dynamical disruption by encounters with other cluster stars, possibly explaining the observed excess in binarity in pre-main sequence, c.f. main sequence, stars [28] (see Mathieu, this volume). Alternatively, the cluster environment may favour binary formation during early pre-main sequence evolution, due to dissipative encounters between stars surrounded by circumstellar discs [29].For the high stellar densities in the core of the Trapezium cluster (5×10^4 pc^{-3}) and typical disc radii (100 A.U.) the encounter timescale is short (2×10^6 years), implying that encounters are indeed important in such regions, possibly generating the observed exponential profiles of the 'silhouette discs' in Orion [30,31]. However, such encounters are too fast, for typical velocity dispersions in young clusters, to produce a significant binary fraction [32–35].

Mass segregation in young clusters.

The stellar content of clusters is now being characterised at increasingly young ages. Recent results for the Orion Trapezium [36] (estimated age a few million years) indicate that the IMF is normal on large scales, despite the conspicuous grouping of massive stars in the central Trapezium (radius ~ 0.05 pc). This prompts the question of whether this mass distribution is primordial or whether it arises through dynamical effects over the cluster lifetime.

Recent simulations by Bonnell (in preparation) have examined this question through N-body simulations, asking how long it takes, starting from an initially flat-topped density profile (Plummer sphere), for the system to dynamically form a 'Trapezium' (defined such that > 5 of the 15 most massive stars are found within a region radius 0.05 pc.) If the IMF is initially uniform then the timescale for 'Trapezium' formation is much longer than the age of the system, whereas if stars are initially assigned radii in inverse ranking with their masses, then $\sim 20\%$ of the simulations have generated a 'Trapezium' within the required time. Thus the mass distribution in the Trapezium *cannot* be the result of two-body relaxation acting on a uniform initial mass and density distribution. Instead the cluster must have formed with massive stars in the middle; it is also probably necessary for these stars to be grouped in a dense sub-cluster at birth, as two body relaxation is not likely to generate a concentrated 'Trapezium' structure within the required time.

Small N Clusters.

Non-hierarchical small N systems dissolve dynamically after of order N crossing times [37]. For compact clusters containing few stars, this timescale is so short that the initial conditions are erased very quickly, so that arguments for their existence are generally indirect. Below we summarize the evidence supporting the postulate that stars are tightly clustered at birth.

Binary formation.

If stars form in small non-hierarchical groups, then one consequence is the generation of binary systems [5]. At least one binary is formed per cluster through purely dynamical effects (i.e. where energy removed from a two-body orbit is transferred to the kinetic energy of a third body), whereas higher binary fractions are incurred if dissipative star-disc interactions provide an additional energy sink [38,39]. (Note that star-disc interactions *are* effective in this context (c.f. the open cluster case above) on account of the lower velocity dispersion within small stellar sub-groups). Although dynamical interactions within small groups are not a unique way to form binary stars, it should be noted that binaries formed in this way have distinct observational characteristics (in terms of mass pairing statistics) compared with binaries formed from the fragmentation of a core into only two pieces [40].

The generation of runaway T Tauri stars (RATTs).

The discovery by the ROSAT All Sky Survey of large numbers of X-ray sources outside classic star forming regions [41], and the evidence from lithium abundances that at least some of these sources are young [42], have prompted a model in which some T Tauri stars are ejected from star forming regions by dynamical interactions within small stellar groups [43]. The ejection velocities that are required in order for the outlying Xray sources to have reached their current locations within their inferred lifetimes (~ 5 km s^{-1}), imply parent mini-clusters of extremely high densities, with mean interstellar separations of ~ 1000 A.U.: the corresponding dissolution timescales are exceedingly short ($< 10^5$ years), so that the observed lack of T Tauri star clustering on these scales does not argue against the existence of such clusters as initial conditions for star formation. Furthermore, a notable characteristic of the RATTs (that they are all Weak Line T Tauri stars) is explicable in this model: the lopping of discs (at a radius of a few A.U.) during close encounters drastically reduces the disc evolution timescale, so that such pared discs are liable to be classified as Weak Line systems within $\sim 10^6$ years of ejection [44].

Tracers in stellar distributions.

Although the postulated mini-clusters dissolve on very short timescales, they would still show up as spatially clustered concentrations of unbound stars over considerably longer timescales. Thus, for example, the dilute, unbound 'Gomez' groups in Taurus [45] may be fossils of formerly bound clusters that were much more compact. Thus it is *not* the observed Gomez groups that will give rise to binary formation and ejection of runaways in their future evolution; instead the Gomez groups and the RATTs may be the low and high velocity tails respectively of the distribution of stars ejected from primordial mini-clusters. Evidently these ideas have to be examined quantitatively, through dynamical modeling combined with kinematic data, both for Taurus and for younger, denser environments like Orion.

Star formation in small N clusters.

The above discussion focused on clusters that had already formed a number of star-disc systems. Earlier stages of cluster evolution, when the stars were actually formed, have not been studied until now, due to numerical problems following the large dynamic range in densities that develop. The introduction of 'sink particles' in SPH simulations (see Bate, this volume) however allows dense clumps of bound gas to be excised from the domain of detailed computation, so that cluster formation calculations can run to completion.

The recent calculations of Bonnell et al [46] start with a marginally unstable sphere of gas, seeded with 10 'mass seeds' comprising a total of 10% of the system mass. During the subsequent evolution, the seeds interact dynamically whilst accreting from the gas flow. The seeds therefore grow by competitive accretion; by the time all the gas has accreted onto the seeds, a large dynamic range in seed masses has developed, even though the initial seeds were identical in mass. The main determinant of final seed mass is the initial position in the cluster: seeds initially in the cluster core start by accreting more rapidly as density gradients develop in the gas, and once they are more massive than their competitors, are then harder to nudge out of the center and so continue rapid growth. Seeds initially at the cluster edge, conversely, grow slowly at first, so that they are easily flung out by interactions with more massive seeds and hence detached from the reservoir of accretable gas.

It would be premature (given the relatively small number of simulations and neglect of effects such as feedback on the cluster gas from outflows and winds) to predict a detailed IMF from this work. Nevertheless, it demonstrates the important principle that the combined effects of cluster dynamics and accretion produce a large dynamic range in resultant stellar masses, without requiring any correspondingly large range in the initial fluctuation spectrum. (Note that in this picture, the issue of what terminates accretion - such as dominates some discussions for the origin of the IMF [47] - is not an issue, because the situation here is *not* one of an isolated protostar, otherwise growing indefinitely from an infinite uniform medium. Instead, the maximum reservoir is set by the finite region of bound gas constituting the initial cluster, whilst the division of mass amongst seeds is set by the process of competitive accretion.) It is interesting that these simulations naturally give rise to mass segregation at birth (as required in the Trapezium), without invoking initial radial gradients in the gas properties and associated Jeans mass.

REFERENCES

1. Miller, G.E. and Scalo, J.M., 1978. PASP 90,506.
2. Hoyle, F., 1953. ApJ 118,513.
3. Larson, R.B., 1978. MNRAS 184,69.

4. Bonnor, W.B., 1956. MNRAS 116,351.
5. Pringle, J.E., 1989, MNRAS 239,361.
6. Lubow, S. and Pringle, J.E., 1993. MNRAS 263,190.
7. Whitworth, A.P. et al ,1994. MNRAS 268,291.
8. Murray, S.D. and Lin, D.N.C., 1989 ApJ., 339, 933
9. Kang, H., Shapiro, P.R, Fall,S.M. and Rees, M.J., 1990. ApJ 363,488.
10. Fall, S.M. and Rees, M.J., 1977. MNRAS 181,37P.
11. Larson, R.B., 1969, MNRAS 145,271.
12. Layzer, D., 1963. ApJ 137,351.
13. Aarseth,S.J., Lin, D.N.C. and Papaloizou, J.C.B., 1988. Ap J 324,288.
14. Murray, S.D. and Lin, D.N.C. 1996, ApJ, 467, 728.
15. Frenk, C.S. and Fall, S.M., 1982. MNRAS 199,565.
16. Elson, R.A.W, Fall, S.M. and Freeman, K.C., 1987. Ap J 323,54.
17. Aarseth, S.J. and Binney, J., 1978. MNRAS 185,227.
18. Lada, C.J., Marguilis, M. and Dearborn, D. 1984. ApJ 285,141.
19. Richer, H.B. and Fahlman, G.G.,1984. ApJ 277,227.
20. Blitz, L. and Shu, F.H., 1980 Ap J 238,148
21. Lizano, S. and Shu, F.H., 1989, ApJ 342,834
22. Lubow, S.H. and Pringle, J.E., 1996, MNRAS, 279, 1251
23. Myers, P.C. and Goodman, A.A., 1988. ApJ., 329,392
24. Casselli, P. and Myers, P.C., 1995. ApJ, 446, 665
25. Turner, J.A. et al, 1995. MNRAS 277,905.
26. Clarke, C.J. and Pringle, J.E., 1996. MNRAS in press.
27. Neufeld, D.A., Lepp, S. and Melnick, G.J., 1995. ApJ. Supp. 100,132.
28. Kroupa, P. ,1995. MNRAS 277,1491.
29. Larson, R.B., 1990, in 'Physical Processes in Fragmentation and Star Formation', R. Capuzzo-Dolcetta et al. (eds), Kluwer,Dordrecht, p.389.
30. Hall, S.M., 1996, MNRAS submitted.
31. McCaughrean, M. and O'Dell, C.R., 1996, AJ 111,1977.
32. Clarke,C.J. and Pringle, J.E., 1991, MNRAS 249,584.
33. Clarke, C.J. and Pringle, J.E., 1993, MNRAS 261,190
34. Heller, C., 1995. ApJ 455,252.
35. Hall, S.M., Clarke, C.J. and Pringle, J.E. 1996. MNRAS 278,303.
36. Hillenbrand, L.A., 1996. Poster: 'Star Formation in Clusters', Wellesley, MA.
37. van Albada, T.S., 1968, Bull. Astron. Inst. Neth., 19,479.
38. McDonald, J.M. and Clarke, C.J., 1993, MNRAS,262,800
39. McDonald, J.M. and Clarke, C.J., 1995, MNRAS 275,671
40. Clarke, C.J., 1996. MNRAS 283,353.
41. Neuhauser, R. et al, 1995. A&A 295,L5.
42. Alcala, J.M. et al, 1996. A&A in press.
43. Sterzik, M.F. and Durisen, R.H., 1995. A&A 304,L9.
44. Armitage, A. and Clarke, C.J., 1996. MNRAS in press.
45. Gomez, M., Hartmann, L., Kenyon, S.J. and Hewett, R., 1993, AJ 105,1927.
46. Bonnell, I., Bate, M., Clarke, C. and Pringle, J., 1996. MNRAS in press.
47. Adams, F.C. and Fatuzzo, M., 1996. Ap J 464,256.

The Initial Mass Function: Now and Then

Harvey B. Richer[†] and Gregory G. Fahlman[†]

[†]*Department of Physics and Astronomy*
University of British Columbia
Vancouver, British Columbia
Canada V6T 1Z4
Email: surname@astro.ubc.ca

Abstract. We examine whether existing data in clusters, both old and young, and in the field of the Galactic disk and halo is consistent with a universal slope for the initial mass function (IMF). The most reasonable statement that can be made at the current time is that there is no strong evidence to support a claim of any real variations in this slope. If the IMF slope is universal then this in itself is remarkable implying that variations in metallicity, gas density or other environmental factors in the star formation process play no part in determining the slope of the mass function.

"Current evidence favours a universal initial mass function, independent of environmental factors that can be expressed by a single power-law of slope $x = 1$ (Salpeter = 1.35) from $100 M_\odot$ down to $1 M_\odot$ and possibly even to $0.1 M_\odot$. [1]"

INTRODUCTION

The mass function (MF) of a stellar population is typically expressed as Ψ, the number of stars / unit mass / unit volume element. The masses of the stars are not observed directly, rather it is their luminosities that are the measurable quantity so that the MF, written in terms of observables, is

$$\Psi(m) = \Theta(M_V) \cdot \frac{dM_V}{dL} \cdot \frac{dL}{dm} \tag{1}$$

where m is the mass, $\Theta(M_V)$ the luminosity function (for illustrative purposes in M_V), $\frac{dM_V}{dL}$ the bolometric correction and $\frac{dL}{dm}$ the mass-luminosity relation. The bolometric correction is usually obtained from theory but can be measured empirically, while the mass-luminosity relation is readily derived for population I from binary systems but currently must be obtained from theory for population II due to a lack of empirical data. Historically, this has been an important difficulty in deriving accurate MFs from luminosity functions for population II samples, but enormous progress has recently been made (see the section below on globular clusters). When the bolometric correction is known, equation 1 reduces to its simplest form

$$\Psi(m) = \Theta(M_V) \cdot \frac{dM_V}{dm} \qquad (2)$$

where it is now explicitly seen that the observed luminosity function must be multiplied only by the *slope* of the M_V-mass relation to obtain the MF. This MF is often assumed to be a power-law so that

$$\Psi(m) \propto m^{-(1+x)}, \qquad (3)$$

x taking on a value of 1.35 in the work of Salpeter [2]. Salpeter derived this slope using Galactic field stars in the mass range $0.5 - 10 M_\odot$.

Moffat, quoted above, was mainly interested in young starbursts when he made his claim of a universal MF slope. We expand his discussion to include much older open clusters, the MF slope in the Galactic disk and halo, and the MF in globular clusters.

YOUNG CLUSTERS, ASSOCIATIONS AND YOUNG FIELD STARS

Massey et al. [3–5] have carried out systematic ground-based work on the MF in young stellar aggregates and in the field both for the Galaxy and in the Magellanic Clouds. Similar research has been done in nearby resolved systems by Hunter et al. [6–9] using *HST*. These studies, together with others, point to a MF slope for all the systems, independent of which galaxy they reside in or the physical conditions in the gas out of which they formed, in the range of $x = 1 - 1.3$ with errors of 10 - 20% in the slope. In these systems it is likely that we are actually seeing the slope of the initial mass function (IMF), at least at the highest mass end, as these objects are too young to have suffered much dynamical evolution. Williams et al. [10] have extended work of this sort to very low masses using infrared observations in the star forming ρ Oph cluster. They find a value of $x = 1.1$ which extends from $\sim 4 M_\odot$ down below $0.1 M_\odot$, although there is some uncertainty in establishing stellar masses from the low resolution infrared spectroscopy used in their analysis.

The one serious divergent data point in the universality of MF slopes for young stars is found in the field of the Galaxy and the Magellanic Clouds where Massey et al. [4] find very steep slopes, up to 4. It is likely that in this case we are *not* seeing the IMF as these stars probably recently escaped from their clusters of origin which are likely to be more efficient at retaining the most massive objects.

OLD OPEN CLUSTERS AND LOW MASS DISK STARS

The most comprehensive work on the MFs in older Galactic clusters is due to Francic [11]. In a sample of eight clusters ranging in age from $\sim 10^8$ to 5×10^9 years, Francic found that the MFs for the 5 youngest systems were all in the range of $x = 1$ with very little scatter. In the three oldest clusters (NGC 6633, NGC 752, M67) the currently observed MF is weighted toward higher mass stars. There is little doubt that this has resulted from the loss of low mass stars through dynamical processes.

Recent efforts at deriving the MF for the disk of the Milky Way have concentrated on improving the theoretical models for low mass stars and on the data sets used to construct the luminosity functions from which the MF is constructed. New models have been presented by Méra et al. [12] who then used the luminosity function derived by Kroupa [13] to obtain a MF for the disk of the Galaxy. This MF is well represented by a power-law with a slope of $x \simeq 0.7$.

THE MASS FUNCTION IN GLOBULAR CLUSTERS

There have been two recent developments which have dramatically altered our view of the MF in Galactic globular clusters. The first of these is observational. New data obtained with the Hubble Space Telescope (*HST*) are both more accurate (as they are less affected by crowding) and penetrate to fainter magnitudes than ground-based data so that the cluster luminosity functions obtained with *HST* are better defined and descend to significantly lower masses. Further, *HST*-derived luminosity functions can be secured in the crowded cores of many Galactic globular clusters so that the effects of mass segregation can be more readily quantified. These points are illustrated in Figure 1 where the color-magnitude diagram (CMD) of M4 obtained from the Las Campanas Observatory [14] is compared with an *HST* cycle 4 CMD [15].

The second major improvement in deriving MFs for globular clusters comes from some very impressive theoretical work. Both the Lyon group in France, led by Gilles Chabrier, and the group in Victoria under Don VandenBerg have been constructing new interior models for metal-poor stars with revised equations of state, and are fitting these to model atmospheres calculated by France Allard and her collaborators. These models, while not yet in final form,

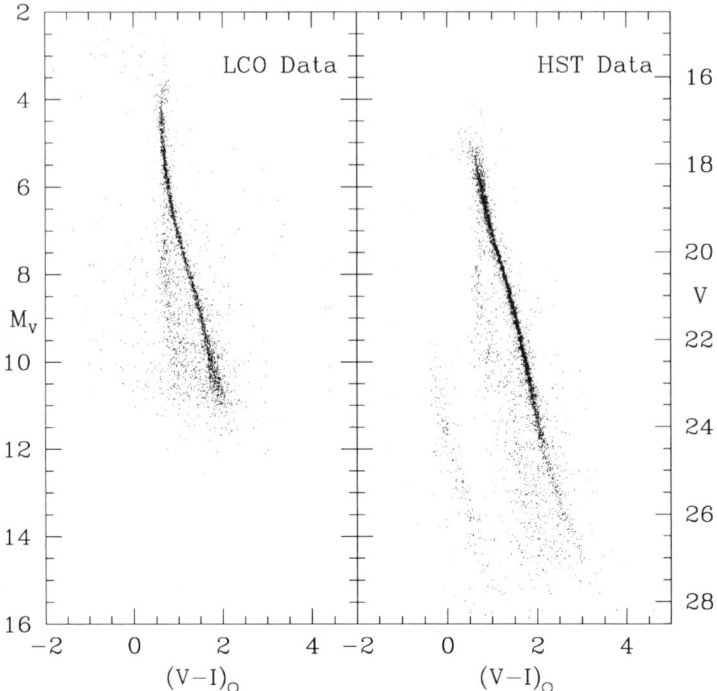

FIGURE 1. A ground-based CMD for the Galactic globular cluster M4 obtained from Las Campanas Observatory (left panel) compared with one secured with *HST*. Note that not only is the main sequence more tightly defined and descends to much fainter objects in the *HST* CMD, but that an extensive white dwarf cooling sequence is also discovered.

are a vast improvement over earlier ones. We illustrate this in Figure 2 where the published models of D'Antona and Mazzitelli [16] are compared with the CMD of M4 in the left panel of the Figure while new VandenBerg-Allard [17] models are shown on the right. It is clear that the older models do a rather poor job of fitting the lower portions of the M4 CMD whereas the more recent models are now an excellent fit all the way to the limit of the data. Significant errors will hence arise into the calculation of MFs constructed from models which are poor fits to the cluster CMDs.

King et al. [18] used the D'Antona and Mazzitelli stellar models together with luminosity functions derived from *HST* observations to produce MFs for 4 globular clusters (47 Tuc, NGC 6397, M15, M30). To these they added the MF of ω Centauri taken from Elson et al. [19] and demonstrated that all of these (with the exception of NGC 6397) appeared to have about the same MF slope. These clusters (NGC 6397 excluded) are not expected to have

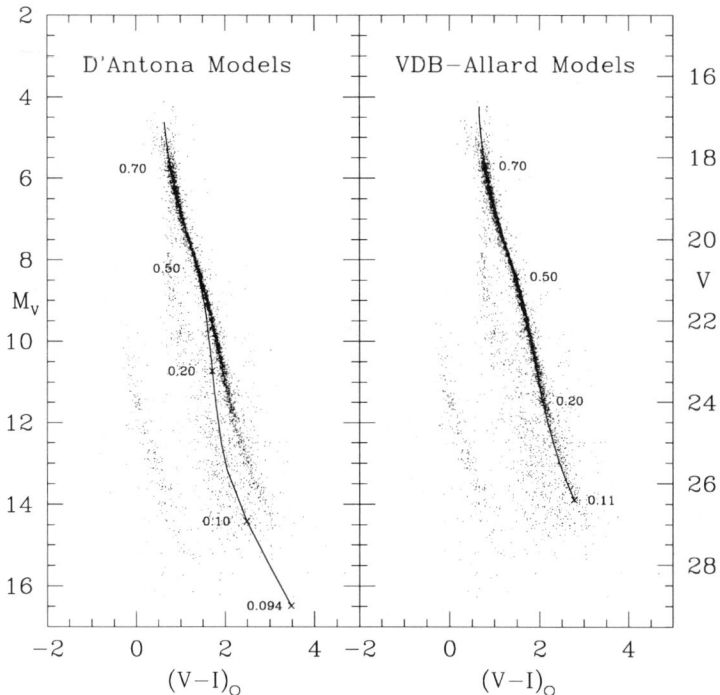

FIGURE 2. D'Antona and Mazzitelli 10 Gyr isochrone for $[m/H] = -1.3$ compared with the CMD of M4 in the left hand panel. The small numbers indicate the stellar masses. In the right panel new interior models calculated by VandenBerg for $[m/H] = -1.0$ are combined with model atmospheres by Allard to produce isochrones which provide a much improved fit to the data.

lost many of their low mass stars as, according to the cluster evolutionary models of Gnedin and Ostriker [20], their destruction rates due to disk and bulge shocks and tidal effects are small. All the data were obtained near the cluster half-mass radii where the local MF well approximates the global one. Below a mass of $0.4 M_\odot$, where the mass segregation effects are small [21], the MF slopes for these globular clusters are in the neighborhood of $x = 0.7$ with a range of about ±0.2. These slopes may be a good representation of the cluster IMF values. The slope for NGC 6397 is near $x = 0.5$. However, what should be kept in mind are the possible errors introduced in using the D'Antona-Mazzitelli models. These produce MF slopes that are steeper than those generated using the newer VandenBerg-Allard mass-luminosity relation. For example, the MF of 47 Tuc, determined using these latter models, has a slope near $x = 0.4$ which is to be compared with $x = 0.8$ using the D'Antona-

Mazzitelli models. Until the theoretical models for low mass population II stars are finalized, this level of uncertainty will remain in the transformation from luminosity functions to MFs and not until then will a clean comparison with MF slopes for population I be possible.

Recent discoveries with *HST* allow for an extension of globular cluster MFs to stars more massive than those at the turnoff. We discuss this with reference to results on M4 [22,15]. The CMD which we obtained for this cluster is displayed in Figure 1 wherein a well-populated white dwarf cooling sequence can be seen. These objects evolved from stars more massive than the current cluster turnoff so they provide the potential of extending the cluster MF to higher masses. Current turnoff stars in a cluster such as M4 ($[m/H] = -1.3$) have masses of $\sim 0.8 M_\odot$ [23]. Because counting faint white dwarfs in the environs of a crowded field of a globular cluster is subject to serious incompleteness, the oldest white dwarfs in M4 for which we are confident of the statistics are 3.7 Gyr. In this cluster, the progenitor of these stars had masses of $0.9 M_\odot$ so the present sample of white dwarfs allow us to extend the MF upward by about $0.1 M_\odot$. The observed main sequence MF of M4 is very flat, with $x \simeq -1$ [24], and the white dwarfs are found to extend this MF at about the same slope. This very flat MF is unlikely to be the cluster IMF as the orbit of M4 keeps it close to the Galactic plane [25] where shocking and tidal effects can efficiently strip low mass stars from it.

The possibility exists, however, to extend the cluster MF to much higher masses. This is discussed in detail in Richer *et al.* [15] and we outline the ideas here. The number of white dwarfs in a cluster brighter than some magnitude depends on four factors: the age of the cluster, the cooling time of the white dwarfs to that magnitude, the dependence of main sequence lifetime on stellar mass ($T_{MS} \propto m^{-\gamma}$), and the cluster IMF ($\Psi(m) \propto m^{-(1+x)}$). The last two factors work in the following way. The larger is x, the steeper the mass function, and the fewer the number of high mass stars that eventually become white dwarfs. On the other hand, the bigger is γ, the shorter the amount of time massive stars spend on the main sequence, and hence the quicker they produce white dwarfs. Under the assumption of no loss of stars from a cluster, the white dwarf luminosity function will then contain information on the cluster IMF for a given value of γ. Very faint white dwarf cooling sequences are required to use this technique efficiently, but such data are possible to obtain with *HST*.

THE MASS FUNCTION OF THE GALACTIC SPHEROID AND HALO

In discussing the MF of the outer environs of the Galaxy away from the disk, a clear distinction between the Galactic spheroid and halo must be kept in mind. If stars are found in a density distribution which varies as $r^{-3.5}$ then it is the spheroid that is being examined. If the halo of the Galaxy has a stellar component associated with it, then these objects will be distributed as r^{-2} so as to account for the flat rotation curve of the Galaxy. It is for this reason that not only should the luminosity function (or MF) of outer Galaxy stars be measured, but, it is critical to examine the density distribution simultaneously so as to be sure which component of the Galaxy is being explored.

With this idea in mind, Richer and Fahlman [26] (hereafter RF) initiated a program using CFHT of determining the MF of stars in the outer Galaxy from stars found *in situ*. The major difficulties with this approach are star/galaxy separation at the faint limits and the assignment of stars to the disk, to the thick disk and to the spheroid or halo. In a single 49$\Box'$ field with a sample of only 31 stars, RF found a very steep MF slope ($x = 3.5$) with the distribution of the stars well represented by $\rho \propto r^{-3.5}$. In a second, similarly-sized field (unpublished), a somewhat flatter MF slope ($x = 2.0$) was measured, and, again, the density distribution fit that expected for the spheroid. These large differences in the derived MF slopes may be suggesting that the outer parts of the Galaxy are quite inhomogeneous as would be expected if it formed from a number of accretion events.

The results of RF were analyzed in some detail by Reid *et al.* [27] who pointed out that the observed MF was likely to have been contaminated by stars in the Galactic thick disk. This could produce an artificial upturn at the low mass end and thus a spuriously rapidly rising slope. Reid *et al.* suggested that all stars redder than $(V - I) = 1.75$ observed by RF actually belong to the thick disk. To avoid the possible inclusion of thick disk objects, we have recomputed the observed MF eliminating all stars below a Z-distance of 5 kpc. Excluding these stars from the original RF sample leaves only 23 objects, while removing them from the second field produces a sample of 45 stars. The MF derived from these very small groups of stars is much flatter with $x \simeq 0.7$, but the error in this slope is large. The objects still fit a spheroid density distribution reasonably well with $\rho \propto r^{-2.9}$.

An alternate approach to deriving the MF for the outer regions of the Galaxy is to study a sample of nearby high proper-motion stars. This technique suffers from the possibility of bias in the selection of the stars which is somewhat difficult to correct for. Further, with such a local sample, it is not possible to solve for the density distribution of the stars. Dahn *et al.* [28] have derived a luminosity function for such objects and we use the new VandenBerg-Allard models to convert this to a MF. This function is displayed in Figure 3 where the approximately linear part of the diagram, from $\log(M/M_\odot) = -0.4$ to -0.8, has a slope of $x = 0.6$ and the MF clearly appears to turn over below a mass of $\sim 0.16 M_\odot$ which is well above the hydrogen-burning limit for metal-poor stars. If this MF represents that of the Galactic halo satisfactorily, with no upturn at lower masses, then very low mass stars and brown dwarfs are unimportant contributors to the total mass budget of the Galaxy. This conclusion is also in agreement with recent results from the microlensing experiments [29].

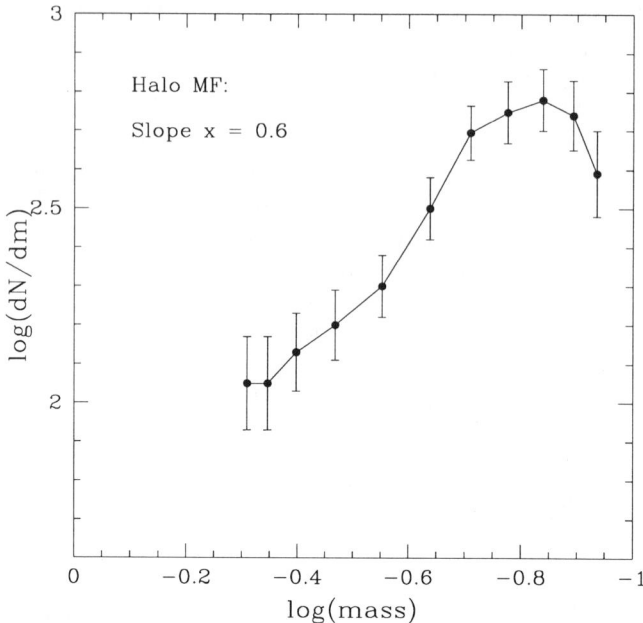

FIGURE 3. Mass function for high-velocity stars in the solar neighborhood. The data are from Dahn *et al.* and the models used to convert the luminosity function into a mass function are from Vandenberg and Allard.

A SUMMARY OF THE MASS FUNCTIONS

To illustrate the similarity of the MFs in extremely different environs, we combine a representative sample of them into a single diagram below (Figure 4). Vertical offsets were applied to each data set to separate the MFs so that only the slopes and the position along the mass axis for the different samples are of interest. The similarity of the slopes in this diagram is remarkable considering the widely different environments under which these stars have formed. There may be some tendency for the MFs involving older stars to have x somewhat smaller, but given the current uncertainty in the modelling, it is premature to attach much weight to this. Also of interest is the mass coverage of each MF, in particular, the manner in which the old open clusters in the Galaxy fill the space between the OB associations and the disk. The average MF slope for all the populations shown is $x = 0.86 \pm 0.23$, somewhat flatter than the Salpeter [2] slope.

Understanding, theoretically, the universality of MF slopes will surely be a challenge. A clue might be found, however, in the precursors to the stars themselves. In regions where star formation is not yet underway, a number of

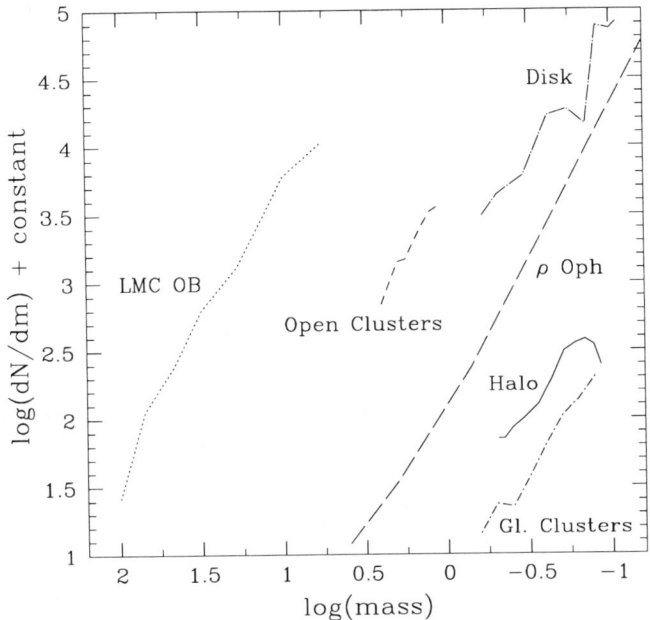

FIGURE 4. Mass functions in a wide variety of environments from extremely young OB associations in the LMC, to a selection of Galactic open clusters and the Galactic disk, through to metal poor globular clusters (M15, M30) and the Galactic halo. Arbitrary offsets have been applied to the data so that only the slopes and the coverage in mass are of interest in this Figure. Sources of the data can be found in the relevant sections in the text.

observations have shown that the spectrum of clump masses are already well fitted by a power law with slope near $x = 1$ [30]. Turbulence in the gas may then be invoked as the cause of this spectrum, but the origin of the turbulence in regions where hot stars do not as yet exist remains a puzzle.

REFERENCES

1. Moffat, A.F.J. *The 1996 INAOE Conference on Starburst Activity in Galaxies*, invited review, *Revista Mexicana de A&A*, in press (1996).
2. Salpeter, E.E., *ApJ*, **121**, 161 (1955).
3. Massey, P., Johnson, K.E., and Degioia-Eastwood, K. *ApJ*, **454**, 151 (1995a).
4. Massey, P., Land, C.C., Degioia-Eastwood, K., and Garmany, C.D. *ApJ*, **438**, 188 (1995b).
5. Massey, P., and Johnson, J. *AJ*, **105**, 980 (1993).

6. Hunter, D.A., Baum, W.A., O'Neil, E.J., Jr., and Lynds, R. *ApJ*, **468**, 633 (1996a).
7. Hunter, D.A., O'Neil, E.J., Jr., Lynds, R., Shaya, E.J., Groth, E.J. and Holtzman, J.A. *ApJ*, **459**, L27 (1996b).
8. Hunter, D.A., Baum, W.A., O'Neil, E.J., Jr., and Lynds, R. *ApJ*, **456**, 174 (1996c).
9. Hunter, D.A., Shaya, E.J., Holtzman, J.A., Light, R.M., O'Neil, E.J., Jr., and Lynds, R. *ApJ*, **448**, 179 (1995).
10. Williams, D.M., Comeron, F., Rieke, G.H., and Rieke, M.J. *ApJ*, **454**, 144 (1995).
11. Francic, S.P. *AJ*, **98**, 888 (1989).
12. Méra, D., Chabrier, G., and Baraffe, I. *ApJ*, **459**, L87 (1996).
13. Kroupa, P. *ApJ*, **453**, 350 (1995).
14. Thompson, I.B., Sivaramakrishnan, A. and Fahlman, G.G. unpublished (1990).
15. Richer, H.B., Fahlman, G.G., Ibata, R.A., Pryor, C., Bell, R.A., Bolte, M., Bond, H.E., Harris, W.E., Hesser, J.E., Holland, S., Ivanans, N., Mandushev, G., Stetson, P.B., VandenBerg, D.A., and Wood, M. submitted to *ApJ* (1997).
16. D'Antona, F. and Mazzitelli, I. *ApJ*, **456**, 329 (1996).
17. VandenBerg, D.A. and Allard, F., private communication (1996).
18. King, I.R., Cool, A.M., and Piotto, G. *Formation of the Galactic Halo....Inside and Out*, eds. H. Morrison and A. Sarajedini, ASP Conference Series, Vol. 92, p. 277 (1996).
19. Elson, R.A.W., Gilmore, G.F., and Santiago, B.X. *AJ*, **110**, 682 (1995).
20. Gnedin, N.Y. and Ostriker, J.P. preprint, (1996).
21. Richer, H.B. and Fahlman, G.G. *The Formation and Evolution of Star Clusters*, ed. K. Janes, ASP Conference Series, Vol. 13, p. 120 (1991).
22. Richer, H.B., Fahlman, G.G., Ibata, R.A., Stetson, P.B., Bell, R.A., Bolte, M., Bond, H.E., Harris, W.E., Hesser, J.E., Mandushev, G., Pryor, C., and VandenBerg, D.A. *ApJ*, **451**, L17 (1995).
23. Bergbusch, P.A. and VandenBerg, D.A. *ApJS*, **81**, 163 (1992).
24. Fahlman, G.G., Richer, H.B., Ibata, R.A., Stetson, P.B., Bell, R.A., Bolte, M., Bond, H.E., Harris, W.E., Hesser, J.E., Mandushev, G., Pryor, C. and VandenBerg, D.A., in preparation (1997).
25. Cudworth, K.M., and Rees, R. *AJ*, **99**, 1491 (1990).
26. Richer, H.B. and Fahlman, G.G. *Nature*, **358**, 383 (1992).
27. Reid, I.N., Yan, L., Majewski, S., Thompson, I. and Smail, I. *AJ*, in press (1996).
28. Dahn, C.C., Liebert, J., Harris, H.C., and Guetter, H.H. *The Bottom of the Main Sequence-and Beyond*, eds. C.G. Tinney (Springer, Berlin), p. 239 (1994).
29. Alcock, C. et al. *ApJ*, **471**, 774 (1996).
30. Stutzki, J. et al., *Fragmentation of Molecular Clouds*, p. 235 (1991).

BINARY MULTIPLICATION: THE FORMATION OF CLOSE BINARIES FROM WIDE ONES

Stephen Watkins, Amardeep Bhattal, Neil Francis
& Anthony Whitworth

*Department of Physics & Astronomy, University of Wales,
Cardiff, CF2 3YB, Wales, U.K.*

Abstract. We hypothesize that the masses of stars and the statistics of binary systems are largely determined by interactions between extended massive protostellar discs in dense protoclusters. We present simulations which show that collisions between extended massive protostellar discs in wide binary systems are very effective at forming smaller binary systems. Therefore one can envisage an hierarchical cascade, which starts with a small-N cluster of predominantly long-period systems, and then by stages multiplies and populates shorter periods. Our simulations indicate that each generation in this proposed hierarchical cascade should approximately double the number of protostellar discs, and produce binaries with periods about an order of magnitude shorter than their parents.

BACKGROUND

We have published SPH simulations of star formation triggered by mildly supersonic collisions between molecular-cloud clumps (Turner et al. 1995, Whitworth et al. 1995). These simulations show the following sequence of events:

1. The colliding clumps produce a shocked layer. Compression in the layer is high, provided the time-scale for radiative cooling is shorter than the time-scale on which the clump is overrun by the shock. For the majority of collisions, the impact parameter is comparable with the clump radius, and so the shocked layer tumbles about an axis perpendicular to the plane containing the collision velocity and the impact parameter.

2. The layer then fragments gravitationally into a network of filaments. At this stage, the layer is still contained by the ram-pressure of the inflowing gas, i.e. not all the clump gas has been shocked. Consequently the filaments are

well spaced (*i.e.* spacing greater than the layer thickness – see Whitworth et al. 1994). Because the layer is tumbling, the filaments are also tumbling.

3. Condensations, $\sim 0.1 - 0.3$ pc in size, form at intervals along the filaments. These condensations continue to grow by accreting material from their parental filament.

4. Because the filaments are tumbling, the material accreting onto the condensations spins them up until they are rotationally unstable. These rotational instabilities break the condensation up into small-N clusters of protostellar discs with diameters $\sim 300 - 1000$ AU, and separations $\sim 3000 - 10000$ AU.

5. Impulsive interactions between these protostellar discs are frequent in the dense protocluster environment. These interactions result in frequent exchanges of matter between protostellar discs, abrupt changes in individual orbits of protostellar discs, and further fragmentation to produce additional protostellar discs.

HYPOTHESIS

However, the SPH code used for the above simulations is unable properly to resolve the internal structure and evolution of the individual protostellar discs. The combination of gravity softening and high shear viscosity (from the standard SPH artificial viscosity prescription) prohibits the inward transport of matter in the disc to form a central star.

In reality, there should be a competition between (i) the internal evolution of individual discs (which transports matter into their central stars), and (ii) impulsive interactions between discs in the dense proto-cluster environment. We hypothesize that this competition is what will determine the masses and binary statistics of newly-formed stars. In order to test this hypothesis, we have developed a new SPH code (Watkins et al. 1996) with a full Navier-Stokes treatment of viscosity and an adaptive gravity-softening.

We present here preliminary results which show that interactions between extended massive protostellar discs in wide binary systems (of the type formed in our earlier simulations) can spawn smaller discs in closer orbits, both with high efficiency, and for very general orbital parameters. Thus it seems likely that there is an hierarchical cascade converting the small number of massive wide binary systems in a small-N cluster (of the type formed in our earlier simulations) into a proliferation of lower-mass, closer systems, and thereby populating the full range of observed binary separations, eccentricities, etc.

SIMULATIONS OF DISC/DISC COLLISIONS

The simulations presented here involve parabolic collisions between two stars of equal mass, each surrounded by a disc of the same mass. The shear viscosity is characterized by a Shakura & Sunyaev $\alpha = 10^{-3}$.

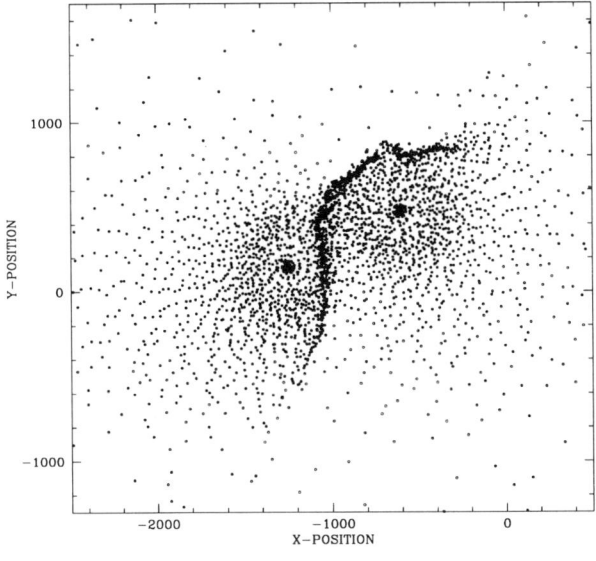

FIGURE 1.

Figure 1 shows an intermediate stage during the interaction of two coplanar discs with their spins anti-parallel, just as the shocked filament in between starts to fragment. Some of the fragments formed here are subsequently destroyed (tidally) or accreted, but many survive. In this simulation 4 additional protostellar discs are created, two of which are unbound.

Figure 2 shows a particle plot of the final frame from a simulation involving two coplanar discs with their spins and orbital angular momenta parallel. In this artificially symmetric case, each of the original primary stars has acquired a secondary with between 10 and 30 % of its own mass, and in an eccentric orbit with semi-major axis between 10 and 20 % of the periastron distance of the two original primary stars.

Non coplanar encounters are no less effective in producing new protostellar discs – but they are harder to visualize from a single particle plot.

CONCLUSIONS

We find that collisions with virtually any orientation of the spins and orbit lead to the efficient creation of new protostellar discs in closer orbits, provided that the periastron of the original collision is less than, or of order, the diameter of the discs involved. Collisions can also be very effective in accelerating the accretion of a circumstellar disc onto its central star. We conclude that it

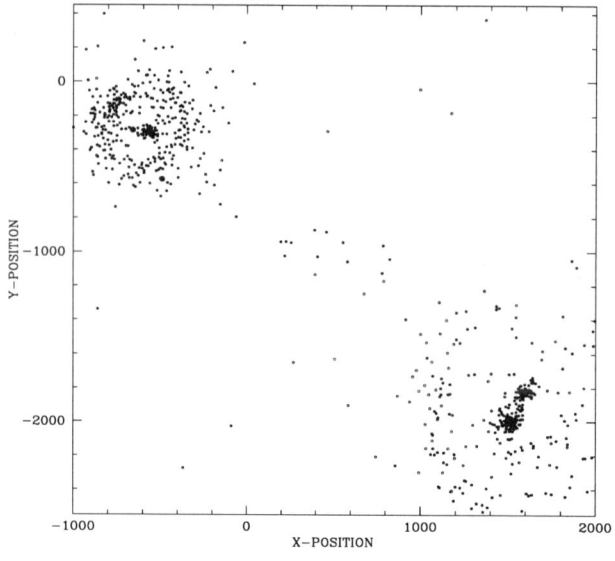

FIGURE 2.

is possible to form close binaries from wide ones, and that the process can be very efficient in a dense protocluster environment.

If this mechanism is to repeat itself in an hierarchical manner, either the time-scale on which impulsive interactions occur must be shorter than, or of order, the time-scale on which an isolated disc dissipates due to internal instabilities; *or* discs must be replenished by accretion from the surrounding medium, in the manner simulated recently by Bate & Bonnell (1996). Both these conditions seem likely to be met in typical protostellar environments.

We speculate that the peak in the binary period distribution near 200 years may be due to the fact that below this separation protostellar discs collide at speeds greater than 10 km s^{-1}; consequently the cooling of post-shock gas is much more efficient, and protostellar discs are dissipated more rapidly.

REFERENCES

1. Bate, M.R. & Bonnell, I.A., 1996, to appear in MNRAS.
2. Turner, J.A. et al., 1995, MNRAS, 277, 705.
3. Watkins, S.J. et al., 1996, A&ASS, 119, 177.
4. Whitworth, A.P. et al., 1994, A&A, 290, 421.
5. Whitworth, A.P. et al., 1995, MNRAS, 277, 727.

Disc Formation in Protobinary Systems

Matthew R. Bate

Max-Planck-Institut für Astronomie, Königstuhl 17, D-69117 Heidelberg, Germany

Abstract. The formation of circumstellar and circumbinary discs from accretion on to young binary stellar systems is investigated. When a protobinary system forms within a collapsing molecular cloud core, gas from the remainder of the cloud falls on to the system, and may form circumstellar discs and/or a circumbinary disc. We investigate how this disc formation depends on the mass ratio of the protobinary and the specific angular momentum of the infalling gas. We find that, in an unequal mass binary, the circumstellar discs may differ dramatically between the two components; in some cases a large circumstellar disc may be formed around the primary, while the secondary has no disc.

INTRODUCTION

Pre-main-sequence binary stellar systems frequently show evidence for circumstellar and/or circumbinary discs that are left-over from the star formation process [1–3]. The presence of such discs is important not only for planet formation, but also for the information it contains about the initial conditions for star formation. A binary's orbit gives an indication of the total angular momentum present in the cloud that formed a binary system. However, the presence or absence of circumstellar and circumbinary discs gives us a method by which we can examine the angular momentum *distribution* of the progenitor cloud. If there is a circumbinary disc, for example, we can determine the specific angular momentum of some of the last material to fall on to the system. Studies of the distributions of gas surrounding young multiple systems may provide important clues about the initial conditions for star formation.

Here, we discuss the formation of circumstellar and/or circumbinary discs in and around a protobinary system via accretion of the gas that remains after the binary's formation. The properties of the discs are found depend both on the mass ratio of the binary and the on specific angular momentum of the infalling gas.

CALCULATIONS

The calculations presented here were performed using a three-dimensional smoothed particle hydrodynamics (SPH) code. The protostars are modelled using sink particles [4].

We assume a protobinary system has formed within a collapsing molecular cloud core, and study how the discs that form in and around the system depend on the mass ratio q of the protobinary and the specific angular momentum of the infalling gas j_{inf}. Accretion of infalling gas also alters the mass ratio and orbit of the binary. These effects are studied elsewhere [5,6].

We define the binary's mass ratio to be $q = M_2/M_1$, where M_1 and M_2 are the masses of the primary and secondary, respectively. The units of specific angular momentum j are defined such that gas in a circular orbit around a mass $M_b = M_1 + M_2$ at a radius equal to the binary's separation a has j of unity (i.e. $j_{\text{circ}} = \sqrt{GM_b a} = 1$). Only circular binary systems are studied.

In the calculations, gas is injected around a binary system, falls on to the system, and is accreted by the binary's components or forms discs around them. The gas is injected far from the binary, with the kinetic energy it would have if it had fallen ballistically from infinity. Throughout each calculation, the infalling gas is injected with the a fixed specific angular momentum j_{inf} about the binary's centre of mass. The gas has an isothermal equation of state and is cold (i.e. the sound speed is much less than the orbital velocity of the stars). The total gas mass that is injected is very much less than the binary's mass M_b.

ACCRETION FLOWS

The discs formed in and around a binary system depend primarily on the specific angular momentum of the infalling gas. As an example, Figure 1 shows the disc formation in systems with mass ratios $q = 0.6$. A general progression is observed with increasing j_{inf}. With most mass ratios, under zero-angular-momentum infall, no discs are resolved around the protostars. When the angular momentum of the gas is increased a circumprimary disc is formed, but there is still no circumsecondary disc. Note that even if no resolved disc is formed around a protostar, it may still accrete significantly via a Bondi-Hoyle-type accretion stream directly on to the star or on to a small, unresolved disc. For still higher j_{inf}, both circumprimary and circumsecondary discs are formed, with the value of j_{inf} above which a circumsecondary disc forms depending on the binary's mass ratio. With angular momenta of $j_{\text{inf}} \approx 1$, two circumstellar discs are formed, and the formation of a circumbinary disc begins. Spiral density waves are produced in the circumbinary material via gravitational torques from the binary. At still higher j_{inf}, a region of exclusion appears near the binary where little gas is present. The infalling gas has too much angular

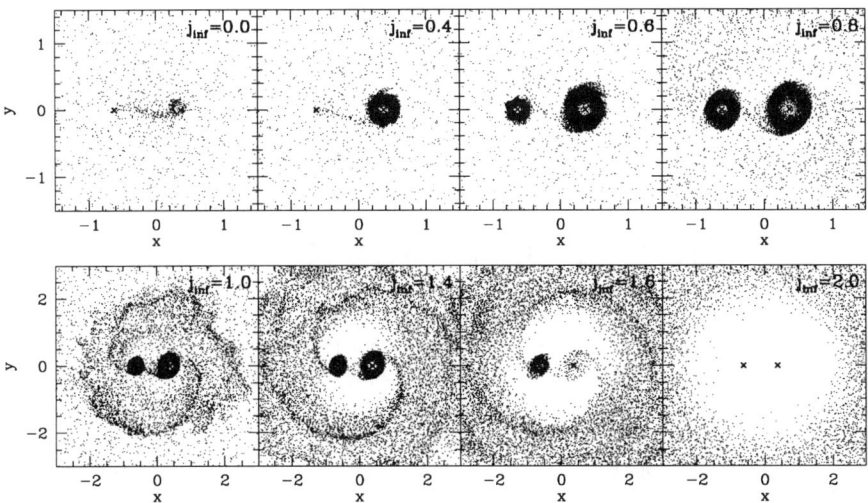

FIGURE 1. The discs formed in and around a protobinary with mass ratio $q = 0.6$ as a function of the specific angular momentum of the infalling gas j_{inf}. The positions of the protostars are marked with crosses, with the primary on the right. Distance is in units of the binary's separation.

momentum to fall directly into this region. However, accretion does proceed on to the circumstellar discs from the circumbinary material via streams of gas which lose angular momentum due to gravitational torques from the binary. The amount of material accreted by the circumstellar discs decreases as more gas goes into the circumbinary disc. Finally, for very high-angular-momentum infall, all the gas goes into a circumbinary disc.

DISC FORMATION CRITERION

For circumstellar-disc formation, the infalling material must have j_{inf} greater than the specific orbital angular momentum of the protostar that captures it so that its excess angular momentum provides spin angular momentum about the protostar. The greater the excess angular momentum of the gas, the larger the radius of the circumstellar disc. Thus, the criterion for circumstellar-disc formation is that

$$j_{\text{inf}} \gtrsim j_* \quad \text{where } j_* = \begin{cases} j_1 & \text{for the primary} \\ j_2 & \text{for the secondary} \end{cases} \quad (1)$$

where j_1 and j_2 are the specific orbital angular momenta of the primary and secondary, respectively. If the infalling gas has $j_{\text{inf}} < j_*$ then, in order to be captured, the gas gains angular momentum when it falls on to the star in a Bondi-Hoyle-type accretion stream.

For the formation of a circumbinary disc, the infalling gas must have j_{inf} roughly equal to, or greater than, the specific angular momentum required to form a circular orbit at the radius of the secondary from the centre of mass of the binary.

OBSERVATIONAL IMPLICATIONS

If a protobinary grows to its final mass via the accretion of low-specific-angular-momentum gas, the primary may have a large circumstellar disc while the secondary is essentially naked. This offers an explanation for the existence of infrared companions to optically visible T-Tauri stars [7]. The optically visible star in this scenario would be the secondary, whereas the embedded object would be the primary viewed through its edge-on circumstellar disc. Note that, if there is still infall on to the system, the secondary may even show significant accretion luminosity due to accretion from its own very small circumstellar disc or even directly from a Bondi-Hoyle-type accretion stream. The infrared companion systems formed by this method are expected to be fairly wide binaries ($\gtrsim$100 AU separation) and there should be no significant circumbinary disc due to the relatively low specific angular momentum of the accreted material.

Alternately, for a binary where a significant circumbinary disc is formed, both stars are expected to form circumstellar discs. However, the masses of these discs may differ considerably and, if the circumstellar discs evolve quickly, the secondary may still be accreting high-angular-momentum material from the circumbinary disc [8] long after the circumprimary disc has been dispersed. In this case the secondary may appear as the infrared companion.

REFERENCES

1. Strom, K. M., Strom, S. E., Edwards, S., Cabrit, S., & Strutskie, M. F. 1989, AJ, 97, 1451
2. Beckwith, S. V. W., Sargent, A. I., Chini, R. S., & Güsten, R. 1990, AJ, 99, 924
3. Dutrey, A., Guilloteau, S., & Simon, M. 1994, A&A, 286, 149
4. Bate, M. R., Bonnell, I. A., & Price, N. M. 1995, MNRAS, 277, 362
5. Bate, M. R. 1997, MNRAS, in press
6. Bate, M. R., & Bonnell, I. A. 1997, MNRAS, in press
7. Zinnecker, H., & Wilking, B. A. 1992, in Binaries as Tracers of Stellar Formation, ed. A. Duquennoy, & M. Mayor (Cambridge: Cambridge Univ. Press), 526
8. Artymowicz, P., & Lubow, S. H. 1996, ApJ, 467, L77

A Near-Infrared/Millimeter Study of Six Southern Hemisphere Star Forming Regions[1]

S. T. Megeath* and Peter K. Sollins[†][2]

*MIT Haystack Observatory
†Swarthmore College

Abstract.
We present observations of six southern hemisphere high mass star forming regions: G333.6-0.2, RCW108, RCW117, RCW122, G351.6-1.3, and NGC 6334 I. K-band observations show extensive nebulosities in all regions and stellar clusters in five of the regions. Maps of $C^{18}O$ ($J = 1 \to 0$) and HC_3N ($J = 15 \to 14$) line emission show extended moderate density gas with embedded high density cores. CO ($J = 1 \to 0$) and SiO ($J = 3 \to 2$) observations show evidence for outflows in five of the detected cores.

The most spectacular examples of outflows are in NGC 6334 I. Our observations show two molecular cores in NGC 6334, both of which contains outflows. Our K-band images show a young cluster embedded in the southern core. Fabry-Perot imaging of this cluster in the Br-γ and H_2 ($1 \to 0$) $S(1)$ lines show the ultracompact H II region NGC 6334 F and evidence for a quadrupolar outflow originating near the H II region. In sharp contrast, the northern core exhibits 70 km s^{-1} broad SiO lines, indicative of an outflow from a massive young stellar object, but it does not contain an embedded cluster. We suggest that outflows from massive stars can precede the appearance of an embedded cluster.

MOTIVATION

Some of the most prodigious examples of recent star formation in our galaxy are the dense, young clusters of high and low mass stars such as those found in the Trapezium region of Orion, in NGC 2024, or around W3 IRS 5 [1–3]. Such clusters can contain hundreds of stars, and the process of star formation in young clusters may play an important role in determining the galactic initial mass function. Despite the growing recognition of the importance of star

---

[1] Based on observations collected at the European Southern Observatory, La Silla, Chile.
[2] Supported under the Haystack Research Experiences for Undergraduates program with funding from the National Science Foundation.

formation in clusters, the process is poorly understood. Clusters are thought to form in massive, turbulently supported, molecular cloud cores. It is not clear what processes control the fragmentation of the cores, the subsequent collapse of the fragments into stars, and the resulting stellar masses.

A better understanding of cluster formation can be obtained through detailed comparisons of the properties and distribution of cluster stars with the structure of the molecular gas. For example, although embedded clusters are typically associated with molecular cores [2], it is not known whether clusters tend to appear in the centers of the cores, distributed throughout the cores, or adjacent to the cores. If star formation in clusters is regulated by ambipolar diffusion and the formation of stars occurs continuously over several million years, we might expect most clusters to be concentrated in the centers of massive cores where ionization by UV radiation from external sources is minimized [4]. In contrast, if clusters form in bursts of star formation, then the clusters should quickly disrupt the surrounding gas. In this latter case, the dense gas detected in the vicinity of most clusters would be in neighboring, dense cores which have not yet formed clusters.

OBSERVATIONAL APPROACH

To address the above questions, we have performed a near–infrared/millimeter study of six southern hemisphere high mass star forming regions. During observing runs from 1992 to 1995, we observed a sample of six regions. The infrared observations were carried out using the facility NICMOS3 camera, IRAC2, on the ESO 2.2 meter, and the millimeter data was collected at the SEST telescope. We obtained the following observations:

1. JHK–band imaging to survey the stellar population in each region.

2. $C^{18}O$ ($J = 1 \to 0$) and HC_3N ($J = 15 \to 14$) maps to locate regions with moderate density (> 1000 cm^{-3}) and high density molecular gas ($> 10^5$ cm^{-3}), respectively.

3. SiO ($J = 3 \to 2$) and CO ($J = 1 \to 0$) observations of the HC_3N peaks to search for outflows.

4. Fabry–Perot line imaging of the Brackett–γ and H_2 S(1) ($1 \to 0$) line emission in three clusters.

We targeted six high mass star forming regions: G333.6-0.2, RCW 108, RCW 117, RCW 122, G351.6-1.2 and NGC 6334 I. These regions are at heliocentric distances of 1.3 to 3 kpc and have total luminosities of $0.5 - 5 \times 10^5$ $L_\odot$. Previous 1300 μm continuum observations, which detect thermal emission from dust grains, showed dense condensation in all of these regions. Since such condensations are likely sites for ongoing star formation, we concentrated our observations on the dust condensations and their immediate environments.

CURRENT RESULTS

- K-band mosaics show bright nebulosities in each region. Clusters are apparent in five of the regions; only in G333.6 is there no clear evidence for a cluster.

- $C^{18}O$ (1 → 0) observations show extended moderate density gas. Except in the case of RCW 117, $C^{18}O$ (1 → 0) emission is detected at every position in each of the maps.

- HC_3N (15 → 14) observations show dense cores in each region. These cores are spatially coincident with the dust condensations detected in previous 1300 μm observations.

- CO (1 → 0) and SiO (3 → 2) observations of the cores show strong detections of broad wings in the NGC 6334 and RCW 122 regions and tentative detections of wings in the RCW 117 and G351.6 regions. We interpret these wings as evidence for outflows driven by massive stars.

- In at least three of the eight detected HC_3N cores, the cores peak on regions of obscuration and are clearly offset from the positions of the clusters. This indicates that the cores are sites of ongoing or future star formation while the neighboring clusters are sites of recent star formation in which most of the dense gas has been cleared.

- Using Fabry–Perot imaging in the H_2 (1 → 0) line, we have detected a quadrupolar outflow in the NGC 6334 I core (see Fig. 1). This outflow may be driven by two luminous infrared sources detected in previous 30 μm mapping [5].

- We have evidence for a powerful outflow in the NGC 6334 I(North) core, but there is no apparent infrared cluster in I(North) (Fig. 1). This result suggests that the formation of massive stars can precede the appearance of an infrared cluster.

REFERENCES

1. McCaughrean, M. J. and Stauffer, J. R. 1994, AJ, 108, 1382
2. Lada, E. A. 1992, ApJ, 393, L25
3. Megeath, S. T., Herter, T. L., Beichman, C., Gautier, N., Hester, J. J., Rayner, J., & Shupe, D. 1996, A&A, 307, 775
4. Bertoldi, F., & McKee, C. F., 1996, in Amazing Light: A Volume Dedicated to Charles Hard Townes on this 80th Birthday, ed. R. Y. Chiao (New York: Springer), 41
5. Harvey, P., & Gately, G. 1983, ApJ, 269, 613

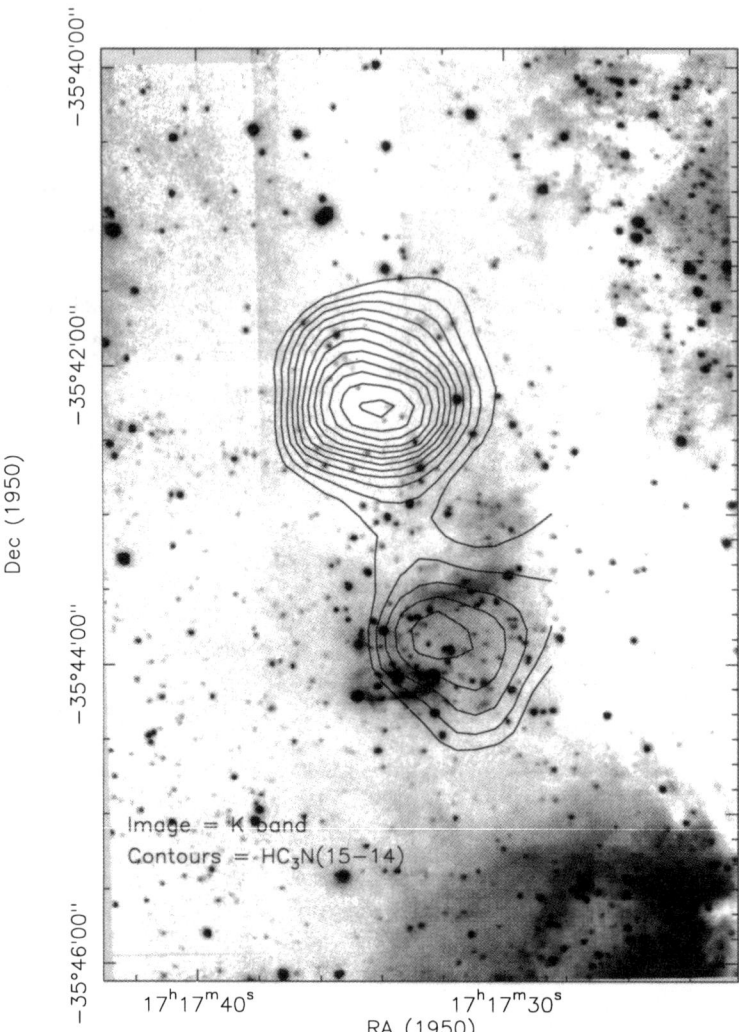

FIGURE 1. The K–band mosaic of the NGC 6334 I cluster. The HC$_3$N map, which covers a $2' \times 4'$ field, has been overlayed. The contours are the velocity integrated T$_{MB}$ of the HC$_3$N $J = 15 \rightarrow 14$ line in levels of 3 to 9 by 0.5 K km s^{-1}. We detect two distinct cores in HC$_3$N. In the southern core, or NGC 6334 I, the K–band mosaic shows an embedded clusters and the ultracompact H II region NGC 6334 F. The northern core, or NGC 6334 I(North), does not show an embedded cluster; however, our SiO and CO spectra show broad wings towards I(North). We take these wings as evidence for an outflow from a massive young stellar object; indicating that massive star formation may occur before the appearance of an embedded cluster. The object(s) producing the outflow in I(North) must be too deeply embedded to stand out in the K–band mosaic. In the lower right hand corner is the NGC 6334 II region; this region is not covered by our HC$_3$N map.

Molecular & Photodissociated Gas in the Massive Star Formation Region NGC 6334

Kathleen E. Kraemer and James M. Jackson

Boston University Astronomy Department
Boston, MA 02215

Abstract. We present mm and FIR wave spectroscopic observations of the southern star formation region NGC 6334. The cloud has been mapped in several transitions of CO, CS and NH_3. The molecular emission shows a complex structure of bubbles and filaments. The bubbles in the CO emission, though devoid of molecular gas, are not empty. Instead, they are filled with photodissociated gas, as shown by the anticorrelation between the CO emission and the [C II] 158 μm emission.

We find an anticorrelation between the 6 cm flux density from the H II regions and the intensity of the CS 7→6 emission and NH_3 (3,3) emission, which are dense gas tracers. We suggest that the continuum sources without dense gas emission have destroyed or dispersed any remnants of the dense gas from which they formed.

Introduction

NGC 6334 is a giant molecular cloud complex/star formation region which contains numerous sites of massive star formation, with inferred stellar types ranging from B9 to O6.5 [2,5,6]. The large range in age and UV field strength among the sources in the cloud makes NGC 6334 an ideal laboratory in which to study the effects of high-mass star formation on the parent molecular cloud. Here we compare the properties of the molecular gas (CO and CS) with those of the photodissociated gas (C^+) and with the 6 cm radio continuum flux.

Gas Morphology: Molecular vs. Photodissociated Gas

We mapped the emission from the CO 2→1, 3→2, ^{13}CO 2→1, CS 3→2, 5→4, and 7→6 transitions in NGC 6334. The CO (2→1) emission from NGC 6334 (Fig. 1a) shows a complex structure of bright filaments which surround dimmer 'bubbles' or holes. The high mass stars are found in the filaments, whereas the bubbles are devoid of such activity. In addition to our molec-

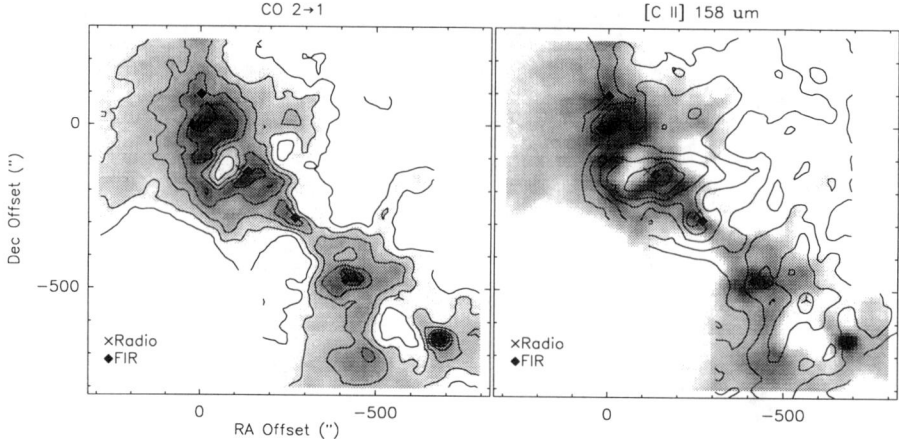

FIGURE 1. (a)CO 2→1 emission from NGC 6334; (b) [C II] 158 μm emission (contours) superposed on the CO 2→1 emission (grayscale). (0,0)=$\alpha = 17^h17^m32.3^s$, $\delta = -35°44'04''$, NGC 6334F

ular data, we have mapped the [C II] 158 μm emission toward NGC 6334. The [C II] emission, which traces photodissociated gas, is *anticorrelated* with the CO emission (Fig. 1b). There is a [C II] emission peak coincident with a CO hole between NGC 6334V and NGC 6334A. The [C II] emission extends east of continuum source NGC 6334D, into the hole in the CO emission ($\Delta\alpha, \Delta\delta \approx -75, -110$). This indicates that at least some of the bubbles in the CO emission contain photodissociated gas. The [C II] 158 μm intensities can be used to estimate minimum atomic hydrogen column densities: $N_H^{min} \approx 6 \times 10^{21}$ cm^{-2} and $\approx 8 \times 10^{21}$ cm^{-2} at the CO hole near NGC 6334 D and the FIR continuum position of D, respectively. The column density of *molecular* hydrogen (assuming a CO/H$_2$ conversion factor of $N_{H_2}/I_{CO\ 1\to0} = 2.2 \times 10^{20}$ cm^{-2} (K kms^{-1})$^{-1}$ [1]): $N_{H_2} \approx 3.7, 8.8$ and 14.1×10^{22} cm^{-2}, for the hole, the continuum position, and the CO emission peak in the ridge of molecular gas. That is, although the column densities of *atomic* hydrogen are comparable at each position, there is much less *molecular* hydrogen in the hole than in the ridge. Excitation models confirm that the molecular hydrogen column density in the hole near NGC 6334D is at least a factor of 4 less than that in the ridge. We conclude that although the hole near NGC 6334D is, indeed a *molecular* cavity, it is filled with *photodissociated* gas.

Dense Gas and Radio Continuum Flux Density

Unlike CO, which requires only modest densities ($n_{crit} \sim 10^3$ cm^{-3}) for collisional excitation, molecular gas ~ 1000 times denser is needed to populate the CS J = 7 level. Thus, the CS J=7→6 transition traces significantly denser gas than the CO lines do. We find CS 7→6 emission (Fig. 2) only at the sites

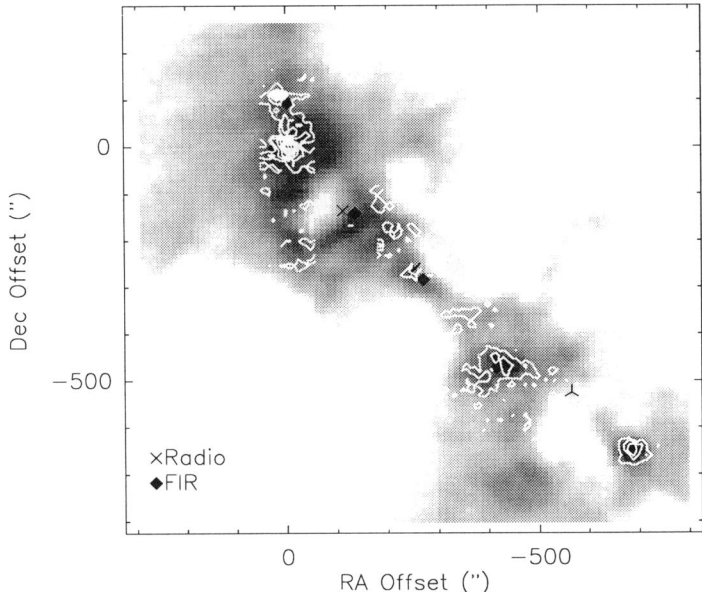

FIGURE 2. CS 7→6 emission (contours) superposed on CO 2→1 emission (grayscale)

of massive star formation in NGC 6334, unlike the CO emission, which is excited throughout the cloud (Fig. 1a).

The brightness of the CS J=7→6 emission is approximately *anticorrelated* with the 6 cm radio continuum flux density (Fig. 3). The CS J=7→6 emission near NGC 6334 C, D and E, which have the largest, brightest H II regions ($d \sim 0.2 - 0.3$ pc, $S_{6\ cm} \sim 10 - 20$ Jy; [6]), is much fainter than the emission near the smaller H II regions. Similarly, the presence of NH_3 (3,3) emission is anticorrelated with 6 cm flux density: NGC 6334 C, D and E have no detected NH_3 (3,3) emission. On the other hand, NGC 6334 V ($S_{3\ cm} \sim 10$ mJy; [3]) and I(N) ($S_{6\ cm} \lesssim 20$ mJy; [6]), which have little or no radio continuum flux, have bright, extended NH_3 (3,3) emission (Fig. 3). Further, the faint CS J=7→6 emission seen near the strong radio sources is offset from the continuum positions.

NGC 6334 C, D and E, which show little evidence for dense gas, have among the earliest spectral types in the cloud (O6.5-O8; [6,4]). We suggest that the intense UV fields from these stars may have destroyed the dense gas which once surrounded the star. The faint CS J=7→6 emission detected is spatially offset from the continuum sources because only clumps of dense gas which are sufficiently far from the stars to survive, though near enough to be heated sufficiently, will radiate detectable CS 7→6 emission.

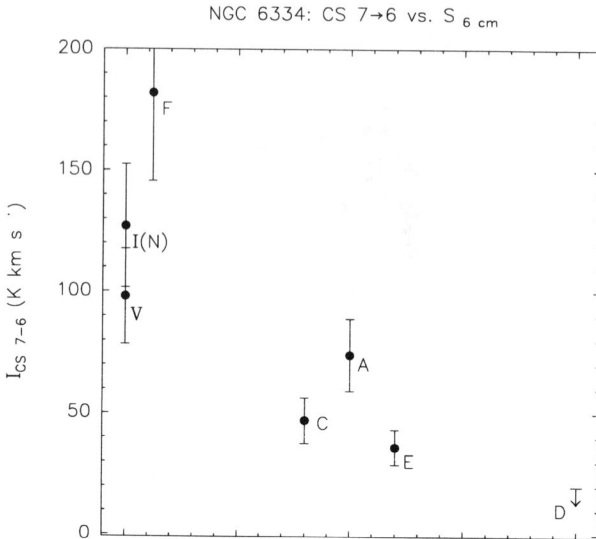

FIGURE 3. CS 7→6 integrated intensity vs. 6 cm radio continuum flux density at each site.

REFERENCES

1. Combes, F. 1991, *Ann. Rev. Astr. Astrop.*, **29**, 195 (1991).
2. Harvey, P. M. & Gatley, I., *Ap. J.*, **269**, 613 (1983).
3. Jackson, J. M. & Kraemer, K. E., in prep (1997).
4. Loughran, L., McBreen, B., Fazio, G. G., Rengarajan, T. N., Maxson, C. W., Serio, S., Sciortino, S. & Ray, T. P., *Ap. J.*, **303**, 629 (1986).
5. McBreen, B., Fazio, G. G., Stier, M. and Wright, E. L., *Ap. J.*, **232**, L183 (1979).
6. Rodríguez, L. F., Cantó, J. & Moran, J. M., *Ap. J.*, **255**, 103 (1982).

A Bright, Young Molecular Outflow near Sharpless 302

Mark Shure[1]

Center for High Angular Resolution Astronomy
Department of Physics & Astronomy
Georgia State University, University Plaza, Atlanta, Georgia 30303

Abstract. A near-infrared bright, optically invisible nebulosity was recently discovered in a dark cloud at the edge of the optical H II region Sharpless 302. Images at 1–2 μm show a bipolar geometry with two lobes separated by a dark lane, while 3–5 μm images show a nearly point source at the center of the dark lane. The red central object is coincident with the coordinates of IRAS 07299-1651, whose far-infrared spectrum implies a luminosity of 1.3×10^4 L_{sun}. ^{12}CO (J = 2–1) maps show an outflow orientation consistent with the near-infrared geometry. The presence of a nearly edge-on dust disk is further evidenced by a deep 9.7 μm silicate absorption. I argue that this represents a young massive protostar formed by the impact of the adjacent H II region on the remnant molecular cloud. An intermediate stage is suggested by a small (0.1 pc) cluster of roughly 20 stellar objects near the IRAS source.

The following three figures and captions provide a summary of this work.

REFERENCES

1. Condon, J. J., Cotton, W. D., Greisen, E. W., Yin, Q. F., Perley, R. A., & Broderick, J. J. 1996, in preparation
2. Price, S. D., Shivanandan, K., Murdock, T. L., & Bowers, P. F. 1983, ApJ, 275, 125
3. Churchwell, E. 1996, private communication
4. Shure, M., Toomey, D. W., Rayner, J., Onaka, P., Denault, A., Stahlberger, W., Watanabe, D., Criez, K., Robertson, L. and Cook, D. 1994, in Infrared Astronomy with Arrays, ed. I. McLean (Dordrecht: Kluwer), 395

---

[1] Visiting astronomer at the Infrared Telescope Facility which is operated by the University of Hawaii under contract to the National Aeronautics and Space Administration.

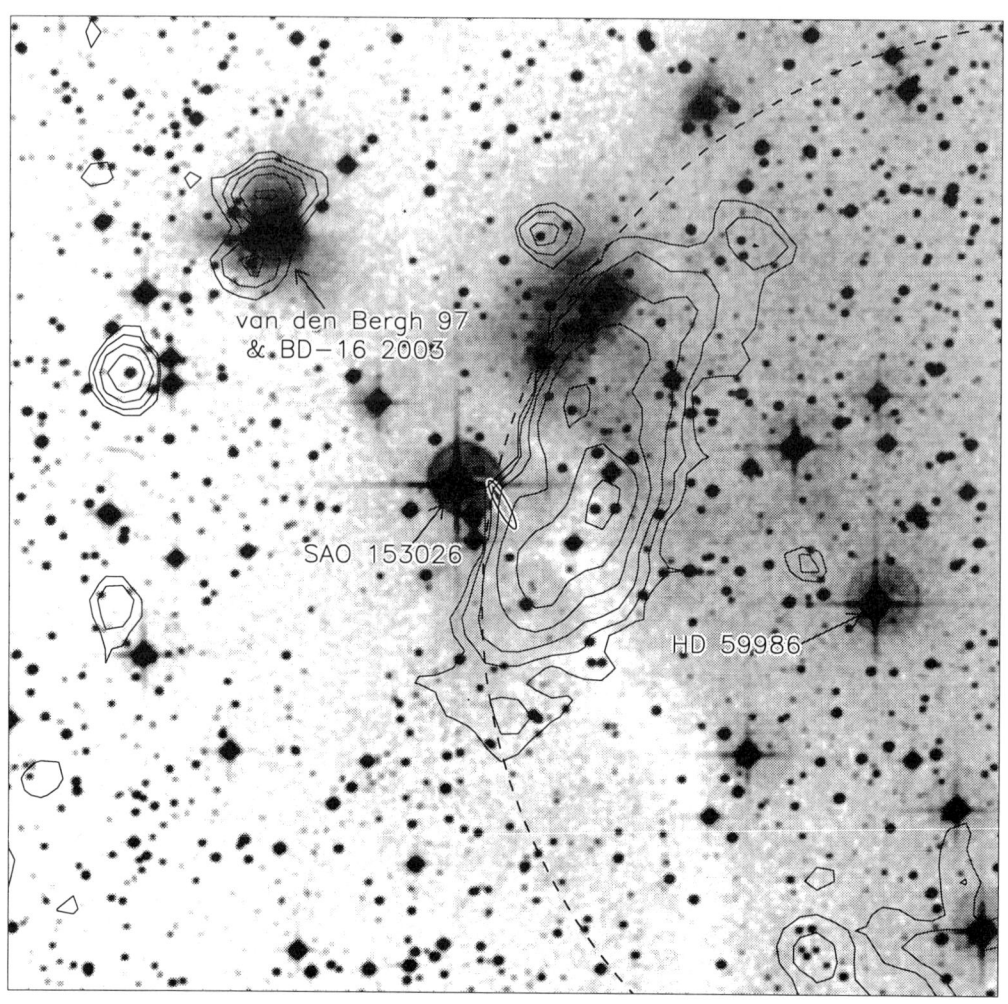

FIGURE 1. SERC Southern Sky Survey grayscale image of optical H II region Sharpless 302 and the reflection nebulosity van den Bergh 97. The field of view is $15' \times 15'$, centered on the coordinates of IRAS 07299-1651 (IRAS error ellipse shown). The solid contours represent 21 cm continuum levels (log-scaled at levels of $-3.0, -2.8, \ldots, -2.0$ in units of Jansky/beam from the NRAO VLA Sky Survey [1]). The dotted line is a circular arc with a radius of 9 arcmin, showing the approximate extent of optical emission from the H II region. The brightest of the ionizing O-stars is HD 59986 (V=9.5, O9.5 V). North is up and east to the left.

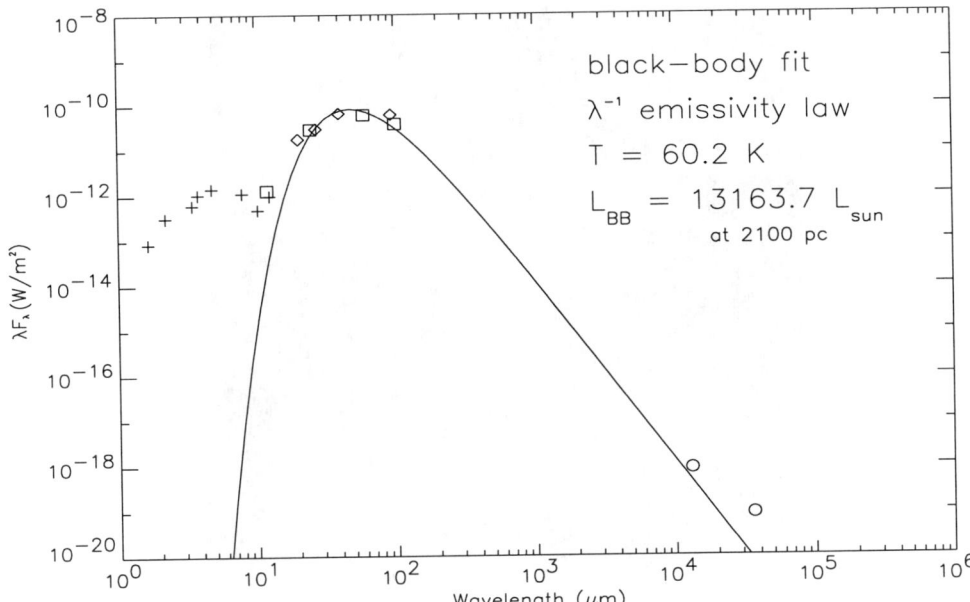

FIGURE 2. Observed fluxes of IRAS 07299-1651. Near- and mid-infrared fluxes (*pluses*) are from this study; 12, 25, 60 and 100 μm fluxes (*squares*) from IRAS Point Source Catalog; 20, 27, 40 and 93 μm fluxes (*diamonds*) from FIRSSE [2] and 1.3 & 3.6 cm fluxes (*circles*) from VLA [3]. The solid curve represents a black-body curve (with λ^{-1} emissivity) derived from a least-squares fit to the data at 20 μm and longer wavelengths.

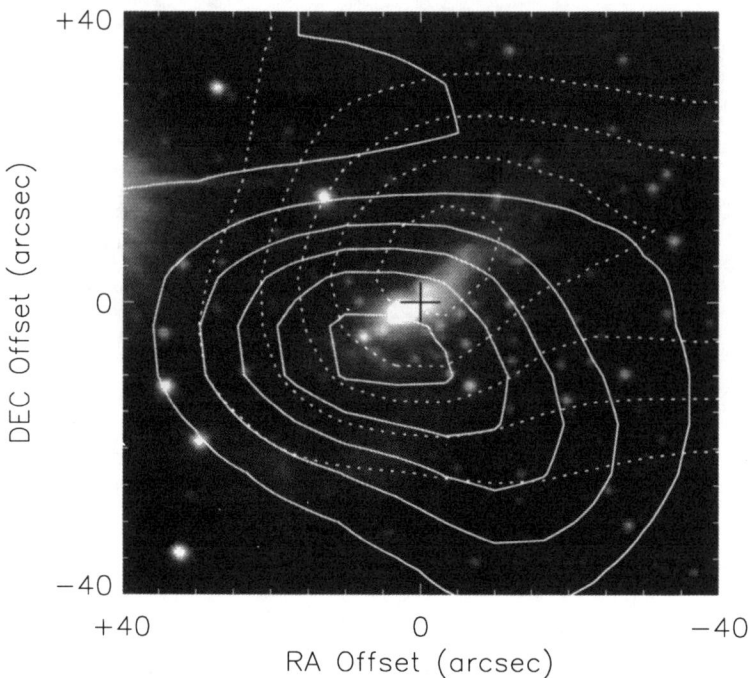

FIGURE 3. NSFCAM K (2.2 μm) image [4] of IRAS 07299-1651 overlaid with ^{12}CO (J = 2–1) integrated line flux contours (at levels of 5, 10, 15, 20 and 25 K). Dotted contours correspond to the red-shifted line wing (22.3 km/s $\leq V_{LSR} \leq$ 32.3 km/s) and the solid contours show the blue-shifted line wing (1.9 km/s $\leq V_{LSR} \leq$ 11.9 km/s). This image covers a field of 80″×80″ centered on the IRAS coordinates indicated by the "+" at $7^h29^m55^s.0$, -16°51′47″ (1950.0)). North is up and east to the left.

Evidence for Core Collapse toward 3 Young Stellar Clusters

J. P. Williams and P. C. Myers

Harvard–Smithsonian Center for Astrophysics
60 Garden Street, Cambridge, MA 02138

Abstract. Observations of CS(2–1) and N_2H^+(1–0) emission toward three young stellar clusters, Serpens NW, NGC1333–IRAS4, and L1251B are discussed. Comparison of the spectra shows that the CS is self-absorbed in many positions, and generally in an asymmetric way (the blue side of the line being of greater intensity than the red) that is characteristic of inward motions onto a warm central core. The predominance for the asymmetry to indicate infall, rather than outflow, in many positions in each core, and in all three sources, suggests that the effect is real, and that large scale core collapse is being observed. A simple one-dimensional model is described and fit to the line profiles resulting in an estimate of the mass infall rate in each case.

INTRODUCTION

The first stage in the star formation prcess is the gravitational collapse of a dense core in a molecular cloud. Although outflows from young protostellar objects have been observed for many years [1], evidence for inward motions associated with core collapse has been harder to find [2], and more controversial [3]. Only in recent years, with more sensitive, higher resolution studies, and more detailed modelling has substantial evidence accumulated [4,5].

The most tractable problem for theoretical studies has been the formation of a single, isolated, low mass star (e.g. [6]) and, consequently, this has been the focus of many observational studies. However, most stars are binaries, and form in clusters. How do observations in these cases compare with similar studies of single, isolated stars, and what does this tell us about star formation in a cluster environment?

The principal difficulty in detecting infall is one of ambiguity: how can one tell apart infall from outflow? An optically thin line would be equally broadened in each case. However, if the core has a temperature gradient, as would be expected for a core with a central protostellar object, then observations of an optically thick line can distinguish between the two possibilities [7]. If the core is static (but has a finite velocity dispersion), then emission from hot gas at the center will be absorbed by cooler gas in the outer parts of the core and the line will be symmetrically self-absorbed. If the core is collapsing, however, the absorbing foreground layer is red-shifted from the perspective of the observer, and the self-absorbtion will be asymmetric: the blue side of the line will be stronger than the red. The reverse would be true for outward motions, and, therefore, infall and outflow can be distinguished.

In practice, we observe both an optically thick and thin line: the thick line to determine the sense of the self-absorbtion, and the thin line to determine the velocity field of the core. The thin line is also used as a check to ensure that any structure in the thick line is indeed due to self-absorbtion and not, for example, multiple density components along the line of sight. Knowledge of the velocity field is essential since core rotatation, for example, can cause additional radiative transfer effects that may reverse the asymmetry in extreme cases [8].

OBSERVATIONS AND RESULTS

We surveyed CS(2–1) and N_2H^+(1–0) emission toward a number of embedded stellar clusters using the Haystack observatory in Winter 95. The three most promising candidates, Serpens NW, NGC1333–IRAS4, and L1251B were then mapped at the FCRAO 14 m telescope in Spring 96. Both lines have similar critical densities, $n \simeq 10^{4-5}$ cm^{-3}, and have brightness temperatures as high as a few Kelvin. The optical depth of the N_2H^+ line can be determined from its hyperfine structure, and is typically less than one. The CS emission is optically thick and is found to be self-absorbed along several lines of sight. A single Nyquist sampled footprint ($250'' \times 300''$) at $25''$ resolution was made toward each source in each line. Due to space limitations, we discuss only the Serpens observations here. The other two clusters have similar (but somewhat less pronounced) features.

The CS spectra are shown in the left panel in Figure 1. Where N_2H^+ emission was detectable, its velocity is shown by a vertical line. Many CS spectra are self-absorbed: visible most clearly as a dip in some spectra, and more often as a red (high velocity) shoulder at the velocity of the N_2H^+ line.

Depending on the degree of asymmetry in the CS line, a single or double gaussian fit has been made, and the velocities compared with the N_2H^+. Considering the double gaussian fits only, we find that the intensity of the blue peak is systematically higher than the red, and that (right panel of Figure 1) there is a velocity offset between the blue gaussian and N_2H^+ velocity that is most prominent around the Class 0 source, S68N [9]. The region of both strongest CS self-absorbtion and velocity asymmetry is extended, but localized. In contrast to the observed, collimated outflow around S68N, it has an amorphous shape. The preponderance of self-absorbed profiles for which the blue peak is greater than the red cannot be explained by either a static or an expanding core, and is strongly suggestive of the presence of large scale, ordered, inward motions.

We define an infall "zone" for each source where $[v_{\rm thin} - v_{\rm thick}]/\Delta v_{\rm thin} > 0.25$, from which we calculate an infall radius, $r_{\rm in}$. CS and N_2H^+ spectra are then averaged over this zone and fit by a simple one-dimensional model [10]. Figure 2 shows the average spectra and model fits for the infall zone around S68N: the fit is narrower than the observations because the radiative transfer is calculated only at the $\tau = 1$ surfaces, which are assumed to be infalling at a speed, $v_{\rm in}$.

From $v_{\rm in}$ and $r_{\rm in}$, we calculate a dynamical mass infall rate, $\dot{M}_{\rm dyn} = 4\pi\rho r_{\rm in}^2 v_{\rm in}$ where ρ is the average density of the infalling gas, here taken to be 10^4 cm^{-3}, which can then be compared to the gravitational infall rate expected for the inside-out collapse of an isothermal sphere [11], $\dot{M}_{\rm grav} = \sigma^3/G$, where σ is the (N_2H^+) velocity dispersion of the gas and G the gravitational constant. These four values for each source are tabulated in Table 1. Given the uncertainties, particularly for $v_{\rm in}$, the agreement between the two mass infall rates to within $\sim 50\%$ is good, and offers further support for the hypothesis that the cores are undergoing gravitational collapse.

REFERENCES

1. Lada, C.J., 1985, A.R.A.A., 23, 267
2. Walker, C.K. et al., 1986, Ap.J., 309, L47
3. Menten, K.M., Serabyn, E., Güsten, R., & Wilson, T.L. 1987, A. A., 177, L57
4. Zhou S., Evans, N.J., Kömpe, C., & Walmsley, C.M. 1993, Ap.J., 404, 232
5. Myers, P.C. et al. 1995, Ap.J., 449, L65
6. Shu, F.H., Adams, F.C., Lizano, S., 1987, A.R.A.A., 25, 23
7. Leung, C.M., & Brown, R.B. 1977, Ap.J., 214, L73
8. Zhou, S. 1995, Ap.J., 442, 685
9. McMullin, J.P. et al. 1994, Ap.J., 424, 222
10. Myers, P.C. et al. 1996, Ap.J., 465, L133
11. Shu, F.H. 1977, Ap.J., 214, 488

TABLE 1. Mass Infall Rates

Source	$v_{\rm in}$ km s^{-1}	$r_{\rm in}$ pc	$\dot{M}_{\rm dyn}$ $10^{-6} M_\odot$ yr^{-1}	$\dot{M}_{\rm grav}$ $10^{-6} M_\odot$ yr^{-1}
Serpens NW	0.03	0.05	4.8	7.8
NGC1333–IRAS4	0.06	0.07	19.0	27.0
L1251B	0.07	0.05	11.0	8.6

Serpens NW

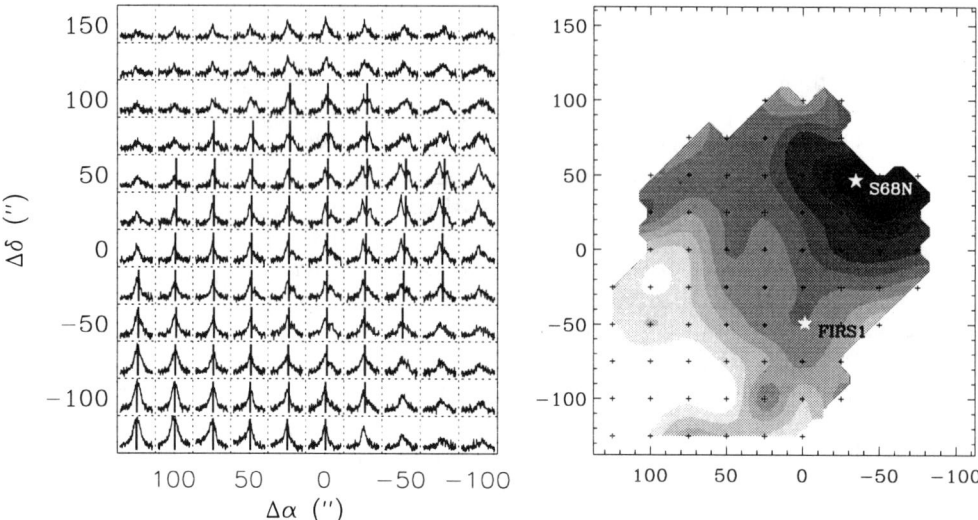

FIGURE 1. CS and N_2H^+ observations toward Serpens. The left panel show the CS spectra with a vertical line superimposed representing the N_2H^+ central velocity (where it was detected). The right panel maps the velocity difference between the blue peak of the CS line and the N_2H^+ velocity. The grayscale ranges from 0, in white, to one N_2H^+ linewidth (= 1.1 km s^{-1}) in black.

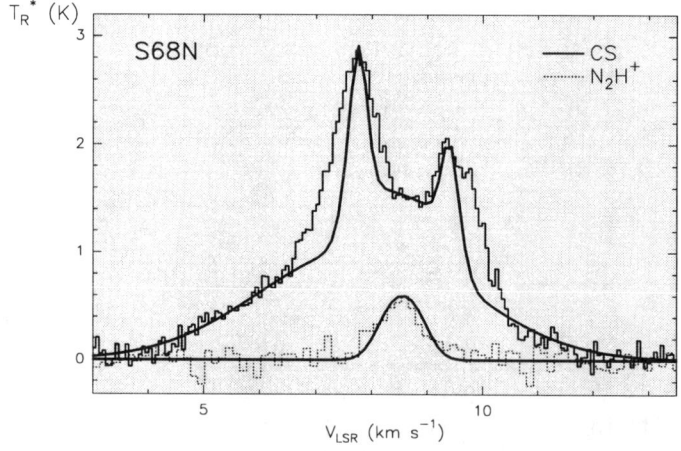

FIGURE 2. Average CS and N_2H^+ spectra around S68N. A model fit, including both outflow and infall is shown. The main difference between the CS and N_2H^+ spectra is the outflow and the optical depth of the infalling gas.

VLA Ammonia (3,3) Observations of Heated and High Velocity Gas in Orion-KL

Jennifer J. Wiseman,*[†] Mary E. Putman,*[‡] and Paul T. P. Ho[†]

*National Radio Astronomy Observatory[1]
520 Edgemont Road, Charlottesville, VA 22903
[†]Harvard-Smithsonian Center for Astrophysics
60 Garden Street, Cambridge, MA 02138
[‡]Department of Astronomy
University of Wisconsin, Madison, WI 53706

Abstract. We present a high resolution VLA mosaic of ammonia (3,3) emission in the central region of OMC-1. Previous maps of (1,1) and (2,2) emission revealed dense filaments of molecular gas fanning out from the core [1]. The (3,3) emission is sensitive to gas heated in regions of outflow and radiation [2] and thus highlights areas where star formation activity is affecting the surrounding molecular filaments. Four fields were combined to cover a $3' \times 3'$ region. The mosaic reveals heated gas near the base of the outflow and also along the eastern filament to distances as far as $2'$ (0.3 pc) from the central source. A high velocity blue-shifted flow is found with its source near the central "hot core". "Hot spots" near the base of the filaments show where this high velocity gas leaves the central region along paths of lower optical depth surrounding the filaments. Thus very dense gas is being heated and accelerated in the OMC-1 outflow. The high spatial resolution of the mosaic reveals detailed heating effects from the outflow and radiation over a large area.

LARGE SCALE MOSAICS OF OMC-1

The Orion-KL core region of the Orion Molecular Cloud (OMC-1) is an active and heavily studied site of high mass star formation. As a regular stage of this process, bipolar outflows and radiation carry energy away from the young sources and into the surrounding dense molecular cloud core. However, detailed studies of the interactions between star formation and the surrounding environment on a large scale have only become feasible with improvements in

---

[1] NRAO is a facility of the National Science Foundation operated under cooperative agreement by Associated Universities, Inc.

receiver and software capabilities. We have recently created such a large scale map in OMC-1 by combining several fields into mosaics of NH_3 (1,1) and (2,2) inversion transition emission covering a large 0.5 parsec region of the OMC-1 ridge surrounding Orion-KL [1]. The maps revealed long filaments of dense molecular gas fanning out from the central core region. Heated regions along the edges of the filaments in the path of the northern lobe of the main OMC-1 bipolar outflow show a direct impact of the outflow on the surrounding environment.

Here we present a similar mosaic of four VLA fields from the central region around Orion-KL, but for this study we have employed the NH_3 (3,3) transition. The (3,3) line is sensitive to heated gas ($\sim > 50$ K) and has also been seen to be excited in regions of warm higher velocity gas within outflows [2]. By mapping the (3,3) line emission at high spatial resolution, the relationship between the dense gas of OMC-1 and the heating and outflows from young sources within the central core can be better understood.

Four fields were observed using the "D" configuration of the VLA. The maps were naturally weighted, yielding a synthesized beamwidth of $7'' \times 6''$. The channels were separated in velocity by 1.2 km s^{-1}. A linear mosaic was created for each channel; the integrated emission map of Figure 1 was then created by combining channels covering the central component of the line transitions. The individual central field was used for kinematical maps of the central core.

RESULTS AND DISCUSSION

Several new findings of Orion-KL activity are made evident in these (3,3) maps. The shades of Figure 1 show the ratio of the (3,3) emission to (1,1) emission in the 4-field mosaic. The contours outline the integrated intensity of the (1,1) line and show the array of molecular filaments based in the central core. It is evident that regions are heated enough to excite (3,3) emission not only in the central core region but also to distances of at least 1' (0.15 pc) north along the eastern (leftmost) molecular filament. Also evident are bright "hot spots" of heated gas around the edge of the core. The (3,3)/(1,1) ratio is saturated within regions of high optical depth in the central core and thus remains at a value of 1 in that region, but as the optical depth drops off around the core edges, hot regions show up as high ratios [3]. These regions show where the outflow and perhaps radiation from the central core region are escaping through regions of lower optical depth around the molecular filaments, heating the gas. Dramatic heating is apparent on the southern edge of the first core to the north of Orion-KL, known as "CS1" [4], in the eastern filament at a declination of $-05°24'15''$. It appears that CS1 is in the path of the outflow and is actively heated by it, as suggested by Wright, Plambeck, and Wilner [5].

We also find a striking lobe of blueshifted ammonia (Figure 2). A strong

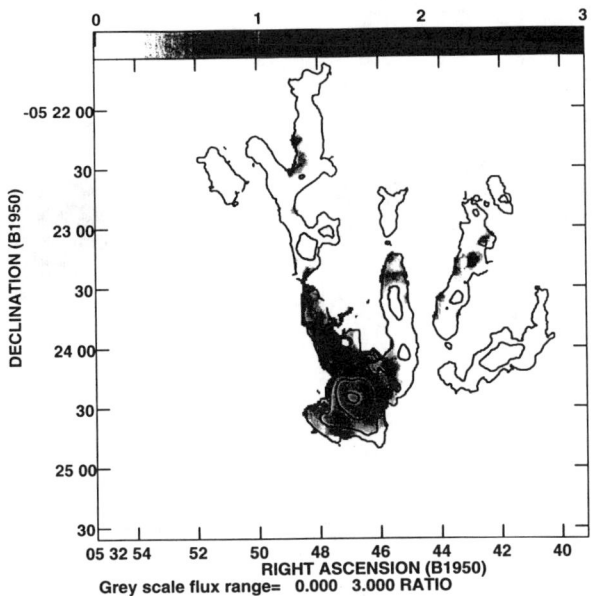

FIGURE 1. A mosaic of OMC-1. The shades portray the ammonia (3,3)/(1,1) integrated intensity ratio, which is related to the temperature and optical depth of the region. The coutours represent the integrated intensity of the (1,1) line. The most heated regions are within the central core (where the ratio saturates to a value of 1) and along the eastern filament to the north, including the edge of "CS1". The brightly shaded "spots" are regions of hot gas where the optical depth of ammonia has fallen at the edge of the denser hot core. These may show the heated paths by which outflowing gas leaves the central core.

blue-shifted feature centered around the 5 km s^{-1} "hot core" source is seen to extend to velocities of -10 km s^{-1}, a 15 km s^{-1} blue-shift. A redshifted counterpart is present but less clear. An overlay of the blueshifted and redshifted gas is seen in Figure 3. The blueshifted gas is similar in structure to that seen in CO J=1-0 with BIMA maps [6]. The outflow axis appears to lie largely along the line of sight with overlapping redshifted and blueshifted lobes. The blueshifted gas, however, extends to the northwest (right) of the redshifted gas in Figure 3, and comparison with Figure 1 reveals that a "hot spot" of high temperature and low optical depth lies at the same position as this western extention of blueshifted gas. Thus these high resolution observations reveal that gas dense enough to excite ammonia ($n = 10^4$ and higher) is being accelerated to high velocity within the outflow, and that the high velocity gas appears to escape from the central core through regions of lower optical depth where dramatic heating results.

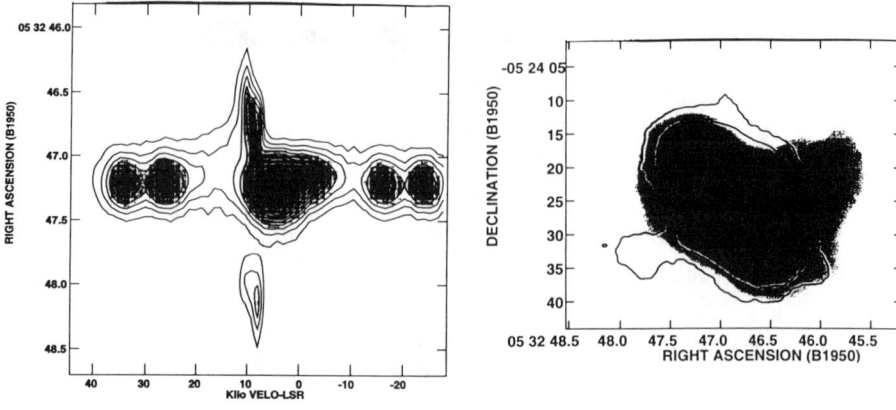

FIGURE 2. A position-velocity cut along the axis of the blueshifted lobe of the OMC-1 outflow, at position angle of 45 degrees west of north. A strong blue-shifted feature based around the 5 km s^{-1} "hot core" region is seen to extend to velocities of -10 km s^{-1}, a 15 km s^{-1} blue-shift. (The two peaks on each spectral side of this central feature are satellite lines from the NH$_3$ hyperfine transition structure.) Quiescent clumps of gas from the 8 km s^{-1} filaments which cross this core region are apparent on either side of the accelerated gas feature.

FIGURE 3. An overlay of the integrated high velocity redshifted gas (in contours) with the high velocity blue-shifted gas (in shades). The two components largely overlap in projection. The blueshifted gas extends to the west (right), where it overlaps with heated regions at the edge of the central core region, possibly where the high velocity gas exits (Figure 1).

REFERENCES

1. Wiseman, J. J., & Ho, Paul T. P. 1996, Nature, 382, 139
2. Bachiller, R., Martín-Pintado, J., and Fuente, A. 1993, ApJ, 417, L45
3. Ho, P. T. P., and Townes, C. H. 1983, ARAA, 21, 239
4. Mundy, L. G., Scoville, N. Z., Baath, L. B., Masson, C. R., and Woody, D. P. 1986, ApJ, 304, L51
5. Wright, M. C. H., Plambeck, R. L., and Wilner, D. J. 1996, ApJ, 469, 216
6. Chernin, L. M., & Wright, M. C. H. 1996, ApJ, 467, 676

High-Spatial Resolution Imaging of the NGC 2024 Molecular Ridge

M. W. Pound (UC Berkeley), R. Gruendl (U. Ill.), E. A. Lada (U. Fl.), and L. Mundy (U. Md.)

Abstract. We present ^{13}CO(1-0) and C^{18}O(1-0) images of the NGC 2024 molecular cloud made with the Berkeley-Illinois-Maryland Association (BIMA) array.

OBSERVATIONS

The interferometric data were taken in 1994 and 1995 with the BIMA array in Hat Creek, CA. We observed 5 fields along the NGC 2024 ridge, observing simultaneously the 110 GHz transition of ^{13}CO(1-0) and the 109 GHz transition of C^{18}O(1-0). We used a mosaicing technique in which each field is observed in succession such that the time for a full cycle over all fields is less than the time for the longest baseline to rotate into an new u-v cell. This "dwell time" is typically 2 to 4 minutes per field. Phase and amplitude were calibrated by observing the quasar 0530+135 at approximately 20 minute intervals. In addition, we obtained the calibrated data for two extra fields in C^{18}O(1-0) from previous work with BIMA by Wilson et al. (1995).

The single dish data were taken with the NRAO 12 m at Kitt Peak in November 1994 and May 1996. The half-power beam width of the NRAO 12 m at these frequencies is about 57″, and the single dish spectra were spaced on a 30″ grid. The interferometric and single dish maps were combined (Figures 2 and 3) using a non-linear maximum entropy technique which simultaneously deconvolves adjacent pointings and also allows for the inclusion of single dish data (Cornwell 1988). The single dish data give the "zero spacing" information missing from the interferometer data. Since the NGC 2024 molecular ridge has significant extended emission, the zero spacing data are critical to proper interpretation of the maps.

RESULTS

The molecular gas is concentrated in a north-south ridge with several emission peaks roughly 10″ (0.02 pc) in size. The typical ratio of integrated intensity I(^{13}CO)/I(C^{18}O) in the ridge is about 3 (Figure 1). A similar ridge with several intense peaks is seen in $\lambda=1.3$ mm dust continuum maps made by Mezger et al. (1988). The dust continuum peaks are also seen in the BIMA maps at $\lambda=2.7$ mm of Wilson et al. (1995). (The $\lambda=2.7$ mm continuum sources are sufficiently weak that they do not contaminate the spectral line maps; we have not subtracted them from our maps). The $C^{18}O$ emission peaks do not, in general, coincide with the dust peaks (the stars in the figures). If, as asserted by Mezger et al., the continuum peaks are isothermal protostars embedded in cold (T$\sim$ 20 K), dense envelopes, the $C^{18}O$ could be depleted in these regions, "frozen out" onto dust grains. However, in this case the $C^{18}O$ emission should surround the dust peaks, with a minimum in the center. We do not observe this; rather, the most intense continuum peaks appear at the edges of the $C^{18}O$ clumps. So the continuum peaks may be embedded young stars rather than protostars, in which case the dust is much warmer than 20 K and the H_2 column density proportionally lower. This is certainly true for FIR 5 and FIR 6 which have associated outflows and thus must be evolved past the protostellar phase. In this case, the $C^{18}O$ emission we see is the remnant of the parent cloud. Finally, we note that opacity may play a role in the apparent morphology of the $C^{18}O$ emission. We will address this in a future paper.

REFERENCES

1. Wilson, T., Mehringer, D., & Dickel, H. 1995, A&A, 303, 840
2. Mezger, P., Chini, R., Kreysa, E., Wink, J., and Salter, C. 1988, A&A, 191, 44
3. Cornwell, T. J., 1988, A&A, 202, 316

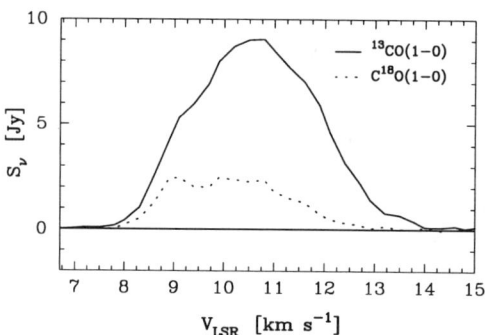

FIGURE 1. Two typical spectra. The relatively high ratio of peak fluxes, $T(^{13}CO)/T(C^{18}O) \sim 3$ indicate that ^{13}CO is optically thick in most regions of this cloud.

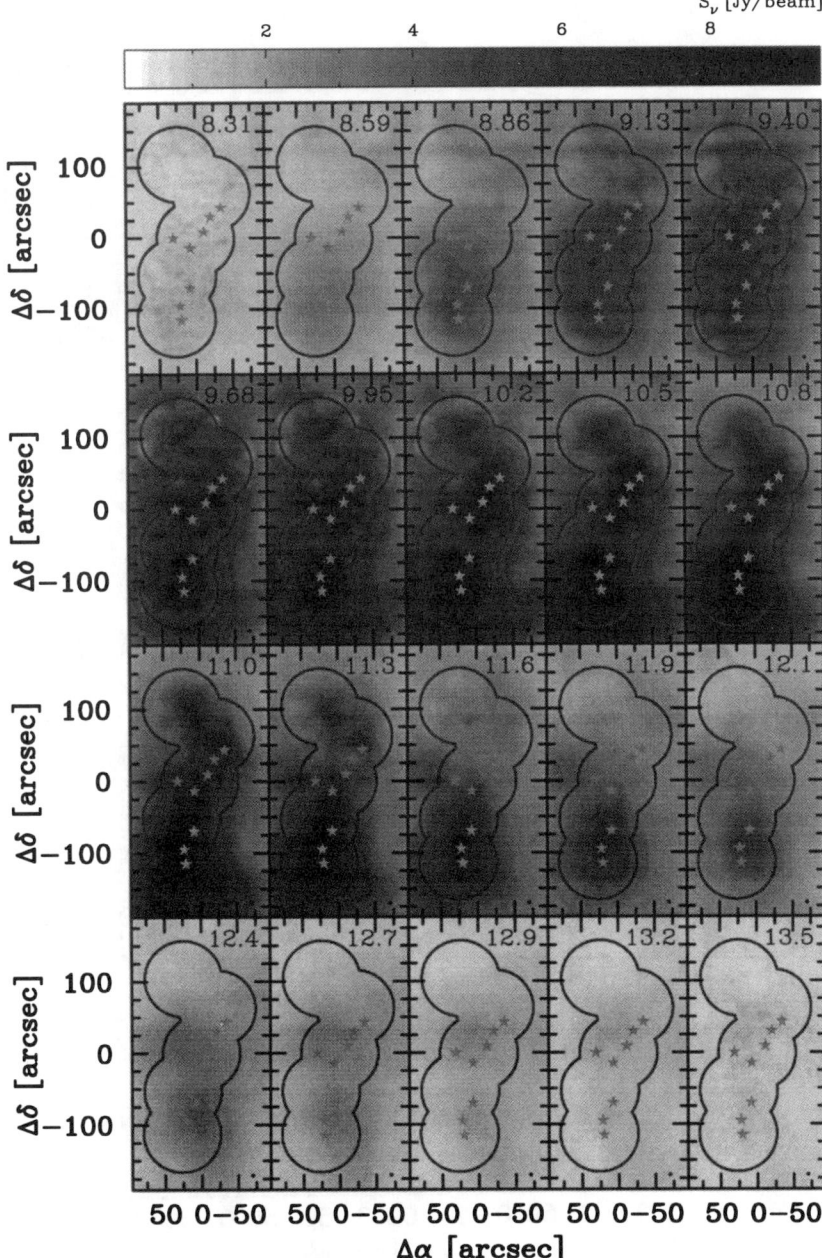

FIGURE 2. ^{13}CO(1-0) channel maps made from combined interferometric and single dish data. The BIMA fields (106″ primary beam patterns) are indicated by open circles. Channel velocity is indicated in the upper right of each panel. The eight far-infrared sources are indicated with stars. The BIMA synthesized beam (lower right corner of each panel) is 5″, corresponding to 0.012 pc.

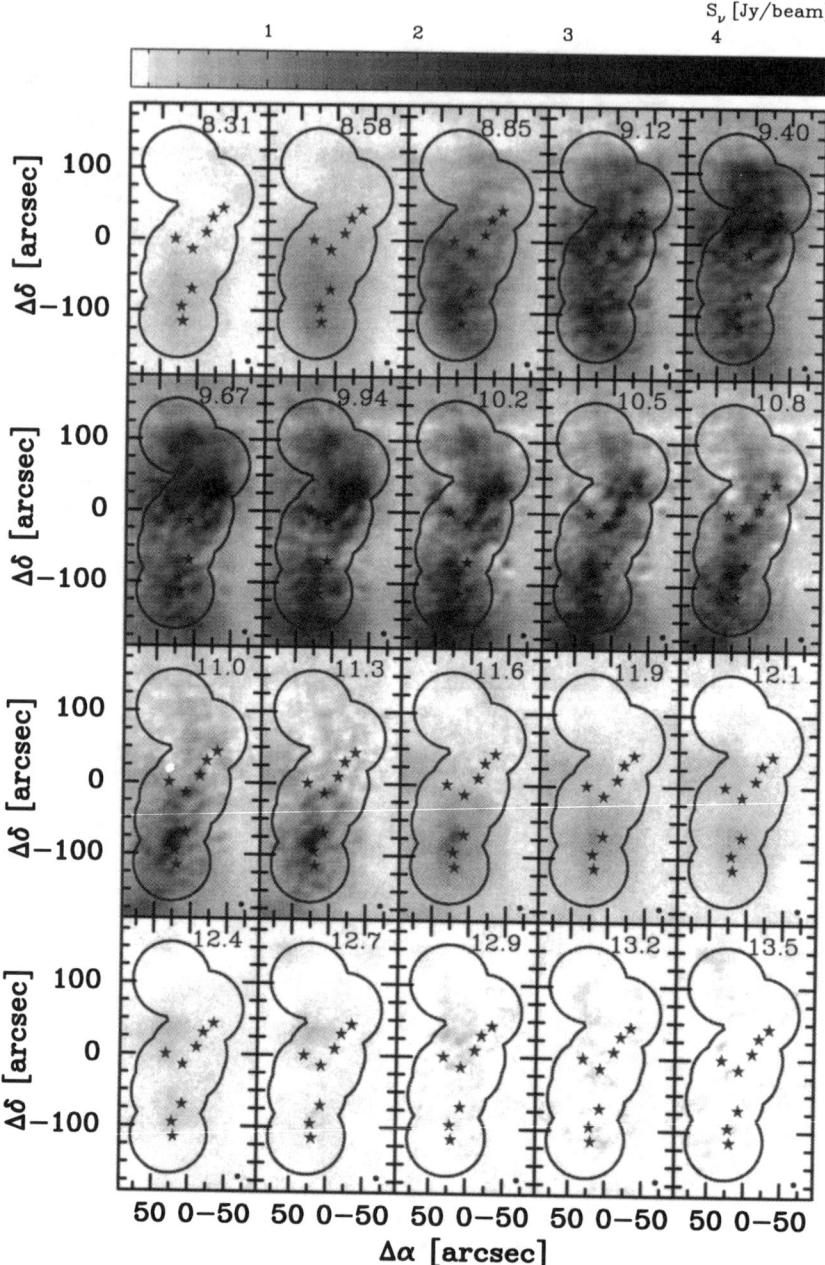

FIGURE 3. $C^{18}O(1-0)$ channel maps made with combined data as in previous Figure. Note the $C^{18}O$ peaks do not in general coincide with the FIR sources. Interestingly, the FIR sources all seem to lie to the west of the molecular ridge. The synthesized beam is 7″, corresponding to 0.016 pc.

Organization of the Magnetic Field in W3(OH) on Fine Spatial Scales

E. E. Bloemhof[*]

[*] *California Institute of Technology*
Pasadena, California 91125

Abstract. Multi-epoch spectral-line VLBI measurements of H_2O maser proper motions have provided kinematic information in several regions of recent star formation on unprecedented spatial scales. Recently, the first measurements of OH maser proper motions have been made in the source W3(OH), and these have provided detailed information about the distribution of the magnetic field strength on fine spatial scales, in addition to the usual three-dimensional kinematics probed by H_2O masers. On a two-dimensional map of the source, the measured field strengths show large variations over small distances, with no apparent pattern. I present here a regularity in the magnetic field distribution for some of the spots that takes the form of a strong correlation of field strength with maser spot z-velocity. This correlation constitutes progress towards a simple understanding of the magnetic field distribution, and might be useful as an additional constraint on the kinematic modelling of maser-emitting sources.

INTRODUCTION

Spectral-line very-long-baseline interferometry (VLBI) has produced some very exciting results when applied to the study of the kinematics of regions of current or recent star formation, using maser-emitting gas as high-intensity markers. The extremely high resolution of the technique permits the study of spatial scales down to the sub-tenth-arc second level, and correspondingly high resolution is achieved in radial and (in multi-epoch proper motion experiments) transverse velocities. A number of H_2O maser sources have been studied in this way, yielding information about source kinematics and distance.

Little work has been done as yet with other maser species. However, an interesting byproduct of synthesis mapping of OH maser emission is the local magnetic field strength in individual maser-emitting gas clumps. Since the magnetic field plays a pivotal role in star formation, these measurements are of great interest in providing a unique opportunity to unravel the detailed field structure in the immediate vicinity of a very young object.

KINEMATICS AND MAGNETIC FIELD DISTRIBUTION OF W3(OH)

The ultracompact HII region W3(OH) is the brightest OH maser source, and the only source to date in which OH maser proper motions have been measured [1]. OH maser emission is strongly circularly polarized, and the maps of left-circularly- and right-circularly-polarized emission may be combined by assuming the velocity shift seen between them is caused by Zeeman splitting in a magnetic field (this appears to be an excellent assumption). The result is a map of 23 "Zeeman-corrected" maser-emitting spots with well-defined radial ("z") velocities to accompany the transverse proper velocities, and a set of local values for the total magnetic field strength derived from the magnitude of the Zeeman splitting (Figure 1). An inspection of Figure 1 shows immediately that the magnetic field strength exhibits sharp variations over small regions of the sky.

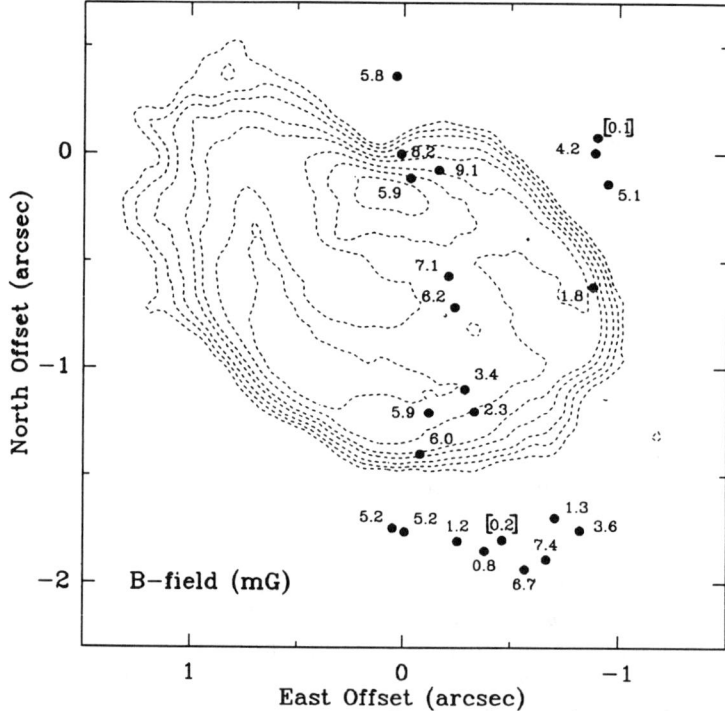

FIGURE 1. Map of individual OH maser spot positions distributed over W3(OH), with corresponding measurements of the local magnetic field strength in milliGauss. Contours are 15 GHz radio continuum emission from the underlying ultracompact HII region (reproduced from reference [1]).

B VERSUS V_Z CORRELATION

A regularity in the apparently random pattern of magnetic field strengths is found by examining the 9 maser spots lying off the southern edge of the HII region. The spatial variations in field are as extreme here as anywhere in the source. However, a plot of field strength versus maser spot z-velocity (Figure 2) reveals a clear trend. More formally, the correlation coefficient is 0.78 which, for 9 spots, implies that there is a relation between the magnetic field and the z-velocity at roughly the 99% confidence level.

This correlation may arise rather naturally from a smooth variation of magnetic field in the z-direction within the source, since the z-velocity for many simple flows is a function of the radial position within the flow. This line of argument indicates that magnetic field measurements might be used to provide an indirect indication of the maser clumps' radial positions, which are an important component of maser source structure not normally measurable.

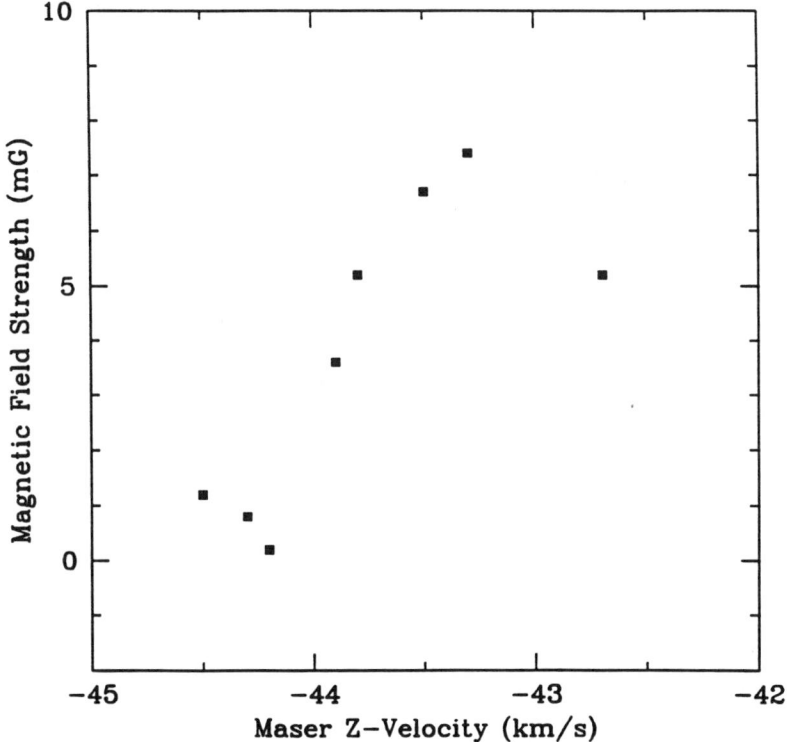

FIGURE 2. Plot of local magnetic field strength versus radial (z-) velocity of maser features in W3(OH), including only the 9 maser spots lying off the southern edge of the ultracompact HII region in Figure 1.

CONCLUSIONS

This preliminary indication of regular structure in the magnetic field distribution of so young an object may provide useful information about the detailed role played by the magnetic field in the process of star formation, a role that is expected to be vitally important. Many OH maser sources suitable for VLBI study exist, and it will be interesting to see if regular field structure is common.

No clear correlation of B and v_z is found for the maser spots located against the HII region in projection or for those lying off the northern edge of the HII region. In the cometary bow-shock model of W3(OH) [2], the southern maser spots for which the regular magnetic field structure is found are very near the stagnation point, or leading edge, of the cometary flow around the HII region; it is tempting to speculate that contact with the bow shock and flow around it have complicated an initially smooth magnetic field structure. Perturbation of maser gas densities would be one obvious mechanism, for magnetic flux frozen to the gas.

A more subtle future application of the results presented here is constraining source modelling for maser proper motion experiments. A large element of uncertainty in such modelling is introduced by the fact that the radial positions of individual maser spots are unknown; but if the source magnetic field is spatially smooth in three dimensions, a crude model of that field and the requirement of continuity of magnetic field values might significantly simplify the solution of the full three-dimensional source kinematics.

ACKNOWLEDGMENTS

I thank Mark Reid, Jim Moran, and Carl Gwinn for helpful discussions on the topics presented here.

REFERENCES

1. Bloemhof, E. E., Reid, M. J., & Moran, J. M. 1992, *ApJ* **397**, 500.
2. Bloemhof, E. E. 1993, *ApJ* **406**, L75.

Observational Evidence for the Present-Day Formation of Globular Clusters

Luis C. Ho* and Alexei V. Filippenko[†]

*Harvard-Smithsonian Center for Astrophysics
[†]Department of Astronomy, University of California at Berkeley

Abstract. We present evidence that some of the compact, luminous, young star clusters recently discovered through images taken with the *Hubble Space Telescope (HST)* have masses comparable to those of old Galactic globular clusters. Three "super star clusters" in the centers of the nearby dwarf galaxies NGC 1569, NGC 1705, and NGC 4214 have been observed with the HIRES echelle spectrograph on the Keck 10 m telescope to measure the velocity dispersion of the stars. The velocity dispersion was measured successfully for the clusters in NGC 1569 and NGC 1705, which, when combined with the size estimates obtained from *HST* images, imply that the clusters have very large dynamical masses. The masses, mass densities, and predicted mass-to-light ratios (at $t = 10$–15 Gyr) of these two 10-Myr old clusters closely resemble those of the majority of evolved Galactic globular clusters. We interpret the results as evidence that these objects are genuinely young globular clusters.

THE UBIQUITY OF "SUPER STAR CLUSTERS"

Images from the *Hubble Space Telescope (HST)* have revealed a class of luminous, compact, young star clusters in a variety of extragalactic environments ranging from interacting galaxies (e.g., NGC 1275, [4]; the "Antennae," [10]; the "Atoms for Peace" galaxy, [11]) to classical starburst systems (e.g., M82, [7]), dwarf irregulars, and circumnuclear rings in barred galaxies (see [1] for a review). Because of their high luminosities (M_V up to –15 mag), compact sizes (half-light radii ≤ 5 pc), and blue colors, there has been speculation that these "super star clusters" may be present-day counterparts of young globular clusters, although some believe that they are more closely related to open clusters (e.g., [9]). The most pertinent piece of information needed to test the young globular cluster hypothesis is the mass of the clusters. On average, globular clusters are much more massive than open clusters, and, because they are so compact, they are also denser. Up to now, however, no direct, model-independent mass estimates exist for super star clusters. Masses quoted in the literature invariably rely on population synthesis calculations to convert the observed luminosities to masses, and the uniqueness and reliability of the models are difficult to judge given their dependence on a large number of poorly constrained parameters. The most straightforward way to obtain a robust mass estimate is through dynamical measurements.

MEASUREMENT OF VELOCITY DISPERSIONS

Using the HIRES spectrograph on the Keck 10 m telescope, we acquired high-dispersion (echelle) optical spectra of one of the two clusters in NGC 1569 (NGC 1569-A; [6]), and of the single dominant cluster in NGC 1705 (NGC 1705-1; [6]) and in NGC 4214 (NGC 4214-1; [5]). All three clusters have size measurements or estimates from recent *HST* imaging studies.

Because of the very young age of NGC 4214-1 (4–5 Myr; [5]), we did not detect stellar absorption lines suitable for velocity dispersion measurements. The other two clusters, on the other hand, have ages of 10–20 Myr, and hence cool supergiants contribute significantly to their integrated light at visual wavelengths. The availability of numerous sharp absorption lines allowed us to measure the velocity dispersions of the clusters using a conventional cross-correlation technique [8]. Some examples of the data and the method are show in Figure 1. The line-of-sight velocity dispersions of NGC 1569-A and NGC 1705-1 are 15.7 ± 1.5 km s^{-1} and 11.4 ± 1.5 km s^{-1}, respectively.

YOUNG GLOBULAR CLUSTERS

From the relatively large velocity dispersions and small physical sizes of the clusters (half-light radii $\sim$1–2 pc), one can infer that the clusters must be gravitationally bound, since their dynamical crossing time ($\sim 10^5$ yr) is about 100 times smaller than their age. Application of the virial theorem then implies that the clusters in NGC 1569 and NGC 1705 have dynamical masses of $\sim 3\times 10^5$ and $\sim 8\times 10^4$ $M_\odot$, and mass densities of $\sim 1\times 10^4$ and $\sim 3\times 10^4$ $M_\odot$ pc^{-3}, respectively. As discussed in greater detail by Ho & Filippenko [2,3], these values are in fact lower limits. These characteristics are very similar to those of evolved globular clusters in the Milky Way. Indeed, even the predicted mass-to-light ratios, 1–5 $(M/L_V)_\odot$, after aging the clusters by 10–15 Gyr, seem to be in good agreement with those of old globular clusters. We therefore believe that these observations provide the first conclusive evidence that at least some super star clusters are genuine, *young* globular clusters.

REFERENCES

1. Ho, L. C. 1996, Rev. Mex. Astr. Astrofis., in press
2. Ho, L. C., & Filippenko, A. V. 1996a, ApJ, 466, L83
3. Ho, L. C., & Filippenko, A. V. 1996b, ApJ, in press (Dec. 1 issue)
4. Holtzman, J. A., et al. 1992, AJ, 103, 691
5. Leitherer, C., Vacca, W. D., Conti, P. S., Filippenko, A. V., Robert, C., & Sargent, W. L. W. 1996, ApJ, 465, 717
6. O'Connell, R. W., Gallagher, J. S., & Hunter, D. A. 1994, ApJ, 433, 65
7. O'Connell, R. W., Gallagher, J. S., Hunter, D. A., & Colley, W. N. 1995, ApJ, 446, L1
8. Tonry, J., & Davis, M. 1979, AJ, 84, 1511
9. van den Bergh, S. 1995, Nature, 374, 215
10. Whitmore, B. C., & Schweizer, F. 1995, AJ, 109, 960
11. Whitmore, B. C., Schweizer, F., Leitherer, C., Borne, K., & Robert, C. 1993, AJ, 106, 1354

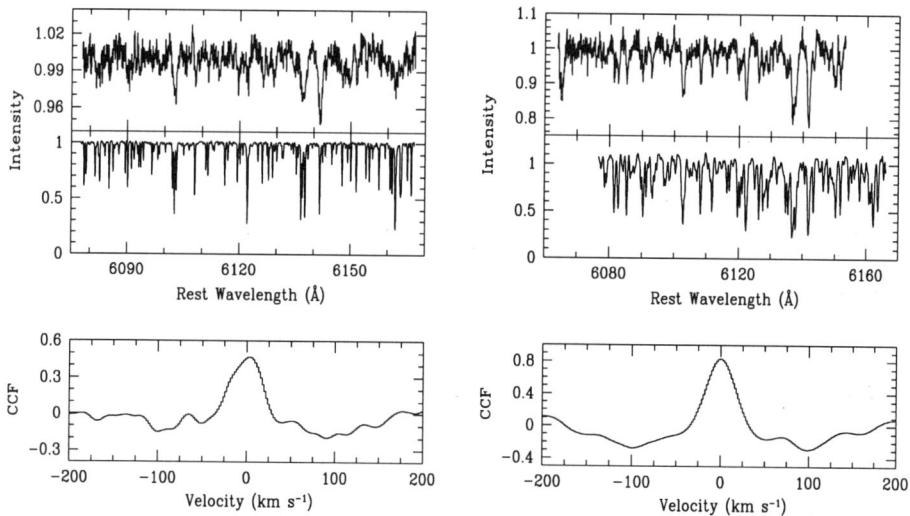

FIGURE 1. Measurement of the stellar velocity dispersions in cluster NGC 1569-A (*left*; [2]) and in NGC 1705-1 (*right*; [3]). The top panel in each case shows the cluster spectrum, the middle panel the template star used to derive the velocity dispersion, and the bottom panel the cross-correlation function between the cluster and star. The width of the main velocity peak of the cross-correlation function is related to the velocity dispersion.

Galactic Star Formation

Some High Class (or High Mass) Neighborhoods – the Sites of the Most Massive Stars in the Milky Way and our Neighbors

Roberta M. Humphreys

University of Minnesota

Abstract. The properties of a few star forming regions famous for their large number of very luminous and very massive stars are briefly described. Two different star formation structures are observed.

WHAT DO WE MEAN BY VERY MASSIVE OR MOST MASSIVE STARS?

For very good astrophysical reasons, massive stars are usually defined as stars ≥ 10 M$\odot$. For the purpose of this paper I want to change your perspective a bit from massive to very massive stars.

Massive stars in our galaxy are found in stellar associations and young clusters. A comparison of two H-R diagrams (Figs.1 and 2) in Blaha and Humphreys [1] clearly illustrates not only the much greater number of massive stars in recognized associations, but also that the most luminous/most massive ones are only in the associations. These diagrams represent the stellar population out to about 3 kpc from the Sun and the stars are relatively bright (HD and BD stars); therefore it is unlikely that a large number of luminous, hot stars has been missed unless they are embedded in dust clouds.

The empirical upper luminosity boundary (see [2] and [3]) provides a defining mass-luminosity boundary in the upper HR diagram. For the hot stars we observe an envelope of declining luminosity with decreasing temperature which turns over and becomes essentially constant near $M_{bol} = -9.5$ to -10 mag for temperatures less than about 9000°K. This luminosity limit for the cooler stars corresponds to an initial mass of 40-50 M$\odot$. Below this initial mass stars become red supergiants; above it their evolution is quite different including episodes of high mass loss as luminous blue variables.

For the purposes of this paper I am going to define the most massive stars as those above the cool star upper luminosity, brighter than Mbol ≅ -10 mag and with initial masses >50 M⊙ .

WHERE ARE THESE STARS FOUND?

The Milky Way

In the Milky Way these stars are found together in some especially prominent star forming regions. Two groupings with resolved stars and MK spectral classifications stand out; these are Cyg OB2 and the Carina Nebula (NGC 3372) with the clusters Tr 14 and Tr 16/Cr 228.

One of the keys to recognizing regions of massive star formation is the presence of the rare O3 type stars [4]. As the hottest and earliest spectral class, this group includes several of the most luminous stars known. Cyg OB2 has two of the 17 known O3-type stars in our galaxy and 9 stars with luminosities brighter than Mbol = -10 mag. One of its members is the visually brightest star in our region of the Galaxy.

The Carina Nebula is even more spectacular with 6 O3-type stars, 36 stars hotter than spectral type B0, 3 WN stars and Eta Carina, the most luminous star known in the Milky Way, with a probable initial mass above 100 M⊙ (see [5]). The surrounding association (Car OB1) contains a large population of evolved blue and red supergiants, and a ring of outer clusters with ages of $\cong 10 \times 10^6$ yrs. The inner association contains the younger clusters Bochum 10 and 11 with ages of 3 to 7×10^6 yrs.[6]. At the core of the association is the giant HII region NGC 3372 with the very young clusters Tr 14 and Tr 16/Cr 228 with ages $\leq 3 \times 10^6$ ([7], [8], [9], [10], [11]). All of the O3 stars, the Wolf-Rayet stars and η Car are all members of these clusters. Another young cluster Tr 15 is usually considered to be behind the HII region but still a member of the Carina association [7]. Thus star formation in this region has been essentially continuous over several million years.

Is star formation still occurring in the Carina Nebula and are massive stars still forming there? Infrared and molecular surveys have revealed few obvious embedded sources, including one embedded IRAS source near Tr 16. However, recent deep JHK imaging by Megeath et al [12] shows a number of very reddened sources; the first clear evidence of continuing star formation in the Carina Nebula. Since the CO survey by Grabelsky et al [13] shows that the Carina Nebula is part of a giant molecular cloud, we expect that massive stars will continue to form there.

The more distant cluster (7 kpc) and HII region NGC 3603 resembles the core of 30 Dor in the LMC, but without the surrounding large association of evolved stars, also observed in the Carina region. Like R136, its central object known as HD 97950 is not resolved by normal groundbased imaging.

However, recent speckle observations by Hoffman et al [14] and FOC images with HST by Drissen et al [15] reveal that the central region may be even more compact than R136 with 28 objects in a 0.2×0.2 pc region. The latter's FOS spectroscopy of 14 stars in HD 97950 reveals 6 new O3 stars, plus 5 other early O-type stars and 3 WN stars which are also the visually brightest members. They describe NGC 3603 as the 'densest concentration of very massive stars known in the galaxy'.

The core of NGC 3603 is largely coeval but a group of evolved red supergiants is located 4 arcmin to the north and pre-main sequence stars are found 2 arcmin to the south. Star formation in the region of NGC 3603 therefore appears to have followed the sequential pattern observed in many other spiral arm regions.

The Large Magellanic Cloud

The 30 Doradus/R136a region in the Large Magellanic Cloud is often described as our 'local starburst' because it most closely resembles the regions of active star formation in 'starburst' galaxies. It is indeed the most spectacular concentration of hot, massive stars close enough that we can resolve individual stars. Walborn [16] lists 40 known O3-type stars in 30 Dor and Parker and Garmany [17] catalogued over 2400 massive stars in the 30 Dor region. Observations with HST have revealed the complexity of the region and the degree of merged images (see [18] Using the pre-costar PC, Malamuth and Heap [19] identified 200 stars within 3 arcsec of the R136a core and 800 stars in the 35 35 arcsec region. R136a is resolved into at least 8 stars.

Like the Carina nebula region, 30 Dor is an example of a region with multiple episodes of star formation. The surrounding population of evolved blue and red supergiants is 10 to 20 $\times 10^6$ yrs. old. These evolved massive stars are the remnants of the initial population and probably provided the trigger for the current starburst in the R136 central cluster.

An infrared survey by Hyland et al. [20] identified 4 candidate protostars with probable masses of 15-20 M$\odot$ located in dense nebular knots. They concluded that both the knots and the star formation are caused by compression of interacting mass-loss winds from the young hot stars in the central cluster. Thus the youngest stars in 30 Dor are now forming in luminous arcs outside the core. This is a second mode of star formation that differs from the standard scenarios in dense molecular clouds. Interestingly, Walborn and Parker [21] have identified another HII region, N11 in the LMC, a somewhat older star forming region, that exhibits this same morphology as 30 Dor.

The Local Group - M33

The nearby spiral M33 is distinguished by the prominence of its star forming regions - the large stellar associations in its spiral arms plus several giant HII regions. The two largest HII regions NGC 604 and NGC 595 contain many WR/Of-type stars [22] implying the existence of a young recently formed population in the central cluster but the presence of evolved blue and red supergiants also indicates a prior star formation epoch like what we see in the Carina and 30 Dor regions (see [23] and [24]). Hunter et al. [24] also note that the density of stars in NGC 604 is comparable to typical OB associations in the Milky Way and comment that the "formation of a large number of massive stars does not necessarily entail a high concentration."

Stellar Density and Structure

The concentrations of massive stars in several of the star forming regions discussed in this paper are summarized in the table below.

In the large star forming regions - the associations - where star formation has been occurring for several million years, the densities are comparable. But in the compact cores of two regions where the most recent formation has occurred, the densities are much, much higher and interestingly similar to each other.

Even though I have described only a few regions of massive star formation in this brief paper, two different formation structures are apparent. Almost all of these regions have a halo/core structure with several epochs or even continuous star formation over many millions of years - that is a large association of evolved massive stars with one or more embedded clusters with a range of

TABLE 1. Concentrations of Massive Stars in Some Selected Associations

Region	Area (pc^2)	Density (stars/pc^2)	Reference
Milky Way – OB	—	0.02	Massey et al. [10]
LMC – OB	—	0.02	Massey et al. [10]
Cyg OB2	409	0.04	this paper
Tr 14, 16, Cr 228	350	0.11	this paper
30 Dor	8700	0.05	Hunter et al. [24]
			Parker & Garmany [17]
R 136	69	1.8	Hunter et al. [24]
NGC 604	10500	0.02	Hunter et al. [24]
cores (based on available HST data)			
R 136a (8 stars)	130	0.063	this paper
NGC 3603	300	0.045	this paper

ages including at the center one or more very young clusters or a very compact core or cluster. The older evolved massive star population is presumably the trigger for the on-going star formation perhaps involving mechanisms similar to that proposed by Hyland et al. [20] as well as the familiar scenarios involving winds and supernovae shock waves. The second formation structure is what we commonly call sequential star formation. Although NGC 3603 is the only region in this paper displaying this scenario, sequential formation is very common among the associations in the spiral arms of the Milky Way.

REFERENCES

1. Blaha C., and Humphreys R. M.., *AJ* **98**, 1598 (1989).
2. Humphreys R.M. and Davidson K., *ApJ* **232**, 409 (1978).
3. Humphreys R.M. and Davidson K., *Science* **223**, 243 (1984)
4. Walborn N.R., *ApJ* **167**, L31 (1971).
5. Davidson K. and Humphreys R.M., *ARA&A* **35**, in press (1997).
6. Fitzgerald P.M. and Mehta S., *MNRAS* **228**, 545 (1987).
7. Walborn N.R., *Rev. Mex. Astrofis. Ser. Conf.* **2**, 51 (1995).
8. Feinstein A., *AJ* **87**, 1012 (1982).
9. Feinstein A., *Rev. Mex. Astrofis. Ser. Conf.* **2**, 57 (1995).
10. Massey P. and Johnson J. *AJ* **105**, 980 (1993).
11. Penny L.R., Gies D.R., Hartkopf W.I., Mason B.D. and Turner N.H., *PASP* **105**, 588 (1993).
12. Megeath S.T., Cox P., Bronfman L. and Roelfsema P.R., *A&A* **305**, 296 (1996).
13. Grabelsky D.A., Cohen R.S., Bronfman L. and Thaddeus P., *ApJ* **331**, 181 (1988).
14. Hofmann, K-H. Seggewiss W. and Weigelt G., *A&A* **300**, 403 (1995).
15. Drissen L., Moffat A.F.J., Walborn N.R. and Shara M.M., *AJ* **110**, 2235 (1995).
16. Walborn N.R., *The MK Process at 50 Years*, San Francisco: ASP Conference Series, 1994, pp. 84-92.
17. Parker J.W. and Garmany C.D. *AJ* **106**, 1471 (1993).
18. Campbell B., Hunter D.A., Holtzman J.A., Lauer T.R., Shaya E.J., Code A., Faber S.M., Groth E.J., Light R.M. and Lynds R., *AJ* **104**, 1721 (1992).
19. Malamuth E.M. and Heap S.R. *AJ* **107**, 1054 (1994).
20. Hyland A.R., Straw S., Jones T.J. and Gatley I. *MNRAS* **257**, 391 (1992).
21. Walborn N.R. and Parker J.M. *ApJ* **399**, L87 (1992).
22. Drissen L., Moffat A.F.J. and Shara M.M., *AJ* **105**, 1400 (1993).
23. Wilson, C.D. and Matthews, B.C. *ApJ* **445**, 125 (1995).
24. Hunter D.A., Baum W.A., O'Neil E.J. and Lynds R., *ApJ* **456**, 174 (1996).

Observations of the Extragalactic Initial Mass Function and Modes of Star Formation

Sara R. Heap

Laboratory for Astronomy & Solar Physics
Goddard Space Flight Center, Greenbelt MD

Abstract. I review recent measurements of the Initial Mass Function in other galaxies for three modes of star formation: stars formed in isolation, in associations, and in starburst clusters. While the observed differences in the IMF are mild, they are enough to rule out a universal IMF.

INTRODUCTION

The Initial Mass Function (IMF) is the mass distribution of stars at the time of birth. The IMF has a form similar to that shown in Figure 1 (Padoan et al. 1996). It has an upper mass limit (M_u), a most frequent mass (M_{peak}), and a lower mass limit (M_l). In most cases, the peak mass is less than the mass of the sun. If we follow Scalo (1986) in defining $\xi(\log M)$ as the number of stars born per unit logarithmic mass interval per unit area (kpc^2) per unit time (Myr), then the slope of the high-mass tail is given by:

$$\Gamma \equiv d\log\xi(\log M)/d\log M. \tag{1}$$

The IMF slope in the solar neighborhood (Salpeter 1955) is $\Gamma = -1.35$. This standard IMF is easily "visualized" by Zinnecker's (1995) handy formula: halve the mass, triple the number of stars. For example, if you have a plate or CCD image showing 10 stars with initial masses between 40 and 80 $M_\odot$, then there will also be 30 stars with masses between 20 and 40 $M_\odot$, 100 with masses between 10 and 20 $M_\odot$, and so on down to near the brown-dwarf limit (0.075 - 0.15 $M_\odot$) with 300,000 stars.

Until direct measurement of stellar mass via microlensing observations comes of age (Gould 1996), the mass of a star has to be estimated indirectly from its luminosity, which effectively limits us to studying the high-mass tail of the IMF in other galaxies. The procedure for determining the mass of a

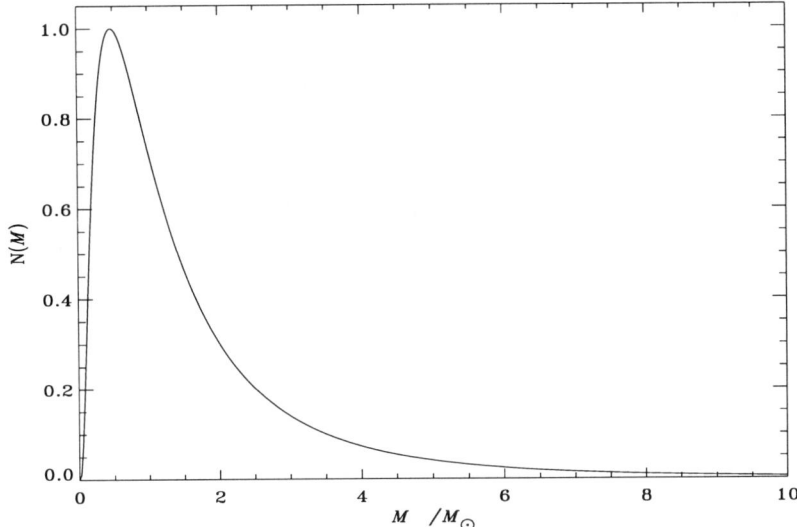

FIGURE 1. General Form of the IMF (predicted for T=10 K, $\overline{N} = 100$ cm^{-3}, and $\sigma_v = 5$ km s^{-1}). From Padoan et al. (1996).

star going into the IMF is to put the star on the HR diagram (HRD) on which evolutionary tracks are also plotted. Then read off the present-day mass from the track that goes through the T_{eff}- L_{bol} position of the star, and follow the track back to its origin to get its initial mass. (The initial mass of a star may be a lot higher than its current mass because of mass-loss.) Determining the IMF is then a matter of counting the number of stars on the HR diagram in different mass intervals.

In practice, star counts are not so simple. There are severe crowding problems because of the rapidly increasing number of lower-mass stars. High spatial resolution, such as is afforded by the Hubble Space Telescope or ground-based telescopes with adaptive optics, is required. There is also the problem that the (B-V)$_o$ color of hot, massive stars is totally insensitive to effective temperature, so spectra are needed to determine their effective temperatures and bolometric corrections. Since spectroscopy is such a time-consuming project, spectral surveys have usually been limited to stars brighter than V≈14. For OB stars in the Magellanic Clouds, this magnitude limit translates to a mass limit of about 40 $M_\odot$. See Massey et al. (1995a) for a thorough discussion of observational sources of error.

There are theoretical sources of error as well. The main problem is that evolutionary tracks of massive stars are quite sensitive to mass-loss, and the mass-loss rates (de Jager et al. 1988) assumed by most evolutionary models are serious underestimates (Lamers & Cassinelli 1966). Meynet et al.'s (1994) most recent grid of evolutionary tracks now multiplies the "standard" mass-loss rate by a factor of two. This correction factor may not be the right number, but at least it shows the (sometimes dramatic) effect of mass-loss on stellar evolution.

In the following sections, I review recent studies of the extragalactic IMF for field stars (§2), stars in OB associations (§3), and stars in starburst clusters (§4). For the most part, these studies have concentrated on the high-mass tail of the IMF because of crowding and sensitivity problems mentioned earlier. In each mode of star formation, I will discuss only direct observations of the IMF (IMF by star counts), since this has to be understood before we can infer the IMF of unresolved systems.

FIELD STARS

Massive stars born in the field? This goes against the grain. The conventional picture of star formation is that stars form as a group from a giant molecular cloud. Nevertheless, Massey et al. (1995a,b) have found more than a thousand such field stars in the Magellanic Clouds and the Galaxy. They argue that these field stars are too far away from the nearest cluster to have been born there. They must have been born in isolation. Not only is the field capable of producing massive stars, but it produces stars just as massive as any found in clusters. In both the Magellanic Clouds and the Galaxy, there are several bona fide field stars of ≈ 100 $M_\odot$.

There is, however, some question about how frequently very massive field stars occur. Optical color-magnitude diagrams of field stars in the Magellanic Clouds (Massey et al. 1995a) yield a very steep IMF: $\Gamma = -4.1 \pm 0.2$ for field stars in the LMC, and $\Gamma = -3.7 \pm 0.5$, in the SMC. Ultraviolet color-magnitude diagrams give a very different picture. Using images from the Ultraviolet Imaging Telescope, Hill et al. (1994) obtain an IMF slope, $\Gamma = -1.74 \pm 0.3$, for field stars in the LMC, which is not much different from what they derive for associations, $\Gamma = -1.08 \pm 0.2$. Cornett et al. (1994) derive similar results for the SMC. Which set of results is right? If I had to choose right now, I would bet on the ultraviolet surveys, since they should be complete down to ≈ 7 $M_\odot$, giving a broader baseline in mass over which to estimate Γ. Still, the discrepancy between optical and ultraviolet results is an important one and should be resolved.

With the superb resolution of the HST, it is now possible to construct color-magnitude diagrams for stars in the Magellanic Clouds that go down to a solar mass or less. A nice example is Gallagher et al.'s (1996) CMD of a quiescent field in the LMC constructed from WFPC2 V- and I-band images. Such images demonstrate the potential of HST for obtaining extragalactic initial mass functions, and we can expect exciting results in the future.

STARS IN OB ASSOCIATIONS

There have been numerous IMF studies of clusters in the Magellanic Clouds and other nearby galaxies. Table 1 gives a non-exhaustive list of the more recent studies. (For a summary of studies prior to 1994, see Maeder & Conti 1994.) By and large, the derived IMF slopes are similar to one another and are consistent with a Salpeter IMF.

TABLE 1. Some Recent Studies of the Extragalactic IMF

OB Association	Γ	Mass Range	Reference
GALAXY			
Cyg OB2	-0.9 ±0.2	7-110	Massey et al. 1995b
Tr 14/16	-1.0 ±0.2	7-120	"
NGC 6611	-0.7 ±0.2	7-75	"
LMC			
30 Doradus	-1.5 ±0.2	12-120	Parker & Garmany 1993
6 Lucke-Hodge assoc.	-1.1 ±0.2	15-40	Hill et al. 1994
4 Lucke-Hodge assoc.	-1.4 ±0.3	10-85:	Massey et al. 1995b
Lucke-Hodge 9	-1.6 ±0.1	7-55	Parker et al. 1992
Lucke-Hodge 10	-1.1 ±0.1	7-85	"
SMC			
NGC 346	-1.4±0.1	7-70	Massey et al. 1995b
M33			
NGC 595 ($Z=0.36Z_\odot$)	-1.50 ±0.12	10-45	Malumuth et al. 1996
CC 93 (($Z=1.38Z_\odot$)	-0.92 ±0.11	10-45	"

Figure 2 summarizes the results on the upper mass limit. The figure compares the most massive stars found in clusters in the Galaxy and Magellanic Clouds as a function of cluster age. The theoretical relation between M_u and cluster age is also shown (Meynet et al. 1994). It simply reflects the fact that massive stars evolve more rapidly than less massive stars. The observed data (Massey et al. 1995a,b) generally conform to this relation, with clusters younger than 3 Myr having stars up to 120 $M_\odot$ in mass. Perhaps, clusters younger than 1 Myr have even more massive stars, but it is questionable whether we would see them, since such clusters might still be embedded in a cocoon (c.f. Bernesconi & Maeder 1996).

The net result of these studies is that the populations of massive stars in the Galaxy and in the Magellanic Clouds are not noticeably different. This finding was unexpected since theories of star formation including the effects of radiative forces on dust grains suggest that $M_u \sim Z^{-\frac{1}{2}}$ (Shields and Tinsley 1976). If the most massive stars in the Galaxy have initial masses of 60 $M_\odot$ (Z=0.020), then there should be stars as massive as 95-135 $M_\odot$ in the LMC (Z=0.004-0.008) or even 220 $M_\odot$ in the SMC (Z=0.015). No such extremely

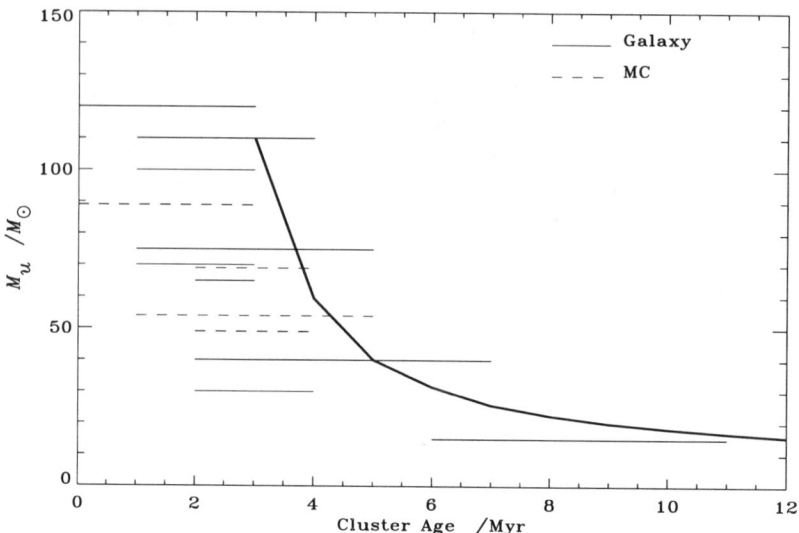

FIGURE 2. Relation between cluster age and upper mass limit for clusters in the Galaxy (solid line) and Magellanic Clouds (dashed line) from Massey et al. (1995b). The theoretical relation (Meynet et al. 1994) is also shown (bold).

massive stars have been identified in the Magellanic Clouds. By the same token, no supressed upper mass limit has been seen in OB associations in M31 having a metallicity twice the solar metallicity (Bresolin et al. 1996). To be fair, there have been reports of a few extremely massive stars based on the results of spectroscopic analysis, but still there is no clear trend in the upper mass limit with metallicity (Pauldrach et al. 1994, de Koter et al. 1997, Taresch et al. 1997).

Evidently, massive star formation proceeds independently of metallicity, or more probably, the role of metallicity is masked by other factors. Wolfire & Cassinelli (1987) argue that it is not only the amount of dust that counts but also its size distribution. Only if the dust is depleted in large grains can accretion continue past 100 $M_\odot$ in regions of low metallicity. Such an alteration in the size distribution could come about through "pre-conditioning" of the dust by turbulence. In fact, turbulence is considered a prime factor in the formation of massive stars (these proceedings).

STARS IN STARBURST CLUSTERS

Here, I will discuss the case of the ionizing star cluster, R136, which is at the heart of the 30 Doradus complex in the LMC. At the nucleus of R136 is the very compact cluster, known as R136a. Walborn (1991) has called R136 the Rosetta Stone for starburst research. Zinnecker (1996) says that "the IMF in R136 is the million-dollar question of extragalactic star formation". R136 earns these accolades because it is the best example of a starburst cluster that can be resolved into individual stars. In the next three sections, I will describe how R136 has been used to investigate the IMF in starbursts.

Low-Mass Stars in Starbursts?

By any definition of "starburst," globular clusters must have been born as magnificent starburst clusters, presumably with an ample number of very massive stars; but now that globular clusters are old, the only stars that are left are low-mass stars. This means that if starburst clusters are the true antecedents of globular clusters, they must contain not only very massive, luminous stars but also low-mass stars.

Do starburst clusters contain low-mass stars? R136a offers the best opportunity to find out. According to Hunter et al. (1995) the answer is yes: HST images show that R136a is fully populated down to the limit of detection, about 2 $M_\odot$. Moreover, the IMF slope between 2.8 and 15 $M_\odot$ is remarkably normal ($\Gamma = -1.22 \pm 0.06$); hence, the lower mass limit must be less than 2.8 $M_\odot$. Some of the low-mass pre-main sequence stars are as young as 1 Myr, which is to say, younger than the cluster itself, which means that low-mass stars can still go on forming despite the hostile environment set up by the massive stars at the core. Their presence in a starburst does not imply that they formed before the very massive stars.

Massive Stars in Starbursts

The first glimpse of the IMF in R136 came from early HST pictures (Campbell et al. 1992, Heap et al. 1992), which showed an incredible concentration of luminous stars near the center – but no super-massive star (1000 $M_\odot$ or more) that had been posited to exist in R136a a decade earlier. A more detailed analysis (Malumuth & Heap 1994) concluded that the most massive stars in R136a had initial masses of $110 M_\odot$ — no more massive than those found in the field or OB associations. Their conclusion is supported by recent spectroscopic analyses (de Koter et al. 1997) that find that the most massive star in R136a, the Wolf-Rayet star star known as R136a1, has an initial mass slightly higher than 120 $M_\odot$ (Figure 3).

What *is* unusual about R136 is the apparent mass segregation. Except for in the nucleus, R136 has a normal distribution of stellar masses ($\Gamma = -1.82 \pm 0.4$); but in the central cluster, R136a, where the concentration of stars is at its highest, it has a very top-heavy IMF ($\Gamma = -0.90 \pm 0.4$ for $r \leq 0.8$ pc, $\Gamma = 0.0$ for $r \leq 0.25$ pc). Since the relaxation time is about the same as the age of

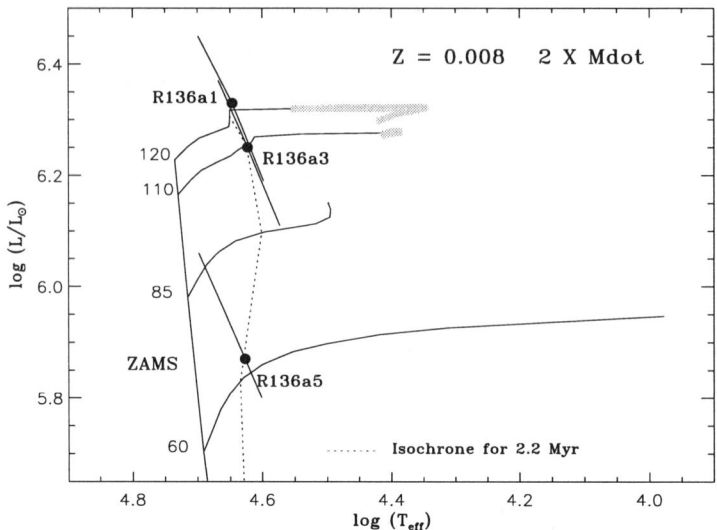

FIGURE 3. HR Diagram for Three Stars in R136a. R136a1 and R136a3 are Wolf-Rayet stars; R136a5 is an Of/WR star. The error bars correspond to adopting a T_{eff} of 40 or 50 kK. (The real errors may be smaller.) The evolutionary tracks are plotted only through the main sequence phase. The shaded gray lines show where the surface hydogen abundance decreases below X=0.4, which defines the start of the Wolf-Rayet phase. Isochrones are shown for the ZAMS and for 2.2 Myr (dotted line). From de Koter et al. (1997).

the cluster itself (2.2 Myr), it is possible that the observed mass segregation may be the result of dynamical evolution. More probably, it is the result of processes operating during star formation. According to Larson (1991), massive stars should preferentially form near the cluster center where the gas density is highest and the relative velocity between an accreting star and the ambient medium is the lowest, so accretion can proceed most efficiently. The concentration of massive stars toward the center should be further accentuated through a feedback mechanism involving dynamical friction.

Wolf-Rayet Stars in Starbursts

The integrated spectrum of R136a resembles those of Wolf-Rayet galaxies in showing emission lines characteristic of WR stars (Heap et al. 1992, Vacca et al. 1995). In fact, early HST images identified the three brightest stars in R136a as WR stars by their strong emission of HeII $\lambda 4686$ (Campbell et al. 1992). Since it was thought that all WR stars were in a highly evolved phase that takes about 3 million years to get to (Maeder & Conti 1994), most investigators assigned an age of 3 Myr to R136a. HST spectra, however, show that the three WR stars in R136a are not highly evolved stars at all, but rather are very massive stars on the main sequence (de Koter et al. 1997). Their positions on the HR diagram (Figure 3) clearly place them in the core hydrogen-burning phase; and their normal (solar) helium abundances are consistent with an early phase of evolution. What gives them their WR appearance is their extraordinarily high rates of mass-loss — two to three times higher than assumed even in the Meynet et al. $2\times\dot{M}$ evolutionary models.

Just because the three WR stars in R136a are main sequence stars does not mean that most WR stars are not highly evolved stars. WR stars that are hotter than their ZAMS temperatures or that have highly altered surface abundances are clearly near the end of their lifecycle. However, it *does* mean that when we see a starburst region or W-R galaxy whose spectrum resembles the integrated spectrum of R136a, we can admit as a possibility that the starburst region or galaxy is simply a scaled-up version of R136a.

Such an interpretation runs counter to the prevailing view of Wolf-Rayet galaxies (Conti 1996), in which the HeII emission lines in their spectra are produced by highly evolved WR stars. Since the evolved WR phase is only about a tenth as long as the main sequence phase, to produce detectable HeII emission requires that the stars be formed in an instantaneous burst about 3 million years ago, and that the IMF must be unusually top-heavy (Meynet 1995). These strict requirements can be greatly relaxed if the WR stars in Wolf-Rayet galaxies are main sequence stars, as in R136a. Continuous star formation could replenish the very massive stars that produce the Wolf-Rayet spectral features. And fewer massive stars would be needed, because WR main sequence stars last much longer, so the IMF does not need to be so top-heavy.

Although the results are not as definitive as one would like, it is clear that the IMF is not universal. There is much for us observers to do, like answering the following questions:
– Does the massive star population in the field really differ significantly from that in OB associations?
– What are the roles of dust and turbulence in determining the IMF?
– What is the proper prescription for mass-loss to be used by evolutionary models?
– What is the lower mass limit in starbursts like R136?

When we have answered these questions, we can get on to the *real* questions of star formation.

REFERENCES

Bernasconi P. & Maeder A. 1996, A&A 307, 829
Bresolin, F., Kennicutt R.C., Stetson P. 1995, AJ 112, 1009
Campbell B. & the WF/PC team 1992, AJ 104, 1721

Conti P. 1996, in *From Stars to Galaxies* (ASP Conf. Series, Vol. 98), eds. C. Leitherer et al.
Cornett R.H., Hill J.K., Bohlin R.C. et al. 1994, ApJ 430, L117
de Jager, C. Niewenhuijzen H., van der Hucht K.A. 1988, A&A 72, 259
de Koter A., Heap S.R. Hubeny I. 1997, ApJ in press (March 10 issue)
Gallagher J.S. Mould J.R., de Feijter E. et al. 1996, ApJ 466, 732
Gould A. 1996, PASP 108, 465
Heap S.R. et al. 1992, in *Sceince with the HST*, ed. P. Benvenuti & E. Schreier (Garching: ESO), p. 347
Hill J.K., Isensee J.E., Cornett R.H. et al. 1994, ApJ 425, 122
Hunter D.A., Shaya E.J., Holtzman J.A. et al. 1995, ApJ 448, 179
Lamers H. & Cassnelli J.P. 1996, in *From Stars to Galaxies* (ASP Conf. Series, Vol 98), eds. C. Leitherer et al., p. 162
Larson R.B. 1991, in *Fragmentation of Molecular Clouds and Star Formation*, eds. E. Falgarone et al. (Kluwer: Dordrecht), p. 261
Maeder A. & Conti P. 1994, Ann Rev Ast & Ap 32, 227
Malumuth E.M. & Heap S.R. 1994, AJ 107, 1054
Malumuth E.M., Waller W.H. & Parker J.W. 1996, AJ 111, 1128
Massey P. Lang C.C, deGoioia-Eastwood K., Garmany C. 1995a, ApJ 438, 188
Massey P., Johnson K., DeGioia-Eastwood K. 1995b, ApJ 454, 151
Meynet G. 1995, A&A 298, 767
Meynet G., Schaller G., Schaerer D. & Charbonnel C. 1994, A&AS 103, 97
Padoan, P., Jones B.J.T., Nordlund Å. 1996, MNRAS, in press
Parker J.W., Garmany C., Massey P., Walborn N. 1992, AJ 103, 1205
Pauldrach A.W., Kudritzki R-P., Puls, J. et al. 1994, A&A 283, 525
Rieke G.H., Loken L., Rieke M.J., Tamblyn P. 1993, ApJ 413, 99
Salpeter E.E. 1955, ApJ 121, 161
Scalo, J.M. 1986, Fund. Cosmic Phys. 11, 1
Schaerer D. 1996, ApJ 467, L17
Schaerer D. 1996, in *Active Galactic Nuclei, Dense Stellar Systems, and their Environments*, eds. S.A. Lamb, J.J Perry (ASP Conf. Series)
Shields G.A. & Tinsley B.M. 1976, ApJ 203, 66
Vacca W.D. & Conti P. 1992, ApJ 401, 543
Vacca W.D., Robert C., Leitherr C. & Conti P. 1995, ApJ 444, 647
Walborn N.D. 1991, in *Massive stars in Starbursts*, ed. C. Leitherer et al.
Wolfire N.G. & Cassinelli J.P. 1987, ApJ, 319, 850
Zinnecker H. 1996, in *The Interplay between Massive Star Formation, the ISM, and Galaxy Evolution*, eds. D. Kunth et al., Ed. Frontieres, p. 249

Controlling Factors for Global Star Formation

Evan D. Skillman

Astronomy Department
University of Minnesota
116 Church St. SE
Minneapolis, MN 55455

Abstract.
Here I attempt to address the question: What do we know, or can we know, about the controlling factors for global star formation? First, I open with a very brief review of measurements of current star formation rates. While absolute estimates of current star formation rates carry a significant degree of uncertainty, the comparison of current star formation rates has proven to be a very valuable guide in understanding the star formation histories of galaxies. I then review the observational evidence for and against the concept of star formation thresholds. I conclude that the concept appears to be a very useful one and enjoys a large range of applicability. However, there are certain troubling aspects, and future work in this area could prove to be quite promising.

STAR FORMATION RATES

Before discussing controlling factors of star formation, it is a good idea to briefly review how star formation rates (SFRs) are measured. The most widely used measurement of the current SFR of a galaxy is its Hα emission. A detailed discussion of the pros and cons of using Hα emission to measure SFRs can be found in Kennicutt [1]. The principle behind this measurement is simple enough. By counting the Balmer line photons, one can infer the total rate of Lyman continuum photons produced by the present stellar population (see Osterbrock [2] for the case of a single star). Assuming an initial mass function (IMF), the inferred Lyman continuum photon rate can be converted into estimate of the total number of massive stars present, which, under the assumption of a constant SFR over the last few 10^7 yr, can then be converted into an estimate of the current SFR.

There are, of course, uncertainties associated with this method. One worries about the effects of dust both through the destruction of the Lyman continuum photons, before they can produce the Balmer emission, and by the extinction of the Hα photons. There is also some concern that the Lyman continuum photons are not absorbed locally (near to their production site), and thus are not converted to measurable Balmer photons (e.g., Ferguson et al. [3]). Additionally, in the presence of a strong underlying continuum of A and B type stars, the Balmer emission will be decreased as it is swallowed up in the stellar Balmer absorption lines (e.g., see Waller et al., these proceedings). These effects would all cause an underestimate of the true current SFR.

The required assumptions are also a source of uncertainty. The uniformity of the IMF is a very controversial topic (cf. [4,5]). Also, a global burst of star formation of very short duration will lead to an overestimate of the current SFR at early times and an underestimate at later times. Luckily, for large galaxies, the Hα flux is coming from a large number of sites of current star formation; as a quantity averaged over many "cells" of star formation, this measurement becomes more robust against the frailties of these assumptions.

A second, and perhaps even more secure measure, is that of the Hα equivalent width, EW(Hα). Kennicutt [1] argued that this observed quantity could be interpreted as a measure of the ratio of the present SFR to the past average SFR (since the underlying stellar continuum at $\sim \lambda 6600$ Å represents the integrated light of stars produced over the last several billion years).

Many advances in our understanding of the evolution of galaxies can be attributed to the interpretation of the Hα fluxes as a measure of the current star formation rate. Kennicutt [1], Gallagher et al. [6], and Sandage [7] have used Hα fluxes in reconstructing star formation rate histories of galaxies. Kennicutt et al. [8] have revisited the calibration of the ratio of current SFR to past average SFR (the "birthrate parameter", Scalo [4]) and found: (1) that the Hubble sequence is chiefly a sequence in birthrate parameter; and (2) that, with the proper treatment of the recycling of the gas, there is sufficient gas in most disks to continue forming stars at the present (average) rate for another Hubble time.

Larson & Tinsley [9] pointed out that interacting galaxies and merging galaxies showed enhanced current star formation rates relative to isolated galaxies. This type of study is fraught with statistical biases, and many of these were addressed by the study of Kennicutt et al. [10]; from a large sample of galaxies they were able to establish a clear offset (although with large dispersion) between the current star formation rates of interacting galaxies and normal galaxies (see also Mihos et al., in these proceedings).

While Hα fluxes are clearly of value in measuring the relative and absolute current star formation rates of galaxies, the question remains, how accurate is this measure. An anonymous astronomer once said that relative SFRs derived from Hα are good to within a factor of 2 while absolute SFRs derived from Hα are good to within a factor of 3. This opinion, when presented to the audience at the conference, drew no sharp criticism, so perhaps the anonymous astronomer is not far from the mark.

Nonetheless, there is a great deal of effort being put into the calibration of Hα as a star formation measure. Current efforts are concentrating on the comparison of different measures of SFRs for individual galaxies and for conversions for individual stars and isolated star forming regions (see, e.g., Hunter [11,12]; Devereux et al. [13,14]; Oey & Massey [15]; Scowen et al. in these proceedings).

One very interesting recent result is the study by Devereux & Hameed [16] (and in these proceedings) of the far infrared fluxes for galaxies in the Nearby Galaxies Catalog. Using the far infrared flux as a measure of the current SFR, they find no trend in normalized current SFR (the ratio of infrared to blue luminosity) as a function of morphological type. They argue that a significant population of early type galaxies with large current SFRs have been overlooked in previous studies, and that their inclusion removes the trend in SFR with morphological type (as seen by e.g., Kennicutt et al. [8]). The situation will be greatly clarified when spatially resolved Hα imaging of the galaxies in the Devereux & Hameed sample are available (there is some concern that enhanced nuclear far infrared emission may skew the estimated SFRs).

STAR FORMATION THRESHOLDS

In the last decade, one idea which has gained considerable attention is that galaxies have star formation thresholds. This concept probably represents the greatest single advance in understanding factors which control global star formation. This idea appears to have its origins in early observations of the Magellanic Clouds. McGee & Milton [17] and, later, Davies et al. [18] noted a strong correlation between regions of high HI column density and high surface brightness HII regions. Gallagher & Hunter [19] proposed that a surface density threshold for star formation may be universal for dwarf irregular galaxies and guessed that the threshold was about 5×10^{20} atoms cm^{-2}. In 1987, I compared the HI distributions (from synthesis observations obtained with a uniform linear resolution of 500 pc) with Hα observations of seven nearby dwarf irregular galaxies and found a very strong correlation between the presence of HI in excess of a column density of 10^{21} atoms cm^{-2} and the presence of adjacent HII regions [20]. Sometimes a large fraction of the gas was associated with star formation, but in other cases only a small fraction of the HI appeared to involved in forming massive stars.

An excellent example of this is seen in IC 1613 (Figure 1) where the HI distribution (Lake & Skillman [21]) shows a single strong peak above 10^{21} atoms cm^{-2} and almost all of the bright H II regions are clustered around that HI peak while the rest of the H II regions are associated with HI above 5×10^{20} atoms cm^{-2} (Hodge, Lee, & Gurwell [22]). Note that the HI associated with the high column density peaks is only a small fraction of the total HI in IC 1613. When I showed this figure to a physicist in our department, he said,

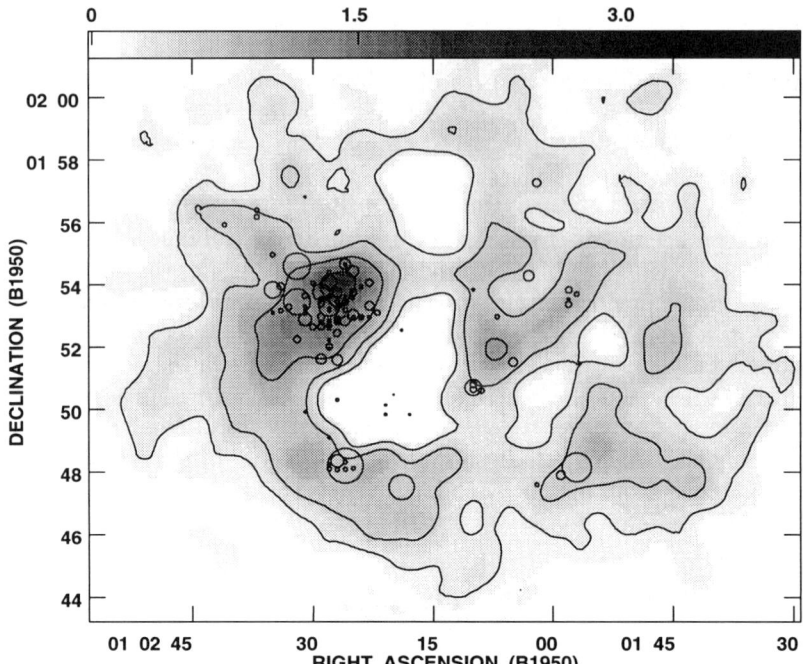

FIGURE 1. The neutral hydrogen column density in the Local Group dwarf irregular galaxy IC 1613 compared with the distribution of H II regions. The HI is shown in greyscale with contours at 0.25, 0.5, 1, and 2×10^{21} atoms cm^{-2}. The image was constructed from VLA observations with a circular beamsize of 1 arcmin (FWHM) which corresponds to 210 pc at the distance of IC 1613. The H II region positions and sizes, which are represented as circles, were taken from Hodge et al. (1990).

"it looks like some sort of threshold effect." If it's clear enough for a physicist to see, there must be something to it.

I think that one of the dangers in inspecting the idea of star formation thresholds is to expect too much. That is, one's natural tendency is to look for a perfect coincidence between gas column densities above a threshold value and the presence of HII regions. Most galaxies fail this test. There are, occasionally, regions of high HI column density without associated HII regions, and HII regions without accompanying high column density HI (although sometimes the offensive HII regions turn out to be planetary nebulae). Given the simplicity of the threshold concept, I think it is much more worthwhile to look for regions of general agreement (or disagreement) rather than to try to fine tune a match to derive the optimal value of the threshold.

I think that the impressive success of Kennicutt's study [23] of star formation thresholds in spirals is due, in part, to the fact that the study concentrated on radial averages. By concentrating on radially averaged properties, Kennicutt was able to pull out the most important pattern, undistracted by small anomalies. Kennicutt's study is now so well known that I don't feel that extensive review is warranted; on the other hand, this review would be lacking without at least a brief overview.

Kennicutt [23] combined radial measurements of the Hα distributions in galaxies with radially averaged total gas surface densities (Σ_g = HI + H$_2$ inferred from CO) and found the following:

- In regions of high gas surface density, the current SFR is well represented by a power law relationship (a "Schmidt" law [24]) with SFR = $a\,\Sigma_g^N$ where N = 1.3 $\pm$ 0.3.
- At lower densities, this relationship fails, and the SFR drops precipitously with decreasing Σ_g.
- The "threshold" values of Σ_g vary over a factor of 10 between galaxies.
- The Toomre disk stability criterion for gravitational instability [25] can be used to predict critical values of the surface gas density (Σ_c). These values show a remarkable ability to predict the limits of star formation in the (predominantly high surface brightness) spiral galaxies studied by Kennicutt.

(As a reminder, the Toomre disk stability criterion holds that a disk is stable against gravitational collapse if:

$$Q \equiv \frac{v_s \kappa}{\pi G \Sigma} > 1 \qquad (1)$$

where v_s is the sound speed and κ is the epicyclic frequency - see Binney & Tremaine [26] §5.3.1 and §6.2.3 for further discussion.)

- By comparing the ratio of Σ_g to Σ_c at the radial edge of the HII region distributions, Kennicutt solved for a value of $\alpha = \frac{\Sigma_g}{\Sigma_c} \approx 0.7$, for the empirical value of the critical threshold for star formation.

Again, the fantastic success of Kennicutt [23] is well known, and doesn't bear extensive reviewing. In the last few years, the focus of related research has been to test the theory in different environments. Below is a partial list of some of those experiments:

- van der Hulst et al. [27] observed a small collection of large, low surface brightness spirals. Although these galaxies all had normal HI contents when measured relative to their masses or luminosities, all were found to have lower HI surface densities relative to high surface brightness spiral galaxies. Their HI surface densities were found to lie at or below the critical densities calculated from the Toomre criterion across the extent of their disks.
- Taylor et al. [28] obtained VLA HI synthesis observations of five HII galaxies and calculated their values of Σ_c. These were compared to optical broadband R and Hα imaging of the galaxies. The results of this study are shown in Figure 2. In all cases, the central HI surface densities exceeded Σ_c, and the radius where the HI surface density fell below Σ_c roughly corresponded to the optical diameter of the galaxy.

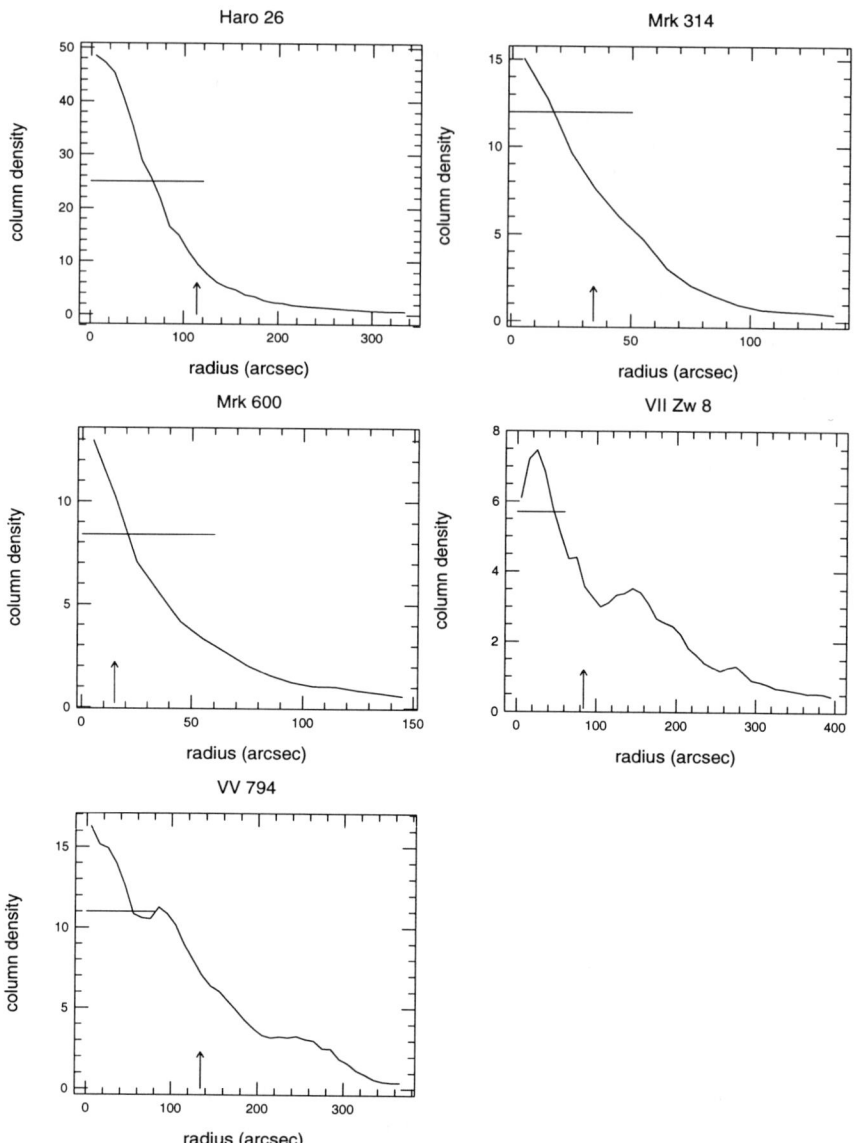

FIGURE 2. Radially averaged HI column density profiles for 5 HII galaxies (in units of 10^{20} atoms cm^{-2}). The horizontal line indicates the value of the threshold density, and its length along the x-axis is the size of the region of solid body rotation. The arrow along the x-axis indicates the isophotal radius, R_{25}, of the 25 mag arcsec^{-2} surface brightness level (from Taylor et al. 1994).

• van Zee and collaborators [29,30] observed ten isolated dwarf galaxies which were divided into (6) low surface brightness dwarf galaxies and (4) "normal" gas-rich dwarf galaxies based on their M(HI)/L(B) ratios. HI distributions and rotation curves were obtained from VLA synthesis observations, and radial values of α ($\equiv \Sigma_g/\Sigma_c$) were derived (see Figure 3). In all cases, values of α were found to be below 1 throughout the disks. Interestingly, *local* peaks in the HI surface densities approached the critical value, and these are associated with sites of active star formation as determined from Hα observations.

At this point, one might find the evidence overwhelming that Kennicutt's star formation threshold is a satisfactory and full explanation of the conditions necessary for star formation in disk galaxies. After uttering such a sentence, one must be struck but how wrong that must be! How can a prescription possibly work when it appears to ignore much of the physics of the interstellar medium which dominates the discussions of star formation?

First off, the Toomre criterion was formulated for the case of a thin, rotating disk. As the observational tests have been performed in different environments (e.g., dwarf galaxies, which have thicker disks and more nearly solid body rotation when compared to spiral galaxies), why should one expect the simple theory to work? After reading the discussions in Binney and Tremaine, I am impressed with the flexibility of the criterion, and, while perhaps expecting the exact value of the criterion to change somewhat, it seems that an instability criterion has applications over a wide range of physical characteristics.

There are, of course, nagging doubts in excess of the intuitive ones. For example:

• Kennicutt [23] pointed out that both M33 and NGC 2403 had regions of "sub critical" star formation in their inner disks. He speculated that perhaps the standard L(CO)-to-M(H$_2$) conversion may be underestimating the H$_2$ contents in the central parts of these galaxies.

• Wilson, Scoville, & Rice [31] showed that the new CO observations of M33 by Wilson & Scoville [32], while raising the H$_2$ mass estimates, did not raise them above the critical threshold level.

• Thornley & Wilson [33] used new CO observations of NGC 2403 to argue that this sub critical star formation was also not due to unseen molecular gas.

• Hunter & Plummer [34] studied the nearby dwarf irregular galaxy Sextans A and found values of α hovering between 0.2 and 0.3 throughout most of the radial extent of the galaxy, while Sextans A is seen to be experiencing vigorous star formation.

Taken together, these observations indicate that while the critical threshold criteria is a very useful concept, in may not be universally applicable (or there may be a "second parameter" to which the value of α is sensitive). Certainly there are other physical ideas which lead to a type of threshold criteria (e.g., a threshold for molecular cloud formation, see review in Skillman [35]). Elmegreen [36] has distinguished between two different gas column density

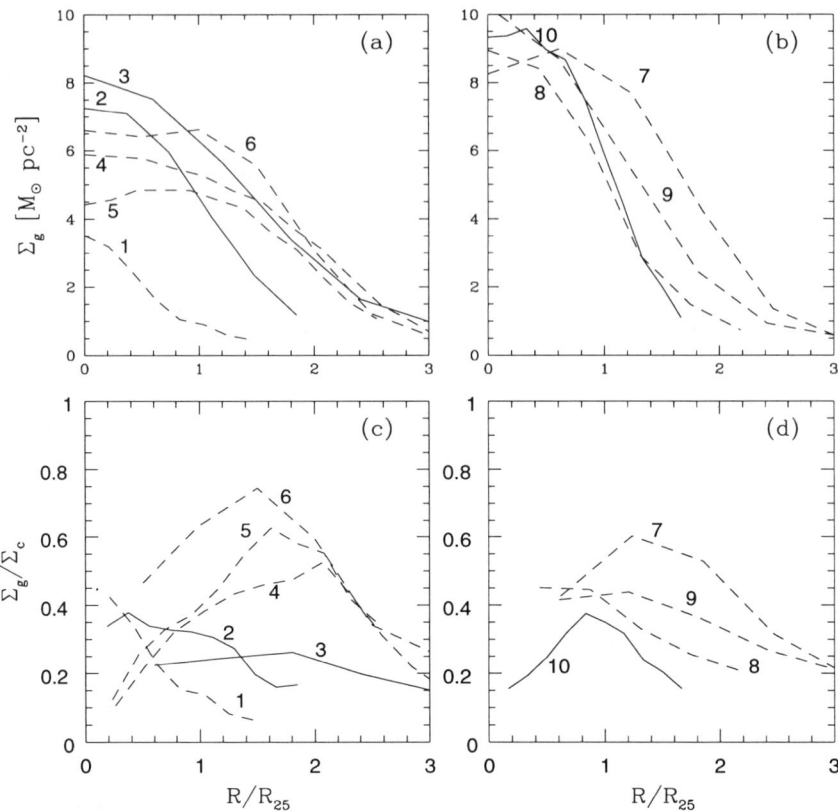

FIGURE 3. The gas surface density distribution of dwarf galaxies. (a) The LSBDGs. (b) The "normal" dwarfs. In panels (c) and (d) the ratio between the gas surface density and the Toomre instability threshold density for the LSBDGs and the "normal" dwarfs, respectively, are shown. Each system is numbered: 1 – UGCA 20; 2 – UGC 2684; 3 – UGC 3174; 4 – UGC 5716; 5 – UGC 7178; 6 – UGC 11820; 7 – UGC 191; 8 – UGC 634; 9 – UGC 891; 10 – UGC 5764. The dashed lines denote galaxies with optical scale lengths greater that 1.5 kpc. The solid lines denote galaxies with smaller optical scale lengths. Note that *all* of the galaxies have gas densities lower than the predicted instability threshold, but that the intrinsically larger systems do have slightly higher ratios of Σ_g/Σ_c.

criteria, that following from the Toomre criterion, and a pressure support criterion [37]. Hunter & Plummer [34] found that this criterion may be in better agreement with the observations in Sextans A, but were wary to draw far reaching conclusions from this individual galaxy.

WHAT'S NEXT?

It is probably obvious that I would like to see more work in the vein of the Hunter & Plummer study of Sextans A. I think that the observational tools are available to give us a better picture of the physical environments of star forming regions. By comparing the physical conditions in various regions, it should be possible to isolate parameters (e.g., gas metallicity, gas velocity dispersion, disk surface mass density, ambient stellar radiation field) to better test the available theoretical alternatives.

In my presentation at the conference, I finished with a presentation of some recent HST observations of color-magnitude diagrams of nearby dwarf irregular galaxies. I wished to emphasize the point that it is now possible to produce very detailed star formation histories of dwarf galaxies, at least over the last billion years. Studies of this type will help us to better understand how we are to interpret the growing compilations of current star formation rates (and the measurements of star formation rates in galaxies at high redshift which are now possible).

I would like to thank Salman Hameed, Rob Kennicutt, Chris Taylor, and Liese van Zee for valuable input into the preparation of this review.

REFERENCES

1. Kennicutt, R. C., Jr. 1983, ApJ, 272, 54
2. Osterbrock, D. 1989, Astrophysics of Gaseous Nebulae and Active Galactic Nuclei, University Science Books
3. Ferguson, A. M. N., Wyse, R. F. G., Gallagher, J. S. III, & Hunter, D. A. 1996, AJ, 111, 2265
4. Scalo, J. M. 1986, Fund. Cos. Phys., 11, 1
5. Massey, P. 1993, in Massive Stars: Their Lives in the Interstellar Medium, eds. J. P. Cassinelli & E. B. Churchwell, ASP Conf. Ser., 35, 168
6. Gallagher, J. S. III, Hunter, D. A., & Tutukov, A. V. 1984, ApJ, 284, 544
7. Sandage, A. 1986, A&A, 161, 89
8. Kennicutt, R. C., Jr., Tamblyn, P., & Congdon, C. W. 1994, ApJ, 435, 22
9. Larson, R. B., & Tinsley, B. M., 1978, ApJ, 219, 46
10. Kennicutt, R. C., Jr., Keel, W. C., van der Hulst, J. M., Hummel, E., & Roettiger, K. A. 1987, AJ, 93, 1011
11. Hunter, D. A. 1994, AJ, 107, 565
12. Hunter, D. A. 1994, AJ, 108, 1658

13. Devereux, N. A., & Scowen, P. A. 1994, AJ, 108, 1244
14. Devereux, N. A., Price, R., Wells, L. A., & Durice, N. 1994, AJ, 108, 1667
15. Oey, M. S., & Massey, P. 1995, ApJ, 452, 210
16. Devereux, N., & Hameed, S. 1997, AJ, submitted
17. McGee, R. X., & Milton, J. A., 1966, Aust. J. Phys., 19, 343
18. Davies, R. D., Eliot, K. H., & Meaburn, J. 1976, Mem. RAS, 81, 89
19. Gallagher, J. S. III, & Hunter, D. A. 1984, ARA&A, 22, 37
20. Skillman, E. D. 1987, in Star Formation in Galaxies, ed. C. J. Lonsdale Persson, NASA CP-2466, 263
21. Lake, G., & Skillman, E. D. 1989, AJ, 98, 1274
22. Hodge, P. W., Lee, M. G., & Gurwell, M. 1990, PASP, 102, 1245
23. Kennicutt, R. C., Jr. 1989, ApJ, 344, 685
24. Schmidt, M. 1959, ApJ, 129, 243
25. Toomre, A., 1964, ApJ, 139, 1217
26. Binney, J., & Tremaine, S. 1987, Galactic Dynamics, (Princeton: Princeton University Press)
27. van der Hulst, J. M., Skillman, E. D., Smith, T. R., Bothun, G. D., McGaugh, S. S., & de Block, W. J. G. 1993, AJ, 106, 548
28. Taylor, C. J., Brinks, E., Pogge, R. W., & Skillman, E. D. 1994, AJ, 971
29. van Zee, L. 1996, Ph.D. thesis, Cornell University
30. van Zee, L., Haynes, M. P., Salzer, J. J., & Broeils, A. H. 1997, AJ, submitted
31. Wilson, C. D., Scoville, N., & Rice, W. 1991, AJ, 101, 1293
32. Wilson, C. D., & Scoville, N. 1989, ApJ, 347, 743
33. Thornley, M. D., & Wilson, C D. 1995, ApJ, 447, 616
34. Hunter, D. A., & Plummer, J. D. 1996, ApJ, 462, 732
35. Skillman, E. D. 1994, in Violent Star Formation, From 30 Doradus to QSOs, ed. G. Tenorio-Tagle, (Cambridge: Cambridge University Press), 168
36. Elmegreen, B. G. 1995, Rev. Mex. A&ASC, 3, 55.
37. Elmegreen, B. G., & Parravano, A. 1994, ApJ, 435, L121

Star Formation Rates, Efficiencies and Initial Mass Functions in Spiral Galaxies

Fernando de Pablos and Jordi Cepa

Instituto de Astrofísica de Canarias, E-38200, La Laguna, Tenerife, Spain

Abstract. Using broad band photometry, a new method is derived to evaluate relative (arm with respect to the interarm disk) star formation rates and relative star formation efficiencies together with spiral arm amplitudes, as a function of the galactocentric radius. The classical method to obtain star formation rates from Hα photometry is discussed, and a new method is inferred to diagnose the possible presence of biassed star formation due to different initial mass functions in the arms and in the interarm disk. As an example, these methods are applied to the spiral galaxy NGC 4321. Evidence is obtained of massive star formation triggered in the spiral arms in a way consistent with the spiral density wave theory, and star formation biassed towards a larger fraction of massive stars in the arms than in the interarm disk, related to and probably caused by the density wave system.

APPLICATION OF THE METHOD

We can evaluate the relative star formation rate of intermediate mass stars as a function of the galactocentric radius as

$$\text{RSFR}^{Measured} = \frac{SFR^{Arm}}{SFR^{Disk}} = \frac{B^{Arm} - B^{Disk} A_\star}{B^{Disk\star} - B^{Disk}} \quad (1)$$

where

$$A_\star = \frac{I^{Arm}}{I^{Disk}} K \quad (2)$$

is the amplitude of the stellar density wave and K is the light contributed to the I Johnson passband coming from the old population of the disk. Its value is ~ 0.85 (Schweizer 1976).

$$\text{RSFR} = \frac{\text{RSFR}^{Measured}}{\chi_B} \quad (3)$$

The proportionality function χ_B handles the dependence of the IMF with (r, θ) in the sense that a larger fraction of massive stars can be formed in the arm than in the interarm region (Güsten & Mezger 1982). If $\chi_B > 1$ then evidence for biased star formation is present. The relative star formation efficiency can be derived dividing the relative star formation rate by $A_\star$.

Relative Massive Star Formation Rates

The relative star formation rate of massive stars as a function of the galactocentric radius can be derived as

$$\text{RSFR}_{OB}^{Measured} = \frac{SFR_{OB}^{Arm}}{SFR_{OB}^{Disk}} = \frac{H\alpha^{Arm}}{H\alpha^{Disk\star}} \quad (4)$$

where

$$\text{RSFR}_{OB} = \text{RSFR}_{OB}^{Measured} Q / \chi_{H\alpha} \quad (5)$$

and $0.7 \leq Q \leq 1.1$ depending on the boundary conditions and the temperature of the HII regions belonging to the arm or the interam regions. $\chi_{H\alpha}$ has an analogous meaning to that of χ_B.

If no correction for possible biassed star formation is applied then,

$$\frac{\text{RSFR}^{Measured}}{\text{RSFR}_{OB}^{Measured}} = Q \frac{\chi_B}{\chi_{H\alpha}} \quad (6)$$

If there is no biassed star formation, $\chi_B/\chi_{H\alpha}=1$. Otherwise this quantity would be less than 1.0. So if the ratio given in equation 6 is less than 0.7, then, evidence for biassed star formation is present.

RESULTS

We present the results of the method previously described when applied to the grand design spiral galaxy NGC4321.

Triggered Star Formation

In both the arms of NGC4321 the qualitative behavior of the measured RSFR and RSFR$_{OB}$ as a function of the radius is notably similar, with a larger SFR of more massive stars in the arms than in the interarm disk (figures 1 and 2). In figure 3 the arm amplitudes of both arms as a function of the radius are represented. We assume that corotation is located at $\sim$110–120″ where dips in RSFR, RSFR$_{OB}$ and arm amplitude can be seen in both

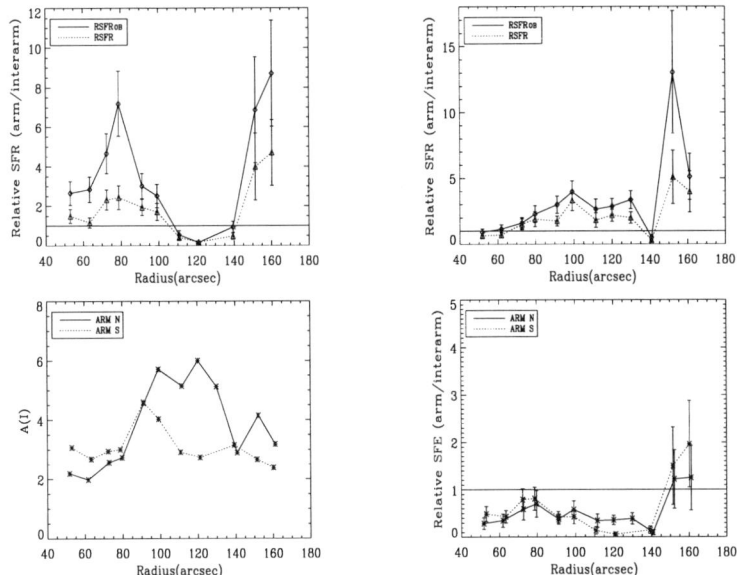

FIGURE 1 (top left): Relative massive star formation rate and relative intermediate mass star formation rate as a function of the galactocentric radius for the southern arm of NGC4321.
FIGURE 2 (top right): The same as figure 2 for the northern arm.
FIGURE 3 (bottom left): Arm amplitudes of NGC 4321 as a function of the galactocentric radius.
FIGURE 4 (bottom right): Relative intermediate mass star formation efficiency for the spiral arms of NGC 4321 as a function of the galactocentric radius.

arms. The corotation radius determined in this way is consistent with that of Sempere et al. (1995) and that of Elmegreen, Elmegreen & Seiden (1989). The relative star formation efficiency of intermediate mass stars and of massive stars are represented in figures 4 and 5 respectively. The arms are in general less efficient, or as efficient, forming intermediate stars than the interarm. However, the southern arm is more efficient forming massive stars than the interarm disk by a factor up to ~2.5–3.5, indicating that density waves are triggering massive star formation. In the northern arm this parameter is in general lower than one, except in a localized region after corotation. del Río (1995) found the same results using different techniques. Moreover, triggering should show up as a non-linear relation between arm amplitude and relative star formation of massive stars, and viceversa: if no triggering of star formation is taking place, a linear relation should be expected between the RSFR$_{OB}$, and the arm amplitude. From figure 6 it appears that there is no evidence of correlation between the RSFR$_{OB}$ and the southern arm amplitude, again leading to the presence of some sort of triggering mechanism. However, there is a remarkable linear relation between the points where no triggering is present (with a RSFE of massive stars less than one), which are, in a way consistent with the spiral density wave theory, situated around the corotation radius. In the northern arm, avoiding the points beyond 140″ (where there is some star formation triggering), we observe a linear relation between RSFR$_{OB}$ and the

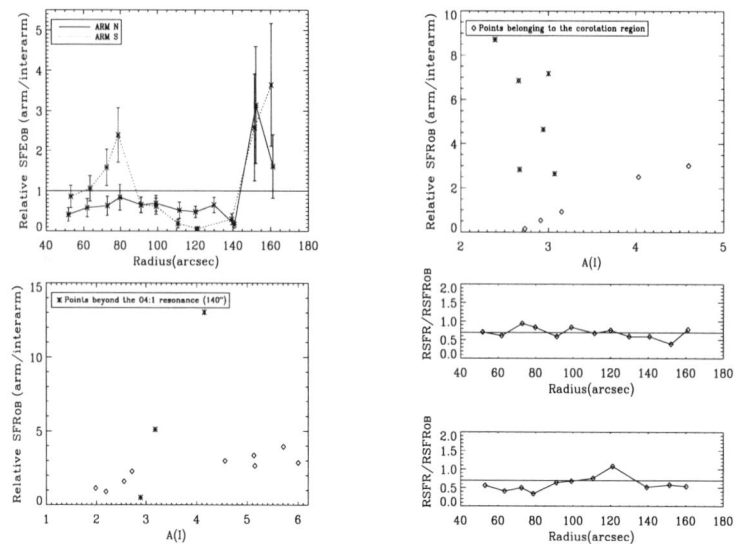

FIGURE 5 (top left): Relative massive star formation efficiency for the spiral arms of NGC 4321 as a function of the galactocentric radius.
FIGURE 6 (top right): Relative massive star formation rate as a function of the arm amplitude for the southern arm of NGC4321.
FIGURE 7 (bottom left): The same as figure 6 for the northern arm.
FIGURE 8 (bottom right): Ratio of the relative (arm/interarm) intermediate mass star formation rate of stars over that of massive stars as a function of the galactocentric radius for the northern (upper diagram) and southern (lower diagram) arms of NGC4321.

northern arm's amplitude (figure 7). This linear correlation is pointing to the absence of star formation triggering in the northern arm. This difference in behavior between both arms might be due to the interaction with NGC4322.

Biassed Star Formation (BSF)

In figure 8 we have plotted the measured ratio RSFR/RSFR$_{OB}$ as a function of the galactocentric radius for both the arms of NGC4321. As pointed out in previous sections, values of RSFR/RSFR$_{OB}$ lower than 0.7 constitute evidence for a different IMF in the arms with respect to the interarm disk. So, the presence of BSF would be mainly situated in the zone where there is triggered massive star formation (*i.e.* RSFE$_{OB}$>1). This suggests that density waves cause a change in the IMF in the arms.

REFERENCES

1. Schweizer, F. 1976,ApJS,31,313
2. Gusten, R., & Mezger, P.G. 1982,Vistas in Astron,26,159
3. Sempere, M.J., García-Burillo, S., Combes, F., & Knapen, J.H. 1995,A&A,296,45
4. Elmegreen, B.G., Elmegreen, D.M., & Seiden, P.E. 1989,ApJ,343,602
5. del Río, M.S. 1995,Ph.D.thesis,Universidad de La Laguna

The History of the Star Formation Rate in the Local Disk

Helio J. Rocha-Pinto and Walter J. Maciel

Instituto Astronômico e Geofísico da USP
Av. Miguel Stefano 4200, CEP 04301-904, São Paulo SP, Brazil

Abstract. We present a method to obtain the history of the local star formation rate (SFR) by using the G dwarf metallicity distribution. The method is tested by simulations and applied to the solar neighborhood. We find strong evidences of two events of star formation: a big burst 8 Gyr ago, and a lull 2-3 Gyr ago. We also find evidences of a present star formation burst. These events agree with some previous determinations of the history of the local SFR.

INTRODUCTION

Previous attempts to recover the history of the star formation rate (SFR) in the solar neighborhood show that the ratio of the present SFR to the mean past SFR is almost unity [1]. This has led many authors to conclude that the SFR has been roughly constant, or slightly decreasing, over the lifetime of the Galaxy. However, several independent evidences suggest that the solar neighborhood has been experiencing star formation bursts. These evidences come from studies relative to stellar age distributions, chromospheric emission in late-type dwarfs, distribution of Li abundances, present-day mass function, stellar kinematics and the white dwarf luminosity function (see [2], for a review).

The occurrence of bursts is likely to put signatures in the metallicity distribution, in the sense that *there will be more stars with the metallicity corresponding to the time of the burst*. Our work consists mainly in recovering the SFR history from these possible signatures in the metallicity distribution of G dwarfs.

A BRIEF DESCRIPTION OF OUR METHOD

The method we have developed to recover the SFR history associates the number of stars with metallicity between $Z(t)$ and $Z(t+\Delta t)$ with the number

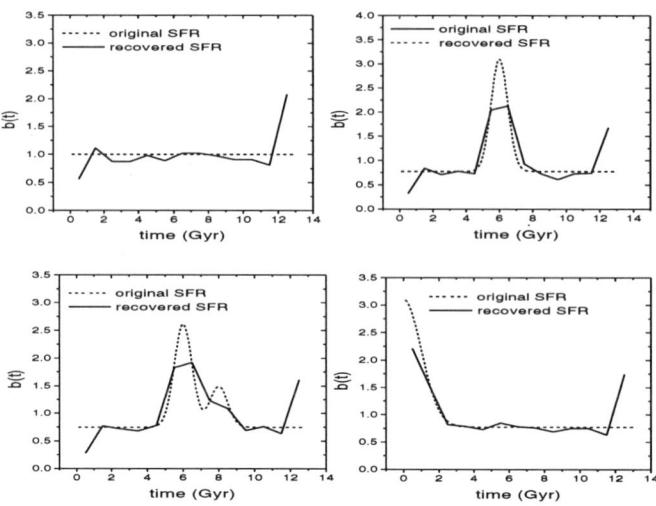

FIGURE 1. Example illustrating the recuperation of the SFR from the metallicity distribution in a fictitious system which experiences irregular SFR.

of stars ever born in the time interval $(t, t + \Delta t)$, were $Z(t)$ is obtained from an empirical age – metallicity relation (AMR).

Due to the cosmic abundance scatter, the metallicity z of a star is often different from the mean metallicity Z of the gas out of which it formed. The method incorporates corrections that correlate the observed metallicity distribution $N(z)$ with a mean metallicity distribution at birth $N(Z)$, for which the counting procedure can be applied. We also have used the Sommer-Larsen factors [3] to correct the SFR history for scale height effects.

TESTING THE METHOD WITH SIMULATED BURSTS

We test the method by simulating the chemical evolution of a system experiencing an irregular SFR. We first derive the age-metallicity relation and the metallicity distribution for such system. After that, we apply both functions to our method to see if it can successfully recover the original star formation rate. Several input SFR histories were used. Typical results are shown in fig.1.

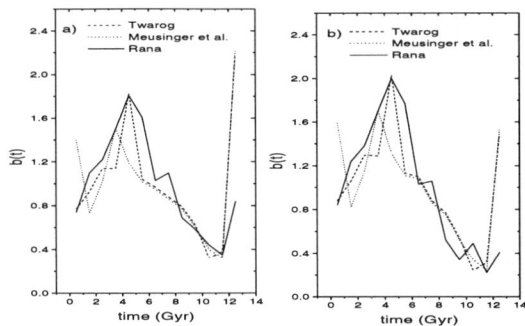

FIGURE 2. Star formation history for the solar neighborhood, extracted from the metallicity distribution of long-lived stars, using three empirical AMRs: a) using the whole G dwarf sample; b) using only stars with lifetime greater than the disk age.

THE LOCAL STAR FORMATION HISTORY

We apply the method to the solar neighborhood. We adopt our recent G dwarf metallicity distribution [4] and three AMRs [5–7] from the literature. The derived SFRs are shown in fig.2. Fig. 2a shows the SFRs derived from all stars in the G dwarf sample. Fig. 2b shows the same functions derived after taking into account only stars with lifetimes greater than the disk age, adopted as 12 Gyr. These long-lived stars were identified by means of a relation between [Fe/H] and $(b-y)$ at the turnoff [8]. Note that there is a good agreement in the derived SFRs, in the sense that all SFRs show a peak at $t = 3$ to 7 Gyr. Also, we can identify a period with weak star formation at $t = 9$ to 12 Gyr. The age of these events varies from one SFR to the other, depending on the AMR adopted. The results also indicate that the solar neighborhood is presently experiencing a strong star formation episode. However our simulations indicate that the method suffers for overestimating of the number of stars in the last temporal bin (see fig 1). Thus, we are not sure how much of this star formation activity is real.

The main trends of our derived SFRs are also present in other works. In fig.3, we show a comparison with a mean SFR from the histories shown in fig.2b, and other SFR histories from the literature. It can be seen that there is a major star formation era at $t = 3$ to 9 Gyr, and a substantial decrease in the star formation activity between 10 to 12 Gyr. Also, some SFR history determinations [9,10] indicate a star formation enhancement in the present time.

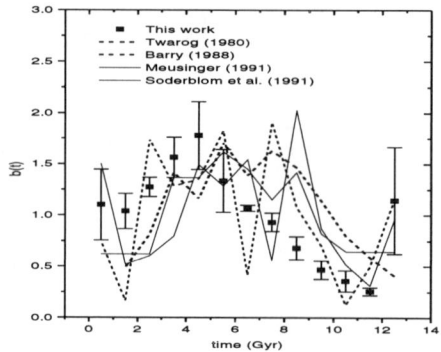

FIGURE 3. Comparison between our mean star formation history for the solar neighborhood with other determinations based on different methods.

ACKNOWLEDGEMENTS

This work was supported by CNPq and FAPESP.

REFERENCES

1. Scalo, J.M., *Fund. Cosm. Phys.* **11**, 1 (1986).
2. Majewski, S.R., *Ann. Rev. Ast. Astrophys.* **31**, 575 (1993).
3. Sommer-Larsen, J., *Month. Not. Roy. Ast. Soc.* **249**, 368 (1991).
4. Rocha-Pinto, H.J., and Maciel, W.J., *Month. Not. Roy. Ast. Soc.* **279**, 447 (1996).
5. Twarog, B.A., *Astrophys. J.* **242**, 242 (1980).
6. Meusinger, H., Reimann, H.-G., and Stecklum, B., *Ast. Astrophys.* **245**, 57 (1991).
7. Rana, N.C., *Ann. Rev. Ast. Astrophys.* **29**, 129 (1991).
8. Rocha-Pinto, H.J., and Maciel, W.J., *submitted.*
9. Barry, D.C., *Astrophys. J.* **334**, 446 (1988).
10. Soderblom, D.R., Duncan, D.K., Johnson, D.R.H., *Astrophys. J.* **375**, 722 (1991).
11. Meusinger, H., *Astrophys. Spac. Sci.* **182**, 19 (1991).

Percolating Star Formation in Barred Spirals

A. S. Mott*, P. Alexander* and J. P. Sleath[†]

*Cavendish Laboratory, Madingley Rd., Cambridge, CB3 0HE, UK
[†]Department of Physics and Astronomy, University of Wales Cardiff,
P.O. Box 913, Cardiff, CF2 3YB, Wales, UK

Abstract. The Stochastic Self-Propagating Star Formation (SSPSF) model [1], for self-regulating star formation based on a percolation process mitigated by supernova shocks, has been very successful in producing quantitative agreement with observations [2] when a Spiral Density Wave (SDW) has been a part of the imposed potential. It has been noted that many Spirals have a significant central bar. In this paper we extend the model to examine whether a barred potential alone is sufficient to produce long-lived spirals in the presence of propagating star formation.

The SSPSF model is based upon a numerical scheme in which test particles (gas-clouds and stars) move and interact in an imposed potential. We follow 30,000 cloud particles, all of which are capable of developing associated stellar clusters. The clouds are treated as a collisional system which accrete material from a cylindrically symmmetric low-density ISM. Clouds may either undergo spontaneous star formation, or stimulated star formation if hit by a supernova shock. Using a realistic galactic potential, including the SDW, we have managed to generate a wide range of spiral galaxy morphologies, predict the cluster formation rate for the Milky Way (although there are a large number of input parameters to the model, these are all determined by Galactic observations), and predict a form for the Schmidt Law, in agreement with observational constraints.

In order to model a realistic galaxy we consider the functional form for the potential of Allen & Santillán [3]. This potential consists of three steady state axisymmetric terms:

(i) a central bulge

$$\phi_1(r,z) = \frac{-M_1}{\sqrt{r^2 + z^2 + b_1^2}} \qquad (1)$$

(ii) a disc component

$$\phi_2(r,z) = \frac{-M_2}{\sqrt{r^2 + \left[a_2 + (z^2 + b_2^2)^{1/2}\right]^2}} \qquad (2)$$

(iii) a halo component

$$\phi_3(R) = -\frac{M_3 R^{1.02}}{a_3^{2.02} \chi} - \frac{M_3}{1.02 a_3} \left[-\frac{1.02}{\chi} + \ln \chi\right]_R^{\{100\ kpc\}} . \qquad (3)$$

In these expressions R and χ are defined by

$$R = \sqrt{r^2 + z^2}, \quad \chi = \left\{1 + \left(\frac{R}{a_3}\right)^{1.02}\right\}. \qquad (4)$$

A fourth term permits us to introduce a non-axisymmetric potential, for example that representing a SDW

$$\phi_4 = \frac{-Ar^2}{(a_4^2 + r^2)^2} \cos(n\theta - n\Omega_p t + \Psi(r)) \qquad (5)$$

in which

$$\Psi(r) = \frac{\ln\left[1 + \left(\frac{r}{r_0}\right)^p\right]}{p \tan i_0}. \qquad (6)$$

Note that there is a domain $\Psi = 0$ in which this potential is bar-like, but this is confined to within $r = 1kpc$, ie within the central bulge not modelled in our simulations of the *disk*. In order to model a bar-like potential we consider a simple perturbation, which replaces the SDW potential (equation 5):

$$\phi_4 = \frac{-Ar^2}{(a_4^2 + r^2)^2} \cos(n\theta - n\Omega_p t). \qquad (7)$$

This potential is bar-like throughout the domain of the simulation and has the advantage of eliminating three of the input parameters required by the SDW form.

In addition to using a bar-like potential, it is also necessary to consider the orbital population. Nearly-circular **loop** orbits dominate axisymmetric and SDW potentials; for the bar potential we expect a new, smaller, but significant, population of highly elongated **box** orbits [4]. We characterise the type of orbit of a particle by a dimensionless ratio formed from its energy and angular momentum. In order to produce a realistic distribution of orbits we have an extended relaxation period at the start of the simulation during which

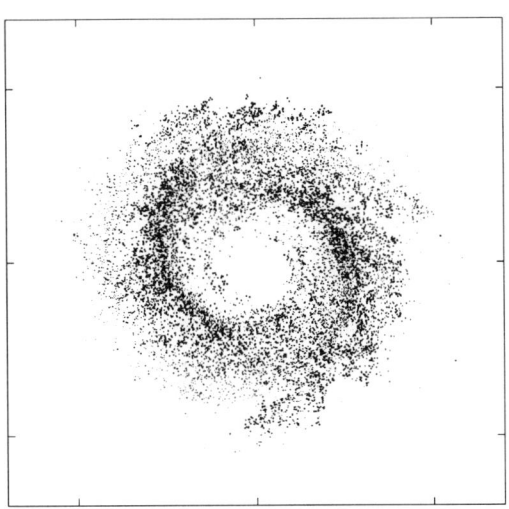

FIGURE 1. The pattern of the galactic disk after $1.8Byr$ simulation time with an imposed SDW perturbing potential. The picture covers a radius of $12.5kpc$. The larger points correspond to the younger, and therefore brighter, stellar associations.

all the test particles are treated as a strongly collisional gas. The collisions are sufficent to scatter particles from loop orbits into box orbits. During this initial relaxation period the star formation processes are kept inactive. The simulation proper is seeded by selecting a fraction of the test particles to represent stars, with an appropriate age distribution, and the rest being cloud particles.

The simulation follows the star particles until they reach an age of $10Myr$ at which point the most massive (O/B) stars will undergo supernovae. These SNR induce star formation in any cloud particle within their radii with a probability proportional to the mass of the cloud. The SNR grow as $t^{2/5}$ and are followed until they are deemed to be too weak to induce star formation (after another $10Myr$). Besides this *stimulated* star formation, the clouds may undergo spontaneous star formation, again with a probability proportional to their masses, but this is typically six orders of magnitude smaller than the probability of stimulated star formation. The clouds accrete mass from the ISM with a density modelled on the Galactic HI distribution. Clouds undergoing star formation have their mass (needed for future star formation) reduced dramatically, thus regulating the percolation process.

Figure 1 shows the structure resulting from a simulation including a SDW potential (equation 5), whereas in figure 2 the structure resulting from a typical simulation in which a bar potential (equation 7) was used is shown. Both simulations have been allowed to run until the galaxies have settled down into

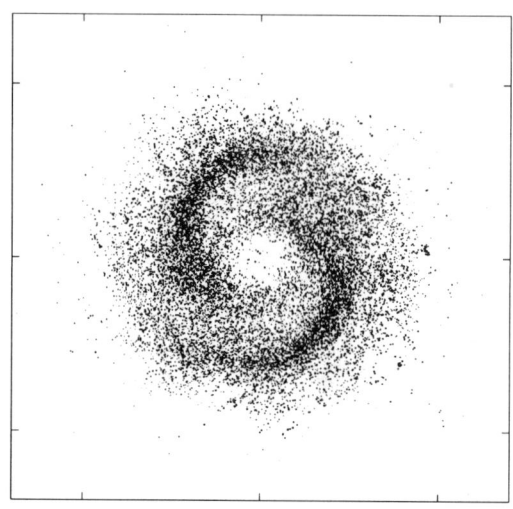

FIGURE 2. As Figure 1, except that a bar potential has been used instead of a SDW.

a steady state in terms of spiral structure and have a steady time-averaged star formation rate. The bar potential together with SSPSF is successful at inducing spiral patterns in the host galaxy, assisted by the interplay between loop and box orbits. However, there do appear to be qualitative differences between figures 1 and 2; the spiral arms are rather brighter and wider when driven by a bar.

Summary

We find that a perturbing bar potential is sufficient to cause spiral structure, but we have yet to clarify any quantitative differences that may exist between SDW and bar simulation galaxies.

REFERENCES

1. Sleath, J. P., *A New Model of the Structure of Spiral Galaxies based on Propagating Star Formation.*, Cambridge: PhD Thesis, 1995.
2. Sleath, J. P. & Alexander, P., 1995a, MNRAS, 275, 507.
3. Alan, C. & Santillán, A., 1991, Rev. Mex. Astron. Astrofís., 22, 255.
4. Binney, J. & Tremaine S., *Galactic Dynamics*, New Jersey: Princeton University Press, 1987, ch. 3, pp. 126-153.

Turbulent Fragmentation of Interstellar Gas and the Stellar Mass Spectrum

Valery Khersonsky

Department of Physics and Astronomy
University of Pittsburgh
Pittsburgh, PA 15217

Abstract. I discuss a connection between the mass spectrum of interstellar gas clouds and the mass spectrum of formed stars. Turbulent processes create the hierarchy of interstellar clouds with mass spectrum $\sim M^{-3/2}$. This mass spectrum is directly related to the mass spectrum of stellar population of young stellar cluster. This relationship is explored in terms of *the partial effectiveness of star formation*, i.e., the number of stars of mass m formed in a cloud of mass M per unit interval of stellar mass. This function transfers the statistical properties of the interstellar cloud population to the statistical properties of the stellar population. It is shown that newly formed stars are not homogeneously distributed in the interval of allowed stellar masses. Their distribution is peaked around some specific mass (depending on cloud mass) and decreases rapidly outside the vicinity of this mass.

THE MASS SPECTRUM OF MOLECULAR CLOUDS

Turbulent fragmentation of highly compressible interstellar gas is the result of interactions of stochastic shock waves (mergers, amplification and dissipation of shocks). In a regime of stationary supersonic turbulence the distribution function of shocks in velocities is

$$f(v) \propto v^{-3}, \qquad (1)$$

[1] where v is a typical velocity behind the shock front. The mass spectrum of fragments can be obtained from the velocity spectrum using the transformation law:

$$\left|\frac{d\mathcal{N}(M)}{dM}\right| \equiv f(M) = 3Df(v)\frac{dv}{dM} = \propto \frac{1}{v^3}\frac{dv}{dM}. \qquad (2)$$

The relation between v and M can be derived from empirical scaling relations between the velocity dispersion, $\Delta v \sim v$, the gas density ρ, the mass of the fragment, M, and the size-scale, l,

$$v \sim l^\alpha, \quad \rho \sim l^{-\beta}, \quad M \sim \rho l^3 \sim l^{3-\beta}, \tag{3}$$

with $0.38 \leq \alpha \leq 0.6$ and $1.0 \leq \beta \leq 1.4$ (see, for example, [2]). The upper end of the hierarchy (largest scales) can be identified as regions of enhanced averaged density in our Galaxy and nearby spiral galaxies [3] which appear as very large HI complexes, giant molecular clouds, HII regions, etc. The lower end of the observed hierarchy corresponds to dense cores of small mass ($\sim$ a few $M_\odot$) found in a variety of nearby clouds with typical sizes over a tenth of a parsec.

From these equations one can derive that

$$l \propto M^{1/(3-\beta)}, \quad v \propto M^{\alpha/(3-\beta)}, \quad \frac{dv}{dM} \propto M^{(\alpha+\beta-3)/(3-\beta)}. \tag{4}$$

Therefore, if $\alpha \approx 0.5$ and $\beta \approx 1$,

$$f(M) \propto M^{(\beta-2\alpha-3)/(3-\beta)} = 3DM^{-3/2}. \tag{5}$$

The empirical mass spectrum is $f(M) = 3Dd\mathcal{N}(M)/dM = 3D\mathcal{N}_0 M^{-\gamma}$, with $\gamma \approx 1.4\text{-}1.7$ in the interval of cloud mass from $M_{c,min} \sim 1\, M_\odot$ to $M_{c,max} \sim 10^5 - 10^6\, M_\odot$, ([4,6] and references therein; [7].)

The basic features of turbulent fragmentation are as follows:

1. Interstellar turbulence forms the hierarchy of interstellar clouds of different scales (see scaling relations).

2. Gravity *may couple* and, therefore, *amplify* such scales which approximately satisfy the virial conditions appropriate for a given temperature, magnetic field strength, and external pressure.

3. If gravitational effects are strong enough, then further evolution of such a structure can be related to the development of gravitational instability and collapse. **However,** *the initial conditions for gravitational instability on each level of structure hierarchy are determined by the properties of the turbulent velocity field and differ significantly from those which are considered in the analysis of Jeans instability in a quiescent uniform medium.*

TRANSFER OF PROPERTIES OF CLOUD ENSEMBLE TO THE STELLAR POPULATION

The Partial Effectiveness of Star Formation or Transfer Function, $S(m, M)$, can be defined as the number of stars of mass m per unit interval of stellar masses in a stellar cluster formed in a cloud of mass M.

It has been shown [8] that the maximum mass of a star formed in a cloud of mass M is $m_{max}(M) \approx pM^\varsigma$, with $p = 3D0.33$ and $\varsigma = 3D0.43$. Therefore, the minimum mass of a cloud which forms a star of mass m is $M_{min}(m) = 3D(m/p)^{1/\varsigma} = 3DPm^\lambda$ with $\lambda = 3D1/\varsigma = 3D2.33$ and $P = 3D1/p^\lambda = 3D13.2$.

The number of stars of mass m per unit stellar mass, $dn(m,M)/dm$ formed in $d\mathcal{N}(M)$ clouds of mass M in the interval $[M, M+dM]$ is $S(m,M)d\mathcal{N}(M) = 3DS(m,M)f(M)dM$.

The integral equation for $S(m,M)$ expresses the fact that the total number of stars of mass m per unit interval of stellar mass formed in clouds of all masses larger than $M_{min}(m)$ is determined by the equation [5]

$$\phi(m) \equiv \frac{dn(m)}{dm} = 3D \int_{M_{min}(m)}^{\infty} S(m,M)f(M)dM. \tag{6}$$

The solution of the integral equation for the function $S(m,M)$ can be obtained for any arbitrary finite and differentiable function $\phi(m)$. This solution is

$$S(m,y) = 3D - \frac{\varsigma}{\mathcal{N}_0 p^\nu}[g(m,M)]^{\nu+1}\frac{d\phi(t)}{dt}\Big|_{t \to g(m,M)}, \tag{7}$$

where $\nu = 3D(\gamma-1)/\varsigma$, and

$$g(m,M) = 3D\frac{m}{\{1 - [M_{min}(m)/M]^{(\gamma-1)}\}^{1/\nu}} = 3D\frac{m}{\{1 - [m/m_{max}(M)]^\nu\}^{1/\nu}}. \tag{8}$$

STAR FORMATION IN THE ORION B GMC

(a) *The mass spectrum of interstellar clumps* is taken from the CS 2→1 molecular line survey [7], $f(M) = 3D\mathcal{N}_0 M^{-\gamma}$, with $\mathcal{N}_0 \approx 220\ M_\odot^{-1}$ and $\gamma \approx 1.6$. The survey includes clumps in the interval 8-700$M_\odot$. To extend this mass spectrum to larger masses we use the same slope (see the review by [6]).

(b) *The mass spectrum of stars* in the same region is taken to be of the form

$$\phi(m) = 3D\frac{dn(m)}{dm} = 3Dn_0 \exp\left[-q\left(\frac{m_*}{m}\right)^\nu\right]\left(\frac{m_*}{m}\right)^\eta, \tag{9}$$

with $\nu = 3D(\gamma-1)/\varsigma \approx 1.4$ and $\eta \approx 2.7$ from a 2.2 micron survey of the same region [9]. The position of the maximum for this distribution is not known. To be more specific I assume that $m_m = 3D1\ M_\odot$. Then $qm_*^\nu \approx 1.93$. Data from the 2.2 micron survey (Lada et al. 1991b) can be used again to determine the normalization constant n_0. The estimate is $n_0 m_*^\eta = 3D3.4 \times 10^3$.

(c) *The $S(m,M)$ function* is

$$S(m,M) = 3D\varsigma\frac{n_0}{\mathcal{N}_0}\left(\frac{m_*}{p}\right)^\nu \left(\frac{m_*}{g(m,M)}\right)^{\eta-\nu} \exp\left\{-q\left(\frac{m_*}{g(m,M)}\right)^\nu\right\}\left[\eta - \nu q\left(\frac{m_*}{g(m,M)}\right)^\nu\right]. \tag{10}$$

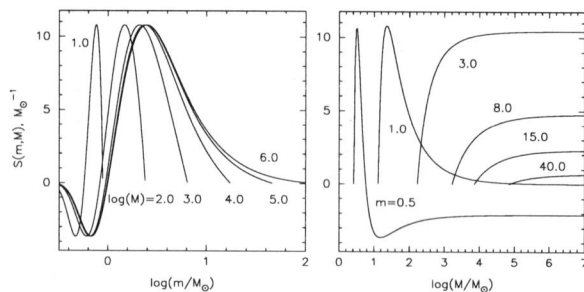

FIGURE 1. Partial effectiveness of star formation in the Orion OB1 association.

The function $S(m, M)$ is strongly dependent on stellar mass. An interstellar cloud clump of mass M **forms stars** $(S(m, M) > 0)$ **with masses mostly in the vicinity of**

$$m = 3D m_{max}(M)\left\{1 + r(\sqrt{\eta} - \sqrt{\nu})\right\}^{-1/\nu}, \quad r = 3D\frac{1}{q}\left(\frac{m_{max}(M)}{m_*}\right)^{\nu}\frac{\sqrt{\eta}}{\nu}.$$

Star formation is suppressed $(S(m, M) < 0)$ **in the vicinity o stellar masses** $m = 3D m_{max}(M)\left\{1 + r(\sqrt{\eta} + \sqrt{\nu})\right\}^{-1/\nu}$ probably due to heating of small mass and low opacity cores by radiation from newly born high mass stars. Small cores can become gravitationally unbound before they form protostellar cores. Radiation and stellar winds from high mass stars can destroy small mass cores and remove them from the star formation process.

REFERENCES

1. Ogul'chanskij Ya.Yu., *Kinematics and Physics of Selestial Bodies*, **8**, 3 (1992).
2. Larson R.B., *MNRAS*, bf 194, 809 (1981).
3. Elmegreen B.G., & Elmegreen D.M., *Ap. J.* **267**, 31 (1983).
4. Solomon P.M., Rivolo A.R., Barret J., & Yahil A., *Ap. J.*, **319**, 730 (1987).
5. Khersonsky V.K. *ApJ*, February 10, 1997
6. Blitz L., in The Physics of Star Formation and Early Stellar Evolution, Lada, C.J., & Kylafis, N.D., (eds.), Kluwer, p. 3 (1991).
7. Lada E.A., Bally, J., & Stark, A.A., *Ap. J.*, **368**, 432 (1991a).
8. Larson R.B., *MNRAS*, **200**, 159 (1982).
9. Lada E.A., DePoy D.L., Evans N.J., II, & Gatley I., *ApJ*, **371**, 171 (1991)

MHD Turbulence and Scaling Laws in Molecular Clouds

Taoling Xie

Laboratory for Millimeter-wave Astronomy, Department of Astronomy, University of Maryland, College Park, MD 20742; email: tao@astro.umd.edu

Abstract. We proposed recently that the Standard Scaling Laws observed in molecular clouds are a manifestation of well-developed MHD turbulence with the energy distribution or Alfvénic fluctuating magnetic field becoming scale-independent (energy spectrum $E_k \propto k^{-1}$). We discuss here some known scaling laws in MHD turbulence and their possible relevance to molecular clouds. It seems that the Iroshnikov-Kraichnan law matches the Standard Scaling Laws if density is added into the analysis and virial equilibrium is assumed. We argue, however, that a more attractive alternative is the inverse turbulent cascade (from small to large scales) which is observed to occur in helical MHD turbulent systems (with tangled or multiply-connected magnetic field lines) given the fact that a lot of energy sources inject energy in the interstellar medium on small scales.

Nature is much more imaginative than we are—E.N.Parker 1979

OBSERVED SCALING LAWS IN MOLECULAR CLOUDS

A major finding of molecular clouds is that they contain a significant internal kinetic energy in the form of non-thermal motions (referred to as turbulence), which is often found to be comparable to the gravitational energy of the clouds, suggestive of virial equilibrium. The nature of the turbulent motions is not yet clear, but one additional interesting finding is the so-called velocity dispersion-size correlation, i.e., $\sigma_v \propto l^\alpha$. Although controversial, most observational and theoretical studies favored $\alpha \sim 0.5$. Together, observations indicate $\sigma_v \propto l^{0.5} \propto \rho^{-0.5}$ (where ρ is the mean density of the cloud over a size l), which are referred to as Standard Scaling Laws by Xie (1997a) [11].

Apparently, the Standard Scaling Laws are not consistent with the Kolmogorov law for incompressible turbulence. For a long time, these scaling laws have been taken as evidence for a constant general magnetic field B_0 in molecular clouds over a large range of size scales [8]. This notion has recently been challenged by Xie [11], who emphasized that the relevant quantity is the

Alfvénic fluctuating magnetic field B instead of the general field B_0. With this modification, arguments parallel to Myers & Goodman (1988) [8] led to the conclusion that the same scaling laws imply a scale-independent Alfvénic B, or a k^{-1} energy spectrum, which is related to the cascade properties of MHD turbulence.

SOME KNOWN SCALING LAWS IN TURBULENCE

By taking on turbulence as a research topic, unfortunately, one is stepping in the murky water of non-linear processes which defy useful rigorous mathematical approaches at the moment. Historically, fortunately, scaling relations, largely based on heuristic energetics arguments of certain arbitration, have afforded significant physical insight into a lot of problems in turbulence. Well-known examples include the Kolmogorov-Obukhov law ($E_k \propto k^{-5/3}$) for hydrodynamic incompressible eddy turbulence and the Iroshnikov-Kraichnan law ($E_k \propto k^{-3/2}$) for incompressible non-helical 3-D MHD turbulence. The common thread to both is the assumption of an existence of inertial range in size scales (or k) where energy is assumed to be injected on the largest scales and the energy transfer rate, $\epsilon = \frac{\rho v_l^2}{\tau}$, from large to small scales l (equivalently from small to large $k \simeq 1/l$, referred to as direct cascade) is held constant, i.e.,

$$\epsilon = \frac{\rho v_l^2}{\tau} = const. \qquad (1)$$

Note that for incompressible fluid, $\rho = 1$.

For hydrodynamic eddy turbulence, energy exchange is assumed to occur only between two eddies with comparable sizes and thus τ is just the eddy turn-over time $\tau_l = l/v_l$, leading to the Kolmogorov law, $v_l = (\epsilon/\rho)^{1/3} l^{1/3}$ o r $E_k \propto \rho(\epsilon/\rho)^{2/3} k^{-5/3}$.

For MHD turbulence, large scale magnetic field enables energy transfer between different scales and τ is different in that the wave interaction time $\tau_A = l/v_A$ (where $v_A = B_0/(4\pi\rho)^{1/2}$ is the Alfvén velocity) should be smaller than $\tau_l = l/v_l$ and many interactions are needed to cause significant distortion on wave amplitudes. The suggestion that $\tau = \tau_l^2/\tau_A = \frac{lv_A}{v_l^2}$ [2] led to the Iroshnikov-Kraichnan law (hereafter I-K law), $v_l = (\epsilon v_A/\rho)^{1/4} l^{1/4}$ or $E_k \propto \rho(\epsilon v_A/\rho)^{1/2} k^{-3/2}$.

Neither the Kolmogorov law nor the I-K law for incompressible fluid ($\rho = 1$) can possibly be consistent with the Standard Scaling Laws, because one of them is $\rho \propto l^{-1}$, which is intimately tied up to the rest of relationships.

The compressibility of molecular clouds might change the situation somewhat, but it is not clear physically how the compressibility should be included. The simplest, naive and straight-forward way might be to allow ρ to vary in the above scaling relations (for a related discussion see, e.g., Fleck 1996 [3]). Assuming that a $\rho \propto 1/l$ relation is established by a separate physical mechanism such as the virial equilibrium, then it is easy to demonstrate that the Kolmogorov law results in $v_l \propto (\epsilon)^{1/3} l^{2/3}$ or $E_k \propto (\epsilon)^{2/3} k^{-4/3}$, inconsistent with the Standard Scaling Laws, while the I-K law results in $v_l = (\epsilon v_A)^{1/4} l^{1/2}$ or $E_k \propto (\epsilon v_A)^{1/2} k^{-1}$, which agree with the Standard Scaling Laws provided that v_A is scale-independent.[1]

A somewhat less well-known case believed to develop a k^{-1} spectrum in a robust manner is MHD turbulence with non-zero magnetic helicity, pioneered by Pouquet et al (1976) [9]. Magnetic helicity density is defined as $H = A \cdot B$, where A is the vector potential of B satisfying $B = \nabla \times A$, which measures the topological structure of the field lines such as the tangling, kinking or connectivity [1]. One major difference between this case and the non-helical MHD turbulence is that inverse cascade (cascade from small to large scales) and considerable self-organized structures on large scales are observed to occur in 3-D case leading to the k^{-1} energy spectrum [2] in an inertial range. Since the k^{-1} energy spectrum corresponds to energy density equipartition between interacting wave packets at different scales, Xie (1997a) [11] argued that it can be understood as a result of effective energy cascade as energy accumulates on larger scales due to lack of significant dissipation there [2]. A formal argument, however, is given in terms of a constant magnetic helicity cascade rate [2], in analogy to energy cascade arguments for the Kolmogorov and I-K laws, as follows. Define a constant magnetic helicity cascade rate,

$$\eta = H_l/\tau = const., \qquad (2)$$

where $H_l \sim B_l^2 l$. Biskamp (1993) suggested $\tau = l/v_l$, which leads to $B_l = const.$ or $E_k \propto k^{-1}$. We feel, however, that this form of τ resembles that for the pure hydrodynamic turbulence case and fails to include the effects of a global magnetic field on the energy exchange. Heuristically, assuming that the general magnetic field is dominantly important, i.e., $\tau = l/v_A$, we obtain $B_l = (\eta/v_A)^{1/2}$ or $E_k \propto v_A^{-1} k^{-1}$. Allowing a varying ρ for compressibility, this would lead to $v_l \propto \rho^{-1/2}$ and $E_k \propto k^{-1}$, consistent with the Standard Scaling Laws provided that v_A is scale-independent.

DISCUSSION

¿From the view point of predicted scaling laws, pure hydrodynamic turbulence is simply not an attractive option for explaining the supersonic motions in molecular clouds. Taking into account of compressibility in the most straight-forward manner for MHD turbulence, it seems that both the I-K

---

[1] Both observations and theory show a field-density scaling in the form $B \propto n^{0.5}$ for the density range $10^2 - 10^9$ cm^{-3} [7,5], which implies a scale-independent v_A.

analysis and helicity analysis lead to scalings consistent with the observed Standard Scaling Laws, provided that v_A is scale-independent and a $\rho \propto l^{-1}$ relation is established by a separate physical mechanism such as the virial equilibrium.

In view of gravitation, shear or differential rotation as well as many other possible forces, certain degree of tangling, kinking and even knotting would almost appear to be entirely plausible in molecular clouds. So the whole idea of helical MHD turbulence and inverse cascade seems particularly attractive for further investigation, as kinetic energy is often injected on very small scales in the interstellar medium through physical processes such as stellar winds and one of the principal problems is how this energy can be channeled to larger scales [6,10]. This also opens up the possibility that turbulence in star-less dark clouds might get a considerable portion of its energy from MHD disturbances caused by streaming cosmic rays on very small scales through resonant interactions with the general magnetic field. As recently suggested and discussed [12], the effective dissipation of this energy into heat might just limit the turbulent energy density at the thermal level, leading to a scale-independent turbulent velocity on small scales, which might explain the reported "velocity coherence" in small dense cores [4].

We thank Amitava Bhattacharjee for a useful suggestion and John Wang and John Scalo for discussions. This research is supported in part by the NSF grant AST9314847 to the Laboratory for Millimeter-wave Astronomy at the University of Maryland.

REFERENCES

1. Berger, M. & Field, G. 1984, J.Fluid Mech. 147, 133
2. Biskamp, D. 1993, Nonlinear Magnetohydrodynamics, Cambridge University Press
3. Fleck, R.C. 1996, ApJ, 458, 739
4. Goodman, A. et al 1997, submitted to ApJ.
5. Heiles, C., Goodman, A.A., McKee, C., & Zweibel, E. 1993, Protostars & Planets III, eds.E.H.Levy & J.I.Lunine (Tucson:Univ. Arizona Press), 279
6. Henriksen, R.N. & Turner, B.E. 1984, ApJ, 287, 200
7. Mouschovias, Th. 1991, The Physics of Star Formation and Early Stellar Evolution, eds.C.J.Lada & N.D.Kylafis, 61
8. Myers, P.C., & Goodman, A.A. 1988, ApJ, 329, 392
9. Pouquet, A., Frisch, U. & Léorat, J. 1976, J. Fluid Mech., 77, 321
10. Shu, F.H., Adams, F.C.,& Lizano, S. 1987, ARAA, 25, 23
11. Xie, T. 1997a, ApJ Letters, 475, L139.
12. Xie, T. 1997b, preprint

Star Formation at the Intermediate Distances: Gravitational Collapse in Massive Cores

Qizhou Zhang, Paul T. P. Ho

Harvard-Smithsonian Center for Astrophysics
60 Garden Street
Cambridge, MA 02138

Abstract. We present spectroscopic evidence of gravitational collapse in massive molecular cloud cores toward the W51 HII region complex. The NH_3 (J,K)=(2,2) and (3,3) lines obtained at $\sim 1''$ angular resolution exhibit inverse P-Cygni or asymmetric profiles, suggesting inward motions of gas toward the central star. The infall motions appear to be localized to individual dense cores, contrary to the suggestion of overall collapse. The two collapsing cores are embedded in a more extended cloud of 1.6pc. If both cores are formed out of this cloud through fragmentation, it seems that both cores are now under collapse again in making the final product: massive OB stars.

BACKGROUND

Massive stars are the key ingredient in galaxies. They dominate the observed radiation from distant galaxies across most of the spectra from UV to radio frequencies. Once they are born, radiation and winds from OB stars dramatically affect the chemistry and evolution of the natal cloud cores. The death of massive stars ejects back into the interstellar medium the metal-enriched material which fuels the next generation of star formation.

In this presentation, we provide spectroscopic evidence of molecular core collapse leading to the formation of massive OB stars. The candidate source is W51, an HII region complex in the Sagittarius spiral arm about 8 kpc away [6]. Rudolph et al. (1990) proposed, based on their HCO^+ observations, an overall collapse involving clouds of $1'$ scale (2.4pc) and 40,000 $M_\odot$. To test this idea, we present $1''$ resolution studies of this region in the NH_3 (2,2) and (3,3) lines.

RESULTS

Distribution of continuum and dense molecular gas

In Figure 1, we present the 1.3cm continuum map and the integrated line emission of the NH_3 (2,2) inversion transition obtained with the VLA in the C-configuration.

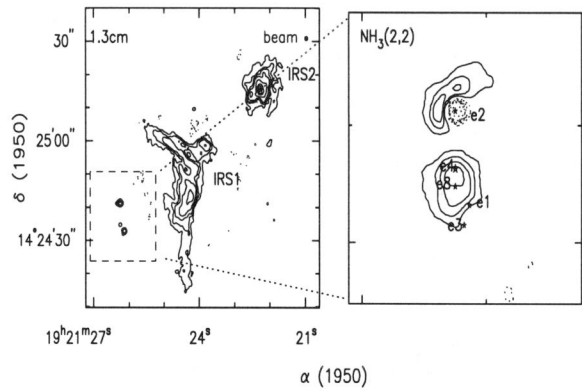

FIGURE 1. The 1.3 cm continuum and integrated flux (from 52 km s^{-1} to 62 km s^{-1}) of the NH_3 (2,2) line. The contours for the continuum are plotted at ± 0.024, 0.036, 0.060, 0.12, 0.24, 0.36 and 0.48 Jy/beam. The contours for the integrated emission are plotted at ± 0.06, 0.12, 0.18, 0.24, 0.30, 0.48 and 0.66 Jy/beam×km s^{-1}. The star denotes the positions of the continuum sources. The synthesized beam is indicated at the upper-right corner of the continuum map.

The 1.3cm continuum arises mostly from the free-free emission of gas ionized by the UV radiation from newly born OB stars. Associated with the molecular gas in the W51e1/e2 region is a cluster of ultra-compact HII regions. Within 5″ (0.2pc) of the projected separation, there are five early B-type stars.

Two dense cores are identified in NH_3 in this region. The high angular resolutions improve the association of the molecular cores with the HII regions. The southern core in the e1/e2 region, which was thought to be associated with e1 [5,3], is centered on W51e8. There is no dense gas detected toward the bright HII region e1 at the sensitivity of 4mJy (1σ).

Kinematics

Figure 2 shows the line spectra of the (2,2) and (3,3) inversion transitions toward the positions of W51e2 and e8 in comparison with the CH_3CN (8-7) [7]. The optically thin line of CH_3CN shows a symmetric line profile and defines the core systemic velocity. The optically thick NH_3 lines exhibit inverse P-Cygni or asymmetric profiles in the main hyperfine component.

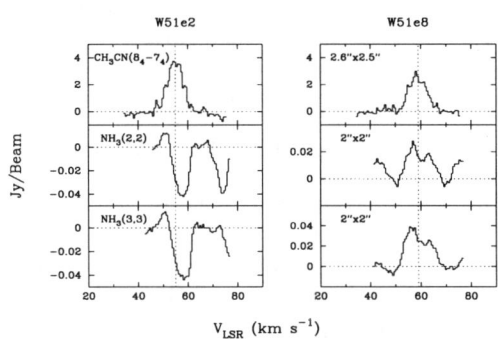

FIGURE 2. The NH_3 spectra at the position of the w51e2 and e8 HII regions.

In the e2 core, the NH_3 lines appear both in emission (52 km s^{-1}) and absorption (58 km s^{-1}). Compared with the cloud systemic velocity of 55 km s^{-1}, the red-shifted absorbing gas in front of the bright continuum source and the blue-shifted emitting gas in the back of the continuum source indicate inward motions of material toward the star.

The line spectra at the position of e8 also suggest infall in the e8 core. The optically thick NH_3 (2,2) and (3,3) lines show self-absorption at the systemic velocity defined by the CH_3CN ($8_4 - 7_4$) line. Both NH_3 lines are doubly peaked with the blue-shifted peak stronger than the red-shifted one. These evidences, *i.e.*, symmetry in the optically thin line and asymmetry in optically thick lines are consistent with infall signatures of a centrally condensed cloud (see [8]).

DISCUSSION

Overall collapse versus localized collapse

The widespread absorption in HCO$^+$ led Ruldolph et al. (1990) to suggest that the W51 e1/e2 and IRS1 region are engaged in a large scale collapse involving 40,000 $M_\odot$ over 1' scale (2.4pc). Observations by others have indicated that the HCO$^+$ absorption is due to a cold foreground cloud that is not related to the W51 dense cores [1,2,4].

Our observations in NH$_3$ do not show widespread absorption features at 62 km s^{-1} as seen in HCO$^+$. Instead, the red-shifted absorption appears at 58 km s^{-1} and is localized only to the bright HII region e2. In addition, the blue- and red-shifted features in NH$_3$ which signify the infall are shifted symmetrically in velocity with respect to the optically thin line. These evidences indicate that the infall is localized to individual cores. Although our observations do not exclude the possibility of large scale fragmentation that form the dense cores, the collapse of individual dense cores seems to be the mode that leads to the formation of the observed OB stars.

Fragmentation, collapse and formation of OB cluster

Compared with the average stellar density in HII regions in the Galaxy, there is an over abundance of OB stars in W51. In C^{18}O, Zhang et al. (1996, in preparation) identified a cloud extending 40″ in the north-south and 15″ in the east-west. It is possible that all the HII regions in the e1/e2 cluster are physically associated with this cloud. If this is the case, the massive stars must have formed synchronously within short periods of time ($\ll 10^5$ years). Otherwise, the UV radiation and winds from the earlier generation will disperse the parental cloud and prevent the formation of the next generation of stars.

It has been proposed that fragmentation is the dominant mode during the collapse of large molecular clouds. It is possible that the e2 and e8 dense cores are formed through the fragmentation of the extended cloud seen in C^{18}O. Dense cores continue to collapse after they are formed. It appears that the infall motion observed in the dense cores may represent the second stage of collapse which leads to the formation of OB stars.

REFERENCES

1. Cox, M. J., Scott, P. F., Andersson, M., and Russell, A. P. G. 1987, MNRAS, 226, 703.
2. Goldader, J. D., and Wynn-Williams, C. G. 1994, ApJ, 433, 164
3. Ho, P T. P., Genzel, R., and Das, A. 1983, ApJ, 266, 596
4. Ho P. T. P. and Young L. M. 1996, ApJ, in Press
5. Rudolph A., Welch W. J., Palmer P., and Dubrulle R. 1990, ApJ, 363, 528
6. Schneps M. H., Lane A. P., Downes D., Moran J. M., Genzel R., and Reid M. J. 1981, ApJ, 249, 124
7. Zhang, Q., Ho P. T. P. and Ohashi, N. 1996, in preparation
8. Zhou, Evans, Kömpe and Walmsley 1993 ApJ, 404, 232

Origin of the Mass in Massive Star Outflows

Ed Churchwell

Department of Astronomy, University of Wisconsin
475 N. Charter St., Madison, WI 53706

Abstract. The molecular outflows associated with massive star formation regions often contain more mass than the central stars that presumably drive the outflows. This paper deals with the origin of the mass. It is shown that both accumulated stellar winds and entrained interstellar matter in bipolar jets are unlikely to be able to account for the observed outflow masses. It is proposed here that the outflow masses are the result of infalling matter that has been diverted outward into bipolar jets and some important consequences of this are explored.

INTRODUCTION

The outflows found in massive star formation regions are often substantially more massive than that of the central star. Table 1 lists several massive outflows believed to be associated with massive stars along with outflow rates, outflow masses, and dynamical ages of the outflows. One can see that the masses are in the range of tens of solar masses. What is the origin of the mass in massive star outflows? Three possibilities come to mind: 1) accumulated stellar winds; 2) mass loaded stellar winds or bipolar jets; and 3) infalling matter (accretion) which has been diverted into bipolar outflows. In the following, I will examine each of these and show that accumulated stellar winds and matter entrained in jets or highly collimated outflows cannot account for the observed masses. In the absence of other explanations, I assume that deflection of infalling matter into bipolar outflows is the origin of the observed mass and explore some consequences of this

ACCUMULATED STELLAR WINDS

For stellar winds to supply an outflow of 50 solar masses over a typical period of 10^4 yr, the average mass loss rate would have to be about 0.005 solar masses per year. This is about two orders of magnitude greater than the highest mass loss rate observed toward O-stars and WR stars. A mass loss rate of this magnitude, if it is radiation driven, cannot be greater than $\frac{L_*/c}{v_\infty}$ where L_* is the luminosity of the star, c is the speed of light, and v_∞ is the wind terminal velocity. This requires that the wind terminal velocities must be in the range 0.4 to 4 km s^{-1} for $L_* = 10^{5-6}$ solar luminosities. Such velocities are several times lower than observed molecular outflow velocities. It therefore appears unlikely that the outflows are the result of accumulated stellar winds.

ENTRAINED ISM IN STELLAR BIPOLAR JETS

Let us now consider entrained interstellar matter (ISM) in stellar bipolar jets. The issue here is how much ISM can be entrained in a bipolar stellar jet. The entrainment rate per unit area is

$$\dot{M} = \varepsilon \rho_o c_o$$

(Cantó and Raga 1991) where ε is the entrainment efficiency ($<<1$), ρ_o is the ambient ISM density, and c_o is the speed of sound in the ambient ISM. For T = 20-50 K and $n_{H_2} \approx 10^5$ cm^{-3},

$$\dot{M} = \varepsilon(0.9-1.4)x10^{-14} \text{ gm cm}^{-2} \text{ s}^{-1}$$

If we take the central flow in G75.78NE as an example (see Table 1), we have a flow radius of 0.69 pc and an age of 3.7×10^4 yr. To estimate its area, we assume it to have an approximate cone shape. For a flow mach number in the range 10-20, the jet opening angle is expected to be $\theta = 2/\sigma \approx 0.04 \, rad = 2.3 \, deg$ from experimental results for a "mixing layer limited" entrainment process (see fig. 2 in Cantó and Raga 1991). This gives a total surface area (both lobes) of about $1.2x10^{36}$ cm^{-2} and an entrained mass

$$M_{flow} = \dot{M} A_{flow} \tau_{flow} \approx \varepsilon(6-10)M_c$$

where A_{flow} is the area of the turbulent mixing layer of the jet and τ_{flow} is the dynamical lifetime of the outflow. For any realistic value of ε, the total mass of entrained ISM would be $M_{flow} < 1M_o$. One could increase this by assuming that the stellar mass loss occurs not as a jet but as large puffs of matter with much larger opening angles. For example, if $\theta = 30 \, deg$, $M_{flow} \approx \varepsilon 50 M_o$. In the "mixing layer limited" entrainment process, $\varepsilon < 0.089$ (Cantó and Raga 1991) and even in the large opening angle case, one can only entrain a few solar masses in outflows; certainly nothing like the several tens of solar masses observed. We, therefore, conclude that neither accumulated stellar wind material nor entrained ISM in stellar jets appear to be capable of providing the mass observed in the outflows from young OB stars.

INFALLING MASS DIVERTED INTO BIPOLAR OUTFLOWS

Although the deflection of infalling matter into bipolar outflows is not understood, in the absence of any other viable mechanism we will assume that this is the primary source of the matter in massive star outflows and examine some implications of this in the following.

First, it implies that massive YSOs must go through a very rapid accretion phase with a mass accretion rate

$$\dot{M}_{acc} > \dot{M}_{flow} \approx 10^{-2} - 10^{-3} \, M_\odot yr^{-1}$$

$\dot{M}_{acc}$ must be larger than $\dot{M}_{flow}$ since some of the accreting matter ultimately ends up on the central star. $\dot{M}_{acc}$ also places restrictions on the density structure of the accreting matter. Since the infall rate cannot exceed the free-fall rate, if gravity is the only inward force, we have

$$\rho(r) \geq \frac{\dot{M}_{acc}}{4\pi r^2}\left(\frac{r}{2GM_*}\right)^{1/2} = \frac{\dot{M}_{acc}}{4\pi r^{3/2}(2GM_*)^{1/2}}$$

where r is the distance from the central star and M_* is the mass of the central YSO.

For example, for $\dot{M}_{acc} = 10^{-2}$ $M_\odot$ yr^{-1} toward an O5 ZAMS star ($M_* = 40$ $M_\odot$; Straizys and Kuriliene 1981), the density of infalling matter at 1000 AU would be $\geq 10^8$ cm^{-3}. This requirement on the density of infalling matter would provide a natural explanation for the very high densities detected in the immediate vicinity of very compact millimeter sources such as G9.62 E and F (Hofner et al. 1995) and others.

Second, if the outflow has achieved its maximum mass (i.e. old and near the end of the bipolar outflow phase), then the outflow mass plus stellar mass gives the minimum mass that a molecular cloud core must have to form a star of a given mass. For example, the central outflow in the G75.78NE region is apparently driven by an early B-star (M_* about 10 $M_\odot$) and has an outflow mass of about 58 $M_\odot$. Thus the fraction of the infalling matter actually incorporated into the star is only about 15% and about 85% is deflected back outward into the outflow. In this sense, the efficiency of getting matter onto the central star is only about 15% and one concludes that a cloud core with a mass at least 6-7 times that of the final star is required to form the star. This may also be part of the reason why massive stars are so rare; typical cloud cores generally do not achieve such large masses.

Third, accretion rates of 10^{-2} to 10^{-3} $M_\odot$ yr^{-1} will result in most of the stellar UV radiation being absorbed close to the central star, thus delaying the formation of a detectable UC HII region until after this phase of star formation. For example, a density greater than 10^8 cm^{-3} of in-falling matter toward an O5 star would result in a Strömgren sphere of radius ≤ 100 AU. This implies that the massive outflows observed to date are likely to be associated with YSOs that have not yet formed a detectable UC HII region. Those that are associated with a UC HII region, such as G5.89, are probably no longer being driven but are relics of an earlier rapid accretion phase. This has the further implication that detection of massive YSOs in their rapid accretion phase or earlier can only be achieved either via their molecular outflows or their thermal dust emission at FIR to mm wavelengths or high excitation molecular line probes, but not from radio free-free emission.

REFERENCES

Acord, J. M., Walmsley, C. M., Churchwell, E. 1997, Ap. J., in press
Cantó, J., Raga, A. C. 1991, Ap. J., 372, 646
Garden, R. P., Hayashi, M., Gatley, I., Hasegawa, T., Kaifu, N. 1991, Ap. J., 374, 540
Hofner, P., Kurtz, S., Churchwell, E., Walmsley, C. M., Cesaroni, R. 1995, Ap. J., 460, 359.
Hunter, T. R., Phillips, T. G., Menten, K. M. 1997, Ap. J., submitted
Hunter, T. R., Taylor, G. B., Felli, M, Tofani, G. 1994, A & A, 284, 215
Shepherd, D. S., Churchwell, E., Wilner, D. J. 1997, Ap. J., in press
Shepherd, D. S., Churchwell, E. 1996, Ap. J., 471, in press
Straizys, V., Kuriliene, G. 1981, Ast. & Sp. Sci., 80, 353.

TABLE 1
Molecular Outflows Associated with OB Stars

Source	Sp. Type	Stellar Mass solar masses	log(outflow rate) solar masses/yr	Outflow Mass solar masses	Dyn. Age 10^4 yr	Notes	Refs.
G5.89-0.39	O6.5	28	-1.59	77	0.3	1,2	1
G45.12+0.13A			-2.35	178	11	1,2	2
G45.12+0.13B			-2.51	117	22	1,2	2
G45.12+0.13C	O5.5	35	-2.82	68	7	1,2	2
G45.12+0.13D			-2.85	72	15	1,2	2
G75.78NE-C	B0	16	-2.80	58	3.7	2	3
G75.78NE-N			-3.45	19	5.3	2	3
G75.78NE-E			-3.36	20.7	4.7	2	3
G98.04+1.45	O6.5	28	-3.36	40	9.1	1	4
G111.25-0.77	B0.5	13	-3.36	16	3.7	1	4
G173.58+2.45	B2.5	7	-3.22	32	5.3	1	4
G192.16-3.82	B2.5	7	-3.67	58	27	1	4
W75N	B1-B2	11,8	-2.92	48	4.4	3	5
DR21	O4-O5	>40	-1.2	~3000	5	3	6

Notes:
1. Single dish results
2. Interferometer results
3. Multiple sources present

References:
1. Acord et al. 1997
2. Hunter et al. 1997
3. Shepherd et al. 1997
4. Shepherd and Churchwell 1996
5. Hunter et al. 1994
6. Garden et al. 1991

Star Formation in Translucent Clouds

Thomas Hearty*, Loris Magnani*, Jean-Pierre Caillault* Ralph Neuhäuser[†], J.H.M.M. Schmitt[†], & John Stauffer[‡]

*Department of Physics & Astronomy, University of Georgia, Athens, GA 30602-2451
[†] Max-Planck-Inst. für Extraterrestrische Physik, 85740 Garching, Germany
[‡] Harvard/Smithsonian Center for Astrophysics, Cambridge, MA 02138

Abstract.
The star-formation capability of three low-extinction translucent molecular clouds (TMCs) at high Galactic latitude is investigated. In an attempt to identify possible PMS stars in and around the clouds we have analyzed *ROSAT* All-Sky Survey and PSPC pointed observations of these TMCs. Follow-up optical spectroscopy (conducted with the 1.5-m Fred Lawrence Whipple Observatory telescope) of the stellar candidates with $m_V < 15.5$ was performed in order to identify standard signatures of pre-main-sequence (PMS) stars. We have found one dozen X-ray bright, lithium rich stars near clouds MBM7 and MBM55. However, all of these may be part of a slightly older stellar population that has retained its lithium. The X-ray sources projected onto cloud MBM40 show no signs of youth and are likely part of the X-ray field-star population.

INTRODUCTION

The TMCs investigated in this paper represent a broad array of the low-mass, low visual extinction TMCs most often identified at high Galactic latitude. Since each of these clouds contains at least one dense core similar to the cores found in star-forming dark clouds, the possibility of star-formation must be examined. The three clouds in our sample represent different types of TMCs; MBM7, located at $l = 150°.4$ and $b = -38°.1$, is a compact, yet gravitationally unbound, cloud that appears to be part of an HI arc associated with several other high-latitude molecular clouds; MBM40 ($l = 37°.6$ and $b = 44°.7$) is an isolated compact gravitationally bound molecular cloud; MBM55 ($l = 89°.2$ and $b = -40°.9$) differs from the other two in that it is embedded in a region at high Galactic latitude containing a large filamentary network of HI within which several other molecular clouds are embedded.

X-RAY AND OPTICAL OBSERVATIONS

X-ray observations of star-forming regions have revealed a large population of previously undetected PMS stars (e.g., Neuhäuser et al. 1995a). Thus, we used the *ROSAT* Position Sensitive Proportional Counter (PSPC) to locate X-ray active stars in the direction of the three TMCs included in this study. The primary method was to investigate the *ROSAT* All-Sky Survey (RASS) data in large regions around the respective clouds to identify any X-ray stars in the vicinity of these clouds since the true extent of the molecular gas is not well known and T Tauri stars can often be displaced several parsecs from their parent clouds (e.g., Neuhäuser et al. 1995b). Several possible mechanisms for the genesis of this newly identified population are discussed by Feigelson (1996). However, caution must be taken when identifying PMS stars several degrees from a molecular cloud since many of these objects my be part of the foreground population of slightly older ($\sim 10^8$ yr) stars predicted by Briceño et al. (1997) that still have T Tauri characteristics. The secondary method was to study three deep *ROSAT* Pointed Observations of the well studied cores of each of the TMCs described above. Finding young stars spatially coincident with the molecular cores would present the most compelling evidence that star formation is ongoing in these low-mass clouds.

We investigated the RASS data in a 55 deg^2 area around MBM7, a 23 deg^2 area around MBM40 and an 82 deg^2 area around MBM55 and found 60, 31, and 222 sources, respectively, above a maximum likelihood detection threshold of 8 in the direction of the three clouds. An additional 22, 14, and 9 sources with a S/N > 2.5 were detected in the three pointed observations of high column density regions. We cross-referenced all of these sources with the SIMBAD[1] and NED[1] databases to remove all non-PMS stars and extragalactic objects; 31 sources were removed via this procedure. An additional 139 sources were eliminated from our list of objects for further study since their high log(f_X/f_V) ratio (> 0.0) precludes them from being stars (Stocke et al. 1991). Typically, each of the 188 remaining X-ray sources had only one possibly-stellar optical counterpart. Many of these stars were found to have X-ray hardness ratios similar to T Tauri stars (e.g., Neuhäuser et al. 1995a) and were therefore considered as candidates for follow-up optical spectroscopy to look for other indicators of youth.

The optical spectra were obtained from September 29 through October 2 of 1995 using the FAST spectrograph with the 600 lines mm^{-1} grating of the 1.5 m Fred Lawrence Whipple Observatory telescope. The 1".5 slit size provided a reciprocal dispersion of ~ 0.75 Å pixel^{-1}. The wavelength range of the spectra ($\sim$ 5500-7500 Å) was selected to detect two indicators of possible youth (Hα

---

[1] This research has made use of the SIMBAD database, operated at CDS, Strasbourg, France and the NASA/IPAC Extragalactic Database (NED) which is operated by the Jet Propulsion Laboratory, Caltech, under contract with the National Aeronautics and space Administration.

emission & Li I λ6708 absorption) and to determine spectral types. We were able to observe all of the optical candidates brighter than $m_V \sim 15.5$ mag (140 possible optical counterparts to 117 X-ray sources) except for a small (~ 6 deg^2) region east of MBM55 where our limiting magnitude was about 13 mag; however, this left only four stars with $13 \leq m_V \leq 15.5$ unobserved.

In the direction of MBM40 we detected four dMe stars which are likely part of the field dMe star population. Another two dMe stars were detected in the direction of MBM7 and twelve in the direction of MBM55. None of the dMe stars observed showed evidence for a large lithium abundance. Thirteen K stars and one G star in direction of MBM55 have Hα in emission; two of the K stars also have weak lithium absorption. A total of twelve F-K stars (4 in the direction of MBM7 and 8 in the direction of MBM55) show lithium absorption.

DISCUSSION

Although the Li I absorption and X-ray activity (and Hα emission for two of the K stars) indicate that these stars are young objects, these properties are still present in stars as old as the Hyades (Briceño et al. 1997). If these stars formed in MBM7 or MBM55, they should have ages similar to the lifetimes of the clouds (a few $\times 10^6$ yr), otherwise they are likely to be part of the foreground population of young stars with ages $\sim 10^8$ yr predicted by Briceño et al. (1997). Unfortunately, since we do not know the distances to these stars, we cannot place them on the HR diagram to estimate their ages from a comparison with theoretical isochrones. However, we can plot the stars on a W(Li) vs. T_{eff} diagram and determine less accurate relative ages by comparing them with different age groups of stars. In Figure 1 we compare our stars (filled triangles) with stars in Taurus (diamonds), the Pleiades (squares), and the Hyades (circles). The measured W(Li) for most of these stars is consistent with their being roughly the age of the Pleiades ($\sim 10^8$ yr) or the Hyades ($\sim 8 \times 10^8$ yr). However, two stars (one in the direction of MBM7 and one in the direction of MBM55) with $T_{eff} \sim 6000$ K have a larger W(Li) than similar spectral type stars in the Pleiades and therefore *may* be young enough to be associated with these clouds. The uncertainties in spectral type of about two subtypes translate to uncertainties in T_{eff} around 160 K and the uncertainty in W(Li) is around 0.03 Å (the standard deviation of three different measurements).

Although there is little distinction in W(Li) between T Tauri and Pleiades age stars hotter than around 5000 K and the stars in our sample cannot be placed on an HR diagram to determine more accurate ages, we conclude that, at best, MBM7 and 55 have each formed one star. In addition, there is no evidence for star formation in MBM40. Thus, the star formation rate of low-extinction TMCs is substantially lower than for dark clouds or even higher extinction TMCs.

REFERENCES

1. Briceño, C, Hartmann, L., Stauffer, J, Gagné, M, Stern, R., & Caillault, J.-P. *AJ*, in press, (1997).
2. Feigelson, E., *ApJ* **468**, 306, (1996).
3. Neuhäuser, R., Sterzik, M.F., Schmitt, J.H.M.M, Wichmann, R., & Krautter, J., *A&A* **297**, 391 (1995a).
4. Neuhäuser, R., Sterzik, M.F., Torres, G., & Martin, E.L. *A&A* **299**, L13 (1995b).
5. Soderblom, D.R., Oey, M.S., Johnson, D.R.H., & Stone, R.P.S. *AJ* **99**, 595 (1990).
6. Stocke, J.T., Morris, S.L., Gioia, I,M., Maccacaro, T., Schild, R., Wolter, A., Fleming, T., & Henry, J.P. *ApJS* **76**, 813 (1991).

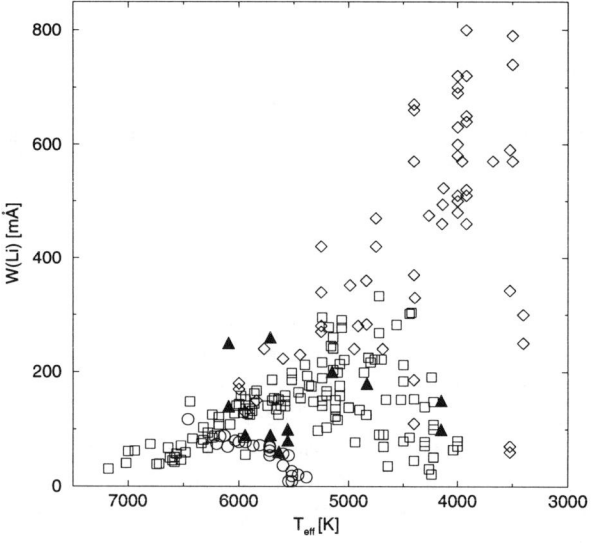

FIGURE 1. The W(Li) vs. T_{eff} diagram of the twelve stars in which we detected lithium (filled triangles) plotted with stars in Taurus (diamonds), the Pleiades (squares), and the Hyades (circles). The data for the Taurus and Pleiades stars follow Briceño et al (1997) and references therein and the Hyades data are taken from Soderblom et al. (1990).

Toward a Better Understanding of the IR Spectral Energy Distributions of H II Regions

David Leisawitz*, Seth W. Digel**, and Margaret M. Hanson[†]

*NASA/GSFC, Code 631, Greenbelt, MD 20771
**Hughes STX, NASA/GSFC, Code 631, Greenbelt, MD 20771
[†]Univ. of Arizona, Steward Observatory, Tucson, AZ 85721

Abstract. Infrared spectral energy distributions (SEDs) of several isolated star–forming regions were derived from observations made by the Diffuse Infrared Background Experiment (*DIRBE*) on NASA's Cosmic Background Explorer (*COBE*) [1]. Recently obtained near–IR photometric observations of five regions will be used in conjunction with 2 μm spectroscopy to locate and classify visually obscured O and early–B stars. Our goal is to learn in some detail how the stellar population and the local interstellar environment affect the IR SED.

MOTIVATION

The infrared emission from a region of massive star formation is affected primarily by two factors: the stellar population and the distribution of interstellar matter around the stars. Of particular importance are the numbers and types of O stars present and the solid angle subtended by molecular clouds as viewed from the stars [1]. Our goal is to characterize the stellar contents of very young, dust–enshrouded OB clusters and to relate that information to the mid–infrared and radio properties of massive star–forming (H II) regions.

Massive stars, especially those still located within or in close proximity to their parental molecular clouds, dominate the interstellar dust heating, and thus the IR SEDs, of normal spiral galaxies. Primordial galaxies, or those at great distances, are unresolvable yet they will soon be studied at infrared wavelengths by space missions such as *ISO*, *WIRE* and *SIRTF*. We

---

[1] NASA's Goddard Space Flight Center is responsible for the design, development, and operation of the *COBE*, and preparation of the data sets. Scientific guidance is provided by the COBE Science Working Group. The data are archived at and distributed from the National Space Science Data Center. For more information, see the *COBE* Home Page at http://www.gsfc.nasa.gov/astro/cobe/cobe_home.html.

hope that an improved understanding of Galactic H II regions will lead to better interpretations of the limited observations obtainable in the cases of remote galaxies.

APPROACH

Traditionally, UBV photometry and MK spectral classification are used to determine the stellar masses, initial mass functions (IMF), and ages of OB clusters. The OB stellar populations in "extended, low-density" and some blister–type H II regions can be studied this way (*e.g.*, [2]), however, unfortunately, the conventional technique cannot be applied to the early stages of star formation because very young stars are always obscured by their parental, visually opaque molecular clouds.

Recently one of us (MMH) demonstrated that near–infrared imaging and spectroscopic observations can be used to locate and classify O stars concealed by many magnitudes of visual extinction [3]. The technique was successfully applied to the heavily reddened ($A_V \sim 10$) OB cluster in M17 [4]. We are using this new method to determine the massive star populations in several young, still–embedded OB clusters.

Five massive star–forming regions (see Table 1) were selected for study according to the criteria that they be well–isolated and luminous in the infrared, small in angular extent, within a few kpc, and at least partially obscured at visible wavelengths. The first two requirements were imposed so that confusion of the target with other discrete or diffuse sources would not pose a problem at the resolution of the *DIRBE* instrument. *IRAS Sky Survey Atlas* (12, 25, 60 and 100 μm) images [5] and optical wavelength images extracted from the *Digitized Sky Survey* [6] were used to guide the selection. The angular size was restricted in order to limit the number of CCD fields needed to cover the cluster. Four of the embedded OB clusters excite blister–type H II regions; one is more deeply embedded in a compact H II region.

TABLE 1. Selected Regions, IR Fluxes and Luminosities

Name	Galactic coordinates		Distance	$\int F_\nu d\nu$	Luminosity
	ℓ (deg)	b (deg)	(kpc)	($10^{-5} nW\ cm^{-2}$)	($10^5 L_\odot$)
S 90	63.1	+0.5	4.0	17 ± 1.7	8.5
Cep A	109.9	+2.1	1.2	9.4 ± 4.4	.41
S 187	126.7	-0.8	1.1	6.9 ± 2.5	0.25
W5 NE	137.6	+1.1	2.3	8.4 ± 1.7	1.3
S 287	218.1	-0.4	3.2	3.3 ± 1.2	1.0

RESULTS

Thus far we have obtained J, H, and K_s band images of each region and derived SEDs extending from the near–IR to 240 μm from the *DIRBE* data. The *DIRBE* has ten photometric bands ranging in wavelength from 1.25 to 240 μm and a $0°.7 \times 0°.7$ instantaneous field of view at all wavelengths [7]. All of the selected clusters and their surrounding H II regions are *DIRBE* point sources.

The average emission detected at several reference positions surrounding each source was considered to be representative of the background and subtracted from the intensity seen in the direction of the source. Preliminary SEDs are shown in Figure 1. The SED can be seen to vary in shape from region to region. Integrated source fluxes and luminosities are presented in Table 1. The quoted uncertainty includes both statistical errors and an estimate of the background uncertainty based on the dispersion of the intensities in the reference positions.

The J, H, and K_s band images were obtained using the NICMOS 256×256 pixel infrared imager on the Steward Observatory's 2.3 m Bok Reflector telescope. The near–IR images will be used to search for O star candidates which will become targets for 2 μm spectroscopy. The spectroscopic observations will allow us to classify the massive stars and deduce their Lyman continuum and bolometric luminosities.

The *DIRBE* and near–IR observations, supplemented by others at a variety of wavelengths, will be used to look for manifestations of the stellar population and the local interstellar environment in the IR SED.

REFERENCES

1. Leisawitz, D. & Hauser, M.G. 1988, ApJ, 332, 954
2. Massey, P., Johnson, E. & DeGioia-Eastwood, K., 1995, ApJ, 454, 151
3. Hanson, M.M., Conti, P.S. & Rieke, M.J. 1996, ApJS, 107, 281
4. Hanson, M.M. & Conti, P.S. 1995, ApJ, 448, L45
5. Wheelock, S.L., *et al.* 1994, *IRAS Sky Survey Atlas Explanatory Supplement*, JPL Publication 94–11 (Pasadena: JPL)
6. Postman, M. *et al.* 1996, in preparation
7. Hauser, M. G., Kelsall, T., Leisawitz, D., & Weiland, J. 1995, *COBE Diffuse Infrared Background Experiment Explanatory Supplement*, Version 2.0, COBE Ref. Pub. No. 95–A (Greenbelt, MD: NASA/GSFC)

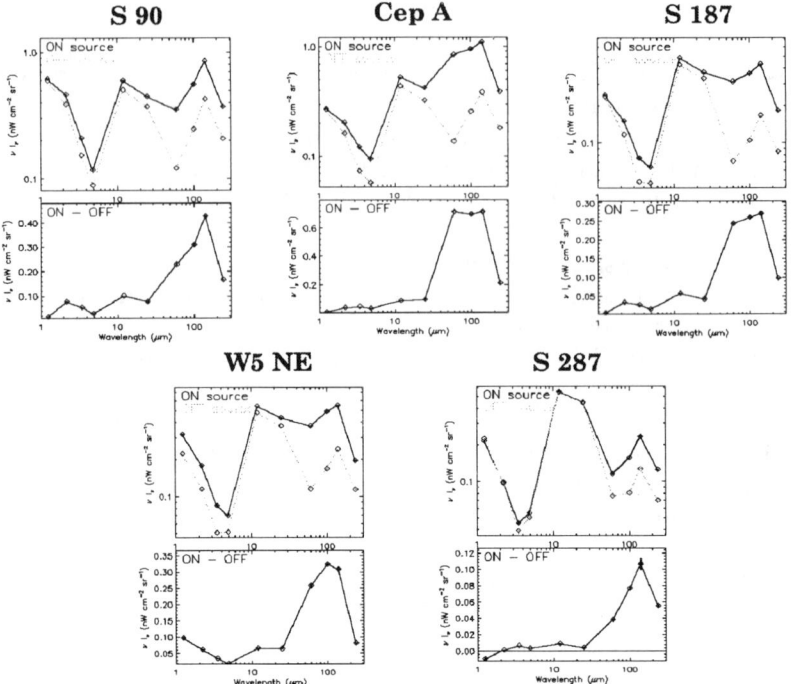

FIGURE 1. *DIRBE* spectral energy distributions of the five regions listed in Table 1. The top portion of each plot shows the SED observed in the direction of the source ("ON") and the average SED seen in the nearby reference positions ("OFF"). The SED of the source ("ON - OFF") is shown in the lower half of each plot. The SED of the compact H II region Cep A differs from those of the other regions which are of blister type, particularly in the 60 to 140 μm wavelength range. There is a greater concentration of dust close to the OB stars in the compact case, and hence a warmer spectrum is seen.

The Spectacular Ionized Interstellar Medium of NGC 55

Annette M. N. Ferguson[1,2], Rosemary F. G. Wyse[1,2]
J. S. Gallagher[3]

[1] *The Johns Hopkins University, Baltimore, MD 21218*
[2] *Guest Observer at Cerro Tololo InterAmerican Observatory*
[3] *University of Wisconsin, Madison, WI 53706*

Abstract. We present deep Hα+[NII], [SII] ($\lambda\lambda$ 6716,6731 Å) and [OII] ($\lambda\lambda$ 3726,3729 Å) images of the highly inclined, actively star–forming SBm galaxy NGC 55, located in the nearby Sculptor Group. Due to its proximity, NGC 55 provides a unique opportunity to study the disk–halo interface in a late–type galaxy with unprecedented spatial resolution. Our images reveal a spectacular variety of ionized gas features, ranging from giant HII region complexes, to supergiant filamentary and shell features, to patches of very faint diffuse emission. Many of these features protrude well above the plane of the galaxy, including a very faint fragmented shell of emission which is projected at 2.6 kpc above the disk. Emission–line ratios ([OII]/Hα+[NII], [SII]/Hα+[NII]) increase smoothly in value from the cores of HII regions, through the haloes of HII regions, into the diffuse ionized gas. Such a continuous trend is predicted by models in which the diffuse gas is photoionized by distant OB stars.

INTRODUCTION

The structure of the disk–halo interface in disk galaxies is likely to be influenced, if not determined, by various feedback effects from massive star formation, such as supernova explosions, stellar winds and photoionization. Many theories have been proposed by which star formation in the disk causes continuous and/or sporadic transfer of mass and energy from the disk to the halo, with the possible return of halo material to the disk (e.g. [1], [2]). Such a cycle could regulate the star formation process in disks, leading to low overall star-formation efficiency, as well as provide a means by which metal–enriched gas could be distributed over large regions of galaxies.

We present here a study of the disk-halo interface in the highly inclined SBm galaxy NGC 55, located in the nearby Sculptor Group. These data were obtained in the course of a large study to map the distribution of star formation in the outer parts of nearby disk galaxies (Ferguson, Ph.D. thesis project). A full account of the present work is given in [3].

OBSERVATIONS

Images of NGC 55 were obtained in the Hα+[NII], [OII] and [SII] emission lines, using the CTIO 1.5 m + Tek 2048 x 2048 CCD during September 1994 and 1995. At f/7.5, the pixel scale was 0.44", corresponding to physical size of $\sim$ 4 pc at the distance adopted here for NGC 55 of 1.6 Mpc. Reductions were carried out using standard procedures. The average sensitivity of the Hα+[NII] continuum–subtracted image, taken to be 1σ of the sky background, is 1.59×10^{-17} erg s^{-1} cm^{-2} arcsec^{-2}, corresponding to an emission measure of 7.8 pc cm^{-6} (for an assumed electron temperature of 10^4K).

MORPHOLOGY OF THE IONIZED GAS

Figure 1 shows our Hα+[NII] continuum–subtracted mosaiced image of the entire galaxy. A spectacular variety of ionized gas features are visible, ranging from bright HII complexes, to loops and plumes protruding out the plane of the galaxy, to faint diffuse emission. The most prominent features are the two large HII complexes located in the central regions of NGC 55, which are clearly the dominant sites of recent massive star formation. The Hα+[NII] luminosities of these features are estimated to be 9 and 3 $\times$ 10^{39} erg s^{-1} (uncorrected for extinction) and they each span more than a kpc in diameter. Their individual luminosities are comparable to that of the 30 Dor complex in the LMC. These two HII complexes are surrounded by a striking network of ionized gas features. Some of these features extend considerable distances above the disk plane and may be examples of galactic 'chimneys', through which hot gas is theorised to be vented into the halo. Such structures are predicted to arise as a result of correlated Type II SNe explosions in OB associations, leading to the production of an expanding superbubble which can break through the disk (e.g. [2]). In our images, several of these chimneys appear to be capped with clumps of relatively bright emission, possibly from gas swept up by the expanding bubble and now directly ionized by the OB association located in the disk below. The cap which sits atop one of these features is particularly interesting, since it is strong in the [OII]–continuum image and may actually be due to *in situ* star formation at 1.5 kpc above the plane of the disk. On the North side of the galaxy, we have detected several patches of extremely faint Hα emission (EM $<$ 5 pc cm^{-6}) lying at projected distances of up to 2.6 kpc above the plane. These clumps of emission appear to trace out a fragmented shell, which has possibly undergone Raleigh–Taylor instability. Throughout the disk of the galaxy, we observe a spatial correlation between bright diffuse ionized gas and HII regions; this correlation has been observed now in several galaxies (e.g. [4], [5]) and lends support for a model in which the gas is photoionized by Lyman continuum photons which leak out of individual HII regions.

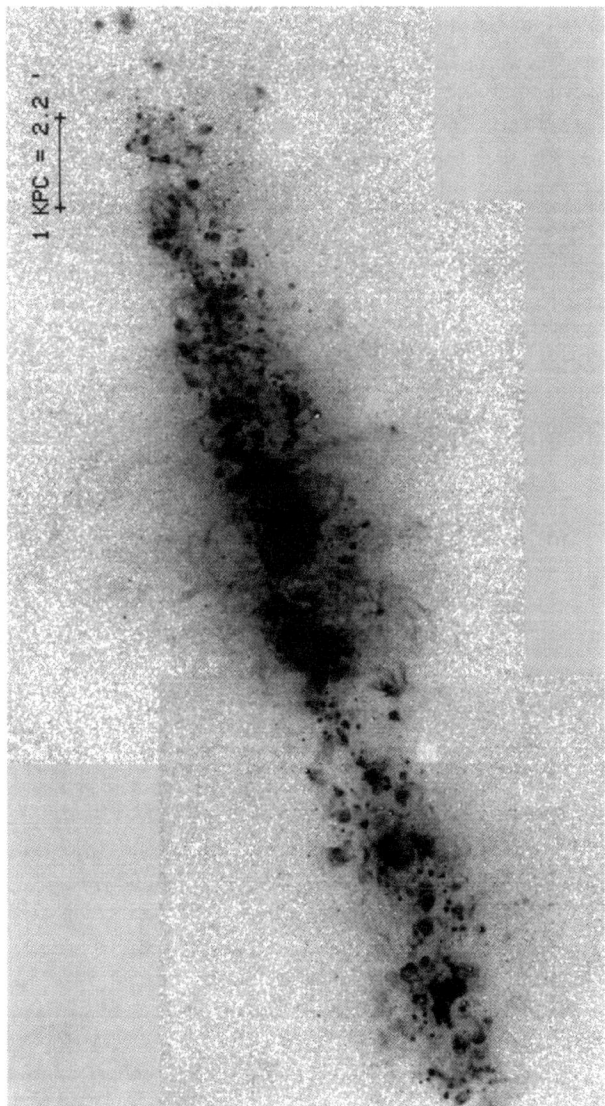

FIGURE 1. An Hα continuum–subtracted mosaic of NGC 55. North is to the left and East is to the top.

LINE RATIOS AND IONIZATION MECHANISM OF THE GAS

Constraints on the ionization mechanism of the extra–planar gas are provided from our data by the values of the line ratios of [OII]/Hα+[NII] and [SII]/Hα+[NII]. Our data allow us to study the variation of emission line ratios over wide ranges of surface brightnesses and of morphology, from bright HII region cores to faint DIG features. Small aperture (2″) photometry was carried out at different locations along various diffuse ionized gas features as well as in the core and halo regions of several bright HII regions. Our results can be summarized as follows:

- There is a smooth increase in both the [OII]/Hα+[NII] and [SII]/Hα+[NII] ratios with decreasing Hα+[NII] surface brightness, as one proceeds from bright HII region cores, through haloes of HII regions and into the diffuse ionized gas. The mean [OII]/Hα+[NII] and [SII]/Hα+[NII] ratios measured for the diffuse ionized features are found to be a factor of $\sim$ 2 higher than the mean found towards HII regions. The smooth sequence in ionization state between HII regions and diffuse ionized gas lends strong support for photoionization of the diffuse gas by a dilute radiation field, such as that produced by massive stars that are located at considerable distances (eg. [6]).

- Different morphological classes of diffuse ionized gas structures (e.g. loops, chimneys, patches of diffuse emission) cannot be distinguished from each other purely on the basis of the line ratios presented here, consistent with all having a common ionization mechanism.

- The [OII]/Hα+[NII] ratio in ionized gas features increases smoothly with projected height above the plane. The [SII]/Hα+[NII] ratio also shows a trend with height above the plane, although there is more scatter. This seems to be due both to low Hα+[NII] surface brightness features in the plane having a high [SII]/Hα+[NII] ratio, as well as some bright extra–planar features which have low [SII]/Hα+[NII] ratios.

REFERENCES

1. Shapiro, P. R. & Field, G. B. 1976, ApJ, 205, 762
2. Norman, C. A. & Ikeuchi, I. 1989, ApJ, 345, 372
3. Ferguson, A. M. N., Wyse, R. F. G. & Gallagher, J. S., 1996, AJ, in press.
4. Walterbos, R. A. M. & Braun, R. 1994, ApJ, 431, 156
5. Ferguson, A. M. N., Wyse, R. F. G., Gallagher, J. S. & Hunter, D. A. 1996, AJ, 111, 2265
6. Domgörgen, H. & Mathis, J. S. 1994, ApJ, 428, 647

Far-Ultraviolet Emission from NGC 4038/4039, "The Antennae": Massive Star Formation in Compact Clusters

Susan G. Neff [1], J. E. Hollis [2], J. K. Hill[2], M. N. Fanelli [2], D. A. Smith [3], A. M. Smith [1], T. P. Stecher [1], R. C. Bohlin [4], R. W. O'Connell [5], and M. S. Roberts [6]

[1] *NASA/GSFC, LASP - Code 681, Greenbelt, Maryland, 20771*
[2] *HSTX/GSFC, LASP - Code 681, Greenbelt, Maryland, 20771*
[3] *NRC/GSFC, LASP - Code 681, Greenbelt, Maryland, 20771*
[4] *STScI, San Martin Drive, Baltimore, Maryland, 21218*
[5] *University of Virginia, Astronomy Department, Charlottesville, Virginia, 22903*
[6] *NRAO, Edgemont Road, Charlottesville, Virginia, 22901*

Abstract.
Far-Ultraviolet (FUV) emission from NGC4038/4039 traces recent massive star formation in the system. The FUV is used to determine the massive star content and to identify compact regions of recent star formation. FUV measurements are combined with ground-based observations and starburst models to determine ages, mass functions, and current star formation rates.

INTRODUCTION

We present the first arcsecond resolution far-ultraviolet (FUV, $\lambda \sim 1500$Å) image of the nearby spectacular galaxy merger NGC4038/ 4039 (figure 1). The observations were obtained by the Ultraviolet Imaging Telescope (UIT) in March 1995, during the Astro-2 mission on Space Shuttle *Endeavour*.

Significant FUV emission is observed in this merger, despite the presence of extensive dust and molecular gas in the system. The total directly observed FUV flux of the pair is $\sim 2.9 \times 10^{-13}$ergs cm^{-2}sec^{-1}Å^{-1}, or $m_{FUV} \sim 10.3$. At 19.8 Mpc, this corresponds to $M_{FUV} \sim -21.2$, and is typical of large spiral galaxies observed by UIT. The observed FUV emission is equivalent to that from $\sim$21,600 early-O dwarfs.

The FUV emission effectively traces recent massive star formation. Star formation occurs mainly along two looping structures within the galaxies' disk remnants (figure 1) and is dominated by compact clusters. 28 separate starburst regions in the system contribute about 60% of the directly observed FUV emission. Individual knots have directly observed FUV luminosities equivalent to $\sim 100 - 1300$ type O 3-6 dwarfs, or to ~ 2-24 30 Dor's.

Ground-based U, V, I, and K band images are used with the FUV to determine individual knots colors. A range of extinction values are found to be present ($A_V = 1.4' - 2.3$). Dust in NGC4038/9 has properties similar that in 30 Dor; Galactic reddening laws do not work.

Further comparisons with models determine properties of individual starbursts. Ages of the FUV-bright knots are estimated to be 2.5–3.6 Myr. The best-fit Initial Mass Function appears to be weighted towards massive stars, with a slope of $\Gamma \sim -1.0$ and extending to 80-100 $M_\odot$. Each individual starburst knot in NGC4038/9 contains at least 6,000–65,000$M_\odot$ in young stars.

These results show that star formation can be extreme in galaxy mergers, generally thought to have been much more common in the early universe. A full understanding of galaxy evolution will require detailed studies of the rest-UV properties of galaxies in both the local and the distant universe.

Details of this work will be published in the *Astronomical Journal*.

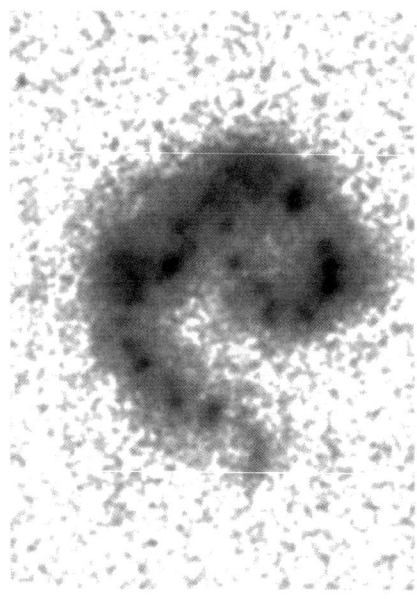

FIGURE 1. Far-UV image of the merging pair NGC4038/4039. Massive star formation is seen to occur predominately in compact knots, each with Far-UV luminosity equivalent to ~ 2-24 that of 30 Dor. The field is about 4 arcmin, the resolution is ~ 2.5 arcsec.

M31 vs. M33: Different Modes of Star Formation

Paul Hodge, Ted Wyder and Knut Olsen

Astronomy Department, University of Washington
Seattle, WA 98195

Abstract. Among the various differences in star-formation modes between various types of galaxies, the differences in HII region luminosity functions are conspicuous. A comparison of those of M31 and M33, using new data, illustrates these differences, which are best understood in terms of different stellar clustering mass functions for the galaxies.

INTRODUCTION

As has been known for many years (e.g., see Baade 1963), the emission nebulae in the Sb galaxy M31 are conspicuously different from those in Sc galaxies like M33. A quantitative measure of one of these differences was pointed out by Kennicutt et al (1989), who showed that the luminosity function of M31 and other Sb galaxies has a different form from that of M33 and other Sc galaxies. The slope is steeper, less linear and the maximum H-alpha luminosity reached (for a given total population) is smaller. Other differences (especially as to structure and morphology) are more subtle and have, until now, not been characterized quantitatively. The reasons for these various differences are largely unknown. Because they are potentially important ingredients in the explanation of the differences in star formation modes in different Hubble types, we are using CCD data to establish quantitative methods of multivariate analysis that will enable us to understand them better.

COMPARING HII REGION POPULATIONS QUANTITATIVELY

Star formation regions in galaxies of different kinds respond in different ways to the properties of their host galaxies. A limited number of studies have been carried out so far to establish quantitative comparisons of HII region populations in different kinds of galaxies. The primary kinds of comparisons have been of HII region luminosity functions (Kennicutt et al, 1989) and size distributions (van den Bergh 1981, Hodge 1987). Spatial distributions are also relevant (e.g., Hodge and Kennicutt 1987, Rozas et al, 1996). We have begun a program of quantitative comparisons of the HII region populations of two rather different galaxies, both sufficiently close that we can examine their star-formation regions in considerable detail: M31 and M33. We use ground-based narrow band images for the gas, and HST and ground-based images for the stars. The accompanying material illustrates an example of our results so far.

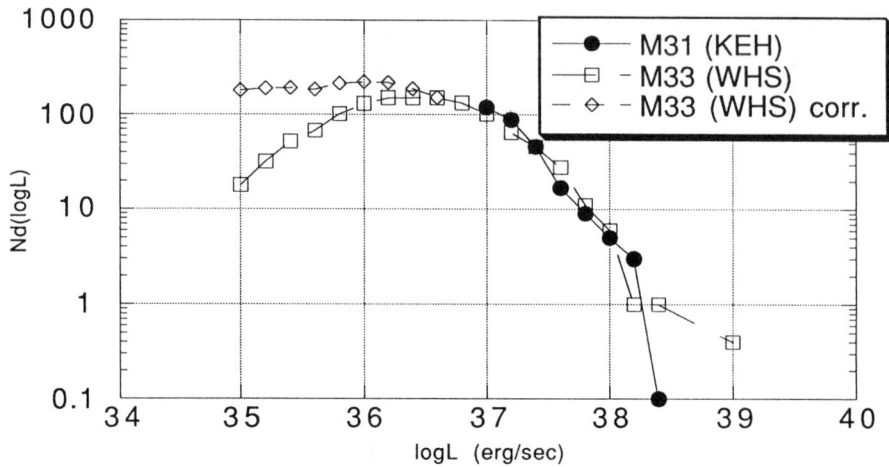

Figure 1. The Kennicutt et al (1989) luminosity function for M31 compared with the new results for M33 obtained by Wyder, Hodge and Skelton (1997).

A NEW HII REGION LUMINOSITY FUNCTION FOR M33

Using the deep Hα CCD KPNO 4-m survey of M33 obtained by Blair, Gordon, Kirschner and Long, we have measured the luminosities of cataloged HII regions in M33 and have added an additional 1200 newly-recognized emission regions (Wyder, Hodge and Skelton 1997). These new data are unique in that they reach to unusually-faint limits. Another unusual feature is that we are able quantitatively to correct the faint end of the luminosity function, using two different kinds of measurements of completeness. The result, shown in Figure 1, is a function that shows a much less pronounced and fainter turn-over than seen in previous diagrams, indicating that the balance between increasing numbers of stars and decreasing temperatures, as one looks at fainter objects, is more equal than previously thought. We are presently working on modeling this portion of the diagram in terms of both gas region and stellar evolution.

The conspicuous difference with the luminosity function for M31, however, is at the bright end. M31's bright HII regions, in spite of the larger mass and total star formation rate of M31, are fainter by an order of magnitude than those in M33, indicating a gross difference in the distribution of the masses of star-forming clouds in the two galaxies.

REFERENCES

1. Baade, W, *Evolution of Stars and Galaxies* (Harvard U. Press, Cambridge) (1963).
2. Kennicutt, R. C., Edgar, B. and Hodge, P., *ApJ* **337**, 761 (1989).
3. Hodge, P., *PASP* **99**, 915 (1987).
4. Hodge, P. and Kennicutt, R. C., *PASP* **99**, 915 (1987).
5. Rozas, M., Knapen, J. and Beckman, J. *A & A*, (1986) in press.
6. van den Bergh, S. *PASP* **86**, 1464 (1981).
7. Wyder, T., Hodge, P. and Skelton, B., (1997), in preparation.

*Star Formation
History in Spirals*

Cosmic Star Formation History

Piero Madau

Space Telescope Science Institute, 3700 San Martin Drive, Baltimore, MD 21218

Abstract. I review some recent progress made in our understanding of galaxy evolution and the cosmic history of star formation. Like bookends, the results obtained from deep ground-based spectroscopy and from the Hubble Deep Field imaging survey put brackets around the intermediate redshift interval, $1 < z < 2$, where starbirth probably peaked at a rate 10 times higher than today. The steady decline observed since $z \sim 1$ is largely associated with late-type galaxies. At $z \gtrsim 2.5$, the Lyman-break selected objects may represent the precursors of present-day spheroids, but appear, on average, quite underluminous relative to the expectations of the standard early-and-rapidly forming picture for spheroidal systems. The observed ultraviolet light density accounts for the bulk of the metals seen today in "normal" massive galaxies.

INTRODUCTION

The knowledge of the star formation rate (SFR) throughout the universe as a function of space and time is one of the primary goal of galaxy formation and evolution studies. Key questions to be answered are as follows: How can we identify high-z galaxies in deep CCD surveys? Are they forming stars more rapidly than quiescent spirals at $z \sim 0$? Are they obscured by dust in analogy with luminous IRAS starbursts? Is there a characteristic epoch of star and element formation in galaxies? How does the distribution of SFR evolve with redshift? Do spheroids form early and rapidly? What is the origin of disk galaxies?

Two complementary approaches can be used to shed light on these questions. One is to study the resolved stellar populations of the Milky Way and nearby galaxies and infer their evolutionary history from fossil records – examples of this approach are nuclear cosmochronology, the color-magnitude diagram of globular clusters, the cooling sequence of white dwarfs. The other is to systematically observe galaxies at increasing cosmological lookback times, and reconstruct the history of stellar birthrate *directly*. In this talk I will review the broad picture that has recently emerged from the "direct" method, focusing on what can be learned from integrated quantities over the entire population, rather than from a detailed study of morphologically-distinct samples whose physical significance remains unclear. I will show how the combination of HST deep imaging and ground-based spectroscopy offers now an exciting first glimpse to the history of the conversion of neutral gas into stars in field galaxies.

In the following, I will make use of the fact that the UV-continuum emission from a galaxy with significant ongoing star formation is dominated by short-lived massive stars, and is therefore nearly independent of the galaxy history. In all the transformation from UV luminosity to SFR, a Salpeter IMF including stars in the $0.1 < M < 125\,M_\odot$ range with solar metallicity will be assumed. For a Scalo IMF – less rich in massive stars – in the same mass range the conversion factor is about 3 times larger. All magnitudes will be given in the AB system, and a flat cosmology with $q_0 = 0.5$ and $H_0 = 50$ km s^{-1} Mpc^{-1} will be adopted.

THE LOCAL UNIVERSE

The Universitad Complutense de Madrid (UCM) objective-prism survey for Hα-emitting objects [1] provides an ideal tool for studying the properties of star-forming galaxies at low redshift. The sample consists of about 250 sources in 500 deg^2 ($z < 0.045$) with EW(Hα + [N II]) > 10Å, and is dominated by intermediate- to low-luminosity late-type galaxies. The line emission comes largely from the nuclear regions, and has been corrected for reddening using the Balmer decrement. We can then use case B recombination theory to relate the Hα line luminosity to the rate of production of ionizing photons, and population synthesis models [2] to estimate the instantaneous SFR, $L(H\alpha) = 3 \times 10^{41} \times$ SFR ergs s^{-1}, where SFR is in units of $M_\odot$ yr^{-1}. A Schechter function with $\alpha = -1.3 \pm 0.2$, SFR$^* = 4.7 \pm 0.8\,M_\odot$ yr^{-1}, and $\phi^* = 10^{-3.2\pm0.2}$ Mpc^{-3}, shown in Figure 1, provides a good fit to the present-epoch "SFR function" – which describes the number of star-forming galaxies as a function of their ongoing SFR – in the range between 0.1 and 10 $M_\odot$ yr^{-1}. Integrating over all luminosities, the total SFR density is

$$\dot{\rho}_* = 10^{-2.4\pm0.2}\,M_\odot\,\text{yr}^{-1}\,\text{Mpc}^{-3}.$$

A similar value can be derived starting from the observed luminosity function (LF) in the B-band [3], applying a luminosity-weighted average color of $\langle 2800 - 4400 \rangle_{AB} = 2.05$ mag [4] to the B-band luminosity density, and then converting the UV flux into an instantaneous SFR.

SURVEYS TO $Z \sim 1$

The recent completion of several comprehensive ground-based redshift surveys [5–7] has significantly improved our undestanding of the evolution of field galaxies to $z \sim 1$. A similar trend is seen by the various groups, namely the rapid evolution – largely driven by late-type galaxies – of the LF with lookback time. In particular, from the marked increase with redshift of the comoving luminosity density at 2800Å observed in the Canada-France Redshift Survey (CFRS) of 730 I-selected galaxies with $17.5 < I_{AB} < 22.5$ [5], the total (integrated over all luminosities) rate of star formation per unit volume is

$$\dot{\rho}_* = 10^{-1.3 \pm 0.15} \left(\frac{1+z}{1.875}\right)^{3.9 \pm 0.75} M_\odot \, yr^{-1} \, Mpc^{-3}$$

over the range $0 < z < 1$. The distribution of SFR in the interval $0.75 < z < 1$ is shown in Figure 1. It has been derived from the rest-frame B-band LF at that epoch [5], by applying a luminosity-weighted average color of $\langle 2800 - 4400 \rangle_{AB} = 1.3$ mag [4]. A Schechter function with $\alpha = -1.28$, SFR$^* = 6.2 \, M_\odot \, yr^{-1}$, and $\phi^* = 10^{-2.17} \, Mpc^{-3}$ provides a good fit to the SFR function in the $2.5 - 30 \, M_\odot \, yr^{-1}$ range.

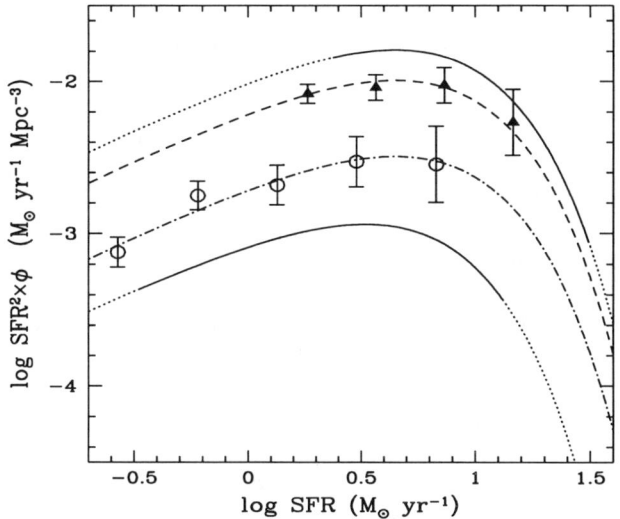

FIGURE 1. Distribution of star formation rates at different redshifts. The solid line represents the Schechter function fitted to the data points, and the dotted line its extrapolation. *Top curve*: SFR density at $0.75 < z < 1$. *Bottom curve*: SFR density today. The dashed line depicts the same $0.75 < z < 1$ Schechter function with ϕ^* lower by a factor of 1.6. This provides a good fit to the HDF points *(filled triangles)* at $\langle z \rangle = 2.75$. *Dot-dashed line*: same for blue dropouts.

A comparison with the present-day distribution shows the sign of a strong density evolution – a rapid increase in ϕ^*. If fitted by an exponential, the ten-fold increase in the volume-averaged SFR over the past 8 Gyr implies an e-folding time of $\tau_{\rm SFR} = 3.2^{+0.8}_{-0.5}$ Gyr. The evolution is strongly differential with color, with the LF of galaxies redder than a present-day Sbc showing very little change with cosmic time. A morphological analysis shows that one can identify disk-dominated galaxies (with bulge/total luminosity < 0.5) with the blue population whose LF is evolving rapidly, and that galaxy disks at $\langle z \rangle = 0.7$ have mean rest-frame central surface brightness ~ 1.6 mag brighter than their local counterpart [8].

LYMAN-BREAK GALAXIES

Recently, the study of field galaxies has been pushed to much earlier lookback times by the development of a broad-band color technique for identifying galaxies at $z \sim 3$ [9,10]. At this redshift, the Lyman-continuum break at 912Å – which arises from a combination of the intrinsic discontinuity in the spectra of hot stars, the opacity of the galaxy to its own ionizing radiation, and the ubiquitous effect of H I absorption in the intergalactic medium along the line of sight [11] – passes through the U-band, resulting in an unmistakable color signature for distant star-forming galaxies: a very red $U - B$ color, combined with colors in longer-wavelength filters that are much bluer. Recent deep spectroscopy with the W. M. Keck telescope has shown the high efficiency of such color selection. Steidel and collaborators [12,13] have confirmed the redshifts of ~ 100 galaxies at $2.9 < z < 3.4$ in four high-latitude fields, and established that the technique is > 90% reliable in going from color-selected "UV dropouts" to confirmed high-z galaxies, the only contaminants being halo subdwarfs.

The UV-continuum fluxes of Lyman-break galaxies imply star formation rates in the range 4–100 $M_\odot$ yr^{-1}, and a SFR density of

$$\dot{\rho}_*(4-100) = 10^{-2.13}\,M_\odot\,{\rm yr}^{-1}\,{\rm Mpc}^{-3}$$

at $\langle z \rangle = 3.25$. This is 6 times higher than the rate of starbirth today, but 3 times lower than the corresponding value at $z \sim 1$ *over the same range of luminosities*, and show the existence of a peak at intermediate redshifts in the volume-averaged SFR in bright galaxies.

Few pieces of evidence may support the interpretation that such galaxies are bulges and spheroids seen in early formation: a) their volume density and rest-frame optical luminosities derived from K-band photometry are comparable to those of bright galaxies today; and b) *HST*-WFPC2 images show the presence of compact, relaxed "cores" which are few kpc in size, in analogy with the bulges and cores of luminous galaxies at the present epoch [14].

THE HUBBLE DEEP FIELD

The Hubble Deep Field (HDF) imaging survey has been specifically designed with the application and generalization of the UV dropout technique in mind. With its unprecedented depth, reaching 5-σ limiting AB magnitudes of roughly 27.7, 28.6, 29.0, and 28.4 in the U_{300}, B_{450}, V_{606}, and I_{814} bandpasses [15], and four-filter strategy in order to detect Lyman-break galaxies at various redshifts, the HDF is now a key testing ground for models of galaxy evolution.

Galaxy Counts There are about 3,000 galaxies in the HDF, corresponding to 2×10^6 deg^{-2} to $V \sim 29$. In all four bands, the slope α of the differential galaxy counts, $\log N(m) = \alpha m$, flattens at faint magnitudes, e.g., from $\alpha = 0.45$ in the interval $21 < B < 25$ to $\alpha = 0.17$ for $25 < B < 29$. This feature cannot be due to the reddening of distant sources as their Lyman break gets redshifted into the blue passband, since the fraction of Lyman-break galaxies at $B \sim 25$ is only of order 10% (cf [16]). Moreover, an absorption-induced loss of sources could not explain the similar flattening of the number-magnitude relation observed in the V and I bands [17]. Rather, the change of slope suggests a decline in the surface density of luminous galaxies beyond $z \sim 1$.

Since, for $\alpha < 0.4$, the extragalactic background light (EBL) is dominated by object at the bright end of the luminosity range, the flattening of the number counts has the interesting consequences that the galaxies that produce $\sim 60\%$ of the blue EBL have $B < 24.5$. They are thus bright enough to be identified in spectroscopic surveys, and are known to have median redshift $\langle z \rangle = 0.6$ [5]. The quite general conclusion is that there is no evidence in the number-magnitude relation for a large amount of star formation at high redshift. Note that these considerations do not constrain the *rate* of starbirth at early epochs, only the total (integrated over cosmic time) amount of stars – hence background light – being produced. The most direct way to track the evolution of the SFR density at early epochs is through a census of the HDF dropouts.

Ultraviolet Dropouts New photometric criteria for robustly selecting Lyman-break galaxies have been developed based on the HDF color system. By simulating colors for an extremely wide range of model galaxy spectra, the criteria have been tuned up to provide what appear to be largely uncontaminated samples of star-forming galaxies at high redshifts [18]. I have further refined them after the many redshift measurements with Keck.

The U_{300} passband – which is bluer than the standard ground-based U filter – permits the identification of star-forming galaxies in the interval $2 < z < 3.5$. Galaxies in this redshift range predominantly occupy the top left portion of the $U_{300} - B_{450}$ vs. $B_{450} - I_{814}$ color-color diagram because of the attenuation by the intergalactic medium and intrinsic extinction. Galaxies at lower redshift can have similar $U - B$ colors, but they are typically either old or dusty, and are therefore red in $B - I$ as well. About 100 ultraviolet dropouts can be identified in the HDF which are brighter than $B = 27$, about 25% of the total. Of these, 17 have spectroscopically confirmed redshift in the range $2.2 < z < 3.4$ [19,20]. Note that, out of the ~ 60 galaxies in the HDF with known redshifts $z < 2$ [21], no low-redshift interlopers have been found among the high-redshift sample. The color-selection is illustrated in Figure 2. The UBI criteria isolate objects that have relatively blue colors in the optical, but a sharp drop into the UV. A "plume" of reddened high-z galaxies is clearly seen in the data.

Assuming the redshift interval $2 < z < 3.5$ has been uniformly probed, I have plotted in Figure 1 the SFR function of the U dropouts in the HDF. From the observed V magnitudes of our sample, a "directly-observed" SFR density at $\langle z \rangle = 2.75$ of $\dot{\rho}_* = 10^{-1.66}\,M_\odot\,\mathrm{yr}^{-1}\,\mathrm{Mpc}^{-3}$ is computed. As only a short segment of the LF can be determined, however, it is dangerous to fit the usual

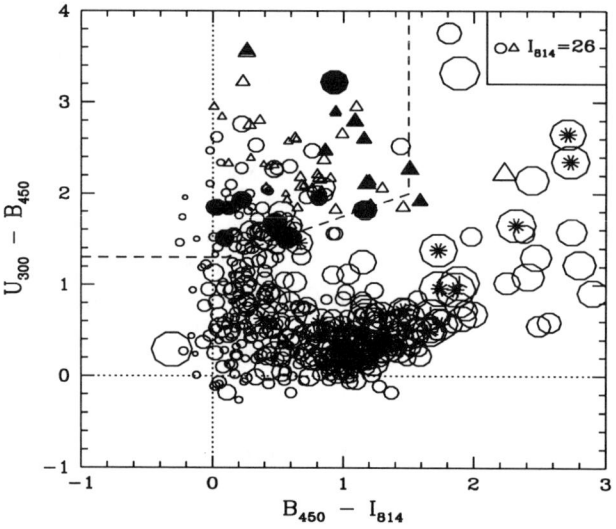

FIGURE 2. Color-color plot of galaxies in the HDF with $B < 27$. Objects undetected in U (with $S/N < 1$) are plotted as triangles at the 1σ lower limits to their $U - B$ colors. Symbols size scales with the I mag of the object. The dashed lines outline the selection region within which we identify candidate $2 < z < 3.5$ objects. Galaxies with spectroscopically confirmed redshifts within this range are marked as solid symbols. Galaxies with confirmed redshifts lower than $z = 2$ are marked as asterisks.

Schechter function to get a better estimate of the total density of starbirth. What I have done instead is to show in Figure 1 that the $z \sim 1$ Schechter function with a normalization ϕ^* *lower* by a factor of 1.6 provides a good fit to the HDF points. The integration of this function over all luminosities yields a star formation density about 1.45 larger than the directly-observed one. In the following, I will adopt the intermediate value

$$\dot{\rho}_* = 10^{-1.57} \, M_\odot \, yr^{-1} \, Mpc^{-3}$$

as our best determination of the SFR density at $\langle z \rangle = 2.75$, and assign to it an uncertainty of ± 0.15 in the log. A comparison between the comoving space density of the HDF U dropouts brighter than $V = 25.5$ and that derived from ground-based statistic [13] ($R < 25.5$) at redshift $\langle z \rangle = 3.25$, shows good agreement, after accounting for the fact that one probes ~ 0.3 mag fainter in the galaxy LF at the HDF average redshift. The combination of the ground-based Lyman-break galaxy survey with the U dropout sample provides a better sketch of the distribution of SFR for galaxies at $z \approx 3$ [22].

Blue Dropouts   In analogous way, the B_{450} passband allows the selection of candidate star-forming galaxies in the interval $3.5 < z < 4.5$. We have identified ~ 20 B dropouts to $V < 28$ in the $B_{450} - V_{606}$ vs. $V_{606} - I_{814}$ plane [18]. The brightest one has been tentatively confirmed to be at $z = 4.02$, consistent with the photometric predictions [23]. From the observed I magnitudes of our sample, a SFR density of

$$\dot{\rho}_* = 10^{-2.06 \pm 0.2} \, M_\odot \, yr^{-1} \, Mpc^{-3}$$

is obtained at $\langle z \rangle = 4.0$, well below the $\langle z \rangle = 2.75$ value. The error bar reflects the uncertainties present in the volume normalization and in the color selection given the lack of spectroscopic confirmations.

COSMIC METAL PRODUCTION

The (rest-frame) radiation flux below 3000 Å is a very good measurement of the instantaneous ejection rate of heavy elements ($Z \geq 6$), since both are directly related to the number of massive stars [24]: the same stars with $m > 10 \, M_\odot$ that manufacture and return most of the metals to the ISM as Type II supernovae also dominate the UV light. Contrary to the conversion from UV to SFR, which is a sensitive function of the IMF slope, the UV-to-metal conversion efficiency is fairly insensitive to the assumed IMF, since the increased metal yield [25] from high mass stars is compensated for by a similar increase in the production of UV photons.

The rate of ejection of newly synthesized material per unit comoving volume as a function of redshift, $\dot{\rho}_Z$, is shown in Figure 3, together with a sketch of the cosmic star formation history in galaxies [18]. When combined with the ground-based data, the HDF results appear consistent with the existence of a peak in the universal metal production rate in the redshift range $1 < z < 2$, in agreement with inferences from quasar absorption systems [26]. The plot suggests that, while the conversion of gas into stars must have been extremely efficient at intermediate redshifts, and galaxies have largely exhausted their reservoirs of cold gas at the present-epoch, there must be a mechanism which prevents the gas within virialized dark matter halos from radiatively cooling and turning into stars at $z \gtrsim 3$.

We may at this stage try to establish a cosmic timetable for the production of metals in relatively bright galaxies, keeping in mind the inherent uncertainties associated with the estimates given above. The mass density of heavy elements observed today in "normal" massive galaxies is

$$\rho_Z(0) = Z_* \rho_B(0) \langle \frac{M}{L_B} \rangle = 6 \pm 3 \times 10^6 (\frac{Z_*}{Z_\odot})\, M_\odot\, \text{Mpc}^{-3}$$

where $\rho_B(0)$ is the local blue light density, and $\langle M/L_B \rangle \approx 3$ is the mean mass-to-blue light ratio of visible matter in solar units. Although a baryonic mass several times larger than the luminous mass may be present in the Galactic halo, metal-rich halo material would be mixed into and over-enrich the disk. Hence, if a substantial amount of metals are missing from our census, they are most likely hidden in the intergalactic gas.

If we define two characteristic epochs of star and element formation in galaxies, z_* and z_Z, as the redshifts by which half of the current stellar and metal content of galaxies was formed, then a straightforward integration of the curve plotted in Figure 3, together with the fact that most of the stars in the inner luminous parts of galaxies are metal rich, imply $z_* \lesssim z_Z \approx 1$, or in other words that a significant fraction of the current metal content of galaxies was formed relatively late by late-type systems, on a timescale of about 8 Gyr. (Note that, contrary to the measured number densities of objects and rates of star formation, the total metal mass density produced is independent of the assumed cosmology.)

This suggests the possibility that we may be observing in the redshift range $z = 0 - 1$ the conversion into stars of gaseous galactic disks. Pure H I disks may be assembled at some higher redshift, and disk gas continuously replenished as a result of ongoing infall from the surrounding hot halo. From stellar population studies we also know that about half of the present-day stars – hence metals – are contained into spheroidal systems, i.e., elliptical galaxies and spiral galaxy bulges, and that these formed early and rapidly (see, however, [27]), experiencing a bright starburst phase at high-z. Where are these protospheroids?

The space density of bright ellipticals today is $\phi(> L_*) \approx 4.5 \times 10^{-4}\, \text{Mpc}^{-3}$ [6]. If a significant fraction of their stellar population formed in a single burst of duration 1 Gyr early in the history of the universe, a comparable number of galaxies should be observed at high-z while forming stars at rates in excess of about 100 $M_\odot$ yr^{-1}. ¿From the Lyman-break galaxy sample, however, the space density of high star-forming galaxies at $z \sim 3$ is about 50 times lower [13]. Hence there is a serious deficit of very bright objects relative to the expectations of the standard early-and-rapidly-forming picture for spheroidal systems. At the star formation density levels inferred from the HDF images, about 1/3 of the observed mass density of metals at $z = 0$ would have been formed during the "spheroid epoch" at $z \gtrsim 1.5$.

Finally, it is only fair to point out that all the values derived above should be considered as lower limits, as newly formed stars which are completely hidden by dust would not contribute to the $H\alpha$ or UV luminosity. It is a fact that, at the present-epoch, we are underestimating the SFR density by about a factor of 2, as a "typical" optically-selected spirals emits 30% of its energy in the FIR region [28]. On the other hand, since the metals we observe being formed are a substantial fraction of the entire metal content of massive galaxies, it appears that – on average – star formation regions remain largely unobscured by dust throughout much of galaxy formation. The opposite can be true only if galaxies eject a large amount of heavy elements in the intergalactic medium.

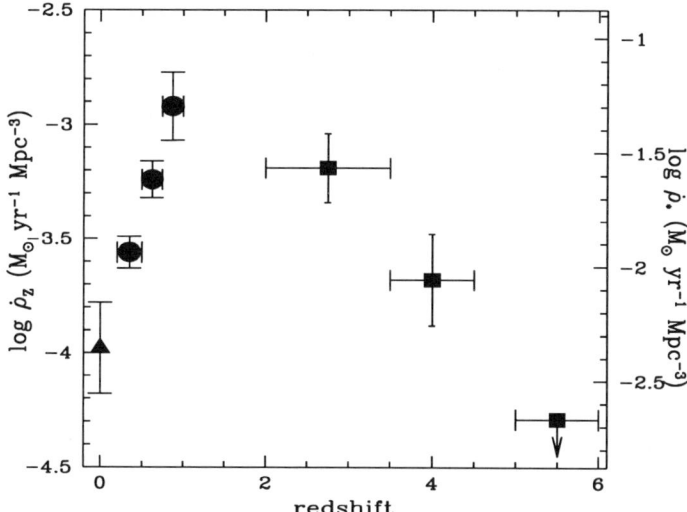

FIGURE 3. Element and star formation history of the universe. The upper limit at $\langle z \rangle = 5.5$ reflects the paucity of V dropouts in the HDF.

REFERENCES

1. Gallego, J., Zamorano, J., Aragón-Salamanca, A., & Rego, M. 1995, ApJ, 455, L1
2. Bruzual, A. G., & Charlot, S. 1993, ApJ, 405, 538
3. Loveday, J., Peterson, B. A., Efstathiou, G., & Maddox, S. J. 1992, ApJ, 390, 338
4. Lilly, S. J., Le Fèvre, O., Hammer, F., & Crampton, D. 1996, ApJ, 460, L1
5. Lilly, S. J., Tresse, L., Hammer, F., Crampton, D., & Le Fèvre, O. 1995, ApJ, 455, 108
6. Ellis, R. S., Colles, M. M., Broadhurst, T. J., Heyl, J. S., Glazebrook, K. 1996, MNRAS, 280, 235
7. Cowie, L. L., Songaila, A., Hu, E. M., & Cohen, J. G. 1996, AJ, 112, 839
8. Schade, D., Lilly, S. J., Le Fèvre, O., Hammer, F., & Crampton, D. 1996, ApJ, 464, 79
9. Steidel, C. C., & Hamilton, D. 1992, AJ, 104, 941
10. Steidel, C. C., Pettini, M., & Hamilton, D. 1992, AJ, 110, 2519
11. Madau, P. 1995, ApJ, 441, 18
12. Steidel, C. C., Giavalisco, M., Pettini, M., Dickinson, M. E., & Adelberger, K. 1996, ApJ, 462, L17
13. Steidel, C. C., et al. 1996, in preparation
14. Giavalisco, M., Steidel, C. C., & Macchetto, F. D. 1996, ApJ, 470, 189
15. Williams, R. E., et al. 1996, AJ, 112, 1335
16. Guhathakurta, P., Tyson, J. A., & Majewski, S. R. 1990, ApJ, 357, L9
17. Pozzetti, L., Madau, P., Ferguson, H. C., & Zamorani, G. 1996, in preparation
18. Madau, P., Ferguson, H. C., Dickinson, M. E., Giavalisco, M., Steidel, C. C., & Fruchter, A. 1996, MNRAS, 283, 1388
19. Steidel, C. C., Giavalisco, M., Dickinson, M. E., & Adelberger, K. 1996, AJ, in press
20. Lowenthal, J. D., et al. 1996, preprint
21. Cohen, J. G., Cowie, L. L., Hogg, D. W., Songaila, A., Blandford, R., Hu, E. M., & Snopbell, P. 1996, ApJ, in press
22. Dickinson, M. E., et al. 1996, in preparation
23. Dickinson, M. E., private communication
24. Songaila, A., Cowie, L. L., & Lilly, S. J. 1990, ApJ, 348, 371
25. Woosley, S. E., & Weaver, T. A. 1995, ApJS, 101, 181
26. Pei, Y. C., & Fall, S. M. 1995, ApJ, 454, 69
27. Worthey, G. 1996, this volume
28. Soifer, B. T., & Neugebauer, G. 1991, AJ, 101, 354

THE STAR FORMATION HISTORY OF LOW REDSHIFT SPIRAL GALAXIES

Robert W. O'Connell

Astronomy Department, University of Virginia
Charlottesville, VA 22903-0818

Abstract.
Recent results on the star formation histories of nearby spiral galaxies do not conform to the classical view of how such systems evolve. Bulges have a range of light-weighted ages and are probably strongly influenced by the inner disks. Disks probably do not have a smoothly monotonic, exponential star forming history but do show a remarkable degree of regulation over time scales with a ratio of 1000:1. UV observations are especially valuable in probing dynamical control of star formation in disks on time scales of 50-500 Myr.

INTRODUCTION

The star forming history of nearby spiral galaxies is a subject which probably hasn't received the attention it deserves. One problem is intrinsic: spirals contain two distinct but interacting components, a bulge and a disk, with very different dynamical properties and star formation histories (SFH's). They are correspondingly more complex than elliptical or pure-disk systems. Another difficulty is technical: because nearby spirals subtend large angles on the sky, it is hard to obtain global information on them, especially spectroscopy. In this sense, it is actually easier to study high redshift systems. Only recently has global spectroscopy become available for a large sample of nearby spirals (e.g. [1]). A final impediment is more sociological. Many astronomers tend to regard spirals as a "solved problem", apparently because there are robust methods of estimating the *current* star formation rate (SFR) in spirals. But this is not at all the same thing as determining the star formation *history* over a Hubble time.

Consequently, information is actually growing faster about spirals at high redshifts than about their low redshift counterparts. Even so, I cannot hope to provide a comprehensive review of low redshift studies, especially those of the Galaxy itself. For that subject, let me just refer to the fine reviews in [2–5]. Instead, I will give here a selective review of some work which bears on the viability of our basic prejudices concerning the evolutionary history of other large spiral galaxies.

PROBING THE STAR FORMATION HISTORY OF SPIRALS

The classical picture of spiral evolution derives largely from the seminal paper of Eggen et al. [6] and the early consensus on stellar populations developed at the Vatican Conference [7]. Spiral bulges were thought to have formed rapidly, early, and nearly synchronously, independent of their environments. After the bulge stabilized, little gas remained to fuel later star formation (much may have been expelled by winds). Disks were imagined to have formed later, under more quiescent circumstances in which significant amounts of gas survived for long periods. This was thought to be smoothly consumed in a process with a time scale which varied from system to system but which was usually taken to have an monotonically decreasing, exponential time dependence (e.g. [8]). In this picture, bulges may influence disks, especially through control of the rate of gas consumption in the disk [9], but disks should not have significant influence on bulges.

How well does the evidence for spiral systems which are too distant for the study of individual stars agree with the classical picture? Though structure and kinematics provide some insights into history, the most useful information is found in the integrated spectral energy distribution (SED) of the galaxies. But this is less easy to extract than is often supposed. The SED of a single generation of stars is only logarithmically sensitive to its age. That is, conventional measures (Q) of the SED such as colors or line indices tend to scale as $Q \sim a + b \log t$, where a and b are constants. This means that we should imagine the history of a galaxy as a set of bins of *constant size in $\delta \log t$* stretching from the present to a lookback time of $\sim$15 Gyr. The SFH is then described by the mass of stars formed in each bin. The $\log t$ binning implies that much less information will be available on the details of the early SFH than on more recent times. The size of the bins, and hence the detail which can be discerned in the SFH, is governed by the amount of stellar population information (not equivalent to spectral information) present in our SED measures. That in turn is determined by the number of data points, the spectral resolution, the wavelength baseline covered, and the photometric precision of the measures [10].

With this perspective, it is evident that the most ubiquitous type of SED information available, namely broad-band colors, can provide only limited constraints on the SFH. Two colors, e.g. $U - B$ and $B - V$, yield at most only two logarithmic resolution cells over the last 15 Gyr. Many different types of histories can be consistent with a given pair of broad-band colors. That is, broad band colors are often degenerate with respect to the SFH. Examples of color degeneracy for SFH's characterized by smooth functions, bursts, and quenching have been discussed in [11–13].

One of the first studies to characterize SFH's of spirals this way was that of Gallagher et al. [14]. They analyzed galaxy histories on three very different timescales. The mean SFR over a Hubble time was determined from the total mass of the galaxies. The mean SFR over the past few Gyr was estimated from the integrated B-band ($\lambda 4400$) luminosity. And finally, the mean SFR over the lifetime of massive ionizing stars was probed by the strength of the Hα emission line. Since the ionizing flux from a population with a standard IMF declines by a factor of 10 after only 5 Myr, Hα effectively probes the instantaneous current SF rate. Gallagher et al. found that the SF rates for irregular galaxies were nearly constant over these three epochs but that spirals seemed to have experienced an early burst of star formation (or contained dark matter). Most other studies of spiral galaxies do not provide much more information on histories than this kind of few-epoch comparison.

HISTORIES OF SPIRAL BULGES

A straightforward prediction of the classical picture is that the bulges of different spirals should have very similar properties (apart from metallicity effects), characteristic of very old populations, and may differ substantially from their surrounding disks. There is considerable debate as to whether this is true of the nearby bulges in which one can study individual stars (e.g. [15]). For other galaxies, the evidence seems to contradict this expectation.

The first discussion of this subject was presented by Morgan and his collaborators [16,17], who had earlier demonstrated that spiral bulges were not pure Pop II systems as had been proposed by Baade in 1944. Morgan found that most small bulge spirals (types $\gtrsim$ Sbc) have strong spectroscopic signatures (F–G spectral types) of nuclear star formation over the past ~ 2 Gyr. The relative youth of the spectra is inversely related to the overall prominence of the bulge (L_{bulge}/L_{tot}). Later quantitative analysis of SED's by Turnrose [18] confirmed this result in late type spirals. An alternative interpretation of these bulges as very old but metal poor systems similar to globular clusters was excluded by high S/N spectra and colors [19–21].

Strictly speaking, the youthful stars found in these studies lie within the volume of the bulges but are not necessarily part of the dynamical bulge itself. It would be important to test their spatial distribution. However, it does seem clear they are a long-lived phenomenon, unrelated to recent starbursts for instance, and that they are not simply the product of normal disk-like star formation extended to the galaxy centers.

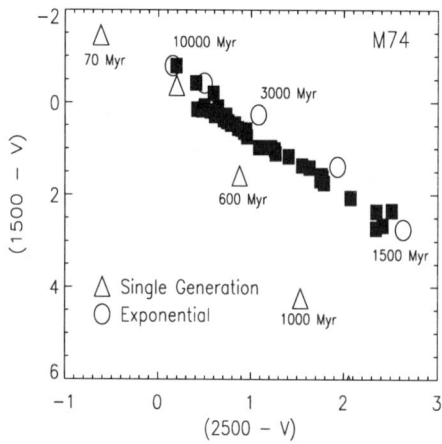

FIGURE 1. Two-color plot for spiral bulges from Peletier & Balcells (1996). In the classical picture, bulges should show a small range of color consistent with large ages. That is clearly not the case here. Synthetic isochrones (lines), labeled by ages in Gyr, are from recent models Vazdekis et al. (1996), and indicate a large age range for the bulges here.

The bulges of earlier type galaxies (Sa–Sb) are outwardly more homogeneous; most were classified as having "K"-type spectra by Morgan, for instance. Photometrically and spectroscopically, however, they exhibit more scatter than do E/S0's, and there are some clear cases of young populations [22–25]. A difficulty is that the effects of dust extinction are particularly important here, where the intrinsic range of SED may be smaller.

Recent multicolor CCD observations of spiral bulges permit an improved treatment of extinction. Peletier & Balcells [26] minimize the dust effects by using special synthetic apertures. They claim that earlier work on the homogeneity of spiral bulges was strongly influenced by extinction, which tended to produce red, "old" colors. They find a significant range in bulge SED properties using several optical/IR colors in a sample of 30 S0–Sbc objects. They also find that the bulge-disk color differential is small in a given galaxy. They conclude that the histories of bulges and disks are not independent and that the mean age differential between bulges and disks is small, $\delta \log t \lesssim 0.1$. The range in bulge colors (see Fig. 1) implies a large range (1–15 Gyr) in *light-weighted* bulge ages. Fagatto et al. [28] have independently examined the problem of extinction in bulges and reach similar conclusions about the range of bulge ages based on intrinsic colors.

The "Morgan effect" therefore suggests that spiral bulges are not uniform structures, that their stars were not produced at a single early epoch, and that bulges and disks probably influence one another. There is no doubt that spiral bulges can form at very early epochs (Dressler, this conference). After formation, however, external circumstances such as mergers or stripping can clearly induce a wide variety of evolutionary histories (e.g. [12,29–32]). The spiral bulges being discussed here may have been influenced more by *internal* processes in which disk material is secularly converted into bulge material by dynamical interactions with the bulge potential well [33]. The moderate starbursts identified in nearby barred spirals (Kenney, this conference) demonstrate that the sheer gas processing rate necessary to build bulges on short time scales (a few Gyr) is available at the present epoch. The resulting changes in galaxy structure, perhaps even in Hubble type, may contribute to the dispersion seen in the Hα-color correlation discussed below. There is also good *kinematic* evidence that in many cases disk material is being converted to bulges. This was reviewed by Kormendy [34], who states, "By type Sc, I do not believe that any galaxies contain true [i.e. classical] bulges."

Could some bulges form late, e.g. at the intermediate redshifts $z \sim 0.5$ suggested if the dating in Fig. 1 is correct and does not result from secular evolution? Probably they could. Evidently, the mild starburst activity involved in current-day bulge building as described by Kenney does not result in galaxies which look very peculiar. The same would be true of initial bulge formation if the central star formation rate were less than ~ 10 M$_\odot$ yr^{-1}. Such systems would have bright centers or inner disks but would not necessarily be seriously disrupted objects like those associated with massive starbursts. The best available cases of earlier ($z > 2$) galaxy formation show surprisingly modest rates of star formation near or below this level (Madau, this conference).

HISTORIES OF SPIRAL DISKS

Ionized Gas and Star Formation Regulation

Is the star formation history of spiral disks the smooth, monotonic function expected in the classic picture? My own view is that the functional form of the SFH in spirals hasn't been determined yet, but that nonetheless emission line data does imply the existence of strong global regulation in spiral disks. There would seem to be good reason to doubt this. One of the remarkable results to emerge from the rapidly increasing volume of digital images of resolved galaxies in the Local Group is the widespread evidence for *strongly discontinuous* star formation histories in the form of bursts or accretion events. This was anticipated nearly 20 years ago by Butcher's [35] study of the LMC field, and the most beautiful recent example is the multiburst Carina system [36]. Most objects showing such discrete episodes of star formation are small galaxies— low luminosity or dwarf. These are easily disturbed by larger neighbors and are more likely to suffer quenching of star formation by expulsion of their

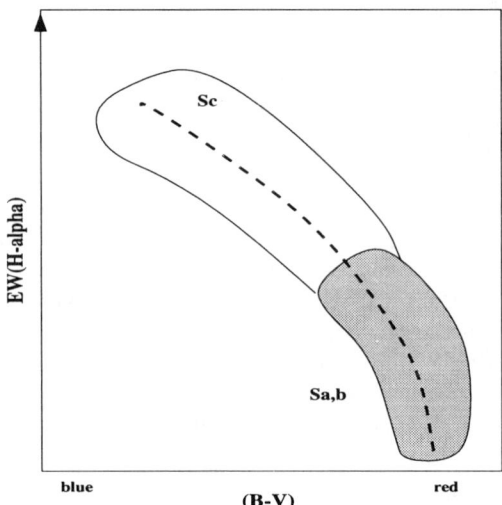

FIGURE 2. Schematic correlation between the global equivalent width of Hα and (B–V) color in spiral galaxies, after Kennicutt, Tamblyn, & Congdon (1994). The dashed line shows a typical fit with an exponentially-declining SFR.

interstellar media following a star forming episode. But size is not the determining factor, since the Milky Way disk itself shows evidence of at least 3 discrete bursts of star formation near the Sun over the past 9 Gyr (reviewed in [5]). Of course, we do not know over what fraction of the total disk such events might remain coherent. The superposition of many incoherent bursts, smoothed over the time resolution of our analysis techniques, may be nicely monotonic.

Evidence for global regulation comes mainly from ionized gas studies. This subject has not changed much since the earlier review by Kennicutt [37], but I approach it from a somewhat different angle here. There is a good correlation between the prominence of Hα or other strong emission lines and the integrated color or spectral type of bright spiral galaxies. This was evident, for instance, from the spectroscopic survey of Humason et al. [38] and from early photoelectric studies [39]. The definitive work on this subject is by Kennicutt and collaborators [40–42]. The composite Hα-(B–V) correlation from these studies is shown in schematic form in Figure 2; both quantities refer primarily to the disks, not bulges, of the galaxies. There is significant scatter within each morphological class and at any color, but the general correlation is quite good.

This correlation has conventionally been interpreted in the context of exponentially-declining star formation histories. The e-folding times needed are in the range 1 Gyr–∞, and the fits are generally good, as suggested in the figure. But this diagram is actually *not* a good test of the *shape* of the star formation history. The Hα equivalent width is a measure of the current SFR compared to the mean past rate, and while (B–V) responds to the mean rate over the past few Gyr, it is quite insensitive to the detailed structure in SFR(t) over that period. In other words, a wide variety of non-exponential SFH's could be consistent with the observed correlation.

The inappropriateness of the monotonically declining exponential SFH is emphasized by another well-established result from the Kennicutt studies. Using several indicators of the mean past SFR, one can compute b, the ratio of the current SFR to the mean past SFR. In many galaxies of type Sbc or later, $b > 1$ [42]—i.e. the disk is presently experiencing enhanced star formation possibly like the irregularities detected in nearby resolved galaxies.

Even if the correlation of Fig. 2 does not establish the shape of the SFH, it does demonstrate a remarkable degree of regulation in the global star formation rates because the two axes in the diagram measure star formation rates over time scales which differ by a ratio of about 1000:1! The correlation implies that the SFR over the past 5 Myr (as measured by Hα) can be used to infer the SFR during the past 500-5000 Myr. The fact that there is any correlation at all implies a strong degree of global regulation in disk SFR's. The correlation cannot be produced unless a large fraction of the disk participates. My impression is that the kind of coherence necessary to explain Fig. 2 is not realistically achieved by existing theoretical models for star formation in disks. If so, then I think the problem merits being called the "regulation mystery".

UV Studies of Disk Histories

The vacuum-UV provides an important opportunity to extend our probes of the SFH to timescales intermediate between those discussed so far (either very short or very long), namely 50-500 Myr, which are the times over which the 1000–3000Å light of a single generation of stars will decay.

UV brightnesses can be used to estimate the mean SFR over the past few 100 Myr [43–45]. Since this is derived from the stellar continuum, rather than from ionized gas, it is an important complement to the emission line methods. Extinction by dust is often cited as an impediment to UV continuum methods, but gaseous ionization, being driven by the Lyman continuum, is comparably sensitive to dust. Good correlations have been found in spiral disks between UV-derived global SFR's and μ, the total gas surface density (H I plus H_2) which show that SFR $\sim \mu^{1.6}$ [46] . The correlations with either atomic or molecular gas alone are less good. In terms of the time dependence of star formation, one can compare the emission line to UV continuum measures to examine changes over the last $\sim$100 Myr [43] or derive UV-optical colors to compare time scales of a few 100 and a few 1000 Myr [44,45].

Of particular interest for the regulation problem are imaging studies where

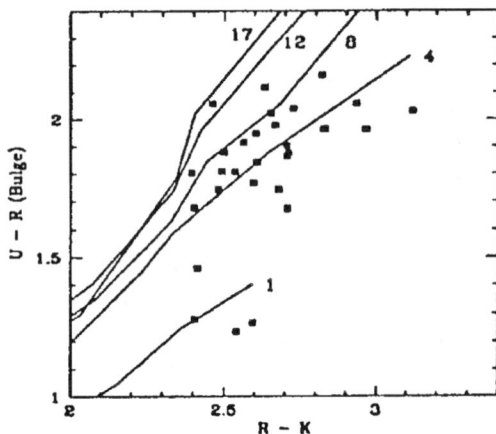

FIGURE 3. UV surface photometry of the Sc M74 obtained by UIT plotted in a UV/optical 2-color diagram, after Cornett (1994). Galaxy data are solid squares, with smaller radii to the right (redder colors) and larger to the left. Triangles and circles are single generation models and exponentially-declining models, labeled by age and e-folding time, respectively.

the UV structure of galaxies can be spatially resolved and gradients can be analyzed. The FOCA and UIT experiments have produced multiband UV continuum maps of a number of prominent spirals, including M33, M51, and M81 [47–50]. From our Astro-1/UIT images, we have studied gradients in the SFH across the face of the Sc I galaxy M74 [51]. Most of the far-UV light does not originate in young, compact H II regions but rather from a more diffuse, presumably older, component. We have compared surface photometry in two UV bands and one optical band to various models (Figure 3). There is a large change in UV color with radius. We believe this is neither a metallicity nor extinction effect, nor does it simply reflect the run of the current SFR or gas density with radius. Instead, it measures changes in the history of the SFR. The entire disk has undergone star formation during the past ~ 500 Myr, but the inner regions have experienced more rapidly declining star formation than the outer regions. If the models are taken at face value, then no part of the disk has experienced a constant SFR (corresponding to an e-folding time of ∞), and the SFR has declined more steeply than an exponential.

What is surprising about the diagram is not that the SFH changes with radius but that it exhibits such a remarkably organized pattern of change. The smooth change with radius removes some of the ambiguity affecting SFH interpretations based on global quantities. For instance, it excludes strong recent bursts of star formation superposed on a strictly old population, which might explain any single point on the M74 locus but can't plausibly reproduce the

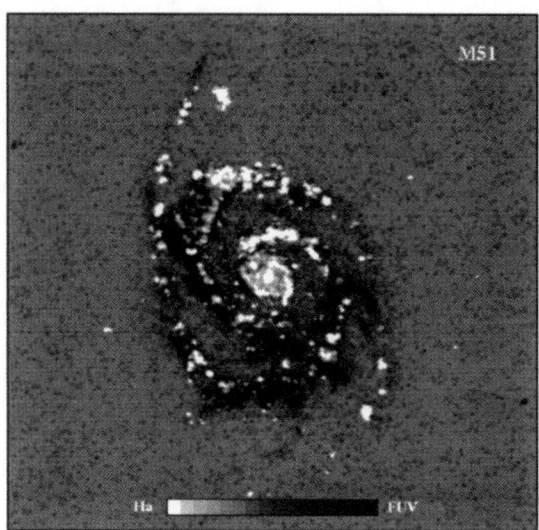

FIGURE 4. An Hα-far-UV difference map of M51 based on Astro-2/UIT images with FWHM ∼ 4″. The grey scale displays the logarithmic difference between the two bands, with lighter areas being relatively brighter in Hα. The perturbing companion galaxy (NGC 5195) at the top of the image is completely absent in the far-UV.

whole well-behaved correlation. Different galaxies, e.g. M33 and M81, present different disk loci in such diagrams [51]. These have not yet been interpreted, but they are probably important clues to the regulation mechanism.

Such UV maps of the SFH over the past ∼ 1000 Myr can be used to relate star formation to dynamics. It is presumably density wave forcing and radiation and gas pressure feedback from young regions that act to produce these examples of star formation regulation. The time scale sampled by the UV is particularly relevant to density wave effects since the period between density wave passages for material in spiral disks is ∼ 50–500 Myr. Using Hα and UV maps, the relationship of star formation on short and intermediate timescales to density waves, molecular and atomic gas, far-IR radiation, and other relevant properties can be examined in detail. A nice example is the nearby Sc spiral M51. Figure 4 shows an Hα/far-UV difference image of M51 derived from our Astro-2/UIT data [52,53]. A similar map with lower resolution was published by the FOCA group [54,55]. One can see how the ∼ 50–100 Myr year old populations are usually spread farther downstream from the density wave than the ∼ 5 Myr-old, Hα bright populations. There are significant differences in the population pattern with position in the galaxy, however, which are presumably related to the tidal interaction which is strongly influencing the density wave structure (e.g. [56]).

If we think of galaxy evolution in Darwinian terms, it may not be surprising that star formation in large spiral disks is well regulated. Because cool interstellar gas at the densities which prevail in spiral galaxies is explosively unstable to star formation in the presence of strong pressure gradients, the thin, relatively quiescent disks we see in current epoch spirals are selected for their survival characteristics. However, I think we don't understand very well *how* the regulation is enforced or how it is to be reconciled with the discontinuous star formation histories revealed in smaller galaxies and even in the Milky Way disk.

ACKNOWLEDGEMENTS

I am grateful to Reynier Peletier for permission to reproduce Fig. 1, to George Becker for producing the map of M51 in Fig. 4 and to Ted Stecher, Bob Cornett, and our other colleagues on the UIT team for continuing support. This work has been supported in part by NASA grants NAGW-2596 and NAG5-700.

REFERENCES

1. Kennicutt, R. C. 1992, ApJ, 388, 310
2. Freeman, K. C. 1987, ARAA, 25, 603
3. Frogel, J. A. 1988, ARAA, 26, 51
4. Gilmore, G., Wyse, R. F. G., & Kuijken, K. 1989, ARAA, 27, 555
5. Majewski, S. R. 1993, ARAA, 31, 575
6. Eggen, O.J., Lynden-Bell, D., & Sandage, A. 1962, ApJ, 136, 748
7. O'Connell, D. J. K. 1958, Stellar Populations (Amsterdam: North Holland)
8. Tinsley, B. M. 1978, in Structure and Properties of Nearby Galaxies (IAU Symposium No. 77), eds. E. Berkhuijsen and R. Wielebinski (Dordrecht: Reidel), 15
9. Roberts, M. S. & Haynes, M. P. 1994, ARAA, 32, 115
10. O'Connell, R. W. 1996, in From Stars to Galaxies: The Impact of Stellar Physics on Galaxy Evolution, eds. C. Leitherer, U. Fritze-von Alvensleben, & J. Huchra (ASP), 3
11. Tinsley, B. M. 1980, ApJ, 241, 41
12. Schweizer, F., & Seitzer, P. 1992, AJ, 104, 1039
13. Fritze-v. Alvensleben, U., & Gerhard, O.E. 1994, A&A, 285, 751
14. Gallagher, J.S., Hunter, D.A., & Tutukov, A.V. 1984, ApJ, 284, 544
15. Habing, H., & Dejonghe, H. 1993, Galactic Bulges (IAU Symposium No. 153), (Dordrecht: Kluwer)
16. Morgan, W. W., & Mayall, N. U. 1957, PASP, 69, 291.
17. Morgan, W. W., & Osterbrock, D. E. 1969, AJ, 74, 515
18. Turnrose, B. E. 1976, ApJ, 210, 33
19. McClure, R. D., Cowley, A. P., & Crampton, D. 1980, ApJ, 236, 112

20. O'Connell, R. W. 1982, ApJ, 257, 89
21. Frogel, J. A. 1985, ApJ, 298, 528
22. Sandage, A. & Visvanathan, N. 1978, ApJ, 223, 707
23. Griersmith, D. 1980, AJ, 85, 1295
24. Veron, P., & Veron-Cetty, M. P. 1985, A&A, 145, 433
25. King, C.R. 1986, PhD Thesis, Yale University
26. Peletier, R. F., & Balcells, M. 1996, AJ, 111, 2238
27. Vazdekis, A., Casuso, E., Peletier, R., Beckman, J. 1996, ApJ, in press
28. Fagatto, F., Corradi, R., & Di Bartolomeo, A. 1996, in From Stars to Galaxies: The Impact of Stellar Physics on Galaxy Evolution, eds. C. Leitherer, U. Fritze-von Alvensleben, & J. Huchra (ASP), 368
29. Dressler, A., & Gunn, J. E. 1982, ApJ, 263, 533
30. Bothun, G., & Dressler, A. 1986, ApJ, 301, 57
31. O'Connell, R. W. 1993, in Nuclei of Normal Galaxies, eds. R. Genzel & A. J. Harris (Dordrecht: Kluwer), 255
32. Caldwell, N., Rose, J., Sharples, R., Ellis, R., & Bower, R. 1993, AJ, 106, 473
33. Larson, R. B. 1990, PASP, 102, 709
34. Kormendy, J. 1993, in [15]
35. Butcher, H.R. 1977, ApJ, 216, 372
36. Smecker-Hane, T. A., Stetson, P. B., Hesser, J. E., & VandenBerg, D. A. 1996, in From Stars to Galaxies: The Impact of Stellar Physics on Galaxy Evolution, eds. C. Leitherer, U. Fritze-von Alvensleben, & J. Huchra (ASP), 328
37. Kennicutt, R. C. 1990, in The Evolution of the Universe of Galaxies, ed. R. G. Kron (ASP), 141
38. Humason, M. L., Mayall, N. U., & Sandage, A. R. 1956, AJ, 61, 97
39. Cohen, J. G. 1976, ApJ 203, 587
40. Kennicutt, R. C. 1983, ApJ, 272, 54
41. Gavazzi, G., Boselli, A., & Kennicutt, R. 1991, AJ, 101, 1207
42. Kennicutt, R. C., Tamblyn, P., & Congdon, C. W. 1994, ApJ, 435, 22
43. Lequeux, J., Maucherat-Joubert, M., Deharveng, J., & Kunth, D. 1981, A&A, 103, 305
44. Donas, J. et al. 1987, A&A, 180, 12
45. Deharveng, J. M., et al. 1994, A&A, 289, 715
46. Buat, V., Deharveng, J. M., & Donas, J. 1989, A&A, 223, 42
47. Landsman, W. B. et al. 1992, ApJ, 401, L83
48. Bohlin, R. C. et al. 1990, ApJ, 352, 55
49. Hill, J. K. et al. 1992, ApJ, 395, L33
50. Reichen, M. et al. 1994, A&AS, 106, 523
51. Cornett, R. H. et al. 1994, ApJ, 426, 553
52. Hill, J. K. et al. 1996, ApJ, in press
53. Marcum, P. M., O'Connell, R. W. et al. 1997, in preparation
54. Bersier, D., Blecha, A., Golay, M., & Martinet, L. 1994, A&A, 286, 37
55. Petit, H., Hua, C. T., Bersier, D., & Courtès, G. 1996, A&A, 309, 446
56. Byrd, G., & Salo, H. 1995, AL&C, 31, 193

Near-IR Discoveries of Groups of Star-Forming Galaxies at $z > 2$

Harry Teplitz, Matthew A. Malkan, and Ian S. McLean

Department of Astronomy & Physics, University of California Los Angeles, LA, CA 90095

Abstract. We report the discovery of several groups of star-forming galaxies at redshifts $2.0 < z < 2.5$. An ongoing program of narrow-band imaging with the Near IR Camera on the Keck I 10-m telescope has shown these galaxies to have strong emission lines at the expected wavelength of redshifted Hα. Optical and IR spectroscopy have confirmed one emission line candidate is at a redshift of z=2.498. This galaxy lies in a field containing two other Hα-emitters, as well as several very red, non-Hα emitting galaxies. Similar emission line galaxies have been observed in three other fields selected for their proximity to QSOs or QSO absorption systems. These galaxies are typically within a projected distance of 250kpc of the QSO line of sight; they are resolved but compact. To date, the program has a detection rate of 1-2 emission line galaxies per square arcminute surveyed.

INTRODUCTION

In this poster, we present the results of an ongoing search for redshifted Hα emission from normal clusters of galaxies at $z > 2$. Using 1% interference filters, we can detect this line emission by comparing the flux in the narrow-band to the continuum measured in the K' filter. An object with Hα emission will appear relatively brighter in the narrow filter, when compared to objects with no emission. We survey fields containing known objects (QSOs or metal absorption-line systems) at redshifts which place Hα in our narrow filters.

Hα suffers less extinction from intrinsic or intervening dust than UV emission lines, allowing us to detect a redder population of galaxies than searches depending on Lyα emission or on the UV-continuum. Further, Hα is directly related to the on-going Star Formation Rate [1].

All observations in this survey were taken at the 10-m Keck I telescope. We obtained images of 19 fields with the Near IR Camera [2], which has a field of view of $38'' \times 38''$. A confirming spectrum of one object, and deep V and I imaging of the field around it, were obtained with the Low Resolution Imaging Spectrometer [3].

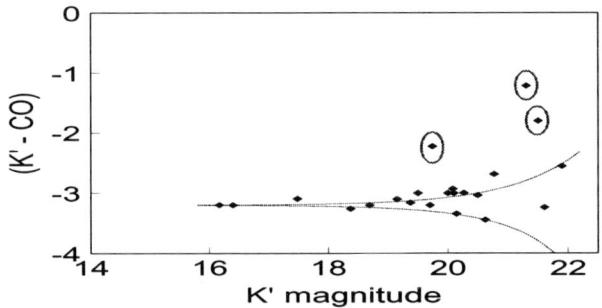

FIGURE 1. Color-Magnitude diagram for the SBS0953 field. The curved lines indicate 1σ errors. The three Hα-emitters are circled

This survey has revealed 8-12 Hα-emitters in small groups of 2 or 3 each. The first object detected [4] proved to lie in a probable cluster of galaxies at z=2.5 [5]. We present here details on that cluster, and then the preliminary conclusions that can be drawn from the entire survey to date. Throughout this poster, we assume $H_0 = 50$ and $q_0 = 0.5$.

MTM0953+549

The quasar SBS0953+549 (z=2.579) was of special interest as a survey target because its spectrum shows a strong absorption system at $z_{abs} = 2.50176 \pm 0.00004$, and weaker systems at 2.49174 and 2.50911 [6]. We obtained narrow-band images, through the 2.30 μm CO(2-0) filter, of two adjacent NIRC fields around this object as well as deep VIJHK photometry of a more extended area. Comparing the broad and narrow-band fluxes reveals three definite Hα-emitting objects (fig.1). The distances of these objects from the QSO line of sight vary from 220kpc to 240kpc, making them likely to be members of a cluster containing the absorber. In addition, this field contains four red $(I - K > 4)$ galaxies that could also be at high redshift, perhaps in the same cluster.

An optical spectrum (fig. 2) of the brightest Hα-emitter (designated by its coordinates as MTM095355+545428) confirms its redshift to be $z = 2.498 \pm 0.001$. The spectrum shows strong, narrow Lyα emission, and also includes substantially weaker UV emission lines of NV, SiIV, CIV, HeII, CIII], SiIII], and OIII]. In addition, interstellar absorption lines (SiII, OI, CII) are definitely present, and some of the emission lines (NV, SiIV, CIV) show distinctive P Cygni profiles; these characteristics are similar to those observed in star-forming galaxies at the present epoch [7]. The presence of stellar and interstellar absorption lines confirms that we are indeed observing on-going star formation. However, the UV emission lines raise the possibility of a non-stellar nuclear component to this galaxy. The flux ratio of the other emission

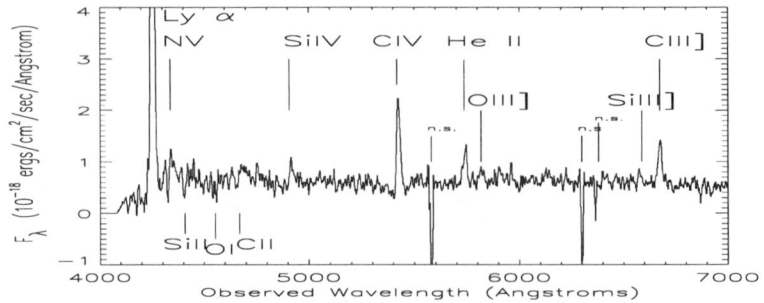

FIGURE 2. LRIS optical spectrum of MTM095355+545428

lines compared to Lyα is weaker than would be expected for a Seyfert 2. In fact, even if all the CIV and HeII were the result of a nonstellar source (which is unlikely given the P Cygni trough near CIV), this source would still contribute less than a third of the observed Lyα flux, and almost none of the UV continuum.

Since we know that we are observing stars in MTM095355+545428, we can use our detailed BVRIJHK Spectral Energy Distribution (fig. 3) to deduce the formation history of this object. Comparing the SED to stellar synthesis models [8], we find this galaxy formed at z=5.5 in a 1 Gyr burst of star formation of which we are seeing the tail end. Similar $z_{formation}$ can be deduced for the other two Hα-emitters in the field. MTM095355+545428 appears to be 0.5–1 L*, and the other two galaxies are at least 5 times less luminous.

SURVEY RESULTS

To date we have surveyed close to 8 square arcminutes to the required depth. We have detected 5 strong Hα-emitting objects ($\Delta m \geq 0.9$), and 5-8 weaker ones. Most of these objects are clustered in groups of two or three. The faintest emission line detection has $f(H\alpha) \sim 5 \times 10^{-17}$ ergs/cm²/sec. The continuum (5σ) detection limits were typically K=22, while the faintest emitting galaxy has K=21.5.

All of the $z > 2$ objects detected are extremely compact. They typically show FWHM$\sim 0.9'' \pm 0.2''$, compared to a usual seeing disk of 0.6''. Even the most extended object is only 1.125''across. These sizes correspond to a projected diameter of 6-9kpc. It is important to point out, however, that the high central surface brightness of these objects favors detection by our technique. While all the objects are compact, they vary in morphology from almost round to noticeably elliptical (b/a ~ 0.5).

The average inferred SFR of detected galaxies is 30 $M_\odot$/year. For example, MTM095355+545428 has SFR=100 $M_\odot$/year, and its companions have 28 and

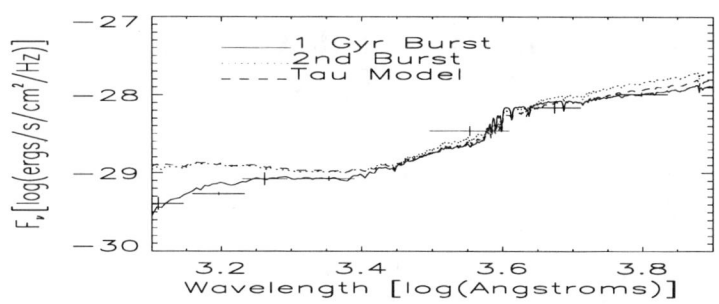

FIGURE 3. SED of MTM095355+545428 compared to Bruzual & Charlot models

72 M$_\odot$/year. Again, this is a bias of the survey which preferentially detects the most vigorous star formation. Nonetheless, this SFR is significantly higher than that observed in the present epoch. It is also significantly higher than the SFR inferred from most optical searches for similar objects (for example Madau in this volume).

The surface density of detections implies that we must be seeing clustering. We detect 1.3-1.6 galaxies/sq. arcmin. in volumes that are only 1% deep in redshift. At the current epoch, the field density is 0.25 galaxies per square arcminute down to L*. Other high-z surveys [9] have found 0.4/sq. arcmin. using the Lyman Limit Dropout technique that admits a much larger redshift range. We would reiterate that our statistics are low. This survey is on-going and we hope to present in the future a 5σ estimate of the galaxy density. Finally, we note that all but one of the objects detected have been in fields containing metal absorption-line systems, not fields at the redshift of the QSO.

REFERENCES

1. Kennicut, R. 1983, ApJ 272, 54
2. Matthews, K., & Soifer, B. T. 1994, in *Infrared Astronomy with Arrays: The Next Generation*, ed I. McLean (Dordrecht:Kluwer), 239
3. Oke, J. B., et al. ,1995, P.A.S.P, 107, 375
4. Malkan, M., Teplitz, H., & McLean, I., 1995, ApJ Letters 448, L5
5. Malkan, M., Teplitz, H., & McLean, I., 1996, ApJ Letters 468, L9
6. Levshakov, S. A. 1992, AJ 104, 950
7. Conti, P. S., Leitherer, C., & Vacca, W. D., 1996, ApJ Letters 461, 87
8. Bruzual, A. G. & Charlot, S., 1993, ApJ 405, 538
9. Steidel, C.C., Giavalisco, M., Pettini, M., Dickinson, M., & Adelberger, K.L., 1996, ApJ Letters 462, L17

The ISO-IRAS Faint Galaxy Survey: Early Results

R.L. Hurt[1], C.J. Lonsdale[1], D.A. Levine[1], H.E. Smith[2], G. Helou[1], D. Van Buren[1], C.A. Beichman[1], S.D. Lord[1], G. Neugebauer[3], M. Moshir[1], B.T. Soifer[3], A. Wehrle[1], C. Cesarsky[4], D. Elbaz[4], U. Klaas[5], R. Laureijs[6], D. Lemke[7], R.G. McMahon[8], R.D. Wolstencroft[9]

[1] *Infrared Processing & Analysis Center, Pasadena, CA, USA*
[2] *University of California at San Diego, San Diego, CA, USA*
[3] *California Institute of Technology, Pasadena, CA, USA*
[4] *Service d'Astrophysique, CEA Saclay, 91191 Gif-sur-Yvette, France*
[5] *Max Planck Institut fur Astronomie, Heidelberg, Germany*
[6] *ESA/ESTEC, Noordwijk, The Netherlands*
[7] *European Southern Observatory, La Silla, Chile*
[8] *Cambridge Institute of Astronomy, Cambridge, England*
[9] *Royal Observatory, Edinburgh, Scotland*

Abstract. We present preliminary results for AGNs and starburst galaxies the ISO-IRAS Faint Galaxy Survey (IIFGS). The goal of the survey is to produce a database of infrared-luminous galaxies at redshifts of about 0.1–1 to help explore the AGN-starburst relationship, study the cosmological evolution of luminous infrared galaxies, and identify possible protogalaxy candidates. The candidate list of ~3700 sources has been extracted from the IRAS Faint Source Survey using criteria selecting for faint, infrared-bright galaxies. The ISO observations will confirm the IRAS detections, yield sensitive 12 & 90 μm fluxes, and provide positions to ~6" accuracy which will allow unambiguous optical identifications. Confirmed sources are being followed up with ground-based observations to determine optical magnitudes and accurate redshifts. In this preliminary phase we have in hand ~100 observed fields and are developing techniques to maximize the sensitivity of the observations. Early results for the ISOCAM 12 μm images indicate we can reliably detect sources as faint as ~0.5 mJy; ~80% of the fields contain at least one source.

INTRODUCTION

One of the many discoveries of IRAS was the population of Luminous Far Infrared Galaxies (LFIRG's) which radiate the bulk of their energy at wavelengths longwards of 10 μm. Such emission is almost certainly powered by warm dust surrounding starbursts and/or AGN [1–3]. However, such samples of IR-bright galaxies are limited by the sensitivity of IRAS and most likely underrepresent such sources at intermediate and large redshifts.

The goal of the ISO-IRAS Faint Galaxy Survey (IIFGS) is to compile a catalog of several hundred LFIRG's at a redshift range of $\sim$0.1–1. Such a sample will allow studies of the mechanisms driving these objects to extend to greater distances and earlier times, and to potentially address issues of evolution. The sample is based on the faintest IRAS detections of potential LFIRG's for which ISO can obtain good photometry and positions. Presented here is a description of and progress report for the IIFGS.

Sources from the IIFGS were selected from IRAS Faint Source Survey (FSS) using criteria to select for distant LFIRG's. The sources were required to have non-stellar colors (i.e. increasing flux density from 12 to 60 μm). To bias the sample towards distant, luminous galaxies, the 60 μm fluxes were constrained to be less than 0.3 Jy (to assure faintness) with ratios of L_{fir}/L_{blue} exceeding 10 (implying $L_{fir} > 10^{11}$ $L_\odot$, using the strong correlation between L_{fir}/L_{blue} and L_{fir}). Blue magnitudes and upper limits were obtained using IPAC's Optical Identification Database, OPTID, which matches FSS sources against digitized sky survey plate data. To further optimize the sample, only candidates at Galactic latitudes exceeding 30 deg were included in the sample to avoid cirrus contamination. The remaining candidates were inspected individually and only those passing a rigorous cirrus rejection procedure were included in the final sample of $\sim$3700 sources.

The IIFGS has been implemented in the ISO mission as a "filler" survey. The IIFGS sources populate many ISO sky bins and are intended to fill in between other observations, helping to optimize the observing efficiency of the satellite. The survey consists of linked short-wavelength ISOCAM and long-wavelength ISOPHOT observations for each object.

The ISOCAM observation uses the LW10 filter (corresponding roughly to the IRAS 12 μm band). The pixel scale of 6"/pixel, provides a total field of view of about 3'x3'. The source is observed with a 2x2 raster using 30" offsets, allowing confirmation of the source in multiple detector pixels. The integration time is 2.1 sec with a total effective exposure time of about 70 sec. The ISOCAM observations are much more sensitive than IRAS and allow the positions of the infrared sources to be determined with 6" accuracy, allowing for unambiguous optical identifications.

The ISOPHOT observation is a chopped PHT22 exposure at 90 μm. Total on-source integration time is 64 sec. In addition to confirming the IRAS detection, it will fill in the spectrum energy distribution between the IRAS wavelengths.

CURRENT STATUS

The IIFGS in presently at an early stage of production. To date, ~150 sources have been observed by ISO; we hope to double that number before the end of the mission. Data from the CAM and PHOT detectors are being evaluated and reduction strategies are being developed for the dataset. The IIFGS data product to be made available to the community shall include positions, fluxes, redshifts, and images of the sources.

The ISOCAM pixels have a characteristic time response to illumination which affects the quality of flat-fielding and coadding. Sample detector responses to an observation are illustrated in Figure 1. The four graphs represent the four detector pixels that are exposed to a single sky position. The horizontal axis represents the time steps for a single pixel (the exposure is 2.1 sec/frame). The actual detector data are indicated with '+' symbols; evident is the slow 'ramp-up' of the background and the presence of the same point source in each of the four raster positions.

By modelling the data with a fit that accounts for both the instantaneous response to illumination and the slower transient ramp, the data may be recti-

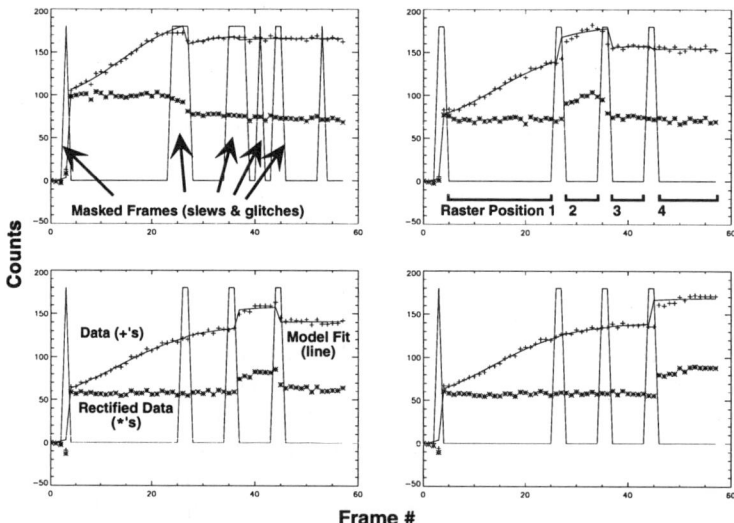

FIGURE 1. ISOCAM Transient Response and Removal. The '+' symbols represent detector responses of an individual pixel for the sequence of frames in the observation. The solid line is the best-fit model of the detector response, and the '*' symbols represent the data with the 'slow' response subtracted out.

FIGURE 2. ISOCAM Images Before and After. These three images present the same ISOCAM field for the IRAS source F12513+7605. The first is the standard pipeline-processed image in which the source is visible but artifacts seriously corrupt the image. The second is the same image after applying the transient model of Fig. 1; the background is flatter and the SNR is improved by a factor of ∼7. In the third image, cosmic ray glitches which have affected processing have been flagged out, and the resulting mosaic shows only the real source.

fied to flatten the response (rectified data are indicated by '*'s). Once rectified, it is possible to construct much-improved flats and more sensitive mosaics of the ISOCAM data. Figure 2 shows a single ISOCAM image, first using the standard data reduction pipeline (with no transient modeling), second after rectifying the dataset, and third after blanking pixels with lingering cosmic ray effects. At this early stage of processing, we have doubled the number of detections from the pipeline processing and improved the sensitivity by factors of five or more.

Complete detection statistics are currently available only for the pipeline-processed data. Of the 115 fields currently reduced, 61 have clear detections. Most of the fluxes range from 1–10 mJy. Inspection of the transient-processed images suggest that 80% of our fields contain at least one source, though this statistic includes serendipitous background sources unrelated to the IRAS FSS detections. Early ISOPHOT results are likewise promising; unambiguous 90 μm detections have been found for many of the brightest ISOCAM detections, although final calibrations are still pending.

Ground-based follow-up for the IIFGS is already underway. Observing time at Lick and Palomar has been allocated to the project for the determination of optical magnitudes and redshifts. Additional proposals are underway for other optical, infrared, and radio observations of the sources observed by ISO.

REFERENCES

1. Mirabel, I.F., & Sanders, D.B. 1996, ARAA, 34, in press
2. Lonsdale, C.J., Smith, H.E., & Lonsdale, C.J. 1993, ApJ, 405, L9
3. Smith, H.E., Lonsdale, C.J., & Lonsdale, C.J. 1996, to be submitted to ApJ

Gas Content and Star Formation Thresholds in the Evolution of Spiral Galaxies

Stacy McGaugh* and Erwin de Blok[†]

*Department of Terrestrial Magnetism, Carnegie Institution of Washington
[†]Kapteyn Astronomical Institute, University of Groningen

Abstract.
The gas mass fraction of spiral galaxies is strongly correlated with the central surface brightness of their disks. There exist many dim galaxies with long gas consumption time scales and $f_g > 0.5$. This resolves the gas consumption paradox.

The surface density of gas follows the optical surface brightness, but does not vary by as large a factor. This is the signature of a critical density threshold for star formation. Such a mechanism seems to be responsible for the slow evolution of dim galaxies.

GAS CONTENT

The fraction of baryonic mass which has been converted from gas into stars is a fundamental measure of the degree of evolution of a galaxy. The gas content of spiral galaxies is a strongly correlated with the optical surface brightness of their disks (Fig. 1). This must be an evolutionary effect indicative of the rate of galaxy evolution in disks of differing surface mass densities [1,2].

A complete analysis describing the details of the derivation of the gas mass fractions is given elsewhere [3]. Gas content increases strongly with decreasing surface brightness. An important consequence of this is the end of the gas consumption paradox. Gas rich galaxies do exist, and are quite common [4].

DISK EVOLUTION

The present epoch gas fraction is a direct chronometer of the star formation history. There is degeneracy in how a given f_g may be reached, but the evolutionary rate and/or age of spiral galaxies must vary systematically with surface brightness to give the observed correlation. Burst and fade scenarios

[5] can be completely ruled out as these should result in low not high f_g for dim galaxies. Indeed, it is difficult to have exponentially declining star formation histories in galaxies with $f_g > 0.5$ unless they are quite young. Such objects can be old if the star formation rate increases rather than decreases with time, but only at the expense of making the mean age of the stars very young (a few Gyr). A roughly constant star formation rate is more plausible, but requires an intermediate age (8 or 10 Gyr rather than 12 or 14 Gyr).

Galaxies with $f_g > 0.5$ have most of their star forming potential in the *future*. These gas rich galaxies are inevitably morphologically late types (Sd & Sm). They could not possibly have experienced an evolution of rapid gas consumption followed by rapid fading. Yet this is precisely the evolution inferred for late types from high redshift data [6]. These results are both sound and utterly contradictory. The nature of the faint blue galaxies therefore remains a mystery.

STAR FORMATION

Why have dim galaxies converted so little of their gas into stars?

A first approximation of the dependence of the star formation rate on gas density is the Schmidt Law:

$$\dot{f}_g \propto \Sigma_g^N. \tag{1}$$

Kennicutt [7] finds $N = 1.3 \pm 0.3$. By this criterion, low surface density disks should evolve slowly, but this alone is not adequate to explain the observations. Gas surface density does follow optical surface brightness, but does not vary by as large a factor [8]. Roughly speaking, Σ_g drops by a factor of 2 for a factor of 5 drop in μ_0. Making the usual assumption that the amount of light traces the global star formation rate averaged over a Hubble time,

$$\frac{\Sigma_{opt}(HSB)/T_H}{\Sigma_{opt}(LSB)/T_H} \approx \frac{\dot{f}_g(\text{HSB})}{\dot{f}_g(\text{LSB})} = \left[\frac{\Sigma_g(\text{HSB})}{\Sigma_g(\text{LSB})}\right]^N \to 5 = 2^N \tag{2}$$

requires a value of $N \approx 2.3$, much larger than observed. Though this is a crude calculation, it is a conservative one: other plausible assumptions require even larger N.

The next order approximation for star formation is based on local gravitational stability in the disk [7]. This leads to a critical density threshold below which star formation is suppressed:

$$\Sigma_c = \frac{\sigma}{3.4G}\frac{V}{R}\left(1 + \frac{R}{V}\frac{dV}{dR}\right)^{1/2}. \tag{3}$$

This provides a good explanation for the evolutionary sluggishness of low surface brightness disks: they are at or below the critical threshold.

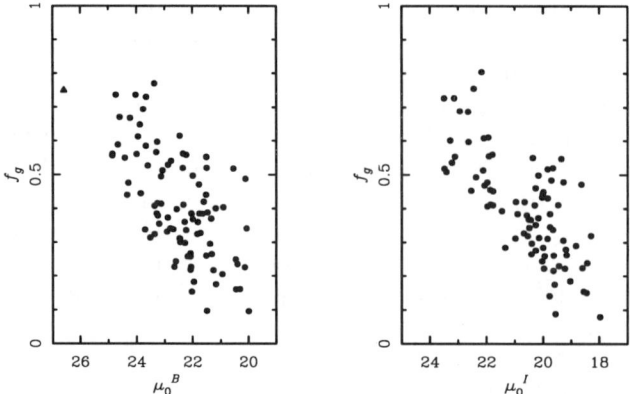

FIGURE 1. The gas mass fraction as a function of disk central surface brightness. The two panels show results derived independently from a) B-band and b) I-band data. These give consistent results. The slopes differ slightly because higher surface brightness galaxies tend to be redder. The data show a strong correlation $[\mathcal{R} = 0.63$ in (a)] with gas fraction decreasing as surface brightness increases. This goes in the sense expected if low densities inhibit star formation.

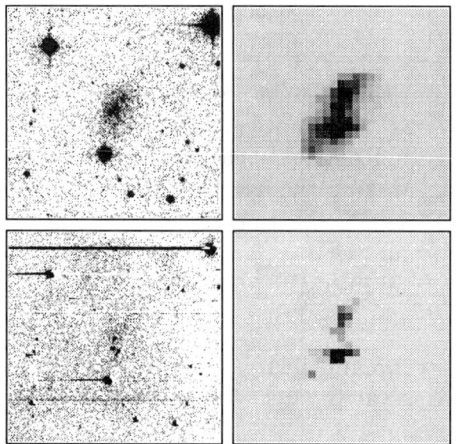

FIGURE 2. The low surface brightness galaxy F563–V1. a) Broad band optical (V-band) image. b) HI gas distribution. c) Continuum subtracted Hα emission. d) Critical density map. The critical density map is constructed by dividing the gas distribution in (b) by Σ_c as given in the text. Only regions near to or exceeding this threshold are shown. There is a reasonable correspondence between these areas which should be forming stars and those which actually are in (c).

We have tested this in a radially averaged way [9] and it works well: lower surface brightness disks are generally further into the critical regime. We can also test this formulation on a point by point basis (Fig. 2). In many cases it works well, there being a reasonable coincidence between regions where $\Sigma_g > \Sigma_c$ and the location of HII regions.

Nevertheless, the agreement between prediction and observation is by no means perfect in all cases. The Σ_c criterion seems especially prone to failure in the solid-body portion of the slowly rising rotation curves of low surface brightness galaxies. This is perhaps not surprising; there is very little shear in these regions. Stability is maintained for lower Q values, and asymmetries can persist in the gas for many dynamical times. Nevertheless, star formation is sometimes observed in regions which should be stable. It seems that further physics is at work — both the Schmidt law and Kennicutt's critical density criteria are operative and directly relevant to the global evolution of galaxies, but they are not the end of the story.

Though further physics is required to understand all the details of star formation, one basic result is clear. Surface mass density is a critical parameter in determining the evolutionary rate of disks. Dim galaxies have consumed little of their gas because of their low surface densities.

GALAXY FORMATION

If surface density determines the evolution of disk galaxies, what determines the surface density? The origin of disk surface density is probably related to the amplitude of the density fluctuation from which a galaxy was born. Lesser density perturbations will expand longer before turn around, and collapse later. This results in a lower final mass concentration and a younger age, as observed. Density begets density.

REFERENCES

1. de Blok, W.J.G., & McGaugh, S.S. 1996, ApJ, 469, L89
2. Mihos, J.C., McGaugh, S.S., & de Blok, W.J.G. 1997, these proceedings
3. McGaugh, S.S., & de Blok, W.J.G. 1997, ApJ, in press
4. McGaugh, S.S. 1996, MNRAS, 280, 337
5. Padoan, P., Jimenez, R., & Antonuccio-Delogu, V. 1997, these proceedings & astro-ph/9609091
6. Madau, P. 1997, these proceedings
7. Kennicutt, R.C. 1989, ApJ, 344, 685
8. de Blok, W.J.G., McGaugh, S.S., & van der Hulst, J.M. 1997, MNRAS, 283, 18
9. van der Hulst, J.M., Skillman, E.D., Smith, T.R., Bothun, G.D., McGaugh, S.S., & de Blok, W.J.G. 1993, AJ, 106, 548

The Evidence for Massive Star Formation in Early-Type Spiral Galaxies

Salman Hameed & Nick Devereux

New Mexico State University
Las Cruces, New Mexico 88003-8001

Abstract. A recent analysis of the IRAS database indicates that the massive star formation rates in early-type (Sa-Sab) spirals are comparable to the massive star formation rates in late-type spirals. $H\alpha$ imaging of some of the infrared luminous early-type spirals reveals two types of galaxies. One type shows clear signs of interaction, whereas the other type appears to host a nuclear starburst. The occurence of nuclear starbursts in early-type spirals may be related to the propensity for such galaxies to also host Seyfert nuclei. The evidence for interactions suggests that early-type spirals are evolving in the current epoch.

INTRODUCTION

Early-type (Sa-Sab) spirals are widely perceived to be the most inert of spiral galaxies. Indeed, the very term "early-type" refers to an early phase of star formation that has long since passed eg. [1]. The perception is based, in large part, on the optical morphology which is dominated by an inert stellar bulge. There are other observational results, however, which suggest that early-type spirals are not as quiescent as once believed. One of the most striking is the fact that early-type spirals are among the most luminous in the population of nearby galaxies (D $\leq$ 40Mpc) when observed in the far-infrared with the Infrared Astronomical Satellite [2]. Adopting the far-infrared luminosity as a measure of the massive star formation rate, one finds that the IRAS data reveals a previously unsuspected population of early-type spirals with star formation rates that rival the most prodigously star forming Sc galaxies.

THE HUBBLE TYPE DEPENDENCE OF THE L(FIR)/L(BLUE) RATIO

Data obtained with the Infrared Astronomical Satellite (IRAS) are used to investigate the dependence of the L(FIR)/L(Blue) ratio on Hubble type for a sample of 1462 galaxies selected from the Nearby Galaxy Catalog [3]. Adopting the far infrared luminosity as a measure of the present day massive star formation rate and the blue luminosity as a measure of the past star formation rate, the L(FIR)/L(Blue) luminosity ratio measures the ratio of present to past star formation in a way that is analogous to $H\alpha$ equivalent widths. Our results (Fig. 1) [2] indicate

a) The mean L(FIR)/L(Blue) luminosity ratios are similar for Sa's, Sb's, and Sc's.

b) The range in L(FIR)/L(Blue) luminosity ratios is as large for the early-type spirals as it is for the late type spirals.

Thus, L(FIR)/L(Blue) luminosity ratios do not support what may be one of the most basic tenets of extragalactic astronomy, that the ratio of present to past star formation is correlated with Hubble type.

$H\alpha$ IMAGING OF LUMINOUS EARLY-TYPE SPIRAL GALAXIES

There is growing evidence to suggest that early-type spiral galaxies are perhaps the most dynamic of all the nearby galaxy systems. The discovery of ripples, shells [4], and counter rotating gas and star disks in some early-type spirals [5] [6] indicate that a major accretion event occured in the past 1-2 Gyr [7]. We have obtained $H\alpha$ images for 5 of the most luminous ($L_{FIR} > 10^{10} L_{\odot}$) Sa-Sab galaxies. The $H\alpha$ images (Fig. 2) suggest two types of early-type spiral galaxies. One type clearly shows signs of disturbance in both the continuum and $H\alpha$ images. The other type includes galaxies that appear to be undisturbed morphologically, but their entire $H\alpha$ emission is radiated from a very small, $\leq 1 kpc$, diameter region centered on the nucleus. The origin of the nuclear $H\alpha$ emission is unclear, but it is most likely a starburst, an active nucleus or a combination of the two.

SUMMARY

The current perception that high mass star formation increases along the Hubble sequence (Sa-Scd) is largely based on measurements of the $H\alpha$ luminosities and equivalent widths by Kennicut & Kent [8]. Their measurements do indicate an increase in the ratio of present to past star formation along the sequence Sa to Sc. Our results, however, indicate that both the massive star formation rate and the present to past star formation ratio are independent of Hubble type, at least for the Sa's, the Sb's and the Sc's. On the other hand, there is still a lingering controversy over the origin of far infrared luminosity. We are therefore obtaining complimentary $H\alpha$ images for a complete, all sky, sample of 57, bright, $m(B) \leq 12$ magnitudes, nearby, $D \leq 40$ Mpc, early type (Sa-Sab) spiral galaxies in order to better understand the difference between the IRAS results and the existing $H\alpha$ measurements.

REFERENCES

1. Sandage, A. 1986, *A & A.* **161**, 89.
2. Devereux, N.A., & Hameed, S.A., 1997, *AJ.* **113**, in press.
3. Tully, R.B., Nearby Galaxies Catalog, 1989 (Cambridge:Cambridge Univ. Press)
4. Schweizer, F., & Seitzer, P., 1988, *ApJ.* **328**, 88.
5. Braun, R., et al. 1992, *Nature* **360**, 442.
6. Merrifield, M.R. & Kuijken, K., 1994, *ApJ.* **432**, 575.
7. Hernquist, L. & Spergel, D.N., 1992, *ApJ.* **399**, L117.
8. Kennicutt, R.C., Jr. & Kent S.M., 1983, *AJ.* **88**, 1094.

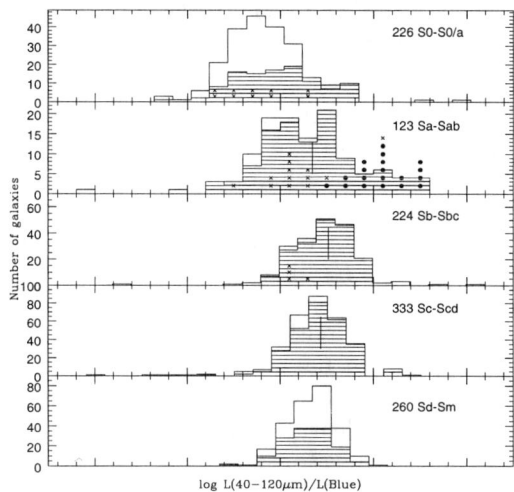

FIGURE 1. Histograms illustrating the Hubble type dependence of L_{fir}/L_{blue} ratios. The hatched histograms identify FIR detections, the unshaded histograms identify upper limits to the FIR flux. The vertical bar identifies the median of the distribution. Solid circles identify Sa-Sab galaxies with $L(40-120) \geq 10^{10} L_\odot$, and crosses identify Sa-Sab galaxies from Kennicutt's sample. Fourteen of Kenicutt's galaxies are classified as either S0-S0/a or Sb-Sbc in the NBG catalog. For details, see [2]

NGC 660

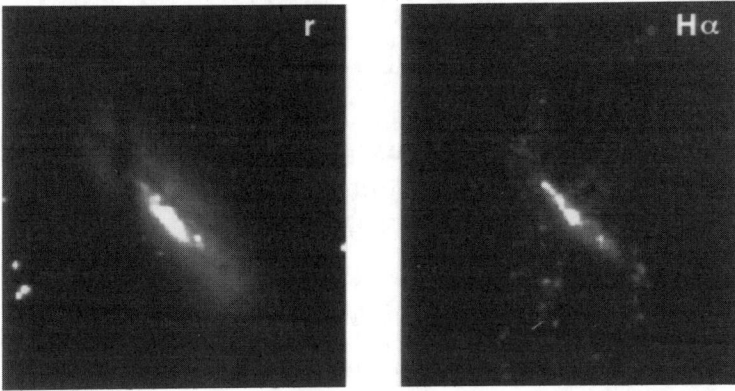

NGC 1022

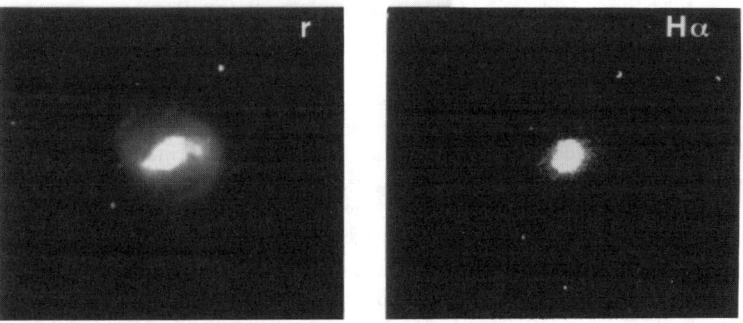

FIGURE 2. The continuum (left) and continuum subtracted $H\alpha$ (right) images of NGC 660 and NGC 1022 illustrate the two types of early-type spirals. NGC 660 shows clear signs of interaction, whereas NGC 1022 appears to host a nuclear starburst.

Hα, Far Infrared & Thermal Radio Continuum Emission within the Late-Type Spiral Galaxy M33

Nick Devereux[1], Neb Duric[2]
and
Paul Scowen[3]

[1] New Mexico State University, Las Cruces, NM 88003
[2] University of New Mexico, Albuquerque, NM 87131
[3] Arizona State University, Tempe, AZ 85287

Abstract. New Hα imaging observations have revealed the morphology of the emission line gas and permitted the first measurement of the total Hα luminosity for the late-type spiral galaxy M33. The total Hα luminosity of M33 is (7.06 $\pm$ 1.40) x $10^6 L\odot$ and is dominated by emission from HII regions.

The Hα image is compared with 6 and 20cm thermal radio continuum images in order to quantify the extinction to HII regions in M33. The extinction is found to be high locally, but low globally. The extinction to the high surface brightness cores of HII regions corresponds to $A_v \sim 1$ magnitude on average with no systematic dependence on radius. However, the extinction correction to the global Hα flux is much lower with $A_v \sim 0.0$ to 0.4 magnitudes. The difference suggests that the extinction is virtually negligible to the low surface brightness Hα emission outside the high surface brightness cores of HII regions.

The Hα image is compared with a high resolution far infrared image, obtained with the Infrared Astronomical Satellite (IRAS), in order to constrain the contribution of O and B stars to the far infrared luminosity of M33. The correspondence between the Hα and far infrared morphology is striking when both images are convolved to a common resolution of 105 arc seconds. The far infrared luminosity, L(FIR), and the Hα luminosity, L(Hα), have been measured at 840 independent locations within M33 and the histogram of L(FIR)/L(Hα) ratios is remarkably similar to that determined for Galactic HII regions. Approximately 70% of the far infrared luminosity of M33 is radiated by dust with temperatures greater than expected for cirrus, but similar to the temperatures measured by IRAS for Galactic and extragalactic HII regions. The results indicate that the majority ($\geq$ 70%) of the far infrared and Hα luminosity of M33 is produced by massive stars.

INTRODUCTION

M33 (NGC 598) is a nearby, 0.84 Mpc, late-type, SA(s)cd, spiral galaxy. With an angular size of approximately one degree, it is one of the half a dozen or so galaxies that was sufficiently well resolved by the Infrared Astronomical Satellite (IRAS) to justify an investigation of the far infrared morphology. The present contribution is one in a series that utilizes high resolution far infrared and Hα imaging to elucidate the origin of the far infrared luminosity in nearby galaxies of large angular size. Galaxies that have been examined to date include M101 [1], M31 [2] and M81 [3].

The motivation for understanding the origin of the far infrared luminosity stems from the potential to use the large database returned by IRAS to quantify massive star formation rates in spiral galaxies. Whereas the Hα emission line luminosity provides an alternative way to measure massive star formation rates, global Hα measurements exist for only 200 or so galaxies. In contrast, the IRAS measured far infrared fluxes for tens of thousands of galaxies. Consequently, the IRAS database would constitute an extremely valuable resource for statistical studies of star formation in galaxies if it could be demonstrated that the far infrared luminosity provides a measure of the massive star formation rate.

Comparison of the far infrared and Hα morphology is a useful way to elucidate the origin of the far infrared emission, having already revealed an important distinction between early and late-type spirals. Whereas the far infrared luminosity of the late-type spiral galaxy M101 can easily be attributed to HII regions [1], the bulges of the early-type spirals M31 and M81 radiate significant *far infrared and Hα emission* that can not be identified with massive stars [2,3].

The present study of M33 is based on higher resolution and more sensitive far infrared images than those discussed previously by [4]. The new far infrared images resulted from application of a high resolution reconstruction algorithm [5] that increased the angular resolution by about a factor of 3 over that of the original IRAS survey data analyzed by [4]. The new high resolution far infrared images are compared with new Hα images of M33 that were obtained recently at Kitt Peak National Observatory. The Hα images are also compared with new 6 and 20cm thermal radio continuum images that were obtained using the Very Large Array and the Westerbork synthesis radio telescopes.

THE DATA

The New Hα Imaging Observations

The new Hα imaging observations were obtained during the night of 1994 September 11 with the Case Western Burrell Schmidt telescope at Kitt Peak

National Observatory. M33 was imaged with a 2048 x 2048 tektronics CCD chip providing a 1 degree field of view that enabled M33 to be imaged with a single pointing of the telescope.

The IRAS HiRes Images

The IRAS 60 and 100μm HiRes images were procured from the Infrared Processing and Analysis Center (IPAC), in Pasadena, CA. Following convolution to a common resolution of 105 arc sec, the 60 and 100μm images of M33 were linearly combined into a 40-120μm far infrared luminosity image. The far infrared luminosity image of M33 is presented in Figure 1 along with the Hα image convolved to the same resolution.

The 6cm Thermal Radio Continuum Image

The radio continuum observations were obtained with the Very Large Array (VLA) and the Westerbork Synthesis Radio Telescope (WSRT). M33 was observed at 6 cm and 20 cm. The VLA and WSRT data were combined into two data sets as described by [6]. The fully cleaned images were smoothed to a common resolution of 10 arc seconds thereby optimizing the sensitivity to structures with linear dimensions $\sim$ 40 pc. The thermal radio sources were then identified on the basis of having flat spectral indices $S_\nu \propto \nu^{-0.1}$ and morphologies similar to those in Hα. The nonthermal sources were identified on the basis of having steeper power-law spectra. The nonthermal sources

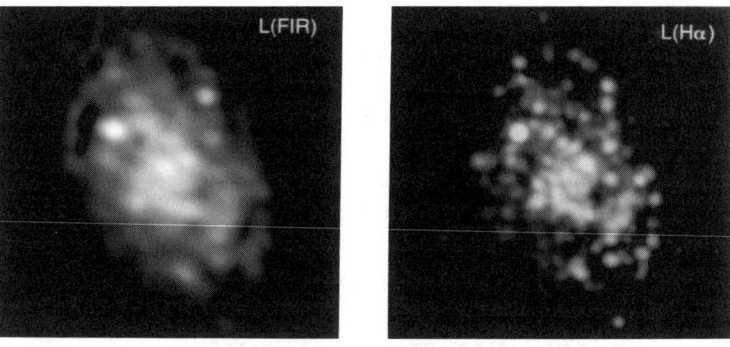

FIGURE 1. Far infrared (left panel) and Hα + [NII] (right panel) images of M33 convolved to a common resolution of 105 arc seconds. The images are stretched logarithmically.

were subtracted from the 6cm radio map. The resulting 6cm thermal radio continuum image is presented in Figure 2 along with the complimentary Hα image convolved to the same 10 arc second resolution.

RESULTS & CONCLUSIONS

A full description of the results and conclusions is presented in [7].

REFERENCES

1. Devereux, N.A., & Scowen, P.A., 1994, AJ, 108, 1244.
2. Devereux, N.A., Price, R., Wells, L.A., & Duric, N., 1994, AJ 108, 1667.
3. Devereux, N.A., Jacoby, G., & Ciardullo, R., 1995, AJ 110, 1115.
4. Rice, W., Boulanger, F., Viallefond, F., Soifer, B.T., Freedman, W.L., 1990, ApJ 358, 418.
5. Aumann, H.H., Fowler, J.W., & Melnick, M., 1990, AJ 99, 1674.
6. Duric, N., Viallefond, F., Goss, W. M., & van der Hulst, J.M. 1993, A&AS 99, 217.
7. Devereux, N.A., Duric, N., & Scowen, P.A., 1997, AJ 113, January issue.

M 33

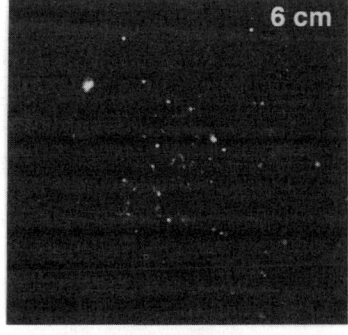 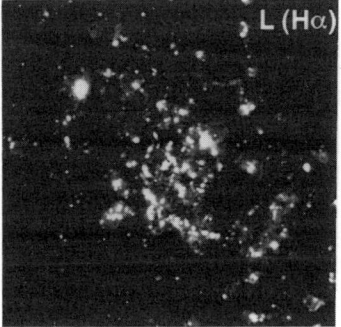

FIGURE 2. 6cm thermal radio continuum (left panel) and Hα + [NII] (right panel) images of M33 convolved to a common resolution of 10 arc seconds. The images are linearly stretched.

Star Formation
History in Ellipticals

Star Formation History of Elliptical Galaxies from Low-Redshift Evidence

Guy Worthey[1]

Astronomy Department, University of Michigan
Ann Arbor, MI 48109-1090

Abstract. Star formation in elliptical galaxies (Es) was and is mostly dominated by mergers and accretions with many suggestive examples seen among local galaxies. Present day star formation in Es is easily measurable in $2/3$ of Es and appears bursty in character. Direct age determinations from integrated light indicate real age scatter. If one assumes the oldest-looking galaxies are a Hubble time old, the light weighted mean ages of the rest spread to 0.5 of a Hubble time, with scatterlings at very young ages. Larger Es and Es in clusters have less age scatter than smaller or field Es. The size trend is clear. The environment trend needs to be rechecked with better data even though it agrees with high redshift field/cluster results.

INTRODUCTION

The most appealing picture of star formation in elliptical galaxies for cosmologists is one in which Es formed very early in the universe and have been quiescent ever since. If such galaxies exist, they potentially measure the curvature parameter q_0 since one could predict their size and luminosity fairly well and use them as standard rulers or candles. This picture is supported by the superficial uniformity of ellipticals in appearance and the existence of scaling relations between observed parameters (stellar velocity dispersion $[\sigma]$, surface brightness, size, luminosity, colors, and line strengths), some of which scale with very small scatter.

The hope for using Es as cosmological tracers is alive and well, but when nearby ellipticals are examined in detail they exhibit a large variety of morphological and kinematic peculiarities and strong evidence for star formation much less than a Hubble time ago, leading to a picture of elliptical formation in which galaxy-galaxy mergers and accretion events are the dominant formation mechanism, and that we still witness the "tail end" of E formation

---

[1] Hubble Fellow

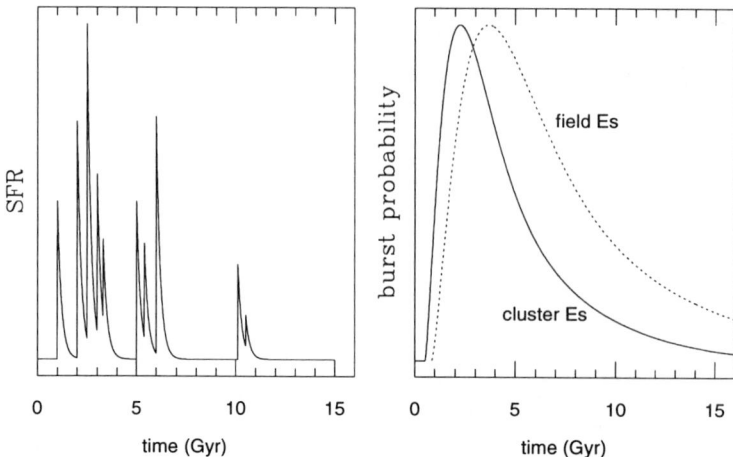

FIGURE 1. A schematic picture of the most probable way that star formation proceeded in E galaxies over the life of the universe. Each galaxy is formed by a series of accretions and mergers. The strengths and times of the star formation bursts are random and have occurred in precursor branches of a merging tree. Bursts are governed by an overall probability distribution in which star formation is more likely to have occurred in the past. The probability curve varies as a function of field/cluster environment and galaxy size.

today in the form of observed galaxy mergers and Es that display current star formation. I think we should view elliptical formation as a *process* rather than an *event*.

Figure 1 does not show the variety of dynamical or morphological changes that could be taking place in the history of an elliptical, but concentrating on star formation alone, a bursting scheme like that of Fig. 1 fits the facts outlined in this article. Note in Fig. 1 a difference in mass-weighted age between cluster and field ellipticals and the bursty star formation with quiescent periods during which ellipticals will look superficially normal if the intraburst period is longer than the stellar population fade time.

This fade time is illustrated in Fig. 2 using the commonly used and easily measured line index Mg_2, which has the additional advantage that it participates in the Mg_2-σ scaling relation (discussed later) so that there is a relatively clear definition of "normal:" the gaussian scatter from this relation. Fig. 2 shows that, if only a few percent by mass of gas is consumed in a starburst, the elliptical galaxy will appear "normal" after less than 0.5 Gyr. Therefore a population of Es will always look "normal" except for a few oddballs caught in the aftermath of starburst. The rest of this article should show that this is a workable scheme.

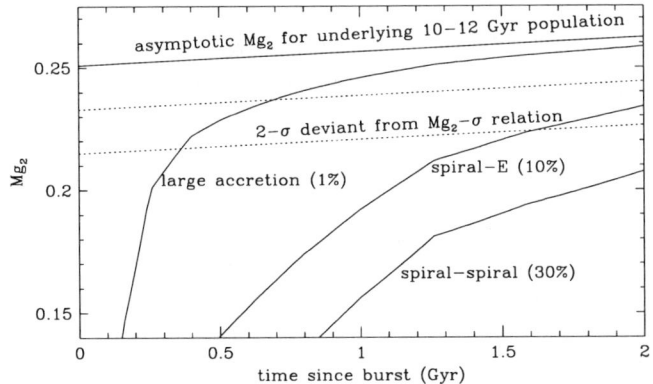

FIGURE 2. The Mg_2 evolution of different fractions by mass (1%, 10%, and 30%) of burst mass / total mass are shown to illustrate how fast an elliptical galaxy will look "normal" after an episode of star formation. Time is counted starting from a burst of star formation that occurred in a passively evolving, 10-Gyr-old "elliptical," whose unperturbed Mg_2 strength is also shown. One-σ and 2-σ increments from the Mg_2-σ relation ($\sigma_{measurement} = \pm 0.018$ mag [Bender et al. 1993]) are shown to indicate when a galaxy would "blend in" with the others. The error in Mg_2 is about 0.01 mag. The models shown in this diagram are a variant of the Worthey (1994) models in which Padova evolution is used instead of VandenBerg.

LOCAL EVIDENCE FOR MERGING

I will differentiate between "merging" and "accretion" events. Accretion events are small, involving less than 10% of the galaxy's mass, while mergers are larger. Each event can also be partially or completely stellar, so that star formation may not result. Theoretical expectations are that gaseous accretions will rapidly, dissipatively sink to galaxy center, forming stars when it becomes dense enough. Stellar accretions survive as kinematic substructure that disappears if there is time to phase mix (e.g. Kormendy 1984; Balcells & Quinn 1990; Balcells 1991). Large, violent mergers are characterized by the ejection of tidal tails and a somewhat protracted series of starbursts as the two galaxies violently relax to an $r^{1/4}$ profile, followed by a steady drizzle of gas that can last as long as a few Gyr (Hibbard & Mihos 1995). The end state of such violent encounters has long been postulated to end in the formation of a bulge-dominated elliptical-like galaxy, with "correct" kinematics and surface brightness profile (e.g. Barnes 1992, Hernquist 1993, Heyl et al. 1994) if a little gas is involved. Note that purely stellar mergers give very large cores that are not observed, so a certain amount of gaseous dissipation is required.

Ongoing disk-disk mergers: Many IRAS galaxies and Arp (1966) peculiar galaxies appear to be disk-disk mergers in progress: still recognizable as two spiral galaxies, but blatantly interacting. Toomre (1977) counted 11 such mergers with NGC numbers. Assuming a constant merger rate and a typical duration of 0.5 Gyr for this morphological phase found that ~250 NGC galaxies would have formed this way over a Hubble time. Compare this number to the fraction of the 6032 (my estimate) galaxies in the original NGC catalog that are Es: about 13% = 780 objects.

Qualitative post-merger signatures: Both morphological and kinematic peculiarities are predicted outcomes of large accretions or mergers. Morphological "fine structure," including ripples, jet features, boxiness, and X structure at large radius was measured by Schweizer et al. (1990) and Schweizer & Seitzer (1992). The galaxies with the most such morphological complexity have systematically bluer colors, weaker metallic line strengths, and stronger $H\beta$ line strength, indicating a connection between recent merger status and the mean age of the stellar populations (the other alternative is that morphologically disturbed galaxies are preferentially metal-poor, but this idea is generally dismissed).

Kinematic peculiarities such as minor axis rotation, cores that rotate counter to the rotation of the galaxy outskirts, and kinematic discontinuities with radius are observed in more than half of all ellipticals (Bender 1996). Many of these peculiarities are dynamically long-lived, and do not constitute evidence for or against *recent* merging, but do argue strongly for merging as a process, since a single radial collapse can not produce such peculiar motions. Kinematic discontinuities are often mirrored by cospatial line strength gradient discontinuities, adding further weight to the merger interpretation (Bender & Surma 1992). In the case of high surface brightness elliptical NGC 1700, Statler et al. (1996) argue from an array of isophotal shape and kinematic data that at least 3 or more stellar subsystems must have merged 2–4 Gyr ago to explain the observed substructure, but the inner regions have phase-mixed to uniformity. Very few galaxies have been studied to this level of detail.

Accretion in Non-Es: A number of galaxies illustrate that multi-episode gaseous accretion/merger events occur. A note of thanks to Kennicutt (1996) for pointing out most of the examples I describe. First, we catch an accretion event in progress in the Milky Way as the Saggitarius dwarf spheroidal is in the process of being tidally disrupted (Ibata et al. 1995) and is strung out over at least ~40 degrees of arc (M. Mateo, private communication). More evidence for the accretion of spheroidal-sized, metal-poor, but not always old galaxies into the Galactic halo comes from young stars on halo orbits, the presence of galactic streams, and age spreads in globular clusters as summarized in Freeman (1996). Some workers now start with the assumption that the entire metal poor halo was built by accretion events over the Galaxy's lifetime, while the bulge may have been built from disk/bar instabilities. It does seem likely that bulges and disks evolved together since their scale lengths are always in nearly the same ratio (Courteau et al. 1996).

Accretion of gas is observed in external spiral galaxies from H I in the form of high velocity clouds of up to $\sim 10^8 M_\odot$ (Kamphius 1993), or in the form of blatant tidal disruption prior to merging as in the case of NGC 3359 and NGC 4565 (Sancisi et al. 1990). HI observations of E galaxies NGC 4472, UGC 7636, NGC 3656, NGC 5128, and NGC 2865, among others, show evidence for ongoing or recent gas accretion (Sancisi 1996). Spiral NGC 4826 (M64) is a rare example of gas-gas counter-rotation in which the sense of rotation switches at a radius of ~ 1 kpc. This galaxy appears to have accreted a gas-rich companion in the recent past, with the inner gas disk the remnant of the original disk (Braun et al. 1992; Rubin 1994; Rix et al. 1995).

Many star-gas counter-rotations are seen. Bertola et al. (1992) list 9 examples of S0 galaxies with gas-star counterrotation. In NGC 4526 (Bettoni et al. 1991) and NGC 3626 (Cirri et al. 1995) the counterrotation extends over the entire disk, with gas masses of $10^8 - 10^9 M_\odot$. More easily seen are polar ring galaxies in which a (usually) spiral galaxy is ringed by gas, stars, and/or dust at a nearly perpendicular angle. From their catalog of (~ 70) polar ring galaxies, Whitmore et al. (1990) estimate that about 5% of S0 galaxies went through a polar ring phase. Even more spectacular than gas-star is star-star counterrotation. The edge-on S0 galaxy NGC 4550 is seen to have two cospatial counterrotating disks with nearly identical masses and scale lengths and with velocity dispersions of 45 and 50 km s^{-1} (Rubin et al. 1992; Rix et al. 1992), indicating two separate epochs of galaxy formation. In Sb NGC 7217 20-30% of the the disk stars are in a retrograde cold disk (Merrifield & Kuijken 1994). In spirals, such oddities are rare probably because spirals are relatively fragile compared to Es.

The evidence from spirals indicates that accretion is a common phenomenon. Further, although galaxies like NGC 4550 must be rare, they reveal a fabulous wealth of possibilities for galaxy formation. Before you knew about 4550, what odds would you have assigned to its existence?

Current E star formation: Most Es have readily measurable nebular emission (Goudfrooij et al. 1994, Gonzàlez 1993) probably indicative of star formation. The Goudfrooij narrow band imaging indicates a variety of morphologies of the ionized gas, from disk-like to nuclear to diffuse and filamentary. Fig. 3 shows reprocessed Gonzàlez (1993) spectroscopic data on Es and Kennicutt (1983) Hα data on spiral galaxies with the same conversion from Hα flux to star formation rate. Fig. 3 shows the fraction of the galaxy assembled over 10 Gyr assuming a constant star formation rate. I corrected roughly for M/L changes as a function of Hubble type, but assumed zero extinction correction. A correction will push the star formation rate to higher values.

There are two points to notice about the elliptical galaxies in relation to spirals. (1) The observed SFR is nonzero in $2/3$ of the Es, but 100 times smaller than late-type spirals. (2) Fig. 3 might almost be an illustration of binomial statistics in which Sc galaxies have large numbers of star formation events and are thus distributed in an almost gaussian way, while E galaxies have small numbers of star formation events and are thus distributed in a way

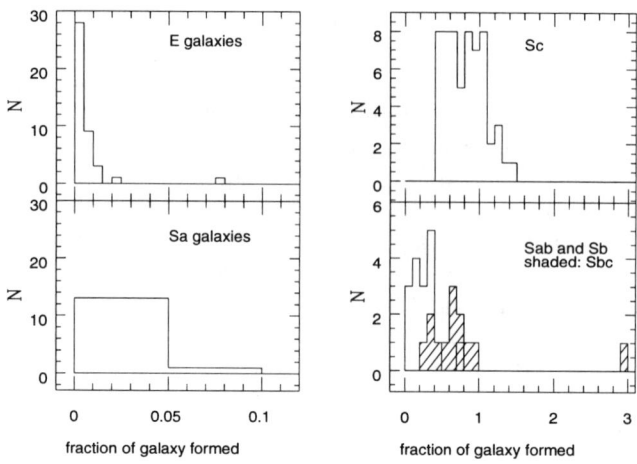

FIGURE 3. Current star formation rates for galaxies of different Hubble types expressed as the fraction of mass that would be assembled over 10 Gyr assuming the current SFR stayed constant. Kennicutt (1983) Hα and Gonzàlez (1993) O[III] data were used to estimate the SFR via the Kennicutt (1983) formula with no correction for extinction. The binning of the Sa data reflects the larger uncertainty in Kennicutt's data.

that resembles Poisson statistics: *fundamentally bursty* in character.

AGE FROM INTEGRATED LIGHT

It is now possible to measure a light-weighted mean age using integrated light indices for old stellar populations. The "light-weighted" part means that young stars, because they are brighter, can heavily influence the mean age that one obtains and that it is easy to obscure older generations of stars. The mean ages are derived from plotting Balmer indices versus hand-picked metal indices that are more metallicity sensitive than average. The spectra from which these measurements are taken need to have good S/N and need to have careful accounting for systematics like instrumental resolution and galaxy velocity dispersion so that the observations transform to the same system as the models.

Such pickiness is needed because easier measurements, such as broad-band colors, D4000, or Mg$_2$ are largely degenerate with age and metallicity along a null spectral change slope of d log(Z) = $-2/3$ d log(age) along which colors and most line strengths stay the same (Worthey 1994). This implies that if one wants a 15% age estimate, one must know the metal abundance to 10%. By picking spectral indices that are preferentially sensitive to age and arraying them against indices that are preferentially sensitive to metal abundance the

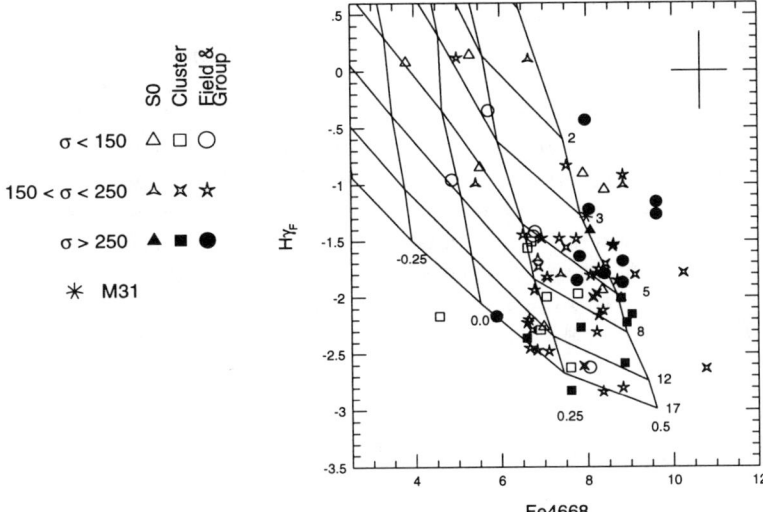

FIGURE 4. Early type galaxies of different size and environment in an age diagnostic diagram; Hγ_F (Worthey & Ottaviani 1997) versus Fe4668 (Worthey et al. 1994). Models (Worthey 1994) are plotted as a grid labeled with [Fe/H] along the bottom and age in Gyr along the right edge. Approximate observational error is indicated. Three galaxies at very young age and with $\sigma < 150$ km s^{-1} are off the scale of this diagram.

age-metal degeneracy is largely broken in a *differential* sense: the age zeropoint is still quite uncertain, but relative mean age changes are readily detectable.

Figure 4 tells us several things. First, the spread in mean age is real in the sense of being well beyond observational error. Regardless of age zeropoint, many galaxies have ages less than half a Hubble time, and several appear very young. At present we can't tell if these ages are the real formation ages of the galaxies or the effect of small, recent bursts of star formation. Second, there is a distinct trend for the younger large galaxies to be more metal rich, a tendency which will be discussed below. Third, there are dependencies on size and probably on field/cluster environment. Smaller galaxies have a larger spread in mean age than the large ones, as well as a tendency to be slightly more metal poor. Field galaxies appear to be more volatile than cluster galaxies (note the lack of large cluster Es younger than 3 Gyr). In this sample there do not appear to be any cluster Es that are both small and young. Note that the sample was not chosen in an intelligent way, and a proper volume-limited sample may show somewhat different trends.

The real scatter of ages tells us that Es have had a fairly complex history, and that they are not quite finished forming. It does not tell us the relative importance of mergers versus accretions.

Aside: Age from abundance ratio variations? Worthey et al. (1992) plotted Mg_2 versus iron indices to find that large Es deviated from solar-neighborhood [Mg/Fe] in the sense of enhanced Mg relative to Fe by about a factor of two. Smaller Es have a nearly solar mixture, so the amount of enrichment from Type II supernovae relative to Type I gets larger in larger Es. Unfortunately, the mechanism for varying this ratio is not known. It could either be a variation in formation timescale or a (mild) variation in upper IMF strength as a function of galaxy size. The implications for E formation are different for those two cases and so, skipping the details, we can't really constrain E formation until we know for sure what mechanism causes the Type II to Type I shift.

SLIPPERINESS OF TIGHT SCALING RELATIONS

As mentioned above, colors and line-strengths usually scale with structural parameters like brightness, size, σ, and combinations of these quantities. The tightest are the Mg_2-σ relation and the fundamental plane (in μ, R, σ space). The interesting thing about these "tightness relations" is the small scatter observed. The small scatter is interesting because significant spread in velocity anisotropy, density structures, and stellar ages could reasonably be expected to raise the scatter to much higher levels than is observed (e.g. Djorgovski et al. 1996). More than one conspiracy must be operating to thus limit the range of E properties.

Empirically, somewhat more than half the scatter from the fundamental plane can be fairly unambiguously tied to stellar population changes because the scatter correlates with color and line strength residuals such that blue colors correspond to high surface brightness, as one would expect from the presence of a younger subpopulation (Prugniel & Simien 1996; Jørgensen & Franx 1996). If that is so, then the tightness of the fundamental plane places a restriction on the age scatter that is allowed for ellipticals. As a first cut, one can take the observed scatter in cluster Es and convert that to a scatter in age or metallicity. Bender et al. (1993) find, analyzing the Mg_2-σ relationship, an allowed age (or metallicity) scatter of 15% RMS at a given σ with a non-gaussian blue tail. This seems to say that $2/3$ of Es were formed in the first $1/3$ of the universe. This interpretation is probably misleading because (1) the scatter does not look one-sided – it looks like there really are galaxies on the redder (older or more metal rich) side of the mean relation, and (2) there appears to be a rough trend for younger Es to be more metal rich (Figure 4; Worthey et al. 1996). Point (1) means that either metallicity plays a significant/dominant role, or that the mean age for Es is substantially younger than that of the universe, and what we are seeing is scatter about a "mean history" of formation. Point (2) implies that younger (bluer) populations will be more metal rich (redder), and this will do a lot to artificially tighten the Mg_2-σ and the fundamental plane relations, allowing more age spread than one might otherwise have guessed.

SUMMARY AND HIGH-REDSHIFT REMARKS

The tightness relations can also be tracked with redshift. If Es are pure passive evolvers, there is a clear prediction for bluer Mg_2 and brighter M/L with redshift. If we think of E formation as a decaying process of continued activity, we can predict a lot less clearly what is going on. In fact, if an ongoing accretion *process* dominates the line strengths and colors, we may see *less* evolution than purely passive. If large bursts of star formation at fairly late times are important we should see *more* than passive evolution. Work on both the fundamental plane and the Mg_2 sigma relation in cluster Es out to redshifts of 0.5 or so (see Dressler, this volume; Bender et al. 1996) indicates that the galaxies identified as Es tend to evolve as passively evolving populations that were formed before $z = 2$.

In the field, however, two different redshift surveys show that the population of red galaxies declines too fast to be consistent with passive evolution of old luminous galaxies with extrememly high confidence (Kauffmann et al. 1997). These high redshift results lend confidence that the Fig. 1 picture is roughly correct. In that picture, field and cluster Es form by some mixture of merging and accretion in a bursty manner which allows lots of time for blue colors to fade and for the galaxy to look "normal" between bursts. Active star formation should be seen in some ellipticals. Merging events and merger remnants should exist. Tight scaling relations are preserved until star formation gets really messy, probably before $z = 2$. The Balmer-metal age-diagnostic diagrams should show a spread in age. The Fig. 1 picture seems to summarize the observed situation fairly well.

REFERENCES

1. Arp, H. C. 1966, Atlas of Peculiar Galaxies, (California Institute of Technology: Pasadena)
2. Balcells, M. 1991, A&A, 249, L9
3. Balcells, M., & Quinn, P. J. 1990, ApJ, 361, 381
4. Barnes, J. 1992, ApJ, 393, 484
5. Bender, R., Burstein, D., & Faber, S. M. 1993, ApJ, 411, 153
6. Bender, R. 1996, in New Light on Galaxy Evolution, ed. R. Bender & R. L. Davies, (Kluwer: Dordrecht), 181
7. Bender, R., & Surma, P. 1992, A&A, 258, 250
8. Bender, R., Ziegler, B., & Bruzual, G. 1996, ApJ, 463, 51
9. Bertola, F., Buson, L. M., & Zeilinger, W. W. 1992, ApJ, 401, L79

10. Braun, R., Walterbos, R. A. M., & Kennicutt, R. C. 1992, Nature, 360, 442
11. Courteau, S., de Jong, R. S., Broeils, A. H. 1996, ApJ, 457, L73
12. Djorgovski, S. G., Pahre, M. A., & de Carvalho, R. R. 1996, in Fresh Views of Elliptical Galaxies, ed. A. Buzzoni, A. Renzini, & A. Serrano, (ASP: San Francisco), ASP Conf. Series 86, 129
13. Freeman, K. C. 1996, in New Light on Galaxy Evolution, ed. R. Bender & R. L. Davies, (Kluwer: Dordrecht), 3
14. Gonzàlez, J. J. 1993, Ph.D. thesis, Univ. California, Santa Cruz
15. Hernquist, L. 1993, ApJ, 409, 548
16. Heyl, J. S., Hernquist, L., & Spergel, D. N. 1994, ApJ, 427, 165
17. Hibbard, J. E., & Mihos, J. C. 1995, AJ, 110, 140
18. Ibata, R. A., Gilmore, G., & Irwin, M. J. 1995, MNRAS, 277, 781
19. Jørgensen, I., & Franx, M. 1996, in Fresh Views of Elliptical Galaxies, ed. A. Buzzoni, A. Renzini, & A. Serrano, (ASP: San Francisco), ASP Conf. Series 86, 139
20. Kamphius, J. 1993, Ph.D. thesis, Univ. Groningen
21. Kauffmann, G., Charlot, S., & White, S. D. M. 1997, MNRAS, in press
22. Kennicutt, R. C., Jr. 1983, ApJ, 272, 54
23. Kennicutt, R. C., Jr. 1996, in New Light on Galaxy Evolution, ed. R. Bender & R. L. Davies, (Kluwer: Dordrecht), 11
24. Kormendy, J. 1984, ApJ, 287, 577
25. Prugniel, R., & Simien, F. 1996, A&A, 309, 749
26. Rix, H.-W., Kennicutt, R. C., Braun, R., & Walterbos, R. A. M. 1995, ApJ, 438, 155
27. Rix, H.-W., Franx, M., Fisher, D., & Illingworth, G. 1992, ApJ, 400, L5
28. Rubin, V. C. 1994, AJ, 107, 173
29. Rubin, V. C., Graham, J., & Kenney, J. 1992, ApJ, 394, L9
30. Sancisi, R., Broeils, A. H., Kamphius, J., & van der Hulst, J. M. 1990, in Dynamics and Interactions of Galaxies, ed. R. Wielen, Springer-Verlag, 304
31. Sancisi, R. 1996, in New Light on Galaxy Evolution, ed. R. Bender & R. L. Davies, (Kluwer: Dordrecht), 143
32. Toomre, A. 1977, in The Evolution of Galaxies and Stellar Populations, ed. B. M. Tinsley & R. B. Larson, (Yale Univ. Observatory: New Haven), 401
33. Schweizer, F., Seitzer, P., Faber, S. M., Burstein, D., Dalle Ore, C. M., & Gonzàlez, J. J. 1990, ApJ, 364, L33
34. Schweizer, F., & Seitzer, P. 1992, AJ, 104, 1039
35. Statler, T. S., Smecker-Hane, T., & Cecil, G. 1996, AJ, 111, 1512
36. Worthey, G. 1994, ApJS, 95, 107
37. Worthey, G., Faber, S. M., & González, J. J. 1992, ApJ, 398, 69
38. Worthey, G., & Ottaviani, D. L. 1997, ApJS, in press
39. Worthey, G., Faber, S. M., González, J. J., & Burstein, D. 1994, ApJS, 94, 687
40. Worthey, G., Trager, S. C., & Faber, S. M. 1996, in Fresh Views of Elliptical Galaxies, ed. A. Buzzoni, A. Renzini, & A. Serrano, (ASP: San Francisco), ASP Conf. Series 86, 203

HST Observations of Distant Clusters: Evidence for Old Ellipticals and Younger S0s

Alan Dressler*

*Carnegie Observatories, 813 Santa Barbara Street
Pasadena, CA 91101

Abstract. The "MORPHS" group has completed the cataloging, parameterization, and morphological classification of ~2000 galaxies in 11 rich clusters from $0.36 < z < 0.56$. Here I discuss the use of the data in regard to the evolution of morphological types in clusters since $z \sim 0.5$. We have, in particular, studied the rest-frame $U - V$ color dispersion of bona-fide ellipticals in high-z clusters, and find the spread to be very small, which suggests an early formation epoch, $z > 3$ for cluster ellipticals. This is consistent with the evolution of the morphology-density relationship, in which we find ellipticals to be as abundant at $z \sim 0.5$ as in clusters today, and already well ensconced in the dense regions. In contrast, S0's are less plentiful and less well-concentrated compared to the present epoch, and spiral galaxies everywhere more abundant. Combined with other spectroscopic and morphological data, these observations suggest that most of these rapidly evolving systems are not likely to become bright ellipticals, which were more likely formed at early epochs. Cluster S0 galaxies, on the other hand, are likely to have been produced in large numbers in the recent past.

INTRODUCTION

In his talk for this conference, Guy Worthey presented evidence that the old paradigm for the early formation of elliptical galaxies has been largely discredited. In particular, he cited analyses of the integrated spectra of present-epoch field and cluster ellipticals that imply an average stellar age of less than 5 Gyr for many of these systems. I will use Hubble Space Telescope observations of distant clusters, for which the lookback time is ~5 Gyr, to argue that, on the contrary, the ellipticals in these high density regions, at least, have stellar ages that imply an early genesis, $z > 3$.

The use of cosmic lookback time has become a valuable tool in studying galaxy evolution. It has long been recognized that very different histories of star formation produce spectral and integrated color characteristics that are essentially indistinguishable by the present epoch. Observations of the state of galaxies at much earlier times help break this degeneracy and offer a clearer picture of the evolutionary path taken by galaxies of various types and masses, and in different environments.

A group of us we call the "MORPHS" — Richard Ellis, Warrick Couch, Ian Smail, Gus Oemler, Harvey Butcher, Ray Sharples, Bianca Poggianti, Amy Barger, and myself, has been using images from the Hubble Space Telescope Wide Field Planetary Camera 2 (WFPC-2) and extensive ground-based photometry and spectroscopy to study the properties of, and galaxy populations in, rich clusters of galaxies at $z \sim 0.5$. Here we report on various results, which should be referred to by the specific papers cited below.

A CONSTRAINT ON THE AGES OF ELLIPTICAL GALAXIES IN DISTANT RICH CLUSTERS FROM THE DISPERSION IN THEIR COLORS

Elliptical galaxies have been conventionally regarded as old galactic systems whose star formation history could be approximated as a single burst that occurred 12–16 Gyr ago. However, in recent years, this simple picture has been challenged from various viewpoints. As Guy Worthey has reported at this conference, numerous cases have been found of ellipticals with intermediate-age stellar populations. Furthermore, dynamical arguments suggest that many peculiarities seen in ellipticals, such as shells and dust-lanes (e.g., Schweizer & Seitzer 1988) are best explained via recent formation from the merger of gas-rich systems. Nevertheless, the small scatter observed for the $(U-V)$ colors of spheroidal galaxies in nearby clusters of galaxies still provides a basic constraint on the history of star formation in dense environments. For example, Bower, Lucey, & Ellis 1992 found a scatter in $U-V$ color of only 0.035 mag for ellipticals in the nearby Coma cluster. A reasonable interpretation of these data, they conclude, is that the stars in these galaxies formed early, $z \gtrsim 3$. However, it is true that admitting a degree of synchronicity would seriously weaken this limit; for example, if all the stars formed within 1 Gyr, they could have all formed much more recently, $z \sim 1$, and still be consistent with the small scatter in $U-V$ color.

In Ellis et al. (1996) we address this ambiguity using high precision rest-frame $(U-V)$ photometry of a large sample of morphologically-selected spheroidal galaxies in three $z \sim 0.5$ clusters which have been observed as part of our HST program. We use our $F555W$ and $F814W$ imaging to determine accurate rest-frame $(U-V)$ colors for spheroidal galaxies in the three clusters Cl0016+16 ($z = 0.55$), Cl0054−27 ($z = 0.56$) and Cl0412−65 ($z = 0.51$). Using these new data we repeat, at a significant look-back time, the color-scatter analysis that has been conducted on present-epoch clusters. Matching aperture sizes, luminosity range and color system to those used for local clusters, we find a small scatter, <0.11 mag rms, for galaxies classified as E or E/S0. After allowing for observational error, we estimate an intrinsic rms scatter in rest frame $U-V$ of only 0.07 mag for these galaxies. This small scatter applies both within each cluster and from cluster to cluster. Furthermore, we do not find any trend for the scatter to increase with decreasing galaxy luminosity beyond that due to expected observational error.

Combined with the Bower et al. result, our data again point to a early formation epoch for the stars in these ellipticals. Again, unless the star formation was well synchronized, it occurred at least 6 Gyr earlier than the epoch of observation. This falls on the same "time line" as the Coma observations, implying an age for both of order $\tau \gtrsim 10$ Gyr. However, by combining the present-epoch and intermediate-redshift observations, we have effectively ruled out the synchronicity argument that allowed for a later formation epoch, since the same more recent epoch of synchronous star formation will not suffice for both. Thus, our result provides a new, stronger constraint on the star formation history of cluster spheroidals prior to $z \simeq 0.5$. Although we cannot rule out the continued production of *some* ellipticals, our results do indicate that the bulk of the stars seen in luminous elliptical cluster galaxies were formed by $z \simeq 3$.

THE MORPHOLOGY-DENSITY RELATION AT HIGH REDSHIFT

A principal goal of our group has been to study the evolution of morphological types in the rich cluster environment. To this end we have morphologically classified 1857 objects brighter than $R < 23.0$ or $I < 23.5$, in the 11 fields, as described in Smail et al. (1996) While the addition of spectroscopically-derived parameters, such as cluster membership or stellar population, is important for understanding the evolutionary state of these populations, photometric/morphological information alone allows a simple and important comparison with the properties of present day clusters. This comparison offers clues as to how clusters of galaxies came to hold their atypical complements of galaxy types.

With a resolution approaching 0.1 arcsecond, our WFPC-2 images show detail at the level of 500 pc in the clusters, equivalent to observing galaxies in the Coma cluster with 1" seeing. While cruder than the resolution usually available for morphological classification of nearby galaxies, it is sufficient for the identification of basic morphological information, and, in particular, is comparable to that presented by Dressler (1980) in the study of galaxy morphology in 55 low-redshift clusters.

Dressler found a strong relation between the fractions of E, S0, and spiral galaxies with the local projected density where they were found, in the well known sense that ellipticals became more prevalent, and spirals less so, in regions of higher surface density. He concluded that, to first order at least, the morphology-density relation is universal, that is, representative of every cluster in the sample, regardless of its global properties.

Dressler's original data have been reanalyzed and are presented in Dressler et al. (1996) along with the morphology-density relation for the HST sample described here. That paper also reviews some of the challenges to the morphology-density relation, for example, the contention by Whitmore, Gilmore, & Jones (1993) that the principal determinant of galaxy type within rich clusters is the radial distance from the cluster center. For our purposes here we simply analyze the morphology-density relation in our $z \sim 0.5$ sample.

Our morphological data for the distant sample has been treated in as similar a fashion as was possible, for example, over the same area and to comparable absolute magnitude limits. We also believe that the morphological classification scheme is comparable to that used in the Dressler study. In Fig. 1 we show the morphology-density relation for the low-redshift Dressler sample, and in Fig. 2 that for the $z \sim 0.5$ sample. The density range encompassed by the more distant sample is shifted by half a dex to higher density, probably reflecting the fact that these clusters are systematically richer than the typical clusters of Dressler's local sample.

Before addressing the question of gradients in Fig. 2, we take note of obvious differences between this and the nearby cluster sample. As is now well known, spirals are greatly overabundant at these high densities compared to present-epoch clusters, but, perhaps surprisingly, the difference seems to made up entirely by a paucity of S0 galaxies rather than an underabundance of *both* S0 and E galaxies. In fact, E galaxies appear to be in even greater abundance! At comparable densities, spirals are a factor of 2 overabundant, S0's are a factor of 2–3 underabundant, and ellipticals are a factor of 1.5 overabundant in the $z \sim 0.5$ sample compared to Dressler's sample of nearby clusters. The paucity of S0 galaxies is particularly noteworthy. As explained in Smail et al. (1996) and Dressler et al. (1996), we have compared the distribution in flattenings of S0 galaxies in our sample with that for the Coma cluster, in an attempt to see if we have misclassified S0 galaxies as ellipticals, particularly for the face-on cases. We conclude that this may, in fact, be the case for a small fraction ($\sim 15\%$) of the S0 population. However, this is of small consequence to the paucity of S0s we find, and, furthermore, our data suggest that this misclassification is roughly the same for both low-z and intermediate-z clusters, for which case no correction is required when comparing the relative number of cluster S0s at the two epochs.

We now ask whether any trend of morphology with density is apparent for the distant sample. From Fig. 2 it appears that a modest relation is present, but it is only for the bins of highest surface density — over the last factor of 5 in surface density. Over this range the spiral fraction plummets and the elliptical fraction rises sharply, but for the lower density zones, over which there is a very noticeable gradient in the nearby clusters, the relationships

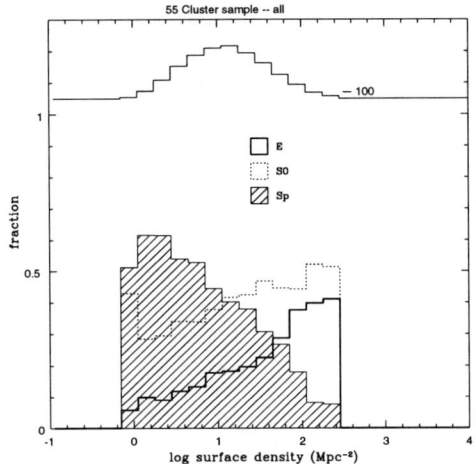

FIGURE 1. The morphology-density relation for the Dressler (1980) 55 cluster low-redshift sample

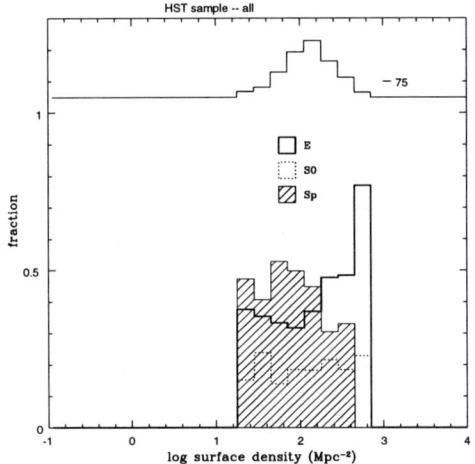

FIGURE 2. The morphology-density relation for 10 clusters at redshifts $0.36 < z < 0.57$.

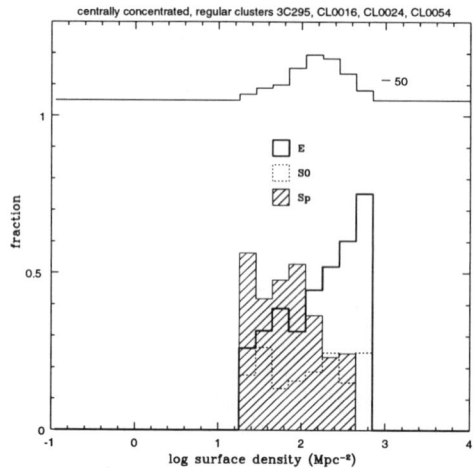

FIGURE 3. The morphology-density relation for 4 high-concentration, regular clusters at intermediate redshift, 3C295, Cl0016+16, Cl0024+16, and Cl0054−27.

are basically flat. However, when the sample is divided by concentration and the degree of regularity, which to a large extent go together, a very different picture emerges. Fig. 3 shows that, for the 4 highest concentration, regular clusters of the $z \sim 0.5$ sample, the morphology-density relation is steep and well defined over the entire density range. In contrast, however, there are no correlations at all for the 4 lowest concentration, irregular clusters. This is a strikingly different result from the situation for present epoch rich clusters, for which Dressler found a strong morphology-density relationship for both irregular and regular clusters. Further discussion of this difference can be found in Dressler et al. (1996).

Perhaps our most important result of this analysis, however, is simply that whether the clusters appear dynamically "mature" or not, the incidence of elliptical galaxies is already very high, and independent of whether they are collected into dense, central regions or not. We suggest, based on this result, that elliptical galaxies predate, and are basically independent of, the virialization of a rich cluster. This is of course, consistent with the Ellis et al. (1996) result described above of early formation of the stars in elliptical galaxies. Furthermore, we find that that the fractional representation of highly asymmetric or disturbed morphologies ($D > 1$, see Smail et al. 1996) with local surface density mirrors the S0 or spiral trend rather than the trend for ellipticals. Together, these three observations suggest that, for the environments of rich clusters at least, elliptical galaxies are not the result of mergers of starforming, gas-rich systems after a redshift $z = 3$. This does not preclude the possibility of dissipationless mergers at $z = 1$, say, when these clusters might have been in the process of amalgamating small groups with lower velocity dispersion, but both the distribution and numbers of ellipticals we have found here, and their photometric and spectral properties, suggest that ellipticals in these regions are not produced by late mergers, or in any process that depended on the dynamical evolution of a rich cluster. Instead, gaseous mergers or coherent collapse at high redshift, or growth of spheroids through dissipationless mergers until later epochs, seems to be the history indicated for ellipticals. It is remarkable, we think, that the environment of proto-clusters of this richness was able to produce such a large population of ellipticals before the identities of the clusters themselves was well established.

The situation for the S0 galaxies seems to be just the opposite. Though the ones we find are, like the ellipticals, red and with little scatter in color, their numbers are so deficient as to suggest that many need to be added since $z = 1$, in order to reach the populations of present-epoch clusters. The source of these S0's seems clear: the overabundance of spirals provides a reservoir of galaxies which may be stripped by ram pressure, tidally harassed (Moore et al. 1995) merged, or subject to strong 2-body gravitational interactions, with the result of producing today's dormant disk galaxies in clusters. Our $z \sim 0.5$ cluster sample includes a significant number of disturbed, distorted morphologies, often with spectroscopic evidence of strong episodes of star formation. These may be the result of mergers, strong interactions, accretions, harassment, or stripping — we are still unable to tell which of these processes are responsible. But, we do know from our morphological classifications that most of these are *disk* systems — they do not seem destined to settle into ellipticals galaxies when their jostling and bursts of star formation have ceased. Though the exact mechanism(s) may be yet unspecified, it seems that at least half of the S0 galaxies in today's clusters have been made by such processes since $z = 0.5$

CONCLUSIONS

We have found a remarkably small scatter of $U - V$ color for galaxies in 10 intermediate-redshift clusters classified as E or E/S0. This suggests an early epoch of formation for the stars of these galaxies, and perhaps of the galaxies themselves. Furthermore, the large number of elliptical galaxies in these clusters, ~40%, suggests that the formation of ellipticals predates cluster virialization. If mergers are responsible for making the ellipticals that now inhabit these rich clusters, they must have been dissipationless, in the "group phase" at $z \sim 1$, or much earlier, $z > 3$, if significant dissipation and star formation were involved. In contrast, the relative paucity of S0's in the intermediate redshift clusters suggests that many of them have indeed been added since $z = 0.5$, by mechanisms that acted on the excessive numbers, compared to today's clusters, of spirals and irregulars.

REFERENCES

1. Schweizer, F. & Seitzer, P. 1988, ApJ, 328, 88.
2. Bower, R.G., Lucey, J.R. & Ellis, R.S. 1991, MNRAS, 254, 601.
3. Ellis, R.S., Smail, I., Dressler, A., Couch. W.J., Oemler, A. Jr., Butcher, H., & Sharples, R.M. 1996, ApJ, in press.
4. Smail, I., Ellis, R.S., Dressler, A., Couch, W.J., Oemler, A. Jr., Sharples, R.M., & Butcher, H. 1996b, Ap.J., in press.
5. Dressler, A. 1980, ApJ, 236, 351.
6. Dressler, A., Oemler, A. Jr., Smail, I., Couch, W.J., Ellis, R.S., Barger, A., Butcher, H., Poggianti, B.M., & Sharples, R.M. 1996, Ap.J., submitted.
7. Whitmore, B.C., Gilmore, D.M., & Jones, C. 1993, ApJ, 407, 489.
8. Moore, B., Katz, N., Lake, G., Dressler, A., & Oemler, A. Jr. 1996, Nature, 379, 613.

The K–band luminosity function of galaxies

J. P. Gardner[1,2], R. M. Sharples[2], C. S. Frenk[2], and B. E. Carrasco[3]

[1] *NASA - GSFC, Code 681, Greenbelt MD 20771*
[2] *University of Durham, Physics Dept., South Road, Durham DH1 3LE, ENGLAND*
[3] *INAOE, Apdo Postal 216 y 51, Puebla, CP 72000, MEXICO*

Abstract.
We present the first determination of the near-infrared K–band luminosity function of field galaxies from a wide field K–selected redshift survey. The best fit Schechter function parameters are $M^* = -23.12 + 5log(h)$, $\alpha = -0.91$, and $\phi^* = 1.66 \times 10^{-2} h^3\ Mpc^{-3}$. We estimate that systematics are no more than $0.1 mag$ in M^* and 0.1 in α, which is comparable to the statistical errors on this measurement.

INTRODUCTION

The luminosity function of galaxies is central to many problems in cosmology, including the interpretation of faint number counts. The near-infrared provides several advantages over the optical for statistical studies of galaxies. The K–corrections and the expected luminosity evolution are smooth and well-understood. The K–band is dominated by near-solar mass stars which make up the bulk of the galaxy. The absolute K magnitude is a measure of the visible mass in a galaxy, and thus the K–band luminosity function is an observational counterpart of the mass function of galaxies. In general, studies of galaxy evolution through number counts, colors, redshift distributions and clustering properties, all require an understanding of the local population of galaxies for interpretation of the faint end data. We have conducted a photometric and spectroscopic survey of galaxies, observed in the near-infrared K band, and the optical B, V, and I bands with linear detectors [1,2], and have obtained spectroscopic redshifts for a sample of 510 galaxies selected in the near-infrared. Our spectroscopy is 90% complete. We present here the K–band luminosity function of field galaxies.

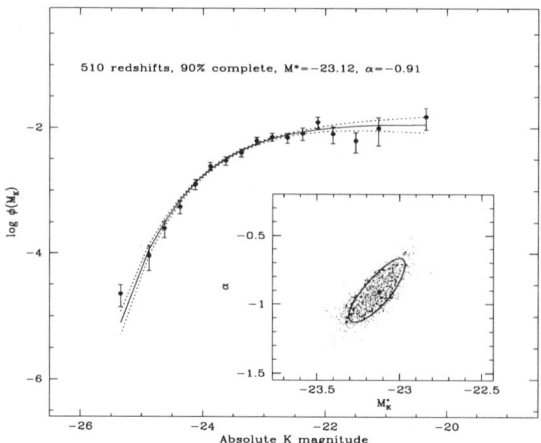

FIGURE 1. The differential K−band luminosity function of galaxies. The solid line is the best fit Schechter function determined using a maximum likelihood method. The dashed lines are the 1σ errors on this fit determined from the error ellipse. Inset are the error ellipse on the Schechter parameter fit to the luminosity function, and the results of 1000 Monte Carlo simulations of our survey parameters. These simulations were binned as 0.03 in M^* and 0.03 in α, and a contour containing 68% (i.e. 1σ) of the points in the binned data is shown as a dashed line.

THE LUMINOSITY FUNCTION

Field galaxy surveys such as this one are typically magnitude limited, and the galaxy distribution has structure along the line–of–sight of the survey. Several methods have been developed to determine the luminosity function from a redshift survey, while avoiding systematics due to clustering. We calculated the luminosity function from our data using the SWML method [3], and plot the results in Figure 1. We used a maximum likelihood method [4] to determine the best fit Schechter [5] function parameters, $M^* = -23.12 + 5log(h)$, and $\alpha = -0.91$, and this function is plotted as a solid line is Figure 1. We used the estimated Schechter parameters within a model of the K−band number counts to determine $\phi^* = 1.66 \times 10^{-2} h^3 \ Mpc^{-3}$. The error ellipse in M^* and α is plotted as an inset figure. We plot in Figure 2 a compilation of the K−band number counts, along with a model prediction based upon the values of M^* and α. We ran extensive Monte Carlo simulations of our survey, to test for systematic errors due to the 10% spectroscopic incompleteness, and the photometric errors. We estimate that the total systematic error in the Schechter function parameters is less than $0.1mag$ in M^* and 0.1 in α.

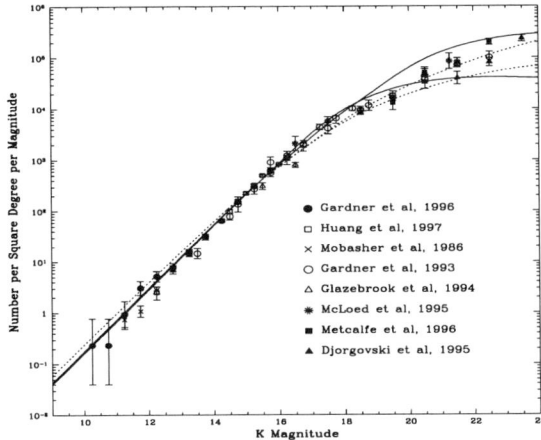

FIGURE 2. The K-band number counts with models based upon our luminosity functions. The solid lines include the effects of passive evolution, the dotted lines are pure K-correction models. The higher line in each case is for $q_0 = 0.02$, while the lower lines are for $q_0 = 0.5$.

DISCUSSION

Previous determinations of the K-band luminosity function were based on K-band photometry of an optically selected redshift survey [6] and a redshift survey of a small number of galaxies selected in the K-band [7]. In Figure 3 we plot our results and the results of those two determinations. Following Glazebrook et al. [7], we apply an aperture correction of -0.30 to their measurement, and a correction of $+0.22 mag$ to the Mobasher et al. [6] measurement to account for their method of calculating K-corrections. In the inset to the figure we plot the error ellipse of our measurement of the luminosity function, and the error ellipse from the Mobasher et al. [6] measurement. Glazebrook et al. [7] fixed α at -1.0, and determined M^*, and this is plotted as an error bar in the inset figure. The errors of the three determinations overlap and are thus consistent at less than the 1σ level. All three determinations are consistent with a flat faint end slope of -1.0, which is approximately the same as has been seen in optical surveys.

There is an excess of faint blue galaxies seen in optical surveys over that predicted by simple models relating local observations to observations at very faint levels [8,9]. Surveys selected in the K band preferentially study normal, massive galaxies. The simple passive-evolution number count models in Figure 2 fit the observed counts well at $K < 20$, and the faint blue galaxy population does not dominate the color distributions until fainter than this

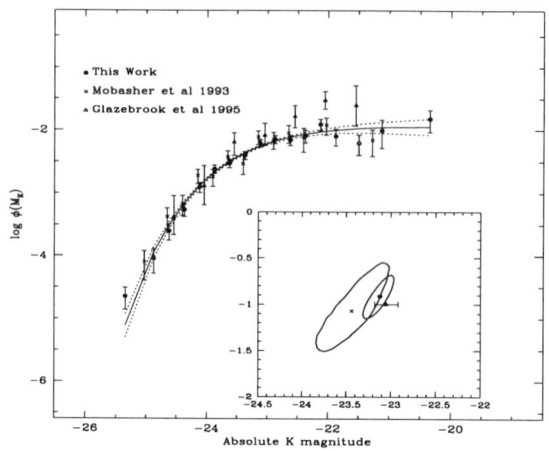

FIGURE 3. Our measurements of the K–band luminosity function from Figure 1 compared to previous measurements [6,7]. Inset are the errors in the Schechter function parameter determinations.

[10]. K–band surveys present a very different picture from optical surveys. Instead of rapid evolution at intermediate or even low redshifts, the galaxies making up the K–band surveys show only passive evolution of the old stellar population to $z \sim 1$, in their counts and colors. To study evolution of the general galaxy population, and to place the faint blue galaxy problem on a statistical footing, theoretical models must first be able to reproduce the local near-infrared properties of galaxies.

REFERENCES

1. Gardner, J. P., Sharples, R. M., Carrasco, B. E., and Frenk, C. S. 1996, MNRAS, 282, L1
2. Baugh, C. M., Gardner, J. P., Frenk, C. S., and Sharples, R. M. 1996, MNRAS, 283, L15
3. Efstathiou, G., Ellis, R. S., and Peterson, B. A. 1988, MNRAS, 232, 431
4. Sandage, A., Tammann, G. A., and Yahil, A., 1979, ApJ, 232, 352
5. Schechter, P. 1976, ApJ, 203, 297
6. Mobasher, B. , Sharples, R. M., Ellis, R. S. 1993, MNRAS, 263, 560
7. Glazebrook, K., Peacock, J. A., Miller, L., and Collins, C. A. 1995, MNRAS, 275, 169
8. Tyson, J. A. 1988, AJ, 96, 1
9. Lilly, S. J., Cowie, L. L., and Gardner, J. P. 1991, ApJ, 369, 79
10. Gardner, J. P. 1995, ApJ, 452, 538

High Redshift Reflection Nebulae: Implications for Galaxy Formation

K. C. Chambers

Institute for Astronomy, 2680 Woodlawn Dr., Honolulu, HI 96822

Abstract. The infrared and optical spatially extended light from high redshift ($z > 1$) radio galaxies is highly polarized, (up to 40% in K band) and can not be dominated by direct starlight. These objects appear to be giant quasar reflection nebulae rather than "galaxies". The extended continuum emission is light from the central regions scattered by dust distributed throughout a 10-30 kpc region. The diverse morphologies (cometary, hourglass, fan, bow-shock) and polarization of these high redshift nebulae are strikingly similar to the appearance of reflection nebulae, jets, and bipolar outflows associated with YSO, although on a vastly larger scale. Based on this similarity and the observed properties of high redshift radio galaxies I propose that the collapse and formation of giant galaxies requires a magnetocentrifugal flow and that disk accretion and bipolar outflow provide the regulating feedback mechanism responsible for the characteristic properties of galaxies. Unique and falsifiable predictions of this hypothesis are discussed in the context of present and future observations of the high redshift universe.

HIGH REDSHIFT RADIO GALAXIES AND REFLECTION NEBULAE

High redshift ($z > 1$) radio galaxies are typically spatially elongated with extended optical and infrared continuum morphologies that are aligned with the axis of the radio source [1,2]. This phenomena is generally referred to in the literature as the "alignment effect" [3]. These objects have no analog at the present cosmic epoch; they only exist at high redshift. HST observations show them to consist of numerous clumps with a remarkable variety of morphologies (e.g. [4–6]) A variety of completely different ideas have been proposed to explain the alignment effect (see references in [7]), but the most attractive explanation is the proposal that a bi-conical beam of light from the AGN is scattered by a diffuse medium and becomes visible [8]. In this picture, the nucleus is either intrinsically anisotropic or a "searchlight" is formed due to obscuration by dust in a disk or torus encircling the nucleus. In the hypothesis of the unification scheme (e.g. [9]) for double lobed radio galaxies and quasars, observers $\lesssim 45°$ of the axis have a clear line of sight to the nucleus and classify such objects as radio quasars, whereas observers $\gtrsim 45°$ from the axis classify them as radio galaxies. Progressing from low redshift to high redshift, the radio galaxies makes a transition at $z \sim 1$ from galaxies where the optical-infrared light is dominated by starlight to objects where the light is dominated by the scattered aligned continuum. Optical polarimetry and spectropolarimetry of the extended continuum of high redshift radio galaxies indicate that the light is scattered, and the scattering medium is dust (e.g. [10,11]). Numerous investigators have argued that any aligned continuum produced in this manner is superimposed on top of an underlying (old) giant elliptical galaxy [12–14], but I find the optical/IR morphological and spectroscopic data show a much more intimate relation with the radio jet axis then can be accounted for by this viewpoint [6,15]. Our recent discovery that the K band light in 8 out of 10 high redshift radio galaxies is highly polarized (20 - 45%) demonstrates the light from these objects is essentially nothing but scattered light, and the objects consist of enormous 10-30 kpc reflection nebulae, without any underlying giant elliptical galaxy [16–18].

MAGNETOCENTRIFUGAL GALAXY FORMATION

Based on the morphological, polarimetric, and kinematical evidence from high redshift radio sources and their associated reflection nebulae, I propose that these objects are protogalaxies in an early stage of formation where the dynamics of the protogalactic gas depend on the balance between gravitational collapse and a giant magnetocentrifugally driven outflow and jet. The timescale of this activity is much less than a Hubble time, and therefore the space density of the remnants is far greater than the observed space density of radio sources at high redshift. Thus all giant galaxies could have undergone such a phase in their past, and I argue that galaxy formation regulated by magnetocentrifugally driven flows (and associated jets) can account for many of the characteristic properties of galaxies today. The basic scheme is analogous to wind mechanisms proposed for YSO's, [19,20] but with the central black hole and its associated accretion disk or torus serving as the central anchor for the magnetic field. This presupposes that the black hole forms in a dense core before the rest of the protogalaxy is formed, but that is precisely what is observed in the case of the high redshift radio galaxies, and nearly every L* galaxy today has a fossil black hole [21]. If a massive black hole and the surrounding material launches a highly supersonic synchrotron radio jet that expands for ~ 100 kpc or more, the sheath of the associated cocoon will be threaded with a well organized magnetic field. The energy in the magnetic field comes from the gravitational binding energy (and rotational energy) of the massive black hole in the core. A substantial fraction of the gas in a bi-conical beam is ionized by the AGN, so the magnetic field becomes frozen in the gas. The protogalaxy collapses and the coldest component of the protogalactic gas settles into a giant Wolfe disk as the magnetic field transfers the angular momentum outward in a giant bipolar wind. As the outermost field lines reach a constant opening angle and become open field lines, they extend throughout the entire protogalaxy. The field lines resist the differential rotation of the forming gas disk and equalize the angular rotation of the disk by sending torsional Alfven waves up and down in the axial direction until the the angular speed of the inner protogalactic disk is roughly constant. If the magnetic field provides some radial (centrifugal) support in this way, while the self-gravity and thermal pressure determine the vertical structure above and below the disk, then stars which form out of the gas and decouple from the magnetic field will populate orbits that maintain some of this structure. In this scheme the primary burst of star formation has a natural trigger when the mass in the disk overcomes the magnetic field and vigorous star formation proceeds throughout the disk. Supernovae, bubbles, and fountains provide enriched high velocity material to upper surfaces of disk, where the magnetic field can lift them into the wind - providing it is still functioning. The enriched gaseous material in the wind cools into grains far from the inital site of stellar nucleosynthesis, and form the observed giant reflection nebulae illuminated by the AGN. If the star formation/wind combination is very efficient, the remaining disk would have little gas. If there were no pre-existing dark halo to dynamically stabilize the disk, it would viralize, "merging" with itself, producing an elliptical galaxy, whereas a spiral could emerge from a less efficient star-formation/wind episode.

Therefore I propose that the characteristic velocity of the protogalactic gas, as established by torsional magnetic Alfven waves produce the feedback mechanism required for the low dispersion in the characteristic properties of galaxies as shown by the existence of the Tully-Fisher relation for spiral galaxies and the Faber-Jackson relation for elliptical galaxies. By identifying the maximum circular velocity V_{max} of the resulting spiral galaxy and the core velocity dispersion $\sigma_c \sqrt{2}$ of the resulting ellipticals (after viralization) with the Alfven velocity $V_A = B/\rho^{1/2}$ of the protogalaxy, then a crude estimate of the required B field before collapse in comes from the magnetic viral theorem $M_{gal} \sim 10^{11} M_\odot (B/10\mu G)(R/100 \text{kpc})^2$, which agrees reasonably with the equipartition magnetic fields in high redshift radio source cocoons.

If magnetocentrifugal flow is the crucial ingredient for galaxy formation, then there should be independent evidence in other observations of the high redshift universe. There may be faint reflection nebulae around quasars and an entire population of "buried" anisotropically emitting radio quiet quasars. Thus I suspect that the features seen in the HST imaging of high z quasar fields [22] are the conical edges of reflection nebulae, not infalling galaxies. Lines of sight through such disks and winds to background quasars could explain the high Faraday rotation seen in damped Lyman alpha systems (Wolfe disks) [23]. If the mechanism is important for normal galaxy formation, then there should be evidence in the deep HST images - and to me many of the objects seen in the deep imaging that have morphologies which "cannot be shoehorned" into any conventional classification system [24,25] are highly reminiscent of the morphologies of the high z radio galaxies with all the morphological features of reflection nebulae: cometary, hourglass, fan, bow-shock, etc. Thus some of the $z > 1$ faint blue objects may actually be highly polarized reflection nebulae, and not direct starlight. Detecting 20% polarization at these faint levels is feasible, but will require a substantial effort with either HST or 10 meter class telescopes. Cool magnetohydrodynamic jets and bipolar outflows at very high redshift could form large scale structure with lower CMB constraints than the thermalized deflagration blast waves of the explosion hypothesis [26], and would leave a unique observational signature.

REFERENCES

1. Chambers, K. C., Miley, G. K., van Breugel, W. J. M. 1987, Nature, 329, 604
2. McCarthy, P., et al., 1987, ApJ, 321, L29 ApJ, 321, L29
3. Chambers, K. C., Miley, G. K., 1989, The Evolution of the Universe of Galaxies, ed. R. G. Kron (San Francisco: Astronomical Society of the Pacific) pp 373-388
4. Miley, G.K., Chambers, K.C., W.J.M. van Breugel, 1992, ApJ 401, L69
5. Longair, M. S., Best, P.N., Rottgering, H.J.A., 1995, MNRAS, 275, L47
6. Chambers, K. C., et al., 1996, ApJ Supp, 106, 247
7. Chambers, K. C., et al., 1996, ApJ Supp, 106, 215
8. Tadhunter, C., et al., 1988, BL Lac Objects, ed. L. Maraschi, p. 79, Springer
9. Barthel, P. D. 1989, ApJ 336, 606
10. di Serego Alighieri, S., et al.,1989 Nature, 341, 307
11. Cimatti, A., et al., 1996, ApJ, in press
12. McCarthy review article
13. Stockton, A., Kellogg, M., and Ridgway, S.E., 1995, ApJ 443, L69
14. Dunlop, J., et. al. Nature 381, 581
15. Chambers, K.C., 1992, First Light in the Universe, Stars or QSO's?, ed. B. Rocca-Volmerange, Paris: Edition Frontiers, p 125
16. Knopp, G. P. & Chambers, K. C. 1996 ApJ Letters, submitted
17. Chambers, K. C. & Knopp G.P. 1996, ApJ Letters, submitted
18. Knopp, G. P. & Chambers, K. C. 1997, ApJ, in prep, 1996a
19. Shu, F., et al, 1995, ApJ, 455, L155
20. Wardle, M & Konigl, 1993, ApJ, 410, 218
21. Kormendy, J. & Richstone, D., ARAA, 1995, 33, 581
22. Cowie, L., Hu, E., McMahon, R., Egamii, E., preprint 1996
23. Kronberg, P.P., & Perry J.J., 1982, ApJ, 263, 518
24. van den Berg et al., 1996, ApJ,
25. Cowie, L., Hu, E., Songaila, A., Nature, 1995
26. Ostriker, J.P., Cowie, L.L., 1981, ApJ, 243, L127

Ultraviolet Imaging Observations of Abell 1795: Further Evidence for Massive Starformation

Eric P. Smith[1], R. C. Bohlin[2], G. D. Bothun[3],
R. W. O'Connell[4], M. S. Roberts[5], S. G. Neff[1],
A. M. Smith[1] & T. P. Stecher[1]

[1] *Laboratory for Astronomy and Solar Physics, NASA/Goddard Space Flight Center*
[2] *Space Telescope Science Institute*
[3] *Department of Physics & Astronomy, University of Oregon*
[4] *Astronomy Department, University of Virginia*
[5] *National Radio Astronomy Observatory*

Abstract. We dicuss observations from the Ultraviolet Imaging Telescope of the Abell 1795 cluster of galaxies. We compare the cD galaxy morphology and photometry of these data with those from existing archival and published data. The addition of a far–UV color helps us to construct and test star formation model scenarios for the sources of UV emission. Models of star formation with rates in the range $\sim 5-20 M_\odot \mathrm{yr}^{-1}$ indicate that the best fitting models are those with continuous star formation or a recent ($\sim$ 4 Myr old) burst superimposed on an old population.

INTRODUCTION

The final fate of the cooling gas seen by X–ray telescopes in clusters of galaxies has long been sought. Therefore, any system which can provide evidence for star formation arising in a system of cooling gas is of great interest. Spectroscopic studies of Abell 1795 in both the ultraviolet and optical reveal an unusual blue continuum [1,2] whose structure has been recently traced in broad band optical imaging [3]. These properties have made it one of the best candidate galaxies for exhibiting star formation arising from a cluster cooling flow.

Hubble Space Telescope observations, improved CCD sensitivity at blue wavelengths and the ultraviolet telescopes on board the Astro Space Shuttle payload have made this galaxy an object of renewed interest [4,5]. These recent observations have revealed regions with colors bluer than normal ($(U-I) \sim 2.1$;

[4]) for an expected old stellar population, near the galaxy nucleus that may be sites of recent star formation.

If the galaxy has undergone recent star formation then it is natural to turn to the ultraviolet (UV) part of the spectrum for observational confirmation. Observations by the Ultraviolet Imaging Telescope (UIT) in its far–UV filter may be used as a valuable diagnostic for stars which are extremely hot ($T_{eff} \gtrsim$ 10000K). At the distance of the cD ($z = 0.0634$) as few as 1200 O stars would be detectable with UIT.

OBSERVATIONAL DATA

The UV data were obtained with the UIT during the Astro–2 Space Shuttle mission of March 1995. The UIT took several exposures using a far–UV filter with λ_c=1520Å; $\Delta\lambda$ = 356Å. We have also assembled a collection of archival space and ground–based data to complement our UIT observations of this cluster. The UV morphology resembles the Hα emission more nearly than the broad band optical emission. Despite this superficial resemblance we conclude from published Lyα [2] fluxes that the bulk of UV light is from a continuum source and not extended emission–line gas. To further investigate the morphology of the UV light we compare its surface brightness profile with those at longer wavelengths. The UV light profile is better fit by a disk galaxy profile (exponential) than by an elliptical ($r^{\frac{1}{4}}$–law) one. There is no evidence for a point source component to the UV surface brightness profile. We also performed fits to an HST F702W profile. The best fit was achieved with a combined point source, exponential disk and $r^{\frac{1}{4}}$-law bulge model. However, the disk central surface brightness for this fit is more than 2 magnitudes brighter than the corresponding characteristic bulge surface brightness (I_e) indicating that the central region ($r < 8''$) of the cD has a predominantly disk-like light distribution. Thus both the UV and visual light from the central portion of the galaxy are better characterized by exponential light distributions than by elliptical galaxy light distributions. Curve of growth analysis reveals the cD galaxy UV light (centered at α =13:48:52.42, δ = +26:35:34.7 J2000) is contained within a circle of radius $\sim 25''$. The aperture magnitude and corresponding flux for the galaxy are $m_{1520} = 15.54 \pm 0.09$ (2.21×10^{-15}erg s^{-1}cm^{-2}Å^{-1}).

STAR FORMATION

To estimate the star formation history or current star formation rate using the UV measurements from UIT we must first determine the nature of the UV light. The bulk of the UV emission in the UIT bandpass comes from a continuum of hot stars which may in turn be responsible for ionizing some of the emission–line gas [6,2,7]. The UIT magnitude listed above with K, and emission–line corrections implies a luminosity at 1500Å, $L_{1500} = 1.9 \times 10^{42}$ erg sec^{-1}. Assuming $H_o = 75$, $q_o = 0.1$, we require 1.8×10^4 O5 stars to match the UIT, K–corrected flux. The implied star formation rate (SFR) is in the range $8 - 23 M_\odot \text{yr}^{-1}$, depending on the IMF model, with the lower bounds corresponding to a Kennicutt IMF [8] and the upper bound corresponding to a Miller-Scalo IMF [9]. This SFR is consistent with the predicted star formation rates one calculates from the Hα luminosity ($5 M_\odot \text{yr}^{-1}$, [10,8]) and is substantially lower than that estimated from recent X–ray analyses ($\sim 300 M_\odot \text{yr}^{-1}$; [11,12]).

To further test the various scenarios for possible star formation histories we use models based upon (a) a single, instantaneous burst of star formation (b) a single, exponentially decaying burst of star formation (c) continuous, constant star formation (d) a single, instantaneous old burst with smaller, younger bursts superimposed. The first three models are described in more detail in [13]. The fourth model is simply a scaled superposition of the models in (a). We have plotted color–color diagrams using the four colors (15-V), (15-U), ($U-V$), ($V-R$) for each set of the above models to compare with the global colors of the cD galaxy (figure 1). The colors have been corrected for K–dimming, but not extinction. Models with exponentially decaying star formation and those in which the central population was created in a single burst of age greater than $\sim$few hundred Myr all have broad-band colors too red to match the cD. Models with some levels of continuous star formation and/or those with recent bursts however can reproduce the colors. The most plausible burst model is one in which an old galaxy ($z_{form} = 5$) experiences a Salpeter IMF burst of age 4Myr and total mass 25% of the old population, within the photometered aperture. The most plausible continuous, constant star formation model which matched the far UV - visible colors is one in which the galaxy has been forming stars at a rate $\sim 5 - 10 M_\odot \text{yr}^{-1}$ over the past 5 Gyr.

Both of these star formation histories are consistent with the UV light coming from an accreted population formed in a cluster cooling flow or from an episode of star formation induced by an interaction with a much smaller galaxy. The cooling flow scenario has the following points as supporting evidence (a) line ratios indicative of shock heated gas and (b) the inner surface brightness profile does not show significant disturbances in the UV or optical bands. However, detailed image analyses of deep U imaging by [14] appears to have revealed, in features too faint to have been detected with the UIT, possible debris associated with an interaction of the cD with two small galaxies. High resolution images in the UV would help further constrain the nature of the UV bright population.

REFERENCES

1. Allen, S. W. 1995, MNRAS, 276, 947
2. Hu, E. M. 1992, ApJ, 391, 608
3. McNamara, B. R., & O'Connell, R. W., ApJ, 393, 579
4. McNamara, B. R., & O'Connell, R. W., AJ, 105, 417
5. Pinkney J., et al., 1996, BAAS, 27, 1443
6. Crawford, C. S., & Fabian, A. C. 1993, MNRAS, 265, 431
7. Nørgaard-Nielsen, H. E.,Jørgensen, H. E., & Hansen, L. 1984, AstrAp, 135, L3
8. Kennicutt, R. C. 1983, ApJ, 272, 54
9. Miller, G. E., & Scalo, J. M. 1979, ApJS, 41, 513
10. van Breugel, W., Heckman,T., & Miley, G. 1984, ApJ, 276, 79
11. Edge, A.C., Stewart, G.C., &Fabian, A.C. 1992, MNRAS, 258, 177
12. Fabian, A.C. 1994, Ann Rev Astr Ap, 32, 277
13. Cornett, R. H., et al. 1994, ApJ, 426, 553
14. McNamara,et al. 1996, ApJL, 466, L9

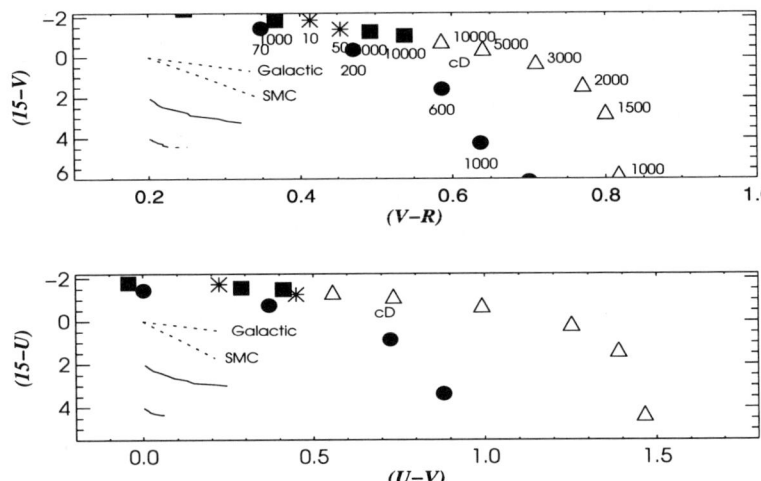

FIGURE 1. Color–color diagrams showing the location of the cD along with several models of star formation histories. Symbols have the following meaning: (•) Single generation, instantaneous bursts of star formation with the age of the burst (in Myr) under the symbol, ($\triangle$) Single generation, exponentially decaying star formation with the decay time (Myr) next to the symbol, ($\square$) Constant, continuous star formation, beginning at the time listed under the symbol, ($*$) Single, 4 Myr old burst added to an old population, ("cD") Abell 1795 measured colors. Tracks indicating the effects of extinction are superimposed. The errors in the cD colors are smaller that the letters used to label the point itself.

Star Formation Associated with the NE Radio Lobe of NGC 5128 (Cen A)

J. A. Graham

Carnegie Institution of Washington
Department of Terrestrial Magnetism
5241 Broad Branch Road NW
Washington, DC 20015

Abstract.
The three clouds of neutral hydrogen around NGC 5128 (Cen A) are most likely surviving remnants of an earlier merger. One of them, NE of the galaxy, appears to be impacted by the adjacent radio jet to the extent that star formation is currently being triggered and loose chains of blue supergiant stars are observed. Some of these stars excite small H II regions whose velocities correspond very closely to that of the H I which in the area. Distinct from these H II regions is the more diffuse, filamentary gas which extends much further along the radio axis and which appears to be excited by shocks associated with the radio jet itself. It is important to realize that the many blue stars are resolved only because NGC 5128 is so very close for a giant radio galaxy. The reported faint blue extensions and plumes in more distant analogs probably have a similar origin.

NGC 5128 (Cen A) is the closest of the giant radio galaxies with a distance about 3.5 Mpc [1]. Its basic structure is that of a giant elliptical galaxy crossed by a dense obscuring lane. There are several indications that a merger event is involved [2,3]. The radio lobes are characteristic, extending over several degrees in the sky. From the early days of radio astronomy, it has been clear that we can observe features in detail here which are also found in more distant and more powerful analogues but which are poorly resolved on an angular scale.

Some optical features extending out towards the NE radio lobe were described by Blanco et al. [4]. These consist of 3 very distinct components: (1) weak, extended gaseous filaments. (2) small, compact H II regions. (3) loose aggregates of blue supergiant stars. They are seen only in the direction of the NE radio lobe with the main concentration appearing 16 arcmin from the galaxy nucleus. No similar features have ever been found in the SW lobe. We now understand why this is so.

Three clouds of neutral hydrogen were discovered a couple of years ago around NGC 5128 by Schiminovich et al. [5]. They are most likely remnant material from the earlier merging event. One of them only, NE of the galaxy, appears to be impacted by the adjacent radio jet to the extent that the cloud is compressed and star formation under rather unusual circumstances has been triggered along its border [6]. The extended gaseous filaments are evidently excited by "autoionization" from photons generated by shocks associated with the jet-cloud interface [7]. The small, compact H II regions on the other hand, are excited by photons from embedded hot stars.

In 1995, I obtained CCD frames of the NE jet region with a large format detector on the du Pont 2.5m telescope at Las Campanas Observatory. The 8 arcmin field of each frame is sufficiently large that an existing photometric sequence to V = 22.5 mag [8] is included. Thus, the frames are well calibrated and, for the first time, magnitudes and colors for the blue supergiant stars can be measured. The brightest blue stars begin at V = 20.0 mag and have average colors close to B−V = 0.0 and U−B = −0.9 mag. Thus they appear similar to the brightest blue stars in the Magellanic Clouds [9]. A surprising number of red stars are clumped along with the blue stars. Among these intrinsically bright stars, stellar evolution is already taking place. The brightest stars will become supernovae and will contribute to the the disruption of the H I cloud. Most of the fainter stars so formed will disperse within the extended ellipsoidal stellar component of NGC 5128 [10].

The loose stellar chains are displaced from the filaments of autoionized gas by about 1 arcmin (approx 1 Kpc at this distance). Examination of the region of the filaments (Figure 1) shows that star formation is going on here too, apparently in smaller clumps which are partially obscured by ambient dust. Are these stellar aggregates predecessors of the more visible and apparently more populous star chains to the west? Will they migrate to the same part of the sky or are they formed with sufficiently small velocities that they will evolve and disperse essentially in place? Radial velocities of the stars themselves when compared with the H I velocities should answer these questions.

Extended emission line and blue continuum features are often seen around more distant radio galaxies [11] and it seems highly probable that at least some of these are similar to the NGC 5128 optical fragments but are at such a distance that their basic structure is beyond resolution, even with the Hubble Space Telescope.

REFERENCES

1. Hui, X., Ford, H.C., Ciardullo, R. & Jacoby, G.H. *Ap. J.*, **414**, 463 (1993)
2. Graham, J.A. *Ap. J.*, **232**, 60 (1979)

3. Malin, D.F., Quinn, P.J. & Graham, J.A. *Ap. J. (Letters)* **272**, L5 (1983)
4. Blanco, V.M., Graham, J.A., Lasker, B.M. & Osmer, P.S. 1975, *Ap. J. (Letters)*, **198**, L63 (1975)
5. Schiminovich, D., van Gorkum, J.H., van der Hulst, J.M. & Kasow, S. *Ap. J. (Letters)*, **423**, L101 (1994)
6. Graham, J.A. *Bull. A.A.S*, **26**, 1504 (1994)
7. Sutherland, R.S., Bicknell, G.V. & Dopita, M.A. 1993, *Ap. J.*, **414**, 510 (1993)
8. Zickgraf, F-J, Humphreys, R.M., Graham, J.A. & Phillips, A. *Pub. A.S.P.*, **102**, 920 (1990)
9. Feast, M.W., Thackeray, A.D. & Wesselink, A.J. *M.N.R.A.S.*, **121**, 337 (1960)
10. Soria, R. et al. *Ap. J.*, **465**, 79 (1996)
11. McCarthy, P.J. *Ann. Rev. A. A.* **31**, 639 (1993)

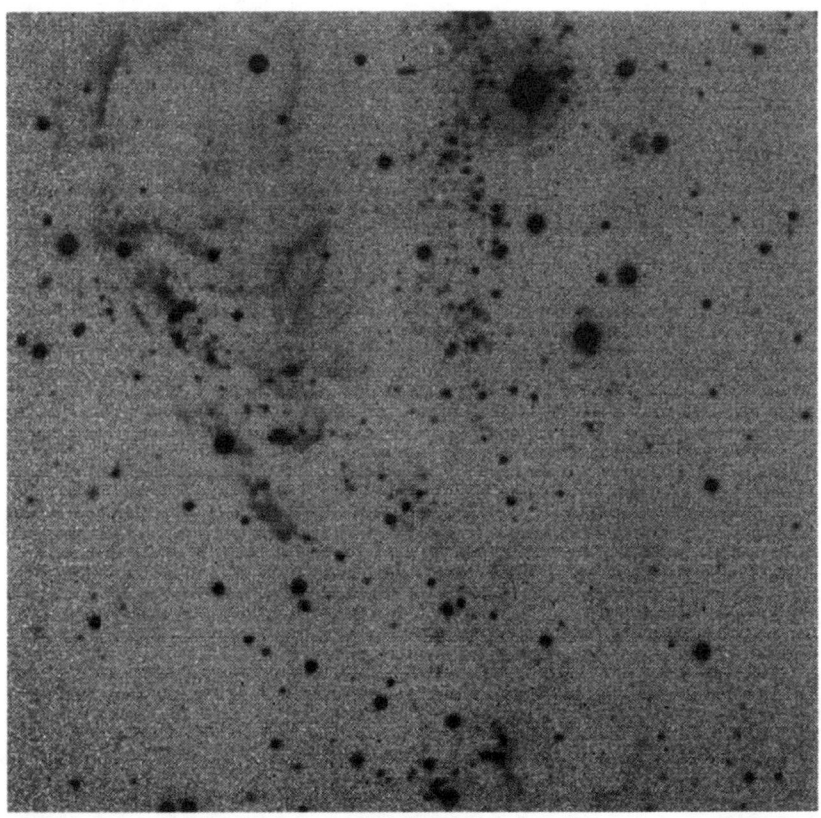

FIGURE 1. CCD frame of a star forming region in NE radio lobe of Cen A showing chain of blue stars (top center) and autoionized gas (center left). An ultraviolet U filter was used with the Las Campanas 2.5m telescope. N is to the top, E to the left.

Star Formation History in Irregulars and Dwarf Spheroidals

Star Formation History of Irregular Galaxies

Despina Hatzidimitriou*

*Physics Department, University of Crete
P.O.Box 2208, Heraklion, Crete, 710 03 Greece

Abstract.
Irregular galaxies are generally less massive and less evolved systems than spirals. They constitute a different sort of system in which to study the processes of star formation and galactic evolution. The star formation history of Irregular Galaxies may be affected by several different factors. However, detailed knowledge of the star-formation history of a significant number of objects is necessary before it becomes possible to determine which parameters are important. In the last few years, there has been significant progress on this front. For the first time, we begin to probe in some detail the star-formation history of a fair number of resolved Irregulars in the Local Group. In the present, these results are summarized.

INTRODUCTION

Irregular galaxies (Irr's), characterized by their lack of organized structure in optical wavelengths, are generally less massive systems than most spiral galaxies. They are gas-rich and have lower metal abundances, being therefore less evolved systems. The star-formation processes in and evolution of Irr's have been the subject of numerous studies and reviews [1] [2] [3] [4] [5] which can lead the reader to the earlier literarure.
Ideally, one would like to address the following fundamental questions:
(i) What are the initial characteristics (for example initial total mass) of the protogalaxy that would lead to a present-day Irr?
(ii) When was the initial epoch of astration in Irr's (if there was such a common epoch)? This is not a trivial problem. Even in the Magellanic Clouds, the closest galaxies to our own, the first epoch of astration is only known to within 3-4 Gyr [6].
(iii) How is star formation (SF) triggered, propagated and regulated in these systems? How are these processes modified as the system evolves? Does a

system with the same initial parameters evolve very differently in different environments?

It should be noted that current star-forming activity in Irr's spans several orders of magnitude without obvious correlation with size, mass and gas content [1]. Therefore, determining the parameters that control SF in these systems is far from straightforward. There are several effects that need to be taken into account:
(i) the stochasticity of star formation in low mass galaxies [7] [8], which may lead to highly variable SFR (over relatively short timescales). Delayed self-regulation of SF can also lead to large oscillations of the SFR ('bursty' mode of SF) [9]
(ii) changes in the gas content, which can result from:
– mass (and metal) loss via galactic winds [10] [11], which may be even more significant in the cases of large star-cluster formation in ralatively small galaxies [12]
– gas accretion or gas loss via galactic interactions (e.g. in the SMC)
– gas infall from outer regions of the galaxy, where the gas is originally passive [13]. In many galaxies the HI distribution is much more extended than the regions of recent SF.
(iii) changes in the internal dynamics, e.g. due to tidal effects, or, Bar formation
(iv) the possible role of dark matter halos in modifying SF activity, e.g. by moderating gas-loss. It has been suggested [14] that episodic SF in low mass galaxies requires both self-regulated quenching of the star-formation rate (SFR) and gas retention, which in turn requires an extended halo, or dark matter. This factor would be more important for the lowest mass Irr's (e.g. the so-called transition objects between dwarf spheroidals and dwarf Irr's).

Some or all of the above effects may be relevant -at varying levels of importance, depending on the total mass of the galaxy and on its environment- in shaping the evolution of Irr's.

Although we are still far from a 'unified theory' for the SF processes and chemical evolution of Irr's, we now begin to have detailed information on the SF history of some Irr's (mostly of dwarf Irr's, which comprise the majority of Irr's in general), owing to the improved quality of ground-based data, the impact of the HST on observations of Local Group galaxies, the improvement of stellar models and the increased sophistication of the techniques applied for the analysis and interpretation of the observations.

The present review will concentrate on recent results on the SF history of nearby resolved systems (for unresolved systems see e.g. [1-4]).

OBSERVATIONAL TOOLS

The SF history of resolved irregulars can be inferred from a combination of observational data, which include:

(i) *Deep colour-magnitude diagrams.* Because most galaxies have complex SF histories, there are numerous interconnected effects that determine the shape of the galactic colour-magnitude diagrams (CMD). The history of a galaxy can therefore be most effectively modeled using numerical simulations, constructing a synthetic CMD from a composition of stellar populations extracted from theoretical stellar evolution tracks, usually using an assumed initial mass function (IMF) [15] [16] [17] [18] [19] [20]. Recently [21], statistical techniques have been applied for the first time to define the likelihoods of the different models. There are several complications with this type of analysis, which include inaccurate knowledge of the distance modulus, of the reddening, and of the age-metallicity relation, possible variability of the IMF, complex selection effects due to crowding, and uncertainties in the stellar models, especially of supergiants. The question of whether the IMF is variable or not is a very important one. For example [22] proposed a time-flattening of the IMF in Irr's, while [11] suggested that one possible explanation of the large variations of [N/O] in Irr's is IMF variability. On the other hand, the numerical models of chemical evolution by [10] rejected any strong dependence of the IMF on metallicity. However, all of the above are based on indirect ways of inferring the IMF. Current observational evidence [23] seems to favour a single universal IMF.

(ii) *Presence of specific types of objects*, such as: RR-Lyrae variables, which are often considered to consist an unambiguously old (15-12 Gyr) population (but see [6]), carbon stars (1-8 Gyr), cepheid variables (<0.1 Gyr), WR stars etc.

(iii) *Star clusters* are a particularly useful diagnostic in close-by systems, as their ages and metallicities can be used to infer the age-metallicity relation and therefore the chemical evolution of their parent galaxy.

(iv) *Metal abundance* distribution of field stars, providing complimentary information on the chemical evolution of the system.

(v) *Element ratios*, e.g. [O/Fe], [N/O], etc. For example, O is expected to be underabundant relative to Fe, when SF proceeds in a small number of widely separated bursts [24]. As mentioned previously, the [N/O] ratio in Irr's shows significant variation [11], which can be either due to inefficient mixing of the ISM (as is the case in the SMC, see below), or due to variable IMF, or due to differential galactic winds.

(vi) *Kinematical and dynamical* studies, to identify possible dynamical parameters that may have affected the SF history of the galaxy, such as the presence of a hot halo (not confirmed observationally in any Irr), Bar formation, signatures of intergalactic interactions, or the presence of dark matter.

(vii) observations of HI, CO, $H\alpha$ emission, HII regions, complexes and supercomplexes of very young stars, to investigate current SF activity and the processes involved.

MAGELLANIC CLOUDS

The proximity of the Magellanic Clouds (the LMC, a 'giant' irregular, and the SMC, a 'dwarf' irregular) allow a detailed study of their star-formation histories. However, it should be kept in mind that the evolution of these galaxies may have been strongly affected by their mutual interactions and their interactions with the MW Galaxy [6]. A first insight into their SFH can be gained from inspecting Table 1, where SFR_o is the current SFR derived from Hα emission measurements [25], $<SFR>$ is the mean SFR derived by dividing the total mass in stars by $\tau = 15$ Gyr, and SFR_i is the 'initial' SFR that can be inferred from the RR-Lyrae content of each galaxy [26], assuming that these types of stars are formed between 15-12 Gyr ago.

It is remarkable that the SFR of 12-15 Gyr ago is very similar to the current SFR in both Clouds. In the mean, the SFR seems to be somewhat higher than these two values, probably indicating that at intermediate ages there may have been some period or periods of increased SFR (this is indeed the case in the LMC, see below).

The initial epoch of astration in the MC's remains uncertain. In the LMC, the oldest star clusters are probably as old as the old Galactic globular clusters, or a few Gyr younger. In the SMC the oldest star cluster is only about 12 Gyr old. Neither Cloud seems to possess a significant field population of blue horizontal branch stars [27] [6]. So, the RR-Lyrae variables are probably the oldest stars (15-12 Gyr) at least in the SMC, although one has to keep in mind that there is no strong observational evidence that can rule out the presence of RR-Lyraes in a population substantially younger than 15-12 Gyr [6].

Large Magellanic Cloud

The SF history of the LMC has been addressed in several recent investigations, based on CMD both from ground-based and from HST observations [16] [18] [28] [29] [30] [31]. A common conclusion of all these studies is that a significant peak ($\simeq$ ×5) on the SF activity occurred about 2-4 Gyr ago,

TABLE 1. Star Formation Rates in the Magellanic Clouds

Galaxy	SFR_i	$<SFR>$	SFR_o
	$M_\odot yr^{-1}$		
SMC	0.05	0.09	0.046
LMC	0.25	0.5	0.26

prior to which the SFR has not been enhanced for several Gyr. This has been suspected for a long time [32]. The pronounced bimodality in the age distribution of star clusters in the LMC [33] [6] also reflects the same event. A detailed study of the earlier SFH depends strongly on knowledge of the age-metallicity relation, which is not well known, exactly because of the absence of star clusters in the age range between 4 and 12-15 Gyr.

Another important -but not unexpected- conclusion that can be drawn from these studies is that populations older than 1-2 Gyr are spatially well mixed out to a radius of 6 Kpc, as can be seen in Table 2, where the slope of the main-sequence luminosity function has been derived from most existing good quality field CMD in the LMC, in the magnitude range $V = 19.3 - 22.0$ (see [30] for details).

All slopes are identical within the errors (with the possible exception of LMC-61), indicating a very similar mixture of intermediate-age populations (contrary to the conclusion of [28]), which is expected given that the dynamical timescale for mixing is of the order of 1 Gyr or less. There are, however, strong variations in the distribution of stars younger than 1 Gyr [6] [31] reflecting the different recent SF histories at different locations.

The chemical evolution of the LMC remains poorly understood [6]. The mean star cluster metallicity jumped by a factor of about 40 during a time when virtually no clusters or significant populations of field stars were forming. It is difficult to produce this chemical enrichment just from supernovae pollution from the small (few %) initial stellar population.

TABLE 2. MS-LF Slope in 11 LMC field regions

Region	Slope	CMD source
LMC-56	0.49±0.06	[18]
LMC-30	0.42±0.08:	[28]
LMC-45	0.48±0.02	[28]
LMC-61	0.57±0.05	[28]
nSL666	0.45±0.03	[30]
n 30 Dor	0.53±0.04	[29]
nNGC1866	0.46±0.03	[16]
nNGC2155	0.53±0.03	[16]
nNGC1783	0.49±0.05:	[16]
nNGC2193	0.48±0.06	[34]
NW Bar	0.48±0.04:	[35]
NW field	0.44±0.07:	[31]

Note: Values followed by ':' signify cases where the LF slope was calculated to a limiting magnitude 1-1.5mag brighter than $V = 22.0$, due to poorer quality data.

Small Magellanic Cloud

No comparably deep photometric studies have been carried out in the SMC to date. The younger populations have been more extensively studied. They are more centrally concentrated than the intermediate-age and old populations, leading to a radial increase of the mean age of the field population [27].

In the main body of the SMC, where most current SF is taking place, there is a plethora ($\simeq$500) of HI shells -identified in a recent high resolution HI map [36]- created by the interactions of the winds of OB stars with the ISM ('OB superbubbles'). The dynamical ages of these shells, which are distributed over an area of about 9 kpc^2, is 5 Myr with a remarkably small dispersion of 2 Myr. Taken at face value, this result infers a global coherent burst of SF over a large area (where SF could not have been self propagated [8] across the region). However, there are several complications in the calculation of these dynamical ages, and one should await for independent measurements of the ages of the underlying OB stars and associations. It should be noted that much more detailed calculations by [37] could not rule out the single burst scenario, although a continuous formation over a much longer period (of 45 Myr or more) is probably more likely. Keeping these reservations in mind, this recent SF event seems to have a profound effect on the ISM of the SMC. The total wind mechanical energy deposited will exceed the whole HI binding energy in $\simeq 10^8$yr (or longer if the dynamical ages are significantly underestimated). So the next close encounter with the LMC, could cost the SMC a large fraction of its current neutral gas, 'turning the Magellanic Stream into a river' [36].

The SF history over the past 2 Gyr in the *outer* regions (at distances larger than $\simeq$2kpc from the center) of the SMC has been studied by [27]. It seems that in most of these regions the SFR has been declining with time (over the past 2 Gyr). There are, however, particular areas which show an enhancement of the SFR, specifically, the region of the 'outer arm', showing a peak of the SFR about 10^8yr ago, and the outer Wing region (with an age of 50 Myr).

The star formation history beyond the past 2 Gyr is much less well known. The horizontal branch and clump morphology can be used as an indication of the 'median' age of the field populations [38] [27]. This 'median' age is of the order of 10 Gyr beyond 2 kpc from the center of the SMC, probably indicating that most stars in the SMC outer areas have been formed about 10 Gyr ago in a major burst or enhanced phase of SF of unknown duration. There are also indications that $\simeq$7% of the stars older than 2 Gyr are older than 10-12 Gyr [27]. Further progress on the SF history of the SMC has to await for high quality deep photometric data. Interestingly, a similar analysis (using the horizontal branch) in one LMC remote field has indicated a median age 2-3 Gyr younger than in the SMC [38].

Intermediate-age (carbon stars) and old populations (RR-Lyraes) seem to be well mixed spatially (at least in the outer regions) in the SMC [39], indicating that they all belong to the same dynamical structure. It must be emphasized at this point that both the youngest (OB stars) and the relatively old stars (carbon stars, planetary nebulae) in the SMC, both in the inner and in the outer regions have very similar velocity dispersions within the observational errors (see Table 2 in [36]), indicating that they share the same kinematics.

A final comment on the history of evolution of the SMC concerns its age-metallicity relation. It was originally thought [33] that for a long period (from 10 to 2 Gyr ago) very little chemical enrichment took place in the SMC, which seemed to require either preferential loss of metals through hot galactic winds, and/or unenriched gas infall from its remote regions. New spectroscopic measurements of the metal abundances of individual giants in a number of star clusters in the SMC [40] led to a revised age-metallicity relation, which does not show the flat distribution of the older version. However, there seems to be a significant metallicity dispersion at the same age in different areas, indicating that the ISM is not well mixed. This is an important result that needs to be taken into account when attempting to model the chemical evolution of a galaxy like the SMC.

OTHER RESOLVED IRREGULARS

Recent Star formation history (last 1 Gyr)

The recent (over past $\simeq 1$ Gyr) SF history has been estimated (see Section 2) for a significant number of other Local Group Irr's. In Table 3 we present

TABLE 3. Star formation history of resolved Irr's over last 1 Gyr

Galaxy	M_B	SFH	SFR $\times 10^{-8} M_\odot yr^{-1} pc^{-2}$	Reference
DDO 210	-9.9	const or gasping		[42] [41]
GR 8	-11.5	$\simeq$const		[43]
Pegasus	-13.1	gasping		[51]
Leo A	-13.3	declining	0.002-1*	[19]
NGC 6822	-15.1	gasping	1-15	[41]
		gasping	1	[44]
WLM	-15.4	const or gasping	0.5	[15]
Sextans B	-15.5	gasping	0.3-3.5	[45] [41]
NGC 2366	-16.3	burst 50Myr		[46]
IC 10	-17.4	burst	5 (Hα)	[47]
NGC 3109	-17.6	gasping	0.3-8.3	[42] [41]
Sextans A	-13.7		0.2(Hα)	[48]
Solar Neighbourhood			0.7	

* The first number corresponds to the current SFR (from Hα) and the second to the recent SFR (1.5-5×10^7yr).

a summary of the general characteristics of the recent SFH in nearby Irr's (mostly dwarf Irr's). Generally, the majority of these galaxies, at least for the last 1 Gyr, appear to have a sort of discontinuous regime of SF, named 'gasping' by [41], with relatively long periods (some 10^8yr?) of moderate activity separated by shorter periods (few 10^7yr) of reduced activity. This could be the result of self-regulation of star formation in these galaxies. The detailed SF history depends strongly on location. Some of the results presented in Table 3 refer to specific areas of the galaxies in question and may not be representative of the SFH of the galaxy as a whole. The timescale for mixing and therefore for wiping out the local differentiations in the newly formed stellar populations is of the order of $\simeq$1 Gyr, i.e. longer than the timescale of the SF history examined by these studies.

The 'bursting' mode of SF (significant increase of SFR over large area) does not seem to be very common. For example, IC 10, which seems to be undergoing an unusual episode of SF, would use up its available gas in under 1 Gyr, at the current rate of SF [47]. Obviously, this sort of star bursting event cannot occur too often, or we wouldn't be observing irregular galaxies at the 'present' time.

Star formation seems to be favoured near the edges of Bars (e.g. LMC, NGC 2366). The younger populations are generally more centrally concentrated than the older populations, probably reflecting the gradual decrease of the gas density in the outer regions below a critical value. Finally, formation of large star clusters in small galaxies (e.g. in NGC 1569, NGC 5253 [49]) may have a significant effect on the ISM of the galaxy, and on its subsequent evolution [1].

Past Star formation history

The past SF history of Irr's has only been studied in some detail very recently and for a handful of objects, as shown in Table 4.

TABLE 4. Star formation history of resolved Irr's over 15 Gyr

Galaxy	M_B	τ_i	SFR(t)	Reference
Pegasus	-13.1	15-12 Gyr	declining over last $\simeq$9 Gyr	[51]
IC1613	-14.8	15-12* Gyr		[52]
NGC 6822	-15.1	15-12 Gyr	const or declining in last 4 Gyr	[20]
LGS3**	-9.2		lower by ×5 in last 1 Gyr	[53]
SMC	-16.7	15-12*Gyr	enhanced $\simeq$10 Gyr ago?	[27]
LMC	-18.2	15-12 Gyr	$\simeq$const + burst $\simeq$2 Gyr ago	[17]

* From the presence of RR-Lyrae variables
** Often thought of as a transition object between a dwarf spheroidal and a dwarf Irr.

The time resolution is much coarser than in the studies of the recent SF history described above. It is important, however, to note that there is a common (to within a few Gyr) epoch of first astration for all galaxies studied. The detailed history of SF seems to be widely different from galaxy to galaxy. The sample is still too small to allow determination of the decisive factors. There are now good quality HST CMD for a few more Irr's in the Local Group [50], reaching below the horizontal branch, which will significantly extend our current knowledge.

Acknowledgements: The author would like to thank A. Aparicio, C. Clarke, C. Gallart, and E. Olszewski for making their results available prior to publication.

REFERENCES

1. Gallagher, J.S. 1996, in ASP Conf. Ser., From Stars to Galaxies: The Impact of Stellar Physics on Galaxy Evolution, ed. C. Leitherer, U. Fritze-von Alvensleben, & J. Huchra (San Francisco: ASP), in press
2. Saha, A. 1995, in IAU Symp. 164, Stellar Populations, ed. P.C. van der Kruit & G. Gilmore (Dordrecht: Kluwer), 175
3. Kennicutt, R.C. in 'Star Formation in Stellar Systems', ed. G. Tenorio-Tagle, M. Prieto, & F. Sanchez, Cambridge: Cambridge U.P., 191
4. Gallagher, J.S., & Hunter, D.A. 1984, ARA&A, 22, 37
5. Hunter, D.A., & Gallagher, J.S. 1986, PASP, 98, 5
6. Olszewski, E.W., Suntzeff, N.B., & Mateo, M. 1996, ARA&A, 34, 511
7. Elmegreen, B., in 'Star Formation in Stellar Systems', ed. G. Tenorio-Tagle, M. Prieto, & F. Sanchez, Cambridge: Cambridge U.P., 381
8. Gerola, H., Seiden, P.E. & Schulman, L.S. 1980, ApJ, 242, 517
9. Parravano, A. 1996, ApJ, 462, 594
10. Matteucci, F., & Tosi, M., 1985, MNRAS, 217, 391
11. Olofsson, K. 1995, A&A, 293, 652
12. O'Connell, R.W., Gallagher, J.S., & Hunter, D.A. 1994, ApJ, 433, 65
13. Firmani, C., & Tutukov, A. 1992, A&A, 264, 37
14. Lin, D.N.C., & Murray, S.D. 1994, in Dwarf Galaxies, ed. G. Meylan & P. Prugniel, ESO Conference and Workshop Proceedings, No 49, 535
15. Ferraro, F.R., Fusi Pecci, F., Tosi, M.,& Buonanno, R. 1989, MNRAS, 241, 433
16. Bertelli, G., Mateo, M., Chiosi, C., & Bressan, A. 1992, ApJS, 388, 400
17. Gallagher, J.S., Mould, J.R., de Feijter, E., Holtzman, J., Stappers, B., Watson, A., Trauger, J., Ballester, G.E., Burrows, C.J., Casertano, S., Clarke, J.T., Crisp, D., Griffiths, R.E., Hester, J.J., Hoessel, J., Krist, J., Matthews, L.D., Scowen, P.A., Stapelfeld, K.R., & Westphal, J.A. 1996, ApJ, 466, 732
18. Vallenari, A., Chiosi, C., Bertelli, G., & Ortolani, S. 1996, A&A, 309, 358
19. Tolstoy, E., 1996, ApJ, 462, 684
20. Gallart, C., Aparicio, A., Bertelli, G., & Chiosi, C. 1996, AJ, 112, 1950

21. Tolstoy, E., & Saha, A., 1996
22. Arimoto, N., & Tarrab, I. 1990, A&A, 228,6
23. Richer, H., 1997, this volume
24. Gilmore, G., & Wyse, R.F.G. 1991, ApJ, 367, L55
25. Kennicutt, R.C., Bresolin, F., Bomans, D.J., Bothun, G.D., Thompson, I.B. 1995, AJ, 109, 594
26. Frogel, J.A. 1984, PASP, 96, 856
27. Gardiner, L.T., & Hatzidimitriou, D. 1992, MNRAS, 257, 195
28. Vallenari, A., Chiosi, C., Bertelli, G., & Ortolani, S. 1996, A&A, 309, 367
29. Elson, R.A.W., Forbes, D.A., & Gilmore, G.F., 1994, PASP, 106, 632
30. Hatzidimitriou, D., & Kontizas, M. 1997, in preparation
31. Westerlund, B.E., Linde, P., & Lynga, G. 1995, A&A, 298, 39
32. Butcher, H.R. 1977, ApJ, 216, 372
33. Da Costa, G.S. 1991, in IAU Symp. 148, The Magellanic Clouds, ed. R. Haynes & D. Milne (Dordrecht: Kluwer), 183
34. Da Costa, G.S., King, C.R., & Mould, J.R. 1987, ApJ, 321, 735
35. Hardy, E., Buonanno, R., Corsi, C.E., Janes, K.A., & Schommer, R.A. 1984, ApJ, 278, 592
36. Staveley-Smith, L., Sault, R.J., Hatzidimitriou, D., Kesteven, M.J., & McConnell, D. 1996, MNRAS, in press
37. Oey, M.S., & Clarke, C.J., 1997, MNRAS submitted
38. Hatzidimitriou, D. 1991, MNRAS, 251, 545
39. Hatzidimitriou, D. 1994, in Astronomy from Wide-Field Imaging, IAU Symp. 161, ed. H.T. MacGillivray et al., (Dordrecht: Kluwer), 489
40. Da Costa, G.S., & Hatzidimitriou, D. 1997, in preparation
41. Marconi, G., Tosi, M., Greggio, L., & Focardi, P. 1995, AJ, 109, 173
42. Greggio, L., Marconi, G., Tosi, M., & Focardi, P. 1993, AJ, 105, 894
43. Tolstoy, E. 1994, in Dwarf Galaxies, ed. G. Meylan & P. Prugniel, ESO Conference and Workshop Proceedings, No 49, 507
44. Gallart, C., Aparicio, A., Bertelli, G., & Chiosi, C. 1996, AJ, in press
45. Tosi, M., Greggio, L., Marconi, G., & Focardi, P. 1991, AJ, 102, 951
46. Aparicio, A., & Gallart, C. 1995, AJ, 110, 2105
47. Massey, P., & Armandroff, T.E. 1995, AJ, 109, 2470
48. Hunter, D.A., & Plummer, J.D. 1996, ApJ, 462, 732
49. Beck, S.C., Turner, J.L., Ho, P.T.P., Lacy, J.H., & Kelly, D.M. 1996, ApJ, 457, 610
50. Skillman, E. 1997, this volume
51. Aparicio, A., Gallart, C., & Bertelli, G. 1996, in preparation
52. Saha, A., Freedman, W.L., Hoessel, J.G., & Mossman, A.E. 1992, AJ, 104, 1072
53. Aparicio, A., 1996, private communication

The Star Formation Histories of Dwarf Spheroidal Galaxies

Tammy A. Smecker-Hane

Dept. of Physics & Astronomy
University of California
Irvine, CA 92697-4575
Email: tsmecker@uci.edu

Abstract. The nine dwarf spheroidal galaxies (dSphs) that are companions of the Galaxy provide excellent laboratories for exploring the physical processes that regulate star formation and galaxy evolution. Although the dSphs are currently devoid of gas, recent investigations of their stellar populations have shown that their star formation histories are surprisingly complex. The Carina dSph is a prime example; it actively formed stars in four distinct episodes over the last 1 to 14 Gyr. Carina retained an interstellar medium over numerous generations of supernovae explosions, which proves that star formation can be self-regulated even in these very low mass galaxies. Explaining the observed star-formation rates and chemical-enrichment histories of the dSphs will be challenging, but in doing so we will gain insight into the balance between star formation and the negative feedback from massive stars, and how galactic winds and massive dark matter halos shape the evolution of galaxies.

I INTRODUCTION

Dwarf galaxies are in theory the simplest galaxies, and thus the best places to study the processes that regulate star formation. If we can thoroughly understand how dwarfs evolve then we have a good start at understanding the evolution of low-metallicity quasar absorption line systems and the early evolution of larger galaxies. In cosmological models that involve hierarchical collapse, the first objects in the Universe to condense out of the Big Bang and begin forming stars are dwarf galaxy-sized fragments. Subsequently, larger galaxies are built up from the growth and mergers of these fragments. We have yet to determine the extent to which the early evolution of our Galaxy is best described as a lumpy, monolithic collapse [1,2] or a merger of isolated fragments [3]. However, the recent discovery of the Sagittarius dwarf spheroidal galaxy, which is being torn apart and accreted onto the Galaxy [4], emphasizes that mergers of dwarf galaxy satellites may be an important

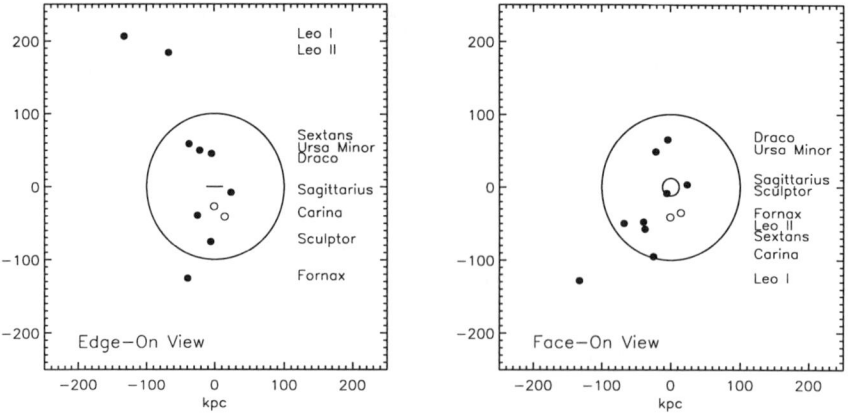

FIGURE 1. The Galaxy and its nearest dwarf galaxy neighbors. Schematically illustrated are the Galactic stellar disk with a radius of 12 kpc, and the Galactic dark matter halo with a radius of 100 kpc. The Large and Small Magellanic Clouds are shown as open symbols, and the nine dSphs are labeled and shown as solid symbols.

and ongoing process in the evolution of the Galaxy. Important clues to the physical sizes and timescales relevant to the formation of the Galactic halo will come from comparing the ages and chemical abundances of Galactic halo and dSphs stars [5,6].

There are many open questions about the evolution of dwarf galaxies:
- What are their star-formation histories?
- How have their internal chemical abundances evolved?
- How have supernovae-driven galactic winds and massive dark matter halos shaped their evolution?
- What physical mechanisms control their evolution and lead to the observed relationships between mass, luminosity, and metallicity?

The eleven dwarf galaxies that are satellites of the Milky Way provide excellent laboratories for studying the processes that regulate star formation, because we can resolve individual stars in them. New technologies make it possible for us to determine the distribution of the ages, chemical abundances, and kinematics of their stars, and accurately determine the evolution of these galaxies. As shown in Figure 1, two of the Galaxy's nearest neighbors are dwarf irregulars, the Large and Small Magellanic Clouds, and the nine others are dSphs. The star-formation histories of the Magellanic Clouds are reviewed by Hatzidimitriou in this volume, and my talk will summarize what we know about the star-formation histories of the dSphs. The spatial extent of the Galactic dark matter halo is controversial but the latest results suggest it extends to at least 100 kpc (see [7] and references therein). Therefore, most of the dSphs are really unique subsystems in the outer Galactic halo. We should keep in mind that the evolution of the inner dSphs may have been effected by their proximity to the Galaxy because of tidal torques or ram-pressure stripping.

The dSphs share a unique morphological class, but they simply may be the low-mass end of the dwarf irregular or dwarf elliptical sequence. DSphs are characterized by their: 1) small sizes with core radii of a few$\times 100$ pc, 2) low surface densities with $\Sigma_V = 1$ to 2×10^{-3} $L_\odot$ pc^{-2}, and 3) intrinsic faintness with total magnitudes ranging from $-9 \leq M_V \leq -14$ [8,9]. The dSphs do not contain HII regions which implies they are not currently forming stars. They contain little or no neutral hydrogen. The upper-limits on the mass in HI are $< 10^4$ $M_\odot$ for most dSphs [10,11]. However, I encourage observers to conduct new searches for HI, because, in some cases, the velocity range searched does not sufficiently bracket the currently-known radial velocity. In addition, we are finding evidence for long quiescent periods in dSphs, and stars as young as 500 Myrs in the Fornax dSph. Thus, there are reasons to suspect that gas in neutral or ionized form may still reside in some dSphs.

A substantial effort has been devoted recently to determining the internal kinematics and the dark matter content of dSphs (see [12–14] for recent reviews). Their central velocity dispersions range from 6 to 12 km/sec where values of 0.5 to 3 km/sec would be expected based on their luminous material alone [15]. This is generally attributed to a dark matter halo, although controversy still exists over this interpretation due to uncertainty about the assumption of dynamical equilibrium. Assuming that dSphs are in equilibrium, mass follows light, and the stellar orbits are isotropic, the estimated total masses of the dSphs range from 10^7 to a few$\times 10^8$ $M_\odot$. However, the dark matter halos are probably more extended than the luminous material so these should be viewed as a lower-limits to the total mass. The derived total mass-to-light ratios decrease from $M/L_V \approx 100$ (Draco, Ursa Minor) to $M/L_V = 5$ (Fornax) with a systematic trend of increasing M/L_V with decreasing L_V (for example, see Figure 8 of [15]). This trend is generally assumed to signify that, in lower mass dSphs, star formation is globally less efficient and a larger fraction of the available gas is lost in supernovae-driven galactic winds.

II STAR FORMATION HISTORIES OF DSPHS

Given the low total mass of a dSph, if it formed stars on a free-fall time (few$\times 10^7$ yr) then supernovae explosions would have quickly driven the interstellar medium out of the dSph leaving it to passively evolve until today. However, pioneering work by Aaronson, Mould and colleagues in the 1980s lead to the discovery of carbon stars in dSphs at levels that implied they contained a significant fraction of intermediate-aged (6 to 10 Gyr old) stars. At the same time, the existence of RR Lyraes in the dSphs proved that they also had stars with ages ≥ 10 Gyr.

We recently have made great progress in quantifying the star formation rates of dSphs, because new high-quantum efficiency CCD cameras on ground-based telescopes and the high-spatial resolution capabilities of the Hubble Space Telescope have given us the ability to obtain high-precision color-magnitude diagrams (CMDs) for the dSphs. We can accurately determine their star formation histories with a temporal resolution of < 1 Gyr by modeling the density of stars in the main-sequence turnoff region (MSTO) of the CMD, which is sensitive to the mix of ages and metallicities of the stars. To overcome the degeneracy of age and metallicity, one can independently determine the distribution of metallicities by obtaining spectra of individual stars. The Ca II triplet lines at 8500 Å are particularly useful for determining metallicities of faint red giants [16,17]. Alternatively, if the evolutionary history is relatively simple, the range of chemical abundances can be estimated from the spread in color of the red giant branch [18], because the color of a red giant is mostly sensitive to metal abundance and only slightly sensitive to age.

Existing CMDs show that each dSphs exhibits a unique and, in most cases, complex history of star formation. **Ursa Minor** and **Sculptor** have predominately old stellar populations with small ranges in age and/or metallicity, similar to Galactic globular clusters [19,20]. **Draco** and **Sextans** have predominantly metal-poor, old populations, but still exhibit small internal dispersions in age and/or metallicity [21,22,16,23]. Although the existing CMDs for Draco and Sextans are not of exceptionally high quality or depth, they show evidence for multiple generations of stars. However, the data is not adequate to quantify the amount or the ages.

High-quality CMDs for the remaining dSphs have been obtained, and they reveal surprisingly complex evolutionary histories. **Leo I** is the dSph in Figure 1 that is the farthest from the Galaxy, and it appears to have the largest fraction of young stars. Ground-based photometry of Leo I revealed a distinct population of 3 Gyr old stars [24]. Subsequent HST imaging has shown that approximately 90% of the stars have ages of ~ 3 Gyr, 10% of the stars have ages of ~ 12 Gyr, and there is little evidence for stars at intermediate ages [25].

HST imaging has also been obtained for **Leo II** [26], and the resulting CMD is shown in Figure 2. The solid line represents a composite of the red giant branch of the globular cluster M2 ([Fe/H]$= -1.6$), and a 10 Gyr old Revised Yale Isochrone with the same metallicity. The average age of Leo II stars is 9 Gyr, although the full range of ages run from 7 to 14 Gyr. Thus Leo II is mostly an intermediate-aged galaxy.

Carina has a remarkably complex star-formation rate. The CMD for Carina obtained with the Cerro Tololo 4-meter telescope [27] is shown in Figure 3a. Four distinct MSTOs are seen which imply that star formation occurred in four distinct phases. The wide gap in the MSTO region suggests a long qui-

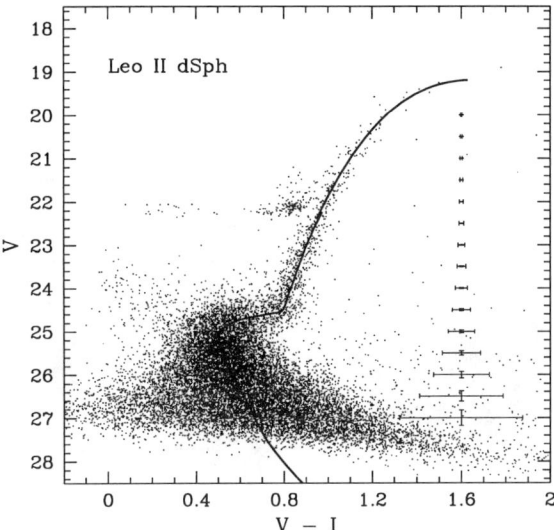

FIGURE 2. The CMD of the Leo II dSph obtained with the Hubble Space Telescope from Mighell & Rich (1996), copyright 1994, American Astronomical Society; reproduced with permission.

escent period lasting from approximately 7 to 10 Gyr ago. The blue horizontal branch and RR Lyrae instability strip represents the older stellar population (ages > 10 Gyr), and the brighter red-clump horizontal branch represents the younger populations (ages < 7 Gyr). New stellar evolutionary tracks [28] have been used to compute models CMDs to determine the star-formation and chemical-evolution histories of Carina. Isochrones for the main sequence and red giant branch for [Fe/H]= −1.84 models are shown in Figure 3b. Figure 3c shows a preliminary model CMD for a star-formation rate in which 2% of the stars formed from 1 to 1.5 Gyr ago, 28% formed from 2.5 to 3.5 Gyr ago, 50% formed 3.5 to 7 Gyr, and 20% formed form 10 to 14 Gyr ago. Figure 3d shows a Monte Carlo simulation of the model CMD with realistic photometric errors applied. With the exceptional quality of our data, we can resolve the width of the subgiant branch of even the oldest "burst" in Carina and find that it lasted ∼ 3 Gyr, which is much longer either the stellar dynamical timescale or the expected cooling timescale of a few$\times 10^8$ yr. By analyzing the density of stars in the real and model CMDs, we can easily determine the best fitting star-formation rate and metallicity distribution. Other researchers are pursuing similar studies on dwarf irregular galaxies and coming up with innovative new ways of quantifying the match between observed and model CMDs (for examples, see [30,31] and references therein).

Another striking feature of the Carina CMD is the narrow range of color

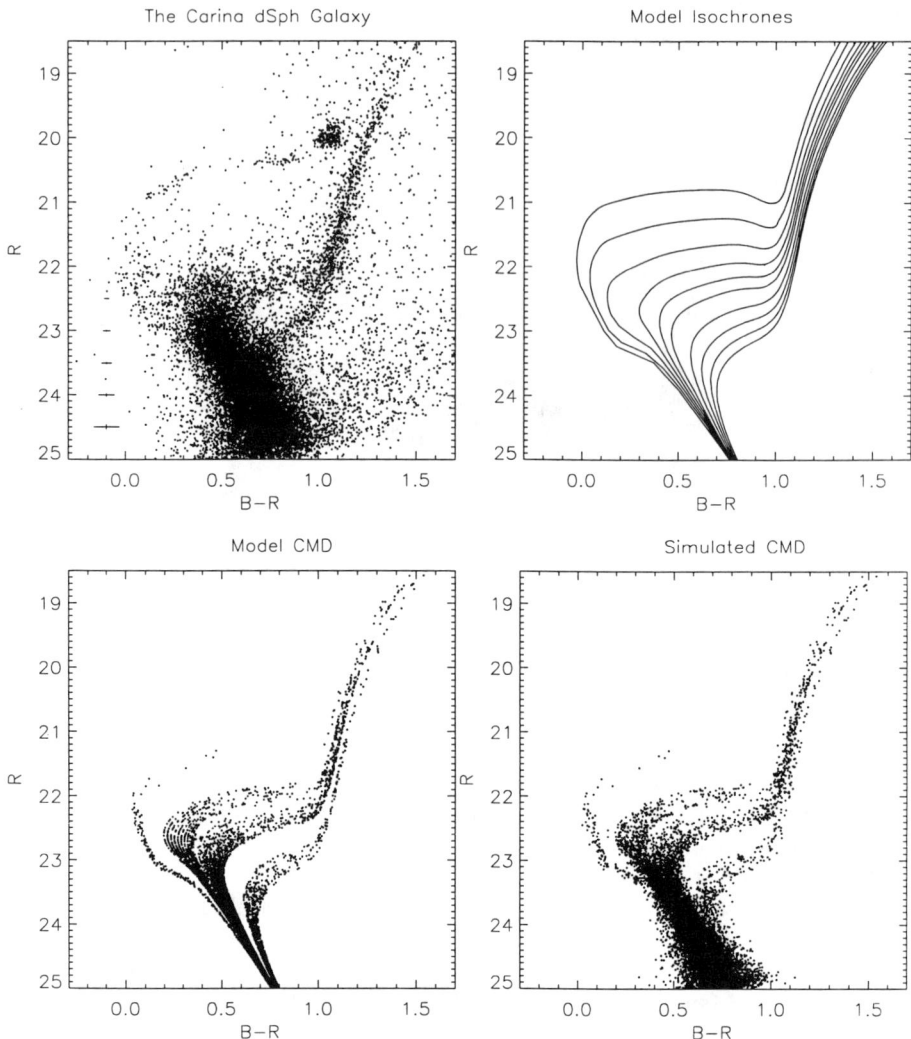

FIGURE 3. *a) upper left* – A CMD for the Carina dwarf spheroidal galaxy [27]. Typical photometric errors are plotted on the left-hand side. *b) upper right* – Theoretical isochrones made from a new set of stellar evolutionary tracks for [Fe/H]= −1.84 [28] for comparison with the Carina dSph. Ages = 1.3, 2, 3, 4, 5, 6, 8, 10, 12, 14 Gyr are shown. *c) lower left* – A model CMD created from the stellar evolutionary models with the star-formation rate described in the text. *d) lower right* – A Monte Carlo simulation of the model CMD with photometric errors applied.

on the red giant branch, which implies the metallicity spread in Carina is ≤ 0.25 dex [29]. If the first episode of star formation enriched the gas to [Fe/H]= −1.84 then the subsequent formation of 4 times more stars would have increased the metallicity to [Fe/H]=−1.1 if the total gas mass remained approximately constant. Therefore, Carina was probably loosing metals, and gas, through galactic winds as it evolved.

It is instructive to quantify what we know about Carina's evolution. If we assume that 20% of the total mass of stars ever formed have an average age of 12 Gyr and 80% have an average age of 5 Gyr, and stars formed with a Salpeter initial-mass function (IMF), then given Carina's total luminosity, $M_V = -9.3$, the masses of stars formed in each burst were approximately 7.3×10^5 $M_\odot$ and 2.9×10^6 $M_\odot$. If we define ν to be the number of supernovae per 50 $M_\odot$ of stars formed, and we assume a Salpter IMF that forms stars with masses 0.1 to 100$M_\odot$, and SNe arise from stars with masses 10 to 100 $M_\odot$, then $\nu = 0.25$. Thus the number of SNe produced in the first and second episodes of star formation were 3.7×10^3 and 1.5×10^4, respectively. The cumulative energy relased by these SNe is clearly enough to unbind the gas if it is released on a short timescale. However, the durations of star formation that we infer from the CMD of Carina tells us that star formation proceeded relatively slowly. For example, if the first episode of star formation in Carina occurred over a period of 3 Gyr then the average star-formation rate was 2.4×10^{-4} $M_\odot$/yr.

What is the critical mass that separates galaxies which will develop a galactic wind from galaxies in which SNe shells can cool, and metals be retained and incorporated into new generations of stars? Dekel & Silk [32] develop an analytical formulation of the critical velocity for dwarf galaxies with dark matter halos. Galactic winds will occur if the velocity is less that a critical velocity,

$$v_c = 123 \ f^{0.5} \ \lambda^{-0.088} \ \epsilon^{0.618} \ \nu^{0.5} \ \tau^{-0.5} \ g^{-0.25} \ n^{-0.0147} \ \text{km/sec}, \qquad (1)$$

where the parameter $f \approx 1$ characterizes the evolution of SNe shells, $\lambda \approx 1$ parameterizes the radiative cooling curve, $\epsilon \approx 1$ is the energy of a typical SNe in units of 10^{51} erg, ν is the number of SNe per 50 $M_\odot$ of stars formed, τ is the ratio of the maximum possible star-formation rate (the mass of gas over the free-fall time) to the star-formation rate, g is the ratio of the gas mass to the total mass, and n is the number density of the hydrogen gas.

Dekel & Silk assumed that $\tau \approx 1$, which would mean that dwarf galaxies efficiently convert their gas into stars on a free-fall time ($\tau_{ff} = 1.9 \times 10^7 \ g^{0.5} \ n^{-0.5}$ yr). However, we have seen from the high-precision CMDs that this is not true for dSphs – star formation proceeded on timescale much longer than the free-fall time. For Carina's first episode of star formation, we can evaluate τ, and $\tau = 2.2 \times 10^3 \ M_7 \ g^{-0.5} \ n^{0.5}$ if M_7 is the initial mass of gas in units of 10^7 $M_\odot$. Assuming that all the metals produced in the first episode of star formation were retained and lead to a metallicity of [Fe/H]= −1.84, then the initial mass of gas in Carina must have been $M_7 \approx 1.4$. Since the global efficiency of star formation is typically 10%, this quantity of gas could reasonably produce the observed mass in stars. The estimated mass of gas is similar to the total mass of $1.1 \times 10^7\,M_\odot$ inferred from the velocity dispersion of Carina [9], although this is a lower-limit to the total mass of the dark matter halo because it is derived assuming the mass follows light. (If the dark matter in dSphs is similar to the dark matter in dwarf irregulars and spirals then its distribution is probably much more extended than the luminous material, and thus the total mass is larger.) With this τ could Carina have been stable during its first episode of star formation? If the other parameters listed above are approximately unity then

$$v_c = 2.6\,\nu^{0.5}\,M_7^{-0.5}\,n^{-0.265}\ \mathrm{km/sec}, \qquad (2)$$

The velocity dispersion of Carina is 6.8 km/sec, and thus Carina may have been marginally stable against the formation of a galactic wind and therefore it only slowly lost its gas. These analytical models are suggestive, but far from comprehensive. It will be very interesting to explore numerical simulations of Carina once we quantify its star-formation rate and chemical enrichment history. In particular, the chemical abundances of dSphs will give us very useful constraints on the inflow/outflow of gas from them.

The more luminous dSphs such as **Fornax** and **Sagittarius** show wide ranges (1 dex or more) in the chemical abundances of their stars [33–37]. The existing CMDs for Sagittarius are not of very high-quality, but they suggest Sagittarius had two distinct episodes of star formation ~ 4 and ~ 10 Gyr ago [38,39]. Preliminary results from a deep imaging study of **Fornax** is shown in Figure 4 [35]. The width in color of the red giant branch shows that metallicities span the range $-2.2 \geq [\mathrm{Fe/H}] \geq -0.7$. Fornax appears to have had a more constant star-formation rate than Carina. Although the star-formation rate dropped to very low levels ~ 2 Gyr ago, stars as young as 500 Myr are seen in Fornax. Therefore, some gas may still remain in Fornax.

III THE PHYSICAL MECHANISMS THAT REGULATE STAR FORMATION IN DSPHS

The long timescales over which stars form in some dSphs tell us that a few massive stars must easily ionize and disturb the interstellar medium (ISM) of dSphs and cause star formation to be self-regulated even in these, the lowest mass, galaxies. Calculations show that massive stars will photoionize surrounding gas and cause holes to expand around the stars prior to their explosion as supernovae [40]. In this way, negative feedback can occur on the timescale upon which massive stars evolve ($10^{6.5}$ to $10^{7.7}$ yrs) and cause the global star formation rate to operate on timescales much slower than the

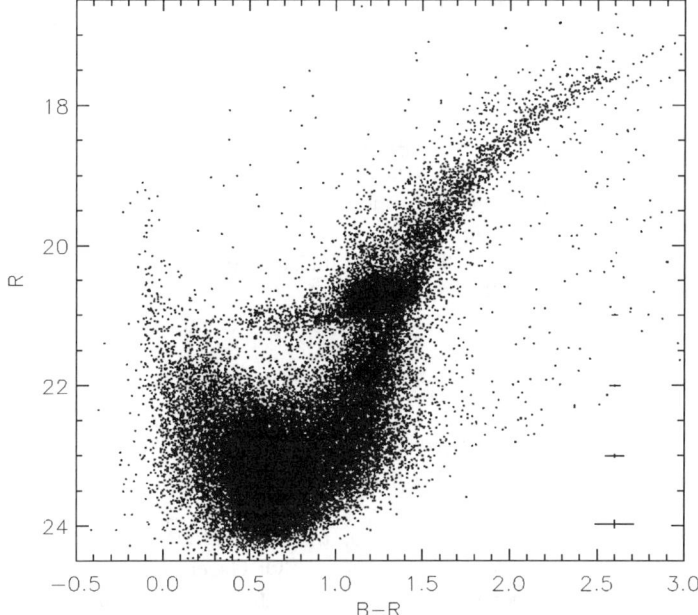

FIGURE 4. A preliminary CMD of the Fornax dwarf spheroidal galaxy [33].

dynamical or cooling timescale. In a small galaxy whose ISM has a disturbed, "swiss cheese"-type structure, the hot metal-enriched gas generated by SNe explosions might easily break out of the galaxy and be lost. However, denser clouds may be compressed and shredded, but not given enough momentum to leave the galaxy. This may explain the multiple bursts and extended star formation activity seen in dSphs, and may be supported by observations of low-mass dwarf irregulars [42,41].

However, the few Gyr quiescent periods seen in Carina, Leo I and possibly Sagittarius, pose a challenge. If gas remained in the dSphs, what kept it from collapsing on a cooling timescale and forming stars within 1 Gyr? The extragalactic UV background may play a role in keeping the gas ionized and creating long dormant periods in dSphs. The decreasing flux of the UV background with time may allow the ISM to recover after a few Gyr. Or gravitational torques as the dSphs orbit around the Galactic halo on timescale of a few Gyr might trigger additional episodes of star formation. Or maybe we will find the only plausible explanation for renewed star formation activity in dSphs is that they accrete fresh gas from clouds wandering around in the outer halo of the Galaxy. Clearly, the observed evolution of the dSphs will be a challenge to explain and lead to new insights into the processes that regulate star formation and galaxy evolution.

REFERENCES

1. Eggen, O., Lynden-Bell, D. & Sandage, A. 1962, APJ, 176, 76
2. Sandage, A. 1990, JRASC, 84, 70
3. Searle, L. & Zinn, R. 1978, ApJ, 225, 357
4. Ibata, R. A., Gilmore, G. and Irwin, M. J. 1994, Nature, 370, 194
5. Unavane, M., Wyse, R. F. G. & Gilmore, G. 1996, MNRAS, 278, 727
6. Mateo, M. 1996, in *Formation of the Galactic Halo...Inside and Out*, eds. H. Morrison & A. Sarajedini, (ASP: San Francisco), p. 434
7. Kochanek, C. 1996, ApJ, 457, 228
8. Caldwell, N., Armandroff, T. E., Seitzer, P., Da Costa, G. S. 1992, AJ, 103, 840
9. Mateo, M., Olszewski, E., Pryor, C., Welch, D. L. & Fischer, P. 1993, AJ, 105, 510
10. Knapp, G. R., Kerr, F. J. & Bowers, P. F. 1978, AJ, 83, 360
11. Mould, J. R., Bothun, G., D., Hall, P. J., Stavely-Smith, L. & Wright, A. E. 1990, APJL, 362, 55
12. Mateo, M. 1994, in *The Proceedings of the ESO/OHP Workshop on Dwarf Galaxies*, eds. G. Meylan and P. Prugniel (ESO: Garching), p. 309
13. Pryor, C. 194, in *The Proceedings of the ESO/OHP Workshop on Dwarf Galaxies*, eds. G. Meylan and P. Prugniel (ESO: Garching), p. 323
14. Gallagher, J. S. & Wyse, R. F. G. 1994, PASP, 106, 1225
15. Vogt, S. S., Mateo, M., Olszewski, E. W. & Keane, M. J. 1995, AJ, 109, 151
16. Suntzeff, N. B., Mateo, M., Terndrup, D. M., Olszewski, E. W., Geisler, D. & Weller, W. 1993, ApJ, 418, 208
17. Da Costa, G. S. & Armandroff, T. E. 1995, AJ, 109, 2533
18. Da Costa, G. S. & Armandroff, T. E. 1990, AJ, 100, 163
19. Olszewski, E. W. & Aaronson, M. 1985, AJ, 90, 2221
20. Da Costa, G. S. 1984, ApJ, 285, 483
21. Lehnert, M. D., Bell, R. A., Hesser, J. E. & Oke, J. B. 1992, ApJ, 395, 466
22. Stetson, P. B. 1996, Baltic Astronomy, in press
23. Mateo, M., Nemec, J., Irwin, M. & McMahon, R. 1991, AJ, 101, 892
24. Lee, M. G., Freedman, W., Mateo, M. & Thompson, I. 1993, 106, 1420
25. Mateo, M., Olszewski, E., Lee, M., Saha, A., Hodge, P., Keane, M., Suntzeff, N., Freedman, W. & Thompson, I. 1994, BAAS, 185, 5103
26. Mighell, K. & Rich, R. M. 1996, AJ, 111, 777
27. Smecker-Hane, T. A., Stetson, P. B., Hesser, J. E. & Vandenberg, D. A. 1997, in preparation
28. VandenBerg, D. , Swenson, F., Rogers, F., Iglesias, C. & Alexander, D. 1996, private communication
29. Smecker-Hane, T. A., Stetson, P. B., Hesser, J. E. & Lehnert, M. D. 1994, AJ, 108, 507
30. Tolstoy, E. & Saha, A. 1996, ApJ, 462, 672
31. Aparicio, A., Gallert, C., Chiosi, C. & Bertelli, G. 1996, ApJL, 469, 97
32. Dekel, A. & Silk, J. 1986, ApJ, 303, 39

33. Buonanno, R., Corsi, C. E., Fusi Pecci, F., Hardy, E. & Zinn, R. 1985, A&A, 145, 97
34. Beauchamp, D., Hardy, E., Suntzeff, N. B. & Zinn, R. 1995, AJ, 109, 1628
35. Smecker-Hane, T. A., Stetson, P. B., & Hesser, J. E. 1997, in preparation
36. Sarajedini, A. & Layden, A. C. 1995, AJ, 109, 1086
37. Whitelock, P.A., Irwin, M. & Catchpole, R.M. 1996, New Astron., 1,17
38. Mateo, M., Udalski, M., Szymański, M., Kalużny, J., Kubiak, M. & Krzemiński, W., 1995, AJ, 109, 588
39. Fahlman, G. G., Mandushev, G., Richer, H. B., Thompson, I. B. & Sivaramakrishnan, A. 1996, ApJL, 459, 65
40. Lin, D. N. C. & Murray, S. D. 1994, in *The Proceedings of the ESO/OHP Workshop on Dwarf Galaxies*, eds. G. Meylan and P. Prugniel (ESO: Garching), p. 535
41. Westpfhal, D. & Puche, D. 1994, in *The Proceedings of the ESO/OHP Workshop on Dwarf Galaxies*, eds. G. Meylan and P. Prugniel (ESO: Garching), p. 295
42. Martin, C. L. & Kennicutt, R. C., Jr. 1995, ApJ, 447, 171

The HI Supergiant Shells in the Large Magellanic Cloud

Sungeun Kim* and Lister Staveley-Smith[†]

*Mount Stromlo and Siding Spring Observatories,
Weston Creek Post Office, ACT2611, Australia
[†]Australian Telescope National Facility,
Epping 76 PO Box, NSW 2121, Australia

Abstract. We present the results of an HI aperture synthesis mosaic of the Large Magellanic Cloud (LMC), made recently with the Australian Telescope Compact Array (ATCA). The resolution of the mosaicked images is $1\rlap{.}'0$ (15 pc, using a distance to the LMC of 50 kpc). In contrast with its appearance at other wavelengths, the LMC is remarkably symmetric in HI on the largest scales, with the bulk of the HI residing in a disk of diameter $8\rlap{.}°4$ (7.3 kpc). Outer spiral structure is clearly seen, though the features appear to be caused by differential rotation, therefore transient in nature. On small to medium scales, the combined action of numerous shells and supershells dominates the structures and motions of the HI in the LMC. Some supergiant shells show good correlation between HI and Hα structures. We compare the results with a new wide-field Hα image.

INTRODUCTION

We have completed a high-resolution HI survey of the Large Magellanic Cloud (LMC) using the Australian Telescope Compact Array (ATCA). The LMC has been studied in neutral hydrogen gas with the Parkes 64m Telescope by McGee and Milton (1966) [2], Rohlfs *et al.* (1984) [3] and Luks and Rohlfs (1992) [4] at a spatial resolution of $14\rlap{.}'9$ (220 pc). The new ATCA survey gives us a much higher spatial resolution, 1' (15 pc). The HI distribution in the LMC can be compared to the distribution of Hα emission, revealed in our images taken with a camera lens piggy-backed on a small telescope at Mount Stromlo and Siding Spring Observatories. The Hα images cover the same area as the HI survey, but at a higher spatial resolution, 20″. These two surveys offer a unique opportunity to study in unprecedented detail the relationship between the ionized phase and atomic phase of the interstellar gas in an irregular galaxy.

OBSERVATIONS

Details of the observing strategy and data reduction, which involve the mosaicking together of 1344 fields, are described by Kim et al. (1997) [5]. The central observing frequency is 1.419 GHz, corresponding to a central heliocentric velocity of 297 km s^{-1}. The velocity coverage is -33 to 627 km s^{-1} and the velocity resolution is 1.6 km s^{-1}.

For the wide-field survey of Hα emission in the LMC, we used a Nikon 30.72 mm $f/5.0$ camera lens with a filter assembly piggy-backed on the 16 inch telescope at Mount Stromlo and Siding Spring Observatories. The detector was a cooled Loral 2148 × 2048 CCD. The 15 μm pixel size corresponded to $20''.63$ on the sky, and the total field size was $11°.7 \times 11°.7$. The Hα filter was centered at 6576 Å with a FWHM of 19 Å. For continuum subtraction, images were taken with a continuum filter at 6620 Å. The exposure times were 4 × 900s in each filter.

RESULTS

Figure 1 shows the HI brightness temperature map of the LMC for each position. The coordinate scale is a J2000 tangent plane centered on $\alpha = 05^h23^m$, $\delta = -69°44'$(J2000). The heliocentric velocity range in which we find HI emission is 190 – 370 km s^{-1}. On large scales, the HI distribution in the LMC is nearly axisymmetric, in contrast with Hα emission. The geometric inclination obtained from the axial ratio (a/b) of the apparent ellipse is 21°.

In the channel maps, we see spiral patterns in the southern part of the LMC. The HI arm near the optical bar appears to extend out from the "B3" stub identified in deep photographic images by de Vaucouleurs and Freeman (1972) [6]. This pattern appears in the velocity range of 215 – 258 km s^{-1}. In the velocity range of 202 – 289 km s^{-1}, the HI gas in the LMC stretches toward the HI bridge between the LMC and the SMC.

On smaller scales, the peak brightness temperature map of the LMC shows filamentary structures interspersed with HI holes and shells. We define HI supergiant shells using the following selection criteria: (1) an HI hole that is larger than 600 pc and visible in at least three channel maps, with each channel integrated over 5 km s^{-1}, and (2) both approaching and receding hemispheres (or one of hemispheres) being visible in position-velocity plots (L-V diagrams). However, Many of the holes are difficult to identify as expanding shells in the L-V diagrams due to relatively low sensitivity and confusion of interlocking shells. The preliminarily defined supergiant shells are marked by ellipses in Figure 1.

In general, we find 3 categories of geometrical correlation between HI and Hα emission. Type I has HII regions residing in the interior of an expanding HI supergiant shell. These interior HII regions might represent the denser

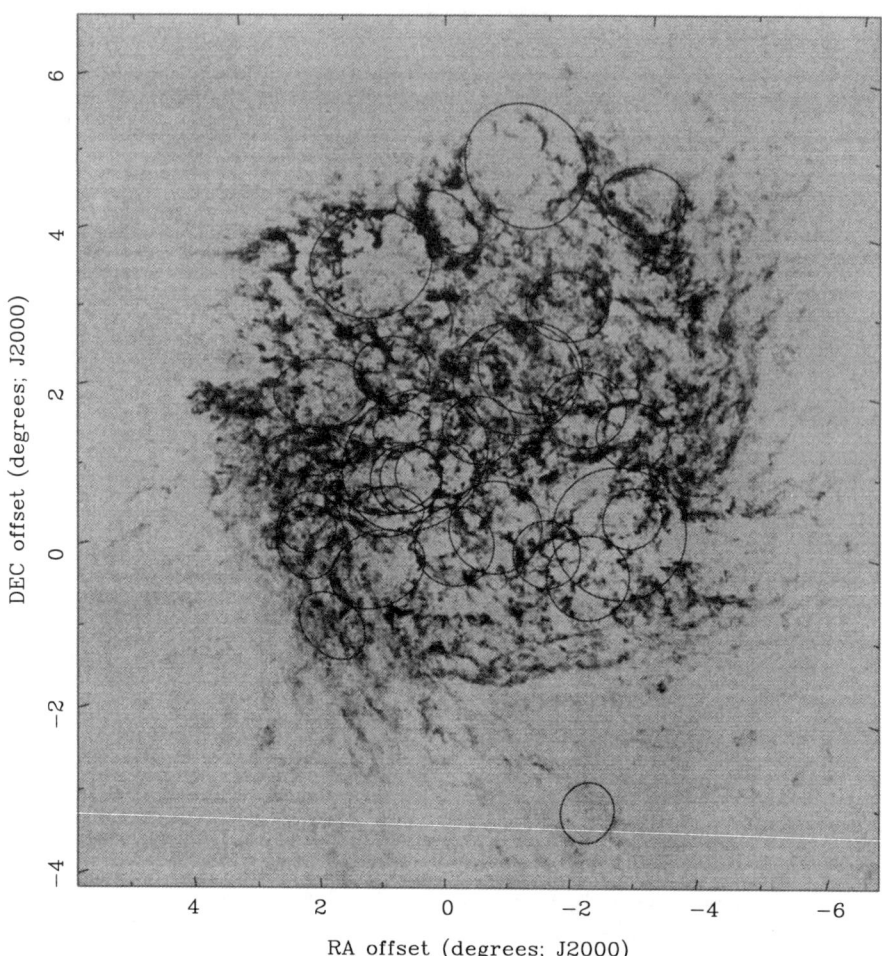

FIGURE 1. The peak 21-cm brightness temperature map (measured at each channel) from the ATCA survey of HI in the LMC. The positions of the preliminarily defined supergiant shells are marked by ellipses.

interstellar clumps that have lagged behind the expanding HI shell. Type II has an HII shell adjacent to the inner wall of an HI shell. This is expected if the ionizing source is inside a supergiant shell and the ionizing front is trapped in the shell. Type III has HII regions distributed outside an HI supergiant shell. This might indicate star formations triggered by the expanding HI shell. Assuming a constant energy input rate without radiative cooling or external pressure, the dynamic age of an expanding shell can be derived from the measured R_s and V_s ([7–9]). The derived ages of these expanding supergiant shells are distributed between 3.5 Myr and 18 Myr.

ACKNOWLEDGMENTS

We would like to thank Neil Killeen for converting CGDISP to accept non-linear coordinates and implement the overlay. We appreciate Richard Gooch and Tom Oosterloo for the support of data cube visualization. We thank Vince McCintyr for helpful suggestions for defining shells. We appreciate Peter Wood, John Dickey and Agris Kalnjas for their comments and lively discussions. We thank You-Hua Chu and Dominik Bomans for pointing out changes which improved the manuscript. Especially Sungeun Kim would like to thank Mount Stromlo Observatory for the financial support for attending this conference.

REFERENCES

1. Meaburn, J., MNRAS, **192**, 365 (1980).
2. McGee, R.X., & Milton, J.A., Aust.J.Phys., **19**, 343 (1966).
3. Rohlfs, K., Kreitschmann, J., Siegman, B.C., & Feitzinger, J.V. A&A, **137**, 343 (1984).
4. Luks, T., & Rohlfs, K. , A&A, **263**, 41 (1992).
5. Kim, S., Staveley-Smith, L., Sault, R.J., Bessell, M., Freeman, K.C., McConnell, D. and Kesteven, M. J., in Preparation (1997).
6. de Vaucouleurs, G. and Freeman, K.C. Vistas in Astronomy, **14**, 163 (1972).
7. Weaver, R., McCray, R., Castor, J., Shapiro, P., Moore, R., ApJ, **218**, 377 (1977).
8. Heiles, C., ApJ, **229**, 533 (1979).
9. Staveley-Smith, L., Sault, R.J., Hatzidimitriou, D., Kesteven, M.J. and McConnell, D., MNRAS (submitted) (1996).

Induced Star Formation and Chemical Enrichment in NGC 5253

Chip Kobulnicky & Evan Skillman

University of Minnesota

Jean-Rene Roy

Universitié Laval

J. R. Walsh & Michael R. Rosa

European Southern Observatory

Abstract. New VLA neutral hydrogen maps for the amorphous starburst galaxy NGC 5253 are presented, along with HST optical and UV spectroscopy of the central HII regions. The data appear to show an infalling HI stream on the SE side of the galaxy which may be responsible for triggering the young burst of star formation. The central HII region exhibits a 300% nitrogen overabundance compared to the surrounding gas, and we propose that recent chemical pollution from massive stars is responsible.

NEUTRAL HYDROGEN KINEMATICS

NGC 5253 is a gas-rich, nearby (D = 4.1 Mpc) amorphous star forming galaxy with a prominent minor-axis dust lane **(Figure 1a)**, which appears to harbor a very young starburst [1]. Previous coarse resolution VLA HI data [2] revealed that the neutral atomic medium appeared to rotate about the minor axis in a "tumbling" manner, while the stellar kinematic studies show little or no rotation. New VLA aperture synthesis spectroscopy at high resolution reveals instead the appearance of two distinct dynamical components.

The main HI component, corresponding morphologically to the stellar light distribution [3], shows little or no rotation, consistent with the stellar kinematics. The second HI component extending SE from the nuclear region appears to be a broad *infalling or outflowing* stream of gas **(Figure 1b)** with a velocity field that merges smoothly with the main body of the galaxy. The good

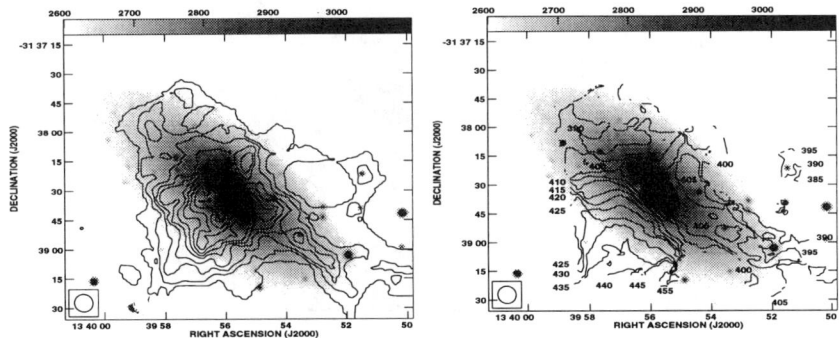

FIGURE 1. The amorphous galaxy NGC 5253. (Scale: $1'' = 20.5$ pc; $M_B = -17.6$) a) Grey: 6450Å continuum in arbitrary units logarithmic units (from [3]). Note the prominent minor axis dust lane. Contours: New VLA HI 21-cm total intensity map. Contours levels are 1, 10, 20, 30, 40, 50, 60, 70, 80, and 90 times 6×10^{19} atoms cm^{-2}. b) Grey: as in a. Contours: HI 21-cm intensity-weighted velocity field diagram. Note the lack of rotation along the major axis, and the well-ordered velocity field SE of the nucleus. We interpret this feature as an in flowing HI stream on the near side of the galaxy.

spatial correspondence of the HI with the prominent dust lane (**Figure 1a**) suggests a physical correspondence, consistent with the idea that most of the gas is on the near side of the galaxy. We prefer to interpret this velocity field as *inflow* from the *near side* of NGC 5253. This infalling HI stream, apparently decelerated as it approaches the nucleus, is likely to be responsible for triggering the young burst of star formation.

CHEMICAL ENRICHMENT FROM MASSIVE STARS

NGC 5253 contains a very young starburst making it an important laboratory for testing theories of nucleosynthesis and dispersal of heavy elements by massive stars in giant HII regions. Optical and ultraviolet spectroscopy obtained with the *Hubble Space Telescope* Faint Object Spectrograph at three locations (**Figure 2**) in the central HII complex confirms an apparent area of enhanced N abundance seen in ground-based studies [4,5].

At two positions near the central HII complex, N appears enhanced by a factor of 3 [$\log(N/O) = -0.85$] compared to a third location 50 pc away (location UV-1) where the N abundance is more typical of metal-poor galaxies [$\log(N/O) = -1.30$]. No other elemental species shows spatial abundance fluctuations, including C (**Figure 3a**) as measured from the [C III] $\lambda\lambda 1907/1909$

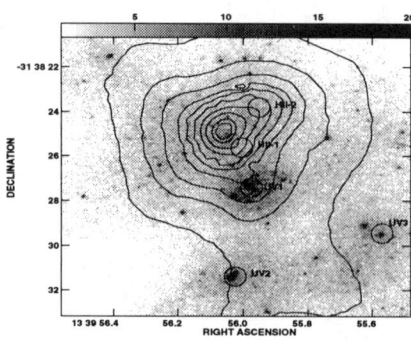

FIGURE 2. Central starburst core of NGC 5253 showing 3 bright UV star clusters. Grey: HST FOC 2200Å continuum [6]. Contours: Ground-based Hα emission. Five HST FOS aperture locations are marked. Position UV-1 is centered on the dominant star cluster, while positions HII-1 and HII-2 sample the regions where the N abundance is enhanced by a factor of three. Note that the Hα peak is centered well away from the obvious ionizing star clusters, suggesting the possibility of a heavily obscured cluster of massive stars.

Å emission lines [7]. Although the elevated N appears near a region likely to have very high extinction ($A_V \cong 8-12$) based on Brackett emission line ratios, the extinction to the optical line–emitting gas is low ($A_V \cong 0.25$) and we rule out reddening uncertainties as the cause of the apparent N enhancement. Comparison to predictions of photoionization models shows that ionization uncertainties are unlikely to be responsible for such an anomalous N/O ratio.

If the N enrichment is due to localized pollution from the winds of WR stars [8], then an accompanying He enrichment of 40% ought to be observed, but we are able to rule out He enhancements at the 2σ level at both locations showing N enhancement (**Figure 3b**). These findings require the existence a N production mechanism that is decoupled from C, O, and He production. Although the high N/He enrichment ratio we find is consistent with that expected from non–Type I planetary nebulae (PNe), the localized nature of the enrichment, large number of PNe required (500) and long timescales (10^9 yrs) required make this explanation highly unlikely. We conclude that the N enrichment must come from either He deficient WR star winds or LBV ejection events which pollute the surrounding ISM on short timescales. If this N-rich ejecta material is incorporated into self-gravitating clumps of molecular gas, and if the 10^6 yr old massive star clusters in starburst galaxies are precursors to globular clusters [10,9], then N–overabundant halo and globular cluster stars in the Galaxy [11,12] may owe their chemical peculiarity to a similar N–enrichment episode early in the history of the Milky Way.

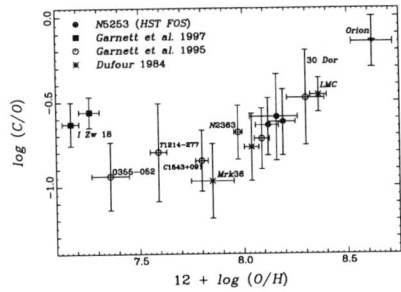
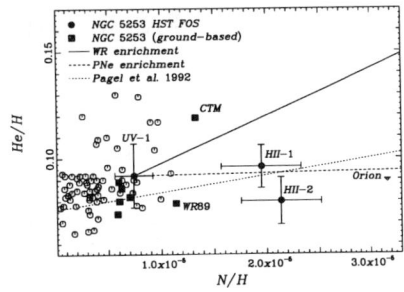

FIGURE 3. 12+log(O/H) versus C/O for metal-poor galaxies with measured C abundances. a) Carbon abundances, measured from the UV C III] 1909Å line, are consistent with each other at all locations, and are typical of other metal-poor galaxies. The lack of a C overabundance compared to systems of similar metallicity enables us to rule out planetary nebulae as the source of the central N enrichment. b) N/H versus He/H in NGC 5253 and 60 other metal-poor galaxies. Positions HII-1 and HII-2 show N abundances three times that of UV-1 and other galaxies of similar metallicity. Other elemental species show no spatial fluctuations. Lines denote the trend expected for enrichment processes dominated by WR star winds and planetary nebulae. The dotted line shows the He–N regression commonly used to derive primordial He abundances.

REFERENCES

1. Beck, S., Turner, J. L., Ho, P. T. P., Lacy, J. H., & Kelly, D. M. 1996, ApJ, 457, 610
2. Kobulnicky, H. A., & Skillman, E. D. 1995, ApJ, 454, L121
3. Martin, C. L., & Kennicutt, R. C. 1995, ApJ, 447, 171
4. Welch, G. A. 1970, ApJ, 161, 821
5. Walsh, J. R., & Roy, J-R. 1989, MNRAS, 239, 297
6. Meurer, G. R., Heckman, T. M., Leitherer, C., Kinney, A., Robert, C., & Garnett, D. R. 1995, AJ, 110, 2665
7. Garnett, D. R., Skillman, E. D., Dufour, R. J., Peimbert, M., Torres–Peimbert, S., Terlevich, E, Terlevich, R. J., & Shields, G. A. 1995a, ApJ, 443, 64
8. Pagel, B. E. J., Terlevich, R. J., & Melnick, J., 1986, PASP, 98, 1005
9. Holtzman, J., et al. 1992, AJ, 103, 691
10. Lutz, D. 1991, A&A, 245, 31
11. Kraft, R. 1994, PASP, 106, 553
12. Laird, J. B. 1985, ApJ, 289, 556

UIT Observations of the SMC

Robert H. Cornett*, Theodore P. Stecher
and the UIT Science Team

Code 681, LASP/GSFC, Greenbelt MD 20771
* Hughes STX Corp.

Abstract.
A mosaic of four UIT far-UV (FUV; 1620Å) images, which covers most of the SMC bar, is presented, with derived stellar and HII region photometry. The UV morphology of the SMC's Bar shows that recent star formation there has left striking features including: a) four concentrations of UV-bright stars spread from northeast to southwest at nearly equal ($\sim$30 arcmin=0.5 kpc) spacings; b) one concentration comprising a well-defined 8-arcmin diameter ring surrounded by a larger Hα ring, suggestive of sequential star formation.

FUV PSF photometry is obtained for 11,306 stars, and FUV photometry is obtained for 42 Hα-selected HII regions, both for the stars and for the total emission contained in the apertures defined by KH [1]. The flux-weighted average ratio of total to stellar FUV flux is 2.15; the stellar FUV luminosity function indicates that most of the excess total flux is due to scattered FUV radiation, rather than faint stars. Both stellar and total emission are well correlated with Hα fluxes measured by KH, and yield FUV/Hα flux ratios that are consistent with models of single-burst clusters with SMC metallicity, ages from 1-5 Myr, and moderate (E(B-V)=0.0-0.1 mag) internal SMC extinction.

UV observations from above the earth's atmosphere are vital for understanding Population I properties of metal-poor, "primitive" galaxies such as the Small Magellanic Cloud (SMC) [2]. Many effects of composition differences appear best, or only, in the UV. Line blanketing strongly affects UV colors; the steep SMC extinction curve ($A_{162}/E(B-V) \sim 16$), widely thought to be due to abundances in SMC dust, is "extreme" only in the UV; and FUV photometry is more effective than optical-band photometry in determining temperatures of hot stars. Here, we present initial results based on a mosaic of four, 40-arcmin diameter, 3-arcsec-resolution, far-UV images, nearly covering the SMC bar, (figure 1) obtained by the Ultraviolet Imaging Telescope (UIT) during the Astro missions. Details of UIT hardware, calibration, operations, and data reduction are in [3], and detailed results of the current study are in [4].

Figure 1 shows that the SMC's FUV emission originates mostly in hot stellar populations which, while not restricted to clusters, are significantly clumped. No diffuse FUV emission is readily apparent. The brightest features are NGC 346 and NGC 330, in Fields 4 and 3, with additional FUV concentrations centered in fields 1 and 2. The bright FUV concentrations, spaced along the Bar at ~0.5kpc intervals, have similar clustering and distribution properties to those evident in wide-field FUV images of the LMC [5].

UIT field 2, near the Bar's center, provides an intriguing instance of what appears to be sequential star formation. A ring of FUV-bright stars dominates the field, and other evidence, including HI shells [6] surrounding the stellar ring, an old supernova remnant 0050-728 [7] at its northern edge, and broad $H\alpha$ linewidths [8] at the ring's center, shows ties between the distribution of hot stars and the gas dynamics.

We have derived far-UV PSF photometry for 11,306 stars and correlated our observations with ground-based stellar photometry. Figure 2 is the $(m(162)-V),V$ color-magnitude diagram for 191 stars from [9]. Discrete symbols are observed stars, uncorrected for reddening, with spectral types. Solid lines and evolutionary tracks show paths of 10, 15, 20, and 40 $M_\odot$ SMC-

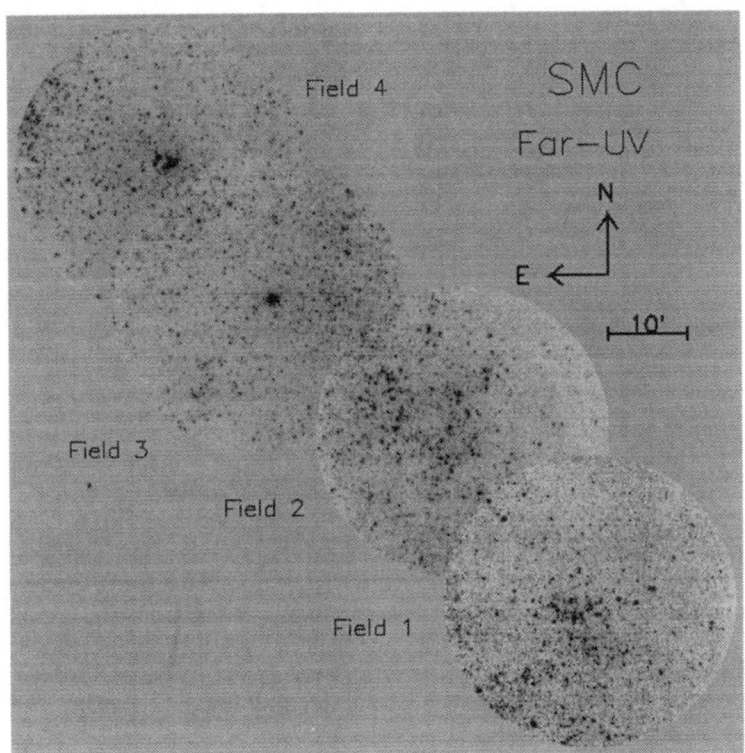

FIGURE 1. FUV Mosaic of the SMC Bar

composition models ([10], [11]) corrected for distance and foreground Galactic extinction. The reddening vector shows a typical large value for the SMC. The tracks show that these stars predominantly have masses 10–20$M_\odot$ and imply ages 10–30 Myr. The general segregation of spectral types by color implies that much of the (m(162)−V) color variation seen in this figure is due to the intrinsic colors of the stars themselves.

We have measured FUV fluxes for 42 HII regions measured at Hα by KH [1], and compared the observed FUV/Hα flux ratios with cluster models. Figure

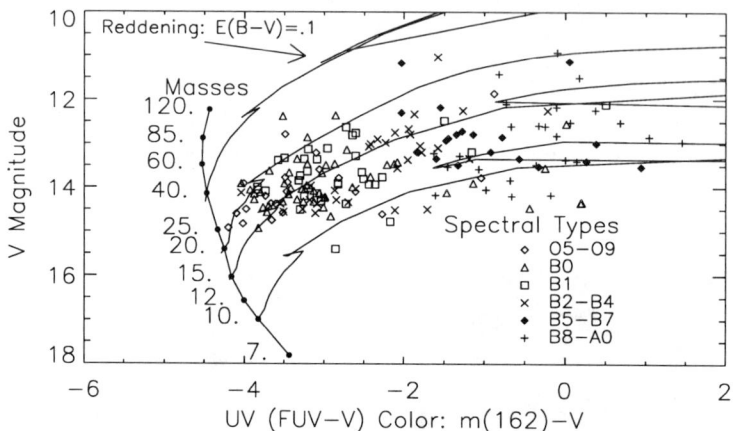

FIGURE 2. SMC (FUV-V,V) Color-Magnitude Diagram

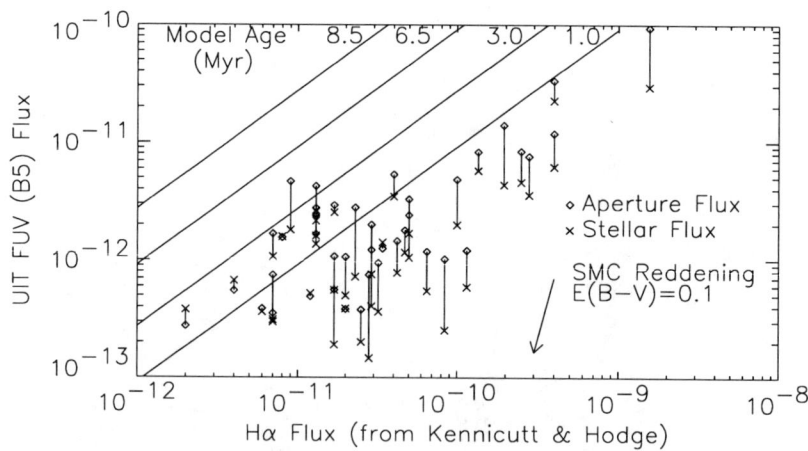

FIGURE 3. (FUV vs. Hα Flux for HII Regions

3 shows these data. Open diamonds are aperture fluxes, and crosses are the sum of stellar fluxes in the apertures, uncorrected for Galactic foreground reddening. The FUV flux is well correlated with Hα and the ratio of FUV aperture to stellar flux is relatively uniform. The ratio measures the relative amounts of "diffuse" light (which includes faint stars and scattered light) and light from bright stars. The flux-weighted average ratio of aperture to stellar flux is 2.15. Two arguments point to a dust-scattering origin for most of the excess FUV aperture flux, however. First, the total-to-stellar FUV flux ratio for the Orion nebula, with few undetected stars, is 2.5 [12]. Second, extrapolating the FUV luminosity function predicts an additional stellar flux contribution of only 22%. In spite of low dust abundance, scattering of FUV radiation is important in the SMC.

The FUV/Hα ratio is known to be a good diagnostic of HII region evolution [13]. We have modelled [14] F(162)/Hα ratio values for SMC-composition clusters with single-burst star formation. The ratio rises monotonically from 0.1 at 1 Myr to 4.3 at 10 Myr; foreground- corrected ratios for various cluster ages are plotted as solid lines in Figure 3. The observed HII regions are evidently no older than $\sim$5 Myr for any internal SMC reddening, and are likely between 1 Myr and 3 Myr in age. This result is consistent with observations of HII regions in galaxies as disparate as NGC 4449 [13] and M81 [15].

REFERENCES

1. Kennicutt, R.C. Jr. & Hodge, P.W. 1986, ApJ 306,130.
2. Cornett, R.H., Hill, J.K., Bohlin, R.C., O'Connell, R.W., Roberts, M.S., Smith, A.M., & Stecher, T.P. 1994, ApJ 425, L117.
3. Stecher et al. 1992, ApJ 395, L1.
4. Cornett et al. 1996, AJ in press.
5. Smith, A.M., Cornett, R.H., and Hill, R.S. 1987, ApJ 320, 609.
6. Staveley-Smith, L., Sault, R.J., Hatzidimitriou, D., Kesteven, M.J., & McConnell, D. 1996, MNRAS in press.
7. Mathewson, D.S., Ford, V.L., Dopita, M.A., Tuohy, I.R., Mills, B.Y., & Turtle, A.J. 1984, ApJ Supp. 55, 189.
8. Okumura, K., 1993, Ph.D. dissertation, University of Paris.
9. Azzopardi, M. & Vigneau, J. 1982, A&A Supp. 50, 291.
10. Kurucz, R.L. 1992, in *The Stellar Populations of Galaxies,* ed. B. Barbuy & A. Renzini (Dordrecht: Kluwer Academic), 225.
11. Charbonnel, C., Meynet, G., Maeder, A., Schaller, G., & Schaerer, D. 1993, A&A Supp. 101, 415.
12. Bohlin, R.C., Hill, J.K., Stecher, T.P., and Witt, A.N. 1982, ApJ 255, 87.
13. Hill, R.S., Home, A.T., Smith, A.M., Bruhweiler, F.C., Cheng, K.-P., Hintzen, P.M., and Oliverson, R.J. 1994, ApJ 430,568.
14. Landsman, W.B. (personal communication).
15. Hill, J.K. et al. 1995, ApJ 438, 182.

UIT Astro-2 Observations of NGC 4449

Robert S. Hill*, Michael N. Fanelli*, Denise A. Smith[†], Ralph C. Bohlin[‡], Susan G. Neff[||], Robert W. O'Connell[¶], Morton S. Roberts[§], Andrew M. Smith[||], Theodore P. Stecher**

*Hughes STX Corp., Code 681, NASA/GSFC, Greenbelt, MD 20771
[†]NRC, Code 681, NASA/GSFC, Greenbelt, MD 20771
[‡]Space Telescope Science Insitute, 3700 San Martin Drive, Baltimore, MD 21218
[||]Code 681, NASA/GSFC, Greenbelt, MD 20771
[¶]University of Virginia, P. O. Box 3818, Charlottesville, VA 22903
[§]National Radio Astronomy Observatory, Edgemont Road, Charlottesville, VA 22903
**Code 680, NASA/GSFC, Greenbelt, MD 20771

Abstract. The bright Magellanic irregular galaxy NGC 4449 was observed by the Ultraviolet Imaging Telescope (UIT) during the Astro-2 Spacelab mission in March, 1995. Far ultraviolet (FUV) images at a spatial resolution of ~ 3" show bright star-forming knots that are consistent with the general optical morphology of the galaxy and are often coincident with bright H II regions. Comparison of FUV with Hα shows that in a few regions, sequential star formation may have occurred over the last few Myr. The bright star forming complexes in NGC 4449 are superposed on a smooth, diffuse FUV background that may be associated with the Hα "froth."

OBSERVATIONS

NGC 4449 is a bright Magellanic irregular galaxy at a distance of 5.4 Mpc [1]. Its absolute FUV magnitude is estimated at -20.2, as compared to -18.8 for the Large Magellanic Cloud (LMC), -19.3 for the Local Group Sc spiral M33, and -21.8 for the SBc spiral M83 [2]. In other words, NGC 4449 is comparable to conspicuous, star-forming spirals in FUV luminosity, and it is brighter than the LMC by a factor of ~ 3.5. This signature of active, ongoing star formation is confirmed by Hα imagery [3], which shows a large number of compact, bright HII regions against a background of loops, filaments, and diffuse emission. The southern bar is a region of particular complexity, with many bright near-point sources and with bright filamentary structure seen in line emission.

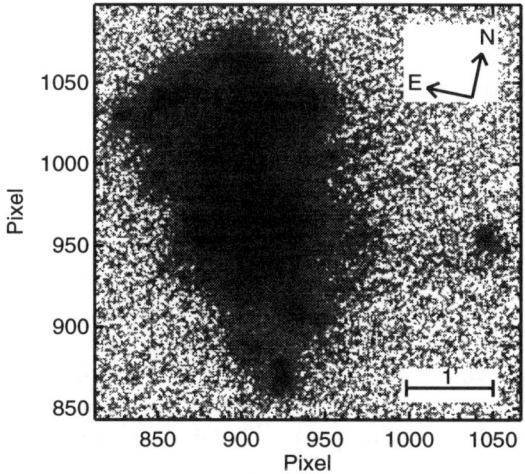

FIGURE 1. UIT FUV Image of NGC 4449 (B1 Filter).

NGC 4449 was observed by the Ultraviolet Imaging Telescope (UIT [4]) during the Astro-2 Space Shuttle mission in March, 1994. UIT observations on the Astro-2 mission were in one of the FUV bands, usually either B1 ($\lambda = 152$ nm, $\Delta\lambda = 35$ nm) or B5 ($\lambda = 162$ nm, $\Delta\lambda = 22$ nm). The two bandpasses are similar; B5 has a longer short-wavelength cutoff to exclude dayglow.

Observations of NGC 4449 were obtained in both B1 and B5. In this paper, we discuss the 900 s B1 image, which is the deepest one (Figure 1), and the 500 s B5 image, which has the best spatial resolution.

NGC 4449 was also observed from the ground using the Goddard Fabry-Perot Imager (GFPI). These observations, made by K.-P. Cheng, R. Oliversen, and P. M. N. Hintzen, are discussed in an earlier paper based on FUV sounding rocket imagery [2]. The GFPI data consist of emission-line images in Hα, Hβ, and nearby, narrow continuum bands.

As a result of analyzing the UIT data, the calibration of observations reported in the sounding rocket paper [2] has been revised. In all bands, the reported fluxes are a factor of ~ 2 too high. Specifically, any single Balmer line flux should be multiplied by 0.6, and any single rocket UV flux should be multiplied by ~ 0.5. The off-band visual continuum measurements are affected in the same way as the line fluxes. In the UV case, the correction is less well defined because of a somewhat uncertain bad-pixel factor, varying from source to source, which is included in the rocket data, but is unnecessary for the UIT data. In summary, the effect on any ratio of two fluxes is $\sim 20\%$. Therefore, the modeled source ages reported in the rocket paper are not significantly affected.

TABLE 1. Photometry of Selected Froth Regions

No.	$F_{H\alpha}$ (erg cm^{-2}s^{-1})	f_{FUV} (erg cm^{-2}s^{-1}Å^{-1})	Equivalent Age (Myr)
1	2.33×10^{-13}	2.28×10^{-14}	4.0
2	1.74×10^{-13}	1.81×10^{-14}	4.2
3	1.00×10^{-13}	1.11×10^{-14}	4.4
4	1.05×10^{-13}	1.33×10^{-14}	4.8

ANALYSIS

The term *froth* has been coined to describe a certain morphology of Hα-emitting material outside of typical H II regions. Froth lacks conspicuous embedded star clusters, and it is distinguished from simple diffuse emission by a complex structure of bright filaments. The froth is thought to be the product of a mixed ionization mechanism, with energy supplied both by ionizing stars and by shocks [3].

A substantial contribution from photoionization is likely. A plausibility argument for this position can be made with a simple application of the FUV image data. We examine the ratio between Hα surface brightness in any given local area and the coincident FUV surface brightness. If the ratio of Hα to FUV is consistent with that expected from an ionizing cluster, then photoionization is a plausible mechanism. Apertures $\sim$ 20" in radius are superposed on 4 regions of diffuse, filamentary Hα emission and extended UV emission. The Hα and FUV fluxes for these regions are given in Table 1, together with the implied ages [2], without extinction corrections or background flux subtraction. The implied FUV surface brightnesses are equivalent to 1 or 2 unreddened early B stars per arcsec^2.

In the case of compact, bright knots of Hα and FUV emission, the age computed from the ratio of the two fluxes has real meaning as the possible age of a coeval cluster of stars. Ages are computed from Hα to FUV (B5) flux ratios in circular apertures. Sources are grouped by proximity and a background surface brightness within NGC 4449 is subtracted from each group. Extinction is computed using the Balmer decrement, assuming that $A_{FUV}/E(B-V) = 4.5$, which is similar to the Orion Nebula extinction curve [5]. Figure 2 shows a map of the compact source ages. Three regions showing a spatial progression of ages are circled. These patterns may indicate sequential star formation of the type discussed by Elmegreen & Lada [6,7]. In this scenario, the typical spatial separation between stellar generations appears to be $\sim$ 100 − 300 pc, a factor of $\sim$ 5 larger than in Elmegreen & Lada's Galactic examples.

REFERENCES

1. Bomans, D. J., Chu, Y.-H., & Hopp, U. 1997, AJ, in press

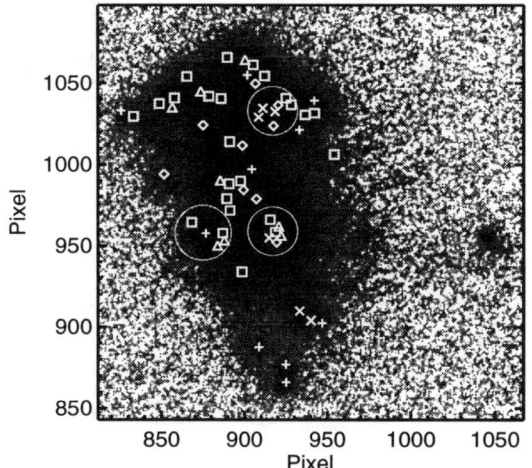

FIGURE 2. Compact Hα + FUV Sources Plotted on B1 Image. Symbols indicate ages in Myr, as follows: triangles, < 1; squares, 1 − 3; diamonds, 3 − 5; plusses, 5 − 7; crosses, > 7.

2. Hill, R. S., Home, A. T., Smith A. M., Bruhweiler, F. C., Cheng, K.-P., Hintzen, P. M. N., & Oliversen, R. J. 1994, ApJ, 430, 568
3. Hunter, D. A. & Gallagher, J. S. 1900, ApJ, 362, 480
4. Stecher, T. P. et al. 1992, ApJ, 395, L1
5. Bohlin, R. C. & Savage, B. D. 1981, ApJ, 249, 109
6. Elmegreen, B. G. & Lada, C. J. 1977, ApJ, 214, 725
7. Home, A. T., private communication

LUMINOSITIES AND STAR FORMATION RATES OF GALAXIES OBSERVED WITH THE ULTRAVIOLET IMAGING TELESCOPE:
A Comparison of Far-UV, Hα, and Far-IR Diagnostics

Michael N. Fanelli *[†], Theodore P. Stecher *,
and the UIT Science Team

* *Laboratory for Astronomy and Solar Physics*
NASA-Goddard Space Flight Center
Greenbelt, Maryland 20771
[†] *Hughes STX Corp, Lanham, MD 20706*

Abstract. During the UIT/Astro Spacelab missions, the *Ultraviolet Imaging Telescope* obtained spatially resolved far-UV ($\lambda\lambda$ 1500 Å) imagery of $\sim$ 35 galaxies exhibiting recent massive star formation. The sample includes disk systems, irregular, dwarf, and blue compact galaxies. The objects span an observed FUV luminosity range from -17 to -22 magnitudes. We estimate global star formation rates by comparing the observed FUV fluxes to the predictions of stellar population models, and compare the FUV-derived astration rates to those derived from Hα and far-IR photometry.

MOTIVATION

Many aspects of galaxy evolution are driven by the spatial distribution and formation history of high-mass (M > 3 M$_\odot$) stellar populations. Massive OB stars emit most of the radiation in the vacuum ultraviolet ($\lambda < 2500$ Å) while cooler, solar-type stars emit minimal radiation at these wavelengths. Therefore UV imagery provides a snapshot of the recently formed stellar populations in galaxies. Prior to 1990, only limited samples of UV photometry were available for galaxies, consisting mostly of integrated fluxes [1–3] and modest resolution images derived from sounding rockets, e.g., [4].

THE DATA

During the Astro/UIT Spacelab missions in December, 1990 and March, 1995 the *Ultraviolet Imaging Telescope* [5] obtained deep, high spatial resolution ($\sim 3''$, FUV ($\lambda\lambda$ 1500 Å) imagery of ~ 35 galaxies exhibiting recent massive star formation. The images cover the full angular extent of each system, most of which have diameters exceeding $5'$. The spatial resolution of the FUV images is an improvement of ≈ 5–20 over previously available data. These data permit determination of both global FUV properties with improved photometric precision, and detailed investigation of galaxian morphology at intermediate (spiral arms, nuclear rings) [6–9], and small (star-forming complexes) [10] scales.

RESULTS

We derive global star formation rates by comparing the observed FUV fluxes to the predictions of stellar population models. For this interation, a simple model was chosen: a power law IMF with slope $= -1.35$, solar abundances, and mass range, $1 < M/M_\odot < 100$. Assuming continuous star formation, the observed flux can be compared to the model FUV luminosity to derive a star formation rate. Astration rates derived from FUV data can be compared to those derived from Hα and far-IR fluxes to explore the utility of star formation rates estimated from FUV data, and the star formation history of these systems (Table 1).

TABLE 1. Star Formation Rates ($M_\odot$ yr^{-1})

Rate	NGC 3310	NGC 4214	NGC 4038/9
$\dot{M}$(FUV)	1.3	0.20	1.6
$\dot{M}$(Hα)	3.7	0.39	6.6
$\dot{M}$(FIR)	8.6	0.50	14.4

- The objects span an <u>observed</u> (uncorrected for internal extinction) FUV luminosity range from -17 to -22 magnitudes (Figure 1).
- The global (FUV–V) colors span a range from -1.3 to $+3.6$, a much larger range than that found using optical bandpasses alone.
- Late Hubble types are bluer in these colors, as expected.
- Massive star formation rates of $0.03 \lesssim \text{SFR} \lesssim 4$ $M_\odot$ yr^{-1} are found based on the <u>observed</u> FUV luminosity.
- For the Sm/Im galaxy NGC 4214 the ratio of FIR/FUV star formation rates is ~ 2.5 indicating that the FUV emission directly traces a significant fraction of the recently formed high-mass stars.
- For the bluest systems, the ratio of FUV/Hα astration rates is found to be comparable.

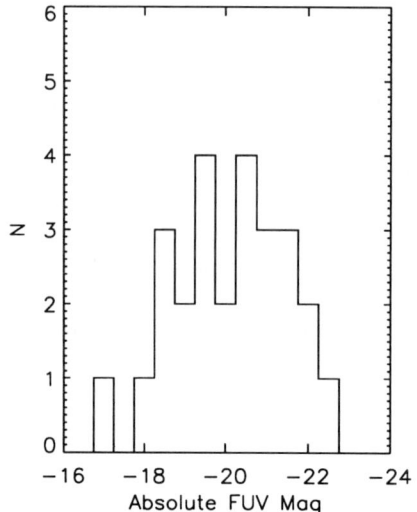 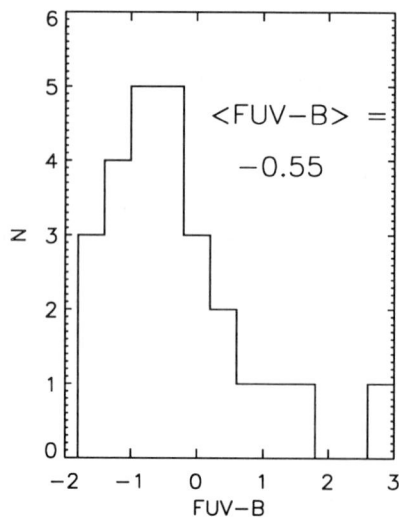

FIGURE 1. The distribution of absolute FUV magnitudes (left) and FUV to optical band colors (right) for the sample of disk galaxies observed by UIT.

- In some dusty, FIR-luminous systems, substantial FUV light is observed, e.g., the merging system NGC 4038/39 (the "Antennae") [10]. Although the FIR / FUV astration rate ratio is ~ 10, the detection of extensive FUV emission indicates that massive star formation can be **directly** probed in these systems, despite the presence of significant extinction.

THE UIT GALAXY ATLAS

We [11] are constructing an Atlas which combines the FUV imagery obtained by the *Ultraviolet Imaging Telescope* and associated optical imagery obtained at ground-based telescopes. Our primary goal is to provide a morphological Atlas of Galaxies extending from far-UV ($\lambda\lambda \sim 1500$ Å) to near-IR ($\lambda\lambda \sim 2.2\mu$) wavelengths. Comparison of the UV images with images at visible/NIR wavelengths will provide critical information on the intensity, spatial pattern, and temporal evolution of the massive stellar populations; the distribution of dust along the arms and central bars of spiral galaxies, and the relationship between recent star formation and the interstellar medium.

REFERENCES

1. Code, A.D., & Welch, G.A. 1982, ApJ, 256, 1
2. Donas, J., Daharveng, J.M., Lage, M., Milliard, B., & Huguein D. 1987, A&A 180, 12
3. Buat, V. 1992, A&A 264, 444
4. Hill, J.K., Bohlin, R.C., & Stecher, T.P. 1984, ApJ 277, 542
5. Stecher, T. P. et al. 1992, ApJ 395, L1
6. Waller, W.H., Stecher, T.P., & the UIT Science Team 1997, this volume
7. O'Connell, R. W., & Marcum, P. 1996, in HST and the High Redshift Universe, eds. N.R. Tanvir, A. Aragon-Salamanca, & J.V. Wall.
8. Smith, D.A., et al. 1996, ApJ 473, L1
9. Fanelli, M. N., et al. 1997, ApJ, in press
10. Neff, S.G., & Stecher, T. P., & the UIT Science Team 1997, this volume
11. Marcum, P., et al. 1997, ApJS, in preparation

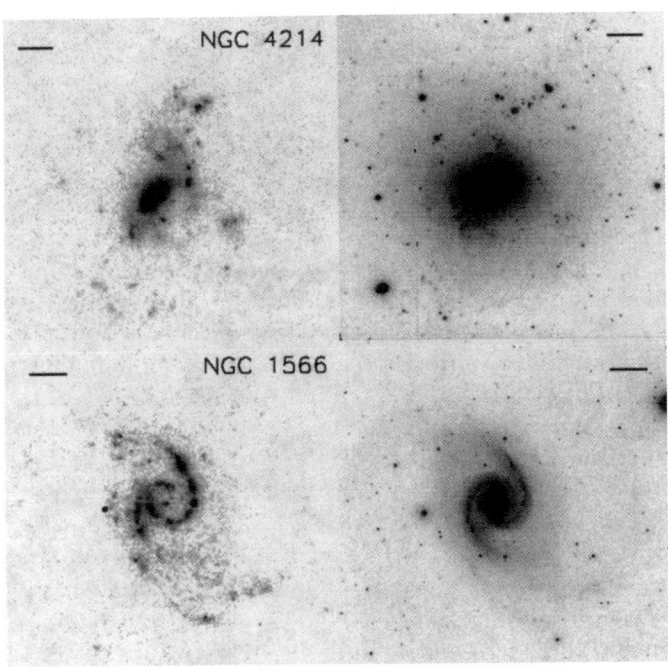

FIGURE 2. A mosaic of far-UV and I-band imagery of two disk galaxies observed by the *Ultraviolet Imaging Telescope*. FUV ($\lambda\lambda \sim 1500$ Å) images from UIT are displayed on the left, ground-based I-band images on the right. All images are displayed in a north up, east left orientation. The horizontal bar indicates an angular scale of $60''$, corresponding to a physical scale of 1.2 kpc for NGC 4214, and 5.1 kpc for NGC 1566.

The Impact of Star Formation on the Interstellar Medium[1]

Crystal L. Martin*[†]

*Space Telescope Science Institute
Baltimore, MD 21286
[†]Hubble Fellow

Abstract. The feedback from star formation on the interstellar medium plays an important role in the formation of galaxies, the regulation of their star formation rate, and the enrichment of the intergalactic medium. The reaction of the interstellar gas to recent star formation can be probed using emission-line radiation from the warm, ionized phase of the interstellar medium. I describe results from new measurements of the physical and kinematic properties of the interstellar gas in 15 nearby, star-forming dwarf galaxies using this technique.

INTRODUCTION

Massive stars shape conditions in the interstellar medium (ISM) through their ionizing radiation, stellar winds, and supernova explosions. Although the mechanical power of the latter is not expected to exceed a few percent of the starburst's bolometric luminosity, as much as 30% of L_{Bol} is emitted as ionizing radiation for up to a few Myrs following the star formation episode. Since the density and temperature structure of the surrounding interstellar gas will affect the propagation of this radiation, the spatial scale of the radiative impact may be coupled to the structure created by the gas dynamical feedback. Hence, it is desirable to study the physical and dynamical properties of the warm ionized medium (WIM), or diffuse ionized gas (DIG) as it is frequently called, in parallel.

My observations have focused on 15 star-forming dwarf galaxies – environments where the feedback is expected to be particularly strong. The luminosity and concentration of the star clusters in most of these galaxies are larger than any star forming region in the Milky Way even though the galaxies are

---

[1] Observations reported here were obtained at the Multiple Mirror Telescope – a joint facility of the University of Arizona and the Smithsonian Institution – and Kitt Peak National Observatory, NOAO, operated by AURA, Inc, under contract with the NSF.

much less luminous than our galaxy. Such clusters may power larger superbubbles. Bubbles that form in dwarfs are also expected to live longer than they might in more massive, metal-rich galaxies, so it may be easier for them to outgrow their host galaxies. The mass-loss from the ensuing galactic outflow could be catastrophic in the weak gravitational field of a low mass dwarf ([6]; [1]; [2]). Here I describe new results concerning the excitation of the DIG and comment on their relation to the gas dynamics in dwarf galaxies.

EXCITATION OF THE DIFFUSE IONIZED GAS

Deep, longslit spectra obtained at the Multiple Mirror Telescope were used to study diagnostic emission-line ratios as a function of position within each galaxy. Narrow-band images obtained at the Steward Observatory 2.3-m telescope were used to strategically select the slit positions.

Spectral Hardness

The intensity ratio of the helium and hydrogen recombination lines from the DIG place an interesting constraint on the hardness of the ionizing spectrum. Figure 1 shows how the nebular He I $\lambda 5876$ / Hα intensity ratio rises as the ionizing spectrum hardens and then saturates at a constant value as hydrogen atoms begin to compete for helium ionizing photons (e.g. [12]). For typical He/H abundance ratios, the break occurs near an effective stellar temperature of 40,000 K. For harder ionizing spectra, the He$^+$ ionization fraction, is approximately equal to the H$^+$ ionization fraction along the sightline through the nebula.

Martin & Kennicutt ([7]) have measured a mean reddening-corrected intensity ratio He I $\lambda 5876$ / H$\alpha \approx 0.041$ across three high surface brightness Magellanic irregular galaxies. Within each galaxy, the ratio is independent of spatial location over a region where the Hα surface brightness decreases by a factor of 100. Referring to Figure 1, we conclude that stars of mass ≥ 40 M$_\odot$ must contribute to the Lyman continuum radiation ionizing the DIG.

Since spectral features from very hot Wolf-Rayet stars are present in our longslit spectra of these galaxies, it is not surprising that $\chi(\text{He})/\chi(\text{H}) \approx 1$ near the giant HII regions, the sites of recent aggregate star formation. The interesting observation is that the relative ionization fraction remains this high a kiloparsec away in the DIG. Hence, the giant HII regions and DIG could share a common source of ionizing photons. As illustrated in Figure 1, the radiation ionizing the DIG in the Milky Way is much softer ([13]; [3]).

Dilution of the Radiation Field

In contrast to the behavior of the He I $\lambda 5876$ / Hα intensity ratio, the overall emission-line spectrum changes significantly across the dwarfs. Previous work had shown that the DIG in Magellanic irregulars was characterized by strong forbidden lines of low-ionization species relative to the HII regions ([4]; [5]). My spectra show that this spectral change is gradual and that the boundary between HII regions and DIG is not well-defined. Details can be found in [8], where I model the spatial changes in the relative density of ionizing photons to gas that will reproduce the spectral gradient. The estimated gradients in ionization parameter are similar to that expected from the geometrical dilution of radiation from a central source. Hence, the leakage of ionizing photons from the HII regions is probably the dominant method of exciting the DIG a kiloparsec away. With the new spatial information, local ionization of the DIG by individual, isolated stars is akward ([7]).

In many of the dwarfs, photoionization by an increasingly dilute stellar radiation field is insufficient to explain the spectrum of the lowest surface brightness emission. I find that the nature of the discrepancy depends in part on the oxygen abundance of the galaxy's HII regions, but in all cases lies in the direction of a small contribution from gas excited by shocks with speeds ranging from 60 to 100 km s^{-1}. The typical signatures of these shocks are relatively high [OIII]$\lambda 5007$ / Hβ from regions with very high [SII]$\lambda\lambda 6717,31$ / Hα and an [OI]$\lambda 6300$ / Hα ratio higher than that in the HII regions. The spectral differences between pure dilution of a stellar radiation field and mixing with

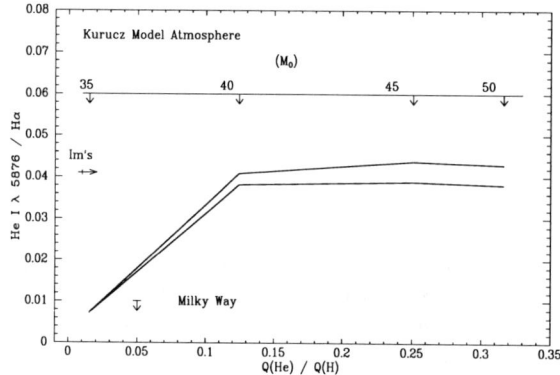

FIGURE 1. Predicted He I $\lambda 5876$ / Hα intensity ratio vs the hardness of the ionizing continuum for the abundance range of the dwarfs. The expression Q(He)/Q(H) denotes the number of photons with energy greater than 24.6 eV to the number with energy greater than 13.6 eV; the same scale is re-expressed in terms of the equivalent stellar mass at the top. Data are from [13] and [8]

shock-excited gas are larger and easier to distinguish in the higher abundance galaxies in my sample.

GALACTIC-SCALE GAS DYNAMICS

The gas kinematics were measured from longslit, echelle spectra obtained at Kitt Peak National Observatory ([11]; [9]; [10]). Atmospheric seeing limits the spatial resolution of the kinematic measurements to $1 - 2''$. To keep the physical resolution comparable to the scale of the largest bubbles formed by individual stars,$\lesssim$ 50 pc; only galaxies closer than 10 Mpc were observed. Almost all of the dwarfs surveyed contain one or more supershells; several contain kiloparsec-scale expanding shells. The expansion velocities are similar to the range of shock speeds inferred from the emission-line spectra. The fate of these superbubbles will be discussed further in a forthcoming paper.

CONCLUDING REMARKS

An intriguing picture is emerging. Many star-forming dwarfs have large cavities supported by the overpressure of very hot, low density gas. Perhaps the topology of these structures allows 10% – 50% of the ionizing photons to escape the HII regions and power the DIG. Many questions remain, however. Why do conditions in the DIG in dwarfs, for example, differ from those in more luminous galaxies like the Milky Way? Addressing issues such as the porosity of the ISM, the role of dust in the transport of ionizing photons, and the ionizing continuum emerging from hot stars may lead to an understanding of how star formation changes the ISM in a wider range of environments.

REFERENCES

1. Dekel, A., & Silk, J. 1986, ApJ, 303, 39.
2. DeYoung, D. S., & Heckman, T. M. 1994, ApJ, 431, 598.
3. Heiles, C., Koo, B.-C., Levenson, N. A., & Reach, W. T. 1996, ApJ, 462, 326
4. Hunter, D. A., & Gallagher, J. S. III 1990, ApJ, 362, 480.
5. Hunter, D. A., & Gallagher, J. S. III 1996, (preprint).
6. Larson, R. B. 1974, MNRAS, 169, 229.
7. Martin, C. L, Kennicutt, R. C. 1997a, ApJ, (submitted).
8. Martin, C. L. 1997b, ApJ.
9. Martin, C. L. 1996a, ApJ, 465, 680.
10. Martin, C. L. 1996, Ph. D. thesis, University of Arizona.
11. Martin, C. L., & Kennicutt, 1995, ApJ, 447, 171.
12. Osterbrock, D. E. 1989, Astrophysics of Gaseous Nebulae and Active Galactic Nuclei (University Science Books: Mill Valley, CA).
13. Reynolds, R. J. & Tufte, S. L. 1995, ApJ, 439, L17.

Dust, gravitational lensing and star formation

Andrew W. Blain

Cavendish Laboratory, Madingley Rd., Cambridge, CB3 0HE, UK

Abstract. Distant ($z \gtrsim 0.6$) dusty star-forming galaxies and their gravitationally lensed images are relatively much brighter in the mm/sub-mm waveband as compared with the optical/near-infrared waveband. Hence the strongly-magnified lensed images of distant galaxies at the Einstein radius of a lensing cluster are expected to be very bright in the mm/sub-mm waveband, and so could introduce significant source confusion noise into observations of the Sunyaev–Zel'dovich (SZ) effect on scales similar to the Einstein radius. The properties of these lensed images, and hence limits to the form of evolution of distant star-forming galaxies, could be inferred from a careful analysis of the confusion noise expected and a set of data on the mm/sub-mm–wave SZ effect.

The flux density–redshift relations of distant star-forming galaxies are predicted to be flat in the mm/sub-mm waveband, due to the large negative K-corrections that arise when the peak of a dust emission spectrum in the rest-frame far-infrared waveband is redshifted towards the mm/sub-mm waveband [1,2]. Hence very steep source counts are expected in the mm/sub-mm waveband for a range of different schemes of galaxy formation [1], and so a significant fraction of faint galaxies are predicted to be gravitationally lensed by intervening galaxies [3]. The fraction of gravitationally-lensed galaxies that could be detected in a deep survey in the mm/sub-mm waveband could therefore be at least two orders of magnitude larger than the fraction detected in a survey in the radio or near-infrared/optical waveband. At some faint flux densities in the mm/sub-mm waveband the surface density of lensed galaxies is predicted to exceed about 10% of the total count, and this fraction can even exceed 50% in some hierarchical models of galaxy formation [4].

Clusters of galaxies are efficient gravitational lenses [5,6], and because distant galaxies are intrinsically bright in the mm/sub-mm waveband, the flux densities of their brightest lensed images in this waveband are predicted to be about an order of magnitude larger than the flux densities of galaxies within the cluster [7]. The appearance and distribution of these lensed images on arcsecond scales are discussed elsewhere [7]. A large number of these images will be detected and resolved using the next generation of large mm/sub-mm–wave interferometer arrays [8], but existing telescopes do not provide the necessary combination of resolution, sensitivity and field of view. However, clusters are currently being observed on arcminute scales in the mm waveband [9,10], in order to detect the Sunyaev–Zel'dovich (SZ) effect [11,12], and a careful analysis of the results could be used to investigate the properties of the lensed images indirectly. The SZ effect describes the modifications made to the spectrum and intensity of the cosmic microwave background radiation (CMBR) by the inverse Compton scattering of CMBR photons in hot gas, such as that bound to clusters of galaxies. The thermal SZ effect, which is due to scattering in gas moving with the Hubble flow, produces a characteristic increment and decrement to the CMBR intensity at frequencies above and below 217 GHz respectively, and is predicted to be most intense at a frequency of about 350 GHz in the sub-mm waveband. The smaller kinematic SZ effect, a Doppler shift in the CMBR intensity, is also present if the scattering takes place in gas which has a peculiar velocity.

Existing observations of the mm-wave SZ effect [9,10] are subject to source

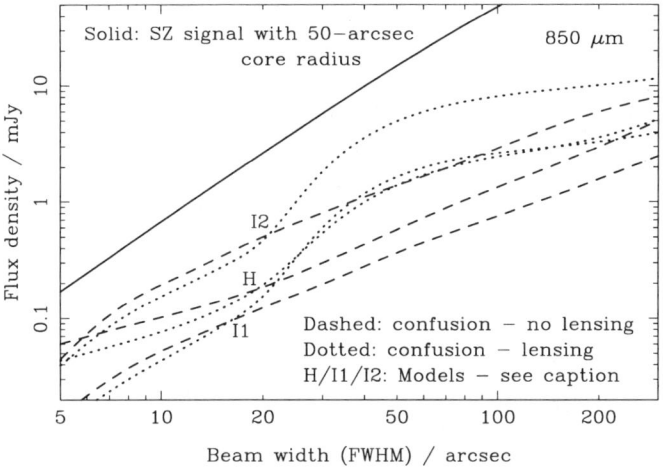

FIGURE 1. The dependence of both the predicted signal of the SZ effect at 850 μm and the expected 1-σ source confusion noise on the width of the observing beam in three different models of galaxy evolution (Blain, in prep.), with and without the effects of gravitational lensing, which modifies the confusion noise significantly. In model I1 the population of *IRAS* galaxies is assumed to undergo pure luminosity evolution of the form $(1+z)^{1.2}$ for $0 \leq z < 5$. In I2 the evolution function takes the form $(1+z)^3$ for $0 \leq z \leq 2$, and is fixed at 27 for $2 < z < 5$. H is a hierarchical clustering model in which galaxies become more numerous and less luminous as redshift increases.

confusion because a different sample of galaxies contribute to the flux density that is detected in each observing beam. This confusion has been estimated elsewhere [13], but without including the effects of gravitational lensing. In the core of a lensing cluster the confusion noise is expected to increase because the strongly-magnified lensed arc images at the Einstein radius, which are brighter and are separated more widely on the sky than unlensed galaxies, fall within the observing beam; however, the total flux density expected within the beam is not increased by lensing. Estimates of the source confusion that is expected in the field of a rich cluster similar to Abell 2218 [5], with and without the effects of lensing, are compared in Figure 1 over a range of observing beam scales and in three different models of galaxy evolution. The estimates were obtained by convolving a series of simulated populations of background galaxies and their lensed images with a Gaussian beam, and finding the variance of the resulting distribution of flux densities (Blain, in prep.). The SZ effect is observed using a differencing method, in which one beam is centred on the core of the cluster and the other is offset. Including the effects of lensing increases the variance of the signal in the on-source beam, and hence the source confusion, on all but the smallest scales. The largest increase, by a factor of about 4, occurs in a 40-arcsec beam (FWHM). On this scale the beam is matched to the Einstein radius of the cluster and the influence of the bright lensed arcs is most pronounced. Estimates of the 1-σ

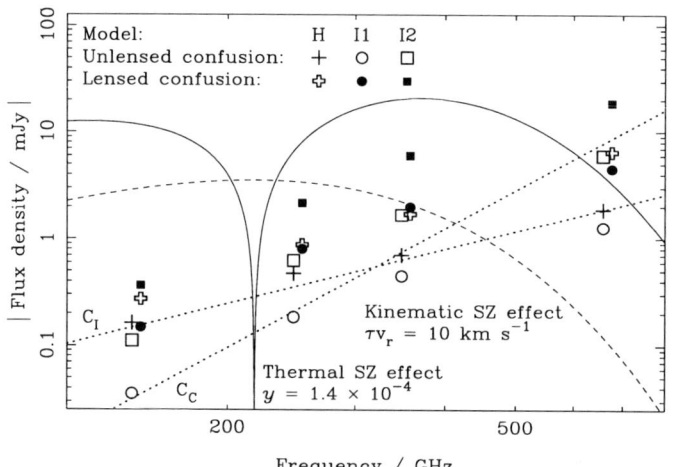

FIGURE 2. The 1-σ source confusion noise predicted for an observation of a cluster similar to Abell 2218 in a 1-arcmin beam with and without the effects of gravitational lensing, in three models of the population of distant galaxies. The predicted SZ signal [12] and estimates of confusion from earlier work are also shown. Estimates of confusion noise [13] due to galaxies and to Galactic cirrus are shown by C_I and C_C respectively.

confusion noise expected in a 1-arcmin beam both in four mm/sub-mm–wave observing bands – 150 GHz, 250 GHz, 850 μm and 450 μm – and in the three galaxy formation models from Figure 1 are compared with the expected flux density due to the SZ effects in Figure 2. The confusion noise is increased significantly by lensing in all four observing bands, and the different predictions for models I1, I2 and H demonstrate that an estimate of the confusion noise in a set of sensitive measurements of the SZ effect in the sub-mm waveband could be used to discriminate between different populations of distant galaxies. In model I2 the source confusion is expected to be so large that high-resolution observations using a large interferometer array [8] would be required in order to subtract bright lensed images and produce an accurate measurement of the SZ effect; for a recent investigation of the effects of gravitational lensing on source subtraction in cm-wave observations of the SZ effect see [14].

Summary

1. The lensed arc images of distant dusty galaxies in the mm/sub-mm waveband should be much brighter than galaxies within the lensing cluster.

2. Gravitational lensing is predicted to increase the source confusion noise due to distant galaxies in observations of the SZ effect in the mm/sub-mm waveband. An estimate of the source confusion could be used to investigate the evolution of dusty star-forming galaxies at large redshifts. Observations with large interferometer arrays may be required in order to subtract confusing sources and measure the SZ effect accurately.

REFERENCES

1. Blain, A. W., & Longair, M. S. 1993, MNRAS, 264, 509
2. Blain, A. W., & Longair, M. S. 1996, MNRAS, 279, 847
3. Peacock, J. A. 1982, MNRAS, 199, 987
4. Blain, A. W. 1997, MNRAS, in press
5. Kneib, J.-P., et al. 1996, ApJ, 471, in press
6. Colley, W. N., Tyson, J. A., & Turner, E. L. 1996, ApJ, 461, L83
7. Blain, A. W. MNRAS, submitted
8. Downes, D. 1994, in Astronomy with millimetre and submillimetre wave interferometry, ed. Wamsteker W. et al. (Dordrecht: Kluwer), 133
9. Wilbanks, T. M., et al. 1994, ApJ, 427, L75
10. Andreani, P., et al. 1996, ApJ, 459, L49
11. Sunyaev, R. A., & Zel'dovich, Ya. B. 1980, ARA&A, 18, 537
12. Rephaeli, Y. 1995, ARA&A, 33, 541
13. Fischer, M. L., & Lange, A. E. 1993, ApJ, 419, 433
14. Loeb, A., & Refregier, A. ApJ, submitted.

Structural Parameters of Hubble Deep Field Galaxies

Marianne Takamiya[1]

The University of Chicago
Department of Astronomy & Astrophysics
5640 South Ellis Ave
Chicago, IL 60637

Abstract. In galaxies, the relation between star formation rate (SFR) and morphological type is controversial [1–4]. See also contribution of Hameed & Devereux in these proceedings. To attempt to understand the physical mechanisms that control the morphology of galaxies (thru measurements of structural parameters) and to provide an objective and quantitative method of measuring the flocculency in galaxies, I have explored a novel technique which bears a simple physical interpretation: Star forming regions within a galaxy have a clumpy distribution and can be measured from a galaxy image as the high-spatial frequency (*hsf*) power [5,6]. This power must be proportional to the number of HII regions within the galaxy. In this contribution, I present a quantitative measurement of the *hsf*-power of 30 distant field galaxies from the Hubble Deep Field (HDF) [7] in the U_{F300W}, B_{F450W}, V_{F606W}, and I_{F814W} bands. Galaxies were taken from the redshift catalog of Cowie [8] and have typically z $\sim$ 0.5 (0.1 < z < 1.4). On average, galaxies with bluer B_{F450W}-I_{F814W} color tend to have higher *hsf*-powers. The *hsf*-power as a function of redshift is presented.

INTRODUCTION

Cosmological tests require a deep understanding of the physical properties of galaxies. Galaxies are good candidates for cosmological studies, however their evolution with cosmological time produce changes in the properties of galaxies (luminosity, color, content of gas, stars, and dust, chemical composition, etc.) that dominate over cosmological effects [9]. While the evolution of galaxies can be approached in several statistical ways, the image analysis of galaxies with HST/WFPC2 opens a new opportunity to explore in detail the structure of distant galaxies.

---

[1] Support for this work was provided by the National Science Foundation through grant numbers GF-1004-95 and GF-1003-96 from the Association of Universities for Research in Astronomy, Inc., under NSF Cooperative Agreement No. AST-8947990.

A simple technique to study the structure of galaxies based on their SFR is considered. Galaxies in stages of high SFR show an enhanced flux originated from extreme Population I stars distributed in a clumpy fashion. Quiescent galaxies on the other hand show a smoother light distribution. In this technique, galaxy images are filtered using a sliding square window with side length $w \simeq 4 + \frac{1}{2} \times R_g$. R_g is the metric radius of the galaxy, which is defined as $R_g = 4 \times R_\eta$. R_η is a fiducial Petrosian metric radius defined at $\eta = \frac{1}{3}$ where the enclosed average surface brightness is 3 times larger than the surface brightness intensity at that radius [10]. It appears that within a galaxy the typical size of HII regions depends on the luminosity of the galaxy [11]. Assuming that the galaxy luminosity is proportional to R_g, w is argued to be a linear function of R_g.

Both R_g and the core radius R_c ($R_c = \frac{1}{2} \times R_\eta$) define the annulus used as the aperture within which the flux of the galaxy is measured. The core flux is excluded from the measurement of the hsf-power since in the central region of the galaxy we cannot differentiate the contribution of the halo from that of the Population I stars. In nearby galaxies the hsf-power normalized to the total flux in the original image (hsf_n) seems to be well correlated with the Hα equivalent width and henceforth, with the current SFR [1,6].

PRELIMINARY RESULTS

I have analysed the structure of 30 HDF galaxies with $20 < I < 24$ and $0.1 < z < 1.4$ and present the results in Figures 1 and 2.

The sample of HDF galaxies span a large range in Hubble types [12] (Fig. 1a). On average the color B_{F450W}-I_{F814W} is related to the Hubble type.

The color B_{F450W}-I_{F814W} increases with redshift (Fig. 1b). This may be explained because at increasingly fainter fluxes, redshifts are preferentially measured from emission lines resulting in a proportionally larger amount of blue galaxies.

In Figure 2, I have plotted the hsf_n-power measured with the V_{F606W} filter as a function of different quantities. The uncertainties in the hsf-power determinations are $\sim 10\%$. The size of the filtering window w scaled by the redshift assuming $H_\circ = 75$ km s^{-1} and $q_\circ = 0.5$ is plotted against hsf_n-power in Figure 2a. No clear correlation exists between the hsf_n-power and R_g. (Note that w scales with R_g.) As seen in Figure 2b, the hsf_n-power shows a good trend with B_{F450W}-I_{F814W} color in the sense that bluer galaxies have on average larger hsf_n-power, as expected. The hsf_n-power dependence on Hubble morphological type is shown in Figure 2c. The trend of hsf_n-power with B_{F450W}-I_{F814W} color along with previous findings of nearby galaxies provide good evidence to incorporate this quantitative and objectively measured parameter (hsf_n-power) in future classifications of galaxies.

Finally, in Figure 2d I present the hsf_n-power as a function of redshift.

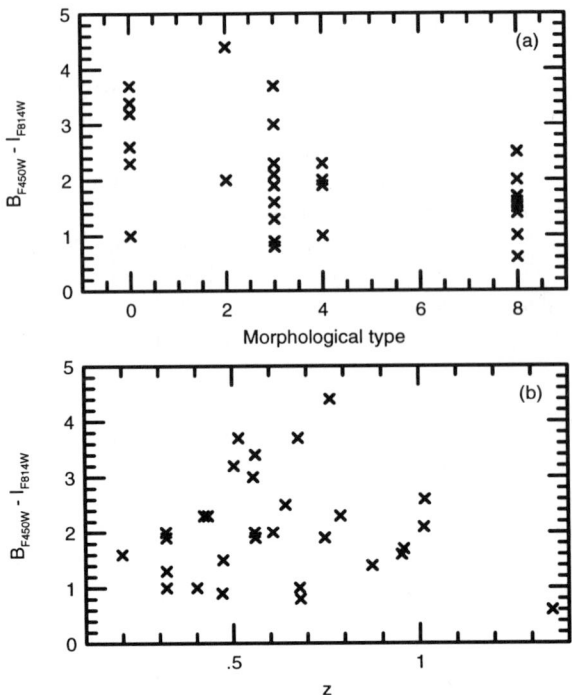

FIGURE 1. B_{F450W}-I_{F814W} color versus redshift (a) and morphological type (b).

REFERENCES

1. Kennicutt, R. C., Tamblyn, P., and Congdon, C. W. 1994, ApJ, 435, 22
2. Devereux, N. A. & Young, J. S. 1991, ApJ, 371, 515
3. Driver, S. P., et al. 1995, ApJ, L449
4. Odewahn, S. C., Windhorst, R. A., Driver, S. P., and Keel, W. C. 1996, 472, L13
5. Isserstedt, J. and Schindler, R. 1986, AA, 167, 11
6. Takamiya, M., 1995, BAAS, 187, 5101
7. Williams, R., et al., 1996, AJ, 112, 1335
8. Cowie, L., 1996, http://www.ifa.hawaii.edu/~cowie/hdf.html
9. Koo, D. C., & Kron, R. G. 1992, ARAA, 30, 613
10. Kron, R.G., 1995, in *The Deep Universe*, ed. Binggeli, B. and Buser, R., (Germany: Springer-Verlag)
11. Elmegreen, D.M., et al., 1994, ApJ, 425, 57
12. van den Bergh, S., et al., 1996, AJ, 112, 359

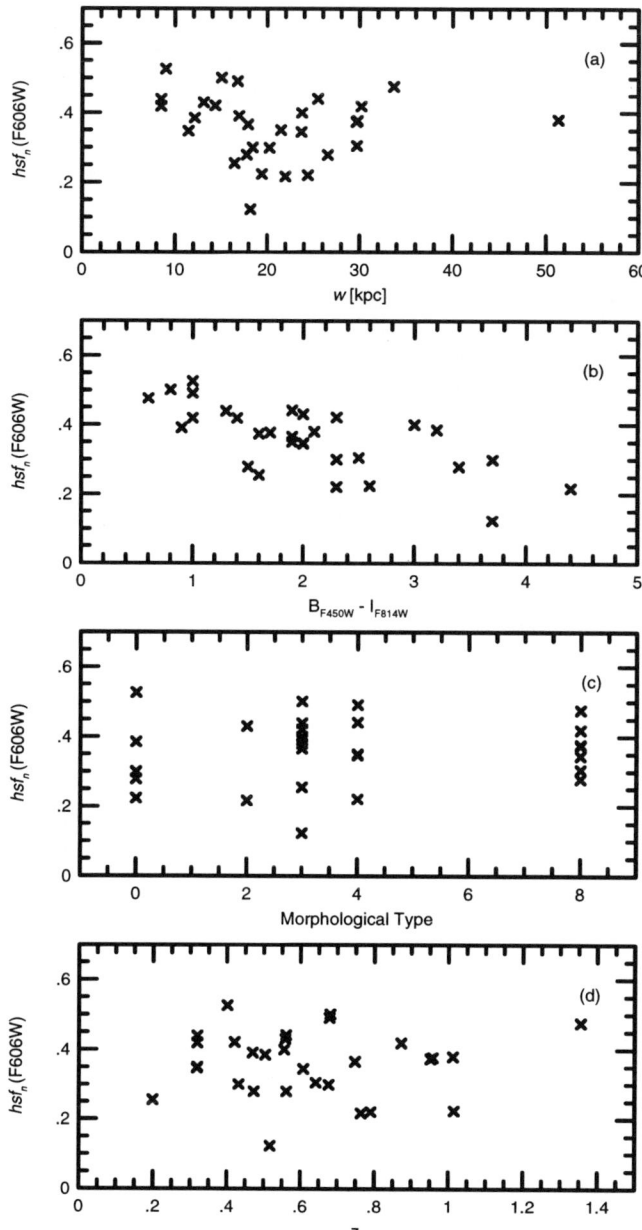

FIGURE 2. Sample of 30 HDF galaxies ($H_\circ = 75$ km s^{-1}, $q_\circ = 0.5$). a) Window size (w) versus hsf_n-power, b) B_{F450W}-I_{F814W} versus hsf_n-power, c) morphological type versus hsf_n-power, d) redshift z versus hsf_n-power.

Summary

Summary: Star Formation Near and Far

Richard B. Larson

Yale Astronomy Department, Box 208101, New Haven, CT 06520-8101

In her historical introduction to this conference, Virginia Trimble reminded us that the study of star formation is a subject with recent origins, and that just fifty years ago it was still considered very speculative to suggest that stars are presently forming from interstellar matter in the Milky Way. In my concluding remarks, I shall try to summarize with a few brief and personal comments where I think we have come in the past fifty years in understanding star formation, as illustrated by the many interesting contributions that we have heard at this meeting. The meeting included a remarkable range of topics and scales, from planets to the universe at large redshifts, and it will not be possible for me to summarize everything of importance that was covered; it will be a challenge even to mention all of the main topics that were discussed.

Let me begin by updating, from a personal perspective, Virginia's historical introduction. The way was prepared for modern studies of star formation by the advances in our understanding of stellar evolution that took place in the 1950s, which made it clear that the Milky Way contains stars and star clusters of all ages and that one cannot avoid the conclusion that stars are presently forming. Dark clouds, now known to be molecular, were discussed during this period as possible stellar birth sites, and the T Tauri stars were identified with some confidence as newly formed pre-main-sequence stars. By the 1960s, when I came on the scene as a graduate student at Caltech, the time was ripe for an attack on the problem of protostellar collapse using the techniques of numerical hydrodynamics that had been developed earlier for nuclear weapons work. I was guided in this direction by my thesis advisor Guido Münch (who, coincidentally, was at the same time also Virginia Trimble's thesis advisor), and I calculated the collapse of a spherical protostar, that being what was then feasible. I initially regarded this calculation as an academic exercise, since theoretical discussions of star formation had already strongly emphasized the importance of magnetic fields and rotation, as well as the possibility of continuing fragmentation. At about the time when I was completing this work, however, another Caltech student, Eric Becklin, discovered an infrared source in Orion which he suggested might be a protostar, and since his observation agreed qualitatively with what I had predicted, we were both very encouraged to believe that protostars like those calculated might actually exist. Theory and observation subsequently made great strides, and the study of star formation was soon no longer purely a matter for speculation.

Almost thirty years have passed since then, and we now possess a great wealth of information on star formation, as we have seen at this conference. The development that has impressed me most is the recent avalanche of high-quality data produced by many modern instruments, among which the Hubble Space Telescope has been especially notable. Many speakers showed spectacular and beautiful photographs of regions of star formation that are unprecedented in the amount of detail they reveal, and I believe that it is no exaggeration to say that observations like these open up a completely new window on the universe. Thanks to them, the science of astronomy, which previously had to be content with fuzzy images and data of generally poorer quality than those of other sciences, can now boast of data whose quality and detail begin to rival those of such Earth-bound sciences as geology, meteorology, and perhaps even biology. As a theorist, I am humbled by the realization that the observations have now far surpassed the limited ability of theory to explain what is being observed, and I believe that this forces a change in the way in which we try to approach an understanding of things. The hope of predicting everything from some kind of "fundamental theory" seems to me a mirage, and I think that instead we now have to focus on the phenomenology and work toward trying to understand it bit by bit. One wouldn't think of trying to predict all of the complex phenomena of geology, for example, from any fundamental theory, and I am sure that star formation, in all its aspects, is no less complicated a subject than the formation of geological landforms.

I have tried to organize my remarks around some of the main questions that one might ask about star formation. Some of these questions were introduced at the opening of the conference by Silk. The first obvious question is "Where do stars form?", and the answer that has been known for many years is that stars form in molecular clouds, or more specifically, in the dense gravitationally bound cloud cores that were discussed here by Myers. Immediately, however, we are confronted with the fact that these clouds are extremely complex in their structure and dynamics, so that we cannot yet provide a good answer to the next obvious question, "How do star-forming clouds evolve?". Molecular clouds appear to be dominated kinematically by a chaotic state of motion often called "turbulence", even though magnetic fields are believed to play an important role and the observed motions may be partly wavelike. The observed turbulent motions contribute importantly to supporting clouds against gravity, at least on the larger scales, and the talk by Ostriker and the posters by Vázquez-Semadeni and collaborators showed that important progress is being made in understanding, with the help of numerical simulations, how MHD turbulence can help to support molecular clouds and account for some of their properties. I think, however, that it has been a myth that molecular clouds are long-lived structures that are maintained close to equilibrium for many dynamical times; such long lifetimes are not supported by the available direct evidence, based on the ages of associated young stars and clusters, which suggests cloud lifetimes that are not much longer than the dynamical time. Most molecular clouds and condensations may therefore be fairly transient structures, like the structures obtained in the numerical simulations which often are more like transient fluctuations than coherent, long-lived "objects".

How is the collapse of star-forming clumps initiated? A widely accepted view has been that self-gravitating clumps gradually separate from a magnetic field by slow ambipolar diffusion until a configuration approximating a singular isothermal sphere is formed, which then begins to collapse from the inside out. However, given the highly turbulent state and the apparently short lifetimes of molecular clouds, it is not clear that there is sufficient time for such a slow ambipolar diffusion process to operate, and it seems likely to me that most star-forming clumps are produced by some more dynamical mechanism like turbulent compression; self-gravitating clumps are indeed found to be produced in this way in the numerical simulations mentioned above. Local dissipation of the turbulent motions that provide part of the support against gravity may then play an important role in initiating the collapse of these clumps, as suggested in the poster by Goodman et al.

How does the collapse subsequently proceed? New results presented at this meeting suggest that inside-out collapse may not occur, even in the standard ambipolar diffusion picture, since the posters by Basu and by Safier et al. both show that in even this case a singular isothermal sphere is never closely approached; instead, magnetic support becomes unimportant at a relatively early stage and a rapid collapse begins from the outside in. Myers presented evidence that in at least one well-studied cloud core, collapse has already begun on a large scale before any central object has formed, contrary to the prediction of the inside-out collapse model. Thus, it appears that revision may be needed to the popular but idealized model in which a self-similar inside-out collapse results in a constant rate of accretion onto a central object; more realistic calculations now indicate that the accretion rate must start out high and then decrease with time. An important implication of this result is that the collapse process cannot really be scale-free, as is often assumed, but instead it must necessarily depend on initial and boundary conditions.

How does rotation affect the collapse? In the simplest case of uniform collapse, rotation should lead to the formation of a disk, whereas in the more realistic case of a centrally condensed collapsing clump, one would expect the formation of a central object surrounded by a disk. Such protostellar disks are indeed now known to be very common in regions of star formation, and it is becoming possible to study their properties systematically, as we heard from Mundy and Sargent. Both inflow and rotation are often observed, and there is evidence for evolution from an early infall-dominated stage to a later rotation-dominated stage. The masses of most of the observed disks are however estimated to be rather small, less than 0.01 solar masses, and it is not yet clear how many of them might form planetary systems like our own. As was reviewed by Hartigan, there is much evidence that material continues to be accreted from these protostellar disks onto their central stars. Theoretical efforts to understand the evolution of such disks have been dominated by the concept of an "accretion disk" and have sought to find transport mechanisms capable of driving an accretion flow. The physics of the various possible transport processes remains poorly understood, however, and we heard from Stone that one of the original ideas, namely that fluid turbulence might drive an accretion flow, does not work because such turbulence tends to transport angular momentum inward rather than outward. The possibility remains that MHD turbulence may serve better to drive an accretion flow, but more work is needed to demonstrate this.

What about the possibility of continuing fragmentation? In principle, one could imagine that a collapsing clump might fragment hierarchically into smaller and smaller objects until fragmentation is stopped by the increase of opacity at a mass less than 0.01 solar masses. However, this clearly is not what happens in most cases, and theory and observation now seem to be converging on an understanding that fragmentation is strongly limited in centrally condensed collapsing clumps and typically produces only two objects, i.e. a binary system. Two stars is the minimum number that could take up any significant fraction of the initial angular momentum, so it is perhaps not surprising that, as Mathieu reminded us, the basic unit of star formation is typically a binary system. The presence of a binary companion clearly prevents the existence of a disk with a size comparable to the binary separation, but it apparently does not prevent the existence of disks with much smaller or much larger sizes. Multiple systems are also commonly observed, especially in the densest star-forming environments, and Whitworth presented numerical simulations illustrating how multiple fragmentation might occur in a realistically complex situation involving protostellar interactions in a star-forming cloud.

Most stars also form in larger groupings or clusters of some kind, so it is important also to understand how star formation might occur in clusters. We heard a whirlwind of interesting theoretical ideas on this subject from Lin and Clarke, but it would be an impossible task to try to summarize them all here. Clearly, much further work and testing will be needed to establish which of these ideas are most useful in helping us to understand the observations. There is beginning to be a rich phenomenology on the subject of cluster formation, and since the processes involved are obviously very complex, being closely related to the evolution of massive molecular cloud cores, I think that theoretical work will have to be guided increasingly by the phenomenology, focusing again on trying to build up an understanding of it bit by bit.

What are the predicted properties of newly formed stars? The prediction that protostars should first be observable as infrared sources, and the good agreement that has been achieved between predicted and observed infrared spectra, have been regarded as triumphs of the theory, although it is not yet clear whether the infrared spectra can be used to diagnose the detailed structure of protostellar envelopes and disks. Theoretical models also predict continuing infall of residual gas onto the central star or disk, but at this meeting we heard more about another phenomenon that was not predicted at all theoretically, namely the jets that all newly formed stars seem to produce at an early stage of their evolution; Mundy and Hester showed HST pictures of the detailed structure of some of these jets. Echoing the words of a famous physicist when confronted with a new particle that did not fit the then-established scheme of things, we might ask "Who ordered jets?" Jets and other outflows are in fact much more conspicuous properties of newly formed stars than is any evidence for continuing gas infall, and for some years they seemed to contradict the theoretical models in this respect. Now it is clear that infall, outflow, and rotation are all present simultaneously in many cases, and that continuing infall effects can indeed be seen when one looks hard enough. Even though we still do not understand in any detail the origin of the jets, we can no longer doubt, after seeing the striking HST picture of the HH30 jet emerging from the center of a protostellar disk, that jets originate from the innermost parts of circumstellar disks close to the central star, and that they are collimated along the rotation axis from the outset. It seems almost certain that the origin and collimation of these jets somehow involve magnetic fields twisted by the rotation of the disk, and that the basic energy source is provided by accretion from the disk onto the central star. The fact that the observed jets are not uniform but contain strings of bright knots, suggesting that they have been emitted in spurts, suggests that the accretion process may itself be sporadic, perhaps related to the FU Orionis flareups that are apparently experienced recurrently by most young stars.

How do newly formed stars influence their birth clouds? Much attention has been devoted to this subject, and some of the effects of star formation on molecular clouds are indeed dramatic. Jets and the associated molecular outflows can blow sizable holes in star-forming clouds, and they may provide a significant energy source for the turbulence in these clouds, although it is not yet clear how much coupling there is between outflows and overall cloud dynamics. Another effect that is clearly of major importance is that the ionizing radiation from newly formed hot stars can rapidly evaporate away the remaining gas in a star-forming cloud, etching away in the process the less dense parts of the cloud and revealing the previously obscured denser parts. The already widely publicized HST photographs of M16 (the "Eagle Nebula") presented here by Hester clearly show both the expected ionized outflows and the intricate structure of the remaining dense parts of the cloud, including a number of small clumps apparently caught in the act of forming stars. Clearly some of the processes of stellar birth are being revealed here, but much remains to be understood about the complex structures that are seen.

My next question is one of broad importance, and it was addressed by several speakers: "What is the stellar Initial Mass Function?" I believe that we have learned two basic facts about the IMF that we can now state with some confidence. First, as we heard in the talks by Kulkarni and Basri, brown dwarfs are clearly not very common objects, even though they can be detected and identified readily with present techniques. Even if the number of stellar objects per unit mass interval remains approximately constant down into the brown dwarf regime, as suggested by Basri, the number per unit logarithmic mass interval then drops strongly below 0.1 solar masses, and the IMF defined in this way peaks at around a few tenths of a solar mass. The total amount of mass per unit logarithmic mass interval peaks at a larger mass around one solar mass, and we can now say definitely that not only is there relatively little mass at the top end of the IMF, there is even less at the bottom end. This implies what seems to me the most basic fact about star formation, namely that there is a characteristic stellar mass of the order of one solar mass, and that most of the mass that forms stars goes into stars with masses of this order. The second basic fact about the IMF, which has long been known, is that at masses above the mass where it peaks it has an approximately power-law tail toward higher masses. As we heard from Heap and Richer, the slope of this power law seems to be very similar in different locations, at least wherever good direct star-count data exist; in fact, most recent studies have yielded results consistent with the original Salpeter law. A possible exception pointed out by Heap is that the IMF appears to be flatter at the center of the 30 Doradus cluster, suggesting that the formation of massive stars has been favored there. Also, as noted by Richer, there is evidence that the IMF for Galactic halo stars may be flatter than the Salpeter law.

What determines the IMF? Those who regularly attend meetings on star formation know that for some years there have been two competing views on the question of what determines the characteristic stellar mass. One view, based on the scale-free inside-out collapse model, is that a forming star continues to grow in mass by accretion at a constant rate until the accretion process is terminated by a wind whose onset is controlled by internal stellar physics; the other, which I have favored, is that stellar masses depend instead on cloud properties, and that the characteristic mass is closely related to the Jeans mass calculated from the typical temperature and total pressure in molecular clouds. I remark here that if the collapse process is not scale-free but depends on initial and boundary conditions, as discussed above, then stellar masses must necessarily depend at least in part on cloud properties. In this case, it is possible that the mass at which the IMF peaks might be different in different circumstances.

On scales larger than those of individual clusters or molecular cloud cores, much additional interesting structure is seen in regions of star formation, as was reviewed by Humphreys. Often several young clusters are seen in the same star-forming region, and a more dispersed population of young stars is also present. Sometimes there is evidence for multiple episodes of star formation, and it is possible that some of the later episodes are triggered by the effects of earlier ones in sweeping up and compressing the remaining gas into expanding shells, as discussed by Surdin. I was particularly impressed again with some of the new data that we saw, especially the spectacular pictures of the giant H II regions 30 Doradus and NGC 604 and their associated star-forming complexes. Here, especially in 30 Doradus, an enormous range of phenomena is observed in the same region of space – massive molecular clouds, infrared protostars, luminous young clusters, ionized flows, windblown bubbles, supernova remnants, hot X-ray emitting gas, and multiple large shells, sometimes interacting with each other – making this region a real happy hunting ground for students of star formation and interstellar matter, as well as a good object lesson in the true complexity of the real world.

Concerning the global history of star formation in galaxies, a question of long standing has been whether star formation rates vary in a smooth and regular way with time, or whether their variation is more discontinuous and characterized by discrete episodes of major star forming activity. The star formation histories of spiral galaxies were reviewed by O'Connell, who noted that the photometric properties of both the disks and the bulges of these systems show well-defined trends suggestive of a smooth evolution regulated by overall galactic properties such as Hubble type. Whatever the detailed mechanisms involved, it would not be surprising if galactic star formation rates were to depend on dynamical timescales and hence on Hubble types in the sense that the galaxies of later Hubble type, which tend to have longer dynamical timescales, turn their gas into stars more slowly, as the observations indicate. The possible controlling factors for star formation were discussed by Skillman, who noted that the occurrence of star formation is always correlated with the presence of a high surface density of gas in galaxies, and that often but not always, significant star formation is observed only where the gas surface density exceeds the critical value required for large-scale gravitational instability to occur in the gas layer. Possible dynamical influences were reviewed by Kenney, who noted that strong spiral waves or bars can play an important role in redistributing the gas in galaxies and causing it to pile up in preferred locations such as nuclear rings. He also noted that since the critical surface density required for gravitational instability is high in the nuclear regions of galaxies, one might expect that if this critical surface density were exceeded as a result of gas inflows, a very high star formation rate, i.e. a starburst, would result.

What causes starbursts in galaxies? As we heard from Hibbard, the most extreme starbursts are found in the central regions of the ultraluminous infrared galaxies, and these galaxies are invariably very peculiar morphologically and are almost certainly strongly interacting or merging systems. As was noted above, the observations suggest that the basic requirement for an extremely high star formation rate is a very high gas surface density. Merger simulations show that sufficiently extreme central concentrations of gas are indeed produced when galaxies merge. High gas surface densities and high star formation rates are sometimes also seen in the regions of overlap between interacting galaxies, and luminous young star clusters seem to be produced in abundance in starburst regions. Heckman discussed the cosmogonical importance of starbursts, and pointed out that they account for a significant fraction of the formation of massive stars in the universe and hence for a significant fraction of the heavy-element production. If nuclear starbursts also produce low-mass stars in the numbers expected for a normal IMF, they must play an important role in the formation of galactic bulges. Outflows from starburst regions may also be important in enriching the surrounding intergalactic medium in heavy elements.

Star formation rates in high-redshift spiral galaxies were reviewed by Madau, who also discussed the overall cosmic history of star formation. The inferred cosmic star formation rate increases strongly with increasing redshift up to at least $z \sim 1$, but this increase does not appear to continue to much higher redshifts, and the evidence for galaxies with $z > 2.5$ suggests lower total star formation rates at these early times (although these early values are strictly lower limits.) The overall cosmic SFR may therefore start out low at very early times, rise to a maximum at a redshift somewhere between 2.5 and 1, and then drop strongly toward smaller redshifts. Integrating this inferred cosmic SFR over time yields roughly the total number of stars presently seen. The striking implication of this result is that we can now observe directly, by looking at these large redshifts, most of the star formation that has ever occurred in the universe, at least in the presently dominant spiral galaxies. It is interesting to note, as Virginia Trimble pointed out, that the apparent peak in the cosmic star formation rate at a redshift of around 2 is reminiscent of the peak in AGN activity that also occurred at about the same time; it is not implausible that these two results could be related, and that both could perhaps reflect an epoch when gas was condensing at a maximum rate into galaxies.

What about the star formation history of elliptical galaxies? We heard contrasting presentations on this subject from Worthey and Dressler. Worthey reviewed studies of the present stellar content of elliptical galaxies, and showed that it is now possible to make some progress with the old problem of separating age and metallicity effects; the mean stellar ages found in this work vary considerably between galaxies, suggesting that many elliptical galaxies have had a long and complex formation history. Dressler disagreed, pointing out that normal-looking elliptical galaxies are seen in about their present numbers in intermediate-redshift clusters, and that these intermediate-redshift elliptical galaxies have red colors indicating that they are already old and therefore must have formed at redshifts significantly greater than 2. Studies of the evolution of the galaxy content of clusters, using Hubble types derived from HST

Appendix A:

Conference Programme

Star Formation, Near and Far

College Park, Maryland
October 14-16, 1996

Monday, 14 October 1996

8:15 Session #1: **Introduction**   Chair: **S. Holt**

    V. Trimble The Fourth Day of Creation: The Search for a Theory of Star Formation

    J. Silk Current Issues in Star Formation

10:00 Coffee (provided outside lecture hall in poster area)

10:30 Session #2: **Close-Up Views of Star Formation** Chair: **R. Brown**

    P. Myers Pre-Stellar Cores and Beginnings of Stellar Collapse

    E. Ostriker Turbulence and Magnetic Fields in Star Formation

    L. Mundy Observations of Circumstellar Disks and Infall

    A. Sargent Evolution of Circumstellar Disks†

12:30 Lunch (provided at the Conference Center)

2:00 Session #3: **Circumstellar Disks in Different Environments**
                                                                Chair: **A. Sargent**

    J. Hester HST Images of Disks and Clusters

    P. Hartigan The Inner Accretion Disk

    J. Stone Theory of Circumstellar Disks

4:00 Tea (provided outside lecture hall in poster area)

4:30 Session #4: **Brown Dwarfs and the Very Low End of the IMF**
                                                               Chair: **S. Maran**

    S. Kulkarni Obs of Brown Dwarfs and the Low End of the IMF

    D. Lin Formation of Low Mass Stars and Brown Dwarfs

    G. Basri Observations of the Substellar Mass Function

6:30 Poster Session (with refreshments -- until 9:30)

Tuesday, 15 October 1996

8:00 Session #5: **Clusters and Multiple Star Systems** Chair: **M. Hauser**

    R. Mathieu   Binaries/Multiple Star Systems

    C. Clarke   Star Formation in Clusters

    H. Richer   IMF in Clusters

    A. Whitworth   Simulations of Dynamically Triggered Star Formation†

10:00 Coffee (provided outside lecture hall in poster area)

10:30 Session #6: **Galactic Star Formation** Chair: **S. Neff**

    R. Humphreys   High Mass Stars in Local Galaxies

    S. Heap   Extragalactic IMF and Star Formation Modes

    E. Skillman   Controlling Factors for Global Star Formation

    F. de Pablos   Efficiency and Biassed Star Formation in Spirals†

12:30 Lunch (provided at the Conference Center)

2:00 Session #7: **Dynamical Processes Influencing Star Formation**
           Chair: **D. Spergel**

    J. Kenney   Dynamical Processes Influencing Star Formation

    J. Hibbard   Mergers, Interactions, and Fueling of Starbursts

    T. Heckman   Starbursts

    V. Surdin   Young Stellar Aggregates Embedded in Expanding Supershells†

4:00 Tea (provided outside lecture hall in poster area)

4:30 Session #8: **Star Formation History in Spirals** Chair: **E. Churchwell**

    R. O'Connell   Star Formation History in Low-z Spirals

    P. Madau   Star Formation History in High-z Spirals

    H. Teplitz   IR Discoveries of Star Forming Galaxies at $2.3<z<2.5$†

7:00 Banquet - **R. Giacconi**

Wednesday, 16 October 1996

8:30 Session #9: **Star Formation History in Ellipticals**   Chair: **E. Dwek**

    G. Worthey   Star Formation History in Low-z Ellipticals

    A. Dressler   Star Formation History in High-z Ellipticals

10:00 Coffee (provided outside lecture hall in poster area)

10:30 Session #10: **Star Formation History in Irregular and Dwarf Spheriodal Galaxies**   Chair: **J. Graham**

    D. Hatzidimitriou Star Formation History in Irregular Galaxies

    T. Smecker-Hane Star Form'n History in Dwarf Spheroidal Galaxies

    S. Kim   HI Mosaic and H-alpha Survey in the LMC†

12:00 Rapporteur - **R. Larson**

1:00 Lunch (provided at the Conference Center)

    End of conference

†contributed paper; other papers are all invited

Appendix B:

List of Attendees

ATTENDEES

Abel, Tom	NCSA	tabel@ncsa.uiuc.edu
Ballesteros-Paredes, Javier	Instituto de Astronomia, UNAM	javier@astroscu.unam.mx
Balsara, Dinshaw	NCSA/University of Illinois	u10956@ncsa.uiuc.edu
Basri, Gibor	University of California-Berkeley	basri@astro.berkeley.edu
Basu, Shantanu	CITA	basu@cita.utoronto.ca
Batchelor, David	NASA/GSFC	batchelor@nssdca.gsfc.nasa.gov
Bate, Matthew	MPI for Astronomy	mbate@mpia-hd.mpg.de
Bennett, Chuck	NASA/GSFC	bennett@stars.gsfc.nasa.gov
Bhatia, Anand	NASA/GSFC	teakb@scfmvs.bitnet
Blain, Andrew	Cavendish Laboratory, Cambridge, UK	awb@mrao.cam.ac.uk
Bloemhof, Eric	Caltech	eeb@astro.caltech.edu
Boldt, Elihu	NASA/GSFC	boldt@lheavx.gsfc.nasa.gov
Borne, Kirk	Hughes STX/NASA/GSFC	borne@xfiles.gsfc.nasa.gov
Brown, Robert L.	NRAO	rbrown@nrao.edu
Byrd, Gene	University of Alabama-Tuscaloosa	byrd@possum.astr.ua.edu
Burrows, Chris	STScI	burrows@stsci.edu
Cahn, Sheri	Arizona State U.	winters.john@al.gc.maricopa.edu
Carkner, Lee	Penn State University	scarkner@astro.psu.edu
Carpenter, Kenneth	NASA/GSFC	hrscarpenter@hrs.gsfc.nasa.gov
Casey, Sean	Hughes STX/NASA/GSFC	sean@irastro.gsfc.nasa.gov
Cepa, Jordi	Inst. de Astrofisica de Canarias	jcn@ll.iac.es
Chambers, Ken	Inst. for Astronomy/U. of Hawaii	chambers@ifa.hawaii.edu
Chen, Xingming	Northwestern University	chen@apollo.astro.nwu.edu
Cheng, Edward	NASA/GSFC	ec@cobi.gsfc.nasa.gov
Cheung, Cynthia	NASA/GSFC	ccheung@nssdc.gsfc.nasa.gov
Churchwell, Ed	University of Wisconsin-Madison	churchwell@astro.wisc.edu
Clarke, Cathie	Institute of Astronomy, Cambridge	cclarke@ast.cam.ac.uk
Cline, Thomas	NASA/GSFC	cline@lheavx.gsfc.nasa.gov
Corcoran, Mike	NASA/GSFC/USRA	corcoran@barnegat.gsfc.nasa.gov

Cornett, Bob	GSFC/IASP/Hughes STX	cornett@uit.gsfc.nasa.gov
Coulson, Iain	Joint Astronomy Centre	imc@jach.hawaii.edu
Cowen, Ron	Science News	
Crannell, Carol	NASA/GSFC	crannell@stars.gsfc.nasa.gov
Currie, Doug	University of Maryland	currie@khaos.umd.edu
Deane, Jim	Institute for Astronomy	deane@galileo.ifa.hawaii.edu
De Pablos, Fernando	Inst. de Astrofisica de Canarias	fpablos@ll.iac.es
Deming, Drake	NASA/GSFC	drake.deming@gsfc.nasa.gov
Devereux, Nick	New Mexico State University	devereux@nmsu.edu
de Winter, Dolf	Universidad Autonoma de Madrid	ditf@astrol.ft.uam.es
Drake, Stephen A.	NASA/GSFC/USRA	drake@lheavx.gsfc.nasa.gov
Dressel, Linda	RJH Scientific, Inc.	dressel@rjhsci.com
Dressler, Alan	Carnegie Observatories	dressler@lynx.ociw.edu
Dwek, Eli	NASA/GSFC	edwek@stars.gsfc.nasa.gov
Fahey, Dick	NASA/GSFC	fahey@stars.gsfc.nasa.gov
Fanelli, Mike	HSTX Corp & NASA/GSFC/LASP	fanelli@kuylym.gsfc.nasa.gov
Feigelson, Eric	Penn State University	edf@astro.psu.edu
Ferguson, Annette	Johns Hopkins University	ferguson@haggis.pha.jhu.edu
Fich, Michel	University of Waterloo	fich@astro.uwaterloo.ca
Gardner, Jonathan P.	NASA/GSFC	gardner@harmony.gsfc.nasa.gov
Garmire, Gordon	Penn State University	garmire@astro.psu.edu
Gezari, Daniel	NASA/GSFC	gezari@stars.gsfc.nasa.gov
Giacconi, Riccardo	ESO	D.Eisenhuth@eso.org
Goodman, Alyssa	Harvard University	agoodman@cfa.harvard.edu
Graber, James S.	Library of Congress	jgra@loc.gov
Grady, Carol	Eureka Scientific	cgrady@mtolympus.ari.net
Graham, John	Carnegie-DTM	graham@jag.ciw.edu
Greyber, Howard D.	Greyber Assoc.	hgreyber@capaccess.org
Gull, Theodore	NASA/GSFC	gull@stars.gsfc.nasa.gov
Hall, Charles	Catholic University	crhall@rosserv.gsfc.nasa.gov
Hameed, Salman	New Mexico State University	shameed@nmsu.edu
Hartigan, Patrick	Rice University	hartigan@sparky.rice.edu
Hatzidimitriou, D.	University of Crete	dh@physics.uch.gr

633

Hauser, Mike	STScI	hauser@stsci.edu
Heap, Sally	NASA/GSFC	hrsheap@stars.gsfc.nasa.gov
Hearty, Thom	University of Georgia	thom@hera.physast.uga.edu
Heckman, Tim	Johns Hopkins University	
Hester, Jeff	Arizona State University	jjh@cygloop.la.asu.edu
Hibbard, John E.	Institute for Astronomy	hibbard@galileo.ifa.hawaii.edu
Hill, Robert S.	Hughes STX/NASA/GSFC	bhill@uit.gsfc.nasa.gov
Hillman, John	NASA/GSFC	u3jjh@lepvax.gsfc.nasa.gov
Hintzen, Paul	NASA/GSFC	hintzen@nevada.edu
Ho, Luis C.	Center for Astrophysics	lho@cfa.harvard.edu
Hodge, Paul	University of Washington	hodge@astro.washington.edu
Holt, Stephen	NASA/GSF	steve.holt@gsfc.nasa.gov
Hubeny, Ivan	NASA/GSFC	hubeny@stars.gsfc.nasa.gov
Humphreys, Roberta	University of Minnesota	roberta@isis.spa.umn.edu
Hurt, Robert L.	IPAC	hurt@ipac.caltech.edu
Imhoff, Catherine	Computer Sciences Corp/IUE Obs.	imhoff@iuegtc.gsfc.nasa.gov
Iping, Rosina	University of Guam	riping@uog9.uog.edu
Jayawardhana, Ray	Harvard University	rayjay@cfa.harvard.edu
Jones, Frank	NASA/GSFC	jones@lheavx.gsfc.nasa.gov
Kane, Brian	Phillips Laboratory	bkane@pldac.plh.af.mil
Kayser, Susan	National Science Foundation	skayser@nsf.gov
Kazanas, Demos	NASA/GSFC	kazanas@lheavx.gsfc.nasa.gov
Kelsall, Thomas	NASA/GSFC	kellsall@stars.gsfc.nasa.gov
Kenney, Jeff	Yale University	kenney@astro.yale.edu
Keohane, Jonathan	NASA/GSFC	jonathan@cassiopeia.gsfc.nasa.gov
Khersonsky, Valery	University of Pittsburgh	vkk@phyast.pitt.edu
Kim, Sungeun	Mount Stromlo Observatory	sek@mso.anu.edu.au
Kimble, Randy	NASA/GSFC	kimble@stars.gsfc.nasa.gov
Klessen, Ralf	Max-Planck-Institute for Astronomy	klessen@mpia-hd.mpg.de
Kobulnicky, Chip	University of Minnesota	chip@astro.spa.umn.edu
Kondo, Yoji	NASA/GSFC	kondo@iue.gsfc.nasa.gov
Kraemer, Kathleen	Boston University	kraemer@fish.bu.edu
Krist, John	STScI	krist@stsci.edu

Kulkarni, S. R.	Caltech	srk@astro.caltech.edu
Larson, Richard B.	Yale University	larson@astro.yale.edu
Leisawitz, David	NASA/GSFC	leisawitz@stars.gsfc.nasa.gov
Leventhal, Marv	University of Maryland	ml@astro.umd.edu
Lin, D.	University of California-Santa Cruz	lin@lick.ucsc.edu
Looney, Leslie	University of Maryland	lwl@astro.umd.edu
Lucas, Phil	Oxford University	p.lucas1@physics.oxford.ac.uk
Madau, Piero	STScI	madau@stsci.edu
Magnani, Loris	University of Georgia	loris@zeus.physast.uga.edu
Magnier, Eugene	University of Washington	gene@astro.washington.edu
Maran, Stephen	NASA/GSFC	hrsmaran@stars.gsfc.nasa.gov
Mardones, Diego	Harvard University	dmardones@cfa.harvard.edu
Marshall, Frank	NASA/GSFC	marshall@lheavx.gsfc.nasa.gov
Martin, Crystal	STScI	cmartin@as.arizona.edu
Mather, John C.	NASA/GSFC	mather@stars.gsfc.nasa.gov
Mathieu, Robert D.	University of Wisconsin-Madison	mathieu@astro.wisc.edu
McGaugh, Stacy	Carnegie-DTM	ssm@dtm.ciw.edu
Megeath, Tom	Haystack Observatory	megeath@newton.haystack.edu
Mendoza, Eduardo	Instituto Nacional de Astrofisica	
Meurer, Gerhardt R.	The Johns Hopkins University	meurer@poutine.pha.jhu.edu
Mihos, Chris	The Johns Hopkins University	hos@pha.jhu.edu
Miller, Kristen	University of Maryland	kadams@astro.umd.edu
Mitchalitsianos, A.	NASA/GSFC	mitchalits@stars.gsfc.nasa.gov
Mott, Andrew	Mullard Radio Astronomy Observatory	asml1@mrao.cam.ac.uk
Mumma, Michael	NASA/GSFC	mmumma@lepvax.gsfc.nasa.gov
Mundy, Lee	University of Maryland	lgm@astro.umd.edu
Murai, Tadayuki	Nagoya University	murai@ibm.net
Myers, Philip C.	Harvard-Smithsonian/CfA	myers@cfa.harvard.edu
Nagar, Neil	University of Maryland	neil@astro.umd.edu
Neff, Susan	NASA/GSFC	neff@stars.gsfc.nasa.gov
Nelson, Charles	University of Nevada-Las Vegas	cnelson@physics.unlv.edu
Niedner, Malcolm	NASA/GSFC	malcolm.niedner@gsfc.nasa.gov

Norman, Colin	The Johns Hopkins University	norman@stsci.edu
O'Connell, Robert	University of Virginia	rwo@virginia.edu
Ohashi, Nagayoshi	Center for Astrophysics	nohashi@cfa.harvard.edu
Olsen, Knut	University of Washington	olsen@astro.washington.edu
Orecchio, Federico	NASA/GSFC	orecchio@achamp.gsfc.nasa.gov
Ostriker, Eve	University of Maryland	ostriker@astro.umd.edu
Ozernoy, Leonid	George Mason University	ozernoy@hubble.gmu.edu
Padoan, Paolo	Theoretical Astrophysics Center	padoan@tac.dk
Petre, Rob	NASA/GSFC	petre@lheavx.gsfc.nasa.gov
Pisarski, Ryszard	NASA/GSFC	ryszard.pisarski@gsfc.nasa.gov
Polidan, Ronald	NASA/GSFC	polidan@aesop.gsfc.nasa.gov
Pound, Marc	University of California-Berkeley	pound@teddi.berkeley.edu
Ramaty, Reuven	NASA/GSFC	ramaty@lheavx.gsfc.nasa.gov
Regan, Mike	University of Maryland	mregan@astro.umd.edu
Richer, Harvey	University of British Columbia	richer@astro.ubc.ca
Riegler, Guenter	NASA Headquarters	griegler@hq.nasa.gov
Rocha-Pinto, Helio	Instituto Astronomico e Geofisico	helio@zeiss.iagusp.usp.br
Rosenbaum, Doris	SMU	rosenbaum@phyvms.physics.smu.edu
Safier, Pedro	University of Maryland	safier@astro.umd.edu
Sargent, Anneila	Caltech	afs@mmstar.caltech.edu
Scowen, Paul A.	Arizona State University	scowen@tycho.la.asu.edu
Shafer, Richard	NASA/GSFC	shafer@stars.gsfc.nasa.gov
Sheth, Kartik	University of Maryland	kartik@astro.umd.edu
Shure, Mark	CHARA/Georgia State University	shure@chara.gsu.edu
Silk, Joe	University of California-Berkeley	silk@pac2.berkeley.edu
Skillman, Evan D.	University of Minnesota	skillman@zon.spa.umn.edu
Smecker-Hane, Tammy	University of California-Irvine	tsmecker@uci.edu
Smith, Eric	LASP/NASA/GSFC	esmith@hubble.gsfc.nasa.gov
Smith, Howard A.	National Air & Space Museum	howard@nasm.edu
Sollins, Peter	Swarthmore College/Haystack Obs.	psollin@sccs.swarthmore.edu
Spergel, David	Princeton University	dns@astro.princeton.edu
Stecher, Ted	NASA/GSFC	stecher@uit.gsfc.nasa.gov

Stecker, Floyd	NASA/GSFC	stecker@lheavx.gsfc.nasa.gov
Stiller, Bertram		bstiller@capaccess.org
Stone, Jim	University of Maryland	jstone@astro.umd.edu
Stone, Robert	NASA/GSFC	
Stringfellow, Guy	CASA/University of Colorado	guy@casa.colorado.edu
Surdin, Vladimir	Sternberg Astronomical Institute	surdin@sai.msu.su
Takamiya, Marianne	University of Chicago	taka@oddjob.uchicago.edu
Teplitz, Harry	UCLA	harry@bnkl01.astro.ucla.edu
Teplitz, Vigdor	SMU	teplitz@phyvms.physics.smu.edu
Teuben, Peter	University of Maryland	teuben@astro.umd.edu
Thornley, Michele D.	University of Maryland	michele@astro.umd.edu
Trasco, John	University of Maryland	jtrasco@astro.umd.edu
Trimble, Virginia	UMCP & UC-Irvine	trimble@astro.umd.edu
Tsuboi, Yohko	Kyoto University	tsuboi@cr.scphys.kyoto-u.ac.jp
Van Steenburg, M. E.	NASA/GSFC	mev@ndadsa.gsfc.nasa.gov
Vázquez-Semadeni, E.	Instituto de Astronomia, UNAM	enro@astroscu.unam.mx
Veilleux, Sylvain	University of Maryland	veilleux@astro.umd.edu
Waller, William H.	Hughes STX & NASA/GSFC/LASP	waller@stars.gsfc.nasa.gov
Walther, Dolores M.	Joint Astronomy Centre/UKIRT	dolores@jach.hawaii.edu
Wang, John C. L.	University of Maryland	jcwang@astro.umd.edu
White, Stephen	University of Maryland	white@astro.umd.edu
Whitworth, Ant	University of Wales	a.whitworth@astro.cf.ac.uk
Williams, Harold	Montgomery College	haroldw@umd5.umd.edu
Williams, Jonathan	Center for Astrophysics	jpw@cfa.harvard.edu
Wilner, David	Center for Astrophysics	dwilner@cfa.harvard.edu
Wilson, Christine	McMaster University	wilson@eccles.physics.mcmaster.ca
Wiseman, Jennifer	National Radio Astronomy Observatory	jwiseman@nrao.edu
Woodgate, Bruce	NASA/GSFC	bruce.woodgate@gsfc.nasa.gov
Wootten, Al	NRAO	awootten@nrao.edu
Worthey, Guy	University of Michigan	worthey@astro.lsa.umich.edu
Wu, Chi-Chao	Computer Sciences Corporation	wu@stsci.edu
Xie, Taoling	University of Maryland	tao@astro.umd.edu
Zhang, Qizhou	Center for Astrophysics	qzhang@cfa.harvard.edu

Appendix C:

Physical Constants

TABLE OF PHYSICAL CONSTANTS

CONSTANT	SYMBOL	MKS		CGS		OTHER
speed of light	c	$3.00 \cdot 10^8$	m/s	$3.00 \cdot 10^{10}$	cm/s	(2.997925)
electron charge	e	$1.60 \cdot 10^{-19}$	coul	$4.80 \cdot 10^{-10}$	esu	
Planck constant	h	$6.63 \cdot 10^{-34}$	J·s	$6.63 \cdot 10^{-27}$	erg·s	
	$\hbar$	$1.05 \cdot 10^{-34}$	J·s	$1.05 \cdot 10^{-27}$	erg·s	
	hc	$1.99 \cdot 10^{-25}$	J·m	$1.99 \cdot 10^{-16}$	erg·cm	
	$\hbar c$	$3.15 \cdot 10^{-26}$	J·m	$3.15 \cdot 10^{-17}$	erg·cm	200 MeV·fm
Boltzmann constant	k	$1.38 \cdot 10^{-23}$	J/K	$1.38 \cdot 10^{-16}$	erg/K	$8.6 \cdot 10^{-5}$ eV/K
	k/h	$2.08 \cdot 10^{10}$	s^{-1}/K	$2.08 \cdot 10^{10}$	s^{-1}/K	
	k/hc	69.5	m^{-1}/K	0.695	cm^{-1}/K	
Gravitational constant	G	$6.67 \cdot 10^{-11}$	$N \cdot m^2/kg^2$	$6.67 \cdot 10^{-8}$	$dy \cdot cm^2/gm^2$	
Gas constant	R	8.314	J/K·mole	$8.31 \cdot 10^7$	erg/K·mole	
Avogadro's number (= R/k)	N	$6.02 \cdot 10^{26}$	amu/kg	$6.02 \cdot 10^{23}$	amu/kg	$6 \cdot 10^{23}$ molecules/mole
electron mass	m_e	$9.11 \cdot 10^{-31}$	kg	$9.11 \cdot 10^{-28}$	gm	0.51 MeV
proton mass	M_p	$1.67 \cdot 10^{-27}$	kg	$1.67 \cdot 10^{-24}$	gm	938 MeV
neutron mass	M_n	$1.67 \cdot 10^{-27}$	kg	$1.67 \cdot 10^{-24}$	gm	939 MeV
pion mass (=270·m_e)	m_π	$2.46 \cdot 10^{-28}$	kg	$2.46 \cdot 10^{-25}$	gm	140 MeV
muon mass (=207·m_e)	m_μ	$1.89 \cdot 10^{-28}$	kg	$1.89 \cdot 10^{-25}$	gm	106 MeV
classical elect radius (=e^2/mc^2)	r_c	$2.82 \cdot 10^{-15}$	m	$2.82 \cdot 10^{-13}$	cm	
Compton wavelength (=h/mc)	λ_c	$2.43 \cdot 10^{-12}$	m	$2.43 \cdot 10^{-10}$	cm	0.02 Å

Quantity	Symbol	Value (SI)	Unit	Value (CGS)	Unit	Other
Thomson cross-section	σ_T	$6.65 \cdot 10^{-29}$	m^2	$6.65 \cdot 10^{-25}$	cm^2	
Planck length ($=\sqrt{\hbar G/c^3}$)	l_{Pl}	$1.61 \cdot 10^{-35}$	m	$1.61 \cdot 10^{-33}$	cm	
Planck time ($=\sqrt{\hbar G/c^5}$)	t_{Pl}	$5.39 \cdot 10^{-44}$	s	$5.39 \cdot 10^{-44}$	s	
Planck density ($=c^5/\hbar G^2$)	ρ_{Pl}	$5.16 \cdot 10^{96}$	kg/m^3	$5.16 \cdot 10^{93}$	gm/cm^3	
Bohr radius ($=\hbar^2/m_e e^2$)	r_B	$0.53 \cdot 10^{-10}$	m	$0.53 \cdot 10^{-8}$	cm	0.5 Å
Fine structure constant ($=e^2/\hbar c$)	α	$7.30 \cdot 10^{-3}$		$7.30 \cdot 10^{-3}$		1/137
Bohr magneton ($=e\hbar/2m_e c$)	μ_B	$9.27 \cdot 10^{-24}$	J/T	$9.27 \cdot 10^{-21}$	erg/gauss	
Nuclear magneton ($=e\hbar/2M_p c$)	μ_N	$5.05 \cdot 10^{-27}$	J/T	$5.05 \cdot 10^{-24}$	erg/gauss	
Permittivity of vacuum	ε_0	$8.85 \cdot 10^{-12}$	fd/m			$1/4\pi\varepsilon_0 = 9.0 \cdot 10^9$
Permeability in vacuum	μ_0	$4\pi \cdot 10^{-7}$	Hen/m			
Stefan-Boltzmann constant	σ	$5.67 \cdot 10^{-8}$	$W/m^2 \cdot K^4$	$5.67 \cdot 10^{-5}$	$erg/s \cdot cm^2 \cdot K^4$	
Rydberg ($=m_e e^4/2\hbar^2$)	R_∞	$2.18 \cdot 10^{-18}$	J	$2.18 \cdot 10^{-11}$	erg	13.6 eV
1 amu		$1.66 \cdot 10^{-27}$	kg	$1.66 \cdot 10^{-24}$	gm	931.5 MeV
1 calorie		4.19	J	$4.19 \cdot 10^7$	erg	
1 year		$3.16 \cdot 10^7$	s	$3.16 \cdot 10^7$	s	
1 atmosphere		$1.01 \cdot 10^5$	N/m^2	$1.01 \cdot 10^6$	$dyne/cm^2$	14.2 lbs/in^2
		$1.01 \cdot 10^5$	Pascal			760 Torr
1 eV		$1.6 \cdot 10^{-19}$	J	$1.6 \cdot 10^{-12}$	erg	11,605 K
		$1.24 \cdot 10^{-6}$	m	$1.24 \cdot 10^{-4}$	cm	
		1	Tesla	10^4	gauss	

ASTROPHYSICAL CONSTANTS

CONSTANT	SYMBOL	MKS	CGS	OTHER
astronomical unit	AU	$1.50 \cdot 10^{11}$ m	$1.50 \cdot 10^{13}$ cm	
	AU/year			4.74 km/s
parsec	pc	$3.09 \cdot 10^{16}$ m	$3.09 \cdot 10^{18}$ cm	3.26 LY
solar mass	$M_\odot$	$1.99 \cdot 10^{24}$ kg	$1.99 \cdot 10^{33}$ gm	
solar luminosity	$L_\odot$	$3.90 \cdot 10^{26}$ J/s	$3.90 \cdot 10^{33}$ erg/s	
solar effective temperature	$T_{eff\odot}$	5780 K	5780 K	
solar radius	$R_\odot$	$6.96 \cdot 10^{8}$ m	$6.96 \cdot 10^{10}$ cm	
Earth radius	$R_\oplus$	$6.38 \cdot 10^{6}$ m	$6.38 \cdot 10^{8}$ cm	
Earth mass	$M_\oplus$	$5.98 \cdot 10^{24}$ kg	$5.98 \cdot 10^{27}$ gm	
Earth density	$\rho_\oplus$	5520 kg/m^3	5.52 gm/cm^3	
Jansky	Jy	$1.0 \cdot 10^{-26}$ W/m$^2 \cdot$Hz	$1 \cdot 10^{-23}$ erg/s$\cdot$cm$^2 \cdot$Hz	
Hubble constant	H_0	$3.24 h \cdot 10^{-18}$ s^{-1}	$3.24 h \cdot 10^{-18}$ s^{-1}	$100 h$ km/s$\cdot$Mpc
critical density (=$3H_0^2/8\pi G$)	ρ_0	$1.88 h^2 \cdot 10^{-26}$ kg/cm^3	$1.88 h^2 \cdot 10^{-29}$ gm/cm^3	
plasma frequency				$8.98\sqrt{n_e(\text{cm}^{-3})}$ kHz/gauss
radian				$57.29578° = 206,265"$
CMB photon density	n_γ	$4.15 \cdot 10^5$ m^{-3}	415 cm^{-3}	

Author Index

A

Abel, T.	329
Alexander, P.	441
van den Ancker, M. E.	189
Anninos, P.	329
Antonuccio-Delogu, V.	315
Aufdenberg, J. P.	291
Avizonis, P.	201

B

Bachiller, R.	113
Ballesteros-Paredes, J.	81
Balsara, D. S.	89
Barranco, J. A.	105
Basri, G.	228
Basu, S.	75
Bate, M. R.	371
Beichman, C. A.	506
Bhattal, A.	367
Bjorkman, K. S.	193
Blain, A. W.	606
Bloemhof, E. E.	399
de Blok, E.	311, 510
Bohlin, R. C.	473, 551, 594
Bonaccini,	201
Borne, K. D.	295
Bothun, G. D.	551
Bushouse, H.	295
Buta, R. J.	283
Byrd, G. G.	283

C

Caillault, J.-P.	461
Carlstrom, J. E.	117
Carrasco, B. E.	543
Cepa, J.	433
Cesarsky, C.	506
Chambers, K. C.	547
Churchwell, E.	457
Clarke, C. J.	347
Clemens, D. P.	137
Colina, L.	295
Cornett, R. H.	590
Coulson, I. M.	209
Crutcher, R. M.	89
Currie, D.	201

D

Dayal, A.	303
Dent, W. R. F.	209
Deutsch, L.	303
Devereux, N.	514, 518
Digel, S. W.	465
Dowling, D.	201
Dressler, A.	535
Duric, N.	518

E

Eikenberry, S.	303
Elbaz, D.	506

F

Fahlman, G. G.	357
Fanelli, M. N.	473, 594, 598
Fazio, G.	303
Feigelson, E. D.	179, 184
Ferguson, A. M. N.	469
Fich, M.	121
Filippenko, A. V.	403
Francis, N.	367
Freeman, T.	283
Frenk, C. S.	543

G

Gallagher, J. S.	469
Gao, Y.	319
Garay, G.	113
Gardner, J. P.	543
Goodman, A. A.	105
Grady, C. A.	193
Graham, J. A.	555
Greyber, H. D.	205
Gruendl, R.	319, 395

H

Hameed, S.	514
Hanson, M. M.	465
Hartigan, P.	153
Hatzidimitriou, D.	561
Hauschildt, P. H.	291
Heap, S. R.	414
Hearty, T.	461
Heckman, T. M.	271
Helou, G.	506
Hester, J. J.	143
Heyer, M. H.	105
Hibbard, J. E.	259
Hill, J. K.	473
Hill, R. S.	594
Hills, R. E.	117
Ho, L. C.	403
Ho, P. T. P.	391, 453
Hodge, P.	475
Hoffmann, W.	303
Hollis, J. E.	473
Hora, J.	303
Hughes, D.	303
Humphreys, R. M.	409
Hurt, R. L.	506
Hwang, C. Y.	319

J

Jackson, J. M.	379
Jayawardhana, R.	303
Jimenez, R.	315
Jogee, S.	247
Jones, B. J. T.	97, 101

K

Kane. B. D.	137
Kenney, J.	247
Khersonsky, V.	445
Kim, S.	582

Kissell, K.	201	Malkan, M. A.	502
Klaas, U.	506	Marcy, G. W.	228
Klessen, R. S.	133	Mardones, D.	109, 113
Kobulnicky, C.	586	Martin, C. L.	602
Koyama, K.	175, 179	Mathieu, R. D.	337
Kraemer, K. E.	379	McCormick, D.	283
		McGaugh, S.	311, 510
		McLean, I. S.	502
		McMahon, R. G.	506

L

		Megeath, S. T.	375
Lada, C. J.	121	Mendoza-Torres, J. E.	197
Lada, E. A.	395	Mihos, C.	311
Larson, R. B.	617	Miller, K.	171
Laureijs, R.	506	Montmerle, T.	179
Lay, O. P.	117	Moshir, M.	506
Leisawitz, D.	465	Moskal', E. V.	279
Lekht, E. E.	197	Mott, A. S.	441
Lemke, D.	506	Mundy, L.	63, 129, 395
Levine, D. A.	506	Murai, T.	241
Lin, D. N. C.	217	Myers, P. C.	41, 109, 113, 387
Lo, K. Y.	319		
Lonsdale, C. J.	506		
Looney, L. W.	129		
Lord, S. D.	506	## N	
Lucas, P.	125	Neff, S. G.	473, 551, 594
Lucas, R. A.	295	Nelson, C. H.	299
		Neugebauer, G.	506
		Neuhäuser, R.	461
		Nordlund, Å.	97, 101

M

		Norman, M. L.	329
Maciel, W. J.	437		
MacKenty, J. W.	299		
Madau, P.	481		
Magnani, L.	461		
Magnier, E. A.	72		

O

O'Connell, R. W.	473, 491, 551, 594
Ohashi, N.	93
Olsen, K.	475
Ostriker, E. C.	51
Ozernoy, L. M.	323

P

de Pablos, F.	433
Padoan, P.	97, 101, 315
Passot, T.	85
Pérez, M. R.	189, 193
Pound, M. W.	395
Pouquet A.	85, 89
Pudritz, R. E.	117
Purcell, G. B.	283
Putman, M. E.	391

R

Regan, M. W.	307
Richer, H. B.	357
Roberts, M. S.	473, 551, 594
Rocha-Pinto, H. J.	437
Roche, P. F.	125
Rosa, M. R.	586
Roy, J.-R.	586

S

Sankrit, R.	291
Schmitt, J. H. M. M.	461
Scowen, P. A.	291, 518
Sharples, R. M.	543
Shaya, E.	201
Sheth, K.	307
Shure, M.	383
Silk, J.	3
Simkin, S. M.	299
Sitko, M. L.	193
Skillman, E. D.	423, 586
Sleath, J. P.	441
Smecker-Hane, T. A.	571
Smith, A. M.	473, 551, 594
Smith, D. A.	473, 594
Smith, E. P.	551
Smith, H. E.	506
Soifer, B. T.	506
Sollins, P. K.	375
Stauffer, J.	461
Staveley-Smith, L.	582
Stecher, T. P.	287, 473, 551, 590, 594, 598
Stone, J. M.	160, 171
Surdin, V. G.	279

T

Tafalla, M.	113
Takamiya, M.	610
Teplitz, H.	502
Trimble, V.	15
Tsuboi, Y.	175

U

Ueno, S.	175

V

Van Buren, D.	506
Vázquez-Semadeni, E.	81, 85
Veilleux, S.	319
Vogel, S. N.	307

W

Waller, W. H.	287
Walsh, J. R.	586
Walther, D. M.	209
Ward-Thompson, D.	117
Watkins, S.	367
Wehrle, A.	506
Welch, W. J.	129
Whitworth, A.	367
Williams, J. P.	387
Wilner, D. J.	105, 109, 113
Wilson, C. D.	117
de Winter, D.	189, 193
Wiseman, J. J.	391
Wolstencroft, R. D.	506
Worthey, G.	525
Wyder, T.	475
Wyse, R. F. G.	469

X

Xie, T.	449

Z

Zhang, Q.	453
Zhang, Y.	329

Subject Index

Page numbers refer to the *first* page of the contribution in which the subject appears

A

Accretion
- disks   63, 117, 143, 153, 160, 171, 337
- into binary systems   63, 337
- onto stars   153, 171, 193, 457
- rates   153, 337, 347
- streams   153, 171, 337, 371

Active Galactic Nuclei   295, 506, 547

Advertisements   15

Angular Momentum
- conservation   217, 617
- specific angular momentum   137, 371
- transport   160, 171, 217

Atomic Line Observations
- C I   193
- [C II]   379
- Ca II   571
- Fe II   193
- Hα 247, 291, 307, 337, 469, 475, 481, 502, 514, 518, 582
- H I   15, 555, 582, 586, 594
- He I   193, 259, 602
- He II   414
- Li I   461
- Mg II   193
- N I   193
- [N II]   469
- [O I]   153
- [O III]   15, 143
- [S II]   143, 153, 291
- Si II   193

B

Binary Stars
- brown dwarfs   228
- circumbinary material   63, 337
- formation   63, 137, 217, 367, 617
- in clusters   347
- pre-main-sequence   63, 129, 337, 367
- spectroscopic   337

Bipolar Outflows   (see Stellar Outflows)

Bok Globules   15, 137

Bonnor-Ebert Sphere   347

Brown Dwarfs   217, 241, 617
- characteristics   228
- doppler surveys   228
- in binaries   228
- in clusters   143, 228
- lithium   228

C

Chemical Evolution   (see Metallicity)

Circumbinary Disks   63, 337, 371

Circumstellar Disks   (see Extrasolar Systems and Vega-Excess Systems)
- models   63, 117, 129, 367, 371
- properties   63, 117, 121, 125, 143, 160, 179, 193, 217, 367
- rotation   63, 93, 121, 153, 160, 171
- sizes   63, 117, 129, 143
- star-disk interactions   160, 171, 337, 347, 371
- turbulence   160

Circumstellar Dust   (see Dust)

Circumstellar Envelopes

               117, 125, 143, 337, 371
Cloud Collapse
        (see Gravitational Infall)
Clumps   (see Molecular Clouds)
Cluster Formation
        (also see Star Clusters)
   embedded clusters   179, 347, 375
   globular clusters   347, 403
   open clusters   347
Color Magnitude Diagram
        357, 414, 491, 561, 571
Comets   193, 209
   (also see Extrasolar Systems
   and Vega-Excess Systems)
Cosmic Rays   15, 160

D

Dark Matter
   baryonic mass   329
   halos   561
Density Waves   283, 371, 433
Disks   (see Circumstellar Disks
      or Galactic Structure: disk)
Dust
   coagulation   153, 193
   continuum emission
      63, 129, 201, 395
   extinction law   15, 121
   grain size distribution   193, 209
   monte carlo modeling   125
   polarization   125, 547
   scattering   121, 125, 547
Dynamical
   collapse   (see Gravitational Infall
          or Instabilities)
   cooling   279, 329
   crossing time   279, 403
   ejection   184
   friction   414

E

Equipartition of Energy   51, 205
Extrasolar Systems   193, 209
   gap clearings   217
   planets   217
   planetesimals   193
   protoplanets   217

F

Fragmentation
   instabilities   15, 217
   in molecular clouds   3, 41, 51,
        133, 217, 347, 445, 453
   rotational   217
   top-down   347
   turbulence driven   445
FU Ori Stars   153

G

Galactic Structures
   bars   247, 311, 491, 561
   chimneys   469
   circumnuclear regions
       247, 271, 514
   disk   491
   H I supershells
       279, 469, 561, 582, 602
   halo   357, 571
   inner Lindblad resonance 247, 287
   rings   247, 287
   rotation   525
   spheroid   357, 525, 535
   spiral arms   283, 287

winds 3, 271, 547, 561, 571, 586, 602
Galaxies
  barred 307, 441, 491
  disk 287, 311, 491, 510, 535
  dwarfs 423, 571, 602
  dwarf spheroidals 571
  ellipticals 525, 535
  flocculent 247
  giant ellipticals 551, 555, 617
  high-z 271, 481, 525, 535, 547
  irregulars 561, 582, 586, 590, 598
  radio galaxies 547, 555
  spirals 247, 475, 510, 514
Galaxy Clusters 525
  cooling flows 551
  evolution 329, 535
  high-z 502, 606
  rich clusters 535
Galaxy Counts 547
Galaxy Evolution 502, 525, 535, 547
  disk-disk interactions 259, 525, 535
  dwarfs 571
  ellipticals 3, 525
  irregulars 561
  spirals 3
Galaxy Luminosity Function 259, 543
Gravitational Infall 41, 93, 371, 617
  (also see Instabilities)
  asymmetric line profiles 41, 63, 109, 113, 387, 453
  inside out collapse 63, 75, 109, 617

Larson-Penston solution 63, 75
mass infall rates 93, 387
prestellar infall 41, 63, 347
velocities 75, 113, 387
Gravitationally Lensed Galaxies 606

H

H II Regions
  diffuse ionized gas 469, 602
  galactic distribution 247, 555, 594
  giant 287, 409, 469, 582
  in LMC and SMC 582, 590
  luminosity function 475
  normal 143, 291, 379, 383, 423, 469, 518
  ultracompact 375, 399, 453, 457
Heirarchical Structure of
  galaxies 571
  molecular clouds 133, 184, 347, 367, 617
Herbig Ae/Be Stars 193
Herbig-Haro Objects 121
Hubble Deep Field 481, 610

I

Infrared Companions 337, 371
  (also see Brown Dwarfs)
Infrared Excesses in young stars 153, 209, 337
Initial Mass Function 3, 41, 143, 217, 271, 315, 347, 357, 414, 423, 561, 571, 617
  globular clusters 357
  measurements 228, 357
  Millar-Scalo Mass Function 101
  minimum stellar mass 143, 228, 241, 414

origins 101, 143, 329, 347
Salpeter Mass Function
    101, 357, 414, 571
  spheroid and halo 357
  substellar 217, 228
  universality 357, 414
  upper mass limit 409, 414, 473
Instabilities
  Balbus-Hawley 51, 160, 171
  Kelvin-Helmholtz 160
  Papaloizou-Pringle 160
  Rayleigh-Taylor
    160, 217, 347, 469
  Richtmyer-Meshkoff 217
  Swing amplification 247
  Toomre Q parameter
    3, 247, 311, 423
Intracluster Medium
  metals in 3, 481
  origins 271
ISO Observations 193, 506

J

Jeans
  mass 3, 15, 101, 217, 347, 617
  length 85

K

Kinematic Viscosity 160

L

Linewidth-Size Relation
    105, 184, 449
  (also see Molecular Clouds:
    velocity structure)
Lithium
  abundances 209

test of youth 228, 461
Low Surface Brightness Galaxies
    311, 315
Luminous IR Galaxies
    259, 319, 506
Lyman-α Forest 271
Lyman Break Galaxies 3, 481

M

Magnetic Fields
  Alfvén Waves 51, 217, 547
  ambipolar diffusion
    3, 41, 51, 75, 89, 347, 617
  dynamical significance
    3, 15, 63, 75, 81, 89, 109,
    160, 205, 347, 399, 547, 617
  field morphology 399
  energy cascade 89, 449
  magnetic Reynolds number 41
  measurements 63, 399
  pressure 89, 449
  pseudo-disk 109, 117
  reconnection 171, 179, 347
  stellar 153, 179
Markarian Galaxies 299
Masers
  OH 399
  water vapor 197, 399
  Zeeman splitting 399
Massive Stars (see OB Stars)
Merger Interactions 247, 259
  galaxy evolution
    295, 319, 525, 535, 571
  galaxy formation 525, 535
  gas-rich systems
    473, 525, 535, 551, 555
  starbursts 3, 259, 295, 319

Metallicities
    chemical evolution   3, 315, 437, 481, 525, 561, 571, 586
    low   414
    mass-metallicity relation   271, 561, 571
    metal-poor galaxies   525, 561, 590
    metal-poor stars   217
    metal production   481

Molecular Clouds
    collisions   3, 15, 323, 367
    cores   97, 133, 143, 189, 387, 367, 375, 391, 617
    definitions   41
    giant molecular associations   247
    large scale structure   85, 97, 113, 184, 189, 347, 391, 395
    mass spectrum   97, 133, 217, 357, 445
    physical characteristics   41
    rotational motions   85, 137, 617
    supercritical core   75
    velocity structure   3, 41, 51, 85, 97, 101, 105, 367, 391, 449

Molecular Depletions   109, 395

Molecular Line Observations
    CO   15, 121, 209, 247, 259, 307, 319, 379, 423
    ^{13}CO   395
    C^{18}O   375, 395
    CH   15
    CN   15
    CS   41, 113, 379, 387
    H$_2$ 121, 337, 375
    HCO+   41, 453
    H$_2$CO   41, 109, 125
    HC$_3$N   41, 109, 375

NH$_2$+   387
NH$_3$ 41, 105, 379, 391, 453
OH   399
SiO   375

N

N-Body Codes   133, 279, 347

Near-infrared Observations
    extinction map   72
    images   72, 125, 153, 189, 201, 283, 303, 307, 375, 502, 543
    spectroscopy   337

O

OB Stars
    associations   15, 143, 189, 201, 279, 383, 409, 453, 457, 465, 473, 561, 571, 598
    atmospheres   291
    birth in clusters   201, 347, 414
    birth in field   414
    O3-type stars   409
    winds   143, 457

Outflows   (see Stellar Outflows or Galactic Stucture: winds)

R

RR Lyrae Stars   561
Radio Continuum Observations   518
    ionized gas   379, 453
Reflection Nebulae
    stellar   15, 93, 121, 143, 383
    high-z   547

S

Self-similar Processes   75, 105

Spectral Energy Distribution
    galaxies   259, 491, 502
    H II regions   465
    infrared sources   383
    T Tauri stars   153
Spectral Veiling   153, 337
Starbursts   3, 247, 259, 617
    at galactic center   299, 323
    dust-shrouded   3, 271
    galaxies   259, 271, 295,
        303, 319, 323, 561, 586
    massive stars   259, 409, 561
    time dependence   315, 414,
        491, 525, 551, 561
Star Clusters   15
    (also see Cluster Formation)
    globular clusters
        259, 271, 357, 414, 571
    open clusters
        143, 189, 347, 357, 375
    super clusters   271, 403
    young globular clusters
                     3, 347, 403
Star Formation
    collision induced   3, 15
    first generation   217, 329
    global regulation
        409, 491, 571, 602
    sequential   409, 594, 617
    stochastic   441, 561
    threshold surface density
                247, 423, 510
    triggered
        15, 143, 323, 409, 433, 555
Star Formation Efficiency
        259, 279, 319, 433
Star Formation History
        423, 535, 551, 617

dwarf spheroidals   571
ellipticals   481, 525, 535
halo   217
irregulars   561, 590, 598
low surface brightness galaxies
                          315
solar neighborhood   437, 461
spiral bulges   217, 491
spiral disks   481, 491
universe   481, 617
Star Formation Rates   3, 259,
    271, 423, 433, 437, 475, 481,
    491, 502, 514, 518, 551, 598
Stellar Evolutionary Tracks   571
Stellar Mass Function
    (see Initial Mass Function)
Stellar Outflows
    bipolar cavities   125
    bow shocks   63, 143
    deflected material   121
    driving mechanisms
        51, 143, 153, 160, 171, 457
    examples
        63, 113, 121, 129, 143, 153
    ionized winds   129
    jets   121, 143, 153, 457, 617
    maser connection   197
    morphologies   93, 121, 383
Sunyaev-Zel'dovich Effect   606
Supersonic Motions on Clouds
    (see Linewidth-Size Relation
             and Molecular Clouds:
                  velocity structure)
Supernovae
    role in metal creation
                (see Metallicity)
    shaping the ISM   441, 469, 571

T

T Tauri Stars
    classical   63, 137, 143, 175, 179, 184, 153, 209, 217, 241, 337, 371, 461
    weak lined   153, 175, 179, 184

Tidal Interactions (also see Mergers)   247, 271, 283, 287, 311, 491

Turbulence
    Burgers   51
    compressible   85, 445, 449, 617
    convective   160
    dissipation timescale   51
    dynamical significance   41, 81, 97, 133, 137, 160, 205, 617
    infall   41
    Kolmogorov Spectrum   51, 184, 449
    magnetohydrodynamic   41, 51, 75, 85, 89, 160, 449, 617
    origins   51, 160
    velocity coherence   89, 105

Two-body Relaxation   347

U

Ultracompact H II Regions (See H II regions)

Ultraluminous Galaxies   3, 259, 271, 295

UV Radiation
    far-UV observations   287, 473, 491
    in galaxies   551, 590, 594, 598
    photodissociation   291, 379
    photoevaporation   143, 189, 201, 271, 291
    photoionization   143, 469, 571, 586, 602

V

Vega-Excess Systems
    comet impacts   193, 209
    CO emission   209
    dust emission   209

Violent Relaxation   347, 525

Virial Equilibrium   81, 279, 347, 403, 449, 547

W

White Dwarfs   357

Wolf-Rayet Stars   409, 414, 457, 586, 602

X

X-ray Observations
    cooling flows   551
    flares   175, 179
    massive stars   323
    young stars   175, 179, 184, 347, 461

Y

Young Stellar Sources
    Class 0   41, 63, 75, 117
    Class I 2   41, 63, 117, 125, 175
    Deuterium-burning   241
    Embedded   63, 72, 113, 129, 143, 179, 189, 201, 217, 241

AIP Conference Proceedings

	Title	L.C. Number	ISBN
No. 360	The Physics of Electronic and Atomic Collisions XIX International Conference (Whistler, Canada, 1995)	95-83671	1-56396-440-6
No. 361	Space Technology and Applications International Forum (Albuquerque, NM 1996)	95-83440	1-56396-568-2
No. 362	Two-Center Effects in Ion-Atom Collisions (Lincoln, NE 1994)	96-83379	1-56396-342-6
No. 363	Phenomena in Ionized Gases XXII ICPIG (Hoboken, NJ, 1995)	96-83294	1-56396-550-X
No. 364	Fast Elementary Processes in Chemical and Biological Systems (Villeneuve d'Ascq, France, 1995)	96-83624	1-56396-564-X
No. 365	Latin-American School of Physics XXX ELAF Group Theory and Its Applications (México City, México, 1995)	96-83489	1-56396-567-4
No. 366	High Velocity Neutron Stars and Gamma-Ray Bursts (La Jolla, CA 1995)	96-84067	1-56396-593-3
No. 367	Micro Bunches Workshop (Upton, NY, 1995)	96-83482	1-56396-555-0
No. 368	Acoustic Particle Velocity Sensors: Design, Performance and Applications (Mystic, CT, 1995)	96-83548	1-56396-549-6
No. 369	Laser Interaction and Related Plasma Phenomena (Osaka, Japan 1995)	96-85009	1-56396-445-7
No. 370	Shock Compression of Condensed Matter-1995 (Seattle, WA 1995)	96-84595	1-56396-566-6
No. 371	Sixth Quantum 1/f Noise and Other Low Frequency Fluctuations in Electronic Devices Symposium (St. Louis, MO, 1994)	96-84200	1-56396-410-4
No. 372	Beam Dynamics and Technology Issues for + - Colliders 9th Advanced ICFA Beam Dynamics Workshop (Montauk, NY, 1995)	96-84189	1-56396-554-2
No. 373	Stress-Induced Phenomena in Metallization (Palo Alto, CA 1995)	96-84949	1-56396-439-2
No. 374	High Energy Solar Physics (Greenbelt, MD 1995)	96-84513	1-56396-542-9
No. 375	Chaotic, Fractal, and Nonlinear Signal Processing (Mystic, CT 1995)	96-85356	1-56396-443-0

Title	L.C. Number	ISBN
No. 376 Chaos and the Changing Nature of Science and Medicine: An Introduction (Mobile, AL 1995)	96-85220	1-56396-442-2
No. 377 Space Charge Dominated Beams and Applications of High Brightness Beams (Bloomington, IN 1995)	96-85165	1-56396-625-7
No. 378 Surfaces, Vacuum, and Their Applications (Cancun, Mexico 1994)	96-85594	1-56396-418-X
No. 379 Physical Origin of Homochirality in Life (Santa Monica, CA 1995)	96-86631	1-56396-507-0
No. 380 Production and Neutralization of Negative Ions and Beams / Production and Application of Light Negative Ions (Upton, NY 1995)	96-86435	1-56396-565-8
No. 381 Atomic Processes in Plasmas (San Francisco, CA 1996)	96-86304	1-56396-552-6
No. 382 Solar Wind Eight (Dana Point, CA 1995)	96-86447	1-56396-551-8
No. 383 Workshop on the Earth's Trapped Particle Environment (Taos, NM 1994)	96-86619	1-56396-540-2
No. 384 Gamma-Ray Bursts (Huntsville, AL 1995)	96-79458	1-56396-685-9
No. 385 Robotic Exploration Close to the Sun: Scientific Basis (Marlboro, MA 1996)	96-79560	1-56396-618-2
No. 386 Spectral Line Shapes, Volume 9 13th ICSLS (Firenze, Italy 1996)		1-56396-656-5
No. 387 Space Technology and Applications International Forum (Albuquerque, NM 1997)	96-80254	1-56396-679-4 (Case set) 1-56396-691-3 (Paper set)
No. 388 Resonance Ionization Spectroscopy 1996 Eighth International Symposium (State College, PA 1996)	96-80324	1-56396-611-5
No. 389 X-Ray and Inner-Shell Processes 17th International Conference (Hamburg, Germany 1996)	96-80388	1-56396-563-1
No. 390 Beam Instrumentation Proceedings of the Seventh Workshop (Argonne, IL 1996)	97-70568	1-56396-612-3
No. 391 Computational Accelerator Physics (Williamsburg, VA 1996)	97-70181	1-56396-671-9
No. 393 Star Formation Near and Far Seventh Astrophysics Conference (College Park, MD 1996)	97-71978	1-56396-678-6